T0353892

Computational Quantum Chemistry

Computational Quantum Chemistry

Second Edition

Ram Yatan Prasad and Pranita

CRC Press is an imprint of the
Taylor & Francis Group, an **informa** business

Second edition published 2021
by CRC Press
6000 Broken Sound Parkway NW, Suite 300, Boca Raton, FL 33487-2742

and by CRC Press
2 Park Square, Milton Park, Abingdon, Oxon, OX14 4RN

First edition published by Foundation Books, an imprint of Cambridge University Press
Second edition published by CRC Press 2021

CRC Press is an imprint of Taylor & Francis Group, LLC

Library of Congress Cataloging-in-Publication Data
Names: Prasad, Ram Yatan, author. | Pranita, author.
Title: Computational quantum chemistry / authored by Prof (Dr.) Ram Yatan Prasad and Dr. Pranita.
Description: Second edition. | Boca Raton : CRC Press, 2021. |
Includes bibliographical references and index.
Identifiers: LCCN 2020044935 | ISBN 9780367679699 (hardback) | ISBN 9781003133605 (ebook)
Subjects: LCSH: Quantum chemistry—Data processing. |
Quantum chemistry—Mathematics—Data processing.
Classification: LCC QD462.6.D38 P73 2021 | DDC 541/.28—dc23
LC record available at https://lccn.loc.gov/2020044935

ISBN: 978-0-367-67969-9 (hbk)
ISBN: 978-0-367-67970-5 (pbk)
ISBN: 978-1-003-13360-5 (ebk)

Typeset in Times
by codeMantra

Ram Yatan Prasad dedicates this book to his parents, wife, children, Dr. Rishav Raj (MBBS, MD), AIIMS, BBSR, Priyanshi, and his beloved grandson Aaryaman Raj, whose contributions in the accomplishment of writing this book cannot be forgotten. He also dedicates this book to his teachers Late Dr. B.P. Gyani, emeritus professor of chemistry; Dr. J.C. Ghosh, scholar of physical chemistry, Patna University; and Bharat Ratna Prof. C.N.R. Rao for his contribution in the field of chemistry.

Pranita dedicates this book to the sacred memory of her beloved mother, Dr. Kamla Prasad.

Contents

Foreword

Computational Quantum Chemistry written by Dr. Prasad and Dr. Pranita is an outcome of diligent effort. The book provides a useful presentation of quantum mechanics and quantum chemistry. It deals with the subject in detail starting with black-body radiation and ending with density functional theory. All the chapters are written in a clear and simple style.

The chapters are arranged in such a manner that each chapter leads to the next. The two chapters on mathematical techniques and quantum mechanical operators are important core topics in the book that help the reader to understand many aspects of physical sciences, electrical engineering, and so on.

The book will prove useful to readers at all levels as it contains several solved and unsolved problems at the end of each chapter which should equip the reader to learn better.

C.N.R. Rao

Preface

Computational Quantum Chemistry provides a mode of description for microscopic systems which has proved supremely successful. Why is quantum mechanics so important? It is important because the theory has led to a large number of practical applications besides providing explanations for various types of physical phenomenon.

It is a fact that students find quantum mechanics tough for two reasons, one conceptual and other technical. The authors have tried to give all the essentials of the subject and some of the trimmings. The style of writing in this book is simple and graspable.

We owe special thanks to Bharat Ratna Prof. C.N.R. Rao for kindly writing the Foreword for this book.

We also express our sincere thanks to Sri N.N. Sinha, IAS and former principal secretary to governor, who encouraged us to write such a book.

Our thanks are also due to Dr. K.K. Nag, former vice chancellor, Ranchi University, for his affection and encouragement.

We thank our reviewers for their suggestions on content.

Lastly, we thank our colleagues at CRC Press, Renu Upadhayay (Commissioning Editor), and Jyotsna for ensuring the quality of publications.

Ram Yatan Prasad
Pranita

Authors

Ram Yatan Prasad, PhD, DSc (India), DSc (hc) Colombo, is a professor of chemistry and former vice chancellor of S.K.M University, Jharkhand, India, and a life fellow of Indian Chemical Society and other societies of repute. He has been two times pro-vice chancellor in Bihar and Jharkhand. To his credit, he has been a member of editorial board in national journals. He has received an Outstanding Service Award from the Governor of Jharkhand. He has been chairperson of World Academy of Sciences, Paris, France, in the international conference and has received an Appreciation Award. He has more than 46 years of experience in teaching quantum chemistry, spectroscopy, and thermodynamics at the postgraduate level. Dr. Prasad is a prolific author of chemistry and has published many research papers in reputed journals to his credit.

Pranita, PhD, DSc (hc) Sri Lanka, FICS, is an assistant professor of chemistry at Vinoba Bhave University, India. She has published many research papers in national and international journals to her credit. Her area of interest is thermodynamics of liquid state. She has 12 years of teaching experience in quantum chemistry, statistics, and liquid state.

1

Quantum Theory

In the eighteenth century, it was believed that light consisted of small particles as postulated earlier by the famous scientist Newton in his corpuscular theory. Various experimental studies were carried out later, which supported the wave theory of light. In the later part of the nineteenth century, a large number of experimental facts/observations came to light, which could not be explained by wave theory. One of the important experimental studies was performed by various scientists on black-body radiation, which led to the development of quantum theory. The pioneers among the scientists in the field of black-body radiation were Stefan–Boltzmann, Rayleigh–Jeans, Wein, and Max Planck. A brief discussion on black-body radiation and distribution of energy in temperature radiation is presented in the subsequent sections.

1.1 Black-Body Radiation

A black body is defined as a perfect absorber and emitter of radiation. Actually, it absorbs all the radiations falling on it, without reflecting any. Experiments have shown that all hot bodies when perfectly blackened give rise to a thermal spectrum that depends only on the temperature and not on the chemical composition or the material the body is made of.

In practice, black-body radiation is studied by making a cavity, which is insulated so that the only energy that can be absorbed is the energy supplied in the form of heat to enhance the temperature of the cavity. The cavity is evacuated and a small hole is made on one side of the apparatus through which the radiation can pass. One approximation suggests that the radiation coming from the hole in the cavity becomes a better sample radiation inside the cavity. In short, we can say that a black body, therefore, comprises the following characteristics:

- It must emit radiation of the same character and intensity as it absorbs.
- It must emit radiation of nearly all wavelengths, and at a particular temperature must emit more radiation than a body with less power of absorbing radiation.
- It must absorb radiations of all wavelengths completely.

The intensity of the radiation emitted from the hole is then studied as a function of wavelength at various temperatures of the cavity. On the basis of the experimental results of the black-body radiation, a graph between energy distribution E_λ and wavelength λ is plotted at various temperatures. It is shown in Figure 1.1.

From the plotted graph, which is based on the experimental data, the following conclusions are drawn:

- The radiant energy E_λ varies with wavelength λ and has maximum value of wavelength, λ_{max}.
- λ_{max} is inversely proportional to the absolute temperature, T, of the radiator, i.e., $\lambda_{max} \propto 1/T$. It has been found that $\lambda_{max}T = 0.287\,cm\ ^\circ C$. The quantitative verification is shown by the data of Lummer and Pringsheim in Table 1.1.

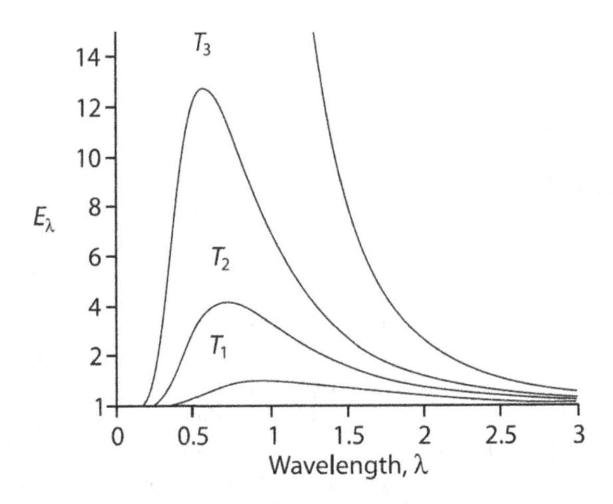

FIGURE 1.1 Distribution of energy in black-body radiation at temperatures $T_3 > T_2 > T_1$.

TABLE 1.1

The Temperature Variation in the Wavelength of Intense Black-Body Radiation

T, K	$\lambda_{max} \times 10^4$ cm	$\lambda_{max} \cdot T$
621.2	4.53	0.2814
908.0	3.28	0.2980
1094.5	2.71	0.2966
1259.0	2.35	0.2959
1460.4	2.04	0.2978
1646.0	1.78	0.2928

The data of $\lambda_{max} \cdot T$ is in close agreement with 0.287. The difference may be approximately 1%.

1.2 Wien's Radiation Law

Wien deduced an expression for the radiation law on the basis of thermodynamics, which is in the following form:

$$E_\lambda = \frac{\alpha}{\lambda^5} e^{\frac{-\beta}{\lambda T}} \tag{1.1}$$

where E_λ is the energy of radiation or absorption and α and β are the constants independent of λ and T.

The position of maximum on the $E_\lambda - \lambda$ curve is found by differentiating the above equation/expression with respect to λ and equating to zero.

$$\because \quad E_\lambda = \frac{\alpha}{\lambda^5} \cdot e^{-\beta/\lambda T} = \left(\alpha\lambda^{-5}\right)\left(e^{-\beta/\lambda T}\right)$$

$$\therefore \quad \frac{dE_\lambda}{d\lambda} = -5\alpha\lambda^{-6} e^{-\beta/\lambda T} + \left(\frac{\alpha}{\lambda^5}\right)\left(e^{-\beta/\lambda T}\right)\left(-\beta/T - 1/\lambda^2\right)$$

$$= e^{-\beta/\lambda T}\left[-5\alpha\lambda^{-6} + \left(\alpha\lambda^{-5}\right)\left(\frac{\beta}{\lambda^2 T}\right)\right]$$

$$= e^{-\beta/\lambda T}\left[-5\alpha\lambda^{-6} + \frac{\alpha\beta}{T\lambda}\lambda^{-6}\right]$$

$$= e^{-\beta/\lambda T}\left[\frac{\beta}{\lambda T} - 5\right]\alpha/\lambda^{6}$$

But, at maximum, $\frac{dE_\lambda}{d\lambda} = 0$

$$\therefore \quad e^{-\beta/\lambda_{\max}T}\left[\frac{\beta}{\lambda_{\max}\cdot T} - 5\right]\frac{\alpha}{\lambda_{\max}^{6}} = 0$$

But, $e^{-\beta/\lambda_{\max}T} \neq 0$ and $\frac{\alpha}{\lambda_{\max}^{6}} \neq 0$

$$\therefore \quad \beta/\lambda_{\max}\cdot T - 5 = 0$$

or $\frac{\beta}{\lambda_{\max}T} = 5$

$$\text{or } \lambda_{\max}T = \beta/5 \tag{1.2}$$

If $\beta = 1.433$ cm. deg, then $\beta/5 = 0.2866$, which is in close agreement with the experimental value.

Wien's expression, however, fails to account for the experimental results obtained at low frequencies and high temperatures, i.e., it holds true only at low temperatures and short wavelengths.

1.3 Rayleigh–Jeans Law

In 1900, Lord Rayleigh applied the classical law of equipartition of energy to the electromagnetic vibrations. Since matter and radiation remain in equilibrium under the applied condition, the energy dE of the radiation at a given frequency υ is equal to the number of black-body oscillators consisting of frequencies between υ and $\upsilon + d\upsilon$ multiplied by the average energy $\bar{\in}$ of an oscillator. On the basis of the general theorem of elasticity, we know that the number of oscillators per unit volume is

$$L = \frac{8\pi v^3}{3c^3} = \frac{8\pi}{3\lambda^3} \tag{1.3}$$

and on differentiation, we get

$$dL = \frac{8\pi v^2 dv}{c^3} = \frac{-8\pi\, d\lambda}{\lambda^4} \tag{1.4}$$

having frequencies between υ and $\upsilon + d\upsilon$. The radiation energy in equilibrium will therefore be

$$dE = dL\cdot\bar{\in} = \frac{8\pi v^2\,\bar{\in}\,dv}{c^3} = \frac{-8\pi\,\bar{\in}\,d\lambda}{\lambda^4}$$

and the energy per unit of frequency will be

$$E_v = \frac{dE}{dv} = \frac{8\pi\,\bar{\in}\,v^2}{c^3} \tag{1.5}$$

and per unit of wavelength will be

$$E_\lambda = \frac{-\mathrm{d}E}{\mathrm{d}\lambda} = \frac{8\pi\,\overline{\in}}{\lambda^4} \tag{1.6}$$

But according to the equipartition principle of energy based on the classical mechanics, the average energy of an oscillator $\overline{\in}$ is given by

$$\overline{\in} = kT \qquad (\text{where } \mathrm{k} = R/N = \text{Boltzmann constant})$$

Hence, Eqs. (1.5) and (1.6) become

$$E_\nu = \frac{8\pi\,\nu^2 \mathrm{k}T}{c^3} \tag{1.7}$$

and

$$E_\lambda = \frac{8\pi\,kT}{\lambda^4}. \tag{1.8}$$

Equations (1.7) and (1.8) are called the *Rayleigh–Jeans radiation law.* These are in excellent quantitative agreement with the experimental observations for low frequencies (or large wavelengths) and high temperatures, thereby leading to the ultraviolet 'catastrophe'.

1.4 Planck's Radiation Law

In 1900, Max Planck could give the correct form of the thermal–radiation law on the basis of the following assumptions:

- The energy of an oscillator does not vary continuously but only by definite amount; thus, there is a finite number of energy levels

$$\in_0, \in_1, \in_2 \ldots$$

- The energies are an integral number of the smallest amount, i.e.,

$$\in_0 = 0,\ \in_1 = \in,\ \in_2 = 2\in,\ \in_3 = 3\in \ldots$$

- The smallest amount of energy (here $\in_1 = \in$) is proportional to the frequency υ of the oscillator, i.e., $\in \propto \upsilon$ or $\in = h\upsilon$, where h = Planck's constant ($h = 6.626 \times 10^{-34}$J s), the dimension of which is energy/frequency = energy × time = action constant.

This packet of energy equal to hv is called quantum.

Let $\in_0$ be the lowest energy of the number of oscillators, $n_0 e^{-\in_1/kT}$ the number of oscillators with energy $\in$ in excess of $\in_0$ on the basis of the classical distribution law. Thus,

$$n_1 = n_0 e^{-\in_1/kT} = n_0 e^{-\in_1/kT}$$

$$n_2 = n_0 e^{-\in_2/kT} = n_0 e^{-\in_2/kT}$$

and the total number of oscillators is

$$n = n_0 + n_1 + n_2 + n_3 + \ldots$$
$$= \left(n_0 + n_0 e^{-\in/kT} + n_0 e^{-2\in/kT} + n_0 e^{-3\in/kT} + \ldots\right)$$
$$= n_0\left(1 + e^{-hv/kT} + e^{-2hv/kT} + e^{-3hv/kT} + \ldots\right)$$
$$= n_0\left(1 + x + x^2 + x^3 + \ldots\right), \quad \text{where } x = e^{-hv/kT}$$

Now, the total energy of n oscillators is clearly

$$E = \in_0 n_0 + \in_1 n_1 + \in_2 n_2 + \in_3 n_3 + \ldots$$
$$= \in_0 n_0 + \in n_0 e^{-\in/kT} + 2\in e^{-2\in/kT} + \ldots$$
$$= 0 + hv\, n_0 e^{-hv/kT} + 2hv\, n_0 e^{-2hv/kT} + 3hve^{-3hv/kT} + \ldots$$
$$= n_0 hv\left(e^{-hv/kT} + 2e^{-2hv/kT} + 3e^{-3hv/kT} + \ldots\right)\left(\text{where, } x = e^{-hv/kT}\right)$$
$$= n_0 hve^{-hv/kT}\left(1 + 2x + 3x^2 + \ldots\right)$$

∴ The average energy of an oscillator

$$= \bar{\in} = E/n = \frac{n_0 hve^{-hv/kT}\left(1 + 2x + 3x^2 + \ldots\right)}{n_0\left(1 + x + x^2 + x^3 + \ldots\right)}$$
$$= \frac{hv\, e^{-hv/kT}\left(1 - x^2\right)^{-2}}{(1-x)^{-1}}$$
$$= \frac{hv \cdot x}{(1-x)}, \quad \text{where } x = e^{-hv/kT}$$
$$= \frac{hv}{\left(\frac{1}{x} - 1\right)} = \frac{hv}{e^{+hv/kT} - 1}$$

$$\therefore \quad \bar{\in} = \frac{hv}{\left(e^{hv/kT} - 1\right)} \tag{1.9}$$

$$\therefore \quad E_v = \frac{8\pi\, v^2}{c^3} \cdot \frac{hv}{e^{hv/kT} - 1} \tag{1.10}$$

$$\therefore \quad E_\lambda = \frac{8\pi\, hc}{\lambda^5} \cdot \frac{1}{e^{hc/\lambda\, kT} - 1} \tag{1.11}$$

Equations (1.10) and (1.11) represent *Planck's radiation law* and account quantitatively for the variation of E_λ with λ over the whole range of wavelength and T.

Now, we shall impose some conditions on the equation related to Planck's radiation law.

I. When $h\nu << kT$, then

$$e^{h\nu/kT} = 1 + \frac{h\nu/kT}{1!} + \frac{\left(\frac{h\nu}{kT}\right)^2}{2!} + \ldots$$

$$\approx 1 + \left(\frac{h\nu}{kT}\right) \text{ on neglecting the higher power terms}$$

$$\therefore \quad \overline{\in} = \frac{h\nu}{1 + e^{h\nu/kT} - 1} = kT \tag{1.12}$$

This result is in accordance with the classical equipartition principle.

II. When $h\nu >> kT$, $e^{h\nu/kT} >> 1$

$$\therefore \quad \overline{\in} \approx \frac{h\nu}{e^{+h\nu/kT} - 1} \approx \frac{h\nu}{e^{h\nu/kT}} \tag{1.13}$$

It became different from kT as predicted classically. This is why the ultraviolet catastrophe arises.

III. At low frequencies and high temperature, when $h\nu/kT << 1$

$$\therefore \quad e^{h\nu/kT} \approx 1 + h\nu/kT$$

$$\therefore \quad \overline{\in} \approx \frac{h\nu}{e^{h\nu/kT} - 1} \approx kT$$

and Eq. (1.10) becomes

$$E_\nu = \frac{8\pi\, \nu^2 kT}{c^3} \tag{1.14}$$

This is the *Rayleigh–Jeans law* and is valid under the above-mentioned conditions.

IV. At high frequency (low λ) and low T, $\lambda\, kT << h\, c$.

$$\therefore \quad e^{hc/\lambda kT} \gg 1$$

and equation $E_\lambda = \frac{8\pi\, hc}{\lambda^5} \cdot \frac{1}{e^{hc/\lambda\, kT} - 1}$ becomes

$$E_\lambda = \frac{8\pi hc}{\lambda^5} \cdot e^{hc/\lambda\, kT} \tag{1.15}$$

which is *Wien's law* with

$$hc/k = \beta = 1.433 \text{ cm } ^\circ\text{C} \tag{1.16}$$

Substituting the value of c and k in Eq. (1.16), we get the value of h. This value has been found to be 6.6×10^{-34} Js. Thus, we see that Planck's radiation law successfully explains the distribution of energy among the various spectral regions of temperature radiation and it puts forth the first piece of evidence for the existence of energy quanta. Thus, Planck proposed *Planck's quantum theory.*

1.5 Quantum Theory

In 1900, Max Planck devised a theory to account for the emission of the black-body radiation from hot bodies. According to this theory "the radiation/energy is emitted by a hot black-body in quanta, each of which has an energy equal to h√, where h is the Planck's constant and √ is the frequency of radiation". Mathematically, it is expressed as $E = h\upsilon$.

It is a fact that the quantum theory has revolutionised the classical ideas and many other phenomena such as photoelectric effect, Compton effect, atomic spectra, etc. have been successfully explained by it. Now we shall try to explain the above-mentioned properties one by one.

1.6 Photoelectric Effect

The photoelectric effect was first discovered in 1887 by Hertz, and it relates to the fact that metals (also semi-conductors) emit electrons when exposed to light. Wave theory was unable to explain this effect. What is photoelectric effect? When light of suitable frequency falls on a thoroughly/scrupulously clean metal surface in an evacuated vessel, electrons are ejected. This phenomenon is called the *photoelectric effect.*

Experiment: The experimental diagram for the demonstration of photoelectric effect is shown in Figure 1.2.

In this experimental set-up, C is an evacuated chamber in which K is functioning as a cathode, made of the metal under investigation, and P is a plate, functioning as an anode. The metal K is joined to the negative pole of the battery B and P is connected to the positive pole of the battery. G is a galvanometer joined as shown in the figure.

When light falls on the metal K from the light source S, photoelectrons are ejected from the surface of the metal, which are attracted by P, and the galvanometer shows deflection, which thereby means that the current is flowing.

An external potential V is applied to arrest the flow of electrons having mass m and velocity v. Under this condition, we can write

$$\frac{1}{2}mv^2 = eV \tag{1.17}$$

where V is the external potential applied to arrest the flow of electrons and e is the electronic charge.

Equation (1.17) can be written as

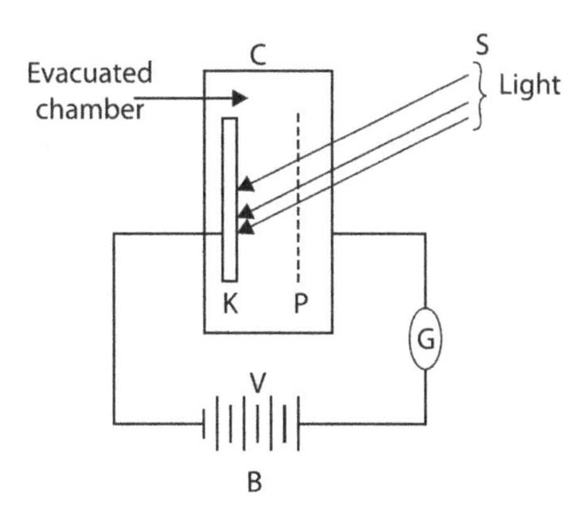

FIGURE 1.2 Schematic diagram for depicting photoelectric effect. (*Note*: C, evacuated chamber; K, cathode made of the metal under investigation; P, plate (anode); G, galvanometer; and B, battery.)

$$mv^2 = 2eV$$
$$v^2 = \frac{2eV}{m}$$

$$\therefore \quad v = \left(\frac{2eV}{m}\right)^{1/2} \tag{1.18}$$

The experimental observations have been mentioned below:

- No electrons are emitted from the metal surface unless the frequency of the incident light approaches a minimum frequency v_0, characteristic of the metal. This minimum frequency has been termed as *threshold frequency*.
- Strength of the photoelectric current, i.e., the number of electrons emitted per second is proportional to the intensity of the incident light.
- The kinetic energy of photoelectrons is directly proportional to the frequency υ of the incident radiation but independent of its intensity.

In 1905, Albert Einstein applied the quantum theory to explain the photoelectric effect. He postulated that the radiant energy is quantised, i.e., radiation consists of particles or quanta (named photon by Lewis in 1926) with an energy hυ.

When the light is incident on a metallic surface, the electron absorbs a photon of energy hυ. For the ejection from the metal surface, it uses up a minimum amount of energy w having minimum frequency υ_0 called *work function* of the metal. The remaining energy absorbed by the photon appears as kinetic energy, $\frac{1}{2}mv^2$, of the electron, where m is the mass of the electron and v is the velocity of electron.

From the above statement, we can write that

$$E = w + \frac{1}{2}mv^2 \tag{1.19}$$

where w is the work function of the metal, E is the total energy absorbed by the metal surface, and $\frac{1}{2}mv^2$ is the kinetic energy of the electron.

But according to Planck's quantum theory,

$$E = hv$$

where h is Planck's constant; υ is the frequency of radiation; and w is the work function.

$$hv_0 = \text{minimum energy required to eject the electron}$$

where υ_0 is the threshold frequency.

On this basis, Eq. (1.19) takes the form

$$hv = hv_0 + \frac{1}{2}mv^2 \tag{1.20}$$

$$= hv_0 + eV \tag{1.21}$$

It should be noted that the number of electrons ejected per second from the surface of the metal indicates the intensity of light striking the metal surface.

Equation (1.20) can be written as

$$\frac{1}{2}mv^2 = hv - hv_0$$

which clearly suggests that the kinetic energy of the photoelectron is directly proportional to the frequency of radiation and independent of its intensity. Thus, we see that all the observations of the experiment are explained on the basis of the quantum theory.

Equation (1.21) can be written as

$$hv = hv_0 + eV$$

$$\text{or} \quad v = v_0 + \left(\frac{e}{h}\right)V$$

$$\text{or} \quad v = \left(\frac{e}{h}\right)V + v_0 \tag{1.22}$$

This equation is of the form $y = mx + c$, and if we plot a graph between υ and V, a straight line is obtained (Figure 1.3).

The intercept PO gives the value of threshold frequency υ_0. The value of slope = tan θ = RS/PS = PT/RT = PT/OW = (e/h)

$$\therefore \quad h = e/\tan\theta = e/\text{PT/OW} = \frac{e.\text{OW}}{\text{PT}}$$

$$= 6.55 \times 10^{-34} \text{ J.s}$$

$$= \text{Planck's constant}$$

which is in excellent agreement with the value of h (Planck's constant) obtained by other methods. It is to be noted that for all the metals under investigation, h has been found to be the same but the value threshold frequency υ_0 varies from metal to metal.

It has been mentioned that the value of the threshold frequency is obtained from the intercept in Y-axis, and hence the work function, $w = h\upsilon_0$, can be found out. The value of λ_0 (threshold wavelength) and the work function w for a few metals at 20°C are given in Table 1.2.

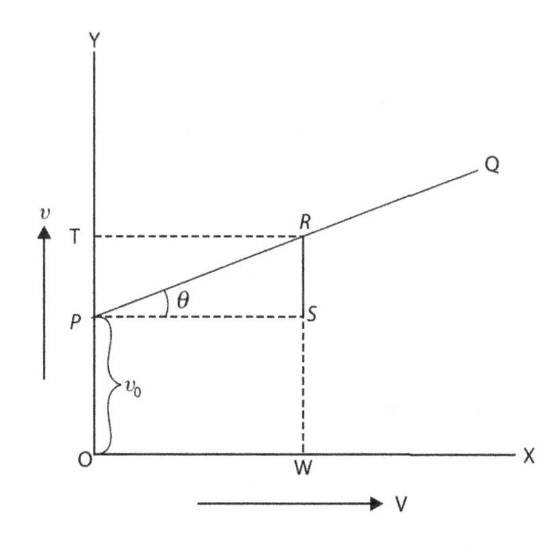

FIGURE 1.3 υ–V graph.

TABLE 1.2

Values of λ_0 and w for a Few Metals at 20°C

Metals	Threshold Wavelength (λ_0)	w in erg × 10^{12}
Li	5400	3.55
Na	5000	3.90
K	5500	3.57
Rb	5700	3.35
Cs	6600	2.98

From Eq. (1.20), it is clear that when KE = 0, $V = 0$, which means $\upsilon = \upsilon_0$, i.e., the electron will not move. When $\upsilon > \upsilon_0$, $1/2mv^2 > 0$, the electron will move but when $\upsilon < \upsilon_0$, $1/2mv^2 < 0$, no electron will be ejected.

1.7 Compton Effect

When a photon collides with an electron, it leaves some energy with the electron, and due to this, the radiation is scattered and its wavelength is increased (or frequency is decreased). This effect is called the *Compton effect*, which was discovered in 1923 by Compton.

Compton found that when various substances are exposed to the X-rays, the wavelength of the scattered radiation is larger than the original wavelength. In other words, we can say that when substances are exposed to the X-rays, the frequency of radiation is decreased with respect to the original frequency. He also suggested that the change in wavelength ($\Delta\lambda$) or frequency (Δv) is independent of the nature of the substance. $\Delta\lambda$ of $\Delta\upsilon$ is always a positive value, which is determined by the scattering angle (the angle between the directions of scattered and original radiations).

Now, we would derive an expression for the Compton effect for which Compton suggested the following:

- Light be considered as a beam of photon energy $h\upsilon$ (it means that Compton applied the quantum theory for explaining the Compton effect).
- The interaction between a photon and an electron be regarded as an elastic collision and the laws of conservation of energy and momentum be obeyed.

Keeping in view the above facts, let us assume that the photon of energy $h\upsilon$ collides with an electron (the energy and momentum are taken to be zero before collision).

Energy before collision = $h\upsilon$

Scattering angle of Photon = θ

Angle of recoil of electron = ϕ

The collision between X-rays and electron is shown in Figure 1.4.

From the figure, it is clear that the resolve part of $h\upsilon/c$ along the X-axis will be

$$\frac{hv'}{c}\cos\theta \tag{1.23}$$

$$\text{and resolve part of } hv'/c \text{ along Y-axis} = \frac{hv'}{c}\sin\theta \tag{1.24}$$

The recoil electron has a momentum of mv.

$$\therefore \text{ The resolve part of } mv \text{ along X-axis} = mv\cos\phi \tag{1.25}$$

$$\therefore \text{ The resolve part along Y-axis} = mv\sin\phi \tag{1.26}$$

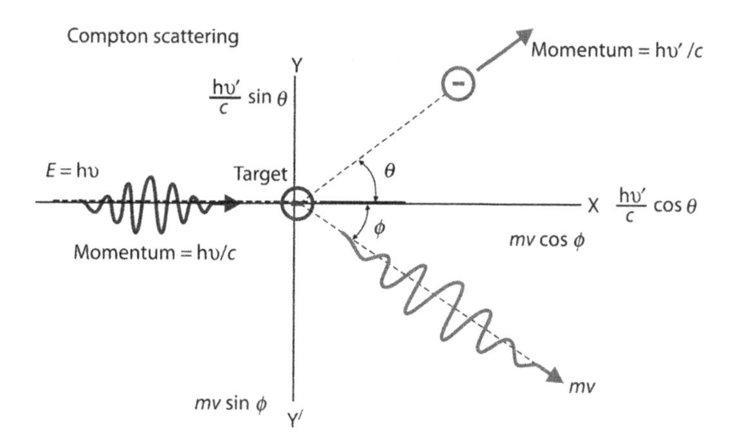

FIGURE 1.4 The Compton effect.

From the above, we can write that

$$\frac{hv'}{c} \cdot \sin\ \theta = mv\ \sin\ \phi \tag{1.27}$$

$$\frac{hv}{c} = \frac{hv'}{c} \cdot \cos\ \theta + mv\ cos\ \phi \tag{1.28}$$

$$\text{or} \left(\frac{hv'}{c}\ \sin\ \theta\right)^2 = (mv\ \sin\ \phi)^2$$

$$\text{and} \left(\frac{hv}{c} - \frac{hv'}{c} \cos\theta\right)^2 = (mv\ \cos\ \phi)^2$$

$$\text{or} \frac{hv'^2}{c}\ \sin^2\ \theta = m^2v^2\ \sin^2\ \phi$$

$$\text{and} \frac{h^2v^2}{c^2} + \frac{h^2v'^2}{c^2}\ \left(\sin^2\ \theta + \cos^2\ \theta\right) - 2\frac{hv}{c}\left(\frac{hv'}{c}\right)\cos\theta = m^2v^2\ \left(\sin^2\ \phi + \cos^2\phi\right)$$

On adding, we get

$$\frac{h^2v^2}{c^2} + \frac{h^2v'^2}{c^2}\left(\sin^2\theta + \cos^2\theta\right) - 2\frac{hv}{c}\left(\frac{hv'}{c}\right)\cos\ \theta = m^2v^2\left(\sin^2\phi + \cos^2\phi\right)$$

$$\text{or} \frac{h^2v^2}{c^2} + \frac{h^2v'^2}{c^2} - 2\frac{h^2v'^2}{c^2}\cos\theta = m^2v^2$$

$$\text{or} \frac{h^2v^2}{c^2} + \frac{h^2v^2}{c^2} - 2\frac{h^2v^2}{c^2}\cos\theta = m^2v^2\ \left[\because vv' \approx v^2\right]$$

$$\text{or } \frac{h^2\nu^2}{c^2} - 2\frac{h^2\nu^2}{c^2}\cos\theta = m^2v^2$$

$$\text{or } \frac{h^2\nu^2}{c^2} - (1-\cos\theta) = 2m^2v^2 \cdot \frac{1}{2} = 2m \cdot \frac{1}{2}mv^2$$

$$= 2m(h\nu - h\nu') = 2hm(\nu - \nu')$$

$$\text{or } (\nu - \nu') = \Delta\nu = \frac{2h^2\nu^2}{2hmc^2}(1-\cos\theta)$$

$$\text{or } \Delta\nu = \frac{h\nu^2}{mc^2}(1-\cos\theta) \tag{1.29}$$

$$\text{or } \left(\frac{c}{\lambda} - \frac{c}{\lambda'}\right) = \frac{h}{m\lambda^2}(1-\cos\theta)$$

$$\text{or } \frac{c(\lambda'-\lambda)}{\lambda\lambda'} = \frac{h}{m\lambda^2}(1-\cos\theta)$$

$$\text{or } \frac{c\Delta\lambda}{\lambda^2} = \frac{h}{m\lambda^2}(1-\cos\theta) \qquad [\because \lambda\lambda' \approx \lambda^2]$$

$$\therefore \quad \Delta\lambda = \frac{h}{mc}(1-\cos\theta) \tag{1.30}$$

when $\theta = 0°$, then $\Delta\lambda = 0$, but when $\theta = 180°$, then $\Delta\lambda = (2h/mc)$.

It is to be noted that $\Delta\nu$ or $\Delta\lambda$ is known as the *Compton shift* and Eq. (1.30) has been confirmed experimentally by measuring the value of λ' and scattering angle θ. By knowing the value of λ' and θ, we can find out the value of Planck's constant (h). The value of h obtained from Eq. (1.30) is in excellent agreement with values obtained from other methods.

1.8 Atomic Hydrogen Spectra

Gaseous atom when heated to a high temperature emits radiation. These radiations, when dispersed through a prism spectrograph, record four main lines on a photographic plate in the visible region.

These four lines are H_α, H_β, H_γ, and H_δ having wavelengths 656.28, 486.13, 434.05, and 410.17 nm, respectively. In the near-ultraviolet region, a number of other lines are also observed. These lines together with the four lines form a series which is called the *Balmer series*, named after the Swiss scientist Balmer who discovered it in 1885. The four important spectral lines are shown in Figure 1.5. Investigation of the hydrogen spectrum in the far-ultraviolet and infrared regions revealed the existence of a few other series of lines called the *Lyman* (ultraviolet region), *Paschen, Brackett*, and *Pfund* (infrared region) series, named after the scientists who investigated them.

Balmer gave an empirical relation to explain the existence of four lines, which is as follows:

$$\lambda = k\frac{n^2}{n^2-4} \tag{1.31}$$

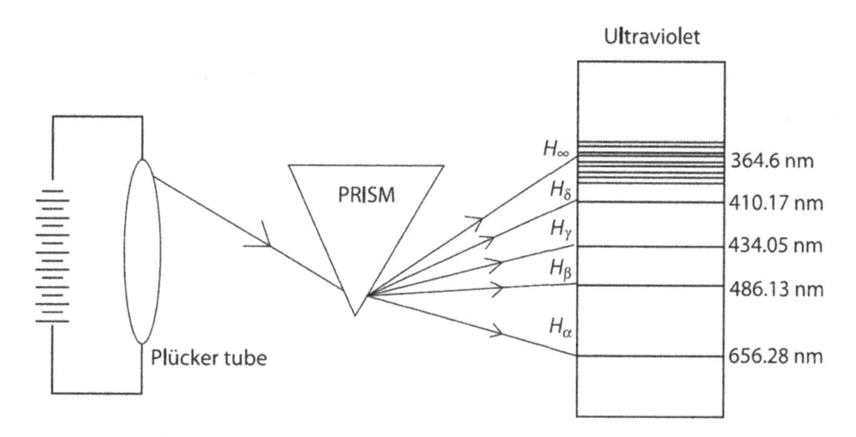

FIGURE 1.5 Schematic representation of atomic hydrogen spectrum from a gas discharge tube. (*Note*: The Plucker tube consists of molecular hydrogen gas at low pressure. The lines illustrated in the figure belong to the Balmer series.)

where λ is the wavelength; k is the constant; and n is an integer, 3 for H_α, 4 for H_β, 5 for H_γ, and 6 for H_δ.

The agreement between the calculated and observed values of λ is very close. It is to be noted that the spectral lines are denoted by wave number ($\bar{\nu}$), and it is equal to $1/A$, i.e., $\bar{\nu} = 1/\lambda$. Therefore, Eq. (1.31) is transformed in terms of wave number. The transformed equation becomes

$$\frac{1}{\lambda} = \frac{n^2 - 4}{kn^2}$$

$$\text{or} \quad \frac{1}{\lambda} = \frac{1}{k}\left(\frac{n^2 - 4}{n^2}\right) = \frac{1}{k}\left(1 - 4/n^2\right)$$

$$\text{or} \quad \bar{\nu} = \frac{4}{k}\left(\frac{1}{4} - \frac{1}{n^2}\right) = \frac{4}{k}\left(\frac{1}{2^2} - \frac{1}{n^2}\right)$$

$$\text{or} \quad \bar{\nu} = \mathrm{R_H}\left(\frac{1}{2^2} - \frac{1}{n^2}\right) \tag{1.32}$$

where $\mathrm{R_H}$ is the Rydberg constant $= 4/k = 109677.6\,\mathrm{cm}^{-1}$ and n = a whole number > 3.

Other workers generalised Eq. (1.32), which is expressed as

$$\bar{\nu} = \mathrm{R_H}\left(\frac{1}{m^2} - \frac{1}{n^2}\right) \tag{1.33}$$

where m is a constant for a particular series and the value of n gives the successive lines in each series. Each spectral series of lines has been mentioned in Table 1.3.

It is amazing that the formulae describing the hydrogen spectrum are controlled by two integers. It is a fact that integers are rather special to humans as we use them in counting. Actually, integers play an important role in quantum theory too. Now, one has to search an ideal model of an atom, which can explain the observed spectral lines satisfactorily.

The credit goes to Niels Bohr who successfully explained the observed spectral lines by putting forward a model of an atom.

TABLE 1.3

Spectral Series and the Values of m and n (Line Series in the Hydrogen Spectrum)

Spectral Series	m	n	Spectral Region
Lyman (1906)	1	2, 3, 4, 5, 6 …	Ultraviolet
Balmer (1895)	2	3, 4, 5, 6 …	Visible
Paschen (1908)	3	4, 5, 6, 7 …	Infrared
Brackett (1922)	4	5, 6, 7, 8 …	Infrared
Pfund (1925)	5	6, 7, 8 …	Infrared

1.9 The Bohr Model

In 1913, Niels Bohr (the Danish Physicist) took a bold step by applying Planck's quantum theory for the quantization of energy to observable like angular momentum and postulated a model for the atom which are as follows:

1. An electron in an atom can move in certain circular orbits around the nucleus under the influence of Coulomb attraction between them. These orbits are called *allowed stationary states.*
2. The allowed orbits of radius r are only those for which the orbital angular momentum is an integral multiple of $(h/2\pi)$, i.e.,

$$mvr = nh/2\pi \tag{1.34}$$

where, mvr = angular momentum

$$n = \text{an integer}$$

$$= 1,2,3,\ldots$$

$$m = \text{mass of electron}$$

$$v = \text{velocity of an electron}$$

3. An electron moving in such an allowed orbit does not radiate/absorb electromagnetic energy, despite the fact that the motion is with constant acceleration. Therefore, in allowed stationary states/orbits, the total energy remains constant.
4. An electron jumps from the orbit having higher energy E to the orbit having lower energy E', emitting only one photon (quantum) of energy hv so that

$$E - E' = h\nu \tag{1.35}$$

It is to be noted that the first postulate confirms the presence/existence of the atomic nucleus. The second postulate introduces the quantisation; it is important to note the difference between Bohr's quantisation of the orbital angular momentum of an atomic electron moving under the influence of Coulomb's force ($\propto 1/r^2$) and Planck's quantisation of energy of the oscillator. The third postulate prevents the instability of an atom. The fourth postulate is actually Einstein's postulate combined with the law of conservation energy.

1.9.1 Energy of an Electron Revolving around the Nucleus in a Permitted Orbit

Let us assume that an electron moves around the nucleus in an allowed orbit of radius r (Figure 1.6). Also suppose that the mass of electron is m, velocity is v, and nuclear charge is +Ze, where Z is the atomic

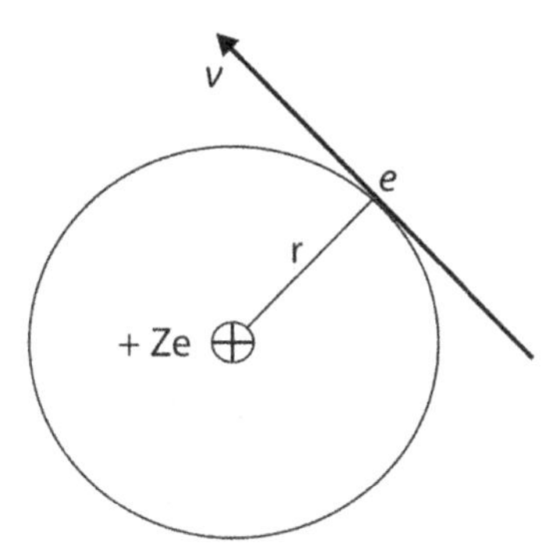

FIGURE 1.6 Movement of electron in the allowed orbit of radius *r*.

number and e is the electronic charge. It is clear that there will be a force of attraction between the nucleus and the electron, and the value of force of attraction will be given by

$$\frac{Ze.e}{r^2} = \frac{Ze^2}{r^2} \tag{1.36}$$

$$\text{The potential energy } \mathrm{d}V = \frac{Ze^2}{r^2} \cdot \mathrm{d}r \tag{1.37}$$

$$\text{or } \mathrm{d}V = \frac{Ze^2}{r^2} \mathrm{d}r$$

where dV is the small change in potential energy. Integrating the above equation, we get

$$\int \mathrm{d}V = \int Ze^2/r^2 \ \mathrm{d}r$$

$$\text{or} \quad V = -Ze^2/r + c \tag{1.38}$$

when $r = \infty$, then $V = 0$ and hence $c = 0$. Substituting the value of $c = 0$ in Eq. (1.38), we get

$$V = -Ze^2/r \tag{1.39}$$

$$\text{The kinetic energy} = \frac{1}{2}mv^2 \tag{1.40}$$

$\therefore$ Total energy = KE + PE

$$\text{or} \quad E = \frac{1}{2}mv^2 + \left(-Ze^2/r\right)$$

$$\text{or} \quad E = \frac{1}{2}mv^2 - Ze^2/r \tag{1.41}$$

1.9.2 Velocity of an Electron

We know that in the stationary state, centrifugal force experienced by an electron equals the force of attraction between the nucleus and the electron.

$$\text{or}\quad \left(\frac{mv^2}{r}\right) = Ze^2/r^2$$

$$\text{or}\quad mv^2 = Ze^2/r$$

$$\text{or}\quad \frac{1}{2}mv^2 = Ze^2/2r \tag{1.42}$$

Substituting the value of 1/2 mv^2 in Eq. (1.41), we get

$$E = Ze^2/2r - Ze^2/r = -Ze^2/2r = -\frac{1}{2}mv^2 \tag{1.43}$$

$$\text{Now according to Bohr's postulate,}\quad mvr = \frac{nh}{2\pi} \tag{1.44}$$

and from Eq. (1.40),

$$\frac{1}{2}mv^2 = Ze^2/2r$$

$$\text{or}\quad mv^2 = Ze^2/r \tag{1.45}$$

From Eqs. (1.44) and (1.45), we can write

$$\frac{mv^2}{mvr} = \frac{Ze^2}{r}\cdot\frac{2\pi}{nh}$$

$$\text{or}\quad v = \frac{2\pi\ Ze^2}{nh} \tag{1.46}$$

Thus, the velocity v of an electron can be determined.

Equation (1.43) is $E_n = -1/2\ mv^2$, where E_n is the energy of electron within nth state

$$\therefore\quad E_n = -\frac{1}{2}mv^2 = -\frac{1}{2}m\left(\frac{2\pi\ Ze^2}{nh}\right)^2 \quad (\text{by putting the value of } v)$$

$$\text{or}\quad E_n = -\frac{2\pi^2 Z^2 e^4 m}{n^2 h^2} \tag{1.47}$$

where n is the quantum number.

1.9.3 Radius of the Orbit

The radius of the orbit can also be found out easily.

$$\because\quad mvr = nh/2\pi$$

$$\text{or} \quad mr\left(\frac{2\pi\ Ze^2}{nh}\right) = \frac{nh}{2\pi}$$

$$\text{or} \quad r\left(\frac{2\pi\ Ze^2 m}{nh}\right) = \frac{nh}{2\pi}$$

$$\therefore \quad r = \frac{(nh)(nh)}{(2\pi)(2\pi\ Ze^2 m)}$$

$$\therefore \quad r = \frac{n^2h^2}{4\pi\ Ze^2 m} \tag{1.48}$$

Now we shall find out the value of wave number $\overline{v}$. We know from Bohr's postulate that the energy radiated or absorbed from an atom must be a quantum of energy, which will be seen in the form of spectral lines having frequency υ. Let us suppose that the electron jumps from orbit 2 to orbit 1 and that the corresponding energies are E_2 and E_1, respectively. Replacing the value of $n = 2$ in Eq. (1.47), we get

$$E_2 = \frac{-2\pi^2 mz^2 e^4}{n_z^2 h^2} \tag{1.49}$$

$$\text{and} \quad E_1 = \frac{-2\pi^2 mz^2 e^4}{n_1^2 h^2} \tag{1.50}$$

$$\therefore \quad E_2 - E_1 = \Delta E = \frac{-2\pi^2 mz^2 e^4}{n_z^2 h^2} + \frac{2\pi^2 mz^2 e^4}{n_1^2 h^2}$$

$$\text{or} \quad \Delta E = \frac{2\pi^2 mz^2 e^4}{h^2}\left(\frac{1}{n_1^2} - \frac{1}{n_2^2}\right)$$

But $\Delta E = h\upsilon$

$$\therefore \quad hv = \frac{2\pi^2 mz^2 e^4}{h^2}\left(\frac{1}{n_1^2} - \frac{1}{n_2^2}\right)$$

$$\text{or} \quad h\overline{v}c = \frac{2\pi^2 mz^2 e^4}{h^2}\left(\frac{1}{n_1^2} - \frac{1}{n_2^2}\right)$$

$$\text{or} \quad \overline{v} = \frac{2\pi^2 mz^2 e^4}{h^3 c}\left(\frac{1}{n_1^2} - \frac{1}{n_2^2}\right) \tag{1.51}$$

For hydrogen, $Z = 1$

$$\therefore \quad \overline{v} = \frac{2\pi^2 me^4}{h^3 c}\left(\frac{1}{n_1^2} - \frac{1}{n_2^2}\right)$$

$$\text{or} \quad \overline{v} = \mathrm{R_H}\left(\frac{1}{n_1^2} - \frac{1}{n_2^2}\right) \tag{1.52}$$

where $\mathrm{R_H} = \dfrac{2\pi^2 me^4}{h^3 c}$ = Rydberg constant.

Equation (1.52) is known as the *Ritz combination formula*. This formula may be compared with Balmer's empirical Eq. (1.32). On comparison, we find that $n_1 = m$ and $n_2 = n$. When the values of m, π, e, h, and c are substituted in $R_H = (2\pi^2 me^4/h^3c)$, R_H is found to be 109677.6 cm^{-1}. This result is in excellent agreement with the Rydberg constant calculated by Balmer. This indicates that Bohr's model of an atom is correct.

1.9.4 Shortcoming of Bohr's Model

Sommerfeld and others proposed a concept that the electron does not revolve around the nucleus in circular orbits but it revolves in elliptical path, and hence, postulate 2 of Bohr was proved wrong.

BIBLIOGRAPHY

Atkins, P. and R. Friedman. 2007. *Molecular Quantum Mechanics*. New York: Oxford University Press
Pauling, L. and E.B. Wilson. 1935. *Introduction to Quantum Mechanics*. New York: McGraw-Hill Book Co. Inc.
Pilar, F.L. 1990. *Elementary Quantum Chemistry*. New York: McGraw-Hill
Rastogi, R.P. 1986. *An Introduction to Quantum Mechanics of Chemical Systems*. New Delhi: Oxford and IBH Publishing Co.
Stevens, B. 1962. *Atomic Structure and Valency*. London: Chapman and Hall Ltd.

Solved Problems

Problem 1. Calculate the average energy of an oscillator with frequency 10^{14}Hz at 1000k (Boltzmann constant $k = 1.38 \times 10^{-23}$JK^{-1}).

Solution: Given that frequency = 10^{14}Hz = 10^{14}s^{-1}, $k = 1.38 \times 10^{-23}$J K^{-1}, and $T = 1000$ K, find the average energy, $\bar{\epsilon} = ?$

We know that average energy

$$\bar{\epsilon}_H = \frac{h\nu}{e^{h\nu/kT} - 1}$$

$$\text{But } \frac{h\nu}{kT} = \frac{(6.627\times10^{-34}\ \text{J s})(10^{14}\ \text{s}^{-1})}{(1.38\times10^{-23}\ \text{J K}^{-1})(1000\ \text{K})}$$

$$= \frac{6.627\times10^{-20}}{1.38\times10^{-23}\ \text{J K}^{-1}} = \frac{6.627\times10^{-20}}{1.38\times10^{-20}} = \frac{6.627}{1.38} = 4.8$$

$$\therefore\quad \bar{\epsilon} = \frac{(6.627\times10^{-34})(10^{14})}{e^{4.8}-1} = \frac{6.627\times10^{-20}}{e^{4.8}-1}$$

$$= \frac{6.627\times10^{-20}}{121.5-1} = \frac{6.627\times10^{-20}}{120.5}$$

$$= 0.05499\times10^{-20}$$

$$= 5.499\times10^{-22}\,\text{J} = 5.5\times10^{-22}\ \text{J}$$

Answer.

Problem 2. Show that Planck's radiation becomes identical with the Rayleigh–Jeans law if the size of the energy quantum is allowed to vanish or if the temperature is too high.

Solution: Planck's radiation law is $E_\nu = \frac{8\pi\nu^2}{c^3} \cdot \frac{h\nu}{e^{h\nu/kT} - 1}$.

At low frequency and high temperature

$$\frac{hv}{kT} \ll 1$$

$$\therefore \quad \bar{\in} = \frac{hv}{e^{hv/kT} - 1} = \frac{hv}{1 + hv/kT - 1} = kT$$

$$\therefore \quad E_v = \frac{8\pi\ v^2}{c^3} \cdot kT$$

which is the Rayleigh–Jeans equation. This can also be proved by considering the λ-containing equation given by Planck.

Hence,

$$E_\lambda \mathrm{d}\lambda = \frac{-8\pi hc}{\lambda^5} \cdot \frac{\mathrm{d}\lambda}{e^{hc/\lambda kT} - 1}$$

$$= \frac{-8\pi hc}{\lambda^5} \cdot \frac{\mathrm{d}\lambda}{\left(1 + \dfrac{hc}{\lambda\ kT} + \ldots\right)} = \frac{-8\pi hc}{\lambda^5} \cdot \frac{\mathrm{d}\lambda}{hc/\lambda\ kT}$$

$$\therefore \quad E_\lambda = \frac{-8\pi hc}{\lambda^5} \cdot \frac{\pi\ kT}{hc} = \frac{-8\pi kT}{\lambda^4} = \frac{c}{\lambda^4} \qquad [\text{where} \quad c = (-8\pi kT)]$$

which is the *Rayleigh–Jeans law.* Answer.

Problem 3. Starting from Planck's distribution law for the energy density, derive Wien's displacement law, $\lambda_{max} = c/T$, where c is a constant and λ_{max} is the maximum wavelength at a given temperature.

Solution: Planck's distribution law is

$$E_\lambda = \frac{8\pi hc}{\lambda^5} \cdot \frac{1}{e^{hc/\lambda kT} - 1}$$

$$= \left(\frac{8\pi hc}{\lambda^5}\right) \cdot \left(\frac{1}{e^{hc/\lambda kT} - 1}\right)$$

$$= \left(\frac{\alpha}{\lambda^5}\right)\left(e^{hc/\lambda kT} - 1\right)^{-1} \qquad (\text{where } \alpha = 8\pi hc)$$

On differentiation with respect to λ, we get

$$\left(\frac{\mathrm{d}E_\lambda}{\mathrm{d}\lambda}\right) = -5\lambda^{-6} \cdot \alpha\left(\frac{1}{e^{\frac{hc}{\lambda kT}} - 1}\right) + \left(\frac{\alpha}{\lambda^5}\right)\left(e^{\frac{hc}{\lambda kT}} - 1\right)^{-2} (-1)\left(e^{hc/\lambda kT}\right)\left(\frac{hc}{kT} \cdot \frac{1}{\lambda^2}\right)(-1)$$

$$= \left(\frac{-5\alpha}{\lambda^6}\right)\left(\frac{1}{e^{hc/\lambda\ kT} - 1}\right) + \left(\frac{\alpha}{\lambda^7}\right)\frac{1}{\left(e^{hc/\lambda kT} - 1\right)^2}\left(e^{hc/\lambda kT}\right)\left(\frac{hc}{kT}\right)$$

$$= \frac{-5\alpha}{\lambda^6\left(e^{hc/\lambda kT} - 1\right)} + \frac{\alpha}{\lambda^7}\left(\frac{hc}{kT}\right)\frac{e^{hc/\lambda kT}}{\left(e^{hc/\lambda kT} - 1\right)^2}$$

Assuming that $hc/\lambda kT >> 1$, thus in the above equation, 1 is neglected.

$$\frac{dE_\lambda}{d\lambda} = \frac{-5\alpha}{\lambda^6 e^{hc/\lambda kT}} + \frac{\alpha}{\lambda^7} \cdot \frac{hc}{kT} \cdot \frac{e^{hc/\lambda kT}}{\left(e^{hc/\lambda kT}\right)^2}$$

$$= \frac{-5\alpha}{\lambda^6 e^{hc/\lambda kT}} + \frac{\alpha}{\lambda^7} \cdot \frac{hc}{kT} \cdot \frac{1}{e^{hc/\lambda kT}}$$

$$= \frac{\alpha}{e^{hc/\lambda kT}} \left[\frac{-5}{\lambda^6} + \frac{hc}{\lambda^7 kT} \right]$$

At maxima, $\frac{dE_\lambda}{d\lambda} = 0$

$$\therefore \quad \frac{\alpha}{e^{hc/\lambda kT}} \left[\frac{-5}{\lambda^6} + \frac{hc}{\lambda^7 kT} \right] = 0$$

$$\text{But} \quad \frac{\alpha}{e^{hc/\lambda kT}} = \frac{8\lambda\, hc}{e^{hc/\lambda kT}} \neq 0$$

$$\therefore \quad \frac{-5}{\lambda^6} + \frac{hc}{\lambda^7 kT} = 0$$

$$\text{or} \quad \frac{5}{\lambda^6} = \frac{hc}{\lambda^7 kT}$$

$$\text{or} \quad 5 = \frac{hc}{\lambda\, kT}$$

$$\therefore \quad 5\lambda\, kT = hc$$

$$\text{or} \quad \lambda \cdot T = \frac{hc}{5k} = C$$

But we have considered $\lambda = \lambda_{\max}$

$$\therefore \quad \lambda_{\max} T = C$$

$$\therefore \quad \lambda_{\max} = C/T, \text{ where, } C = \frac{hc}{5k} = 2.88 \times 10^{-3} \text{ mK}$$

This is Wien's law. Answer.

Problem 4. Calculate $\lambda_{\max}$ at 727°C and with the use of $\lambda_{\max}$, find Planck's constant.

Solution: Given that $t = 727$°C

$$\therefore \quad T = (727 + 273) \text{ K} = 1000 \text{ K}$$

$$\lambda_{\max} = ?$$

We know from Wien's law that $\lambda_{max} = C/T$

$$= \frac{2.880 \times 10^{-3} \text{ mK}}{1000} \quad (\because C = 2.880 \times 10^{-3} \text{ mK})$$

$$= 2.880 \times 10^{-6} \text{ mK}$$

$$= 2880 \times 10^{-9} \text{ mK} = 2880 \text{ nm}$$

We also know that $\lambda_{max} = \dfrac{hc}{5kT}$

or 2880×10^{-9}mK

$$= \frac{h \times 3.0 \times 10^{8} \text{ m s}^{-1}}{5 \times 1.38 \times 10^{-23} \cdot 1000 \text{ J K}^{-1}}$$

$$\therefore \quad h = \frac{2880 \times 10^{-9} \text{ m K} \times 5 \times 1.38 \times 10^{-23} \cdot 1000 \text{ J K}^{-1}}{3.0 \times 10^{8} \text{ m s}^{-1}}$$

$$= \frac{2.880 \times 5 \times 1.38 \times 10^{-26}}{3.0 \times 10^{8}}$$

$$= \frac{2.880 \times 5 \times 1.38 \times 10^{-34} \text{ J s}}{3}$$

$$= 6.624 \times 10^{-34} \text{ J s}$$

$$\therefore \quad h = 6.624 \times 10^{-34} \text{ J s}$$ Answer.

Problem 5. The energy radiated by the sun shows a peak at 480 nm. Assuming that the sun behaves as a black body, calculate the temperature of its surface.

Solution: Given that $\lambda_{max} = 480$ nm

$$T = ?$$

$$\frac{hc}{5k} = 2.88 \times 10^{-3} \text{ mK}$$

$$T = \frac{hc}{5k\ \lambda_{max}}$$

But, $\dfrac{hc}{5k} = 2.88 \times 10^{-3}$ mK

$$\therefore \quad T = \frac{2.88 \times 10^{-3} \text{mk}}{480 \text{ nm}}$$

$$= \frac{2.88 \times 10^{-3} \text{ mK}}{480 \times 10^{-9} \text{ m}}$$

$$= \frac{2.88}{480} \times 10^6 \text{ K}$$

$$= 0.006 \times 10^6 \text{ K}$$ Answer.

$$= 6 \times 10^3 \text{ K} = 6000 \text{ K}$$

Problem 6. Given that the work function for sodium metal is 1.82 eV, calculate the threshold frequency for sodium.

Solution: Given that

$$w = 1.82 \text{ eV}$$

$$= (1.82 \text{ eV})\left(\frac{1.602 \times 10^{-19} \text{ J}}{\text{eV}}\right)$$

$$= 2.92 \times 10^{-19} \text{ J}$$

But we know that $w = h\upsilon_0$

$$\therefore \; \nu_0 = \frac{w}{h} = \frac{2.92 \times 10^{-19} \text{ J}}{6.627 \times 10^{-34} \text{ J}}$$

$$= 4.40 \times 10^{14} \text{ Hz}$$ Answer.

Problem 7. When Li is irradiated with light, the kinetic energy of the ejected electron is 2.935×10^{-19} J for $\lambda = 300$ nm and 1.280×10^{-19} J for $\lambda = 400$ nm. Find (a) Planck's constant, (b) the threshold frequency, and (c) the work function of Li from this data.

Solution: Given data are $\lambda_1 = 300$ nm, $KE_1 = 2.935 \times 10^{-19}$ J, $KE_2 = 1.280 \times 10^{-19}$ J, and $\lambda_2 = 400$ nm

$$h = ?; \; \nu_0 = ?; \; w = ?$$

a. we know that

$$KE_1 = h\nu_1 - h\nu_0$$

$$KE_2 = h\nu_2 - h\nu_0$$

$$\therefore \; KE_1 - KE_2 = h(\nu_1 - \nu_2) = hc\left(\frac{1}{\lambda_1} - \frac{1}{\lambda_2}\right)$$

$$\text{or} \; \left(2.935 \times 10^{-19} - 1.280 \times 10^{-19}\right) J = hc\left(\frac{1}{300} - \frac{1}{400}\right)$$

$$\text{or} \; 1.655 \times 10^{-19} \; J = \frac{\text{h}.3.0 \times 10^8 \text{ ms}^{-1}}{10^{-19}} m \left(\frac{1}{300} - \frac{1}{400}\right)$$

$$h(3 \times 10^8 \times 10^9)\left(\frac{1}{1200}\right)$$

$$h\left(\frac{3 \times 10^{15}}{12}\right)$$

$$h\left(0.25 \times 10^{15}\right) = \left(25 \times 10^{13}\right)h$$

$$\therefore \quad h = \frac{1.655 \times 10^{19}\ \text{J}}{25 \times 10^{13}\ \text{s}^{-1}} = 0.0662 \times 10^{-32}\,\text{J s}$$

$$= 6.62 \times 10^{34}\,\text{J s Answer}$$

b. Now, $KE_1 = h\upsilon_1 - h\upsilon_0 = h(\upsilon_1 - \upsilon_0)$

$$\text{or} \quad 2.935 \times 10^{-19}\ J = 6.62 \times 10^{-34}\ \text{J s}\left(\frac{c}{\lambda_1} - \nu_0\right)$$

$$\text{or} \quad \frac{2.935 \times 10^{-19}\ \text{J}}{6.62 \times 10^{-34}\ \text{J s}} = \left(\frac{3 \times 10^{8}\ \text{m s}^{-1}}{300 \times 10^{-9}\ \text{m}} - \nu_0\right)$$

$$\text{or} \quad 0.4433 \times 10^{15} = \left(\frac{10^{8}\ \text{m s}^{-1}}{100 \times 10^{-9}\ \text{m}} - \nu_0\right)$$

$$\text{or} \quad 0.4433 \times 10^{15} = \left(\frac{10^{8}}{10^{-7}} - \nu_0\right)$$

$$= \left(10^{15} - \nu_0\right)\ \text{Hz}$$

$$\text{or} \quad \nu_0 = \left(10^{15} - 0.4433 \times 10^{15}\right)\ \text{Hz}$$

$$= 10^{15}(1 - 0.4433)\ \text{Hz}$$

$$= 0.5567 \times 10^{15}\ \text{Hz} = 5.567 \times 10^{14}\ \text{Hz} \qquad \text{Answer.}$$

c. Work function $= w = h\upsilon_0$

$$= \left(6.627 \times 10^{-34}\ \text{J s}\right)\left(5.567 \times 10^{14}\right)\ \text{s}^{-1}$$

$$= 6.627 \times 5.567 \times 10^{-20}\ \text{J}$$

$$= 36.8925 \times 10^{-20}\ \text{J}$$

$$= \frac{36.8925 \times 10^{-20}}{1.602 \times 10^{-19}}\ \text{eV} \quad \left[\because\ 1\ \text{eV} = 1.602 \times 10^{-19}\ \text{J}\right]$$

$$= 23.209 \times 10^{-1}\ \text{eV} = 2.3209\ \text{eV Answer}$$

Problem 8. The threshold wavelength of tungsten is 2300 A. What wavelength light must be used for the electrons to be ejected with a maximum energy of 1.5 eV?

Solution: Given that $E = 1.5\text{eV} = 1.5 \times 1.602 \times 10^{-19}\text{J}$ and $\lambda_0 = 2300\ \text{A} = 23 \times 10^{-8}\text{m}$

$$\lambda = ?$$

Kinetic energy = KE = $h(\upsilon - \upsilon_0)$

$$E = hc\left(\frac{1}{\lambda} - \frac{1}{\lambda_0}\right)$$

$$\text{or} \quad \frac{E}{hc} = \frac{1}{\lambda} - \frac{1}{\lambda_0}$$

$$\therefore \frac{1}{\lambda} = \frac{E}{hc} + \frac{1}{\lambda_0}$$

$$= \frac{1.5 \times 1.602 \times 10^{-19}\ \text{J}}{\left(6.627 \times 10^{-34}\ \text{J s}\right)\left(3 \times 10^{8}\ \text{m s}^{-1}\right)} + \frac{1}{23 \times 10^{-8}\ \text{m}}$$

$$= \frac{2.403 \times 10^{-19}\ \text{J}}{19.881 \times 10^{-26}\,\text{J m}} + \frac{1}{23 \times 10^{-8}\ \text{m}}$$

$$\text{or} \quad \frac{1}{\lambda} = 0.1208 \times 10^{-7}\ \text{m}^{-1} + \frac{1}{23 \times 10^{-8}\ \text{m}}$$

$$= 0.1208 \times 10^{-7}\ \text{m}^{-1} + 0.0434 \times 10^{8}\ \text{m}^{-1}$$

$$= 10^{7}(0.1208 + 0.434)\ \text{m}^{-1} = 0.5548 \times 10^{7}\ \text{m}^{-1}$$

$$\therefore \quad \lambda = \frac{1}{0.5548 \times 10^{7}\ \text{m}^{-1}} = 1.802 \times 10^{-7}\ \text{m}$$

Answer.

Problem 9. Calculate the number of photons from the given line of mercury ($\lambda = 4961 \times 10^{-10}$m) required to perform 1 J of work.

Solution: Given that $\lambda = 4961 \times 10^{-10}$m, $c = 3 \times 10^{8}$m s^{-1}, and $h = 6.627 \times 10^{-34}$J s

$$n = \text{number of photon} = ?$$

Since the energy of a photon is $h\upsilon$

$$\therefore \quad \text{Energy of } n \text{ photon} = nh\nu$$

$$\therefore \quad nh\nu = 1\ \text{J}$$

$$\text{or} \quad \frac{nhc}{\lambda} = 1\ \text{J}$$

$$\therefore \quad n = \frac{1\text{J}}{hc/\lambda} = \frac{\lambda \times 1\ \text{J}}{hc}$$

$$= \frac{4961 \times 10^{-10}\ \text{m} \times 1\ \text{J}}{6.627 \times 10^{-34}\ \text{J s} \times 3 \times 10^{8}\ \text{m s}^{-1}}$$

$$= \frac{4961 \times 10^{-10}}{6.627 \times 3 \times 10^{-26}} = 249.5 \times 10^{16} = 2945 \times 10^{15}$$

Answer.

Problem 10. Calculate the energy of a quantum of light of wavelength $\lambda = 5.4 \times 10^{-5}$cm and express it in eV. If such a light falls on calcium with a work function 1.9 eV, find the retarding potential to stop the most energetic electrons.

Solution: Given that $\lambda = 5.4 \times 10^{-5}$cm and $c = 3 \times 10^{10}$cm.s^{-1}, the energy of one quantum of light

$$= h\nu = \frac{hc}{\lambda} = \frac{6.62 \times 10^{-27} \times 10^{10} \times 3 \text{ cm s}^{-1}}{5.4 \times 10^{-5} \text{ cm}} \text{ ergs}$$

$$= \frac{3.677 \times 10^{-12}}{1.6 \times 10^{-12}} \text{ eV} = 2.298 \text{ eV}$$

Now let V = stopping potential

$$w = 1.9 \text{ eV} = 1.9 \times 1.6 \times 10^{-12} \text{ ergs}$$

Energy of electron = $h\upsilon - h\upsilon_0 = Ve$

$$\therefore \quad \text{eV} = h\nu - w$$

$$V = \frac{h\nu - w}{e}$$

$$= \frac{2.298 \times 1.6 \times 10^{-12} - 1.9 \times 1.6 \times 10^{-12}}{4.8 \times 10^{-10}}$$

$$= \frac{1.6 \times 10^{-2}}{4.8 \times 10^{-10}}(2.298 - 1.9)$$

$$= \frac{1}{3} \times 10^{-2} \times 0.398 \times 300 \text{ Volts}$$

$$\therefore \quad V = 0.398 \text{ Volt}$$

Answer.

Problem 11. A photon of energy 1.02 MeV is scattered through 90° by a free electron. Calculate the energy of photon and electron after interaction.

Solution: Given that energy of photon = 1.02 MeV and $\theta = 90°$, we know that

$$\Delta\lambda = \left(\lambda^1 - \lambda\right) = \frac{h}{mc}(1 - \cos\theta)$$

$$= \frac{6.627 \times 10^{-34} \text{ J s}}{9.1 \times 10^{-31} \times 3 \times 10^{8}}(1 - \cos 90°) = 2.42 \times 10^{-12} \text{ m}$$

Change in the energy of photon = ΔE

$$\therefore \quad \Delta E = h \cdot \Delta\nu$$

$$= \frac{6.627 \times 10^{-34} \times 3 \times 10^{8} \text{ m}}{2.42 \times 10^{-12}} = 0.51 \text{ MeV}$$

The energy is transferred to the free electron which in turn will be the KE of the recoil electron.

$\therefore$ Energy of photon after interaction

$= \left(1.02 - 0.51\right)$ MeV

$= 0.51$ MeV

Answer.

Problem 12. In the Compton experiment, a beam of X-rays is scattered by loosely bound electrons at 45° to the direction of the beam. The wavelength of the scattered X-rays is 0.22 A. What is the wavelength of the incident beam? (Given *h*/*mc* = 0.02426 Å.)

Solution: It is known to us that

$$\theta = 45°$$

$$\lambda' = 0.22 \text{ Å} = 0.22 \times 10^{-10} \text{ m}$$

$$h/mc = 0.02426 \text{ Å} = 2.426 \times 10^{-12} \text{m}$$

We know that

$$\lambda' - \lambda = h/mc(1 - \cos\theta)$$

$$\therefore \lambda = \lambda' - h/mc(1 - \cos\theta)$$

$$= \left(0.22 \times 10^{-10} \text{ m}\right) - 2.426 \times 10^{-12} \text{ m } (1 - \cos 45°)$$

$$= 0.22 \times 10^{-10} \text{ m} - 2.426 \times 10^{-12} \text{ m} \left(1 - \frac{1}{1.414}\right)$$

$$= 0.22 \times 10^{-10} \text{ m} - 2.426 \times 10^{-12} \times 0.293 \text{ m}$$

$$= 0.22 \times 10^{-10} \text{ m} - 0.71 \times 10^{-12} \text{ m}$$

$$= \left(0.22 \times 10^{-10} - 0.0071\right) \times 10^{-10} \text{ m}$$

$$= (0.22 - 0.0071) \times 10^{-10} \text{ m}$$

$$\therefore \quad \lambda = 0.213 \times 10^{-10} \text{ m}$$

Answer.

Problem 13. Calculate the wavelengths of the first few lines in the visible region of the hydrogen atomic spectrum using Balmer's formula.

Solution: Balmer's formula is $\bar{v} = 109{,}680\left(\frac{1}{2^2} - \frac{1}{n^2}\right)$,

where RH = 109,680

The first line will be obtained by substituting $n = 3$

$$\therefore \quad v = 109{,}680\left(\frac{1}{2^2} - \frac{1}{3^2}\right) \text{cm}^{-1}$$

$$= 109{,}680 \left(\frac{1}{4} - \frac{1}{9}\right) \text{cm}^{-1}$$

$$= 109{,}680 \left(\frac{9-4}{36}\right) \text{cm}^{-1}$$

$$= 109{,}680 \times \frac{5}{36} \text{cm}^{-1}$$

$$\bar{v} = \frac{1}{109{,}680 \times 5} \times 36 \text{ cm} = 0.0000656 \text{ cm}$$

$$= 6.56 \times 10^{-5} \text{ cm} = \frac{6.56 \times 10^{-5} \text{ nm}}{100 \times 10^{-9}}$$

$$= 656 \text{ nm}$$

The next line is obtained by substituting $n = 4$

$$\therefore \quad \bar{v} = 109{,}680\left(\frac{1}{2^2} - \frac{1}{4^2}\right) \text{cm}^{-}$$

$$= 109{,}680\left(\frac{1}{4} - \frac{1}{16}\right) \text{cm}^{-1}$$

$$= 109{,}680\left(\frac{4-1}{16}\right) \text{cm}^{-1}$$

$$= \frac{109{,}680 \times 3}{16} \text{cm}^{-1}$$

$$\therefore \quad \lambda = \frac{1}{v} = \frac{16}{109{,}680 \times 3} \text{cm}$$

$$= \frac{16 \times 10^{-2} \times 10^{+9}}{109{,}680 \times 3} \text{nm} = 0.0000486 \times 10^{7} \text{ nm}$$

$$= 4.86 \times 10^{-5} \times 10^{7} \text{ nm}$$

$$= 4.86 \times 10^{2} \text{ nm} = 486 \text{ nm}$$

Problem 14. Calculate the wavelength of the second line in the Paschen series and show that this line lies in the near infrared.

Solution: In case of Paschen, $m = 3$ and $n = 4,5,6 \ldots$
Thus, the second line in the Paschen series is given by $m = 3$ and $n = 5$.
We know that

$$\bar{v} = \text{R}_\text{H}\left(\frac{1}{m^2} - \frac{1}{m^2}\right)$$

$$= 109{,}680\left(\frac{1}{3^2} - \frac{1}{5^2}\right)$$

$$= 109{,}680\left(\frac{1}{9} - \frac{1}{25}\right) \text{cm}^{-1}$$

$$= 109{,}680\left(\frac{25-9}{9 \times 25}\right) \text{cm}^{-1}$$

$$= \frac{109{,}680 \times 16}{225} = 7799.46 \text{ cm}^{-1}$$

$$= 7.79946 \times 10^{-3} \text{ cm}^{-1}$$

$$\therefore \quad \lambda = \frac{1}{\nu} = \frac{1}{7.799 \times 10^{-3}} \text{ cm}$$

$$= 0.1282 \times 10^{3} \text{ cm} = 1.282 \times 10^{4} \text{ cm}$$

which belongs to the near-infrared region. Answer.

Questions on Concepts

1.
   a. What do you mean by black-body radiation?
   b. What are the qualities of a black body?
   c. What conclusions do you draw from the E_λ–λ graph?
2. Wien's radiation law is expressed as

   $W_\lambda = \frac{\alpha}{\lambda^5} e^{-\beta/\lambda T}$, where a and β are constants independent of λ and T.

   Differentiate this equation with respect to λ and find λ_{max}.
3. Prove that $\lambda_{max} \propto 1/T$.
4. The Rayleigh–Jeans radiation law is expressed as $E_\nu = \frac{8\pi\nu^2 kT}{c^3}$.

   How will you arrive at this equation?
5.
   a. What are the assumptions of Planck's radiation law?
   b. Prove that $E_\nu = \frac{8\pi\nu^2}{c^3} \cdot \frac{h\nu}{e^{h\nu/kT} - 1}$, where the terms have their usual significance.
6.
   a. Improving the condition that $h\nu >> kT$ and $e^{h\nu/kT} >> 1$ in Planck's radiation law, arrive at ultraviolet catastrophe.
   b. When $h\nu/kT << 1$, prove that $\bar{\epsilon} = kT$.
7. What is photoelectric effect? How did Einstein use the quantum theory to explain this effect?
8. What is threshold frequency? How is this frequency related to the frequency of incident radiation?
9.
   a. What observations are recorded from photoelectric effect?
   b. How will you find out Planck's constant and threshold frequency from Einstein photoelectric equation?
   c. What do you mean by stopping potential?
10.
    a. What do you mean by the Compton effect?
    b. Prove that $\Delta\lambda = h/mc(1 - \cos\theta)$.
11. What is the Compton effect? How is the Compton wavelength related to the incident wavelength?
12.
    a. Write down Balmer's empirical formula, which explains the existence of four lines.
    b. How will you arrive at $\bar{\nu} = R_H\left(\frac{1}{m^2} - \frac{1}{n^2}\right)$ from Balmer's empirical relation?

13.
   a. Write down Bohr's model of an atom after the use of Planck's quantum theory.
   b. Derive the Ritz combination formula.
14. Calculate the energy of the photon corresponding to $\lambda = 140\,\text{nm}$. [Ans. 1.42×10^{18}J]
15. A photon of energy 10 eV falls on molybdenum, the work function of which is 4.15 eV. Find the stopping potential (V). [Ans. 5.85 Volts]
16. The threshold wavelength for a metal is 3800×10^{-10}m. Calculate the maximum kinetic energy of the photoelectrons ejected when UV light of wavelength 2500×10^{-10}m falls on it. [Ans. 2.725×10^{-19}J, ≈ 1.7 eV]
17. Prove that in photoelectric effect from a metal surface, the maximum velocity of the photoelectrons is related to stopping potential by the equation $v_{max} = 5.923 \times 10^5 \times \sqrt{V_0}$, where v_{max} is in m/s and V_0 in volts. [Ans. $5.923 \times 10^5 \times \sqrt{V_0}\,\text{ms}^{-1}$]
18. Show that the maximum recoil energy of a free electron of rest mass m_0 when struck by a photon having wavelength λ is given by

$$E_{max} = \frac{2m_0c^2\lambda_c^2}{\lambda^2 + 2\lambda_e\lambda}$$

where λ_e is the Compton wavelength of the electron.
19. A beam of X-rays is scattered by loosely bound electron at 45° from the direction of the beam. The wavelength of the scattered X-rays is 0.22 Å. What is the wavelength of X-ray in the direct beam?
   [Hint: use $\lambda = \lambda' - h/mc(1 - \cos\theta)$] [Ans. 0.2129 Å.]
20. In the Compton experiment, a beam of X-rays having $\lambda = 0.0558$ nm is scattered through an angle of 45°. What is the wavelength of the scattered beam? [Ans. 0.565 nm]
21. X-rays having $\lambda = 0.085$ Å are scattered by carbon. At what angle will the scattered photons have a wavelength of 0.09 A? [Ans. 37.5°]
22. What is the wavelength (eV) of the line emitted when the atom comes back to its ground state? [1 eV = 1.6×10^{-12} ergs] [Ans. 1218 A]
23. The line of greatest wavelength of a particular series in the atomic spectrum of hydrogen is at 6563 Å. What is this series? $R_H = 10978\,\text{cm}^{-1}$. [Ans. Balmer's series]
24. A beam of electrons is used to bombard gaseous H atom. What is the minimum energy the electrons must have if the first member of the Balmer series corresponding to a transition from $n = 3$ to $n = 2$ state is to be emitted? [Ans. 2.98×10^{-19} J]

2

Wave–Particle Duality

In Chapter 1, we discussed about the black-body radiation, wherein we learnt about Wien's law, Rayleigh–Jeans law, and Planck's radiation law. In general, Wien's law and Rayleigh–Jeans radiation law can be arrived at using Planck's quantum hypothesis or Planck's radiation equation. Actually, Wien's law and Rayleigh–Jeans law are special cases of Planck's radiation law. We also studied the photoelectric effect, Compton effect, and atomic spectra in the first chapter. The photoelectric effect and Compton effect are vividly indicative of the corpuscular (particle) nature of visible light; X-radiation, interference, and diffraction phenomena are evidence of its wave nature. Therefore, it can be concluded that the motion of photons clearly obeys special laws in which corpuscular and wave characteristics are combined. In the same chapter, we dealt with the atomic spectrum of hydrogen. Analysis of hydrogen atom spectrum should convince us that light is emitted or absorbed in energy quanta, i.e., in particles, which we know as photons. It does not mean that we must revive the old controversy of the corpuscular versus the undulatory theories of light. On the contrary, we now know that both theories are part of a more general theory. In some instances, one theory applies, and in others, we must invoke the alternative.

Einstein explained the photoelectric effect by Planck's quantum theory, and he also handled the specific heat of solids considering the said theory. In this chapter, we shall focus on wave–particle duality and Heisenberg's uncertainty principle and their verification by experiments.

2.1 Dual Nature of Electron/de Broglie Wave

Keeping in view that light has dual nature, in 1924 de Broglie suggested that a small particle like electron would behave as both a particle and a wave. This is called the dual nature electron or matter wave or de Broglie wave.

We know that the energy of a photon can be written in terms of Planck's quantum theory, i.e.,

$$E = hv \tag{2.1}$$

But according to Einstein's theory of relativity, energy can be expressed as

$$E = mc^2 \tag{2.2}$$

where m is the mass of electron, c the velocity of light, h Planck's constant, and υ the frequency of radiation.

Equating Eqs. (2.1) and (2.2), we can write

$$hv = mc^2 \tag{2.3}$$

$$\text{or} \frac{hv}{c} = mc$$

Also since

$$\therefore \quad \frac{h}{\lambda} = mc$$

$$\therefore \quad \lambda = \frac{h}{mc} \tag{2.4}$$

If the particle of mass m possesses velocity v, then Eq. (2.4) can be rewritten as

$$\lambda = \frac{h}{mv} = \frac{h}{p} \tag{2.5}$$

where p is the momentum of the particle (here electron).

Equation (2.5) is called the *de Broglie equation.* It is clear from Eq. (2.5) that the wavelength of the particle is inversely proportional to its momentum, i.e.,

$$\lambda \propto \frac{1}{p}$$

Equation (2.5) clearly suggests the dual nature of electron, that is, corpuscular and wave characteristics since the equation involves mass m and wave length λ. de-Broglie also suggested that the idea of duality should be extended not only to radiation but also to all micro particles.

One important conclusion of Eq. (2.5) is that there exists stationary states, i.e., states of constant momentum as proposed by Bohr, which follow from the existence of line spectra.

Now think of the de Broglie equation $\lambda = h/mv$ and suppose that the velocity v of an electron depends on the potential drop V through which it passes.

If V = potential and e = electronic charge, v = velocity of the particle, then the kinetic energy $\frac{1}{2}mv^2$ will be equal to energy eV.

$$\therefore \quad eV = \frac{1}{2}mv^2 \tag{2.6}$$

$$\text{or } 2eV = mv^2$$

$$\text{or } mv^2 = 2eV$$

$$\therefore \quad v^2 = \frac{2eV}{m}$$

$$\therefore \quad v = \left(\frac{2eV}{m}\right)^{1/2} \tag{2.7}$$

Now $\lambda = h/mv = \dfrac{h}{m\left(\frac{2eV}{m}\right)^{1/2}}$,

on replacing the value of v in the de Broglie equation

$$\therefore \quad \lambda = \frac{h}{\left(\frac{2eV\ m^2}{m}\right)^{1/2}} = \frac{h}{\sqrt{2emV}} \tag{2.8}$$

From the known values of Planck's constant, charge, and mass, it is found that $\frac{\text{h}}{\sqrt{me}} \approx 10^{-8}$, as calculated below:

$$\frac{h}{\sqrt{me}} = \frac{6.626 \times 10^{-27}\ \text{ergs.sec}}{\sqrt{9.11 \times 10^{-28} g \times 4.8 \times 10^{-10}\ \text{e.s.u}}}$$

$$= \frac{6.626 \times 10^{-27}}{\sqrt{43.728 \times 10^{-38}}}$$
$$= \frac{6.626 \times 10^{-27}}{6.6127 \times 10^{-19}} \approx 1.002 \text{ Å}$$
$$= 1 \times 10^{-8} \text{ cm}$$

$$\therefore \quad \lambda = \frac{1}{\sqrt{2mV}} = \sqrt{\frac{1}{2V}} \times 10^{-8} \text{ cm}$$
$$= \sqrt{\frac{1 \times 300}{2V}} \times 10^{-8} \text{ cm}$$
$$= \sqrt{\frac{150}{V}} \times 10^{-8} \text{ cm} \tag{2.9}$$

where V is now the potential in volts after multiplying it with 300, which is a conversion factor from e.s.u to volt. The wave nature of the electron was shown experimentally by Davisson and Germer through the diffraction of electrons by crystal lattice. The results were similar, as observed in diffraction of X-rays by crystals.

2.2 Davisson and Germer's Experiment

It has been known for many years that a crystal can act as a diffraction grating for X-rays, because the spacing of atoms or molecules in a crystal has been found to be of the same order as the wavelength of the rays. Atom or molecule layers, therefore, act as reflecting planes and generate diffraction effects.

In 1927, Davisson and Germer experimentally showed that a stream of fast moving electrons is diffracted by crystals. An outline of the experimental set-up used by them is shown in Figure 2.1. In the figure, A is a tungsten filament, B is a nickel crystal, C is a Faraday cage, and G is a galvanometer.

A beam of electrons emitted from the heated tungsten filament passes through a known fall potential, and after this, the electrons are allowed to impinge on a single nickel crystal B. Finally, they are reflected into the opening of the Faraday cage. As the angle of incidence of the fast moving electrons on crystal was fixed, the accelerating potential varied gradually. The intensity of the reflected electrons is determined by the deflection observed in the galvanometer.

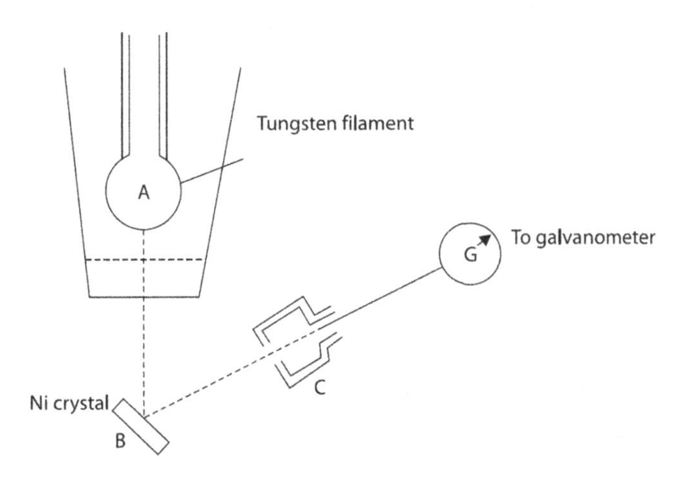

FIGURE 2.1 Diffraction of electrons by a crystal.

It was noted that the variation of intensity was irregular. When intensity versus square root of accelerating voltage ($\sqrt{V}$) graph was plotted (Figure 2.2), a series of almost equally spaced maxima was found. This type of behaviour was expected for a wave motion, such as X-rays, reflected from a crystal acting as diffraction grating. Since the wavelength is increased, the reflection maxima should be equally spaced when plotted against $\frac{1}{\lambda}$.

The results obtained are in agreement with equation $\lambda = h/\sqrt{2emV}$, which demands that the wavelength of the electron wave should be inversely proportional to the square root of the applied potential V, i.e., $\lambda \alpha \frac{1}{\sqrt{V}}$.

Davisson and Germer did another series of experiments in which they took a single crystal of nickel and exposed the (111) plane of the crystal. A beam of electrons was allowed to impinge on it at normal incidence. The crystal was rotated and the position of the Faraday cylinder was changed for the determination of intensities at various angles. At potential difference from 40 to 48 V, the curve was found to be smooth, whereas at 54 V, a distinct maximum spur appeared on the curve at 50° (Figure 2.3).

The selective reflection at 54 V of electrons at an angle 50° can be attributed to constructive interference. The bump at 54 volts offers evidence for the existence of electron waves. In case of Ni crystal (cubic), the edge length is 3.72 Å. Therefore, for the (111) plane the interplanar distance

$$\mathrm{d}_{hkl} = \frac{a}{\sqrt{h^2+k^2+l^2}} = \frac{3.72}{\sqrt{1^2+1^2+1^2}} = \frac{3.72}{\sqrt{3}} \quad [\mathrm{d}_{hkl} = \mathrm{d}_{111}]$$

$$= 2.1478 \text{ Å} \approx 2.15 \text{ Å}$$

$$= \text{Interplanar spacing}$$

Now we shall apply Bragg's law to calculate λ for first-order reflection, i.e., $n\lambda = D \sin\theta$, where D is the interplanar spacing

$$\text{or } \lambda = D \sin\theta \quad [\because \; n = 1]$$

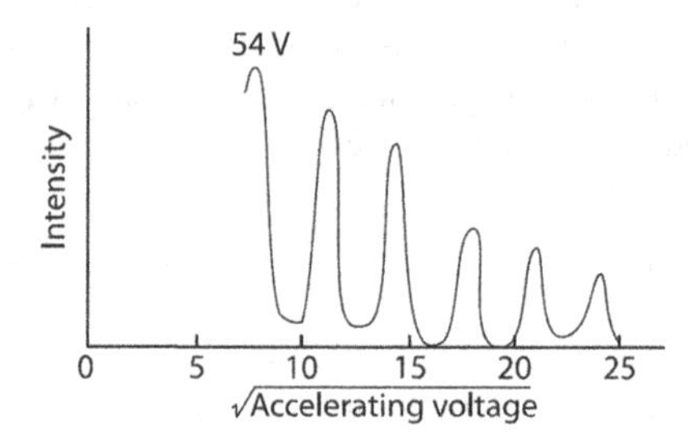

FIGURE 2.2 Intensity of electron diffraction vs $\sqrt{V}$ graph

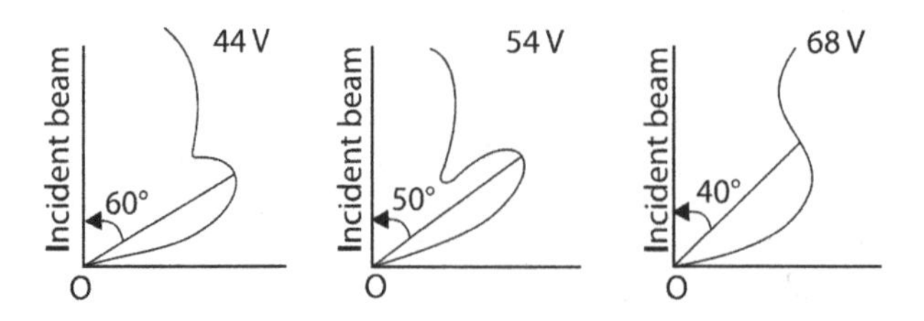

FIGURE 2.3 Intensity vs voltage graph.

$$\text{or } \lambda = 2.15 \sin 50° = 2.15 \times 0.776$$
$$= 1.668 \text{ Å} \approx 1.67 \text{ Å} \tag{2.10}$$

Further, we shall verify it from the de Broglie equation containing voltage, i.e.,

$$\lambda = \sqrt{\frac{150}{V}} \text{ Å}$$

Here $V = 54\,\text{V}$

$$\therefore \quad \lambda = \sqrt{\frac{150}{54}} \text{ Å} = 1.66 \text{ Å} \tag{2.11}$$

The values of λ in Eqs. (2.10) and (2.11) are in close agreement. This evidence shows that electrons have wave characteristics.

2.3 Quantisation of Angular Momentum

In Bohr's theory, it was arbitrarily assumed that the angular momentum of an electron is quantised, i.e., $mvr = nh/2\pi$.

This relation is easily derived on the basis that electrons have wave characteristics. For this purpose, wave motion of the electron is depicted in Figure 2.4.

If λ is the wavelength of the electron, the circumference of the orbit $2\pi r$ must be an integral multiplication of λ, i.e.,

$$2\pi r = n\lambda \tag{2.12}$$

But according to de Broglie's equation,

$$\lambda = \frac{h}{mv}$$

$$\therefore \quad 2\pi r = \frac{nh}{mv}$$

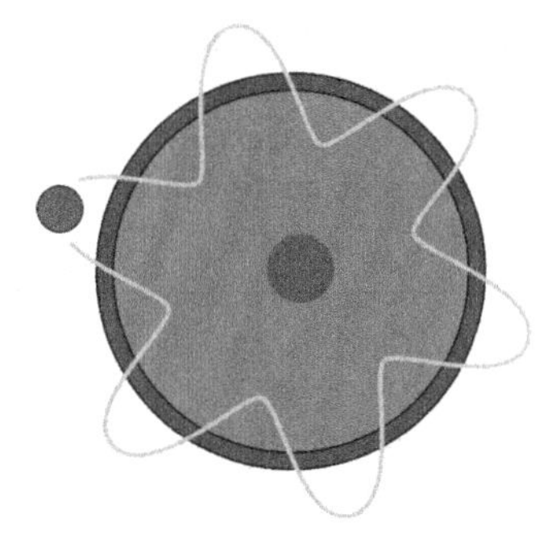

FIGURE 2.4 Wave motion of electron.

$$\text{or} \quad mvr = \frac{nh}{2\pi} \tag{2.13}$$

Thus, the angular momentum is quantised. In other words, we can say that the wave concept of electrons provides a simple interpretation of quantisation of angular momentum.

2.4 Heisenberg's Uncertainty Principle

A wave extends through space, and it is difficult to locate the exact position of an electron behaving as a wave at any given point of time. The dual nature of an electron as a particle and as a wave presents difficulty in locating the exact position and momentum at the same time. This difficulty was overcome by Heisenberg's uncertainty principle. In 1927, Heisenberg proposed his principle, according to which 'It is not possible to specify both the position and the momentum of a particle simultaneously with infinite precision'.

It is important to note that the measurement of particle position leads to loss of knowledge about particle momentum. Heisenberg's uncertainty principle may also be stated as follows.

The product of the simultaneous uncertainties in position (Δx) and momentum (Δp) of a particle can never be smaller than the value of Planck's constant h divided by 4π, i.e.,

$$\Delta x \cdot \Delta p \geq h/4\pi \tag{2.14}$$

$$\Rightarrow \Delta x \cdot m\Delta v \geq h/4\pi \Rightarrow \Delta x \cdot \Delta v \geq h/4\pi n \tag{2.15}$$

or in other words, we can state that the product of uncertainty in position (Δx) and uncertainty in momentum (Δp) of a particle is equal to or greater than $h/4\pi$, where Δx is the measure of the uncertainty in the variable x or position, Δp is the measure of the uncertainty in the variable p or momentum, and Δv is the measure of uncertainty in velocity.

The uncertainty principle may also be stated as follows. 'It is impossible to determine simultaneously the position and momentum of a particle with any desired accuracy'.

This statement will be clear by taking the following example.

Let us suppose that we want to locate the position of an electron with uncertainty $\Delta x = 10^{-13}$ m. The uncertainty in momentum will be calculated by using

$$\begin{aligned}\Delta p &= \frac{h}{4\pi\Delta x}\\ &= \frac{6.627\times10^{-34}\ \text{J s}}{= 4\pi\times10^{-13}\ \text{m}}\\ &= \frac{6.627\times10^{-34}\ \text{J s}}{12.57\times10^{-13}\ \text{m}} = 0.527\times10^{-21}\ \text{kg m s}^{-1}\end{aligned}$$

This value is negligible with regard to a macroparticle, but it is not negligible in systems having electrons and microscopic particles. It may be said from the values of Δx and Δp that if Δx is determined with more uncertainty, Δp will be measured with less uncertainty or Δp will be determined with accuracy. We can also find the uncertainty relation between energy and time.

2.5 Phase Velocity

If a plane simple harmonic wave moves in x-direction, it is mathematically represented by the equation as given below:

$$y = A\sin(\omega t - x/v) \quad \text{and} \quad \omega = 2\pi v \tag{2.16}$$

where A is the amplitude, v is the velocity of propagation of the wave in x-direction, and ω is the angular frequency, which is equal to $2\pi v$ (v = frequency of the wave).

The wave velocity is also known as phase velocity. The reason for this is given subsequently.

In Eq. (2.16), the phase ϕ of the wave at position x and time t is

$$\phi(x,t) = \omega(t - x/v) \tag{2.17}$$

Differentiating this equation with respect to t, we get

$$\frac{d\phi}{dt} = \omega\left(1 - \frac{1}{v}\frac{dx}{dt}\right)$$

For a point of constant phase $\frac{d\phi}{dt} = 0$

$$\therefore \quad 1 - \frac{1}{v}\left(\frac{dx}{dt}\right)_\phi = 0$$

$$\text{or} \quad \frac{1}{v}\left(\frac{dx}{dt}\right)_\phi = 1$$

$$\therefore \quad \left(\frac{dx}{dt}\right)_\phi = v$$

$\left(\frac{dx}{dt}\right)_\phi$ is the velocity with which a displacement of a given phase moves forward. Therefore, this quantity is termed as the phase velocity and it is normally denoted by v_ρ or ν_ϕ.

2.6 Group Velocity

Plane waves of different wavelengths move simultaneously in the same direction along a straight line through a medium in which the phase velocity $\nu_p = w/k$, where k is the propagation constant or phase constant or propagation number of the wave. $\nu_p = w/k$ of a wave depends on the wavelengths that the successive groups of the waves generate. These wave groups are also known as wave packets. Each wave group moves with a velocity v_g, called the group velocity. It should be noted that group velocity is different from the phase velocity of a wave.

2.7 Uncertainty Relation between Energy and Time

The uncertainty relation between energy and time is expressed as

$$\Delta E \cdot \Delta t \geq h/4\pi$$

This relation may be stated as follows: 'It is impossible to determine simultaneously the energy and time of a particle with any desired accuracy' or 'The product of uncertainty in energy (ΔE) is equal to or greater than $h/4\pi$', where ΔE is the uncertainty in the variable E and Δt is the uncertainty in the variable t.

The uncertainty relation between energy and time can be obtained by taking into consideration the motion of a wave packet. A particle in motion is indicated by a wave packet. The group velocity of the wave packet is equal to the velocity of the particle. Suppose that is:

v_g = group velocity of the wave packet

v_g = velocity of the particle along x-axis

Let Δx be the distance travelled by the wave packet in time Δt

$$\therefore \quad \Delta t = \frac{\Delta x}{v_g} = \frac{\Delta x}{v_x} \tag{2.18}$$

The kinetic energy of the particle is

$$E = \frac{p_x^2}{2m} \tag{2.19}$$

The differential of this equation gives uncertainty in E, i.e., ΔE, and the RHS of this equation becomes

$$\Delta E = \frac{2p_x \Delta p_x}{2m} = \frac{p_x}{m} \cdot \Delta p_x = \frac{mv_x}{m} \cdot \Delta p_x \tag{2.20}$$

$$\therefore \quad \Delta E \cdot \Delta t = \frac{mv_x}{m} \cdot \Delta p_x \cdot \frac{\Delta x}{v_x} = \Delta x \cdot \Delta p_x \geq h/4\pi$$

$$\therefore \quad \Delta E \cdot \Delta t \geq h/4\pi \tag{2.21}$$

This is the uncertainty relation between energy and time.

2.8 Experimental Evidence of Heisenberg's Uncertainty Principle

To confirm the correctness of Heisenberg's uncertainty principle, some experiments were carried out. These are described in the following sections.

2.8.1 Diffraction of Electrons through a Slit

Let us suppose that a particle of mass m moves along x-direction with a velocity v and consequently with momentum p. The particle passing through a slit of width Δx will undergo diffraction.

The error in the determination of the position of the particle is Δx. We shall now assume that the particle incident on the screen is corresponding to the position of the first minimum of the diffraction pattern. The particle's path difference = BF = $\lambda/2$, where λ = wavelength. It is obtained by drawing a perpendicular from A to BD. Now DE = Δp = uncertainty in momentum (Figure 2.5).

From the figure, it is clear that ∠DBE = α = ∠FAB (in Δ FAB)

$$\therefore \quad \sin \alpha = \frac{\Delta p}{p} \quad [\text{from } \Delta\text{DBE}] \tag{2.22}$$

$$\text{and from } \Delta \text{ BAF, } \sin \alpha = \frac{\lambda}{2\Delta x} = \frac{\text{BF}}{\text{AB}} \tag{2.23}$$

From Eqs. (2.22) and (2.23), we have

$$\frac{\Delta p}{p} = \frac{\lambda}{2\Delta x}$$

$$\text{or} \quad \Delta x \cdot \Delta p \approx \frac{p\lambda}{2} = h/2 \tag{2.24}$$

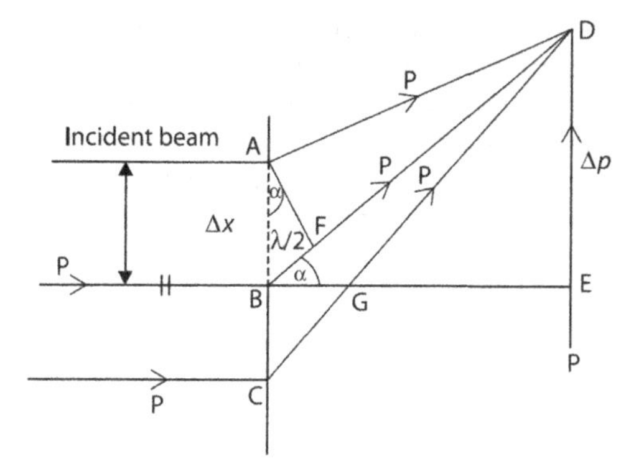

FIGURE 2.5 Thought experiment.

The probability of an electron approaching the centre of the pattern is the greatest. Hence, Eq. (2.24) gives the maximum uncertainty in Δp for a given value of uncertainty in position Δx.

Therefore, Eq. (2.24) is inconsistent with the uncertainty relation.

$$\therefore \quad \Delta x \cdot \Delta p \geq h/4\pi$$

Thus, Heisenberg's uncertainty principle is correct.

2.8.2 Gamma Ray Microscope Thought Experiment

This thought experiment was first proposed by Heisenberg. It is usually called γ-ray microscope experiment. Let us suppose that we want to locate the position of the electron by means of a microscope. It is known to us that the radius of an atom is of the order of 10^{-10}m. Therefore, to determine the position of an electron with an uncertainty of approximately 10% of the radius of an atom, we must use radiation having a wavelength of the order of 10^{-12}m, i.e., 0.01 Å. It clearly gives an idea that the electron must be illuminated with γ-rays having wavelength 0.01 Å (Figure 2.6).

Now, let υ be the frequency radiation and λ be the wavelength of incident γ-rays. Then, the momentum of the incident rays photon will be $= \dfrac{hv}{c} = h/\lambda$

It should be kept in mind that at least one photon should be scattered by the electron into the microscope to keep the electron visible. In this process, the frequency and wavelength of the scattered photon are varied and the electron suffers a Compton recoil because of momentum gain.

Let 2α be the angle subtended at the electron by the diameter AB of the instrument's aperture. The scattered photon may enter the microscope along the surface of the cone, the semivertical angle of which is α. Suppose that the photon enters the microscope along OA.

Let υ' be the frequency of scattered photon and λ' be the wavelength of scattered photon.

$$\therefore \quad \text{Momentum of scattered photon} = \frac{hv'}{c} = h/\lambda'$$

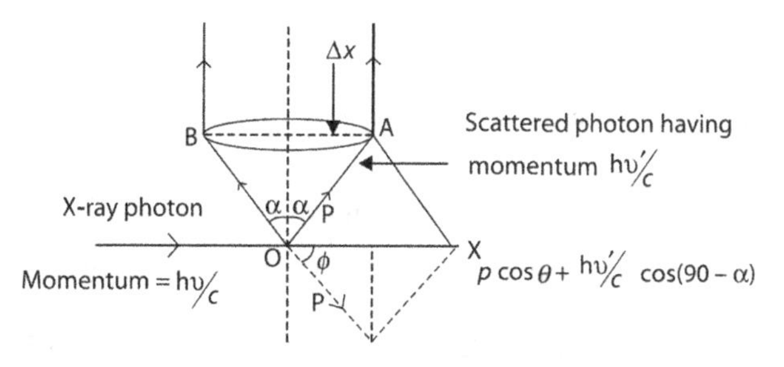

FIGURE 2.6 Diagrammatic operation of gamma ray microscope.

It is to be noted that the image of the electron formed by the microscope will be a diffraction pattern possessing a central bright disc surrounded by alternate dark and bright rings.

Since the electron may occupy position within the central bright disc, the uncertainty in the position of the electron is the diameter of the central disc. Then, the uncertainty in position is Δx.

In optics, we have learnt that the resolving power (RP) of an optical instrument is the distance between the peak intensity and the first minimum of the diffraction pattern and its related equation is

$$\text{RP} = \frac{\lambda'}{2 \sin \alpha}$$

In this case, $\text{RP} = \Delta x/2$

$$\therefore \quad \frac{\Delta x}{2} = \frac{\lambda'}{2 \sin \alpha}$$

$$\therefore \quad \Delta x = \frac{\lambda'}{\sin \alpha}$$

$$= \text{uncertainty in the position of the electron} \qquad (2.25)$$

Further, let θ be the direction of recoil and p be the gain of momentum by the electron. Resolving the momenta along OX, we shall get

$$\frac{hv}{c} = hv'/c \cos(90 - \alpha) + p \cos \theta$$

or $h/\lambda = h/\lambda' \sin \alpha + p \cos \theta$

$$\therefore \quad p \cos \theta = h/\lambda - h/\lambda' \sin \alpha \qquad (2.26)$$

Here, $p \cos \theta$ is the x-component of momentum p.

In Eq. (2.26), h/λ is accurately known and the scattered photon can enter the microscope along any other direction making an angle less than α, and the second term of the RHS of Eq. (2.26) represents uncertainty in momentum.

$$\therefore \quad h/\lambda' \sin \alpha = \Delta p \qquad (2.27)$$

From Eqs. (2.25) and (2.27), we get

$$\Delta x \cdot \Delta p = \frac{\lambda'}{2 \sin \alpha} \cdot h/\lambda' \sin \alpha$$
$$\approx h/2$$

$$\therefore \quad \Delta x \cdot \Delta p \approx h/2 \qquad (2.28)$$

Since Δp is maximum, this is consistent with $\Delta x \cdot \Delta p \geq h/4\pi$. This is Heisenberg's uncertainty principle.

2.8.3 Physical Significance of Uncertainty Principle

The following represent the physical significance of Heisenberg's uncertainty principle.

- From this principle, it is clear that an electron cannot exist within the nuclei of the atom.
- It is valid for microparticles.

- If one can measure the uncertainty in position of the microparticle accurately, he/she cannot measure the uncertainty in momentum (Δp) accurately and simultaneously.
- It helps to know the limit to the accuracy with which one can measure the frequency of the radiation emitted by an atom.

BIBLIOGRAPHY

Ball, D.W. 2003. *Physical Chemistry*. Pacific Grove, CA: Thomson Learning Inc.
Landau, L.D. and E.M. Lifshitz. 1977. *Quantum Mechanics*. New York: Pergamon Press.
Mortimer, R.G. 2008. *Physical Chemistry*. San Diego, CA: Academic Press.
Peleg, Y., R. Pnini and E. Zaarur. 2004. *Theory and Problems of Quantum Mechanics*. New Delhi: McGraw-Hill.
Schiff, L.I. 1955. *Quantum Mechanics*. New York: McGraw-Hill.

Solved Problems

Problem 1. An electron is accelerated by applying a potential difference of 1000 V. Calculate the de Broglie wavelength associated with it.

Solution: We know the λ–V relation, which is

$$\therefore \lambda = \sqrt{\frac{150}{V}} \times 10^{-8} \text{ cm}$$

$$= \sqrt{\frac{150}{1000}} \times 10^{-8} \text{ cm} \; [\because V = 1000 \text{ V}]$$

$$= \frac{3.873}{10} \times 10^{-8} \text{ cm}$$

$$= \frac{3.873}{10 \times 100} \times 10^{-8} \text{ m}$$

$$= 3.873 \times 10^{-11} \text{ m} = 38.73 \times 10^{-12} \text{ m}$$

Answer.

Problem 2. Calculate the de Broglie wavelength of an electron ($m = 9.1 \times 10^{-31}$kg) moving at 1% speed of light.

Solution: We know that

$$\lambda = \frac{h}{mv} = \frac{6.627 \times 10^{-34}}{9.1 \times 10^{-31} \times 3 \times 10^{6} \text{ m s}^{-1}} \qquad \left[1\% \text{ of } c = 3 \times 10^{6} \text{ m s}^{-1}\right]$$

$$= \frac{6.627 \times 10^{-3} \times 10^{-6}}{27.3} \text{ m}$$

$$= 0.243 \times 10^{-9} \text{ m} = 2.43 \times 10^{-10} \text{ m}$$

$$\therefore \quad \lambda = 2.43 \times 10^{-10} \text{ m}$$

Answer.

Problem 3. Through what potential must an electron initially at rest fall in order for λ to be 10^{-10}m or 0.1 nm?

Solution: Given that $\lambda = 10^{-10}$m, V =?
The momentum associated with this wavelength is

$$p = \frac{h}{\lambda} = \frac{6.627 \times 10^{-34} \text{ J s}}{10^{-10} \text{ m}}$$

$$\text{or} \quad p = \frac{6.627 \times 10^{-34} \text{ kg m}^2 \text{ s}^{-2} \text{ s}}{10^{-10} \text{ m}}$$

$$\text{or} \quad p = 6.627 \times 10^{-24} \text{ kg m s}^{-1}$$

$$\therefore \quad E = \frac{1}{2}mv^2 = p^2/2m = \frac{\left(6.627 \times 10^{-24} \text{ kg m s}^{-1}\right)^2}{2 \times 9.11 \times 10^{-31} \text{ kg}}$$

$$= \frac{(6.627)^2\left(10^{-48}\right)\text{kg}^2 \text{ m}^2 \text{ s}^{-2}}{18.22 \times 10^{-31} \text{ kg}}$$

$$= \frac{43.917. \times 10^{-48} \text{ kg}^2 \text{ m}^2 \text{ s}^{-2}}{1822 \times 10^{-31} \text{ kg}}$$

$$= 2.41 \times 10^{-17} \text{ kg m}^2 \text{ s}^{-2}$$

$$= 2.41 \times 10^{-17} \text{ J}$$

$$= \frac{2.41 \times 10^{-17} \text{ kg}^2 \text{ m}^2 \text{ s}^{-2}}{1.602 \times 10^{-19} \text{ J}} \text{ eV} = 1.50 \times 10^2 \text{ eV}$$

$$\therefore \quad E = 1.50 \times 10^2 \text{ eV} = 150 \text{ eV}$$

$\therefore$ Voltage applied $= 150$ eV Answer.

Problem 4. Calculate the de Broglie wavelength of a 60 kg man walking at the rate of 100 m min^{-1}.

Solution: Given that $m = 60\,\text{kg}$, $v = \frac{100}{60}\text{m s}^{-1}$

$$\therefore \quad \lambda = \frac{h}{mv} = \frac{6.627 \times 10^{-34} \text{ J s}}{60 \text{ kg} \times 100 \text{ m min}^{-1}} = \frac{6.627 \times 10^{-34} \text{ kg m}^2 \text{ s}^{-2}\text{s}}{60 \text{ kg} \times \frac{100}{60} \text{m s}^{-1}}$$

$\therefore \quad \lambda = 6.627 \times 10^{-36}$ m Answer.

Problem 5. An electron beam emerges from an accelerator with a KE of 100 eV. What is its de Broglie wavelength?

Solution: Given that KE = 100 eV, λ =?

$$\therefore \quad V = 100 \text{ V}$$

$$\therefore \quad \lambda = \frac{12.26}{\sqrt{V}} \text{ Å}$$

$= \frac{12.26}{\sqrt{100}} \text{ Å} = \frac{12.26}{10} \text{ Å} = 1.226 \text{ Å}$ Answer.

Problem 6. What is the wavelength of an H^+ ion having mass 1.7×10^{-27} kg moving with a velocity equal to $\frac{1}{100^{th}}$ of the velocity of light?

Solution: Given that $m = 1.7 \times 10^{-2}$kg

$$v = \frac{c}{100} = \frac{3\times10^{8}\ \text{m s}^{-1}}{100}$$

$$= 3\times10^{6}\ \text{m s}^{-1}$$

$$\therefore\ \ \lambda = \frac{\text{h}}{mv} = \frac{6.627\times10^{-34}\ \text{kg m}^{2}\ \text{s}^{-2}\ \text{s}}{1.7\times10^{-27}\ \text{kg}\times3\times10^{6}\ \text{m s}^{-1}}$$

$$= \frac{6.627}{17\times3}10^{-13}\ \text{m}$$

$$= \frac{6.627}{51}\times10^{-13}\ \text{m} = 1.3\times10^{-13}\ \text{m}$$ Answer.

Problem 7. The most rapidly moving valence electron in metallic sodium at absolute zero temperature has a KE of 3 eV. Show that its de Broglie wavelength is 7 Å.

Solution: Given that KE = 3 eV = $3 \times 1.6 \times 10^{-19}$ J, and $m = 9.1 \times 10^{-31}$ kg, λ =?
We know that momentum $p = (2mK.E)^{1/2}$

$$\therefore\ \ P = \left(2\times9.1\times10^{-31}\ \text{kg}\times3\times1.6\times10^{-19}\ \text{J}\right)^{1/2}$$

$$= (18.2\times4.8)^{1/2}\cdot\left(10^{-31}\times10^{-19}\right)^{1/2}\ \text{m s}^{-1}$$

$$= (87.36)^{1/2}\left(10^{-50}\right)^{1/2}\ \text{m s}^{-1}$$

$$= 9.346\times10^{-25}\ \text{m s}^{-1}$$

$$\because\ \ \lambda = \frac{h}{p} = \frac{6.627\times10^{-34}}{9.346\times10^{-25}} = 0.709\times10^{-9}\ \text{m}$$

$$= 7.09\times10^{-10}\,\text{m} = 7.09\ \text{Å} \approx 7\text{Å}$$ Answer.

Problem 8. The uncertainties in the position and velocity of a particle are 10^{-10} m and 5.27×10^{-24} m s^{-1}, respectively. Calculate the mass of the particle.

Solution: Given that $\Delta x = 10^{-10}$ m and $\Delta v = 5.27 \times 10^{-24}$ m s^{-1}, m =?
From Heisenberg's uncertainty principle, we know that

$$\Delta x\cdot\Delta v\cdot m \geq \frac{\text{h}}{4\pi}$$

$$\text{or}\ \ \Delta x\cdot\Delta vm \geq \frac{h}{4\pi}$$

$$\therefore\ \ m = \frac{h}{4\pi\cdot\Delta x\cdot\Delta v}$$

Substituting the given data in the above equation, we get

$$m = \frac{6.627 \times 10^{-34} \text{ kg m}^2 \text{ s}^{-2} \text{ s}}{4 \times 3.14 \times 10^{-10} \text{ m} \times 5.27 \times 10^{-24} \text{ m s}^{-1}}$$

$$= \frac{6.627 \times 10^{-34} \text{ kg m}^2 \text{ s}^{-2} \text{ s}}{4 \times 3.14 \times 5.27 \times 10^{-24} \text{ m}^2 \text{ s}^{-1}} = 0.1 \text{ kg}$$

Answer.

Problem 9. On the basis of Heisenberg's uncertainty principle, show that an electron cannot exist in the nucleus.

Solution: The radius of the nucleus is 10^{-14} m. Now if the electron were to exist within the nucleus, then the maximum uncertainty in its position would be 10^{-14} m.

$$\because \quad \Delta x = 10^{-14} \text{ m}$$

We know that

$$\Delta x \cdot \Delta v \cdot m \geq \frac{h}{4\pi}$$

$$\because \quad \Delta x \cdot \Delta v = \frac{h}{4\pi m}$$

$$\text{or } \Delta v = \frac{h}{4\pi m \Delta x}$$

$$= \frac{6.627 \times 10^{-34} \text{ kg m}^2 \text{ s}^{-1}}{4 \times 3.14 \times 9.11 \times 10^{-31} \text{ kg} \times 10^{-14} \text{ m}}$$

$$= \frac{6.627}{4 \times 3.14 \times 9.11} \times 10^{11} \text{ m s}^{-1}$$

$$= \frac{6.627}{114.42} \times 10^{11} \text{ m s}^{-1}$$

$$= 0.0579 \times 10^{11} \text{ m s}^{-1} = 5.79 \times 10^{9} \text{ m s}^{-1}$$

$\therefore$ We see that $\Delta v > c$ (velocity of light = 3×10^8 m s^{-1}).

$\therefore$ It is not possible.

Problem 10. An electron is moving with a velocity of 10^6 m s^{-1}. Suppose its position can be measured to 0.001 nm or 1% of the magnitude of a typical atomic radius. Compare the uncertainty in its momentum with the momentum of the electron itself.

Solution: Given that $\Delta x \approx 0.001$ nm = 1×10^{-12} m, we know that

$$\Delta x . \Delta p \approx \hbar \approx 1.0545 \times 10^{-34} \text{ J s}$$

$$\therefore \quad \Delta p \approx \frac{1.054 \times 10^{-34} \text{ J s}}{\Delta x} = \frac{1.054 \times 10^{-34} \text{ J s}}{10^{-12} \text{ m}}$$

$$\approx 1.06 \times 10^{-22} \text{ kg m s}^{-1}$$

The momentum is

$$p = mv = (9.11 \times 10^{-31}\ \text{kg})(10^{6}\ \text{m s}^{-1})$$
$$= 9.11 \times 10^{-31} \times 10^{6}\ \text{kg m s}^{-1}$$
$$= 0.911 \times 10^{-24}\ \text{kg m s}^{-1}$$
$$\approx 1 \times 10^{-24}\ \text{kg m s}^{-1}$$

Thus, it is clear that the uncertainty in momentum of the electron is about 100 times as great as momentum itself. In other words,

$$\Delta p \gg p$$ Answer.

Problem 11. Calculate the width of the spectral line, which results when a system in excited stage of lifetime 10^{-10}s makes a transition to the ground state.

Solution: From the uncertainty principle,

$$\Delta E \cdot \Delta t \approx \hbar$$

But $E = \underline{h}v$ $\therefore$ $\Delta E = h\Delta \upsilon$

$$\therefore \quad (h\Delta v)(\Delta t) = \hbar$$

$$\Delta v = \frac{\hbar}{h\Delta t} = \frac{1}{2\pi\Delta t}$$

According to question,

$$\Delta t \approx t$$

$$\therefore \quad \Delta v = \frac{1}{2\pi t} = \frac{1}{2(3.14) \times 10^{-10}\ \text{s}} \approx 1.6 \times 10^{9}\ \text{Hz}$$
$$= \frac{1.6 \times 10^{9}\ \text{s}^{-1}}{3 \times 10^{10}\ \text{cm s}^{-1}} = 0.053\ \text{cm}^{-1}$$

$\therefore$ The spectral line width = 0.053 cm s^{-1}, which is equal to the uncertainty in frequency. Answer.

Problem 12. Calculate the uncertainty in the position of a baseball (0.14 kg) thrown at 90 mph if we measure its velocity to a millionth of 1%.

Solution: Given that ν = 90 mph = 40 m s^{-1}

$$\therefore \quad \text{momentum} = p = mv$$
$$= (0.14\ \text{kg})(40\ \text{m s}^{-1}) = 5.6\ \text{kg m s}^{-1}$$

millionth of 1% of 5.6 kg m s^{-1} = 5.6×10^{-8}kg m s^{-1} = Δp.

$\therefore$ The uncertainty in the position of the baseball = $\Delta x = \dfrac{\text{h}}{\Delta p}$

$$= \frac{6.627 \times 10^{-34}\ \text{J s}}{5.6 \times 10^{-8}\ \text{kg m s}^{-1}}$$

$$= \frac{6.627 \times 10^{-26}\ \text{kg m}^{2}\ \text{s}^{-1}}{5.6\ \text{kg m s}^{-1}} = 1.2 \times 10^{-2}\ \text{m}$$

This is a completely inconsequential distance. Answer.

Problem 13. What is the uncertainty in momentum if we wish to locate an electron within an atom so that Δx is nearly 50 pm?

Solution: Given that $\Delta x = 50$ pm $= 50 \times 10^{-12}$ m, we know that

$$\Delta p \cdot \Delta x = \frac{h}{4\pi}$$

$$\therefore \quad \Delta p = \frac{h}{4\pi\Delta x} = \frac{6.627 \times 10^{-34} \text{ J s}}{12.56 \times 50 \times 10^{-12} \text{ m}}$$

$$= \frac{6.627 \times 10^{-34} \text{ J s}}{628 \times 10^{-12} \text{ m}}$$

$$= 0.0105 \times 10^{-22} \text{ kg m s}^{-1} = 1.05 \times 10^{-24} \text{ kg m s}^{-1}$$

$$\text{New } \Delta v = \frac{\Delta p}{m} = \frac{1.05 \times 10^{-24} \text{ kg m s}^{-1}}{9.11 \times 10^{-31} \text{ kg}} = 0.1097 \times 10^{7} \text{ m s}^{-1}$$

This is a very large uncertainty in the velocity. Answer.

Questions on Concepts

1. Derive the de Broglie equation or show that wavelength of a particle is inversely proportional to its momentum.
2. Show that for a microparticle $\lambda = \frac{h}{\sqrt{2emV}}$, where the terms have their usual significance.
3. Show that λ of a microparticle is equal to $\sqrt{\frac{150}{V}} \times 10^{-8}$ cm, where $V =$ potential in volts.
4. What does the experiment by Davisson and Germer prove? Describe the experiment in detail.
5. How will you show the quantisation of angular momentum on the basis of the de Broglie equation?
6.
   a. What is Heisenberg's uncertainty principle?
   b. What do you mean by phase velocity?
   c. Define group velocity.
7. Derive the uncertainty relation between energy and time.
8. Give the experimental evidence of Heisenberg's uncertainty principle.
9.
   a. Describe the gamma ray microscope thought experiment.
   b. Give the physical significance of uncertainty principle.
10. Calculate the de Broglie wavelength of an electron travelling with a velocity $3c/5$ ($c =$ velocity light).
11. Find the wavelength of the waves having energy equal to 1 MeV. [Ans. 0.0087 A]
12. An electron microscope uses 1.25 keV electrons. Find its ultimate resolving power on the assumption that this is equal to the wavelength of the electron, given that
    $m = 9.11 \times 10^{-31}$ kg $= 9.11 \times 10^{-28}$ g
    $e = 4.8 \times 10^{-10}$ e.s.u

$h = 6.627 \times 10^{-27}$ erg.sec [Ans. RP = 0.351 Å]

13. Find the energy of the neutron in eV when its de Broglie wavelength is 1 Å. [Ans. 8.12×10^{-2} eV]
14. Compute the de Broglie wavelength of 10^{11} keV neutron:
    Mass of neutron = 1.675×10^{-27}kg [Ans. $\lambda = 2.86 \times 10^{-15}$ m]
15. Calculate the de Broglie wavelength for an electron which has been accelerated through a potential difference of 28.8 V given that
    $h = 6.62 \times 10^{-27}$ erg.s
    $m = 9.1 \times 10^{-28}$ g
    $e = 4.8 \times 10^{-10}$ e.s.u [Ans. 2.4 Å]
16. If the uncertainty in position of an electron is 4×10^{-10}m, calculate its uncertainty in momentum. [Ans. $\Delta p = 1.65 \times 10^{-24}$kg m s]
17. Using the uncertainty relation $\Delta E \cdot \Delta t = \frac{h}{2\pi}$, calculate the time required for the atomic system to retain the rotation energy for line wavelength 6000 Å and width 10^{-4} Å. [Ans. 1.9×10^{-8} s]
18. An electron is confined to a box length 10^{-9}m. Calculate the minimum uncertainty in its velocity. [Ans. $\Delta \nu = 7.39 \times 10^{5}$ m/s]
19. An electron moving in the x-direction is accelerated by a potential difference of 10 ± 0.01 kV. Calculate the uncertainty in position. [Ans. $\Delta x = 1.928$ nm]
20. Suppose the position of the electron moving around the proton in the Bohr model of H atom can be located to within 0.01 nm, calculate the uncertainty in velocity (Δv) and comment on your result. [Ans. $\Delta \nu = 1.2 \times 10^{7}$m s^{-1}]
22. A cricket ball weighing 100 g is to be located within 1 nm. What is the uncertainty in its velocity. [Ans. $\Delta \nu = 5.27 \times 10^{-25}$m s^{-1}]

3

Mathematical Techniques

This chapter is intended to provide the students of chemistry with the mathematics they need in the course of study of quantum chemistry or quantum mechanics. The students will face difficulties in understanding the theoretical portion of quantum mechanics without the knowledge of mathematics. This chapter will help in developing a strong mathematics background for those who will continue into the mathematics of advanced theoretical chemistry. It is to be noticed that the subject of differential equations, for example, is no longer a series of trick solutions of abstract but the solutions and general properties of the differential equations, which the students will most frequently encounter in the description of our real chemical world. The other topics such as matrices, determinants, and vectors are equally important in quantum mechanics.

3.1 Differential Equations

An equation involving the dependent variable and independent variable and also the derivatives of the dependent variable is called a differential equation.

For example:

$$dy = (x + \sin x)dx \tag{3.1}$$

$$\frac{d^4x}{dt^4} + \frac{d^2x}{dt^2} + \left(\frac{dx}{dt}\right)^5 = e^t \tag{3.2}$$

$$y = \sqrt{x}\frac{dy}{dx} + k/dy/dx \tag{3.3}$$

$$k\frac{d^2y}{dx^2} = \left[1 + \left(\frac{dy}{dx}\right)^2\right]^2 \tag{3.4}$$

3.1.1 Ordinary Differential Equation

A differential equation comprising derivatives with respect to a single independent variable is known as an ordinary differential equation.

For example:

Equations (3.1–3.4) are the examples of ordinary differential equations.

3.1.2 Partial Differential Equation

A differential equation consisting of partial derivatives with respect to more than one independent variables is known as a partial differential equation.

For example:

i. $\frac{\partial^2 u}{\partial x^2} + \frac{\partial^2 u}{\partial y^2} + \frac{\partial^2 u}{\partial z^2} = 0$

ii. $\frac{\partial z}{\partial x} + \frac{\partial z}{\partial y} = c$

iii. $x\frac{\partial z}{\partial x} + y\frac{\partial z}{\partial y} = 2z$

iv. $\frac{\partial^2 y}{\partial t^2} = c^2 \frac{\partial^2 y}{\partial x^2}$

3.1.3 Order and Degree of a Differential Equation

3.1.3.1 Order

Order of a differential equation is defined as the order of the highest derivative in the differential equation.

For example:

i. $\left(\frac{dy}{dx}\right)^3 + \left(\frac{dy}{dx}\right)^2 + 3x = 0$ is the differential equation of the first order because the highest derivative of y, with respect to x is $\frac{dy}{dx}$.

ii. $\frac{d^2y}{dx^2} + 9y = e^x$

Here the order is 2 since the highest derivative of y, with respect to x, is $\frac{d^2y}{dx^2}$.

3.1.3.2 Degree

It is defined as the degree of the highest derivative which occurs in it, after the differential equation has been made free from fractions and radicals as far as the derivative is concerned.

For example:

i. $\left(\frac{dy}{dx}\right)^2 - 5\left(\frac{dy}{dx}\right) + 6y = 0$

Here the degree is 2 because $\left(\frac{dy}{dx}\right)$ has been squared.

ii. $y = x\left(\frac{dy}{dx}\right) + \left(\frac{dy}{dx}\right)^3$ is of the first order and the third degree.

iii. $\left(\frac{dy}{dx}\right)^2 - 10\frac{dy}{dx} + 21y = 0$ is of the first order and the second degree.

iv. $\frac{d^2y}{dx^2} - 7\frac{dy}{dx} + 10y = e^x$ is of the second order and the first degree.

3.1.4 Linear and Non-Linear Differential Equation

A differential equation is called linear if (a) every dependent variable and every derivative involved occur in the first degree only and (b) no products of dependent variables and/or derivatives occur.

For example:

i. $dy = (x + \sin x)dx$ and

ii. $\frac{\partial^2 u}{\partial x^2} + \frac{\partial^2 u}{\partial y^2} + \frac{\partial^2 u}{\partial z^2} = 0$

These are linear differential equations, but the following are the examples of non-linear differential equations:

iii. $\frac{d^4x}{dt^4}+\frac{d^2x}{dt^2}+\left(\frac{dx}{dt}\right)^5=e^t$

iv. $\frac{\partial^2 v}{dt^2}=k\left(\frac{d^3v}{dx^3}\right)^x$

3.1.5 General Solution, Particular Solution, and Arbitrary Constants

3.1.5.1 General Solution

A solution of a differential equation having independent arbitrary constant equal in number to the order of the differential equation is known as its general solution.

3.1.5.2 Particular Solution

A solution which is found out by giving particular values to the arbitrary constants in the general solution is termed as a particular solution or a particular integral.

3.1.5.3 Arbitrary Constants

The solution of a differential equation may consist of as many arbitrary constants as is the order of the differential equation.

For example:

$$y=x\frac{dy}{dx}+\frac{a}{dy/dx}$$

We shall observe that the general solution of the above equation is $y = mx + a/m$, where m is the arbitrary constant. By giving m a particular value, we shall get a particular solution.

3.1.6 Differential Equation of the First Order and the First Degree

3.1.6.1 Worked Out Examples

Example 1

Solve by simple substitution

$$(x+y-3)dx+(x+y+4)dy=0$$

Solution: ***Let us try the substitution***

$$z=x+y$$

Then $dz = dx + dy$.

Now the above equation takes the form

$$(z-3)dx+(z+4)dy=0$$

$$\Rightarrow(z-3)dx+(z+4)(dz-dx)=0$$

$$\Rightarrow zdx-3dx+zdz-zdx+4dz-4dx=0$$

$$\Rightarrow -7dx+zdz+4dz=0$$

On integration, we get

$$\Rightarrow -7\int dx + \int z dz + 4\int dz = 0$$

$$\Rightarrow -7x + z^2/2 + 4z = c'$$

$$\Rightarrow -14x + z^2 + 8z = c$$

$$\Rightarrow -14x + (x+y)^2 + 8(x+y) = c \quad [\text{By substituting } z = x + y]$$

$$\Rightarrow (x+y)^2 - 6x + 8y = c$$

$$\Rightarrow x^2 + 2xy + y^2 - 6x + 8y = c$$

Answer.

Example 2

Solve: $\frac{dy}{dx} + 1 = e^{x-y}$

Solution: ***Let x – y = z.***

Differentiating with respect to x, we have

$$1 - \frac{dy}{dx} = \frac{dz}{dx}$$

$$\therefore \quad 1 - \frac{dz}{dx} = dy/dx$$

Putting the value of $\frac{dy}{dx'}$, we get

$$1 - \frac{dz}{dx} + 1 = e^z$$

$$\text{or } \frac{dz}{dx} = 2 - e^z$$

$$\text{or } dx = \frac{dz}{2 - e^z}$$

On integration, we get

$$\int dx = \int \frac{dz}{2 - e^z}$$

$$\text{or } \int dx = \int \frac{e^{-z} dz}{2e^{-z} - 1}$$

Now put $2e^{-z} - 1 = t$ so that

$$-2e^{-z} dz = dt$$

$$\therefore \quad \int dx = \int \frac{dt}{-2t}$$

$$\text{or } x = -\frac{1}{2}\log t + c$$

$$\text{or } x - c = -\frac{1}{2}\log t$$

$$\text{or } 2c - 2x = \log t$$

$$\therefore t = e^{2c-2x}$$

$$\text{or } e^{2c-2x} = 2e^{-z} - 1 = 2e^{y-x} - 1$$

$$\text{or } e^{2c}\cdot e^{-2x} = 2e^{y}\cdot e^{-x} - 1$$

$$\text{or } e^{2c}\cdot e^{-2x}\cdot e^{2x} = 2e^{y}\cdot e^{-x}\cdot e^{2x} - e^{2x}$$

$$\text{or } e^{2c} = 2e^{y}\cdot e^{x} - e^{2x}$$

$$\text{or } e^{2c} = 2e^{x+y} - e^{2x}$$

$$\text{or } 2e^{x+y} = e^{2c} + e^{2x} = e^{2x} + \alpha\left[\text{where } \alpha = e^{2c}\right]$$ Answer.

Example 3

Solve: $(2x + y)^2 dx - 2dy = 0$

Solution*: Let 2x + y = z.*

On differentiation, we get

$$2dx + dy = dz$$

$$\therefore \quad dy = (dz - 2dx)$$

Substituting these values in the above expression, we get

$$z^2 dx - 2(dz - 2dx) = 0$$

$$\Rightarrow z^2 dx - 2dz + 4dx = 0$$

$$\Rightarrow (z^2 + 4)dx = 2z$$

$$\Rightarrow dx = \frac{2dz}{z^2 + 4}$$

On integration, we get

$$\int dx = 2\int \frac{dz}{z^2 + 4} = 2\int \frac{dz}{z^2 + 2^2}$$

$$\Rightarrow x = \frac{2}{2}\tan^{-1}\frac{z}{2} + c\left[\therefore \int \frac{dx}{x^2 + a^2} = \frac{1}{a}\tan^{-1} x/a\right]$$

$$= \tan^{-1} z/2 + c$$

$$\therefore \quad x = \tan^{-1}\left(\frac{2x + y}{2}\right) + c \quad [\text{on substituting } z = 2x + y]$$ Answer.

Example 4

Solve: $(x^2 + y^2)\,dx - 2xy\,dy = 0$

Solution: ***This equation may be expressed as*** $dy/dx = \dfrac{x^2 + y^2}{2xy}$.

[It is of the form $\dfrac{dy}{dx} = \dfrac{f(x,y)}{\phi(x,y)}$, where $f(x, y)$ and $\phi(x, y)$ are homogeneous functions of x, y and of the same degree, and hence, it is said to be homogeneous. Every homogeneous equation of the above type may/can be solved by putting $y = vx$, so that $\dfrac{dy}{dx} = v + x\ dv/dx$.]

Since the above equation is homogeneous,
Let $y = vx$

$$dy = v dx + x dv$$

Substituting these values for y and dy in the given equation, we get

$$\left(x^2 + v^2x^2\right)dx - 2vx^2\left(v dx + x dv\right) = 0$$

Collecting coefficients of dx and dv, we have

$$\left(x^2 + v^2x^2 - 2v^2x^2\right)dx - 2vx^3\ dv = 0$$

or $x^2(1 - v^2)dx - 2vx^3dv = 0$

Division by $x^2(1 - v^2)$ results

$$\frac{x^2\left(1 - v^2\right)dx}{x^3\left(1 - v^2\right)} - \frac{2vx^3dv}{x^3\left(1 - v^2\right)} = 0$$

$$\Rightarrow dx/x - \frac{2v dv}{1 - v^2} = 0$$

Integrating, we get

$$\int \frac{dx}{x} - \int \frac{2v dv}{1 - v^2} = 0$$

$$\Rightarrow \ln x - \left[-\ln\left(1 - v^2\right)\right] = c$$

$$\Rightarrow \ln x + \ln\left(1 - v^2\right) = \ln c$$

$$\Rightarrow x\left(1 - v^2\right) = c$$

Replacing v by its equal y/x, we get

$$x\left(1 - y^2/x^2\right) = c$$

$$\text{or } \frac{x\left(x^2 - y^2\right)}{x^2} = c$$

$$\text{or } x^2 - y^2 = xc$$

Answer.

Example 5

Solve: $(x^3 + y^3)\,\mathrm{d}x - xy^2\mathrm{d}y = 0$.

Solution: ***The above equation may be written as***

$$\frac{\mathrm{d}y}{\mathrm{d}x} = \frac{x^3 + y^3}{xy^2} \tag{3.5}$$

Since this equation is homogeneous, we shall put

$$y = vx$$

$$\therefore \quad \frac{\mathrm{d}y}{\mathrm{d}x} = v + x\frac{\mathrm{d}v}{\mathrm{d}x} \tag{3.6}$$

$\therefore$ Equation (3.5) becomes

$$\frac{\mathrm{d}y}{\mathrm{d}x} = \frac{x^3 + v^3x^3}{xx^2v^2} = \frac{x^3\left(1+v^3\right)}{x^3\left(v^2\right)}$$

$$= \frac{\left(v^3+1\right)}{v^2}$$

$$\Rightarrow v + \frac{x\mathrm{d}v}{\mathrm{d}x} = \frac{v^3+1}{v^2}$$

$$\Rightarrow \frac{x\mathrm{d}v}{\mathrm{d}x} = \frac{v^3+1}{v^2} - v = \frac{v^3+1-v^3}{v^2}$$

$$= \frac{1}{v^2}$$

$$\Rightarrow \frac{x\mathrm{d}v}{\mathrm{d}x} = \frac{1}{v^2}$$

$$\Rightarrow v^2\mathrm{d}v = \mathrm{d}x/x$$

Integrating both the sides, we get

$$\int v^2\mathrm{d}v = \int \mathrm{d}x/x$$

$$\Rightarrow \frac{v^3}{3} = \ln x + c \tag{3.7}$$

Substituting the value of v in Eq. (3.7), we get

$$\frac{\left(\frac{y}{x}\right)^3}{3} = \ln x + c$$

$$\frac{y^3}{3x^3} = \ln x + c \qquad \text{Answer.}$$

Example 6

Solve: $x\mathrm{d}y - y\mathrm{d}x - \sqrt{x^2+y^2}\ \mathrm{d}x = 0.$

Solution: ***From the given differential equation, we can write as***

$$x\mathrm{d}y = \left(y + \sqrt{x^2+y^2}\right)\mathrm{d}x$$

$$\therefore \quad \frac{\mathrm{d}y}{\mathrm{d}x} = \left(\frac{y+\sqrt{x^2+y^2}}{x}\right) \tag{3.8}$$

Equation (3.8) is a homogeneous equation; therefore, we shall write $y = vx$. On differentiation, we have

$$\frac{\mathrm{d}y}{\mathrm{d}x} = v + x\frac{\mathrm{d}v}{\mathrm{d}x}$$

Hence, the above problem reduces to

$$v + \frac{x\mathrm{d}v}{\mathrm{d}x} = \frac{vx + \sqrt{x^2 + v^2x^2}}{x}$$

$$= \frac{vx + x\sqrt{1+v^2}}{x}$$

$$\therefore \quad v + \frac{x\mathrm{d}v}{\mathrm{d}x} = \left(v + \sqrt{1+v^2}\right)$$

$$\therefore \quad x\frac{\mathrm{d}v}{\mathrm{d}x} = \sqrt{1+v^2}$$

$$\text{or} \quad \frac{\mathrm{d}v}{\sqrt{1+v^2}} = \frac{\mathrm{d}x}{x} \tag{3.9}$$

Integrating Eq. (3.9), we get

$$\int \frac{\mathrm{d}v}{\sqrt{1+v^2}} = \int \mathrm{d}x/x$$

$$\Rightarrow \log\left(v + \sqrt{1+v^2}\right) = \log x + \log c = \log xc$$

$$\Rightarrow v + \sqrt{1+v^2} = xc$$

$$\Rightarrow y/x + \sqrt{1 + y^2/x^2} = xc$$

$$\Rightarrow \frac{y + \sqrt{x^2+y^2}}{x} = xc$$

$$\Rightarrow y + \sqrt{x^2+y^2} = x^2c$$

3.1.7 Linear Differential Equation

The general solution of differential equations of the form $\frac{\mathrm{d}y}{\mathrm{d}x} + Py = Q$, where P and Q are functions of x or constants but not of y, will be

$$y \times [\text{integrating factor(I.F.)}] = \int Q(\text{I.F})\ \mathrm{d}x$$

Now we shall cite some examples and solve them.

Example 1

Solve: $\frac{\mathrm{d}y}{\mathrm{d}x} + \frac{y}{x} = x^3$.

Solution: ***This equation is of the form***

$$\frac{\mathrm{d}y}{\mathrm{d}x} + Py = Q$$

where $P = \frac{1}{x}$ and $Q = x^3$.

In such cases, integrating factor should be found out

$$\text{Integrating factor} = \int_e P\mathrm{d}x$$

$$\Rightarrow \int_e \frac{1}{x}\mathrm{d}x = \int_e \mathrm{d}x/x$$

$$\Rightarrow e^{\log x} = x$$

$\therefore$ The general solution of such equation will be

$$y(\text{I.F}) = \int Q(\text{I.F})\mathrm{d}x = x^3$$

$$\Rightarrow yx = \int x^3(x)\mathrm{d}x$$

$$\therefore \quad yx = \int x^4\mathrm{d}x = x^5/5 + c$$

$$\therefore \quad yx = x^5/5 + c$$

Answer.

Example 2

Solve: $\frac{\mathrm{d}y}{\mathrm{d}x} + \frac{2}{x}y = x^2 + 2$.

Solution: ***It is of the form*** $\boldsymbol{\frac{dy}{dx} + Py = Q}$***, where*** $\boldsymbol{P = 2/x}$ ***and*** $\boldsymbol{Q = x^2 + 2}$

$$\therefore \quad \text{I.F.} = e^{\int P\mathrm{d}x} = e^{\int \frac{2}{x}\mathrm{d}x} = e^{2\int \mathrm{d}x/x}$$

$$\Rightarrow e^{\ln x^2} = x^2$$

∴ The general solution will be

$$y(\text{I.F.}) = \int Q(\text{I.F.})\text{d}x$$

$$\Rightarrow y\left(x^2\right) = \int \left(x^2+2\right)x^2 \text{ d}x$$

$$\Rightarrow yx^2 = \int \left(x^4+2x^2\right)\text{d}x$$

$$= \int x^4 dx + 2\int x^2 \text{d}x$$

$$= \frac{x^5}{5} + \frac{2x^3}{3} + c$$

$$\therefore \quad x^2 y = x^5/5 + \frac{2}{3}x^3 + c \qquad \text{Answer.}$$

Example 3

Solve: $\dfrac{\text{d}x}{\text{d}y} + 3/yx = 2y$

Solution: ***The equation is of the form*** $\boldsymbol{\dfrac{dx}{dy} + Px = Q}$.

Here $P = 3/y$ and $Q = 2y$.

Therefore, integrating factor

$$\text{I.F.} = e^{\int P\text{d}y} = e^{\int 3/y \text{ d}y} = e^{3\int \text{d}y/y}$$

$$= e^{\ln y^3} = y^3$$

∴ The general solution will be

$$x(\text{I.F.}) = \int Q(\text{I.F.})\text{d}y$$

$$\Rightarrow xy^3 = \int 2y\cdot\left(y^3\right)\text{d}y$$

$$\Rightarrow xy^3 = 2\int y^4 \text{d}y$$

$$= \frac{2y^5}{5} + c$$

$$\therefore \quad x = \frac{2y^5}{5y^3} + c/y^3$$

$$\therefore \quad x = \frac{2}{5}y^2 + cy^{-3} \qquad \text{Answer.}$$

3.1.8 Equation of the Type $\dfrac{\text{d}y}{\text{d}x} + Py = Qy^n$

We shall understand the method of getting the general solution by solving some examples.

Example 1

Solve: $\frac{dy}{dx} + \frac{y}{x} = 3x^2y^2$.

Solution: ***This equation will be reduced to the form*** $\frac{dy}{dx} + Py = Qy^n$.

Dividing the whole equation by y^2, we get

$$\frac{1}{y^2}\frac{dy}{dx} + \frac{y}{xy^2} = \frac{3x^2y^2}{y^2}$$

$$\text{or } y^{-2}\frac{dy}{dx} + \frac{y^{-1}}{x} = 3x^2$$

Now put $y^{-1} = v$, and on differentiation, we get

$$-y^{-2}\frac{dy}{dx} = \frac{dv}{dx}$$

Putting these values, we get

$$\frac{-dv}{dx} + \frac{v}{x} = 3x^2$$

$$\Rightarrow \frac{dv}{dx} - \frac{v}{x} = -3x^2$$

It is of the form $\frac{dy}{dx} + Py = Q$, where $y^{-1} = v$, $P = \frac{-1}{x}$, $Q = -3x^2$.

$\therefore$ Integrating factor will be

$$e^{\int P dx} = e^{\int -1/x^{dx}} = e^{-\ln x} = e^{\ln x^{-1}} = x^{-1}$$

$\therefore$ The general solution will be equal to

$$vx^{-1} = \int Q(\text{I.F})dx \quad (\because v = y^{-1})$$

$$\Rightarrow y^{-1}x^{-1} = \int (-3x^2)(x^{-1})dx$$

$$\Rightarrow y^{-1}x^{-1} = \int \frac{-3x^2}{x}dx$$

$$\Rightarrow y^{-1}x^{-1} = -3\int x dx$$

$$\Rightarrow y^{-1}x^{-1} = \frac{-3x^2}{2} + c$$

$$\therefore \quad \frac{1}{xy} = \frac{-3x^2}{2} + c$$

which is the required solution. Answer.

Example 2

Solve: $dy + ydx = 2xy^2e^x dx$.

Solution: ***The above equation may be written as***

$$\frac{dy}{dx} + y = 2xy^2e^x$$

It is of the form $\frac{dy}{dx} + Py = Qy^n$. Dividing both sides by y^2, we get

$$y^{-2}\frac{dy}{dx} + \frac{y}{y^2} = 2xe^x$$

$$\Rightarrow y^{-2}\frac{dy}{dx} + \frac{1}{y} = 2xe^x \tag{3.10}$$

Now substitute $y^{-1} = v$

$$\therefore \quad -y^{-2}\frac{dy}{dx} = \frac{dv}{dx}$$

Putting this value in Eq. (3.10), the equation takes the form

$$-\frac{dv}{dx} + v = 2xe^x$$

$$\Rightarrow \frac{dv}{dx} - v = -2xe^x$$

where $P = -1$ and $Q = -2xe^x$.

$\therefore$ The integrating factor will be equal to $e^{\int P dx} = e^{\int -I dx} = e^{-x}$.

$\therefore$ The general solution will be

$$v(\text{I.F}) = \int Q(\text{I.F.})dx$$

$$\Rightarrow ve^{-x} = \int -2xe^x \cdot e^{-x}dx$$

$$\Rightarrow y^{-1}e^{-x} = -\int 2xdx$$

$$\Rightarrow y^{-1}e^{-x} = -2\int xdx = -2\frac{x^2}{2} + c$$

$$\therefore \quad y^{-1}e^{-x} = (c - x^2)$$

$$\therefore \quad ye^x(c - x^2) = 1$$ Answer.

3.1.9 Linear Differential Equation with Constant Coefficient/ Second-Order Differential Equation with Constant Coefficient

In such differential equations, first of all, we find the *auxiliary equation*. We write $\frac{d^2y}{dx^2} = m^2$, and the coefficient of y is also taken to frame the auxiliary equation.

After getting the auxiliary equation, we find the roots of the equation. The roots may be of the following types:

- Case (i): The roots may be real and unequal.
- Case (ii): The roots may be real and equal.
- Case (iii): The roots may be complex conjugate.

After knowing the values of roots, we then find the general solution of the differential equation. We shall discuss each case one by one by taking examples.

Case (i): The roots may be real and unequal.

Example 1

Solve: $\frac{d^2y}{dx^2} - 5\frac{dy}{dx} + 6y = 0.$

Solution: ***The auxiliary equation is***

$$m^2 - 5m + 6 = 0$$

$$\Rightarrow m^2 - 3m - 2m + 6 = 0$$

$$\Rightarrow m(m-3) - 2(m-3) = 0$$

$$\Rightarrow (m-2)(m-3) = 0$$

$\therefore$ $m = 2$ and $m = 3$. These are the roots of the equation.

The general solution is given by

$$y = c_1 e^{2x} + c_2 e^{3x} \qquad \text{Answer.}$$

Example 2

Solve: $\frac{d^2y}{dx^2} - 3\frac{dy}{dx} - 4y = 0.$

Solution: ***The auxiliary equation is***

$$m^2 - 3m - 4 = 0$$

$$\Rightarrow m^2 + m - 4m - 4 = 0$$

$$\Rightarrow m(m+1) - 4(m+1) = 0$$

$$\Rightarrow (m+1)(m-4) = 0$$

$\therefore m = -1$ and $m = 4$

Hence, the general solution will be

$$y = c_1 e^{-x} + c_2 e^{4x} \qquad \text{Answer.}$$

Example 3

Solve: $\dfrac{d^3y}{dx^3}+6\dfrac{d^2y}{dx^2}+3\dfrac{dy}{dx}-10y=0$

Solution: ***The auxiliary equation is***

$$m^3+6m^2+3m-10=0$$

$$\Rightarrow (m-1)(m+2)(m+5)=0$$

$\therefore$ The roots are $m = 1$, $m = -2$, and $m = -5$

$\therefore$ The general solution is

$$y=c_1e^{x}+c_2e^{-2x}+c_3e^{-5x}$$ Answer.

Case (ii): The roots may be real and equal.

Example 1

Solve: $\dfrac{d^2y}{dx^2}-6\dfrac{dy}{dx}+9y=0.$

Solution: ***The auxiliary equation is***

$$m^2-6m+9=0$$

$$\Rightarrow m^2-3m-3m+9=0$$

$$\Rightarrow m(m-3)-3(m-3)=0$$

$$\Rightarrow (m-3)(m-3)=0$$

$$\therefore m=3,\ 3$$

Therefore, the general solution is

$$y=(c_1+c_2x)e^{3x}$$ Answer.

Example 2

Solve: $\dfrac{d^3y}{dx^3}-3\dfrac{dy}{dx}+2y=0.$

Solution: ***The auxiliary equation is***

$$m^3-3m+2=0$$

$$\Rightarrow (m+2)(m-1)^2=0$$

$$\Rightarrow m=-2 \text{ and } m=1,\ 1.$$

$\therefore$ The general solution is $y = c_1e^{-2x} + (c_2 + c_3x)\ e^{x}$ Answer.

Case (iii): The roots may be complex conjugate.

Let $m_1 = \alpha + i\beta$ and $m_2 = \alpha - i\beta$ be the two roots of an auxiliary equation. Then, the general solution will be

$$y = c_1 e^{(\alpha+i\beta)x} + c_2 e^{(\alpha-i\beta)x}$$

$$\Rightarrow c_1 e^{\alpha x} \cdot e^{i\beta x} + c_2 e^{\alpha x} e^{-i\beta x}$$

$$\Rightarrow e^{\alpha x}\left[c_1 e^{i\beta x} + c_2 e^{-i\beta x}\right]$$

$$\Rightarrow e^{\alpha x}\left[c_1 e^{i\beta x} + c_2 e^{-i\beta x}\right]$$

$$\Rightarrow e^{\alpha x}\left[c_1(\cos\beta x + i\sin\beta x) + c_2(\cos\beta x - i\sin\beta x)\right]$$

$$\Rightarrow e^{\alpha x}\left[c_1 \cos\beta_x + c_1 i\sin\beta x + c_2\cos\beta x - c_2 i\sin\beta x\right]$$

$$\Rightarrow e^{\alpha x}\left[(c_1 + c_2)\cos\beta x + (c_1 - c_2)i\sin\beta x\right]$$

$$\Rightarrow e^{\alpha x}\left[A\cos\beta x + B\sin\beta x\right]$$

where $(c_1 + c_2) = \text{A}$ and $i(c_1 - c_2) = \text{B}$.

Example 1

Solve: $\dfrac{d^2y}{dx^2} + 4y = 0.$

Solution: ***The auxiliary equation is***

$$m^2 + 4 = 0$$

$$\Rightarrow m^2 = -4 = 4i^2$$

$$\therefore m = \pm 2i$$

Hence, the real root is zero (i.e., $\alpha = 0$), and the imaginary part is $\pm i^2$.

$\therefore$ The general solution is

$$y = e^{0x}\left[A\cos 2x + B\sin 2x\right]$$

$$= \left[A\cos 2x + B\sin 2x\right] \qquad \text{Answer.}$$

Example 2

Solve: $\dfrac{d^2y}{dx^2} + \dfrac{dy}{dx} + y = 0.$

Solution: ***The auxiliary equation is*** $\boldsymbol{m^2 + m + 1 = 0}$

$$\therefore m = \frac{-1 \pm \sqrt{1 - 4.1.1}}{2} = \frac{-1 \pm \sqrt{3i^2}}{2}$$

$$= \frac{-1}{2} \pm \frac{\sqrt{3}}{2}i$$

Hence, the real part of the root is $\frac{-1}{2}$, and the imaginary part of the root is $\frac{\sqrt{3}}{2}$.

$\therefore$ The general solution is

$$y = e^{-x/2}\left[c_1 \cos\frac{\sqrt{3}}{2}x + c_2 \sin\frac{\sqrt{3}}{2}x\right] \qquad \text{Answer.}$$

Example 3

Solve: $\frac{d^4y}{dx^4} - a^4y = 0.$

Solution: ***The auxiliary equation is***

$$m^4 - a^4 = 0$$

$$\Rightarrow (m^2 - a^2)(m^2 + a^2) = 0$$

$$\therefore m^2 = a^2 \therefore m = \pm a$$

$$\text{or } m^2 + a^2 = 0$$

$$\text{or } m^2 = -a^2$$

$$\therefore m = \pm ai$$

Hence, the general solution will be

$$y = c_1e^{ax} + c_2e^{-ax} + (c_3 \cos ax + c_4 \sin ax) \qquad \text{Answer.}$$

3.1.10 Solving Differential Equations by Power Series

The method of solution will be clear by taking examples.

Example 1.

Solve:

$$\frac{d^2y}{dx^2} - xy = 0 \tag{3.11}$$

Solution: ***Let us assume that there exists a solution having the form***

$$y = c_0 + c_1x + c_2x^2 + c_3x^3 + c_4x^4 + c_5x^5 + \cdots \tag{3.12}$$

where c_0, c_1, c_2, c_3, c_4, ... are constants.

Differentiating this equation with respect to x, we get

$$\frac{dy}{dx} = c_1 + 2c_2x + 3c_3x^2 + 4c_4x^3 + 5c_5x^4 + \cdots \tag{3.13}$$

and

$$\frac{d^2y}{dx^2} = 2c_2 + 6c_3x + 12c_4x^2 + 20c_5x^3 + \cdots \tag{3.14}$$

Putting the values of $\frac{d^2y}{dx^2}$ in Eq. (3.11), we obtain

$$2c_2 + 6c_3x + 12c_4x^2 + 20c_5x^3 + \cdots - x\left(c_0 + c_1x + c_2x^2 + c_3x^3 + c_4x^4 + \cdots\right) = 0$$

$$\Rightarrow 2c_2 + 6c_3x + 12c_4x^2 + 20c_5x^3 + \cdots - c_0x - c_1x^2 - c_2x^3 - c_3x^4 - c_4x^5 + \cdots = 0$$

$$\Rightarrow 2c_2 + (6c_3 - c_0)x + (12c_4 - c_1)x^2 + (20c_5 - c_2)x^3 + \cdots$$

$$\Rightarrow 2c_2 + (3.2c_3 - c_0)x + (4.3c_4 - c_1)x^2 + (5.4c_5 - c_2)x^3$$

$$+ (6.5c_6 - c_3)x^4 + \cdots + \left[n(n-1)c_n - c_{n-3}\right]x^{n-2} + \cdots = 0 \tag{3.15}$$

Equation (3.15) is an identity. Hence, equating coefficients of powers of x to zero, we get

$$2c_2 = 0, 3.2c_3 = c_0, 4.3c_4 - c_1 = 0, 5.4c_5 - c_2 = 0$$

$$6.5c_6 - c_3 = 0, n(n-1)c_n - c_{n-3} = 0$$

From these, we can write

$$c_2 = 0, c_3 = \frac{c_0}{2.3}, c_4 = \frac{c_1}{3.4}, c_5 = 0, c_6 = \frac{c_3}{5.6}$$

Hence, the complete solution is

$$y = c_0 + c_1x + \frac{c_0}{6}x^3 + \frac{c_1}{12}x^4 + \ldots$$

$$= c_0 + c_1x + \frac{c_0x^3}{6} + \frac{c_1}{12}x^4 + \ldots$$

$$= c_0\left(1 + x^3/6\right) + c_1x\left(1 + \frac{x^3}{12}\right) + \ldots$$

Answer.

Example 2

Solve: $\frac{d^2y}{dx^2} - x\frac{dy}{dx} - 2y = 0.$

Solution: ***Let us assume that there exists a solution having the form***

$$y = c_0 + c_1x + c_2x^2 + c_3x^3 + c_4x^4 + .$$

$$\frac{dy}{dx} = c_1 + 2c_2x + 3c_3x^2 + 4c_4x^3 + .$$

$$\frac{d^2y}{dx^2} = 2c_2 + 6c_3x + 12c_4x^2 + .$$

Substituting the values of y, $\frac{dy}{dx}$ and $\frac{d^2y}{dx^2}$ in the above equation, we get

$$2c_2 + 6c_3x + 12c_4x^2 + \ldots$$

$$- x\left(c_1 + 2c_2x + 3c_3x^2 + 4c_4x^3 + \ldots\right)$$

$$- 2\left(c_0 + c_1x + c_2x^2 + c_3x^3 + c_4x^4 + \ldots\right) = 0$$

$$\Rightarrow 2c_2 + 6c_3x + 12c_4x^2 + \ldots$$
$$- c_1x - 2c_2x^2 - 3c_3x^3 + \ldots$$
$$- 2c_0 - 2c_1x - 2c_2x^2 - 2c_3x^3 + \ldots = 0$$
$$\Rightarrow (2c_2 - 2c_0) + (6c_3 - 3c_1)x + (12c_4 - 4c_2)x^2 + = 0$$

Now equating the coefficient of powers of x to zero, we get

$$2c_2 - 2c_0 = 0 \quad \therefore \quad c_2 = c_0$$

$$6c_3 - 3c_1 = 0 \quad \therefore \quad 2c_3 - c_1 = 0 \quad \therefore \quad c_3 = \frac{c_1}{2}$$

$$12c_4 - 4c_2 = 0 \quad \therefore \quad 3c_4 - c_2 = 0$$

$$\therefore \quad c_4 = \frac{c_2}{3} = \frac{c_0}{3}$$

The solution of the above differential equation is

$$y = c_0 + c_1x + c_2x^2 + c_3x^3 + \ldots$$

$$\text{or } y = c_0 + c_1x + c_0x^2 + \frac{c_1}{2}x^3 + \frac{c_0}{3}x^4 + \ldots$$

$$\text{or } y = c_0\left(1 + x^2 + \frac{1}{3}x^4 + \ldots\right) + c_1\left(x + x^3/2 + \ldots\right)$$

This is the desired solution. Answer.

3.2 Matrices

An arrangement of mathematical elements into rows and columns according to algebraic rules, in order to solve a set of linear equations, is called a matrix.

It can be defined in another way in the following manner.

A matrix is a rectangular array of numbers or elements, the operation of which is governed according to algebraic rules.

or

A matrix may be defined as an ordered array of elements (real or complex), the operation of which is subject to certain rules.

The elements are arranged in rows and columns. An example of an arbitrary matrix A is presented as follows:

$$A = \downarrow \begin{matrix} \text{C} \\ \text{O} \\ \text{L} \\ \text{U} \\ \text{M} \\ \text{N} \end{matrix} \begin{pmatrix} a_{11} & a_{12} & a_{13} & a_{1n} \\ a_{21} & a_{22} & a_{23} & a_{2n} \\ a_{m1} & a_{m2} & --- & a_{mn} \end{pmatrix} \text{—ROW}$$

wherein a_{ij}, ($i = 1,2,3, \ldots, m$ and $j = 1,2,3,4 \ldots, n$), are real or complex numbers. We may denote the set of real number by F. Keep in mind that the index i identifies rows and index j identifies columns.

The matrix has m rows, and each row comprises n numbers, i.e., the matrix has n columns. Due to this reason, we shall call it an $m \times n$ matrix over F. Thus, $m \times n$ matrix over F is a matrix, which consists of m rows and n columns and each element $\in F$. The individual numbers in the array are known as components or elements of the matrix. The systems of enclosing the numbers constituting the matrix are the following:

$$[\,] \quad (\,) \quad \|\,\|$$
$$(1) \quad (2) \quad (3)$$

We shall use the second one.

The concept of matrix helps us to consider the solution of simultaneous equation. For example, let us consider the following three simultaneous linear equations consisting of four unknowns x, y, z, and t.

$$x - 2y + 3z + 2t = 0$$

$$2x + y + 4z + 5t = 0$$

$$3x + y - z + t = 0$$

The system of coefficients in the above set of equation written in order in which they occur determines the 3×4 matrix.

$$\begin{pmatrix} 1 & -2 & 3 & 2 \\ 2 & 1 & 4 & 5 \\ 3 & 1 & -1 & 1 \end{pmatrix}$$

It has three rows and four columns. With the help of matrices, we can express the result in more compact form. It must be remembered that the manipulations of quantum mechanical equations frequently need some algebraic and geometrical operations. Fortunately, most of the manipulations can be put into a very elegant and compact form by the use of matrices. Indeed, it is the matrix form of quantum mechanics, which lends itself to computational procedures on high-speed electronic computers. Here, we are going to learn about the types of matrices.

3.2.1 Types of Matrices

3.2.1.1 Rectangular Matrix

If the number of rows (m) is not equal to the number of columns (n) in a matrix, i.e., $m \neq n$, then it is called a rectangular matrix.

For example: $\begin{pmatrix} 2 & 3 \\ 3 & 2 \\ 4 & 1 \end{pmatrix}$.

This matrix consists of three rows and two columns, i.e., $m \neq n$; therefore, it is a rectangular matrix.

3.2.1.2 Square Matrix

If in an $m \times n$ matrix $m = n$, then it will constitute a square matrix. In such matrix, the number of rows remains equal to the number of columns.

For example: $\begin{pmatrix} 1 & 0 \\ 0 & 1 \end{pmatrix}$.

It is a 2×2 matrix, i.e., $m = n$; hence, it is an example of a square matrix.

3.2.1.3 Non-Singular and Singular Matrices

A square matrix A is said to be non-singular when $(A) \neq 0$ and singular when $(A) = 0$.

3.2.1.4 Unit Matrix

A square matrix having each of its diagonal element 1 and each off-diagonal element zero is called a unit matrix.

In other way, we can say that a square matrix $A = (a_{ij})$ is a unit matrix provided $a_{ij} = 1$ for $i = j$ and $a_{ij} = 0$ for $i \neq j$.

For example: $\begin{pmatrix} 1 & 0 & 0 \\ 0 & 1 & 0 \\ 0 & 0 & 1 \end{pmatrix}$.

It is a unit matrix of order 3×3 as each of its diagonal element is 1 and each off-diagonal element is zero. We may have unit matrix of different orders.

For example: $\begin{pmatrix} 1 & 0 \\ 0 & 1 \end{pmatrix} \begin{pmatrix} 1 & 0 & 0 \\ 0 & 1 & 0 \\ 0 & 1 & 0 \end{pmatrix} \begin{pmatrix} 1 & 0 & 0 & 0 \\ 0 & 1 & 0 & 0 \\ 0 & 0 & 1 & 0 \\ 0 & 0 & 0 & 1 \end{pmatrix}$.

2×2 matrix 3×3 matrix 4×4 matrix

3.2.1.5 Null Matrix or Zero Matrix

A matrix, all elements of which are zero, is called a zero matrix or a null matrix.

For example: $\begin{pmatrix} 0 & 0 & 0 \\ 0 & 0 & 0 \end{pmatrix}$ is a zero matrix because all its elements are zero.

3.2.1.6 Row Matrix

A matrix which consists of only one row and any number of columns is known as a row matrix or a row vector.

or

A matrix which has n elements arranged in one row only is known as a row matrix or a row vector. For example, $(a_1, a_2, a_3 \ldots\ldots a_n)$ and (2, 1, 6, 3, 8) are row matrices. The first one is $1 \times n$ matrix, and the second is 1×5 matrix.

3.2.1.7 Column Matrix

A matrix having n elements arranged in one column only is called a column matrix or a column vector. It is $n \times 1$ matrix, when n = any value.

For example: $\begin{pmatrix} a_1 \\ a_2 \\ a_3 \\ \cdot \\ \cdot \\ a_n \end{pmatrix}$ is a column matrix.

Another example of a column matrix is $\begin{pmatrix} 5 \\ 8 \\ 7 \end{pmatrix}$, which is a 3×1 column matrix.

3.2.1.8 Diagonal Matrix

A square matrix in which all elements, except the diagonal elements, are zero is called a diagonal matrix, i.e., $a_{ij} = 0$ for all $i = j$.

For example: $\begin{pmatrix} 5 & 0 & 0 \\ 0 & 6 & 0 \\ 0 & 0 & 7 \end{pmatrix}$ is a diagonal 3×3 matrix.

3.2.1.9 Scalar Matrix

A diagonal matrix is said to be a scalar matrix when its diagonal elements are same or equal.

For example: $\begin{pmatrix} 3 & 0 & 0 \\ 0 & 3 & 0 \\ 0 & 0 & 3 \end{pmatrix}$ is a scalar matrix.

3.2.2 Operation of Matrices

In this section, we shall learn addition, subtraction, and multiplication of two matrices. We shall handle them one by one.

3.2.2.1 Addition of Two Matrices

The addition of matrices of the same order $(m \times n)$ is defined as a matrix, and the elements of which are equal to the sum of the corresponding elements in the original matrices.

This will be clear by taking an example.

$$\text{Let } (A) = \begin{pmatrix} a_{11} & a_{12} & a_{13} \\ a_{21} & a_{22} & a_{23} \end{pmatrix}$$

$$(B) = \begin{pmatrix} b_{11} & b_{12} & b_{13} \\ b_{21} & b_{22} & b_{23} \end{pmatrix}$$

$$(A)+(B) = \begin{pmatrix} a_{11} & a_{12} & a_{13} \\ a_{21} & a_{22} & a_{23} \end{pmatrix} + \begin{pmatrix} b_{11} & b_{12} & b_{13} \\ b_{21} & b_{22} & b_{23} \end{pmatrix}$$

$$\Rightarrow \begin{pmatrix} a_{11}+b_{11} & a_{12}+b_{12} & a_{13}+b_{13} \\ a_{21}+b_{21} & a_{22}+b_{22} & a_{23}+b_{23} \end{pmatrix}$$

The addition operation will be clearer when we shall consider some numerical examples.

Example 1

$$\begin{pmatrix} 1 & 0 \\ 0 & 0 \end{pmatrix} + \begin{pmatrix} -1 & 0 \\ 0 & -1 \end{pmatrix}$$

Solution: $\begin{pmatrix} 1 & 0 \\ 0 & 0 \end{pmatrix} + \begin{pmatrix} -1 & 0 \\ 0 & -1 \end{pmatrix}$

$$\Rightarrow \begin{pmatrix} 1-1 & 0+0 \\ 0+0 & 0-1 \end{pmatrix} = \begin{pmatrix} 0 & 0 \\ 0 & -1 \end{pmatrix}$$ Answer.

Example 2

$$\begin{pmatrix} 6 & 0 & 1 \\ -3 & 1 & -4 \\ 2 & 0 & 5 \\ 1 & 1 & 1 \end{pmatrix} + \begin{pmatrix} -7 & 1 & 0 \\ 4 & 3 & 2 \\ 0 & 1 & -3 \\ 0 & -2 & 1 \end{pmatrix}$$

Solution: $\begin{pmatrix} 6 & 0 & 1 \\ -3 & 1 & -4 \\ 2 & 0 & 5 \\ 1 & 1 & 1 \end{pmatrix} + \begin{pmatrix} -7 & 1 & 0 \\ 4 & 3 & 2 \\ 0 & 1 & -3 \\ 0 & -2 & 1 \end{pmatrix}$ Answer.

$$\Rightarrow \begin{pmatrix} 6+(-7) & 0+1 & 1+0 \\ -3+4 & 1+3 & -4+2 \\ 2+0 & 0+1 & 5-3 \\ 1+0 & 1-2 & 1+1 \end{pmatrix}$$

$$\Rightarrow \begin{pmatrix} -1 & 1 & 1 \\ 1 & 4 & -2 \\ 2 & 1 & 2 \\ 1 & -1 & 2 \end{pmatrix}$$

3.2.2.2 Subtraction of Two Matrices

If A and B are two matrices $(m \times n)$, then we define subtraction as $A - B = A + (-B)$.

The difference between A and B is obtained by subtracting from each element of A the corresponding element of B. We take an example to understand this.

Example 1

If $(A) = \begin{pmatrix} 2 & 7 \\ 9 & 8 \end{pmatrix}, (B) = \begin{pmatrix} 1 & 2 \\ 0 & 3 \end{pmatrix}$ Find, $(A) - (B)$

Solution: $(A - B) = \begin{pmatrix} 2 & 7 \\ 9 & 8 \end{pmatrix} + \begin{pmatrix} -1 & -2 \\ -0 & -3 \end{pmatrix}$

$$\Rightarrow \begin{pmatrix} 2-1 & 7-2 \\ 9-0 & 8-3 \end{pmatrix} = \begin{pmatrix} 1 & 5 \\ 9 & 5 \end{pmatrix}$$ Answer.

3.2.2.3 Multiplication of Two Matrices

Two matrices can be multiplied with each other. To perform their multiplication, it is required to multiply each element in the rows of the first matrix by the corresponding elements in the columns of the second matrix. The pairs of numbers are then added to yield the element of the resultant matrix.

If A is an $m \times n$ matrix and B is an $n \times p$ matrix, then AB will be $m \times p$ matrix.

Now we shall cite some numerical examples to understand the process of multiplication.

Example 1

Let $A = \begin{pmatrix} 1 & 2 & 3 & 4 \\ 5 & 4 & 3 & 2 \\ 0 & 1 & 2 & 3 \end{pmatrix}_{3\times4}$ and $B = \begin{pmatrix} 1 & 2 \\ -2 & 0 \\ 3 & -2 \\ -4 & 3 \end{pmatrix}_{4\times2}$. Then, find AB

Solution: ***Since A is 3 × 4 matrix and B is a 4 × 2 matrix, we shall get a 3 × 2 matrix on multiplication.***

$$\therefore \quad AB = \begin{pmatrix} 1 & 2 & 3 & 4 \\ 5 & 4 & 3 & 2 \\ 0 & 1 & 2 & 3 \end{pmatrix} \times \begin{pmatrix} 1 & 2 \\ -2 & 0 \\ 3 & -2 \\ 4 & 3 \end{pmatrix}$$

The resultant matrix will be of the type

$$\begin{pmatrix} c_{11} & c_{12} \\ c_{21} & c_{22} \\ c_{31} & c_{32} \end{pmatrix}$$

Now we shall calculate every element of the matrix as given below:

$$c_{11} = 1\times 1 + 2\times(-2) + 3\times 3 + 4\times(-4) = -10$$

$$c_{21} = 5\times 1 + 4\times(-2) + 3\times 3 + 2\times(-4) = -2$$

$$c_{31} = 0\times 1 + 1\times(-2) + 2\times 3 + 3\times(-4) = -8$$

$$c_{12} = 1\times 2 + 2\times 0 + 3\times 3 + 2\times(-4) = -2$$

$$c_{22} = 5\times 2 + 4\times 0 + 3\times(-2) + 2\times 3 = 10$$

$$c_{32} = 0\times 2 + 1\times 0 + 2\times(-2) + 3\times 3 = 5$$

$$\therefore \quad AB = \begin{pmatrix} -10 & -2 \\ -2 & 10 \\ -8 & 5 \end{pmatrix}$$ Answer.

Example 2

If $A = \begin{pmatrix} 2 & 3 & 4 \\ 1 & 2 & 3 \\ 1 & 1 & 2 \end{pmatrix}$ and $B = \begin{pmatrix} 1 & 3 & 0 \\ -1 & 2 & 1 \\ 0 & 0 & 2 \end{pmatrix}$. Find AB and BA and show that $AB \neq BA$.

Solution: ***Since A is 3 × 3 matrix and B is also 3 × 3 matrix, AB will also be 3 × 3 matrix. The resultant matrix will be of the type***

$$AB = \begin{pmatrix} c_{11} & c_{12} & c_{13} \\ c_{21} & c_{22} & c_{23} \\ c_{31} & c_{32} & c_{33} \end{pmatrix}$$

Now we shall calculate each element of the resultant matrix:

$$c_{11} = 2\times 1 + 3(-1) + 4\times 0 = 2 - 3 + 0 = -1$$

$$c_{12} = 2\times 3 + 3\times 2 + 0\times 4 = 6 + 6 + 0 = 12$$

$$c_{13} = 2\times 0 + 3\times 1 + 4\times 2 = 0 + 3 + 8 = 11$$

$$c_{21} = 1\times1 + 2(-1) + 3\times0 = 1 - 2 + 0 = -1$$

$$c_{22} = 1\times3 + 2\times2 + 3\times0 = 3 + 4 + 0 = 7$$

$$c_{23} = 1\times0 + 2\times1 + 3\times2 = 0 + 2 + 6 = 8$$

$$c_{31} = 1\times1 + 1(-1) + 2\times0 = 1 - 1 + 0 = 0$$

$$c_{33} = 1\times0 + 1\times1 + 2\times2 = 0 + 1 + 4 = 5$$

$$\therefore \quad AB = \begin{pmatrix} -1 & 12 & 11 \\ -1 & 7 & 8 \\ 0 & 5 & 5 \end{pmatrix}$$

Similarly BA may be found out as follows:

$$BA = \begin{pmatrix} 1 & 3 & 0 \\ -1 & 2 & 1 \\ 0 & 0 & 2 \end{pmatrix} \times \begin{pmatrix} 2 & 3 & 4 \\ 1 & 2 & 3 \\ -1 & 1 & 2 \end{pmatrix}$$

$$= \begin{pmatrix} 1\times2+3\times1+0(-1) & 1\times3+3\times2+0\times1 & 1\times4+3\times3+0\times2 \\ -1\times2+2\times1-1\times1 & -1\times3+2\times2+1\times1 & -1\times4+2\times3+1\times2 \\ 0\times2+0\times1-2\times1 & 0\times3+0\times2+2\times1 & 0\times4+0\times3+2\times2 \end{pmatrix}$$

$$= \begin{pmatrix} 5 & 9 & 13 \\ -1 & 2 & 4 \\ -2 & 2 & 4 \end{pmatrix}$$

Thus, we see that $AB \neq BA$. Proved.

Example 3

If A, B, and C are three matrices such that

$$A = (x, y, z),\ B = \begin{pmatrix} a & h & g \\ h & b & f \\ g & f & c \end{pmatrix} \text{ and } C = \begin{pmatrix} x \\ y \\ z \end{pmatrix}, \text{ find ABC}$$

Solution: ***We see that A is 1 × 3 matrix B is 3 × 3 matrix, and C is 3 × 1 matrix. Therefore, associative law holds for A, B, and C.***

Now AB = (ax + hy + gz, hx + by + fz, gx + fy + cz)

$$AB(C) = (ax + hy + gz,\quad hx + by + fz,\quad gx + fy + cz) \times \begin{pmatrix} x \\ y \\ z \end{pmatrix}$$

$$= \{x(ax + hy + gz),\quad y(hx + by + fz),\quad z(gx + fy + cz)\}$$

$$= (ax^2 + hxy + gxz,\quad hxy + by^2 + fyz,\quad gxz + fyz + cz)$$ Answer.

Example 4

$$\underset{1\times 3 \text{ matrix}}{(2,3,5)} \times \underset{3\times 3 \text{ matrix}}{\begin{pmatrix} 3 & 0 & 2 \\ 5 & 8 & 3 \\ 7 & 2 & 0 \end{pmatrix}} \times \underset{3\times 1 \text{ matrix}}{\begin{pmatrix} 3 \\ 3 \\ 3 \end{pmatrix}}$$

Solution:

$$(2,3,5)\times\begin{pmatrix} 3 & 0 & 2 \\ 5 & 8 & 3 \\ 7 & 2 & 0 \end{pmatrix}$$

$$=(2\times 3+3\times 0+5\times 2,\ 2\times 5+3\times 8+5\times 3,\ 2\times 7+3\times 2+5\times 0)$$

$$=(6+0+10,\ 10+24+15,\ 14+6+0)$$

$$=(16,49,20)$$

$$ABC=(16,49,20)\begin{pmatrix} 3 \\ 3 \\ 3 \end{pmatrix}$$

$$=(16\times 3+49\times 3+20\times 3)=(48+147+60)$$

$$=255$$

Answer.

3.2.3 Transpose of a Matrix

It is the matrix which is obtained from a given matrix A by interchanging its rows and columns and is represented by A^t or A^I.

For example, if

$$A=\begin{pmatrix} 1 & 3 \\ 4 & 5 \\ 6 & 8 \end{pmatrix}, \text{ then } A^t=\begin{pmatrix} 1 & 4 & 6 \\ 3 & 5 & 8 \end{pmatrix}$$

Thus, A^t is the transpose of A. It follows that if A is an $m \times n$ matrix, then A^t will be an $n \times m$ matrix. Thus, if $A = (a_{ij})$, then $A^t = (a_{ji})$.

3.2.4 Symmetric Matrix

A square matrix $A = (a_{ij})$ is called a symmetric if $a_{ij} = a_{ji}$, i.e., the (i, j)th element is the same as the (j, i) th element. Thus, it may be said that a square matrix A will be symmetric if it is equal to its transpose A^t or A^I. It means that a symmetric matrix does not alter if we interchange its rows and columns. Therefore, a square matrix A is symmetric if its transpose $A^t = A$.

For example: $A=\begin{pmatrix} a & h & g \\ h & b & f \\ g & f & c \end{pmatrix}$

$$\text{then } A^t=\begin{pmatrix} a & h & g \\ h & b & f \\ g & f & c \end{pmatrix}$$

$\therefore A = A^t \ \therefore A$ is symmetric.

3.2.5 Skew-Symmetric Matrix

A square matrix A is called a skew-symmetric, if $a_{ij} = -a_{ji}$.

Thus, a square matrix A will be a skew matrix if it is equal to transpose A^t with a negative sign, i.e., $A^t = -A$. By definition,

$$a_{ij} = -a_{ij} \; \ldots \; \therefore \; 2a_{ij} = 0 \; \ldots \; \therefore \; a_{ij} = 0$$

Hence, the diagonal elements of a skew-symmetric matrix are always zero.

For example, the matrix $\begin{pmatrix} 0 & a & b \\ -a & 0 & c \\ -b & -c & 0 \end{pmatrix}$ is a skew-symmetric because its transpose matrix is $A^t = \begin{pmatrix} 0 & -a & -b \\ a & 0 & -c \\ b & c & 0 \end{pmatrix}$

If we take the negative sign of A^t, we shall get $\begin{pmatrix} 0 & a & b \\ -a & 0 & c \\ -b & -c & 0 \end{pmatrix}$ which is A

$$\therefore A^t = -A \text{ or } A = -A^t$$

Example 1

If $A = \begin{pmatrix} 1 & 1 \\ x & y \\ x^2 & y^2 \end{pmatrix}$, find the value of AA^t.

Solution: ***Here the transpose matrix*** $A^t = \begin{pmatrix} 1 & x & x^2 \\ 1 & y & y^2 \end{pmatrix}$

$$AA^t = \begin{pmatrix} 1 & 1 \\ x & y \\ x^2 & y^2 \end{pmatrix} \begin{pmatrix} 1 & x & x^2 \\ 1 & x & y^2 \end{pmatrix}$$

$$= \begin{pmatrix} 1\times1+1\times1 & 1\times x+1\times y & 1\times x^2+1\times y^2 \\ x\times1+y\times1 & x\times x+y\times y & x\times x^2+y\times y^2 \\ x^2\times1+y^2\times1 & x^2\times x+y^2\times y & x^2\times x^2+y^2\times y^2 \end{pmatrix}$$

$$= \begin{pmatrix} 2 & x+y & x^2+y^2 \\ x+y & x^2+y^2 & x^3+y^3 \\ x^2+y^2 & x^3+y^3 & x^4+y^4 \end{pmatrix}$$

Evidently, AA^t is a symmetric matrix because by interchanging the rows and columns we get $\begin{pmatrix} 2 & x+y & x^2+y^2 \\ x+y & x^2+y^2 & x^3+y^3 \\ x^2+y^2 & x^3+y^3 & x^4+y^4 \end{pmatrix}$; i.e., $A = A^t$ Answer.

3.2.6 Complex Matrix

If the elements of a matrix are complex numbers, the matrix is known as a complex matrix. Any complex matrix Z can be expressed in the form $x + iy$, where x and y are two real matrices.

For example:

$$Z = \begin{pmatrix} 3+i2 & 4+i3 \\ 5+i4 & 3-i5 \end{pmatrix}$$

$$= \begin{pmatrix} 3 & 4 \\ 5 & 3 \end{pmatrix} + \begin{pmatrix} i2 & i3 \\ i4 & -i5 \end{pmatrix}$$

$$= \begin{pmatrix} \underbrace{\begin{matrix} 3 & 4 \\ 5 & 3 \end{matrix}}_{x} \end{pmatrix} + i \begin{pmatrix} \underbrace{\begin{matrix} 2 & 3 \\ 4 & -5 \end{matrix}}_{y} \end{pmatrix}$$

3.2.7 Complex Conjugate of a Matrix

The matrix obtained by replacing the elements of a complex matrix Z by the corresponding conjugate complex numbers is known as the conjugate of the matrix Z and is represented by $\overline{Z}$.

If $Z = x + iy$, then $\overline{Z} = x - iy$.

3.2.8 Hermitian Matrix

A complex square matrix $Z = Z_{ij}$ is called Hermitian if $Z_{ii} = \overline{Z}_{ii}$, i.e., the (i, j)th element is equal to the imaginary conjugate of (j, i)th element.

Evidently, a matrix Z is Hermitian if and only if it coincides with its conjugate transpose, i.e., $Z = Z^*$, since by definition $Z_{ii} = \overline{Z}_{ii}$, i.e., every diagonal element is equal to its conjugate; therefore, we see that the diagonal elements of a Hermitian matrix are real. This can be illustrated as follows:

Let $Z_{ij} = a + ib$ and then $\overline{Z}_{ii} = a - ib$

Now since $Z_{ii} = \overline{Z}_{ii}$

$$\therefore a + ib = a - ib$$

From this, it follows that

$$2ib = 0 \text{ i.e, } b = 0$$

Therefore, the diagonal elements consist of real parts only. Hence, the diagonal elements are real. It is to be noted that a Hermitian matrix over the field of real numbers is a real symmetric matrix.

For example:

$$\begin{pmatrix} 2 & 2-3i & 3+4i \\ 2+3i & 0 & 4-5i \\ 3-4i & 4+5i & 2 \end{pmatrix} \text{ is Hermitian,}$$

Similarly, $\begin{pmatrix} a & b+ic \\ b-ic & d \end{pmatrix}$ is also a Hermitian matrix.

3.2.9 Skew-Hermitian Matrix

A square matrix Z is said to be skew-Hermitian if

$$Z_{ij} = -\bar{Z}_{ji}$$

i.e., the (i, j)th element is equal to the negative conjugate of (j, i)th element.

Evidently, a matrix Z is skew-Hermitian if $Z^* = -Z$.

From definition, $Z_{ij} = -\bar{Z}_{ji}$

$$\therefore Z_{ij} + \bar{Z}_{ji} = 0$$

Thus, it is clear from this that the diagonal of a skew matrix is either zero or a pure imaginary number. For example, the matrix

$$\begin{pmatrix} i & 3+4i & 4-5i \\ -3+4i & 0 & 3+2i \\ -4-5i & -3+2i & 0 \end{pmatrix} \text{ is skew-Hermitian}$$

3.2.10 Adjoint of a Matrix

Let $A = (a_{ij})$ be any square matrix and $B = (A_{ij})$ is a matrix, the elements of which are the co-factors of the corresponding elements in A, then the transpose of B is called the adjoint of matrix A and it is equal to (A_{ij}). Adjoint of A is generally written as Adj A.

$$\text{Let } A = \begin{pmatrix} a_{11} & a_{12} & a_{13} \\ a_{21} & a_{22} & a_{23} \\ a_{31} & a_{32} & a_{33} \end{pmatrix}$$

$\therefore$ According to the definition,

$$B = \begin{pmatrix} A_{11} & A_{12} & A_{13} \\ A_{21} & A_{22} & A_{23} \\ A_{31} & A_{32} & A_{32} \end{pmatrix}$$

The transpose of B

$$= \begin{pmatrix} A_{11} & A_{21} & A_{31} \\ A_{12} & A_{22} & A_{32} \\ A_{13} & A_{32} & A_{33} \end{pmatrix}$$

and it is called the adjoint of A.

It will be clear by taking an example.

Example 1

Find the adjoint of the matrix $A = \begin{pmatrix} 1 & 2 & 3 \\ 0 & 5 & 0 \\ 2 & 4 & 3 \end{pmatrix}$

Solution: ***It is given that the matrix*** $A = \begin{pmatrix} 1 & 2 & 3 \\ 0 & 5 & 0 \\ 2 & 4 & 3 \end{pmatrix}$.

Co-factors of the elements of the first row of $|A|$ are as follows:

$$\begin{vmatrix} 5 & 0 \\ 4 & 3 \end{vmatrix}, -\begin{vmatrix} 0 & 0 \\ 2 & 4 \end{vmatrix}, \text{ and } \begin{vmatrix} 0 & 5 \\ 2 & 4 \end{vmatrix}, \text{ i.e., 15, 0, and } -10$$

The co-factors of the second row of $|A|$ are as follows:

$$-\begin{vmatrix} 2 & 3 \\ 4 & 3 \end{vmatrix}, \begin{vmatrix} 1 & 3 \\ 2 & 3 \end{vmatrix}, \text{ and } \begin{vmatrix} 1 & 2 \\ 2 & 4 \end{vmatrix}, \text{ i.e., 6, } -3\text{, and 0, respectively.}$$

The co-factors of the elements of the third row of $|A|$ are as follows:

$$-\begin{vmatrix} 2 & 3 \\ 5 & 0 \end{vmatrix}, \begin{vmatrix} 1 & 3 \\ 0 & 0 \end{vmatrix}, \text{ and } \begin{vmatrix} 1 & 2 \\ 0 & 5 \end{vmatrix}, \text{ i.e., } -15\text{, 0, and 5, respectively.}$$

∴ Matrix B will be framed as

$$B = \begin{pmatrix} 15 & 0 & -10 \\ 6 & -3 & 0 \\ -15 & 0 & 5 \end{pmatrix}$$

Hence, adjoint A = transpose of $B = \begin{pmatrix} 15 & 6 & -15 \\ 6 & -3 & 0 \\ -10 & 0 & 5 \end{pmatrix}$ Answer.

3.2.11 Inverse of a Matrix

If the matrices A and B are so related that $AB = BA = I$, then B is known as the inverse of A and is expressed as A^{-1}, i.e., $A^{-1} = B$.

It should be kept in mind that A and B both are square matrices of the same order. It means that non-square matrices cannot possess inverse matrix.

It is important to note that the necessary and sufficient condition for a square matrix A to possess the inverse is that $|A| \neq 0$. Under this condition,

$$B = A^{-1} = \frac{\text{adj } A}{|A|}$$

$$\therefore AA^{-1} = I$$

3.2.12 Orthogonal Matrices

An orthogonal matrix is the one the inverse of which is equal to transpose.

Mathematically, it may be expressed as $A^{-1} = A^t$, where A^t = transpose of matrix A. The name orthogonal is because of the fact that the rows (or columns) of such a matrix behave like orthonormal row (or column) vectors since

$$A^t \cdot A = AA^t = 1$$

3.3 Determinants

A determinant is an algebraic method of solving simultaneous equation in which an expression is written out in a square array. It is generally represented by Δ.

Thus, the determinant of $a_1b_2 - a_2b_1$ is expressed as

$$\begin{vmatrix} a_1 & b_1 \\ a_2 & b_2 \end{vmatrix}$$

This is an example of determinant of order 2.

The value of a determinant of order 2 is equal to the product of the elements along the principal diagonal minus the product of the off-diagonal elements. Keep in mind that in a determinant, the number of rows is equal to the number of columns.

The example of third order determinant is

$$\Delta = \begin{vmatrix} a_1 & b_1 & c_1 \\ a_2 & b_2 & c_2 \\ a_3 & b_3 & c_3 \end{vmatrix}$$

and it is called a determinant of order 3 and its value is obtained by expanding it, i.e.,

$$\Delta = a_1\begin{vmatrix} b_2 & c_2 \\ b_3 & c_3 \end{vmatrix} - b_1\begin{vmatrix} a_2 & c_2 \\ a_3 & c_3 \end{vmatrix} + c_1\begin{vmatrix} a_2 & b_2 \\ a_3 & b_3 \end{vmatrix}$$

This is known as the expansion of the determinant along its first row. In this expansion, each element in the first row is multiplied by that determinant of the second order, which is obtained by suppressing the row and column containing that element. Beginning with the first, the signs of the products are alternately positive and negative.

The value of the above determinant will be

$$\Delta = a_1(b_2c_3 - b_3c_2) - b_1(a_2c_3 - a_3c_2) + c_1(a_2b_3 - a_3b_2)$$

$$= a_1b_2c_3 - a_1b_3c_2 - b_1a_2c_3 + b_1a_3c_2 + a_2b_3c_1 - a_3b_2c_1$$

Similarly, the expansion of determinants of order 4 may be done.

3.3.1 Properties of Determinants

- A determinant remains unchanged if its row is changed into columns and its columns into rows.

$$\Delta = \begin{pmatrix} a_1 & b_1 & c_1 \\ a_2 & b_2 & c_2 \\ a_3 & b_3 & c_3 \end{pmatrix} \text{ and } \Delta^1 = \begin{pmatrix} a_1 & a_2 & a_3 \\ b_1 & b_2 & b_3 \\ c_1 & c_2 & c_3 \end{pmatrix}$$

  The leading term in both determinants is $a_1b_2c_3$. The results in the two determinants are the same. Hence, $\Delta = \Delta^1$.

- Interchange of two rows or two columns changes the sign of a determinant without changing its absolute value.

  Interchange of two rows is equivalent to the interchange of two suffixes, and the interchange of two columns is equivalent to the interchange of two letters. Therefore, in either case, the sign of every term is altered, whereas the absolute value remains unchanged.

A determinant in which two rows or two columns are identical is equal to zero.

If Δ be the determinant in which two rows or two columns are identical, the value of Δ remains unchanged if these two rows or these two columns are interchanged.

But on the basis of point 2 above, Δ changes to $-\Delta$. Therefore, $\Delta = -\Delta$ or $\Delta = 0$.

3.3.2 Minors and Co-factors

The minor of an element is the determinant left after erasing the row and column of a given determinant having the element.

We shall understand this by taking an example.

Let us consider the determinant

$$\Delta = \begin{pmatrix} a_{11} & a_{12} & a_{13} \\ a_{21} & a_{22} & a_{23} \\ a_{31} & a_{32} & a_{33} \end{pmatrix}$$

If we leave the row and column passing through the element a_{ij}, then the second-order determinant thus obtained is known as the minor of the element a_{ij}. We may denote the minor of the element a_{ij} as M_{ij}.

For example:

The minor of the element $a_{11} = \begin{pmatrix} a_{22} & a_{23} \\ a_{32} & a_{33} \end{pmatrix} = M_{11}$

The minor of the element $a_{21} = \begin{pmatrix} a_{12} & a_{13} \\ a_{32} & a_{33} \end{pmatrix} = M_{21}$

The minor of the element $a_{32} = \begin{pmatrix} a_{11} & a_{13} \\ a_{21} & a_{23} \end{pmatrix} = M_{32}$, and so on. In terms of the notation of minors, if we expand Δ along the first row, then

$$\Delta = (-1)^{1+1} a_{11}M_{11} + (-1)^{1+2} a_{12}M_{12} + (-1)^{1+3} a_{13}M_{13}$$

$$= a_{11}M_{11} - a_{12}M_{12} + a_{13}M_{13}$$

The co-factor of an element is defined as the coefficient of that element in the expanded form of the determinant and is, therefore, equal to the corresponding minor with the proper sign attached. For an element which is in the rth row and the sth column, the co-factor = $(-1)^{r+s}$ (the corresponding minor).

3.3.3 Uses of Determinants in Quantum Chemistry

Most students are probably familiar with the use of determinants in solving n-linearly independent equations for unknowns. The two other uses of determinant in quantum chemistry are as follows:

- It is used to represent the product of one particle antisymmetric functions with respect to the full spatial and spin co-ordinates of a pair of indistinguishable fermions. It is to be noted that the antisymmetric property is the consequence of the fundamental property of determinants: a change in sign when rows or columns are interchanged.
- In the determination of roots of secular equations in course of Hückle Molecular Orbital calculations for energy. For this, first of all, carbon atoms are numbered in the conjugated double-bond systems. Then, secular equations and finally secular determinant are set up. The secular determinant is solved to get the roots of the secular equation. Finally, energy levels and molecular orbital are deduced.

3.4 Characteristics Value Problem

We want to find the values of a scalar parameter λ for which there exists $X \neq 0$, satisfying

$$AX = \lambda X \tag{3.16}$$

where A is an nth-order matrix. Such a problem is called a characteristics value problem.

If $X \neq 0$, Eq. (3.16) satisfies for a given scalar λ, when operating an operator A on vector X which is scalar multiple of X.

If $X = 0$, $AX = 0$ has a trivial solution. For vectors $X = 0$, the above equation can be written as

$$AX = \lambda IX \tag{3.17}$$

$$\text{or,} (A - \lambda I)X = 0 \tag{3.18}$$

since $X \neq 0$, $|A - \lambda I| = 0$ if and only if

$$\begin{vmatrix} a_{11} - \lambda & a_{12} & \cdots & a_{1n} \\ a_{21} & a_{22} - \lambda & \cdots & a_{2n} \\ \cdots & \cdots & \cdots & \cdots \\ \cdots & \cdots & \cdots & \cdots \\ a_{n1} & a_{n2} & \cdots & a_{nn} - \lambda \end{vmatrix} = 0 \tag{3.19}$$

Obviously, there is a polynomial in λ. The highest order term in λ comes from the product of changed element. Thus, $(-\lambda)^n$ is the highest order term and $|A - \lambda I|$ is an nth-degree polynomial.

We can write as

$$\begin{aligned} f(\lambda) &= |A - \lambda I| \\ &= (-\lambda)^n + b_{n-1}(-\lambda)^{n-1} + \cdots + b_1(-\lambda) + b_0 \end{aligned} \tag{3.20}$$

where $f(\lambda)$ = characteristics polynomial for A

$|A - \lambda I| = 0$ = characteristics or secular equation for matrix A.

The $f(\lambda) = 0$ has n roots which may be either real or complex number.

The roots of the characteristic equation, which may be represented by λ_i, $i = 1, 2, \ldots, n$ and are known as characteristic values or proper values or latent roots of the matrix A.

The vectors $X \neq 0$ are known as characteristic vectors or Eigen vectors of the matrix A.

We can write the polynomial $f(\lambda)$ in the factored form, using the root $f(\lambda) = 0$, i.e.,

$$f(\lambda) = (\lambda_1 - \lambda)(\lambda_2 - \lambda)(\lambda_n - \lambda) \tag{3.21}$$

The coefficients of different powers can be expressed as

$$b_{n-1} = \sum_{i=1}^{n} \lambda_i = \lambda_1 + \lambda_2 + \cdots + \lambda_n \tag{3.22}$$

$$b_{n-2} = \sum_{j>1}^{n} \lambda_i \lambda_j = \lambda_1\lambda_2 + \cdots + \lambda_1\lambda_n + \lambda_2\lambda_3 + \cdots + \lambda_2\lambda_n + \cdots + \lambda_{n-1}\lambda_n \tag{3.23}$$

$$b_{n-r} = \sum_{\substack{k \gg i \\ i > j}}^{n} \lambda_i \lambda_j \lambda_k, \text{ each term is a product of } r \text{ of this } \lambda_i \text{ and}$$

$$b_0 = \lambda_1 \lambda_2 \lambda_n \tag{3.24}$$

3.5 Similarity Transformation

Let us suppose that X is an Eigen vector of A corresponding to the Eigen value λ and P is an nth-order non-singular matrix. Then, the vector $y = PX$ will not, in general, be an Eigen vector of A corresponding to the Eigen value λ because on multiplying the left hand side of $AX = \lambda X$ by P gives

$$PAX = \lambda PX \tag{3.25}$$

where $AX = \lambda X$ is called a characteristic value problem. A = a given nth-order matrix.

Keep in mind that Eq. (3.25) is not the same as $APX = \lambda PX$.

However, $X = P^{-1}PX$.

Putting this value in Eq. (3.25), we can write

$$PAP^{-1}PX = \lambda PX \tag{3.26}$$

$$\text{or, } PAP^{-1}y = \lambda y \tag{3.27}$$

where y = an Eigen vector of the matrix PAP^{-1}, corresponding to the Eigen value λ.

We have shown that if λ is an Eigen value of A, then λ will also be an Eigen value of PAP^{-1} for any nth-order non-singular matrix P.

If $B = PAP^{-1}$, then $A = P^{-1}BP$ and $X = P^{-1}y$. This shows that any Eigen value of B must also be an Eigen values of A. Hence, the matrices A and B consist of identical sets of Eigen values and are known as similar matrices.

Similarity: If there exists a non-singular matrix P such that $B = PAP^{-1}$, the square matrices A and B are called similar. If $B = PAP^{-1}$, we say that B is obtained by similarity transformation on A. A similarity transformation is a special case of an equivalence transformation. If B is similar to A, B is also called equivalent to A.

Note that if $B = PAP^{-1}$, then $B = R^{-1}AR$, where $R = P^{-1}$. Thus, either PAP^{-1} or $P^{-1}AP$ represents a similarity transformation on A. We shall call B similar to A if $B = PAP^{-1}$ for some P or $B = R^{-1}AR$ for some R. The definitions are equivalent since putting $R = P^{-1}$ converts one form into another.

Example 1

Find the Eigen value and Eigen vector of $A = \begin{bmatrix} 2 & \sqrt{2} \\ \sqrt{2} & 1 \end{bmatrix}$.

Solution: ***The characteristic equation is***

$$|A - \lambda I| = \begin{bmatrix} 2-\lambda & \sqrt{2} \\ \sqrt{2} & 1-\lambda \end{bmatrix}$$

$$= (2-\lambda)(1-\lambda) - 2 = 0$$

$$= 2 - \lambda - 2\lambda + \lambda^2 - 2 = 0$$

$$= \lambda^2 - 3\lambda = 0$$

$$= \lambda(\lambda - 3) = 0$$

$\therefore$ The Eigen value will be $\lambda_1 = 0$ and $\lambda_2 = 3$.

To find the Eigen vectors corresponding to λ_i, we have to solve the set of homogeneous equations $(A - \lambda_i I)x = 0$. Let us put/take $\lambda_1 = 0$ first, and then the set of equations will be

$$2x_1 + \sqrt{2}x_2 = 0$$

$$\text{or } \sqrt{2}x_1 + x_2 = 0$$

$$\therefore x_1 = -\frac{x_2}{\sqrt{2}}$$

If we want to find an Eigen vector of unit length, we must require $x_1^2 + x_2^2 = 1$.

$$\text{Thus } \frac{x_2^2}{2} + x_2^2 = 1 \quad \left[\therefore \quad x_1 = -\frac{x_2}{\sqrt{2}} \right]$$

$$\text{or } \frac{3}{2}x_2^2 = 1$$

$$\therefore x_2 = \sqrt{2/3} \text{ and } x_1 = -1/\sqrt{3}$$

And the Eigen vector of unit length corresponding to λ_1 be

$$u_1 = \left[\frac{-1}{\sqrt{3}}, \sqrt{2}/3 \right]$$

Note carefully that u_1 is not completely specified by the requirement of $|u_1| = 1$. The vector $-u_1$ is also an Eigen vector of length 1. However, u_1 and $-u_1$ are not linearly independent. Only one linearly independent Eigen vector corresponds to λ_1.

For $\lambda_2 = 3$, the set of equations becomes

$$-x_1 + \sqrt{2}x_2 = 0$$

$$\text{or } \sqrt{2}x_1 - 2x_2 = 0$$

$$\text{or } x_1 = \sqrt{2}x_2$$

If $x_1^2 + x_2^2 = 1$, then $3x_2^2 = 1$; from this, we obtain

$$x_2 = \frac{1}{\sqrt{3}} \text{ and } x_1 = \frac{\sqrt{2}}{3} \text{ and } u_2 = \left[\frac{\sqrt{2}}{3}, \frac{1}{\sqrt{3}} \right]$$

Again there is only one linearly independent Eigen vector which corresponds to λ_2.

It can easily be checked that $u_2 u_1 = 0$ which is in agreement with the theory.

Alternatively: Similarity transformation

Let us consider two sets of co-ordinate system, such as $(x_1, x_2, x_3 \ldots x_n)$ and $(x_1', x_2', x_3' \cdots x_n')$, which are taken to be as column vectors. For example,

$$x = \begin{bmatrix} x_1 \\ x_2 \\ x_3 \\ . \\ . \\ x_n \end{bmatrix} \quad \text{and} \quad X' = \begin{bmatrix} x_1' \\ x_2' \\ x_3' \\ . \\ . \\ x_n' \end{bmatrix}$$

These two column vectors may be supposed to be interrelated as follows:

$$X = QX' \tag{3.28}$$

With Q as n-dimensional transformation matrix, i.e.,

$$\begin{bmatrix} x_1 \\ x_2 \\ x_3 \\ \cdot \\ \cdot \\ x_n \end{bmatrix} = Q \begin{bmatrix} x_1' \\ x_2' \\ x_3' \\ \cdot \\ \cdot \\ x_n' \end{bmatrix}$$

It is thus clear that a new type of transformation is found out by matrix A such that $AX = Y$, where $Y = QY'$.

Both X and Y belong to the same system of co-ordinates. If we write $A'X' = Y'$, then an additional transformation will be obtained.

Keep in mind that an established relationship may exist, between A and A'. As $X = QX'$, $Y = QY'$ and $Q^{-1}AQX' = Y'$ also $A'X' = Y'$.

Since Q is a non-singular matrix, multiply by Q^{-1} such that

$$Q^{-1}AQX' = Q^{-1}QY' = Y' \tag{3.29}$$

This kind of transformation is regarded as similarity transformation. The two matrices A and A' will be similar

$$Q^{-1}AQ = X_A$$

$$Q^{-1}BQ = X_B$$

X_A and X_B are numbers which commute so that

$$Q^{-1}X_BQ = Q^{-1}QX_B = X_B$$

$$Q^{-1}X_AQ = Q^{-1}QX_A = X_A$$

3.6 Block Diagonalisation of Matrices

A block diagonal matrix is a block matrix which is a square matrix, having main diagonal blocks square matrix such that the off-diagonal blocks are null matrix.

Let us consider a square matrix A of three dimensions such as

$$A = \begin{bmatrix} a_{11} & a_{12} & a_{13} \\ a_{21} & a_{22} & a_{23} \\ a_{31} & a_{32} & a_{33} \end{bmatrix}$$

It is a 3×3 matrix.

The matrix A is to be transformed into B. The matrix retains its dimension in the course of transformation $A \to B$. This can be achieved through similarity transformation. $Q^{-1}AQ = B$ denotes similarity transformation. This process is repeated so that a point is reached when similarity transformation does not further reduce the dimensions.

The matrix will be rearranged in such a way that the zero elements are symmetrically distributed about the diagonal. The non-zero elements are distributed into blocks along the diagonal.

$$Q^{-1}AQ = Q^{-1}\begin{pmatrix} a_{11} & a_{12} & a_{13} \\ a_{21} & a_{22} & a_{23} \\ a_{31} & a_{32} & a_{33} \end{pmatrix} Q \rightarrow A = \begin{pmatrix} b_{11} & b_{12} & 0 \\ b_{21} & b_{22} & 0 \\ 0 & 0 & b_{33} \end{pmatrix} = B$$

Keep in mind that matrix B is regarded as made up of two smaller matrices B_1 and B_2. Here, we obtain a 2×2 block and a 1×1 block such that

$$\text{and } B_1 = \begin{bmatrix} b_{11} & b_{12} \\ b_{21} & b_{22} \end{bmatrix} \text{and } B_2 = [b_{33}]$$

We see that B_1 is a 2×2 matrix whereas B_2 is a 1×1 matrix.

It is thus clear that we have changed the three-dimensional representation of A to another three-dimensional representation of smaller dimensions B_1 and B_2. Thus, the reduction of the original representation has been affected.

BIBLIOGRAPHY

Anderson, J.M. 1968. *Introduction to Quantum Chemistry*. New York: W.A. Benjamin Inc.

Hameka, H.F. 1967. *Introduction to Quantum Theory*. New York: Harper and Row

Margenau, H. and G.M. Murphy. 1943. *The Mathematics of Physics and Chemistry*. Princeton, NJ: D. Van Nostrand

Miller, K.S. 1935. *Partial Differential Equation in Engineering Problems*. New York: Prentice Hall

Spiegel, M.R. 1971. *Advanced Mathematics for Engineering and Scientists*. New York: McGraw-Hill

Solved Problems

Problem 1. Using matrices $A = \begin{pmatrix} -3 & 6 & 4 \\ 1 & 0 & 2 \end{pmatrix}$ and $B = \begin{pmatrix} 2 & 1 & 1 \\ -6 & 4 & 3 \end{pmatrix}$ form the matrix $C = 3A - 2B$.

Solution: Matrix

$$C = 3\begin{pmatrix} -3 & 6 & 4 \\ 1 & 0 & 2 \end{pmatrix} - 2\begin{pmatrix} 2 & 1 & 1 \\ -6 & 4 & 3 \end{pmatrix}$$

$$= \begin{pmatrix} -9 & 18 & 12 \\ 3 & 0 & 6 \end{pmatrix} - \begin{pmatrix} 4 & 2 & 2 \\ -12 & 8 & 6 \end{pmatrix}$$

$$= \begin{pmatrix} -13 & 16 & 10 \\ 15 & -8 & 0 \end{pmatrix}$$

Problem 2. Find $D = BA$ if $B = \begin{pmatrix} 1 & 2 & 1 \\ 3 & 0 & -1 \\ -1 & -1 & 2 \end{pmatrix}$ and $A = \begin{pmatrix} -3 & 0 & -1 \\ 1 & 4 & 0 \\ 1 & 1 & 1 \end{pmatrix}$.

Solution: Both matrices B and A are 3×3 matrices. Hence, the product will also be a 3×3 matrix. Given that $D = BA$,

$$D = \begin{pmatrix} 1 & 2 & 1 \\ 3 & 0 & -1 \\ -1 & -1 & 2 \end{pmatrix}\begin{pmatrix} -3 & 0 & -1 \\ 1 & 4 & 0 \\ 1 & 1 & 1 \end{pmatrix}$$

$$= \begin{pmatrix} -3+2+1 & 0+8+1 & -1+0+1 \\ -9+0-1 & 0+0-1 & -3+0-1 \\ +3-1+2 & 0-4+2 & 1+0+2 \end{pmatrix}$$

$$= \begin{pmatrix} 0 & 9 & 0 \\ -10 & -1 & -4 \\ 4 & -2 & 3 \end{pmatrix}$$ Answer.

Problem 3. Do the matrices A and B commute if $A = \begin{pmatrix} 2 & 1 \\ 0 & 1 \end{pmatrix}$ and $= \begin{pmatrix} 1 & 1 \\ 0 & 1 \end{pmatrix}$?

Solution: We have to prove that $AB = BA$; otherwise, they will not commute.
We shall first find AB.

$$AB = \begin{pmatrix} 2 & 1 \\ 0 & 1 \end{pmatrix}\begin{pmatrix} 1 & 1 \\ 0 & 1 \end{pmatrix} = \begin{pmatrix} 2 & 3 \\ 0 & 1 \end{pmatrix}$$

$$\text{and then } BA = \begin{pmatrix} 1 & 1 \\ 0 & 1 \end{pmatrix}\begin{pmatrix} 2 & 1 \\ 0 & 1 \end{pmatrix}$$

$$= \begin{pmatrix} 2 & 2 \\ 0 & 1 \end{pmatrix}$$

Thus, $AB \neq BA$. Therefore, A and B will not commute. Answer.

Problem 4. Prove that the product of the matrices

$$\begin{pmatrix} \cos^2\theta & \cos\theta\sin\theta \\ \cos\theta\sin\theta & \sin^2\theta \end{pmatrix} \text{and} \begin{pmatrix} \cos^2\phi & \cos\phi\sin\phi \\ \cos\phi.\sin\phi & \sin^2\phi \end{pmatrix}$$

is a zero matrix when θ and ϕ differ by an odd multiple of $\pi/2$.

Solution: The product of the two given matrices is

$$\begin{pmatrix} \cos^2\theta\cos^2\phi + \cos\theta.\sin\theta.\cos\phi.\sin\phi & \cos^2\theta\cos\phi.\sin\phi + \cos\theta.\sin\theta.\sin^2\phi \\ \cos\theta\sin\theta\cos^2\phi + \sin^2\theta.\cos\phi.\sin\phi & \cos\theta\sin\theta.\cos\phi\sin\phi + \sin^2\theta.\sin^2\phi \end{pmatrix}$$

$$\begin{pmatrix} \cos\theta\cos\phi(\cos\theta\cos\phi + \sin\theta\sin\phi) & \cos\theta\sin\phi(\cos\theta\cos\phi + \sin\theta\sin\phi) \\ \sin\theta\cos\phi(\cos\theta\cos\phi + \sin\theta\sin\phi) & \sin\theta\sin\phi(\cos\theta\cos\phi + \sin\theta\sin\phi) \end{pmatrix}$$

$$\begin{pmatrix} \cos\theta\cos\phi\cos(\theta-\phi) & \cos\theta\sin\phi\cos(\theta-\phi) \\ \sin\theta\cos\phi\cos(\theta-\phi) & \sin\theta\sin\phi\cos(\theta-\phi) \end{pmatrix}$$

$= \begin{pmatrix} 0 & 0 \\ 0 & 0 \end{pmatrix}$ provided $\cos(\theta-\phi) = 0$, i.e., provided $(\theta-\phi)$ is an odd multiple of $\pi/2$. Answer.

Problem 5. If $A = \begin{pmatrix} 2 & 3 \\ 0 & 1 \end{pmatrix}$, $B = \begin{pmatrix} 3 & 4 \\ 2 & 1 \end{pmatrix}$, verify that $(A + B)' = A' + B'$ and $(AB)' = (B'A')$

Solution: Given that

$$A' = \begin{pmatrix} 2 & 3 \\ 0 & 1 \end{pmatrix}' = \begin{pmatrix} 2 & 0 \\ 3 & 1 \end{pmatrix} \text{ and } B' = \begin{pmatrix} 3 & 4 \\ 2 & 1 \end{pmatrix}' = \begin{pmatrix} 3 & 2 \\ 4 & 1 \end{pmatrix}$$

$$\therefore \quad A + B = \begin{pmatrix} 2+3 & 3+4 \\ 0+2 & 1+1 \end{pmatrix} = \begin{pmatrix} 5 & 7 \\ 2 & 2 \end{pmatrix}$$

$$(A+B)' = \begin{pmatrix} 5 & 7 \\ 2 & 2 \end{pmatrix}' = \begin{pmatrix} 5 & 2 \\ 7 & 2 \end{pmatrix}$$

$$\text{and } A' + B' = \begin{pmatrix} 2 & 0 \\ 3 & 1 \end{pmatrix} + \begin{pmatrix} 3 & 2 \\ 4 & 1 \end{pmatrix}$$

$$= \begin{pmatrix} 2+3 & 0+2 \\ 3+4 & 1+1 \end{pmatrix} = \begin{pmatrix} 5 & 2 \\ 7 & 2 \end{pmatrix}$$

$$\therefore = (A+B)' = A' + B'$$

Further,

$$AB = \begin{pmatrix} 2 & 3 \\ 0 & 1 \end{pmatrix}\begin{pmatrix} 3 & 4 \\ 2 & 1 \end{pmatrix}$$

$$= \begin{pmatrix} 2\times3+3\times2 & 2\times4+3\times1 \\ 0\times3+1\times2 & 0\times4+1\times1 \end{pmatrix}$$

$$= \begin{pmatrix} 12 & 11 \\ 2 & 1 \end{pmatrix}$$

$$\therefore (AB)' = \begin{pmatrix} 12 & 2 \\ 11 & 1 \end{pmatrix}$$

Also,

$$B'A' = \begin{pmatrix} 3 & 2 \\ 4 & 1 \end{pmatrix}\begin{pmatrix} 2 & 0 \\ 3 & 1 \end{pmatrix} = \begin{pmatrix} 3\times2+2\times3 & 3\times0+2\times1 \\ 4\times2+1\times3 & 4\times0+1\times1 \end{pmatrix}$$

$$= \begin{pmatrix} 12 & 2 \\ 11 & 1 \end{pmatrix}$$

$$\therefore (AB)' = B'A'$$

Problem 6. Prove that the matrix $\frac{1}{3}\begin{pmatrix} 1 & 2 & 2 \\ 2 & 1 & -2 \\ -2 & 2 & -1 \end{pmatrix}$ is orthogonal.

Solution: Let the given matrix be A.

$$\therefore A = \frac{1}{3}\begin{pmatrix} 1 & 2 & 2 \\ 2 & 1 & -2 \\ -2 & 2 & -1 \end{pmatrix}$$

$$\therefore A' = \frac{1}{3}\begin{pmatrix} 1 & 2 & -2 \\ 2 & 1 & 2 \\ 2 & -2 & -1 \end{pmatrix}$$

We have $AA' = \frac{1}{9}\begin{pmatrix} 1 & 2 & 2 \\ 2 & 1 & -2 \\ -2 & 2 & -1 \end{pmatrix}\begin{pmatrix} 1 & 2 & -2 \\ 2 & 1 & 2 \\ 2 & -2 & -1 \end{pmatrix}$

$$= \frac{1}{9}\begin{pmatrix} 9 & 0 & 0 \\ 0 & 9 & 0 \\ 0 & 0 & 9 \end{pmatrix} = \begin{pmatrix} 1 & 0 & 0 \\ 0 & 1 & 0 \\ 0 & 0 & 1 \end{pmatrix}$$

$\therefore$ The matrix A is orthogonal. Answer.

Problem 7. Find the inverse of the matrix $A = \begin{pmatrix} 1 & 3 & 3 \\ 1 & 4 & 3 \\ 1 & 3 & 4 \end{pmatrix}$.

Solution: Given that $A = \begin{pmatrix} 1 & 3 & 3 \\ 1 & 4 & 3 \\ 1 & 3 & 4 \end{pmatrix} = \begin{pmatrix} 1 & 3 & 0 \\ 0 & 1 & 0 \\ 0 & 0 & 1 \end{pmatrix}$

By applying, $R_2 \to R_2 - R_1$.

$R_3 \to R_3 - R_1 = 1$ on expanding the determinant along the first column.

Since $|A| \neq 0$, matrix A is non-singular and possesses inverse.

Now the co-factors of the elements of the first row of the determinant $|A|$ are

$$\begin{vmatrix} 4 & 3 \\ 3 & 4 \end{vmatrix}, -\begin{vmatrix} 1 & 3 \\ 1 & 4 \end{vmatrix}, \begin{vmatrix} 1 & 4 \\ 1 & 3 \end{vmatrix}, \text{ i.e., } 7, -1, \text{ and } -1, \text{ respectively.}$$

The co-factors of the elements of the second row of $|A|$ are

$$-\begin{vmatrix} 3 & 3 \\ 3 & 4 \end{vmatrix}, -\begin{vmatrix} 1 & 3 \\ 1 & 3 \end{vmatrix}, \begin{vmatrix} 1 & 3 \\ 1 & 4 \end{vmatrix}, \text{ i.e., } -3, 0, \text{ and } 1, \text{ respectively.}$$

$\therefore$ The Adj A = the transpose of the matrix $\begin{pmatrix} 7 & -1 & -1 \\ -3 & 1 & 0 \\ -3 & 0 & 1 \end{pmatrix}$

$$\therefore \quad Adj\ A = \begin{pmatrix} 7 & -3 & -1 \\ -1 & 1 & 0 \\ -1 & 0 & 1 \end{pmatrix}$$

$$\therefore \quad A^{-1} = \frac{1}{|A|}\text{Adj } A = \begin{pmatrix} 7 & -3 & -1 \\ -1 & 1 & 0 \\ -1 & 0 & 1 \end{pmatrix}, \text{ since } |A| = 1$$

Answer.

Problem 8. Solve the following system of equations using matrix method:

$$5x + 2y = 3 \quad 3x + 2y = 5$$

Solution: The given system of equations can be expressed in the form of matrix as

$$AX = B$$

where $A = \begin{pmatrix} 5 & 2 \\ 3 & 2 \end{pmatrix}$, $X = \begin{pmatrix} x \\ y \end{pmatrix}$, $B = \begin{pmatrix} 3 \\ 5 \end{pmatrix}$

We have $|A| = 10 - 6 = 4 \neq 0$; hence, A^{-1} exists.
Co-factors of the elements of the first row of $|A|$ are 2, –3.
Co-factors of the elements of the second row of $|A|$ are –2, 5.

$$\text{Now Adj } A = \begin{pmatrix} 2 & -3 \\ -2 & 5 \end{pmatrix}^t = \begin{pmatrix} 2 & -2 \\ -3 & 5 \end{pmatrix}$$

$$\therefore \quad A^{-1} = \frac{1}{|A|}\text{ Adj } A = \frac{1}{4}\begin{pmatrix} 2 & -2 \\ -3 & 5 \end{pmatrix}$$

$$\therefore \quad X = A^{-1}B = \frac{1}{4}\begin{pmatrix} 2 & -2 \\ -3 & 5 \end{pmatrix}\begin{pmatrix} 3 \\ 5 \end{pmatrix}$$

$$= \begin{pmatrix} -1 \\ 4 \end{pmatrix} \text{or} \begin{pmatrix} x \\ y \end{pmatrix} = \begin{pmatrix} -1 \\ 4 \end{pmatrix}$$

$$\therefore \quad x = -1 \text{ and } y = 4$$

Answer.

Problem 9. Show that $\begin{vmatrix} 1 & \omega & \omega^2 \\ \omega & \omega^2 & 1 \\ \omega^2 & 1 & \omega \end{vmatrix} = 0$ where ω is a complex cube root of unity.

Solution: Given that

$$\Delta = \begin{vmatrix} 1 & \omega & \omega^2 \\ \omega & \omega^2 & 1 \\ \omega^2 & 1 & \omega \end{vmatrix}$$

Applying $c_1 \rightarrow c_1 + c_2 + c_3$, we have

$$\Delta = \begin{vmatrix} 1+\omega+\omega^2 & \omega & \omega^2 \\ 1+\omega+\omega^2 & \omega^2 & 1 \\ 1+\omega+\omega^2 & 1 & \omega \end{vmatrix}$$

$$= \begin{vmatrix} 0 & \omega & \omega^2 \\ 0 & \omega^2 & 1 \\ 0 & 1 & \omega \end{vmatrix} = 0 \quad \left[\therefore \; 1+\omega+\omega^2 = 0\right]$$

because all entries in c_1 are zero. Answer.

Problem 10. Using determinants, solve the system of equations $2x - 2y = 1$, $x + 2y = 2$.

Solution: We have $D = \begin{vmatrix} 2 & -2 \\ 1 & 2 \end{vmatrix} = 4 + 2 = 6 \neq 0$

Also,

$$D_1 = \begin{vmatrix} 2 & -2 \\ 1 & 2 \end{vmatrix} = 2 + 4 = 6$$

$$D_2 = \begin{vmatrix} 2 & 1 \\ 1 & 2 \end{vmatrix} = 4 - 1 = 3$$

$$\therefore \quad x = \frac{D_1}{D} = \frac{6}{6} = 1, \text{ and } y = \frac{D_2}{D} = \frac{3}{6} = 1/2$$

Hence, $x = 1$, $y = 1/2$ Answer.

Questions on Concepts

Solve the following differential equations:

1.

a. $\dfrac{dy}{dx} = \sqrt{1 + x^2 + y^2 + x^2y^2}$ [Ans. $\log\left\{y + \sqrt{1+y^2}\right\} = \dfrac{x - \sqrt{1+x^2}}{2} + \dfrac{1}{2}\log\left(x + \sqrt{1+x^2}\right) + 0$]

b. $\dfrac{dy}{dx} = e^{x+y} + x^2e^2$ [Ans. $e^x + x^3/3 + e^{-y} = k$]

c. $ydx - xdy = xydx$ [Ans. $y = xe^{c-x}$]

d. $x\dfrac{dy}{dx} - y = 2x^2y$ [Ans. $y = cxe^{x^2}$]

e. $e^{x-y}dx + e^{x-y}dy = 0$ [Ans. $e^{2x} + e^{2y} = c$]

2.

a. $(6x^2 - 7y^2)\, dx - 14xy\, dy = 0$ [Ans. $2x^2 - 7xy^2 = c$]

b. $xdy - ydx = \sqrt{x^2 + y^2 dx}$ [Ans. $y + \sqrt{x^2 + y^2} = cx^2$]

c. $(x^2 + y^2)\, dx - 2xy\, dx = 0$ [Ans. $x^2 - y^2 = cx$]

d. $y(x^2 + xy - 2y^2)\, dx + x\,(3y^2 - xy - x^2)\, dy = 0$ [Ans. $2y^2 \ln\,(y^2/x^2) + 2xy + x^2 = cy^2$]

e. $(3x + 2y)dx + 2xdy = 0$ [Ans. $3x^2 + 4xy = c$]

3.

a. $x\left(\frac{dy}{dx}\right) + 2y = x^3$ [Ans. $5yx^2 = x^5 + c$]

b. $\frac{dy}{dx} - 2xy = 2e^{y^2}y$ [Ans. $xe^{-y^2} = y^2 + c$]

c. $\frac{dy}{dx} + \frac{1}{x}y = x^3 - 3$ [Ans. $y = \frac{x^4}{5} - \frac{3x}{2} + cx^{-1}$]

d. $\frac{dy}{dx} + y\cot x = 2\cos x$ [Ans. $y\sin x + \frac{\cos 2x}{2} = c$]

4.

a. $\frac{dy}{dx} + 1 = e^{x-y}$ [Ans. $e^x + y = \frac{e^{2x}}{2} + c$]

b. $\frac{dy}{dx} + \frac{y}{x} = y^2/x^2$ [Ans. $\frac{1}{xy} = \frac{1}{2x^2} + c$]

c. $\left(x^3y^2 + xy\right)dx = dy$ [Ans. $\frac{1}{y} = \left(x^2 - 2\right) + Ce^{-x^2/2}$]

d. $\frac{dy}{dx} + \frac{y}{x} = y^2$ [Ans. $xy\left(\log x + c\right) + 1 = 0$]

e. $x^2\frac{dy}{dx} + xy = y^2$ [Ans. $\frac{1}{xy} = \frac{1}{2x^2} + c$]

f. $y^2\frac{dy}{dx} = x + y^3$ [Ans. $y^3 + x + 1/3 = ce^{3x}$]

5.

a. $\frac{d^2y}{dx^2} - \frac{dy}{dx} - 2y = 0$ [Ans. $y = c_1 e^{-x} + c_2e^{2x}$]

b. $\frac{d^2y}{dx^2} + k^2y = 0$ [Ans. $y = A\cos kx + B\sin kx$]

c. $\frac{d^2y}{dx^2} - 3\frac{dy}{dx} + 2y = 0$ [Ans. $y = c_1e^x + c_2e^{2x}$]

d. $\frac{d^2x}{dt^2} = \omega^2x = 0$

Subject to the initial condition $x(0) = A$ and $dx/dt = 0$ at $t = 0$ [Ans. $x = A\cos\omega t$]

e. $\frac{d^2y}{dx^2} - 7\frac{dy}{dx} + 12y = 0$ [Ans. $y = c_1e^{-3x} + c_2e^{-4x}$]

f. $\frac{d^4y}{dx^2} + k^4y = 0$ [Ans. $y = c_1e^{kx} + c_2e^{-kx} + c_3\cos kx + c_4\sin kx$]

g. $\frac{d^3y}{dx^3} - 3\frac{dy}{dx} + 2y = 0$ [Ans. $y = 0; y = c_1e^{-2x}\left(c_2 + c_3x\right)e^x$]

h. $\frac{d^3y}{dx^3} - 8y = 0$ [Ans. $y = c_1e^{2x} + e^{-x}\left(c_2\cos\sqrt{3x} + c_3x\right)$]

6. Solve the following by power series method.

a. $\frac{d^2y}{dx^2} + xy = 0$ [Ans. $y = c_0\left(1 - \frac{x^3}{3!} + \frac{1.4x^6}{6!} - \frac{1.4.7x^9}{9!}\right) + c_1\left(x - \frac{2x^4}{4!} + \frac{2.5x^7}{7!}\right)$]

b. $\frac{d^2y}{dx^2} - xy = 0$ [Ans. $y = c_0\left(1 + x^3/3! + \cdots\right) + c_1\left(x + 2x^4/4! + \cdots\right) + \frac{2.5.8\cdots(3n-1)c^{3n+1}}{(3n+1)!} + \cdots$]

c. $\dfrac{d^2y}{dx^2} + x\dfrac{dy}{dx} - 2y = 0$ [Ans. $y = c_0\,(1 + x + x/3 + \ldots) + c_1\,(x + x/2 + \ldots) = 0$]

d. $(x-1)\dfrac{d^2y}{dx^2} + y = 0$ [Ans. $y = c_0 + c_1x + c_0x^2/2! + (c_1 + c_0)\,x^3/3! + (3c_0 + 2c_1)\,x^4/4! + \ldots$]

e. $\dfrac{dy}{dx} - 2xy = 0$ [Ans. $y = c_0 \sum_{n=0}^{\infty} x^{2n}/n!$]

7. If $A = \begin{pmatrix} 1 & 0 & -1 \\ -1 & 2 & 0 \\ 0 & 1 & 1 \end{pmatrix}$ and $B = \begin{pmatrix} -1 & 1 & 0 \\ 3 & 0 & 2 \\ 1 & 1 & 1 \end{pmatrix}$ form the matrices $C = 2A - 3B$ and $D = 6B - A$ [Ans. $C = \begin{pmatrix} 5 & -3 & -2 \\ -11 & 4 & -6 \\ -3 & -1 & -1 \end{pmatrix}$ and $D = \begin{pmatrix} -7 & 6 & 1 \\ 19 & -2 & 12 \\ 6 & 5 & 5 \end{pmatrix}$]

8. Examine $AB = BA$, when $A = \begin{pmatrix} -2 & 3 & -1 \\ -1 & 2 & -1 \\ -6 & 9 & -4 \end{pmatrix}$ and $B = \begin{pmatrix} 1 & 3 & 3 \\ 2 & 4 & 3 \\ 3 & 3 & 4 \end{pmatrix}$ [Ans. Yes]

9. Find the inverse of $A = \begin{pmatrix} 1 & 3 & 3 \\ 1 & 4 & 3 \\ 1 & 3 & 4 \end{pmatrix}$ and verify that $A^{-1}A = I_3$

10. If $A = \begin{pmatrix} -4 & -3 & -3 \\ 1 & 0 & 1 \\ 4 & 4 & 3 \end{pmatrix}$ prove that adj $A = A$. [Ans. $|A| \neq 0$]

11. If $A = \begin{pmatrix} 3 & 1 & -1 \\ 0 & 1 & 2 \end{pmatrix}$ find AA' and $A'A$ [Ans. $AA^1 = \begin{pmatrix} 11 & -1 \\ -1 & 5 \end{pmatrix}$ and $\begin{pmatrix} 9 & 3 & -3 \\ 3 & 2 & 1 \\ -3 & 1 & 5 \end{pmatrix}$]

12. If A is a Hermitian matrix, show that iA is skew-Hermitian.
[Hint: Let A be a Hermitian matrix $A^* = A$
we have

$$(iA)^* = \bar{i}A^* \qquad \because (kA)^* = \bar{k}A^*$$

$$= -iA^* \qquad \because \bar{i} = -i$$

$$= -(iA^*) = -(iA) \qquad \because A^* = A$$

Hence, iA is a skew-Hermitian.]

13. Solve the following simultaneous equations by the matrix inverse method.

$$x + y - z = 1$$

$$2x - 3y + z = +$$

$$x + 3z = 0$$

First show that $A^{-1} = \frac{1}{13}\begin{pmatrix} 6 & 3 & 1 \\ 5 & -4 & 3 \\ -2 & -1 & 4 \end{pmatrix}$ and evaluate $x = A^{-1}c$.

[Ans. $x = \frac{24}{13}$, $y = -\frac{19}{13}$ and $z = -\frac{8}{13}$]

14. What do you mean by characteristic value problem?
15. What do you mean by block diagonalisation of matrices? What role does the similarity transformation play in such a procedure?

4

Quantum Mechanical Operators

Before taking into consideration the quantum mechanical operators, it is perhaps the best to clear the meaning of the term *operator* in the mathematical sense. The term operator may be defined in the following manner:

An operator is a symbol, which represents a mathematical operation to be carried out on a specific operand.

or

An operator is a symbol that tells to do something to whatever follows the symbol.

or

An operator may be defined as a short-hand notation for a set of well-defined mathematical operations to be carried out on a function (called the operand).

or

An operator conveys the instruction to carry out operation on a function.

or

An operator is defined as a symbol of mathematical procedure, which alters one function into another.

or

A mathematical entity that acts to transform various functions of the independent variables to different functions of these variables.

It means, operator function = new function

$$\text{e.g.,} \frac{\mathrm{d}}{\mathrm{d}x}(\sin x) = \cos x$$

Here symbol d/dx is an operator that alters the function sin x into first derivative with respect to x, i.e., cos x.

Similarly, the symbol $\left(\sqrt{\ }\right)$ is an operator, the instruction being to take the square root of the quantity, which follows it. For example, if a number, say 25, is kept under operator $\left(\sqrt{\ }\right)$, it alters 25 into its square root, i.e., $\left(\sqrt{25} = 5\right)$. Addition, subtraction, multiplication, division, log, differentiation, integration, etc. are also operators. It is to be noted that the operators always operate on the function written to the right of them and the operator has no physical significance if written alone.

An operator is normally written with a cap sign (^) overhead, e.g., $\hat{A}$ Thus, $\hat{A}$ represents an operator. To make it clearer, if we write $\hat{A}f(x) = g(x)$, it means $\hat{A}$ is an operator operating on the function $f(x)$. $g(x)$ is an another function, which is a new one obtained after the operation of operator A on the function $f(x)$.

If $\hat{A} = \mathrm{d}/\mathrm{d}x$ and $f(x) = ax^2$, then $\hat{A}f(x) = \mathrm{d}/\mathrm{d}x\left(ax^2\right) = 2ax = g(x)$. Similarly, we can cite other examples similar to the above.

4.1 Linear Operator and Non-Linear Operator

An operator is said to be linear if it follows the following mathematical instruction:

$$\text{i.e., } \hat{A}(u+v) = \hat{A}u + \hat{A}v \tag{4.1}$$

If $u = f(x)$ and $v = g(x)$, then

$$\hat{A}[f(x)+g(x)] = \hat{A}f(x) + \hat{A}g(x) \tag{4.2}$$

$$\text{and also } \hat{A}(cu) = c\hat{A}u \tag{4.3}$$

$$\text{or } \hat{A}[cf(x)] = c\hat{A}f(x)$$

where c = a complex number (constant)

$f(x)$ and $g(x)$ = functions

$\hat{A}$ = an operator

In quantum mechanics, only linear operator is used.

Example 1

Prove that d/dx is a linear operator.

Solution: *Let $f(x) = ax^m$ and $g(x) = bx^n$, then* $\frac{d}{dx}[f(x)+g(x)] = \frac{d}{dx}g(x)$

$$= \frac{d}{dx}\left[ax^m + bx^n\right]$$

$$= \left[\frac{d}{dx}(ax^m) + \frac{d}{dx}(bx^n)\right] \tag{4.4}$$

This is of the form $\hat{A}(u + v) = \hat{A}u + \hat{A}v$

Hence, d/dx is a linear operator. Answer.

Non-linear operators are those which do not follow

$$\hat{A}(u+v) = \hat{A}u + \hat{A}v$$

$$\text{i.e., } \hat{A}(u+v) \neq \hat{A}u + \hat{A}v$$

For example: Square root is non-linear because

$$\sqrt{f(x)+g(x)} \neq \sqrt{f(x)} + \sqrt{g(x)} \tag{4.5}$$

Suppose $f(x) = x^2$ and $g(x) = 4x^2$

$$\text{then } \sqrt{x^2 + 4x^2} = \sqrt{5x^2} = x\sqrt{5}$$

$$\text{Now } \sqrt{f(x)+g(x)} = \sqrt{x^2} + \sqrt{4x^2} = x + 2x = 3x$$

$$\therefore \quad \sqrt{f(x)+g(x)} \neq \sqrt{f(x)} + \sqrt{+g(x)}$$

Thus, in case of non-linear operator, the LHS is not equal to the RHS.

We learnt above about the sum of two operators. Now we want to gain knowledge about the product of the two operators. The product of the two operators is defined as the successive application of the operators and is represented by writing the two operator symbols by the side of each other.

Let us write an operator equation

$$\hat{C} = \hat{A}\hat{B} \tag{4.6}$$

Then its equivalent form may be expressed as

$$\begin{aligned} \hat{C}f(x) &= \hat{A}\left(\hat{B}f(x)\right) \\ &= \hat{A}g(x) \end{aligned} \tag{4.7}$$

where $g(x)$ = the function generated by the operation of operator

$$\hat{B} \text{ on } f(x)$$

The operator written next to the symbol for the function (the operator on the right side) always operates first; therefore, operators operate from the right to the left. If we square the operator, it means operating two times with that operator.

The property of operator multiplication is associative, which means that

$$\hat{A}\left(\hat{B}\hat{C}\right) = \left(\hat{A}\hat{B}\right)\hat{C} \tag{4.8}$$

Suppose that the product of the operator $\hat{B}\hat{C} \approx \hat{F}$ and that of $\hat{A}\hat{B}$ is $\hat{G}$, then $\hat{A}\hat{F} = \hat{G}\hat{C}$.

The second property is that operator multiplication and addition are distributive, which means that

$$\hat{A}\left(\hat{B} + \hat{C}\right) = \hat{A}\hat{B} + \hat{A}\hat{C} \tag{4.9}$$

4.2 Commutator

The commutator is mathematically defined as

$$\left[\hat{A}\hat{B}\right] = \hat{A}\hat{B} - \hat{B}\hat{A} \tag{4.10}$$

where $\hat{A}$ and $\hat{B}$ are operators and $\left[\hat{A}, \hat{B}\right]$ = representation of commutator of two operators $\hat{A}$ and $\hat{B}$, i.e., commutator of two operators is denoted by $\left[\hat{A}, \hat{B}\right]$.

If $\hat{A}\hat{B} = \hat{B}\hat{A}$, then the two operators are said to commute, i.e., they are known as commuting operators. But if $\hat{A}\hat{B} \neq \hat{B}\hat{A}$, then the two operators are said not to commute, i.e., they are called the non-commuting operators. We shall cite some examples to clarify the commuting and non-commuting operators in the following discussion.

We have already mentioned that operators operate on functions; therefore, the above can be expressed as

$$\hat{A}\hat{B}f(x) = \hat{B}\hat{A}f(x)$$

for commuting and $\hat{A}\hat{B}f(x) \neq \hat{B}\hat{A}f(x)$ for non-commuting operators.

Example 1

Let $\hat{A} = \sqrt{\ }, \hat{B} = (\ )^2$ and $f(x) = x^2$, then show that $\hat{A}$ and $\hat{B}$ are commuting operators.

Solution: *Given that* $\hat{A} = \sqrt{\ }$, $\hat{B} = (\ )^2$ *and* $f(x) = x^2$

$$\therefore \quad \hat{A}\hat{B}f(x) = \hat{A}\left(x^2\right)^2 = \sqrt{\left(x^2\right)^2} = x^2$$

$$\text{and } \hat{B}\hat{A}f(x) = \hat{B}\sqrt{x^2} = \hat{B}x = (x)^2 = x^2$$

$\therefore$ We see that $\hat{A}\hat{B}f(x) = \hat{B}\hat{A}f(x)$

$\therefore \hat{A}$ and $\hat{B}$ will commute, i.e., $\hat{A}\hat{B}f(x) - \hat{B}\hat{A}f(x) = x^2 - x^2 = 0$

Thus, it is clear that if two operators commute, their commutator vanishes. Answer.

Example 2

Let $\hat{A} = \mathrm{d}/\mathrm{d}x$, $\hat{B} = x$, and $f(x) = x^2$, then show that the two operators are non-commuting.

Solution: *Given that* $\hat{A} = \mathbf{d/dx}$; $\hat{B} = x$ *and* $f(x) = x^2$

$$\therefore \quad \hat{A}\hat{B}f(x) = \frac{\mathrm{d}}{\mathrm{d}x}\left(x.x^2\right) = \frac{\mathrm{d}}{\mathrm{d}x}\left(x^3\right) = 3x^2$$

$$\text{and } \hat{B}\hat{A}f(x) = x\frac{\mathrm{d}}{\mathrm{d}x}\left(x^2\right) = x.2x = 2x^2$$

$$\therefore \quad \hat{A}\hat{B}f(x) \neq \hat{B}\hat{A}f(x)$$

Hence, $\hat{A}$ and $\hat{B}$ are non-commuting. Answer.

Example 3

Find the commutator $\left[x, \frac{\mathrm{d}}{\mathrm{d}x}\right]$

Solution: *Commutator is* $\left[x, \frac{\mathrm{d}}{\mathrm{d}x}\right]$

$$\therefore \left[x, \frac{\mathrm{d}}{\mathrm{d}x}\right]f(x) = x\frac{\mathrm{d}f}{\mathrm{d}x} - \frac{\mathrm{d}(xf)}{\mathrm{d}x}$$

$$= x\frac{\mathrm{d}f}{\mathrm{d}x} - x\frac{\mathrm{d}f}{\mathrm{d}x} - f(x)$$

$$= -f(x)$$

As an operator equation

$[x, \mathrm{d}/\mathrm{d}x] = -\hat{E} = -1$ Answer.

4.2.1 Facts about Commutation

The following facts are very useful in connection with commutation.

- Every operator commutes with itself.
- Commutation of multiplication operators is held with each other.
- A constant multiplication operator commutes with all other operators under consideration.
- Operators operating on different variables commute with each other.
- A derivative operator does not commute with a multiplication having the same independent variable.

4.3 Hermitian Operator

A Hermitian operator is a linear operator ($\hat{A}$), which satisfies the following condition:

$$\int \psi\left(\hat{A}\phi\right)\mathrm{d}_\tau = \int (A\phi)\psi d_\tau \tag{4.11}$$

$$\text{or} \int \psi^*\left(\hat{A}\phi\right)\mathrm{d}_\tau = \int (A\psi)^*\phi \mathrm{d}_\tau \tag{4.12}$$

where ϕ and ψ are square-integrable functions and $\hat{A}$ is defined as a Hermitian operator.

Also, ψ^* = complex conjugate of ψ

$\mathrm{d}\tau$ = volume element

Let $\psi = e^{ix}$

$$\therefore \quad \psi^* = e^{-ix}$$

$$\phi = \sin x$$

$$\hat{A} = \mathrm{d}^2/\mathrm{d}x^2$$

Applying these on Eq. (4.12), we shall get

a.

$$\int e^{ix}\frac{\mathrm{d}^2}{\mathrm{d}x^2}(\sin x)\mathrm{d}x = \int e^{-ix}\frac{\mathrm{d}}{\mathrm{d}x^2}(-\sin x)\mathrm{d}x$$
$$= -\int e^{-ix}\sin x\ \mathrm{d}x \tag{4.13}$$

b. Putting the above values in the RHS of Eq. (4.12), we obtain

$$\int \phi\left(\hat{A}\psi\right)^*\mathrm{d}\tau = \int \sin x\left[\frac{\mathrm{d}^2}{dx^2}\left(e^{ix}\right)^*\right]\mathrm{d}x$$

$$= \int \sin x\left[\frac{\mathrm{d}}{\mathrm{d}x}\left\{\frac{\mathrm{d}}{\mathrm{d}x}\left(e^{ix}\right)^*\right\}\right]\mathrm{d}x$$

$$= \int \sin x\left[i\frac{\mathrm{d}}{\mathrm{d}x}\left(e^{ix}\right)^*\right]\mathrm{d}x$$

$$= \int \sin x\left[\left(i\times ie^{ix}\right)^*\right]\mathrm{d}x$$

$$= -\int \sin x e^{-ix}\mathrm{d}x \tag{4.14}$$

where minus sign in (*ix*) comes from carrying out the operation denoted by the asterisks (*). We find that the two integrals are equal; therefore, operator $\hat{A} = d^2/dx^2$ is Hermitian.

Now we shall consider another example to test an operator by equation

$$\int \psi^* \hat{A} \phi d\tau = \int \phi (A\psi)^* d\tau \tag{4.15}$$

For this, let us consider ψ and ϕ to be square-integrable functions of x and also suppose that the operator $\hat{A} = (id/dx)$. Then, the LHS of Eq. (4.15) becomes, upon integration by parts,

$$\int_{-\infty}^{+\infty} \psi^* \left(i\frac{d\phi}{dx} \right) dx = i\left[\psi^* \phi \right]_{-\infty}^{+\infty} - i\int_{-\infty}^{+\infty} \phi \frac{d\psi^*}{dx} dx = -i\int_{-\infty}^{+\infty} \phi \left(\frac{d\psi^*}{dx} \right) dx \tag{4.16}$$

Since ψ and ϕ are square-integrable, they must vanish at infinity, resulting zero term in Eq. (4.14).

Now we shall write the RHS of Eq. (4.13) as

$$\int_{-\infty}^{+\infty} \phi (id/dx)^* \psi^* dx = -i\int_{-\infty}^{+\infty} \phi \left(\frac{d\psi^*}{dx} \right) \tag{4.17}$$

where minus sign comes from the operation denoted by the asterisks. We found that Eq. (4.16) is equal to Eq. (4.17) and thus concluded that the operator (*id*/d*x*) is Hermitian.

Since the effect of i was to introduce an essential sign reversal, it is apparent that the equality does not result for $\hat{A} = d/dx$. It is obvious that any Hermitian operator having first derivative in any Cartesian co-ordinate must have the factor '*i*'.

4.3.1 Properties of Hermitian Operator

The following are the properties of Hermitian operator.

4.3.1.1 The Eigen Values of a Hermitian Operator Are Real

Proof: Let us suppose that ψ_i is an Eigen function of the Hermitian operator $\hat{A}$ having Eigen value a, i.e.,

$$\hat{A}\psi_i = a\psi_i \tag{4.18}$$

Because $\hat{A}$ is a Hermitian operator, we can write

$$\int \psi_i^* \hat{A} \psi_i d\tau = \int \psi_i \left(\hat{A}\psi_i \right)^* d\tau$$

$$\text{or} \int \psi_i^* a \psi \psi_i d\tau = \int \psi_i (a\psi_i)^* d\tau$$

$$\text{or} \left(a - a^* \right) \int \psi_i^* \psi_i d\tau = 0$$

$$\text{But} \int \psi_i^* \psi_i d\tau \neq 0$$

$$\therefore \quad \left(a - a^* \right) = 0$$

$$\therefore \quad a = a^* \tag{4.19}$$

Thus, Eigen value of a Hermitian operator is real. Q.E.D.

Alt Proof: Let $\hat{A}$ = a Hermitian operator with a square-integrable Eigen function ψ

$$\text{Then} \quad \hat{A}\psi = a\psi \tag{4.20}$$

where a = Eigen value.

Each side of Eq. (4.20) must be expressed as real and imaginary parts. The real part and also the imaginary part must be equal to each other.

Now we shall take the complex conjugate of Eq. (4.20). This causes the imaginary parts to reverse sign but they remain equal. Therefore, we write

$$\hat{A}^*\psi^* = a^*\psi^* \tag{4.21}$$

Now multiply the LHS of Eq. (4.21) by ψ^*, and it is integrated over all spatial variables.

$$\int \psi^*\hat{A}\psi \mathrm{d}\tau = a\int \psi^*\psi \mathrm{d}\tau \tag{4.22}$$

Similarly, we multiply Eq. (4.21) by ψ, and then it is integrated, i.e.,

$$\int \psi\hat{A}^*\psi^* \mathrm{d}\tau = a^*\int \psi\psi^* \mathrm{d}\tau \tag{4.23}$$

Since $\hat{A}$ is a Hermitian operator, the LHS of Eqs (4.22) and (4.23) are equal.

$$\therefore \quad a\int \psi^*\psi \mathrm{d}\tau = a^*\int \psi^*\psi \mathrm{d}\tau$$

Therefore, $(a - a^*)\int \psi^*\psi \mathrm{d}\tau = 0$

But $\int \psi^* \psi \mathrm{d}\tau \neq 0$

$\therefore (a - a^*) = 0$, since $\hat{A}$ is square-integrable.

$\therefore a = a^*$, which proves that Eigen values of a Hermitian operator are real. Q.E.D.

4.3.1.2 Non-Degenerate Eigen Functions of a Hermitian Operator Form an Orthogonal Set

Proof: Let us consider that ψ and ϕ are two square-integrable Eigen functions of the Hermitian operator $\hat{A}$.

$$\therefore \quad \hat{A}\psi = a_1\psi \tag{4.24}$$

$$\text{and} \quad \hat{A}^*\psi^* = a_2\phi^* \tag{4.25}$$

Multiplying Eq. (4.24) from left by ϕ^* and Eq. (4.25) from left by ψ and then on integration yields

$$\int \phi^*\hat{A}\psi \mathrm{d}\tau = a_1\int \phi^*\psi \mathrm{d}\tau \tag{4.26}$$

$$\text{and}\int \psi\hat{A}^*\phi^* \mathrm{d}\tau = a_2\int \psi\phi^* \mathrm{d}\tau \tag{4.27}$$

The LHS of Eqs (4.26) and (4.27) are equal.

$$\therefore \quad a_1 \int \phi^* \psi \mathrm{d}\tau = a_2 \int \psi \phi^* \mathrm{d}\tau$$

$$\therefore \quad (a_1 - a_2) \int \phi^* \psi \mathrm{d}\tau = 0$$

$$\Rightarrow \int \phi^* \psi \mathrm{d}\tau = 0 \left[\text{Since } a_1 \neq a_2 \right]$$

This proves that non-degenerate Eigen function of a Hermitian operator forms an orthogonal set. Q.E.D.

Further, we are going to note down the other properties of Hermitian operator.

- Hermitian operators are linear.
- Two Hermitian operators are not required to commute with each other.
- A Hermitian operator has a set of Eigen functions and also Eigen values.
- Two commuting Hermitian operators can have a set of common Eigen functions.
- The set of Eigen functions of a Hermitian operator makes a complete set for the expansion of functions following the same boundary conditions.

4.3.1.3 If a Hermitian Operator $\hat{A}$ Commutes with an Arbitrary Operator $\hat{B}$, and Ψ_k and Ψ_l Are Two Eigen Functions of $\hat{A}$ with Non-Degenerate Eigen Values, then Bra-Ket Notation, Prove That $< \Psi_k | \hat{B} \mid \Psi_l > = 0$

Proof : Let us suppose that

$$\hat{A}\, \Psi_k = a\, \Psi_k \text{ and } \hat{A}\, \Psi_1 = b\, \Psi_1$$

We know that since

$$\left[\hat{A},\, \hat{B}\, \right] = \hat{A}\, \hat{B} - \hat{B}\, \hat{A} = 0$$

Then we can write

$$< \Psi_k \Big| \left[\hat{A},\, \hat{B}\, \right] \Big| \Psi_i > = < \Psi_k \Big| \hat{A},\, \hat{B} \Big| \Psi_i > - < \Psi_k \Big| \hat{B}\, \hat{A} \Big| \Psi_1 >$$

$$= \; < \hat{A}\, \Psi_k \Big| \hat{B} \Big| \Psi_i > - < \Psi_k \Big| \hat{B} \Big| \hat{A}\, \Psi_1 >$$

$$= < a\, \Psi_k \Big| \hat{B} \Big| \Psi_1 > - < \Psi_k \Big| \hat{B} \Big| B\psi_1 >$$

$$= (a - b) < \Psi_k \Big| \hat{B} \Big| \psi_1 > = 0$$

But (a–b) $\neq$ 0

Therefore, $< \Psi_k \Big| \hat{B} \Big| \Psi_l > = 0$ Q.E.D

4.3.1.4 If Two Hermitian Operators Â and B̂ Possess a Common Eigen Function, Then They Commute

Proof: According to 4.3.1.4, we have seen that $\hat{A}\ \Psi = \text{a}\ \Psi$ and $\hat{B}\ \Psi = \text{b}\ \Psi$. Hence, we can write as

$$\left[\hat{A},\hat{B}\right]\Psi = \hat{A},\hat{B}\,\Psi - \hat{B}\,\hat{A}\,\Psi$$

$$= \hat{A}b\,\Psi - \hat{B}a\,\Psi$$

$$= ba\,\Psi - ab\,\Psi = 0$$

Hence, they commute Q.E.D

4.3.1.5 If Two Hermitian Operators Â and B̂ Commute, Then They Must Have a Common Eigen Function

Proof: According to the above statement, we can write

$$\left[\hat{A}\,\hat{B}\right] = 0$$

Here, we shall assume that $\hat{A}\Psi = a\Psi$ then it is to prove that Ψ is an Eigen function of $\hat{B}$ also. If this is not the fact, then we can find some functions which are Eigen functions of $\hat{A}$ as well as $\hat{B}$. Then we can write

$$= \left[\hat{A},\hat{B}\right]\Psi = \hat{A}\,\hat{B}\Psi - \hat{B}\hat{A}\Psi$$

$$= \hat{A}\left(\hat{B}\Psi\right) - \text{a}(\hat{B}\psi) = 0$$

$$\text{Or}\ \hat{A}(\hat{B}\psi) = a\left(\hat{B}\Psi\right) = 0$$

This gives a clue to understand that $\hat{B}\ \Psi$ is an Eigen function of $\hat{A}$ having Eigen value *a*. This can take place under these two conditions.

a. $\hat{B}\Psi$ is a multiple of Ψ, i.e, $\Psi\hat{B} = b\Psi$; thus, Ψ is an Eigen function of $\hat{B}$ having the Eigen value *b*.
b. $\hat{B}\Psi$ and Ψ are degenerate Eigen function of $\hat{A}$.

At this stage, we want to illustrate (b) with simple example.

Suppose $\hat{A} = \text{d}^2/\text{d}x^2$and $\hat{B} = id/\text{d}x$. The functions sin *kx* and cos *kx* are degenerate Eigen functions of $\hat{A}$ having value $-k^2$. The fact is that they are not Eigen function of $\hat{B}$ even though $[\hat{A}\ ,\hat{B}] = 0$. But the following functions

$$\Psi_1 = \cos kx + i\ \sin kx = e^{ikx}$$

$$\Psi_2 = \cos kx - i\ \sin kx = e^{-ikx}$$

are given Eigen function of $\hat{A}$ as well as $\hat{B}$. Keep in mind that when a particular Eigen value of $\hat{A}$ is *n*-fold degenerate, we can have *n* linear combinations of the degenerate Eigen functions, so that they are Eigen function of $\hat{B}$ too.

4.4 Schmidt Orthogonalisation

Let ψ_1 and ψ_2 are two linearly independent degenerate functions which are normalised but not orthogonal.

Suppose $\phi_1 = \psi_1$ and $\phi_2 = a_1\psi_1 + a_2\psi_2$, and also $\int \psi_1\psi_2 d\tau = S$. The problem now is to find a_1 and a_2 such that

$$\int \phi_1\phi_2 d\tau = 0$$

$$\text{and} \int \phi_2^2 d\tau = 1$$

$$\text{or} \int \psi_1 (a_1\psi_1 + a_2\psi_2) d\tau = 0$$

$$\text{or}\, a_1 \int \psi_1^2 d\tau + a_2 \int \psi_1\psi_2 d\tau = 0$$

$$\text{or}\, a_1 + a_2 S = 0 \qquad \left[\therefore \quad \int \psi_1^2 d\tau = 1 \quad \text{and} \quad \int \psi_1\psi_2 d\tau = S \right]$$

$$\text{or}\, a_1 + a_2 S = 0$$

$$\therefore \quad a_1 = -a_2 S \tag{4.28}$$

$$\text{Again} \int (a_1\psi_1 + a_2\psi_2)^2 d\tau = 1$$

$$\text{or} \int \left(a_1^2\psi_1^2 + a_2^2\psi_2^2 + 2a_1a_2\psi_1\psi_2\right) d\tau = 1$$

$$\text{or}\, a_1^2 \int \psi_1^2 d\tau + a_2^2 \int \psi_2^2 d\tau + 2a_1a_2 \int \psi_1\psi_2\, d\tau = 1$$

$$\text{or}\, a_1^2 + a_2^2 + 2a_1a_2 S = 1 \tag{4.29}$$

Putting the value of a_1 from Eq. (4.28), we get

$$(-a_2 S)^2 + a_2^2 + 2(-a_2 S)(a_2) S = 1$$

$$\Rightarrow a_2^2 S^2 + a_2^2 - 2a_2^2 S^2 = 1$$

$$\Rightarrow a_2^2 - a_2^2 S^2 = 1$$

$$\Rightarrow a_2^2 \left(1 - S^2\right) = 1$$

$$\therefore \quad a_2^2 = \frac{1}{\left(1-S^2\right)} = \left(1-S^2\right)^{-1}$$

$$\therefore \quad a_2 = \left(1-S^2\right)^{-1/2} \tag{4.30}$$

$$\therefore \quad a_1 = -\left(1-S^2\right)^{-1/2} S$$

$$\therefore \quad a_1 = -S(1-S)^{-1/2} \tag{4.31}$$

The required orthogonal pair will be

$$\phi_1 = \psi_1$$

$$\phi_2 = a_1\psi_1 + a_2\psi_2$$

$$= -S(1-S)^{-\frac{1}{2}}\psi_1 + \left(1-S^2\right)^{-1/2}\psi_2$$

$$\text{or}\,\phi_2 = (1-S)^{-1/2}\left(\psi_2 - S\psi_1\right) \tag{4.32}$$

This is known as Schmidt orthogonalisation.

4.5 ∇ and ∇² Operators

We know that an operator can be a vector or a complex quantity. If an operator is a vector, one usually handles it in terms of its component. In rectangular co-ordinates, the del operator is represented as

$$\nabla = i\frac{\partial}{\partial x} + j\frac{\partial}{\partial y} + k\frac{\partial}{\partial z} \tag{4.33}$$

But since i, j, and k are mutually perpendicular, Δ·Δ is expressed as

$$\nabla^2 = \frac{\partial^2}{\partial x^2} + \frac{\partial^2}{\partial y^2} + \frac{\partial^2}{\partial z^2} \tag{4.34}$$

The quantity ∇f, where f is a scalar function, is known as the gradient of f. For example, suppose $f = x^2 + y^2 + z^2$, then the gradient of f is the vector

$$\nabla f = 2xi + 2yi + 2zk \tag{4.35}$$

Since for a scalar function f, the quantity $\frac{\partial f}{\partial x'}\frac{\partial f}{\partial y}$ and $\frac{\partial f}{\partial z}$ are the rates of change of f with respect to distance in the x, y, and z directions, and the gradient of f supplies a means of finding out the rate of change of f with distance in any direction.

4.6 Linear Momentum Operator

The linear momentum operators are px, py, and pz, which may be obtained from the wave equation. For an electron wave, the wave function may be expressed as

$$\psi = Ae^{\pm 2\Pi ix/\lambda} \tag{4.36}$$

Differentiating this equation with respect to x, we get

$$\psi = \pm Ae^{\pm 2\Pi i\, x/\lambda}$$
$$= \pm \frac{2\prod i}{\lambda}\psi \tag{4.37}$$

But we know that $\lambda = h/p_x$ (in the x-direction)

$$\therefore \quad \frac{d\psi}{dx} = \pm \frac{2\pi i}{h/p_x}\psi = \pm \frac{2\pi i \cdot p_x\psi}{h}$$

$$\therefore \quad p_x\psi = \pm \frac{h}{2\pi i}\frac{d\psi}{dx}$$

$$\therefore \quad p_x = \pm \frac{h}{2\pi i}\frac{d}{dx} \tag{4.38}$$

$\therefore$ In the $+x$-direction

$$p_x = +\frac{h}{2\pi i}.\frac{d}{dx} = \frac{i\hbar}{i^2}.\frac{d}{dx} = -i\hbar\frac{d}{dx} \tag{4.39}$$

and p_x in the $-x$-direction is

$$p_x = -\frac{h}{2\pi i}.\frac{d}{dx} = i\hbar.\frac{d}{dx} \tag{4.40}$$

Similarly, $p_y = -\frac{h}{2\pi i}.\frac{d}{dy}$ and $p_z = -\frac{h}{2\pi i}.\frac{d}{dz}$

or $p_y = +i\hbar.\frac{d}{dy}$ and $p_z = +i\hbar.\frac{d}{dz}$

4.6.1 Operators of Every Two Components of the Momentum Commute

Proof: The operators of every two components of the momentum commute, because they are independent from each other, and the order of the differentiation can be altered.

$$[p_x, p_y] = p_x p_y - p_y p_x = 0 \tag{4.41}$$

Thus, we see that the above statement is correct. Q.E.D.

4.6.2 Momentum Components Commute with Unlike Co-Ordinates

Proof: Let us consider the commutator of the momentum component with not like co-ordinate, i.e., with unlike co-ordinate such as

$$[y, p_x]\psi = y(p_x\psi) - p_x(y\psi) = y\left[-i\hbar\frac{\partial\psi}{\partial x} + i\hbar\frac{\partial}{\partial x}(y\psi)\right]$$

$$= i\hbar\left[-y\frac{\partial\psi}{\partial x} + y\frac{\partial\psi}{\partial x}\right] = 0$$

$$\therefore \quad [y, p_x]\,\psi = 0 \qquad \text{Q.E.D}$$

4.6.3 Momentum Components Do Not Commute with Their Relative Co-Ordinates

Proof: Let us consider p_x as the momentum component and its relative co-ordinate x. Now operating with the commutator operator $[x, p_x]$ for the function ψ, we have

$$[x, p_x]\psi = \left[x\frac{\hbar}{i}\frac{\partial\psi}{\partial x} - \frac{\hbar}{i}\frac{\partial(x\psi)}{\partial x}\right]$$

$$= \frac{\hbar}{i}\left[x\frac{\partial\psi}{\partial x} - x\frac{\partial\psi}{\partial x} - \frac{\partial x}{\partial x}\psi\right]$$

$$= \frac{\hbar}{i}(-\psi) = \frac{i\hbar}{i^2}(-\psi) = -i\hbar(-\psi) = i\hbar\psi$$

$$\therefore \quad [x, p_x] = i\,\hbar \tag{4.42}$$

$$\text{Similarly}\,[y, p_y] = i\hbar \quad \text{and} \quad [z, p_z] = i\hbar \qquad \text{Q.E.D.} \tag{4.43}$$

4.7 Angular Momentum Operator or Angular Momentum Vector $(\vec{L})$

Angular momentum operator or angular momentum vector $(\vec{L})$ is very important in quantum mechanics, especially for a rotating system (Figure 4.1). In classical mechanics, if a particle having mass m is rotating around O, then

$$\vec{L} = \vec{r} \times \vec{p}$$

where $\vec{L}$ = angular momentum vector directed outwards from O at right angles to the plane
 $\vec{p}$ = linear momentum
 $\vec{r}$ = distance of the particle from O

In Cartesian co-ordinates, a vector may be expressed as unit vectors and the Cartesian components will be written as

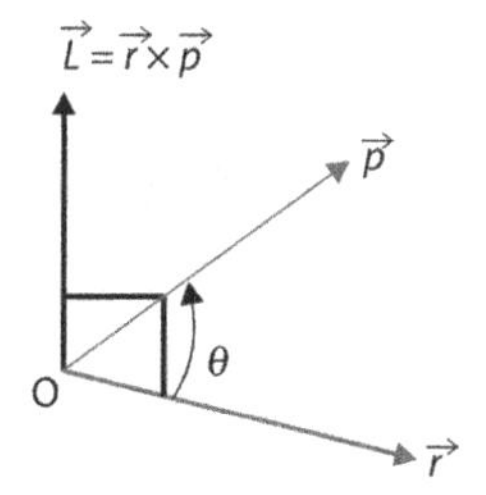

FIGURE 4.1 Representation of angular momentum vector $(\vec{L})$.

$$\left.\begin{aligned} \vec{r} &= \vec{i}x + \vec{j}y + \vec{k}z \\ \vec{p} &= \vec{i}p_x + \vec{j}p_y + \vec{k}p_z \\ \text{and} \quad \vec{L} &= \vec{i}L_x + \vec{j}L_y + \vec{k}L_z \end{aligned}\right\} \tag{4.44}$$

From definition, the vector product is generally expressed as

$$\vec{L} = \vec{r} \times \vec{p} = \begin{vmatrix} \vec{i} & \vec{j} & \vec{k} \\ x & y & z \\ P_x & p_y & p_z \end{vmatrix} \tag{4.45}$$

$$\Rightarrow \vec{i}\left(yp_z - zp_y\right) + \vec{j}\left(zp_x - xp_z\right) + \vec{k}\left(xp_y - yp_z\right) \tag{4.46}$$

Therefore, the Cartesian components of $\vec{L}$ will be obtained by comparing Eqs (4.44) and (4.45), i.e.,

$$L_x = yp_z - zp_y$$

$$L_y = zp_x - xp_z$$

$$L_z = xp_y - yp_x$$

Replacing in terms of quantum mechanical operators, we shall get

$$L_x = \frac{h}{2\pi i}\left(y\frac{\partial}{\partial z} - z\frac{\partial}{\partial y}\right) = \frac{\hbar}{i}\left(y\frac{\partial}{\partial z} - z\frac{\partial}{\partial y}\right) = -i\hbar\left(y\frac{\partial}{\partial z} - z\frac{\partial}{\partial y}\right)$$

Similarly, $L_y = -i\hbar\left(z\frac{\partial}{\partial x} - x\frac{\partial}{\partial z}\right)$ and

$$L_z = i\hbar\left(x\frac{\partial}{\partial y} - y\frac{\partial}{\partial x}\right) \tag{4.47}$$

For square of the angular momentum, we can write

$$L^2 = L_x^2 + L_y^2 + L_z^2$$

$$= \hbar^2\left(y\frac{\partial}{\partial z} - z\frac{\partial}{\partial y}\right)^2 + \hbar^2\left(z\frac{\partial}{\partial x} - x\frac{\partial}{\partial z}\right)^2 + \hbar^2\left(x\frac{\partial}{\partial y} - y\frac{\partial}{\partial x}\right)^2$$

Now we are going to see what are the commuting properties of the operators obtained above.

4.7.1 Operators of the Angular Momentum Components Do Not Commute

a. Prove that $[L_x, L_y] = i\hbar L_z$

Proof: We know that

$$\left[L_x, L_y\right] = \left[L_xL_y - L_y, L_x\right]$$

$$\Rightarrow \left(yp_z - zp_y\right)\left(zp_x - xp_z\right) - \left(zp_x - xp_z\right)\left(yp_z - zp_y\right)$$

$$\Rightarrow \left(yp_z zp_x - yp_z xp_z - zp_y xp_z\right) - \left(zp_x yp_z - zp_x zp_y - xp_z yp_z + xp_z zp_y\right)$$

$$\Rightarrow yp_x p_z.z - yxp_z^2 - z^2 p_x p_y + xzp_y p_z - zyp_x p_z + z^2 p_x p_y + xyp_z^2 - xp_y p_z.z$$

$$\Rightarrow \left(yp_x p_z.z + xzp_y p_z - zyp_x p_z - xzp_y p_z\right)$$

$$\Rightarrow yp_x\left(p_z.z - zp_z\right) + xp_y\left(zp_z - p_z.z\right)$$

$$\Rightarrow yp_x(-i\hbar) + xp_y(i\hbar)$$

$$\Rightarrow i\hbar\left(xp_y - yp_x\right)$$

$$\Rightarrow i\hbar L_z$$

b. Prove that $[LyL_z] = i\hbar L_x$

Proof: We have

$$\left[L_y, L_z\right] = \left[L_y L_z - L_z L_y\right]$$

$$= \left[\left(zp_x - xp_z\right)\left(xp_y - yp_x\right) - \left(xp_y - yp_x\right)\left(zp_x - xp_z\right)\right]$$

$$= \left(zp_x.xp_y - zp_x.yp_x - xp_z - xp_y + xp_z.yp_x\right)$$

$$-\left(xp_y - zp_x + xp_y.xp_z - yp_x.zp_x - yp_x.xp_z\right)$$

$$= \left[zp_x xp_y - zp_x yp_y - xp_z.xp_y + xp_z.yp_x - xp_y.zp_x + xp_y.xp_z + yp_x.zp_x - yp_x.xp_z\right]$$

$$= \left(zp_x.xp_y + xp_z.yp_x - xp_y.zp_x - yp_x.xp_z\right)$$

$$= zp_y\left(p_x x - xp_x\right) + p_z y\left(xp_x - p_x.x\right)$$

$$= zp_y(-i\hbar) + p_z y(i\hbar)$$

$$= i\hbar\left(yp_z - zp_y\right)$$

$$= i\hbar L_x$$

4.7.2 Operators of the Angular Momentum Components Do Commute with the Operator of the Square of the Angular Momentum

Proof: We know that the total angular momentum is given by

$$L^2 = L_x^2 + L_y^2 + L_z^2$$

Let us consider

$$\left[L^2, L_x\right] = \left[\left(L_x^2 + L_y^2 + L_z^2\right)L_x\right]$$

$$= \left[L_x^2, L_x\right] + \left[L_y^2, L_x\right] + \left[L_z^2, L_x\right]$$

But

$$\left[L_x^2, L_x\right] = \left[L_x.L_x, L_x\right]$$
$$= L_x\left[L_x, L_x\right]$$
$$= L_x\left[L_x.L_x - L_x, L_x\right] = 0$$

$$\therefore \quad \left[L_x^2, L_x\right] = 0 \tag{4.48}$$

Again

$$\left[L^2, L_x\right] = \left[L_y^2, L_x\right] + \left[L_z^2, L_x\right]$$
$$= L_y\left[L_y, L_x\right] + \left[L_y, L_x\right]L_y + L_z\left[L_z, L_x\right] + \left[L_z, L_x\right]L_z$$

$$\left[\therefore [a,b,c] = a[b,c] + [a,c]b\right]$$
$$= L_y\left(-i\hbar L_z\right) + \left(-i\hbar L_z\right)L_y + L_z\left(i\hbar L_y\right) + \left(i\hbar L_y\right)L_z$$
$$= -i\hbar L_y L_z - i\hbar L_z L_y + i\hbar L_z L_y + i\hbar L_y L_z = 0$$

It follows, therefore, that L^2 also commutes with L_y and L_z, and hence, we can write

$$\left[L^2, L_x\right] = 0, \ \left[L^2, L_y\right] = 0, \left[L^2, L_z\right] = 0 \tag{4.49}$$

Alternatively: We have the knowledge of the following commutator:

$$\left.\begin{aligned} \left[L_x, L_y\right] &= i\hbar L_z \\ \left[L_z, L_x\right] &= i\hbar L_y \\ \left[L_y, L_z\right] &= i\hbar L_x \end{aligned}\right\} \tag{4.50}$$

We then multiply its first equality on the right and also on left by the operator L_y, giving rise to the following equation:

$$\left(L_x L_y - L_y L_x\right)L_y = i\hbar L_z L_y$$

$$\Rightarrow L_x L_y^2 = L_y L_x L_y + i\hbar L_z L_y$$

$$\text{and } L_y^2 L_x = L_y L_x \quad i\hbar L_y L_z$$

On subtraction, we get

$$\left(L_x L_y^2 - L_y^2 L_x\right) = i\hbar\left(L_z L_y + L_y L_z\right) \tag{4.51}$$

By analogy from $[L_z, L_x] = i\hbar L_y$ after multiplying by L_z, we get

$$\left(L_x L_z^2 - L_z^2 L_x\right) = i\hbar\left(L_z L_y + L_y L_z\right) \tag{4.52}$$

Adding Eqs (4.51) and (4.52), we obtain

$$\left(L_x L_y^2 - L_y^2 L_x\right) + \left[L_x L_z^2 - L_y^2 L_z\right] = 0$$

$$= L_x L_y^2 - L_y^2 L_x + L_x L_z^2 - L_z^2 L_x = 0$$

$$= L_x\left(L_y^2 + L_x^2\right) - L_x\left(L^2 - L_x^2\right) = 0$$

$$= \left(L^2 L_x - L_x^3 - L_x L^2 + L_x^3\right) = 0$$

$$= \left(L^2 L_x - L_x^3 - L_x L^2 + L_x^3\right) = 0$$

$$= \left(L^2 L_x - L_x L^2\right) = 0$$

$$= \left[L^2, L_x\right] = 0 \tag{4.53}$$

Similarly, with the help of two commutators from Eq. (4.52), we can get a similar expression for L_y and L_z, i.e.,

$$\left[L^2, L_y\right] = 0 \text{ and } \left[L^2, L_z\right] = 0 \qquad \text{Q.E.D.}$$

4.7.3 Angular Momentum in Spherical Polar Co-Ordinates

We have already derived that

$$L_x = -i\hbar\left(y\frac{\partial}{\partial z} - z\frac{\partial}{\partial y}\right)$$

$$L_y = -i\hbar\left(z\frac{\partial}{\partial x} - x\frac{\partial}{\partial z}\right)$$

$$\text{and } L_z = -i\hbar\left(x\frac{\partial}{\partial y} - y\frac{\partial}{\partial x}\right)$$

We now want to transform the Cartesian co-ordinates (x, y, z) to spherical co-ordinates (r, θ, ϕ).

We also know that

$$\left.\begin{aligned} x &= r\sin\theta\cos\phi \\ y &= r\sin\theta\cos\phi \\ \text{and } z &= r\cos\theta \end{aligned}\right\} \tag{4.54}$$

$$r^2 = x^2 + y^2 + z^2$$

$$\cos\theta = z/r \quad \text{and} \quad \tan\phi = y/x$$

To do so, we require to express the differential operators in Cartesian co-ordinates in terms of differential operators in spherical co-ordinates, i.e.,

$$\frac{\partial}{\partial x}=\frac{\partial r}{\partial x}\cdot\frac{\partial}{\partial r}+\frac{\partial\theta}{\partial x}\cdot\frac{\partial}{\partial\theta}+\frac{\partial\phi}{\partial x}\cdot\frac{\partial}{\partial\phi}$$

$$\frac{\partial}{\partial y}=\frac{\partial r}{\partial y}\cdot\frac{\partial}{\partial r}+\frac{\partial\theta}{\partial y}\cdot\frac{\partial}{\partial\theta}+\frac{\partial\phi}{\partial y}\cdot\frac{\partial}{\partial\phi}$$

$$\frac{\partial}{\partial z}=\frac{\partial r}{\partial z}\cdot\frac{\partial}{\partial r}+\frac{\partial\theta}{\partial z}\cdot\frac{\partial}{\partial\theta}+\frac{\partial\phi}{\partial z}\cdot\frac{\partial}{\partial\phi}$$

These nine derivatives of r, θ, and ϕ with regard to x, y, and z are easily found from Eq. (4.54). These nine derivatives shown above can be expressed as given follows:

$$\frac{\partial r}{\partial x}=\sin\theta\cos\phi,\frac{\partial r}{\partial y}=\sin\theta\sin\phi,\frac{\partial r}{\partial z}=\cos\theta$$

$$\frac{\partial\theta}{\partial x}=\frac{\cos\theta\cos\phi}{r},\frac{\partial\theta}{\partial y}=\frac{\cos\theta\sin\phi}{r},\frac{\partial\theta}{\partial z}=\frac{-\sin\theta}{r}$$

$$\frac{\partial\phi}{\partial x}=\frac{-\sin\phi}{r\sin\theta},\frac{\partial\phi}{\partial y}=\frac{\cos\phi}{r\sin\theta'}\frac{\partial\phi}{\partial y}=\frac{\cos\phi}{r\sin\theta'}\frac{\partial\phi}{\partial z}=0 \tag{4.55}$$

Now using equation for L_x, L_y, and L_z and Eqs (4.54) and (4.55), one can evaluate the angular momentum operators in spherical co-ordinates as follows:

$$L_x=-i\hbar\left(y\frac{\partial}{\partial z}-z\frac{\partial}{\partial y}\right)$$

$$=-i\hbar\left[r\sin\theta\sin\phi\left(\cos\theta\frac{\partial}{\partial r}+\frac{-\sin\theta}{r}\frac{\partial}{\partial\theta}\right)-r\cos\theta\left(\sin\theta\sin\phi\frac{\partial}{\partial r}+\frac{\cos\theta\sin\phi}{r}\cdot\frac{\partial}{\partial\theta}\right)+\frac{\cos\phi}{r\sin\theta}\cdot\frac{\partial}{\partial\phi}\right] \tag{4.56}$$

$$=i\hbar\left[-r\sin\theta\sin\phi\cos\theta\frac{\partial}{\partial r}+\sin^2\theta\sin\phi\frac{\partial}{\partial\theta}+r\cos\theta\sin\theta\sin\phi\frac{\partial}{\partial r}+\cos^2\theta\sin\phi\frac{\partial}{\partial\theta}+\frac{\cos\theta\sin\phi}{\sin\theta}\cdot\frac{\partial}{\partial\phi}\right]$$

$$=i\hbar\left[\left(\sin^2\theta+\cos^2\theta\right)\sin\phi\frac{\partial}{\partial\theta}+\frac{\cos\theta\sin\phi}{\sin\theta}\cdot\frac{\partial}{\partial\phi}\right]$$

$$=i\hbar\left[\sin\phi\frac{\partial}{\partial\theta}+\cot\theta\cos\phi\frac{\partial}{\partial\phi}\right] \tag{4.57}$$

Similarly, we can prove that

$$L_y=i\hbar\left[-\cos\phi\frac{\partial}{\partial\theta}+\cot\theta\sin\phi\frac{\partial}{\partial\phi}\right]\text{ and }L_z=-i\hbar\frac{\partial}{\partial\phi} \tag{4.58}$$

Since $L^2=L_x^2+L_y^2+L_z^2$

Therefore, squaring the equations for L_x, L_y, and L_z as mentioned above, we shall come to the following results:

$$L^2 = -\hbar^2\left(\frac{\partial}{\partial\theta} + \cot\theta\frac{\partial}{\partial\theta} + \frac{1}{\sin^2\theta}\frac{\partial^2}{\partial\phi^2}\right)$$

$$= -\hbar^2\left(\frac{1}{\sin\theta}\frac{\partial}{\partial\theta} - \sin\theta\frac{\partial}{\partial\theta} + \frac{1}{\sin^2\theta}\frac{\partial^2}{\partial\phi^2}\right) \tag{4.59}$$

In atomic units, $\hbar$ becomes unity and does not appear.

4.7.4 Ladder Operators or Step-Up and Step-Down Operators for Angular Momentum

The ladder operators are defined by two equations, namely

$$L_+ = L_x + iL_y \tag{4.60}$$

$$L_- = L_x - iL_y \tag{4.61}$$

$$\text{or } L_\pm = L_x \pm iL_y \tag{4.62}$$

These operators are also known as *step-up* and *step-down* operators, respectively. These are also named as *raising* and *lowering* operators. Actually, these correspond to no observable property and are not Hermitian. The purpose of devising these operators is that they are useful in formal analysis. We are going to prove useful relations with the help of ladder operators.

Prove that

a. $L_zL_+ = L_+(L_z + 1)$ and
b. $L_zL_- = L_-(L_z - 1)$ in atomic units.

Proof: (a) $L_zL_+ = L_+(L_z + 1)$

We know that $L_+ = L_x + iLy$, therefore, expanding L_+ results in

$$\begin{aligned} L_zL_+ &= L_z\left(L_x + iL_y\right) \\ &= L_zL_x + iL_zL_y \\ &= \left(L_zL_x - L_xL_z\right) + L_xL_z + i\left(L_zL_y - L_yL_z\right) + iL_yL_z \\ &= \left[L_z, L_x\right] + L_xL_z + i\left[L_z, L_y\right] + iL_yL_z \\ &= i\hbar L_y + L_xL_z + i\left(-i\hbar L_x\right) + i\hbar L_yL_z \end{aligned}$$

Recombining terms, we get

$$\begin{aligned} L_zL_+ &= \hbar\left(L_x + iL_y\right) + \left(L_x + iL_y\right).L_z \\ &= \hbar L_+ + L_+.L_z \\ &= L_+\left(\hbar + L_z\right) \\ &= L_+\left(L_z + 1\right) \quad [\therefore \text{ in atomic units } \hbar = 1] \end{aligned}$$

Proof: (b) $L_zL_- = L_-(L_z - 1)$

$$\begin{aligned}
L_zL_- &= L_z\left(L_x - iL_y\right) \\
&= L_zL_x - iL_zL_y \\
&= \left(L_zL_x - L_xL_z\right) + L_xL_z - iL_zL_y \\
&= \left(L_zL_x - L_xL_z\right) + L_xL_z - i\left(L_zL_y - L_yL_z\right) - iL_yL_z \\
&= \left[L_z, L_x\right] + L_xL_z - i\left[L_z, L_y\right] - iL_yL_z \\
&= i\hbar L_y + L_xL_z - i\left(i\hbar L_x\right) - iL_yL_z \\
&= -\hbar\left(L_x - iL_y\right) + L_z\left(L_x - iL_y\right) \\
&= -\hbar\left(L_-\right) + L_-L_z \\
&= L_-\left(L_z - \hbar\right) \\
&= L_-\left(L_x - 1\right) \;\left[\text{in atomic units } \hbar = 1\right]
\end{aligned}$$

c. Prove that $[L^2, L_+] = 0$

Proof: $[L^2, L_+] = [L^2L_+ - L_+L^2]$

$$\begin{aligned}
&\Rightarrow L^2\left(L_x + iL_y\right) - \left(L_x + iL_y\right)L^2 \\
&\Rightarrow L^2L_x + iL^2L_y - L_xL^2 - iL_yL^2 \\
&\Rightarrow \left(L^2L_x - L_xL^2\right) + i\left(L^2L_y - L_yL^2\right) \\
&\Rightarrow \left[L^2, L_x\right] + i\left[L^2, L_y\right] \\
&= 0 + 0 = 0
\end{aligned}$$

d. Prove that $[L^2, L_-] = 0$

Proof: $[L^2, L_-]$

$$\begin{aligned}
&\Rightarrow \left(L^2L_- - L_-L^2\right) \\
&\Rightarrow L^2\left(L_x - iL_y\right) - \left(L_x - iL_y\right)L^2 \\
&\Rightarrow L^2L_x - iL^2L_y - L_xL^2 + iL_yL^2 \\
&\Rightarrow \left(L^2L_x - L_xL^2\right) - i\left(L^2L_y - L_yL^2\right) \\
&\Rightarrow \left[L^2, L_x\right] - i\left[L^2, L_y\right] \\
&\Rightarrow 0 - 0 = 0
\end{aligned}$$

e. Prove that $L^2 = L_+L_- + L_z^2 - \hbar L_z$

Proof: $L_+L_- = (L_x + iL_y)(L_x - iL_y)$

$$= L_x^2 + L_y^2 + iL_yL_x - iL_xL_y$$

$$= L_x^2 + L_y^2 + i\left(L_xL_y - L_yL_x\right)$$

$$= L_x^2 + L_y^2 - i\left(i\hbar L_z\right)$$

$$= L_x^2 + L_y^2 + \hbar L_z$$

$$\text{or } L_+L_- + L_z^2 = L_x^2 + L_y^2 + L_z^2 + \hbar L_z$$

$$\text{or } L_+ \; L_- + L_z^2 = L^2 + \hbar L_z$$

$$\therefore L^2 = L_+L_- + L_z^2 - \hbar L_z$$

Q.E.D

f. Prove that $L^2 = L_-L_+ + L_z^2 + \hbar L_z$

Proof: $L_-L_+ = (L_x - iL_y)(L_x + iL_y)$

$$= L_x^2 + L_y^2 - iL_yL_x + iL_xL_y$$

$$= L_x^2 + L_y^2 - i\left(i\hbar L_z\right)$$

$$= L_x^2 + L_y^2 - \hbar L_z$$

$$\text{or } L_-L_+ + L_z^2 = L_x^2 + L_y^2 + L_z^2 - \hbar L_z$$

$$\text{or } L_-L_+ + L_z^2 = L^2 - \hbar L_z$$

$$\therefore \quad L^2 = L_-L_+ + \quad L_z^2 + \hbar L_z$$

Q.E.D.

4.8 Hamiltonian Operator

It is known to us that Hamiltonian operator is the sum of kinetic and potential energies. It is denoted by (H), i.e.,

$$H = T + V$$

$$= \frac{p^2}{2m} + V\left[\text{K.E.} = \frac{mv^2}{2} \times \frac{m}{m} = \frac{m^2v^2}{2m} = \frac{p^2}{2m}\right]$$

$$\because p = mv$$

In quantum mechanics, Hamiltonian operator is obtained by putting the corresponding operator value in classical formula.

$$p = \frac{h}{2\pi i}\left(\frac{\partial}{\partial x} + \frac{\partial}{\partial y} + \frac{\partial}{\partial z}\right) = \frac{h}{2\pi i}\nabla$$

where ∇ is called the *de*'.

$$\therefore \quad p^2 = \frac{-h^2}{4\pi^2}\nabla^2 \qquad \therefore \quad H = \frac{-h^2\nabla^2}{8\pi^2 m} + V(x,y,z)$$

where ∇^2 is the Laplacian operator.

We can express Schrödinger's equation in the form of Hamiltonian operator, i.e.,

$$\nabla^2\psi + \frac{8\pi^2 m}{h^2}(E - V)\psi = 0$$

$$\text{or} \frac{-h^2\nabla^2\psi}{8\pi^2 m} = E\psi - V\psi$$

$$\text{or} \frac{-h^2}{8\pi^2 m}\nabla^2\psi + V\psi = E\psi$$

$$\text{or} \left(\frac{-h^2\nabla^2}{8\pi^2 m} + V\right)\psi = E\psi$$

$$\text{or } H\psi = E\psi \tag{4.63}$$

where H = Hamiltonian operator or Eigen state
E = Eigen value
ψ = Eigen function

4.9 Commutation Relation of Angular Momentum Operators with Hamiltonian Operators and with Each Other

It is well known that a rotating system experiencing torque maintains E, L_z, and $|L|$ (or equivalently L^2) as fixed motion but it is not the case with L_x or L_y.

Similar situation is expected in quantum mechanics. It is clearly meant that a state function ψ would be an Eigen function for H, L^2, and L_z, but would not have Eigen function for L_x or L_y. This in turn requires that H, L^2, and L_z should commute with each other but L_x and L_y will not commute with all of the above-mentioned H, L^2, and L_z.

Let us first consider H and L^2.$\cdot L^2$ and ∇^2 are related as follows

$$\nabla^2 = \frac{1}{r^2}\frac{\partial}{\partial r}.r^2\frac{\partial}{\partial r} - 1/r^2 L^2 \tag{4.64}$$

Since L^2 does not consist of the variable r, L^2 commutes with any function, which depends on only r. Since L^2 must also commute with itself, it follows from this that L^2 and ∇^2 commute, i.e.,

$$\left[L^2, \nabla^2\right] = 0 \tag{4.65}$$

If V in a Hamiltonian operator is $f(r)$ only, then $[L^2, H] = 0$. This obviously proves that H for hydrogen like ion commutes with L^2.

We have already proved that $[L_z, L^2] = 0$, i.e., L_z and L^2 commute. We also know that L_z does not operate on functions of r; hence, L_z commutes with ∇^2, i.e., $[L_z, \nabla^2] = 0$.

Since for spherically symmetric system $V = f(r)$, and hence, $[L_z, H] = 0$, i.e., L_z and H will commute.

Finally, we can state that H, L^2, and L_z all commute in a system consisting of a spherically symmetric potential.

4.10 Projection Operators

Let us consider that $\hat{A}$ is an observable, and its Eigen vectors constitute a complete set. Then, we can write

$$|\psi\rangle = \sum_a Ca|a\rangle \tag{4.66}$$

If we take the scalar product of $|a'\rangle$ with Eq. (4.66) and the orthonormality condition is employed, then we have

$$\langle a' | \psi\rangle = \sum_a C_a \delta_{aa'} = ca' \tag{4.67}$$

$$\left[\because \ \langle a' | a\rangle = \delta_{a'a}\right]$$

Thus, $C_a = \langle a | \psi\rangle$.

Now we shall substitute this in Eq. (4.66) and we obtain

$$|\psi\rangle = \sum_a |a\rangle\langle a|\psi\rangle \tag{4.68}$$

In this equation, a typical term on the RHS is a vector where length is the scalar product of $|a\rangle$ and $|\psi\rangle$ and whose direction is the same as that of $|a\rangle$; hence it is the projection of $|\psi\rangle$ on $|a\rangle$.

$$\text{The quantity } \hat{P}_a = |a\rangle\langle a| \tag{4.69}$$

is the operator, which carries into effect this projection. It is called the *projection operators.*

Equation (4.68) opines that the sum of all the projection operators $\hat{P}_a$ leaves any vector unaltered. Hence, it is the unit vector which is written as

$$\sum_a |a\rangle\langle a| = \hat{I} \tag{4.70}$$

Remember that the closure relation

$$\sum_a \phi_a(x)\phi^*(x') = \delta(x - x') \tag{4.71}$$

is the Schrödinger representation of Eq. (4.70).

This equation may be viewed as an expansion of the unit operator and will be found of greatest value in the following discussion.

As an example, we notice that it helps us to expand an arbitrary operator $\hat{F}$ as

$$\hat{F} = \hat{I}\hat{F}\hat{I}$$

$$= \sum_{a} |a\rangle\langle a| \hat{F} \sum |a'\rangle\langle a'| \tag{4.72}$$

$$= \sum_{a,a'} |a\rangle\langle a'| \hat{F}aa'$$

where $\hat{F}aa' = \langle a|\hat{F}|a'\rangle$. Thus, we can say that $\hat{F}$ may be written as a linear combination of the operators $|a\rangle\langle a'|$, which are really more general in nature than our projection operators $\hat{P}_a$.

Suppose that $\hat{F}$ is $\hat{A}$ itself, then $|a\rangle\hat{A}|a'\rangle = a\delta aa'$ and the above expansion can simply be expressed as

$$\hat{A} = \sum_{a} |a\rangle\langle a|a\rangle = \sum a\hat{P}_a \tag{4.73}$$

which clearly shows that an observable can be expressed as a weighted sum of projection operators to its own Eigen states, and the weighing factor will be the Eigen value itself. In general, for any function $\hat{A}$, we can write

$$f(\hat{A}) = \sum_{a} f(a)|a\rangle\langle a| = \sum_{a} f(a)\hat{P}_a \tag{4.74}$$

Now, we shall make an important point so that one can understand what a projection operator is.

I. An operator $P \in L(V, V)$ is called a projection operator if $P^2 = P$. Show that $P \in L(V, V)$ is a projection operator if and only if there is an operator $Q \in L(V, V)$ with

a. $P + Q = I$
b. $PQ = QP = 0$

(where I = identity matrix and V = set space)

Solution: $P + Q = 1$ *clearly opines that*

$Q = I - P$. If P is a projection operator, then

$$P(I - P) = (P - P^2) = 0$$

or $p^2 = P$

and similarly for $(1 - P)P$.

Conversely, if $P(I - P) = 0$, then $P - P^2 = 0$, and hence, $P^2 = P$. Hence, P is a projection operator.

II. Here we shall give some characteristic properties of the projection operator. In the valence bond (VB) method, the spatial two electron function $\phi(r_1, r_2)$ is constructed as follows:

$$\phi(r_1, r_2) = N_{VB}\left[a(1)b(2) + b(1)a(2)\right] \tag{4.75}$$

This can also be written as

$$\phi(r_1, r_2) = N_{VB}\left[1 + \frac{b(1)a(2)}{a(1)b(2)}\right]a(1)b(2)$$

$$\text{or,} \phi(r_1, r_2) = N_{VB}\left[1 + P_{12}^{r}\right]a(1)b(2) \tag{4.76}$$

where P_{12}^r is the permutation operator, which interchanges r_1 and r_2. After interchanging/exchanging r_1 and r_2, we have

$$\phi(r_1,r_2) = N_{VB}\left[a(2)b(1) + b(1) + b(2)a(1)\right]$$
$$= N_{VB}\left[\frac{a(2)b(1)}{a(1)b(2)} + 1\right]a(1)b(2)$$
$$= N_{VB}\left[P_{12}^r + 1\right] = N_{VB}\left[1 + P_{12}^r\right]$$

Apart from a constant factor, the operator in square brackets in Eq. (4.76) is a so-called projection operator.

Now, let us square the terms in the bracket; we have

$$\left[1 + P_{12}^r\right]\left[1 + P_{12}^r\right] = 1.1 + 1 \cdot P_{12}^r + 1 \cdot P_{12}^r + 1 \cdot P_{12}^r \cdot P_{12}^r$$
$$= 1 + P_{12}^r + P_{12}^r + 1 \tag{4.77}$$
$$= \left(2 + 2P_{12}^r\right) = 2\left(1 + P_{12}^r\right)$$

We see that a projection operator is an operator that does not change or alter when it is squared. It also proves that $P^2 = P$.

III. *The operator* $\left(1 + P_{12}^r\right)$ *is Hermitian*. Since $1 + P_{12}^r$ does not contain any imaginary part, it is a real quantity, and hence, it is Hermitian.

IV. The projection operator may be used as the following:

We can write as

$$|\psi\rangle = \sum_n \langle w_n|\psi\rangle|w_n\rangle \tag{4.78}$$

where w_n = an Eigen function

And according to the closure relation

$$\sum_n |w_n\rangle\langle w_n| = \hat{I} \tag{4.79}$$

It will be useful to define a projection operator P_n as

$$P_n = |w_n\rangle\langle w_n|,\ P_n^2 = P_n \cdot P_n = P_{n'} \sum_n P_n = \hat{I} \tag{4.80}$$

Now use the basis $\left\{\psi_i^0\right\}$ to define the projection operator (4.80) to get

$$P_k = \left|\psi_k^0\right\rangle\left\langle\psi_k^0\right| \equiv |k\rangle\langle k|, \sum_k P_k = \hat{I} \tag{4.81}$$

The complementary projection operator

$$P_k = \hat{I} = P_n = \hat{I} - |n\rangle\langle n| = \sum_{k \neq n} |k\rangle\langle k| \tag{4.82}$$

This helps one to express the equation

$$\psi_n = \psi_n^0 + \lambda\left(\frac{H'_{1n}}{E_n^0 - E_1^0}\psi_2^0 + \frac{H'_{2n}}{E_n^0 - E_2^0}\psi_2^0 + \dots\right)\text{as}$$

$$|\psi_n\rangle = |n\rangle + \lambda\sum_{k\neq n}\frac{k|H'|n\rangle}{E_n^0 - H_0}|k\rangle = |n\rangle + \lambda k_n H'|n\rangle \tag{4.83}$$

$$\text{Where } k_n = \sum_{k\neq n}\frac{|n\rangle\langle k|}{E_n^0 - H_0} = \frac{P_n}{E_n^0 - H_0} = \frac{1-P_n}{E_n^0 - H_0} \tag{4.84}$$

The inverse operator $\frac{1}{E_n^0 - H_0}$ is well behaved so long it does not operate on $|\psi_n^0\rangle$, since E_n^0 is one of the Eigen values of H_0. The operator P_n makes certain this. In a closely packed form, we may write

$$\begin{aligned} E_n^{(1)} &= |n\rangle H'|n\rangle \\ E_n^{(2)} &= |n\rangle H'k_n H'|n\rangle \\ E_n^{(3)} &= |n\rangle H'k_n \bar{H}' k_n H'|n\rangle \end{aligned} \tag{4.85}$$

$$\begin{aligned} \psi_n^{(1)} &= k_n H'|n\rangle \\ \psi_n^{(2)} &= k_n \bar{H}' k_n H'|n\rangle \end{aligned}$$

At this stage, we have learnt about the projection operator and understood its applications.

4.11 Parity Operator (π *operator)*

Parity operator is defined by the relation

$$\hat{\pi} f(x) = f(-x) \tag{4.86}$$

$$\text{Therefore, } \hat{\pi}^2 f(x) = \hat{\pi} f(-x) = f(x) \tag{4.87}$$

which clearly suggests that Eigen value of $\hat{\pi}^2$ is 1. However, if $f(x)$ be arbitrary function, it will not essentially be an Eigen function of $\hat{\pi}$. On the other hand, if (x) happens to be the Eigen function of $\hat{\pi}$, then the problem arises that what will be the corresponding Eigen value?

In answering the question, let us suppose that

$$\hat{\pi} f(x) = c f(x)$$

$$\hat{\pi}^2 f(x) = c^2 f(x) \tag{4.88}$$

On comparing the equation (4.87) and (4.88), we get $c^2 = 1$; therefore, $c = \pm 1$

Thus, we have

$$\hat{\pi} f(x) = f(x) \tag{4.89}$$

$$\hat{\pi} f(x) = -f(x) \tag{4.90}$$

It is inferred from equations (4.86), (4.89), and (4.90) that f(−x) is equal to either *f*(*x*) or f(−x). Actually, there are even and odd functions. These are called a function of even parity and odd parity respectively with regard to parity operator.

We have read in chapter 7 that the wave functions of a particle in a box of length a ($-a \leq x \leq a$) are found to be either even or odd functions. It is because of the fact the Hamiltonian operator for this system commutes with operator

It can be shown in the following manner:

$$\hat{\pi}\hat{H} = \hat{\pi}(-\hbar^2/2m\ d^2/dx^2) = -\hbar^2/2m\left[d(-x)^2/dx\ d^2/d(-x)^2\right]$$

$$= -\hbar^2/2m\,d^2/dx^2 = \hat{H}$$

$$\text{Therefore, } [\hat{H}\hat{\pi}]\Psi(x) = \hat{H}\hat{\pi}\Psi(x) - \hat{\pi}\hat{H}\ \Psi(x)$$

$$= \hat{H}\Psi(-x) - (\hat{\pi}\hat{H})\Psi(x)$$

$$= \hat{H}\Psi(-x)\ -\ H\Psi(-X) = 0$$

$$\text{Or } [\hat{H}, \hat{\pi}] = 0 \tag{4.91}$$

Thus, the above statement is proved.

BIBLIOGRAPHY

Chadra, A.K. 1989. *Introductory Quantum Chemistry*, 3rd ed. New Delhi: McGraw-Hill.

Dicke, R.H. and J.P. Wittke. 1960. *Introduction to Quantum Mechanics*. Reading, MA: Addison-Wesley Publishing Co.

Edmonds, A.R. 1957. *Angular Momentum in Quantum Mechanics*. Princeton, NJ: Princeton University Press.

Eyring, H., J. Walter and G.E. Kimball. 1944. *Quantum Chemistry*. New York: Wiley.

Greiner, W. 1989. *Quantum Mechanics- An Introduction*, vol. 1. Springer-Verlag.

Roman, P. 1962. *Advanced Quantum Theory*. Reading, MA: Addison-Wesley Publishing Co.

Spolsky, E.V. 1963. *Atomic Physics*, vol. 1,2. Mir Publishing House.

Solved Problems

Problem 1. Write out the operator $\hat{A}^2$ for

(a) $\hat{A} = \left(\frac{d}{dx} + x\right)$; (b) $\left(x\frac{d}{dx}\right)$; (c) $\left(\frac{d}{dx}x\right)$;(d)$\left(\frac{1}{x} + \frac{d}{dx}\right)$

Solution:

a. $\hat{A} = \left(\frac{d}{dx} + x\right)$

$$\therefore\quad \hat{A}^2 = \left(\frac{d}{dx} + x\right)^2$$

Now we shall include an arbitrary function $\psi(x)$ before carrying out the operation

$$\hat{A}\psi^2 = \left(\frac{d}{dx}+x\right)^2 \psi = \left(\frac{d}{dx}+x\right)\left(\frac{d}{dx}+x\right)\psi$$

$$= \left(\frac{d}{dx}+x\right)\left(\frac{d\psi}{dx}+x\psi\right)$$

$$= \frac{d^2}{dx^2}+\frac{d}{dx}(x)\psi + x\frac{d\psi}{dx}+x\frac{d\psi}{dx}+x^2\psi$$

$$= \frac{d^2\psi}{dx^2}+1\psi+2x\frac{d\psi}{dx}+x^2\psi$$

$$= \left(\frac{d^2}{dx^2}+2x\frac{d}{dx}+x^2+1\right)$$

$$\therefore \quad \hat{A} = \frac{d^2}{dx^2}+2x\frac{d}{dx}+x^2+1$$ Answer.

b. $\hat{A} = \left(x\frac{d}{dx}\right)$

$$\therefore \quad \hat{A}^2\psi = \left(x\frac{d}{dx}\right)^2\psi$$

$$= x^2\frac{d^2\psi}{dx^2}+x\frac{d\psi}{dx}$$

$$= \left(x^2\frac{d^2}{dx^2}+x \ d/dx\right)\psi$$

$$\therefore \quad \hat{A} = \left(x^2\frac{d^2}{dx^2}+x \ d/dx\right)$$ Answer.

c. $\hat{A} = \left(\frac{d}{dx}x\right)$

$$\therefore \quad \hat{A} = \left(\frac{d}{dx}x\right)^2$$

$$\hat{A}^2\psi = \left(\frac{d}{dx}x\right)^2\psi = \left(\frac{d}{dx}x\right)\left(\frac{d}{dx}x\right)\psi$$

$$= \left(\frac{d}{dx}x\right)\frac{d}{dx}(x\psi)$$

$$= \left(\frac{d}{dx}x\right)\left(x\frac{d\psi}{dx}+\psi\right)$$

$$= \frac{d}{dx}\left(x^2\frac{d\psi}{dx}+x\psi\right)$$

$$= x^2\frac{d^2\psi}{dx^2}+2x\frac{d\psi}{dx}+\frac{d\psi}{dx}+\psi$$

$$= \left(x^2\frac{d^2}{dx^2}+2x\frac{d}{dx}+\frac{d}{dx}+1\right)\psi$$

$$\therefore \quad \hat{A}^2 = \left(x^2 \frac{\mathrm{d}^2}{\mathrm{d}x^2} + 2x \frac{\mathrm{d}}{\mathrm{d}x} + \frac{\mathrm{d}}{\mathrm{d}x} + 1 \right) \qquad \text{Answer.}$$

d. $\hat{A} = \left(\frac{1}{x} + \frac{\mathrm{d}}{\mathrm{d}x} \right)$

$$\therefore \quad \hat{A}^2 = \left(\frac{1}{x} + \frac{\mathrm{d}}{\mathrm{d}x} \right)^2$$

$$\begin{aligned}
\hat{A}\psi &= \left(\frac{1}{x} + \frac{\mathrm{d}}{\mathrm{d}x} \right)^2 \psi \\
&= \left(\frac{1}{x} + \frac{\mathrm{d}}{\mathrm{d}x} \right)\left(\frac{1}{x} + \frac{\mathrm{d}}{\mathrm{d}x} \right)\psi \\
&= \left[\frac{1}{x^2}\psi + \frac{1}{x}\frac{\mathrm{d}\psi}{\mathrm{d}x} + \frac{\mathrm{d}}{\mathrm{d}x}\left(x^{-1}\psi\right) + \frac{\mathrm{d}^2\psi}{\mathrm{d}x^2} \right] \\
&= \left[\frac{1}{x^2}\psi + \frac{1}{x}\frac{\mathrm{d}\psi}{\mathrm{d}x} + x^{-1}\frac{\mathrm{d}\psi}{\mathrm{d}x} + \psi(-1)x^{-2} + \frac{\mathrm{d}^2\psi}{\mathrm{d}x^2} \right] \\
&= \frac{1}{x^2}\psi + \frac{1}{x}\frac{\mathrm{d}\psi}{\mathrm{d}x} + \frac{1}{x}\frac{\mathrm{d}\psi}{\mathrm{d}x} - \frac{\psi}{x^2} + \frac{\mathrm{d}^2\psi}{dx^2} \\
&= \frac{\mathrm{d}^2\psi}{\mathrm{d}x^2} + \frac{2}{x}\frac{\mathrm{d}\psi}{\mathrm{d}x} \\
&= \left(\frac{\mathrm{d}^2}{\mathrm{d}x^2} + \frac{2}{x}\frac{\mathrm{d}}{\mathrm{d}x} \right)\psi
\end{aligned}$$

$$\therefore \quad \hat{A}^2 = \left(\frac{\mathrm{d}^2}{\mathrm{d}x^2} + 2/x \ d/x \right) \qquad \text{Answer.}$$

Problem 2. Show that

a. $\left(\frac{\mathrm{d}}{\mathrm{d}x} + x \right)\left(\frac{\mathrm{d}}{\mathrm{d}x} - x \right) = \frac{\mathrm{d}^2}{\mathrm{d}x^2} - x^2 + x\frac{\mathrm{d}}{\mathrm{d}x} - 1$

b. $\left(\frac{\mathrm{d}}{\mathrm{d}x} - x \right)\left(\frac{\mathrm{d}}{\mathrm{d}x} + x \right) = \left(\frac{\mathrm{d}^2}{\mathrm{d}x^2} - x\frac{\mathrm{d}}{\mathrm{d}x} - x^2 + 1 \right)$

Solution:

a. $\left(\frac{\mathrm{d}}{\mathrm{d}x} + x \right)\left(\frac{\mathrm{d}}{\mathrm{d}x} - x \right)$

Let $\frac{\mathrm{d}}{\mathrm{d}x} = D$, then we can write

$$= (D + x)(D - x)$$
$$= D(D - x) + x(D - x)$$
$$= D^2 - Dx + xD - x^2$$
$$= \frac{d^2}{dx^2} - 1 + x\frac{d}{dx} - x^2$$
$$= \left(\frac{d^2}{dx^2} + x\frac{d}{dx} - x^2 - 1\right)$$

Answer.

b. $\left(\frac{d}{dx} - x\right)\left(\frac{d}{dx} + x\right)$

Let $D = d/dx$

∴ The above expression may be expressed as

$$(D - x)(D + x)$$
$$= D(D + x) - x(D + x)$$
$$= D^2 + Dx - xD - x^2$$
$$= \frac{d^2}{dx^2} + 1 - x \ \ d/dx - x^2$$
$$= \left(\frac{d^2}{dx^2} - x\frac{d}{dx} - x^2 + 1\right)$$

Answer.

Problem 3. Determine whether the following operators are linear or non-linear.

a. $\hat{A}f(x) = x^2 f(x); (b)\hat{A}f(x) = \text{SQRT}f(x)$ [SQRT: Square Root]

Solution: (a) $\hat{A}f(x) = x^2 f(x)$

The condition of a linear operator is

$$\hat{A}\left[c_1 f_1(x) + c_2 f_2(x)\right] = c_1\hat{A}f_1(x) + c_2\hat{A}f_2(x)$$

$$\text{But}\ \hat{A}\left[c_1 f_1(x) + c_2 f_2(x)\right] = x^2\left[c_1 f_1(x) + c_2 f_2(x)\right]$$

$$= c_1 x^2 f_1(x) + c_2 x^2 f_2(x)$$
$$= c_1\hat{A}f_1(x) + c_2\hat{A}f_2(x)$$

and therefore, x^2 multiplied by a function is a linear operator.
Answer.

b. $\hat{A}f(x) = \text{SQRT}\,f(x)$

$$\hat{A}\left[c_1 f_1(x) + c_2 f_2(x)\right]$$

$$= [c_1 f_1(x) + c_2 f_2(x)]^{1/2}$$
$$\neq c_1 f_1^{1/2}(x) + c_2 f_2^{1/2}(x)$$

and hence, SQRT is a non-linear operator. Answer.

Problem 4. Is d/dx a linear operator?
Is $\sqrt{\ }$ a linear operator?

Solution: An operator is said to be linear if

$$\hat{A}[f(x) + g(x)] = \hat{A}f(x) + \hat{A}g(x) \text{and}$$

$$\hat{A}[cf(x)] = c\hat{A}f(x)$$

where $f(x)$ and $g(x)$ are arbitrary functions.

$$\text{Now} \frac{\mathrm{d}}{\mathrm{d}x}[f(x) + g(x)] = \frac{\mathrm{d}f(x)}{\mathrm{d}x} + \frac{\mathrm{d}g(x)}{\mathrm{d}x}$$

$$\text{and} \left(\frac{\mathrm{d}}{\mathrm{d}x}\right)[cf(x)] = c\frac{\mathrm{d}f(x)}{\mathrm{d}x}$$

We see that d/dx obeys the above equation and so it is a linear operator.

The second problem is that whether $\sqrt{\ }$ is a linear operator, where SQRT becomes an operator then

$$\sqrt{f(x) + g(x)} \neq \sqrt{f(x)} + \sqrt{g(x)}$$

and therefore, $\sqrt{\ }$ does not obey the above equation and is non-linear operator. Answer.

Problem 5. If $\hat{A} = \frac{\mathrm{d}^2}{\mathrm{d}x^2} + 3x\ \mathrm{d}/\mathrm{d}x$ and $f(x) = 4x^3$, then find $\hat{A}f(x)$.

Solution: Given that $\hat{A} = \frac{\mathrm{d}^2}{\mathrm{d}x^2} + 3x\ \mathrm{d}/\mathrm{d}x$ and $f(x) = 4x^3$

$$\hat{A}f(x) = [\mathrm{d}^2/\mathrm{d}x^2 + 3x\ \mathrm{d}/\mathrm{d}x]\, f(x)$$

$$= \left[\frac{\mathrm{d}^2 f(x)}{\mathrm{d}x^2} + 3x\frac{\mathrm{d}f(x)}{\mathrm{d}x}\right]$$

$$= \frac{\mathrm{d}^2}{\mathrm{d}x^2}(4x^3) + 3x\frac{\mathrm{d}}{\mathrm{d}x}(4x^3)$$

$$= 24x + 36x^3$$

Answer.

Problem 6. Show that

a. $[\hat{A}, \hat{B}] = -[\hat{B}, \hat{A}]$

b. $[\hat{A}^2, \hat{B}] = \hat{A}[\hat{A}, \hat{B}] + [\hat{A}, B]\hat{A}$

c. $\left[\hat{A},\left[\hat{B},\hat{C}\right]\right]=\left[\left[\hat{A},\hat{B}\right],\hat{C}\right]+\left[\hat{B},\left[\hat{A},\hat{C}\right]\right]$

Solution:

a. $\left[\hat{A},\hat{B}\right]=-\left[\hat{B},\hat{A}\right]$

$$\therefore \quad \left[\hat{A},\hat{B}\right]=\left[\hat{A}\hat{B}-\hat{B}\hat{A}\right]$$
$$=-\left[\hat{B}\hat{A}-\hat{A}\hat{B}\right]$$
$$=-\left[\hat{B},\hat{A}\right]$$

b. $\left[\hat{A}^2,\hat{B}\right]=\hat{A}\left[\hat{A},\hat{B}\right]+\left[\hat{A},\hat{B}\right]\hat{A}$

We shall consider the RHS for getting the result

$$\text{RHS}=\hat{A}\left[\hat{A},\hat{B}\right]+\left[\hat{A},\hat{B}\right]\hat{A}$$
$$=\hat{A}\left[\hat{A}\hat{B}-\hat{B}\hat{A}\right]+\left[\hat{A}\hat{B}-\hat{B}\hat{A}\right]\hat{A}$$
$$=\hat{A}^2\hat{B}-\hat{B}\hat{A}^2$$
$$=\left[\hat{A}^2,\hat{B}\right]=\text{LHS}$$

QED

c. $\left[\hat{A},\left[\hat{B},\hat{C}\right]\right]=\left[\left[\hat{A},\hat{B}\right],\hat{C}\right]+\left[\hat{B},\left[\hat{A},\hat{C}\right]\right]$

$$\text{LHS}=\left[\hat{A},\left[\hat{B},\hat{C}\right]\right]$$
$$=\hat{A}\left(\hat{B}\hat{C}-\hat{C}\hat{B}\right)-\left(\hat{B}\hat{C}-\hat{C}\hat{B}\right)\hat{A}$$
$$=\hat{A}\hat{B}\hat{C}-\hat{A}\hat{C}\hat{B}-\hat{B}\hat{C}\hat{A}+\hat{C}\hat{B}\hat{A}$$

Now,

$$\text{RHS}=\left[\left[\hat{A},\hat{B}\right],\hat{C}\right]+\left[\hat{B}\left[\hat{A},\hat{C}\right]\right]$$
$$=\left[\left(\hat{A}\hat{B}-\hat{B}\hat{A}\right),\hat{C}\right]+\left[\hat{B},\left[\hat{A}\hat{C}-\hat{C}\hat{A}\right]\right]$$
$$=\hat{A}\hat{B}\hat{C}-\hat{B}\hat{A}\hat{C}-\hat{C}\hat{A}\hat{B}+\hat{C}\hat{B}\hat{A}+\hat{B}\hat{A}\hat{C}-\hat{B}\hat{C}\hat{A}-\hat{A}\hat{C}\hat{B}+\hat{C}\hat{A}\hat{B}$$
$$=\hat{A}\hat{B}\hat{C}-\hat{A}\hat{C}\hat{B}-\hat{B}\hat{C}\hat{A}+\hat{C}\hat{B}\hat{A}$$
$$=\text{LHS}$$

Here, LHS = RHS QED

Problem 7. Consider the function $f(x, y) = x^2 + y^2 + 2xy$. Let $\hat{P} = \left(\frac{\partial}{\partial x}\right)_{yz}$ and $\hat{Q} = \left(\frac{\partial}{\partial y}\right)_{x,z}$ operate first on $f(x, y)$ with $\hat{P}\hat{Q}$ and then with $\hat{Q}\hat{P}$. What would be the result after operating on $f(x, y)$ with $\hat{P}\hat{Q} - \hat{Q}\hat{P}$?

Solution: According to the question $f(x, y) = x^2 + y^2 + 2xy$

Also $\hat{P} = \left(\frac{\partial}{\partial x}\right)_{yz}$ and $\hat{Q} = \left(\frac{\partial}{\partial y}\right)_{x,z}$

Now,

$$\hat{P}f(x,y) = \frac{\partial}{\partial x}\left(x^2 + y^2 + 2xy\right)_{yz}$$
$$= (2x + 2y)$$

and

$$\hat{Q}f(x,y) = \frac{\partial}{\partial x}\left(x^2 + y^2 + 2xy\right)_{x,z}$$
$$= (2y + 2x)$$
$$= (2x + 2y)$$
$$\left(\hat{P}\hat{Q} - \hat{Q}\hat{P}\right) = (2x + 2y)(2x + 2y) - (2y + 2x)(2x + 2y)$$
$$= 0$$

It means $\hat{P}$ and $\hat{Q}$ commute.

We can also see that

$$\hat{P}\hat{Q} = \frac{\partial}{\partial x}\left(\frac{\partial}{\partial y}\right)f(x)$$
$$= \frac{\partial}{\partial x}\left[\frac{\partial}{\partial y}\left(x^2 + y^2 + 2xy\right)\right]$$
$$= \frac{\partial}{\partial x}\left[(2y + 2x)\right] = 2$$
$$\hat{Q}\hat{P} = \frac{\partial}{\partial y}\left[\frac{\partial}{\partial x}\left(x^2 + y^2 + 2xy\right)\right]$$
$$= \frac{\partial}{\partial y}(2x + 2y) = 2$$
$$\therefore \quad \hat{P}\hat{Q} - \hat{Q}\hat{P} = 0$$

$\therefore \hat{P}$ and $\hat{Q}$ commute Q.E.D.

Problem 8. Find the commutator $[x, \mathrm{d}/\mathrm{d}x]$.

Solution: Let us suppose that the commutator operate on an arbitrary differentiable function $f(x)$, i.e.,

$$\left[x,\frac{d}{dx}\right]f(x)=x\frac{df(x)}{dx}-\frac{d\left[xf(x)\right]}{dx}$$

$$=x\frac{df(x)}{dx}-x\frac{df(x)}{dx}-f(x)$$

$$=-f(x)$$

$$\therefore\left[x,\frac{d}{dx}\right]=-1$$

Answer.

Problem 9. Find the commutator $[x^2, d/dx]$.

Solution: $\left[x^2,\frac{d}{dx}\right]f(x)=\frac{x^2 df(x)}{dx}-\frac{d}{dx}x^2 f(x)$

$$\Rightarrow\frac{x^2 df(x)}{dx}-x^2\frac{df(x)}{dx}-2xf(x)$$

$$\Rightarrow -2xf(x)$$

$$\therefore\left[x^2,\frac{d}{dx}\right]=-2x$$

Answer.

Problem 10. Give the three different operator $\hat{A}$, which satisfy $\hat{A}e^x = e^x$.

Solution:

i. $\frac{d}{dx}=\hat{A}$

$$\therefore\quad \hat{A}\ e^x=\frac{d}{dx}\left(e^x\right)=e^x$$

ii. $\frac{d^2}{dx^2}=\hat{A}$

$$\therefore\quad \hat{A}\ e^x=\frac{d^2}{dx^2}\left(e^x\right)=e^x$$

iii. $\hat{A}=\int$

$$\therefore\quad \hat{A}e^x=\int e^x dx=e^x$$

Answer.

Problem 11. Let $\hat{P}=\frac{d}{dx},\hat{Q}=x$ and $f(x)=x^2+2x+1$, show that $\hat{Q}\hat{P}f(x)\neq\hat{P}\hat{Q}f(x)$

Solution: First we shall calculate

$$\hat{P}\hat{Q}f(x) = \frac{d}{dx}x.f(x)$$

$$= \frac{d}{dx}\left[x\left(x^2 + 2x + 1\right)\right]$$

$$= \frac{d}{dx}\left[x^3 + 2x^2 + x\right]$$

$$= 3x^2 + 4x + 1$$

Now,

$$\hat{Q}\hat{P}f(x) = x\frac{d}{dx}\left(x^2 + 2x + 1\right)$$

$$= x(2x + 2)$$

$$= 2x^2 + 2x = 2x(x + 1)$$

$$\therefore \quad \hat{P}\hat{Q} \neq \hat{Q}\hat{P}$$ Answer.

Problem 12. Verify that the operator ∇^2 is linear.

Solution: An operator will be linear if $\hat{A}\left[f(x) + g(x)\right] = \hat{A}f(x) + \hat{A}g(x)$ and

$$\hat{A}\left[cf(x)\right] = c\hat{A}f(x)$$

Suppose $\hat{A} = \nabla^2$

$$\therefore \quad \nabla^2\left[f(x) + g(x)\right] = \nabla^2 f(x) + \nabla^2 g(x)$$

and $\nabla^2\left[cf(x)\right] = cV^2 f(x)$

Hence, ∇^2 is a linear operator. Q.E.D

Problem 13. Which of the following operators are linear?

a. $\hat{A}u = \lambda u,\ \lambda = \text{constant}$
b. $\hat{B}u = u^*$
c. $\hat{C}u = u^2$
d. $\hat{D}u = \dfrac{du}{dx}$
e. $\hat{E}u = 1/u$

Solution:

a. $\hat{A}u = \lambda u$

$$\because \quad \hat{A}(u + v) = \lambda(u + v)\hat{A}u + \hat{A}v$$

and $\hat{A}(cu) = \lambda cu = c\hat{A}u$

Hence, $\hat{A}$ is a linear operator. Answer.

b. $\hat{B}u = u^*$

$$\because \quad \hat{B}(cu) = \lambda cu = c\hat{B}u^* \neq c\hat{B}u$$

Hence, $\hat{B}$ is not a linear operator. Answer.

c. $\hat{C}u = u^2$

$$\because \quad \hat{C} \quad (u+v) = (u+v)^2 = u^2 + 2uv + v^2$$

$$\text{But } \hat{C}u + \hat{C}v = u^2 + v^2 \neq \hat{C}(u+v)$$

$\therefore \quad \hat{C}$ is not linear. Answer.

d. $\hat{D}u = \dfrac{du}{dx}$

$$\because \quad \hat{D} \quad (u+v) = \frac{du}{dx} + \frac{dv}{dx} = \hat{D}u + \hat{D}v$$

$$\hat{D}(cu) = c\frac{du}{dx} = c\hat{D}u$$

$\therefore \hat{D}$ is a linear operator. Answer.

e. $\hat{E}u = 1/u$

$$\hat{E}(cu) = \frac{1}{(cu)} = \left(\frac{1}{c}\right)\hat{E}u$$

$\therefore \hat{E}$ is not a linear operator. Answer.

Problem 14. In the following, show that $f(x)$ is an Eigen function of the operator given. Find the Eigen value.

	$\hat{A}$	$f(x)$
(a)	$\dfrac{d^2}{dx^2}$	$\cos \omega x$
(b)	$\dfrac{d}{dx}$	$e^{i\omega t}$
(c)	$\dfrac{d^2}{dx^2} + 2\dfrac{d}{dx} + 3$	$e^{\alpha x}$

Solution:

a. $\hat{A} = \dfrac{d}{dx^{2'}} f(x) = \cos \omega x$

$$\therefore \quad \hat{A}f(x) = \frac{d^2}{dx^2}[\cos \omega x]$$

$$= \frac{d}{dx}\left[\frac{d}{dx}(\cos \omega x)\right]$$

$$= \frac{d}{dx}[(-\omega \sin \omega x)] = -\omega^2 \cos \omega x$$

$\therefore$ The Eigen value $= -\omega^2$ Answer.

b. $\hat{A} = \frac{d}{dx'} f(x) = e^{i\omega t}$

$$\therefore \quad \hat{A} \ f(x) = \frac{d}{dx}\left(e^{i\omega t}\right) = i\omega e^{i\omega \ t}$$

$\therefore$ Eigen value $= i\omega$ Answer.

c. $\hat{A}f(x) = \frac{d^2}{dx^2} + 2\frac{d}{dx} + 3, f(x) = e^{\alpha x}$

$$\therefore \quad \hat{A}f(x) = \left(\frac{d^2}{dx^2} + 2\frac{d}{dx} + 3\right)e^{\alpha x}$$

$$= \frac{d^2}{dx^2}\left(e^{\alpha x}\right) + 2\frac{d}{dx}\left(e^{\alpha x}\right) + 3e^{\alpha x}$$

$$= \alpha^2 e^{\alpha x} + 3e^{\alpha x}$$

$$= \left(\alpha^2 + 2\alpha + 3\right)e^{\alpha x}$$

$\therefore$ The Eigen value $= \alpha^2 + 2\alpha + 3$ Answer.

Problem 15. Show that the function $\psi = \sin(k_1 x) \sin(k_2 y) \sin(k_3 z)$ is an Eigen function of ∇^2. What is the Eigen value?

Solution: Given that $\psi = \sin(k_1 x) \sin(k_2 y) \sin(k_3 z)$

$$\therefore \quad \frac{d\psi}{dx} = k_1 \cos(k_1 x)\sin(k_2 y)\sin(k_3 z)$$

$$\text{and} \frac{d^2\psi}{dx^2} = -k_1^2 \sin(k_1 x)\sin(k_2 y)\sin(k_3 z) = -k_1^2\psi$$

Similarly, $\frac{d^2\psi}{dy^2} = -k_2^2\psi$

and $\frac{d^2\psi}{dz^2} = -k_3^2\psi$

$$\therefore \quad \nabla^2\psi \ = \frac{d^2\psi}{dx^2} + \frac{d^2\psi}{dy^2} + \frac{d^2\psi}{dz^2} = -\left(k_1^2 + k_2^2 + k_3^2\right)\psi$$

Hence, ψ is the Eigen function of ∇^2 with the Eigen value $-\left(k_1^2 + k_2^2 + k_3^2\right)$ Answer.

Problem 16. Find the operator equal to the commutator $\left[\hat{x}, \hat{P}_x\right]$.

Solution: Let us operate on an arbitrary differentiable function $f(x)$ (or in short f)

$$\left[\hat{x}, \hat{P}_x\right] = \frac{\hbar}{i}\left[x\frac{\partial f}{\partial x} - \frac{\partial}{\partial x}(x, f)\right] \quad \left[\ \therefore \quad \hat{p}_x \quad \frac{h}{i}\frac{d}{dx}\right]$$

$$= \frac{\hbar}{i}\left[x\frac{\partial f}{\partial x} - x\frac{\partial}{\partial x} - f\right]$$

$$= \frac{\hbar}{i}(-f) = -\frac{\hbar}{i}f = i\hbar f$$

$\therefore$ The operator equation is $\left[\hat{x}, \hat{p}_x\right] = i\hbar$ Answer.

Problem 17. Show that two operators $\left[\hat{x}, \hat{p}_x\right]$ are non-commuting.

Solution: According to the question $\hat{x}$ and $\hat{p}_x$ are two operators which may be written as $\left[\hat{x}, \hat{p}_x\right]$. Let us operate on an arbitrary differentiable function $\psi = \hat{A}$ sin kx

$$\therefore \quad \psi = A \sin kx$$

$$\therefore \quad \frac{\partial \psi}{\partial X} = Ak \cos kx$$

$$\therefore \quad \left[\hat{x}, \hat{p}_x\right]\psi = \hat{x}\hat{p}_x\psi - \hat{p}_x \cdot x\psi$$

But $\hat{x}\hat{p}_x\psi = \hat{x}\left[-i\hbar\frac{\mathrm{d}\psi}{\mathrm{d}x}\right] = i\hbar x \cdot Ak \cos k\ x$

and

$$\hat{p}_x \cdot x\psi = i\hbar\frac{\mathrm{d}\psi}{\mathrm{d}x}(xA\sin k_x)$$
$$= -i\hbar\left[A\sin kx + x.Ak\cos kx\right]$$

Thus, $\widehat{x,}\hat{p}_x\psi \neq \hat{p}_x\hat{x}\psi$

$\therefore \hat{x}$ and $\hat{p}_x$ do not commute. Q.E.D

Problem 18. Deduce the commutation relation $\left[\hat{p}_x \cdot \hat{H}\right]\psi = \left[\hat{p}_x - \hat{H}\psi - \hat{H}\hat{p}_x\psi\right] = 0.$

Solution: Let us suppose that

$$\psi = A\sin kx \text{ then}$$

$$\frac{\mathrm{d}\psi}{\mathrm{d}x} = Ak\cos k\ x \text{and}$$

$$\frac{\mathrm{d}^2\psi}{\mathrm{d}x^2} = -k^2 A\sin kx$$

$$\therefore \quad \hat{H}\psi = \frac{-\hbar^2}{2m}\cdot\frac{\mathrm{d}^2\psi}{\mathrm{d}x^2} = \frac{\hbar^2}{2m}\cdot k^2 A\sin kx$$

$$\therefore \quad \left[\hat{p}_x \cdot \hat{H}\right]\psi - i\hbar\frac{\mathrm{d}\psi}{\mathrm{d}x}\left(\frac{\hbar^2}{2m}\cdot k^2 A\sin kx\right)$$
$$= \frac{-\hbar^3 k^3}{2m}A\cos kx$$

$$\therefore \quad \hat{p}_x \cdot \psi = -i\hbar \frac{d\psi}{dx} = -i\hbar \ Ak \cos kx$$

$$\hat{H}\hat{p}_x \cdot \psi = \frac{-h^2}{2m} \cdot \frac{d^2}{dx^2} = (-i\hbar \ Ak \cos kx \ )$$

$$\frac{d^2\psi}{dx^2} = -k^2 A \sin kx$$

$$\therefore \quad \hat{H}\psi = \frac{-\hbar^2}{2m} \cdot \frac{d^2\psi}{dx^2} = \frac{\hbar^2}{2m} \cdot k^2 \quad A \sin kx$$

$$\therefore \quad \left[\hat{p}_x \cdot \hat{H}\right]\psi = -i\hbar \frac{d\psi}{dx}\left(\frac{\hbar^2}{2m} \cdot k^2 A \sin kx\right)$$

$$= \frac{-\hbar^3 k^3}{2m} A \cos kx$$

$$\therefore \quad \hat{p}_x \quad \cdot \psi = -i\hbar \frac{d\psi}{dx} = -i\hbar \ Ak \cos kx$$

$$\hat{H}\hat{p}_x \cdot \psi = \frac{-\hbar^2}{2m} \cdot \frac{d^2}{dx^2} = (-i\hbar \ Ak \cos kx)$$

$$= -i\hbar \ Ak \frac{d^2}{dx^2}(\cos kx)$$

$$= \frac{i\hbar^3 Ak}{2m} \cdot \frac{d}{dx}(-k \sin kx)$$

$$= \frac{i\hbar^3 Ak}{2m} \cos kx$$

$$\therefore \quad \left[p_x, \hat{H}\right] = \left[p_x \cdot \hat{H}\psi - p_x\psi\right] = 0$$

Hence, $\hat{p}_x$ and $\hat{H}$ commute. Answer.

Problem 19. Show that the linear momentum operator $\hat{p}_x$ is Hermitian.

Solution: An operator $\hat{A}$ will be Hermitian if

$$\int \psi_i^* \left(\hat{A}\psi_j\right) d\tau = \int \left(\hat{A}\psi_L\right)^* \psi_j d\tau$$

where ψ_i and ψ_j = Eigen functions of $\hat{A}$.

We know that

$\hat{p}_x = -i\hbar \frac{d}{dx}$ and in the x-direction $dx = d\tau$ and $-\infty < x < \infty$

From the method of integration by parts, we know that if u and V are functions of x, then

$$\int_a^b u dv = [uv]_a^b - \int_a^b v du$$

$$\text{Now, } \int_{-\infty}^{\infty} \psi_i^* \left(-i\hbar \frac{d}{dx}\right) \psi_j \cdot dx$$

$$= -i\hbar \left[\psi_i^* \psi_j\right]_{-\infty}^{\infty} \; i\hbar \int_{\infty}^{\infty} \psi_j \frac{d\psi_i^*}{dx} \cdot dx$$

$= 0$ (because according to the fundamental postulate of quantum mechanics, the wave function ψ_i and ψ_j must vanish at infinity). The second term can be written as

$$\int_{-\infty}^{\infty} \psi_j \left(-i\hbar \frac{d}{dx}\right)^* \psi_i^* dx$$

Hence the operator for $\hat{p}_x$ is Hermitian. Similarly, the operators for $\hat{p}_y$ and $\hat{p}_z$ can also be illustrated to be Hermitian.

QED

Problem 20. A system is described by the Hamiltonian operator $\hat{H} = \frac{d^2}{dx^2} + x^2$, show that $\psi = Axe^{-x^2/2}$ is Eigen function of $\hat{H}$. Find the Eigen value.

Solution: It is clear by inspection that $\psi = \left(Axe^{-x^2/2}\right)$ is well behaved.

Given that $\hat{H} = \frac{-d^2}{dx^2} + x^2$

$$\therefore \quad \frac{-d^2}{dx^2}\left[\left(Axe^{-x^2/2}\right)\right] = -A\frac{d}{dx}\left[1 - x^2\right]e^{-x^2/2}$$

$$= A\left(3 - x^2\right)xe^{-x^2/2}$$

$$\therefore \quad \widehat{H}\ \psi = \left(\frac{d^2}{dx^2} + x^2\right)\left(Axe^{-x^2/2}\right) = 3\left[Axe^{-x^2/2}\right]$$

$\therefore$ Eigen value = 3 Answer.

Questions on Concepts

1. Define operator. What do you mean by linear and non-linear operators? Is SQRT linear?
2. Find out an expression for the following commutators:

a. $x, \left[\frac{d^2}{dx^2}\right]$

b. $x, \left[\frac{d}{dx'} x\right]$

c. $\left[\frac{1}{x} + \frac{d}{dx}\right]^2$

3. Find the commutator $[x^2, \mathrm{d}/\mathrm{d}x]$
4. Show that
   a. the operator $\mathrm{d}/\mathrm{d}x$ is linear operator.
   b. $\frac{\mathrm{d}}{\mathrm{d}x}$ is not Hermitian.
5. Determine whether each of the following operators is linear and whether it is Hermitian.
   a. $\frac{\mathrm{d}^2}{\mathrm{d}x^2}$
   b. $\frac{\mathrm{d}^3}{\mathrm{d}x^3}$
6. Show that the operator $i\left(\frac{\mathrm{d}}{\mathrm{d}x}\right)$ is linear and Hermitian.
7. Prove that any Hermitian operator is linear.
8. Prove that
   a. $L_x = yp_z - Zp_y$
   b. $L_y = zp_x - xp_z$
   c. $L_z = xp_y - yp_x$
9. Prove that
   a. $\left[L_x, L_y\right] = i\hbar L_z$
   b. $\left[L_x, L_y\right] = i\hbar L_x$
   c. $\left[L_x, L_y\right] = i\hbar L_y$
10. Prove that
    a. $[L^2, L_y] = 0$
    b. $[L^2, L_z] = 0$
11. Show that
    a. $L_z, L_+ = L_+(L_z + 1)$
    b. $L_z, L_- = L_-(L_z - 1)$, where the terms have their usual significance.
12.
    a. What do you mean by Hermitian operator?
    b. Prove that $L^2 = L_+L_- + L_z^2 - \hbar L_z$.
13. Prove that the Eigen values of a Hermitian operator are real.
14. Prove that non-degenerate Eigen functions of a Hermitian operator form an orthogonal set.
15.
    a. Find $\left[z^3, \mathrm{d}/\mathrm{d}z\right]$.
    b. Find the square of $\left(\frac{\mathrm{d}}{\mathrm{d}x} + x\right)$.
16.
    a. What do you mean by the commutator of two operators?
    b. Write out the operator $\hat{A}^2$ when $\hat{A}^2 = \frac{\mathrm{d}^1}{\mathrm{d}x^2} - 2x\ d/dx + 1$.
17. Determine whether or not the following pairs of operators commute.

	$\hat{A}$	$\hat{B}$
(a)	$\frac{d}{dx}$	$\frac{d^2}{dx^2}+2\frac{d}{dx}$
(b)	$\frac{\partial}{\partial x}$	$\frac{\partial}{\partial y}$
(c)	x	d/dx

18. When $\hat{A} = d/dx$ and $\hat{B} = x^2$, show that
    a. $\hat{A}^2 f(x) \neq [\hat{A}f(x)]^2$
    b. $\hat{A}\hat{B}f(x) \neq \hat{B}\hat{A}f(x)$
19. Show that e^{ikx} is an Eigen function of the momentum operator, $\hat{p}_x = -i\hbar\frac{\partial}{\partial x}$ what is the Eigen value?
20. Prove that

$$L_x = -i\hbar\left[r\sin\theta\sin\phi\left(\cos\theta\frac{\partial}{\partial r}+\frac{-\sin\theta}{r}\frac{\partial}{\partial\theta}\right)-r\cos\theta\right.$$

$$\left.\left(\sin\theta.\sin\phi.\frac{\partial}{\partial r}+\frac{\cos\theta\sin\theta}{r}.\frac{\partial}{\partial\theta}+\frac{\cos\theta}{r\sin\theta}.\frac{\partial}{\partial\phi}\right)\right]$$

$$L_y = -i\hbar\left[-\cos\theta\frac{\partial}{\partial\theta}+\cot\theta\sin\theta\frac{\partial}{\partial\phi}\right] \text{ and } L_z = i\hbar\frac{\partial}{\partial\phi}$$

21. What do you mean by Hamiltonian operator? Prove that $[L^2, \nabla^2] = 0$.
22. If $\hat{A} = 3x^2$ and $\hat{B} = d/dx$, then show that $\hat{A}\hat{B} \neq \hat{B}\hat{A}$.
23. Show that the function $\sin(k_1x)\cdot\sin(k_2y)$ is an Eigen function of $\nabla^2 = \frac{\partial^2}{\partial x^2}+\frac{\partial^2}{\partial y^2}+\frac{\partial^2}{\partial z^2}$; find the Eigen value.
24. Prove that if two operator $\hat{A}$ and $\hat{B}$ are Hermitian, then $\left(\hat{A}\cdot\hat{B}\right)$ is also Hermitian if and only if $\hat{A}$ and $\hat{B}$ commute.
25. Show that $\frac{h}{2\pi i}x(-d/dx)$ is not Hermitian.
26. Illustrate that the operator $\hat{p}_x$ for linear momentum is Hermitian. What is its physical significance?
27. The Hamiltonian operator a given system is $\hat{H} = -(\hbar^2/2m)d^2/dx^2 + V$ (where V = |constant). The corresponding Eigen functions (not normalised) are $\psi_n = e^{\pm inx}$ (n = 1, 2, 3 ...)

5

Postulates of Quantum Mechanics

A postulate is an idea that is suggested as, or assumed to be, the basis for a theory, argument, or calculation. In quantum mechanics, the postulates concern the atomic and molecular properties, which are quite far removed from everyday experience. Consequently, in this regard, these may be difficult to understand. The important point is that the postulates are justified by their ability to predict. They should also have the ability to correlate experimental facts by their general applicability.

Before discussing the postulates of quantum mechanics, it will be useful to understand the meaning of two important terms: (a) dynamical variable and (b) observable.

First of all, we should know about the dynamical variable. Any property of a system of interest is known as the dynamical variable. Examples are the position r, the energy E, the x-component of linear momentum p_x, and so on.

Generally, any quantity of interest in classical mechanics is a dynamical variable. A very useful dynamical variable, which will be used later, comprises three components of the momentum vectors, which a particle in a system has when it remains at a fixed point.

Now, we shall know about observable. An observable is defined as any dynamical variable that can be measured. It should be kept in mind that in classical mechanics, all dynamical variables are observables, but it is not so in quantum mechanics.

In quantum mechanics, certain fundamental restrictions are imposed on simultaneously measurable variable quantities. To measure the component of the momentum vector, it is essential to make a simultaneous measurement of the position and momentum of the particle. We are aware of the fact that there exists an uncertainty relation for such kind of simultaneous measurement on microscopic particle, and the dynamical variable 'the momentum at a point' is not an observable. With this background in mind, we are going to introduce the basic postulates of quantum mechanics.

5.1 Postulate 1

The state of the quantum mechanical objects is described by a wave function.

or

A quantum mechanical system of n particles is described as fully as possible by a function, $\Psi(x, y, z, t)$ called the wave function, which determines all the measurable quantities of the system, where x, y, z, t are spatial co-ordinates.

The wave function ψ is needed to be a finite, single valued, and continuous function, which becomes zero at infinity. It must have a square integrable, and its first derivative $\mathrm{d}\psi/\mathrm{d}x_i$ must also be continuous. Ψ is physically interpreted by $\psi\psi^*\mathrm{d}x_i$, which is called the probability of finding the particles having co-ordinates lying between x, y, z, t and $x + \mathrm{d}x,\ y + \mathrm{d}y,\ z + \mathrm{d}z,\ t + \mathrm{d}t$. Since each particle must be found somewhere in space, the integrated probability density must be unity, i.e., $\int \psi\psi^*\mathrm{d}\tau = 1$, where $\mathrm{d}\tau$ is the volume element and $(\mathrm{d}x \cdot \mathrm{d}y \cdot \mathrm{d}z) \cdot \psi^*$ is the complex conjugate of Ψ. The integrated probability density equal to 1 also denotes the *normalisation condition.* If $\Psi\Psi^*$ denotes probability, it is reasonable to desire that Ψ should be finite, otherwise there could be an infinite probability of finding the particle. It is further specified that it should be single valued, otherwise there may be two or more probabilities of finding a particle at a particular point at the same time, and this would not be meaningful. It is also problematic to understand how a particle can be at infinity so that $\Psi \rightarrow 0$ at infinity. It is further pointed out that Ψ and $d\psi/dx_i$ should be continuous, which is the restriction imposed by the first postulate. It is not at once clear why this should be such, but the experience says that there is a necessary restriction/condition.

5.2 Postulate 2

Every physical observable is represented by a linear operator.

or

To every observable physical property of a system, there corresponds a linear Hermitian operator and the physical properties of the observable may be inferred from the mathematical properties of the corresponding operator.

It is to be noted that the observable means a quantity, which can be measured experimentally, e.g., position, momentum, energy, etc. It is clear from the second postulate that 'the observables are described by operators and the operators of the observables are Hermitian'. The above postulate consists of some new ideas, which can best be clarified by considering some examples in depth. The first thing is to establish the method of construction of quantum mechanical operators.

The quantum mechanical operators corresponding to the pertinent observables are listed in Table 5.1.

TABLE 5.1

Quantum Mechanical Operators Corresponding to the Observables

Observable		Quantum Mechanical Operator
Position	(x)	x
Position	(r)	r
Momentum	(p_x)	$\frac{h}{2\pi i}\cdot\frac{\partial}{\partial x}$ or $-ih\frac{\partial}{\partial x}$
	(p_y)	$\frac{h}{2\pi i}\cdot\frac{\partial}{\partial x}$ or $-ih\frac{\partial}{\partial y}$
	(p_z)	$\frac{h}{2\pi i}\cdot\frac{\partial}{\partial x}$ or $-ih\frac{\partial}{\partial z}$
	(P)	$-i\hbar\left(i\frac{\partial}{\partial x}+j\frac{\partial}{\partial y}+k\frac{\partial}{\partial z}\right)\nabla$
Kinetic energy	(k_x)	$-\frac{h^2}{8\pi^2 m}\nabla^2$ or $-\frac{h^2}{2m}\cdot\frac{\partial^2}{\partial x^2}$
Total kinetic energy	$(k_x+k_y+k_z)$	$-\frac{h^2}{2m}\left(\frac{\partial^2}{\partial x^2}+\frac{\partial^2}{\partial y^2}+\frac{\partial^2}{\partial x^2}\right)$ or $-\frac{h^2}{8\pi^2 m}\nabla^2$
Potential energy	(V)	V
Total energy	(K.E + P.E)	$H=-\frac{h^2}{8\pi^2 m}\nabla^2+V$
Angular momentum	$I_x = yp_z - zp_y$	$L_x=-ih\left(y\frac{\partial}{\partial z}-z\frac{\partial}{\partial y}\right)$
	$I_y = zp_x - xp_z$	$L_y=-ih\left(z\frac{\partial}{\partial x}-x\frac{\partial}{\partial z}\right)$
	$I_z = xp_y - yp_x$	$L_z=-ih\left(x\frac{\partial}{\partial y}-y\frac{\partial}{\partial x}\right)$

5.2.1 Construction of Quantum Mechanical Operator

We have stated in the previous discussion that the second postulate gives an idea regarding the set up or construction of quantum mechanical operator.

Suppose, for example, that the quantum mechanical operator corresponding to total energy is required.

We know that the total energy E = kinetic energy + potential energy,

$$\text{i.e., } E = T + V \tag{5.1}$$

The kinetic energy, $T = \frac{1}{2}mv^2 = \frac{(mv)^2}{2m} = \frac{p^2}{2m}$

$$E = \frac{p^2}{2m} + V$$

where V = potential energy of the particle, which depends only on its position; therefore, V is a function only of co-ordinates; P = momentum; m = mass of the particle.

To construct the total energy operator, V is left unaltered as it is only the function of the co-ordinates, and the momentum terms have to be replaced by momentum operator.

$$\because\ p_x = -ih\frac{\partial}{\partial x}$$

$$\therefore\ p_x^2 = \left(-ih\frac{\partial}{\partial x}\right)^2 = -\,h^2\frac{\partial^2}{\partial x^2}$$

$$= -\left(\frac{h}{2\pi}\right)^2\frac{\partial^2}{\partial x^2} = -\frac{h^2}{4\pi^2}\cdot\frac{\partial^2}{\partial x^2}$$

$$\therefore\ T_x = \frac{p_x^2}{2m} = -\frac{1}{2m}\cdot\frac{h^2}{4\pi^2}\cdot\frac{\partial^2}{\partial x^2}$$

$$= -\frac{h^2}{8\pi^2 m}\cdot\frac{\partial^2}{\partial x^2} \tag{5.3}$$

$$T_y = -\frac{h^2}{8\pi^2 m}\cdot\frac{\partial^2}{\partial y^2} \tag{5.4}$$

$$T_z = -\frac{h^2}{8\pi^2 m}\cdot\frac{\partial^2}{\partial z^2} \tag{5.5}$$

$$\therefore\ T = T_x + T_y + T_z = -\frac{h^2}{8\pi^2 m}\left(\frac{\partial^2}{\partial x^2} + \frac{\partial^2}{\partial y^2} + \frac{\partial^2}{\partial z^2}\right)$$

$$= -\left(\frac{h^2}{8\pi^2 m}\right)\cdot\nabla^2 \tag{5.6}$$

$$\therefore\ E = \left(-\frac{h^2}{8\pi^2 m}\cdot\nabla^2 + V\right)$$

This particular operator for total energy is known as Hamiltonian operator, which is represented by *H*.

∴ From Eq. (5.7), we can write

$$H = \left(-\frac{h^2}{8\pi^2 m} \cdot \nabla^2 + V \right) = \text{Hamiltonian operator}$$

Thus, we could construct a new operator, *H*, with the help of the second postulate.

Next, suppose we want to construct the quantum mechanical operator L_x (angular momentum operator). For this, we shall take the help of equation:

$$l_x = yp_z - zp_y$$

and the values of p_y and p_z, where l_x, l_y, l_z, are angular momentum.

$$L_x = y\left(-ih\frac{\partial}{\partial z}\right) - z\left(-ih\frac{\partial}{\partial y}\right)$$

$$= -ih\left(y\frac{\partial}{\partial z} - z\frac{\partial}{\partial y}\right)$$

where L_x = angular momentum operator. Similarly, we can find L_y and L_z also.

5.3 Postulate 3

Postulate 3 states that the measurement of a physical observable will give a result that is one of the Eigen values of the corresponding operator for that observable.

This statement immediately raises the possibility of quantisation, because only certain values (Eigen values) are possible; however, it is possible that the set of Eigen values for a specific case may be infinite, continuous, or both, that is, any real number could be a possible result of a measurement.

Let us assume that $\hat{A}$ is an operator operating on the function Ψ_i. As a result of this, a set of Eigen function $\{\Psi_i\}$ with Eigen values $\{a_i\}$ can be expressed as

$$\hat{A}\Psi_i = a_i\Psi_i \tag{5.9}$$

where $\hat{A}$ = an operator, Ψ_i=Eigen function, and $\{a_i\}$=a set of Eigen values.

Now, we shall cite another example, i.e., $\Psi = e^{ax}$. On differentiation with respect to *x*, we get

$$\frac{\partial\Psi}{\partial x} = ae^{ax}$$

where $\partial/\partial x$ is an operator operating on the function Ψ and *a* is the Eigen value.

Now let us consider, for example, the operator equation for momentum *p* in one dimension, i.e.,

$$p\psi_\lambda(x) = -ih\frac{\mathrm{d}}{\mathrm{d}x}\psi_\lambda(x)$$

$$= \frac{h}{2\pi i}\frac{\mathrm{d}}{\mathrm{d}x}\psi_\lambda(x) = p_\lambda\psi_\lambda(x) \tag{5.10}$$

or

$$\frac{d\psi_\lambda(x)}{dx} = \frac{2\pi i}{h} p_\lambda \psi_\lambda(x)$$

or

$$\frac{d\psi_\lambda(x)}{\psi_\lambda dx} = \frac{2\pi i p_\lambda}{h} dx$$

Integrating both the sides, we shall get

$$\int \frac{d\psi_\lambda(x)}{\psi_\lambda dx} = \frac{2\pi i p_\lambda}{h} \int dx$$

$$\text{or } \ln \psi_\lambda(x) = \frac{2\pi i p_\lambda}{h} x + \ln A$$

$$\ln \frac{\psi_\lambda(x)}{A} = \frac{2\pi i p_\lambda}{h} x$$

$$\therefore \frac{\psi_\lambda(x)}{A} = e \frac{2\pi i p_\lambda}{h} x$$

$$\therefore \psi_\lambda(x) = Ae \frac{2\pi i p_\lambda}{h} x \tag{5.11}$$

where A = arbitrary constant.

Now all values of Eigen value p_λ are allowed, which keep the functions $\psi_\lambda(x)$ well-behaved, i.e., finite, continuous, and single valued. If we put $2\pi p_\lambda / h = k$

$$\therefore \ p_\lambda = \frac{kh}{2\pi} \tag{5.12}$$

Then the Eigen function of momentum operator can be written as

$$\psi_\lambda(x) = Ae^{ikx} \text{ with Eigen value, } p_\lambda = \frac{kh}{2\pi}$$

5.4 Postulate 4

If a system is in a state denoted by the wave function ϕ, then the average of a sequence of measurements of an observable which is connected with the operator S is

$$\overline{x} = \frac{\int \phi^* S \phi d\tau}{\int \phi^* \phi d\tau} \tag{5.14}$$

In this equation, ϕ^* is the complex conjugate of ϕ and $d\tau$ = the volume element. The wave function ϕ is not essentially an Eigen function of *S*.

If we assume the expected mean of a series of measurements is *x*, the expectation value of × will be

$$\langle x \rangle \text{ or } \bar{x} = \frac{\int \phi^*(x) \times \phi(x) dx}{\int \phi^*(x)\phi(x) dx} = \int \phi^*(x) \cdot x\phi(x) dx \tag{5.15}$$

This is a simple case because the operator $\hat{A}$ is simply *x* itself, and it leads to an interpretation of the wave function as a function, which determines the probability of finding the system at position *x*.

Let the function *f*(*x*) is defined as

$$f(x) = \phi^*(x)\phi(x) \equiv |\phi(x)|^2 \tag{5.16}$$

In terms of *f*(*x*), Eq. (5.15) takes the form

$$x \text{ or } \bar{x} = \frac{\int xf(x) dx}{\int f(x) dx} \tag{5.17}$$

Actually, this is the usual formula for the average value of *x* if the probability of finding the system between *x* and *x* + d*x* is given by

$$\frac{f(x) dx}{\int f(x) dx} \tag{5.18}$$

We shall usually consider the wave functions ϕ such that they are normalised, that is

$$\int \phi^*(x) \times \phi(x) dx = \int f(x) dx = 1 \tag{5.19}$$

Therefore, we shall be able to omit the denominator in Eq. (5.14) through Eq. (5.18).

Now we are going to find the expectation value of momentum for the above system, i.e.,

$$\langle p_x \rangle \text{ or } \bar{p}_x = \frac{\hbar}{2\pi i} \int_{-\infty}^{+\infty} \phi^* \frac{d\phi}{dx} dx \tag{5.20}$$

and the KE

$$\langle T_x \rangle \text{ or } \bar{T}_x = -\frac{\hbar^2}{8\pi^2 m} \int_{-\infty}^{+\infty} \phi^* \frac{d^2\phi}{dx^2} \cdot dx \tag{5.21}$$

Similarly, we can find expectation value of other observables.

5.5 Postulate 5

The time dependence of a wave function is expressed as:

$$i\hbar\frac{\mathrm{d}\phi}{\mathrm{d}t} = H\phi \tag{5.22}$$

where H=Hamiltonian operator for the system.
Postulate 5 is also called the time-dependent postulate.

This equation may be separated as

$$\phi(x,y,z,t) = \phi(x,y,z,t)\cdot F(t) \tag{5.23}$$

$$\text{or } i\hbar\frac{\mathrm{d}}{\mathrm{d}t}(\phi F) = H(x,y,z)\cdot(\phi F)$$

$$\phi i\hbar\frac{\mathrm{d}}{\mathrm{d}t}F = FH\phi \tag{5.24}$$

The LHS follows because the time-independent wave function ϕ does not depend on time, and hence, commutes with $\mathrm{d}/\mathrm{d}t$. The RHS follows because F does not depend on the co-ordinates, and hence, commutes with the time-independent Hamiltonian H.

Dividing Eq. (5.24) by ϕ, we get

$$\frac{i\hbar\dfrac{\mathrm{d}F(t)}{\mathrm{d}t}}{F(t)} = \frac{H(x,y,z)\phi(x,y,z)}{\phi(x,y,z)} \tag{5.25}$$

It is clear that the LHS depends only on t and the RHS depends only on x, y, z (or r in vector notation). The both sides of Eq. (5.25) remain equal, no matter what values of t are used. It can only be true if each side of the equation is separately equal to a constant. Let us call the constant E so that

$$\frac{i\hbar\dfrac{\mathrm{d}F}{\mathrm{d}t}}{F} = E = \frac{H\phi}{\phi} \tag{5.26}$$

$$\text{or } H\phi = E\phi \tag{5.27}$$

which is a time-independent equation.

According to Eq. (5.26), we have

$$\frac{i\hbar\dfrac{\mathrm{d}F}{\mathrm{d}t}}{F} = E$$

$$i\hbar\frac{\mathrm{d}F}{\mathrm{d}t} = EF \tag{5.28}$$

This is a differential equation, which has a simple solution like

$$F(t) = Ce^{-iE\,t/\hbar} \tag{5.29}$$

where C = an arbitrary constant.

For a particular solution of the time-dependent equation, ϕ_λ, which satisfies $H\phi_\lambda = E_\lambda\phi_\lambda$, we shall get a solution of the time-dependent equation in the following form:

$$\Phi_\lambda = \phi_\lambda\ e^{-iEt/\hbar} \tag{5.30}$$

In the above equation, C has been taken unity since this makes ϕ normalised. If ϕ is normalised

$$\int \Phi_\lambda^* \Phi_\lambda\ d\tau = \int \phi_\lambda^* e^{iE\lambda t/\hbar} \phi_\lambda e^{-iE\lambda t/\hbar} d\tau = \int \Phi_\lambda^* \Phi_\lambda\ d\tau = 1 \tag{5.31}$$

Solutions of the form of Eq. (5.31) are stationary as the observable properties of the system represented by Eq. (5.30) do not change with time.

The stationary state is such a state for which the Hamiltonian H remains independent of time.

5.6 Postulate 6

The wave functions must be antisymmetric (symmetric) with respect to the simultaneous interchange of space and spin co-ordinates of fermions (bosons).

or

The wave function ψ must be antisymmetric (symmetric) for the exchange of identical fermions (bosons).

It must be remembered that a fermion is distinguished by half integral spin quantum number and a boson is characterised by integral spin quantum number. Electrons consist of spin quantum number (1/2), and hence, they are fermions.

Let us suppose that in a two-particle system, if we cannot tell particle 1 and 2, then

$$|\psi(r_1 r_2)|^2 = |\psi(r_2 r_1)|^2 \tag{5.32}$$

Therefore, either

$$\psi(r_1 r_2) = \psi(r_2 r_1) \tag{5.33}$$

$$\text{or } \psi(r_1 r_2) = -\psi(r_2 r_1) \tag{5.34}$$

Neither condition is met if we use the equation.

$$\psi\ (r_1 r_2 \ldots\ldots\ldots\ldots\ldots\ldots r_n, t) = \Psi_1(r_1, t)\Psi_2(r_2, t) ------ \Psi_N(r_N, t) \tag{5.35}$$

Instead of this, we must take $\psi(r_1 r_2) = \frac{1}{\sqrt{2}}[\Psi_1(r_1) \cdot \Psi_2(r_2) + \Psi_2(r_1)\Psi_1(r_2)]$ (5.36)

$$\psi(r_1 r_2) = \frac{1}{\sqrt{2}}[\Psi_1(r_1)\Psi_2(r_2) - \Psi_2(r_1)\Psi_1(r_2)] \tag{5.37}$$

for Eqs. (5.33) and (5.34), respectively. The factor $1/\sqrt{2}$ is the normalisation constant. The + (–) case is known as symmetric (antisymmetric).

In the antisymmetric case, $\psi \rightarrow 0$ when $\Psi_1 = \Psi_2$. It means the particles are in the same state. It also gives a clue to understand that if the particles are characterised by antisymmetric wave function, there will be zero probability that they will coincide, which clearly indicates that they tend to keep away from each other. This does not happen in the case of symmetric situation.

The rule that no two particles can have the same wave function, i.e., they cannot occupy the same quantum state, is called Pauli's exclusion principle.

Pauli suggested that in many atoms, the electrons stack up in higher and higher energy levels rather than collapsing together into the ground state. A study of nuclear structure reveals that the protons and neutrons obey Pauli's principle. It can be shown that this important exclusive characteristic is related to the existence of intrinsic spin for the particles involved, and for such particles, it is essential to use antisymmetric wave functions.

BIBLIOGRAPHY

Levine, I.N. 2000. *Quantum Chemistry*. Singapore: Pearson Education.
Lowe, J.P. and K.A. Peterson. 2006. *Quantum Chemistry*, 3rd ed. London: Academic Press.
McQuarrie, D.A. 2007. *Quantum Chemistry*. New Delhi: Viva Books Pvt Ltd.
Mortimer, R.G. 2008. *Physical Chemistry*. San Diego, CA: Academic Press.

Solved Problems

Problem 1. Construct the operator for the *z*-component of the angular momentum of one particle.

Solution: The *z*-component of angular momentum is $l_z = xp_y - yp_x$
The operator for this component is found as follows:
We know that

$$l_z = xp_y - yp_x$$

Replacing the values of p_y and p_z, we get

$$L_z = x\frac{h}{i}\frac{\partial}{\partial y} - y\frac{h}{i}\frac{\partial}{\partial x}$$

$$= \frac{h}{i}\left(x\frac{\partial}{\partial y} - y\frac{\partial}{\partial x}\right) \qquad \text{Answer.}$$

Problem 2. Find the operator equal to the commutator $[x, p_x]$.

Solution: Let us operate on an arbitrary differentiable function $\Psi(x)$

$$[x, p_x]\psi = \frac{h}{i}\left(x\frac{\partial\psi}{\partial y} - \frac{\partial(x\psi)}{\partial x}\right)$$

$$= \frac{h}{i}\left(x\frac{\partial\psi}{\partial y} - x\frac{\partial\psi}{\partial x} - \psi \times 1\right)$$

$$= \frac{\hbar}{i}(-\psi) = -\frac{\hbar}{i\psi}$$

∴ The operator equation is

$$[x, p_x] = -\frac{\hbar}{i} = i\hbar \qquad \text{Answer.}$$

Problem 3. Show that the formula in $\bar{A} = \dfrac{\int \Psi^* \hat{A}\psi \mathrm{d}q}{\int \Psi^* \psi \mathrm{d}q}$

for the expectation value is unchanged if ψ is replaced by $C\psi$ where C is a constant.

Solution: We know that

$$\bar{A} = \frac{\int \Psi^* \hat{A}\psi \mathrm{d}q}{\int \Psi^* \psi \mathrm{d}q}$$

$$= \frac{\int C^* \Psi^* \hat{A} C\psi \mathrm{d}q}{\int C^* \Psi^* C\psi \mathrm{d}q}$$

$$= \frac{C^* C \int \Psi^* \hat{A}\psi \mathrm{d}q}{C^* C \int \Psi^* \psi \mathrm{d}q}$$

$$\bar{A} = \frac{\int \Psi^* \hat{A}\psi dq}{\int \Psi^* \psi dq}$$

Problem 4. Find the expectation value for the position of a particle in one-dimensional box of length a for $n = 1$.

Solution: The expectation value is given by

$$\langle x \rangle = \frac{2}{a}\int_0^a \left\{\sin\left(\frac{\pi x}{a}\right)\right\} x \sin\left(\frac{\pi x}{a}\right) \mathrm{d}x$$

$$= \frac{2}{a}\int_0^a \left\{\sin\left(\frac{\pi x}{a}\right)\right\}^2 \mathrm{d}x$$

This is a standard integral which can be expressed as

$$= \frac{2}{a}\int_0^a x\{\sin bx\}^2 \mathrm{d}x, \text{ where, } b = \frac{\pi}{a}$$

The value of the standard integral is given by

$$\int_0^a x\sin^2 bx\mathrm{d}x = \frac{x^2}{4} - \frac{\cos^2 bx}{8b^2} - \frac{x\sin 2bx}{4b}$$

Substituting the value of the integral, we get

$$x = \frac{2}{a}\left[\frac{x^2}{4} - \frac{\cos\frac{2\pi x}{a}}{8\left(\frac{\pi}{a}\right)^2} - \frac{x\sin\left(\frac{2\pi x}{a}\right)}{4\left(\frac{\pi}{a}\right)}\right]_0^a$$

$$= \frac{2}{a}\left[\frac{a^2}{4} - \frac{a^2}{8\pi^2} - 0 + \frac{a^2}{8\pi^2}\right]$$

$$= \frac{2}{a}\left[\frac{a^2}{4} - 0\right]$$

$$= \frac{2}{a}\cdot\frac{a^2}{4} = \frac{a}{2}$$

$$\therefore \langle x \rangle = \frac{a}{2}$$ Answer.

This means that the average position is at the middle point of the box length.

Problem 5. For a particle in one-dimensional box in the stationary state with $n = 2$, find the probability that the particle will be found in each of the regions making up thirds of the box.

Solution: For $0 < x < \frac{a}{3}$

$$\text{Probability} = \int_0^{a/3} \Psi^2(x)\text{d}x = \frac{2}{a}\int_0^{a/3} \sin^2\left(\frac{2\pi x}{a}\right)\text{d}x$$

$$= \frac{2}{a}\frac{a}{2\pi}\int_0^{2\pi/3} \sin^2 y\, dy$$

$$= \frac{1}{\pi}\left[\frac{y}{2} - \frac{1}{4}\sin 2y\right]_0^{2\pi/3}$$

$$= \frac{1}{\pi}\left[\frac{\pi}{3} - \frac{1}{4}\sin\frac{4\pi}{3}\right]$$

$$= 0.40225$$

Questions on Concepts

1. Show that if $\Psi(x,t) = \Psi(x)e^{-iEt/h}$ and if $\Psi(x,t)$ are normalised, then $\Psi(x)$ is also normalised.
2. Show that the momentum operator $\left|\frac{\hbar}{i}\frac{\partial}{\partial x}\right|$ gives the correct sign for p_x for a travelling wave given by

$$= e^{ikx}e^{-iE\,t/h}$$

3. For a particle in a hard box of length a, find the expectation value of the quantity p_x^4 for $n = 1$ state.

4. a. What do you mean by a dynamical variable?
   b. Define observable.
   c. Discuss the first postulate of quantum mechanics.
5. How will you show that every physical observable is represented by a linear operator?
6. a. What do you mean by Eigen value? Explain it by giving an example.
   b. State the third postulate of quantum mechanics and show that you have understood this postulate.
7. State and explain the fourth postulate of quantum mechanics.
8. With the help of the fifth postulate of quantum mechanics, prove that

$$F(t) = Ce^{-iEt/\hbar}$$

9. Construct the operator for the x component of the angular momentum of one particle.
10. Show that $[x, p_x] = i\hbar$
11. Show that the expectation value for the position of a particle in one-dimensional box of length a is zero for $n = 1$ state.
12. For a particle in ID box in the stationary state with $n = 2$, find the probability that the particle will be found in each of the regions making up 1/2 of the box.

6

The Schrödinger Equation

The propagation of a periodic disturbance carrying energy is called the wave motion. At any point along the path of the wave motion, a periodic displacement or vibration about a mean position occurs. This may take the form of a displacement of air molecules (e.g., sound waves in air), of water molecules (wave on water), or of elements of a string or wire. The locus of these displacements at any instant is known as *wave*. The wave motion moves forward a distance equal to its wavelength in the time taken for the displacement at any point to undergo a complete cycle about its mean position. Waves in which the displacement or vibration takes place in the direction of propagation of the waves are called *longitudinal waves*, e.g., sound waves. However, the waves in which the vibration or displacement occurs in a plane at right angles to the direction of propagation of the waves are called *transverse waves*. We are discussing the waves and wave motion, and therefore, we want to establish a relationship or an equation of wave motion. In order to do so, it will be convenient to consider the simplest type of wave motion, namely, the vibration of string, and then we shall derive an equation of wave motion.

6.1 Equation of Wave Motion

In deriving the equation for wave motion (e.g., vibration of string), let w = the amplitude of vibration at any point, the co-ordinate of which is x at a time t.

We can write

$$\frac{\partial w}{\partial t} = \frac{\partial w}{\partial x} \cdot \frac{\partial x}{\partial t} \tag{6.1}$$

$$\text{or } \frac{\partial}{\partial t} = \frac{\partial}{\partial x} \cdot \frac{\partial x}{\partial t}$$

Differentiating Eq. (6.1) with respect to t, we shall get

$$\frac{\partial^2 w}{\partial t^2} = \left[\frac{\partial w}{\partial x} \cdot \frac{\partial^2 x}{\partial t^2} + \frac{\partial}{\partial t}\left(\frac{\partial w}{\partial x} \cdot \frac{\partial x}{\partial t}\right)\right]$$

$$= \left[\frac{\partial w}{\partial x} \cdot \frac{\partial^2 x}{\partial t^2} + \left(\frac{\partial}{\partial t} \cdot \frac{\partial x}{\partial t}\right)\left(\frac{\partial w}{\partial x} \cdot \frac{\partial x}{\partial t}\right)\right]$$

$$= \left[\frac{\partial w}{\partial x} \cdot \frac{\partial^2 x}{\partial t^2} + \frac{\partial^2 w}{\partial x^2}\left(\frac{\partial x}{\partial t}\right)^2\right] \tag{6.2}$$

But $\dfrac{\partial x}{\partial t}$ = velocity u of propagation of wave = constant

$$\therefore \frac{\partial^2 x}{\partial t^2} = 0$$

Putting $\frac{\partial^2 x}{\partial t^2} = 0$ in Eq. (6.2), we get

$$\text{or } \frac{\partial^2 w}{\partial t^2} = \frac{\partial^2 w}{\partial x^2} \cdot u^2 \tag{6.3}$$

This is the equation of *wave motion in one dimension.*

Equation (6.3) is a differential equation, and it may be solved by the method of separating the variables, provided u remains constant. Thus, w may be written as

$$w = f(x) \cdot g(t) \tag{6.4}$$

where $f(x)$ = a function of x only
$g(t)$ = a function of time t only

For the motion of standing waves, which occur in a stretched string, it is known to us that $g(t)$ may be expressed as

$$g(t) = A \sin 2\pi vt$$

where υ = frequency of vibration
A = constant, which is maximum amplitude

Therefore, Eq. (6.4) may be expressed as

$$w = f(x) A \sin 2\pi vt \tag{6.5}$$

Differentiating this equation with respect to t, we get

$$\frac{\partial w}{\partial t} = f(x) 2\pi v \; A \cos 2\pi vt$$

Further differentiation gives

$$\begin{aligned} \frac{\partial^2 w}{\partial t^2} &= f(x)(2\pi v)(2\pi v)(-) \; A \cos 2\pi vt \\ &= -4\pi^2 v^2 f(x) A \sin 2\pi vt \\ &= -4\pi^2 v^2 f(x) g(t) \end{aligned} \tag{6.6}$$

From Eq. (6.4), we can write

$$\frac{\partial^2 w}{\partial x^2} = \frac{\partial^2 w}{\partial x^2} \cdot g(t) \tag{6.7}$$

But $\frac{\partial^2 w}{\partial x^2} = \frac{1}{u^2} \frac{\partial^2 w}{\partial t^2}$

$$\therefore \quad \frac{\partial^2 f(x)}{\partial x^2} \cdot g(t) = -\frac{1}{u^2} 4\pi^2 v^2 f(x) g(t)$$

$$\therefore \quad \frac{\partial^2 f(x)}{\partial x^2} = -\frac{1}{u^2} 4\pi^2 v^2 f(x)$$

But $u = \lambda \upsilon$, so we can write

$$\frac{\partial^2 f(x)}{\partial x^2} = -\frac{4\pi^2 v^2 f(x)}{\lambda^2 v^2}$$

$$\text{or } \frac{\partial^2 f(x)}{\partial x^2} = -\frac{4\pi^2}{\lambda^2} \cdot f(x) \tag{6.8}$$

This is the equation for wave motion in one dimension, which may be extended to three dimensions. If $f(x)$ is replaced by $\Psi(x, y, z)$, which is the amplitude function for three dimensions, then Eq. (6.8) takes the form

$$\frac{\partial^2 \Psi}{\partial x^2} + \frac{\partial^2 \Psi}{\partial y^2} + \frac{\partial^2 \Psi}{\partial z^2} = -\frac{4\pi^2}{\lambda^2} \Psi$$

$$\text{or } \nabla^2 \Psi = -\frac{4\pi^2}{\lambda^2} \Psi \tag{6.9}$$

where $\Psi = \Psi(x, y, z)$

$$\nabla^2 = \frac{\partial^2}{\partial x^2} + \frac{\partial^2}{\partial y^2} + \frac{\partial^2}{\partial z^2}$$

∇ is the Laplacian operator.

6.1.1 Time-Independent Schrödinger Equation

We have derived that

$$\nabla^2 \Psi = -\frac{4\pi^2}{\lambda^2} \Psi$$

Schrödinger stated that this equation may be applied to all particles, including electrons, atoms, and photons.

Applying the de Broglie equation, i.e., $\lambda = h/p$

where λ = wavelength
h = Planck's constant
p = momentum of the particle

∴ Equation (6.9) takes the form

$$\nabla^2 \Psi = -4\pi^2 \Psi . \frac{1}{\lambda^2}$$

$$\text{or } \nabla^2 \Psi = -\left(4\pi^2 \Psi\right) \cdot \left(\frac{p}{h}\right)^2$$

$$\text{or } \nabla^2 \Psi = -\frac{4\pi^2 p^2}{h^2} \Psi$$

$$\text{or } \nabla^2 \Psi + \frac{4\pi^2 p^2}{h^2} \Psi = 0 \tag{6.10}$$

We know that the kinetic energy of the particle will be

$$T = \frac{1}{2}mv^2$$

where m = mass of the particle
v = velocity of the particle

$$v^2 = \frac{2T}{m} \tag{6.11}$$

Further, $T = (E - V)$
where E = total energy of the particle
V = potential energy of the particle

$$\therefore v^2 = \frac{2(E - V)}{m}$$

Multiplying both sides by m^2, we get

$$m^2v^2 = \frac{2m^2(E - V)}{m} = 2m(E - V)$$

$$\therefore p^2 = 2m(E - V)$$

Substituting the value of p^2 in Eq. (6.10), we get

$$\nabla^2\Psi + \frac{4\pi^2}{h^2} \times 2m(E - V)\Psi = 0$$

$$\text{or } \nabla^2\Psi + \frac{8\pi^2 m}{h^2}(E - V)\Psi = 0 \tag{6.12}$$

This equation is the *time-independent Schrödinger equation in three dimensions.* It is to be noted that the Schrödinger wave equation is an equation, which describes the properties of particles in terms of wave motion.

Equation (6.12) may also be expressed as follows:

$$\nabla^2\Psi = -\frac{8\pi^2 m}{h^2}(E - V)\Psi$$

$$\text{or } -\frac{h^2}{8\pi^2 m} \cdot \nabla^2\Psi = (E - V)\Psi = E\Psi - V\Psi$$

$$\text{or } -\frac{h^2}{8\pi^2 m} \cdot \nabla^2\Psi + V\Psi = E\Psi$$

$$\text{or } \left(-\frac{h^2}{8\pi^2 m} \cdot \nabla^2 + V\right)\Psi = E\Psi$$

$$\text{or } H\Psi = E\Psi \tag{6.13}$$

where $\left(-\frac{h^2}{8\pi^2 m}.\nabla^2 + V\right) = H$ = Hamiltonian operator

Ψ and E are called the *Eigen function* and *Eigen value*, respectively, and $\nabla^2 = \frac{\partial^2}{\partial x^2} + \frac{\partial^2}{\partial y^2} + \frac{\partial^2}{\partial z^2}$.
Thus, we see that the Schrödinger wave equation may also be represented by Eq. (6.13). It is clear that the solution of Eq. (6.13) gives the value of E for the system and Ψ as Eigen state.

6.1.2 Time-Dependent Schrödinger Equation

The time-dependent Schrodinger equation is taken as one of the postulates, i.e., fundamental assumption of quantum mechanics, and it is often written in the abbreviated form as

$$H\Psi = -\frac{\hbar}{i}\frac{\partial \Psi}{\partial t} \tag{6.14}$$

where H is the Hamiltonian operator, which is the same as used in the time-independent Schrödinger equation and Ψ is the time-dependent wave function.

Here, we shall use a capital psi (Ψ) a time-dependent wave function and a lower case psi(ψ) for a co-ordinate wave function or a co-ordinate factor.

The time-dependent Schrödinger equation can be found out from Eq. (6.14) by separation of variables. For motion in the x-direction, we assume that

$$\Psi(x,t) = \Psi(x)\phi(t) \tag{6.15}$$

For simplicity sake, we can write Eq. (6.15) as

$$\Psi = \Psi\phi \tag{6.16}$$

Substituting Eq. (6.16) into Eq. (6.14), we get

$$H\Psi\phi = -\frac{\hbar}{i}\frac{\partial(\Psi\phi)}{\partial t} \tag{6.17}$$

Since H operates only on ϕ and because $\partial/\partial t$ operates only on ϕ, we may express Eq. (6.17) as

$$\phi H\Psi = \Psi\left(-\frac{\hbar}{i}\frac{\partial \phi}{\partial t}\right) \tag{6.18}$$

Dividing both sides of this equation by $\Psi\phi$, we get

$$\frac{\phi H\Psi}{\Psi\phi} = -\frac{\Psi\left(\frac{\hbar}{i}\frac{\partial \phi}{\partial t}\right)}{\Psi\phi}$$

$$\text{or } \frac{H\Psi}{\Psi} = -\frac{1}{\phi}\frac{\hbar}{i}\frac{\partial \phi}{\partial t} \tag{6.19}$$

The variables x and t are now separated, as x and t are independent variables; each can be fixed, while the other varies. It means each side of the equation must be a constant, which may be represented by E.

$$\therefore \frac{1}{\Psi}H\Psi = E \tag{6.20}$$

$$\text{or } H\Psi = E\Psi \tag{6.21}$$

$$-\frac{1}{\phi}\frac{\hbar}{i}\frac{\partial\phi}{\partial t} = E$$

$$\text{or } -\frac{\hbar}{i}\frac{\partial\phi}{\partial t} = E\phi \tag{6.22}$$

It should be kept in mind that Eq. (6.22) is an ordinary differential equation, and hence, partial differentiation notation is dropped and we shall now write d in place of ∂.

Equation (6.22) can be rearranged as

$$\frac{\mathrm{d}\phi}{\phi} = -\frac{iE}{\hbar}\mathrm{d}t$$

Integrating both the sides, we obtain

$$\int\frac{\mathrm{d}\phi}{\phi} = -\frac{iE}{\hbar}\int\mathrm{d}t$$

$$\text{or } \ln\phi = -\frac{iEt}{\hbar} + \ln A$$

$$\text{or } \ln\phi - \ln A = -\frac{iEt}{\hbar}$$

$$\text{or } \ln\frac{\phi}{A} = -\frac{iEt}{\hbar}$$

$$\therefore \quad \frac{\phi}{A} = e^{-iEt/\hbar}$$

$$\phi = Ae^{-iEt/\hbar} \tag{6.23}$$

where A = constant. We may take the value of $A = 1$, and the complete wave function may be expressed as

$$\Psi(x,t) = \Psi(x)e^{-iEt/\hbar} \tag{6.24}$$

This is the time-dependent Schrödinger equation in one dimension. It may be written in three dimensions as

$$\Psi(x,y,z,t) = \Psi(x,y,z)e^{-iEt/\hbar} \tag{6.25}$$

This equation may also be written in *sine* and *cosine* form as $e^{-iEt/\hbar}$ is of the form $e^{-i\theta}$.

6.1.3 Interpretation of Wave Function, Ψ

The wave function Ψ has been found to be a complex number, which may be represented as

$$\Psi = x + iy$$

where x and y are real, and i is an imaginary quantity. The complex conjugate of Ψ may be written as Ψ^* and represented as follows:

$$\Psi^* = x - iy$$

When Ψ and Ψ^* are multiplied, they give the following equation:

$$\Psi\Psi^* = (x+iy)(x-iy) = x^2 + y^2 \approx \Psi^2$$

It means the product of Ψ and Ψ^* is real and non-negative because x and y are real, and hence, x^2+y^2 will be also real.

In general, $|\Psi|^2$ indicates the probability density. It means $\Psi\Psi^*$ also indicates the probability density, whereas $\Psi^2 d\tau$ or $\Psi\Psi^* d\tau$ denotes the probability of finding the particle in the volume element $d\tau$, which is the product of dx, dy, and dz, i.e., $d\tau = dx \cdot dy \cdot dz$. From the above facts, it is clear that Ψ and Ψ^* alone have no physical meaning, but Ψ^2 or $\Psi\Psi^*$ has.

The equation $H\Psi = E\Psi$ has a number of mathematical solutions; however, the only acceptable one corresponding to various values of E are those that satisfy the following characteristics of Ψ.

1. Ψ must be single valued, and this is required because we want $|\Psi|^2$ to yield an unambiguous probability for finding the particle (e.g., electron) in a certain region.
2. Ψ must be piecewise continuous. In order for $H\Psi$ to be defined everywhere, it is essential that $d^2\Psi/dx^2$ be defined everywhere. For this, $d\Psi/dx$ should be piecewise continuous and also Ψ itself should be continuous.
3. Ψ must be quadratically integrable, i.e., $\int ---- \int |\Psi|^2 d\tau$ over all space must be equal to a finite number of particles. In other words, one may say that $\Psi \to 0$ at the boundaries of the system.

   It should be noted that wave functions satisfying the above conditions are called *well-behaved functions.*
4. The gradient of wave function Ψ must be continuous. It may be mathematically expressed as

$$\text{grad } \Psi(x) \text{ or } \nabla\Psi(x) = -\frac{2m}{\hbar^2}\int [E - V(x)\Psi(x)dx] \tag{6.26}$$

   This has higher degree of continuity than $\Psi(x)$ unless $V(x)$ be suddenly infinite.

6.1.4 Acceptable Wave Function

For functions, which are single-valued, nowhere infinite, continuous, and having piecewise continuous, the first derivative will be called acceptable wave functions. The meaning of the above-described terms is explained in Figure 6.1.

a. Ψ is triple valued at $x = 0$.
b. Ψ is discontinuous at $x = 0$.
c. Ψ approaches infinity.
d. Ψ is continuous and consists of a *cusp* at $x = 0$.

Therefore, the first derivative of Ψ is not continuous at $x = 0$ and is only piecewise continuous.

Other most pertinent general restriction we impose on Ψ is that it should be a normalised function, i.e.,

$$\int_{\text{allspace}} \Psi\Psi^* d\tau = 1$$

This means that $\left|\int_{\text{allspace}} \Psi\Psi^* d\tau = 1\right|$ should not be equal to zero or infinity. When this condition is satisfied, it is said to be *square integrable.*

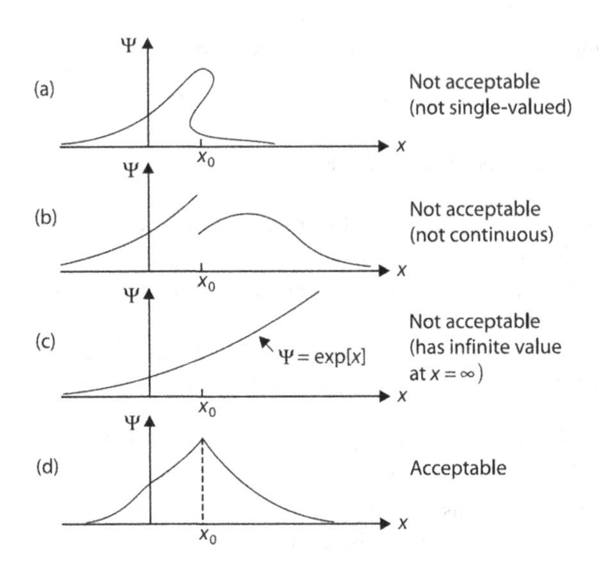

FIGURE 6.1 Acceptable wave functions.

6.2 Normalisation

If any wave function satisfies the Schrödinger equation multiplied by an arbitrary constant, then it will still satisfy the Schrödinger equation and will yield the same value for any expectation value.

If we choose a value of a constant, multiplying a wave function such that

$$\int \Psi\Psi^* d\tau = 1$$

Then, this will be the definition of normalisation, and the wave function Ψ is then said to be normalised.

Generally, Ψ is not normalised. When Ψ is multiplied by a constant A, it gives new function $A\Psi$. Then, the new function $A\Psi$ will also be a solution of the wave equation. The value of A should be selected such that the new function should be normalised. For this condition, the new wave function has to fulfil the following requirement, i.e.,

$$\int \left(A\Psi^*\right)(A\Psi)d\tau = 1$$

$$\text{or} \int A^2\Psi^*\Psi d\tau = 1$$

$$\text{or } A^2 \int \Psi^*\Psi d\tau = 1$$

$$\therefore A^2 = \frac{1}{\int \Psi^*\Psi d\tau}$$

$$A = \left(\frac{1}{\int \Psi^*\Psi d\tau} \right)^{1/2} \tag{6.27}$$

where A is called the *normalising constant.* Multiplying the normalisation constant A with the wave function Ψ, a new wave function $A\Psi$ be obtained, which is also a solution of the wave equation.

6.3 Orthogonality

It is a mathematical characteristic of wave functions that if any two of them are exact solution of the wave equation, then they are said to be *orthogonal.* It means they are independent of one another, and the integral of their product over whole space is equal to zero. It is expressed, in general, as

$$\int_0^a \Psi_n \Psi_m \mathrm{d}\tau = 0 \quad (n \neq m)$$

For specific case in one dimension, we may express it as

$$\int_0^a \Psi_n \Psi_m \mathrm{d}x = 0 \quad (n \neq m)$$

For any value of n and m, we can write

$$\int_0^a \Psi_n \Psi_m \mathrm{d}x = \int_0^a \left[\left(\frac{2}{a}\right)^{1/2} \sin\frac{n\pi}{a}\cdot x\right]\left[\left(\frac{2}{a}\right)^{1/2} \sin\frac{m\pi}{a}\cdot x\right]\mathrm{d}x$$

$$\text{where } \Psi_n = \left(\frac{2}{\mathrm{a}}\right)^{1/2} \sin\frac{n\pi}{a}\cdot x \text{ and } \Psi_m = \left(\frac{2}{a}\right)^{1/2} \sin\frac{m\pi}{a}\cdot x$$

$$= \frac{2}{a}\int_0^a \left(\sin\frac{n\pi}{a}\cdot x\right)\left(\sin\frac{m\pi}{a}\cdot x\right)\mathrm{d}x$$

$\therefore 2\sin\alpha\sin\beta = \cos(\alpha-\beta) - \cos(\alpha+\beta)$; therefore, the integral may now be written as

$$\int_0^a \Psi_n \Psi_m \mathrm{d}x$$

$$\int_0^a \Psi_n \Psi_m \mathrm{d}x = \frac{2}{\mathrm{a}}\int_0^a \frac{1}{2}\left\{\cos[(n-m)\frac{\pi}{a}\cdot x - \cos\left[(n+m)\frac{\pi}{a}\cdot x\right]\right\}\mathrm{d}x$$

$$= \frac{1}{a}\int_0^a \left\{\frac{a}{(n-m)\pi}\cdot\sin\left[(n-m)\frac{\pi}{a}\cdot x\right] - \frac{a}{(n+m)\pi}\cdot\sin\left[(n+m)\frac{\pi}{a}\cdot x\right]\right\}_0^a$$

$$= \frac{1}{a}\left\{\frac{a}{(n-m)\pi}\cdot\sin[(n-m)\pi] - \frac{a}{(n+m)\pi}\cdot\sin[(n+m)\pi]\right\}$$

Since $(n-m)$ and $(n+m)$ must be integers, the *sines* of the angles are all zero, and hence, the value of integral will be zero. Thus,

$$\int_0^a \Psi_n \Psi_m \mathrm{d}x = 0 \text{ and, hence, for general case also } \int \Psi_n \Psi_m \mathrm{d}\tau = 0 \quad (n \neq m)$$

Proof: Let Ψ_m and Ψ_n be the two different solutions of wave equation. Ψ_m and Ψ_n correspond to the energy E_1 and E_2, respectively. Thus, we can write

$$\nabla^2\Psi_m + \frac{2m}{\hbar^2}(E_1 - V)\Psi_m = 0 \tag{6.28}$$

$$\text{and } \nabla^2\Psi_n + \frac{2m}{\hbar^2}(E_2 - V)\Psi_n = 0 \tag{6.29}$$

If Ψ_m and Ψ_n are the complex conjugates of Ψ_m and Ψ_n, respectively, then Eq. (6.28) may take the form

$$\nabla^2\Psi_m^* + \frac{2m}{\hbar^2}(E_1 - V)\Psi_m^* = 0 \tag{6.30}$$

Multiplying Eq. (6.29) by Ψ_n^* and Eq. (6.30) by Ψ_n, we get

$$\Psi_m^*\nabla^2\Psi_n + \frac{2m}{\hbar^2}(E_2 - V)\Psi_m^*\Psi_n = 0 \tag{6.31}$$

$$\text{and } \Psi_n\nabla^2\Psi_m^* \frac{2m}{\hbar^2}(E_1 - V)\Psi_m^*\Psi_n = 0 \tag{6.32}$$

Subtracting Eq. (6.32) from Eq. (6.31), we obtain

$$\Psi_m^*\nabla^2\Psi_n - \Psi_n\nabla^2\Psi_n^* + \frac{2m}{\hbar^2}(E_2 - E_1)\Psi_m^*\Psi_n = 0 \tag{6.33}$$

Integrating Eq. (6.33) over the space variables, we get

$$\iiint\left(\Psi_m^*\nabla^2\Psi_n - \Psi_n\nabla^2\Psi_m^*\right)d\sigma + \frac{2m}{\hbar^2}\left[(E_2 - V - E_1 - V)\Psi_m^*\Psi_n\right]d\sigma = 0$$

It is now possible to change the first integral to a surface integral at infinity by applying Green's theorem.

That is, $\iint\left(\Psi_m^*\nabla^2\Psi_n - \Psi_n\nabla^2\Psi_m^*\right)d\sigma + \frac{2m}{\hbar^2}(E_2 - E_1)\iiint \Psi_m^*\Psi_n\, d\sigma = 0$

As Ψ_m and Ψ_n become zero at infinity, the first term becomes zero and finally we get

$$\iiint \Psi_m^*\Psi_n d\sigma = 0 \quad \text{if } E_1 \neq E_2,$$

Hence, the two wave functions are called *orthogonal*.

6.3.1 Orthonormality

Wave functions which are both normalised and orthogonal are called *orthonormal*, and this phenomenon is called *orthonormality*. Normalisation and orthogonality may be combined to give the orthonormal set of wave functions.

Mathematically, it is expressed as

$$\int \Psi_i\Psi_j d\tau = 1 \quad \text{if } i = j$$

$$\int \Psi_i\Psi_j d\tau = 0 \quad \text{if } i \neq j$$

It should he kept in mind that only Eigen functions, which are exact solutions to the wave equation, constitute the orthogonal pairs.

6.3.2 Eigen Function and Eigen Value

Eigen function is an allowed wave function, enabling a meaningful solution to be obtained from the wave equation. For each Eigen function, there is a fixed energy value for the system.

The word 'Eigen value' is a partial translation of the German word 'Eigenwert', and a full translation is 'characteristic value'. The time-independent Schrödinger equation, $H\Psi = E\Psi$, belongs to a class of equations called the *Eigen value equation.* An Eigen value equation contains on one side an operator operating on a function, and on the other side, a constant, which is called the Eigen value multiplying the same function, which is known as *Eigen function.*

If A = a mathematical operator operating on a function

$f_n = a$ function, then

$$\hat{A}f_n = a_n f_n \tag{6.34}$$

where f_n = Eigen function
a_n = Eigen value

The time-independent Schrödinger equation is the Eigen value equation for the Hamiltonian operator, i.e., $H\Psi = E\Psi$. Here H is an operator operating on the Eigen function Ψ to yield an Eigen value E. The Eigen value of the Hamiltonian operator E is the value of energy of the system and is known as *energy Eigen value.*

To understand better, let us take an example. Suppose $\Psi = e^{ax}$.

Differentiating this equation with respect to x, we get

$$\frac{d\Psi}{dx} = \frac{d}{dx}e^{ax}$$

$$\text{or } \frac{d}{dx}(\Psi) = e^{ax} = a\Psi$$

$\therefore a$ = Eigen value, and the Eigen function here is d/dx.

6.3.3 Degeneracy

We have already stated that for every Eigen function Ψ, there must be a corresponding Eigen value E. It may also happen that for two different Ψ's, there may be same value of energy. These two states (of different Ψ values) are said to be degenerate states, and such phenomenon is called *degeneracy.* It will be more explicit while dealing with the particle in a box problem.

6.4 Transformation of the Laplacian into Spherical Polar Co-Ordinates

The Laplacian

$$\nabla^2 = \frac{\partial^2}{\partial x^2} + \frac{\partial^2}{\partial y^2} + \frac{\partial^2}{\partial z^2}$$

when transformed to spherical polar co-ordinates becomes

$$\nabla^2 = \frac{1}{r^2}\frac{\partial}{\partial r}r^2 \cdot \frac{\partial}{\partial r} + \frac{1}{r^2 \sin\theta}\frac{\partial}{\partial\theta}\sin\theta\frac{\partial}{\partial\theta} + \frac{1}{r^2\sin^2\theta}\cdot\frac{\partial^2}{\partial\phi^2}$$

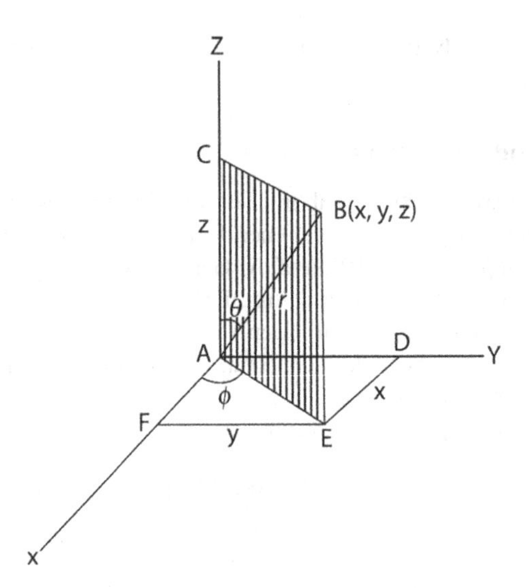

FIGURE 6.2 Cartesian co-ordinates and spherical polar co-ordinates.

$$\text{or } \nabla^2 = \frac{\partial^2}{\partial r^2} + \frac{2}{r}\frac{\partial}{\partial r} + \frac{1}{r^2}\frac{\partial^2}{\partial \theta^2} + \frac{\cos\theta}{r^2 \sin^2\theta}\frac{\partial}{\partial \theta} + \frac{1}{r^2 \sin^2\theta}\frac{\partial^2}{\partial \phi^2}$$

The transformation of Cartesian co-ordinates to spherical polar co-ordinates r, θ, and ϕ may be easily understood from Figure 6.2.

From the figure, it is clear that r = the distance of the point B from the origin at A = radius vector, θ = angle between the radius vector r, and ϕ = angle between the positive direction of the x-axis and the projection of the radius vector on the XY plane. The parameters r, θ, and ϕ locate the position of a point.

Let $B(x, y, z)$ is a point on the radius vector and E be the reflection of B in the XY plane.

Let us draw perpendicular ED on AY, EF on AX, and BC on AZ, then

$$\frac{BC}{AB} = \sin\theta \quad \therefore BC = AB\sin\theta = r\sin\theta$$

$$\frac{AC}{AB} = \cos\theta \quad \therefore \frac{z}{r} = \cos\theta \quad \therefore z = r\cos\theta$$

From the figure, it is obvious that $BC = AE = r\sin\theta$.

$$\therefore \frac{AF}{AE} = \frac{x}{r\sin\theta} = \cos\phi \quad \therefore x = r\sin\theta\cos\phi$$

$$\text{and} \therefore \frac{EF}{AE} = \frac{AD}{AE} = \frac{y}{r\sin\theta} = \sin\phi \quad \therefore y = r\sin\theta\sin\phi$$

Thus, the equations of transformation are

$$x = r\sin\theta\cos\phi$$

$$y = r\sin\theta\sin\phi$$

$$z = r\cos\phi$$

We shall perform this transformation in two steps; the first is given by

$$x = \rho\cos\phi \quad [\text{where } \rho = r\sin\theta]$$

$$y = \rho\sin\phi$$

We have

$$\frac{\partial\Psi}{\partial\rho} = \frac{\partial x}{\partial\rho}\cdot\frac{\partial\Psi}{\partial x} + \frac{\partial y}{\partial\rho}\cdot\frac{\partial\Psi}{\partial y} \tag{6.35}$$

$$\frac{\partial\Psi}{\partial\phi} = \frac{\partial x}{\partial\phi}\cdot\frac{\partial\Psi}{\partial x} + \frac{\partial y}{\partial\phi}\cdot\frac{\partial\Psi}{\partial y} \tag{6.36}$$

$$\text{or } \frac{\partial\Psi}{\partial\rho} = \cos\phi\frac{\partial\Psi}{\partial x} + \sin\phi\frac{\partial\Psi}{\partial y} \tag{6.37}$$

$$\frac{\partial\Psi}{\partial\phi} = -\rho\sin\phi\frac{\partial\Psi}{\partial x} + \rho\cos\phi\frac{\partial\Psi}{\partial y} \tag{6.38}$$

Multiplying Eq. (6.37) by $\cos\phi$, Eq. (6.38) by $-\frac{\sin\phi}{\rho}$, we shall get

$$\cos\phi\frac{\partial\Psi}{\partial\rho} = \cos^2\phi\frac{\partial\Psi}{\partial x} + \sin\phi\cos\phi\frac{\partial\Psi}{\partial y}$$

$$-\frac{\sin\phi}{\rho}\frac{\partial\Psi}{\partial\phi} = \sin^2\phi\frac{\partial\Psi}{\partial x} - \sin\phi\cos\phi\frac{\partial\Psi}{\partial y}$$

On adding, we shall get

$$\begin{aligned} &\cos\phi\frac{\partial\Psi}{\partial\rho} - \frac{\sin\phi}{\rho}\frac{\partial\Psi}{\partial\rho} = \frac{\partial\Psi}{\partial x} \\ &\text{Similarly, } \sin\phi\frac{\partial\Psi}{\partial\rho} + \frac{\cos\phi}{\partial\rho}\frac{\partial\Psi}{\partial\phi} = \frac{\partial\Psi}{\partial y} \end{aligned} \tag{6.39}$$

In the same way, we may obtain the second derivatives, i.e.,

$$\frac{\partial^2\Psi}{\partial x^2} = \cos^2\phi\frac{\partial^2\Psi}{\partial\rho^2} - \frac{2\sin\phi\cos\phi}{\rho}\cdot\frac{\partial^2\Psi}{\partial\rho\cdot\partial\phi} + \frac{\sin^2\phi}{\rho^2}\frac{\partial^2\Psi}{\partial\phi^2} + \frac{\sin^2\phi}{\rho}\frac{\partial\Psi}{\partial\rho} + \frac{2\sin\phi\cos\phi}{\rho^2}\frac{\partial\Psi}{\partial\phi} \tag{6.40}$$

$$\text{and } \frac{\partial^2\Psi}{\partial y^2} = \sin^2\phi\frac{\partial^2\Psi}{\partial\rho^2} + \frac{2\sin\phi\cos\phi}{\rho}\cdot\frac{\partial^2\Psi}{\partial\rho\cdot\partial\phi} + \frac{\cos^2\phi}{\rho^2}\frac{\partial^2\Psi}{\partial\phi^2} + \frac{\cos^2\phi}{\rho}\frac{\partial\Psi}{\partial\rho} - \frac{2\sin\phi\cos\phi}{\rho^2}\frac{\partial\Psi}{\partial\phi} \tag{6.41}$$

On adding Eqs. (6.40) and (6.41), we shall get

$$\frac{\partial^2\Psi}{\partial x^2} + \frac{\partial^2\Psi}{\partial y^2} = \frac{\partial^2\Psi}{\partial\rho^2} + \frac{1}{\rho}\frac{\partial\Psi}{\partial\rho} + \frac{1}{\rho^2}\frac{\partial^2\Psi}{\partial\phi^2} \tag{6.42}$$

The Laplace operator ∇^2 may now be written as follows:

$$\nabla^2\Psi = \frac{\partial^2\Psi}{\partial\rho^2} + \frac{\partial^2\Psi}{\partial z^2} + \frac{1}{\rho}\frac{\partial\Psi}{\partial\rho} + \frac{1}{\rho^2}\frac{\partial^2\Psi}{\partial\phi^2} \tag{6.43}$$

The second step of transformation is

$$z = r\cos\theta$$

$$\rho = r\sin\theta$$

By analogy to Eq. (6.42), we may write as

$$\frac{\partial^2\Psi}{\partial\rho^2} + \frac{\partial^2\Psi}{\partial z^2} = \frac{\partial^2\Psi}{\partial r^2} + \frac{1}{r}\frac{\partial\Psi}{\partial r} + \frac{1}{r^2}\frac{\partial^2\Psi}{\partial r^2} \tag{6.44}$$

Similarly, we can derive from Eqs. (6.35) and (6.36) that

$$\frac{\partial\Psi}{\partial\rho} = \sin\theta\frac{\partial\Psi}{\partial r} + \frac{\cos\theta}{r}\frac{\partial\Psi}{\partial\theta} \tag{6.45}$$

Consequently, we have

$$\frac{1}{\rho}\frac{\partial\Psi}{\partial\rho} = \frac{1}{r}\frac{\partial\Psi}{\partial r} + \frac{\cos\theta}{r^2\sin\theta}\frac{\partial\Psi}{\partial\theta} \tag{6.46}$$

In addition, we have

$$\frac{1}{\rho^2}\frac{\partial^2\Psi}{\partial\phi^2} = \frac{1}{r^2\sin^2\theta}\cdot\frac{\partial^2\Psi}{\partial\phi^2} \tag{6.47}$$

By substitution of Eqs. (6.46) and (6.47) in Eq. (6.44), we shall get

$$\nabla^2\Psi = \frac{\partial^2\Psi}{\partial r^2} + \frac{2}{r}\frac{\partial\Psi}{\partial r} + \frac{1}{r^2}\frac{\partial^2\Psi}{\partial\theta^2} + \frac{\cos\theta}{r^2\sin\theta}\frac{\partial\Psi}{\partial\theta} + \frac{1}{r^2\sin^2\theta}\frac{\partial^2\Psi}{\partial\phi^2} \tag{6.48}$$

$$\text{or } \nabla^2 = \frac{\partial^2}{\partial r^2} + \frac{2}{r}\frac{\partial}{\partial r} + \frac{1}{r^2}\frac{\partial^2}{\partial\theta^2} + \frac{\cos\theta}{r^2\sin\theta}\frac{\partial}{\partial\theta} + \frac{1}{r^2\sin^2\theta}\frac{\partial^2}{\partial\phi^2} \tag{6.49}$$

$$\text{or } \nabla^2 = \frac{1}{r^2}\frac{\partial}{\partial r}\cdot r^2\frac{\partial}{\partial r} + \frac{1}{r^2\sin^2\theta}\cdot\sin\theta\frac{\partial}{\partial\theta} + \frac{1}{r^2\sin^2\theta}\frac{\partial^2}{\partial\phi^2} \tag{6.50}$$

Thus, $\nabla^2\Psi$ or ∇^2 is transformed into polar co-ordinates.

6.5 Ehrenfest's Theorem

Ehrenfest's theorem opines that quantum mechanics provides the same results as classical mechanics for a particle for which the average or expectation values of dynamical quantities are involved.

Proof: Here, we are going to prove the theorem for one-dimensional motion of a particle by illustrating that

$$(a)\ \frac{d\langle x\rangle}{dt} = \frac{\langle P_x\rangle}{m}$$

and (b) $\frac{d\langle P_x\rangle}{m} = \langle F_x\rangle$

First of all, we shall show that

$$(a)\ \frac{d\langle x\rangle}{dt} = \frac{\langle P_x\rangle}{m}$$

Let x = the position co-ordinate of a particle of mass m at time t. The expectation value of x is indicated by

$$\langle x\rangle = \int_{-\infty}^{+\infty} \Psi^*(x,t)\cdot x\Psi(x,t)dx \tag{6.51}$$

Differentiating this equation with respect to time t, we get

$$\frac{d\langle x\rangle}{dt} = \int_{-\infty}^{+\infty} x \frac{d(\Psi\Psi^*)}{dt} dx \tag{6.52}$$

We know that

$$\frac{\partial}{\partial t}\Psi\Psi^* = \frac{i\hbar}{2m}\frac{\partial}{\partial x}\left(\Psi^*\frac{\partial\Psi}{\partial x} - \Psi\frac{\partial\Psi^*}{\partial x}\right) \tag{6.53}$$

Substituting Eq. (6.53) in Eq. (6.52), we get

$$\frac{d\langle x\rangle}{dt} = \frac{i\hbar}{2m}\int_{-\infty}^{+\infty} x\frac{\partial}{\partial x}\left(\Psi^*\frac{\partial\Psi}{\partial x} - \Psi\frac{\partial\Psi^*}{\partial x}\right)dx$$

Integrating the RHS of the above equation by parts, we shall get

$$\frac{d\langle x\rangle}{dt} = \frac{i\hbar}{2m}\left[x\left(\Psi^*\frac{\partial\Psi}{\partial x} - \Psi\frac{\partial\Psi^*}{\partial x}\right)\right]_{-\infty}^{+\infty} - \frac{i\hbar}{2m}\int_{-\infty}^{+\infty}\left(\Psi^*\frac{\partial\Psi}{\partial x} - \Psi\frac{\partial\Psi^*}{\partial x}\right)dx$$

As $x \to$ either ∞ or $-\infty$, Ψ and $\partial\Psi/\partial x \to 0$, and hence, the first term becomes zero. Hence, we have

$$\frac{d\langle x\rangle}{dt} = \frac{i\hbar}{2m}\int_{-\infty}^{+\infty}\left(\Psi^*\frac{\partial\Psi}{\partial x} - \Psi\frac{\partial\Psi^*}{\partial x}\right)dx \tag{6.54}$$

The expectation value of $\langle p_x\rangle$ is given by

$$\langle p_x\rangle = \int_{-\infty}^{+\infty}\Psi^*\frac{h}{i}\frac{\partial\Psi}{\partial x}dx$$

$$\therefore \int_{-\infty}^{+\infty}\Psi^*\frac{\partial\Psi}{\partial x}dx = \frac{i}{h}\langle p_x\rangle \tag{6.55}$$

$$\text{Similarly, } \int_{-\infty}^{+\infty} \Psi^* \frac{\partial \Psi}{\partial x} \mathrm{d}x = \frac{i}{\hbar}\langle p_x \rangle \tag{6.56}$$

Substituting these values in Eq. (6.54), we get

$$\begin{aligned} \frac{\mathrm{d}\langle x \rangle}{\mathrm{d}t} &= -\frac{i\hbar}{2m}\left(\frac{i}{\hbar}\langle p_x \rangle + \frac{i}{\hbar}\langle p_x \rangle\right) \\ &= \frac{\langle p_x \rangle}{m} \qquad \text{Q.E.D} \end{aligned} \tag{6.57}$$

(b) To prove that $\frac{\mathrm{d}\langle p_x \rangle}{\mathrm{d}t} = \langle F_x \rangle$

The expectation value of momentum $\langle p_x \rangle$ is expressed as

$$\langle p_x \rangle = \int_{-\infty}^{+\infty} \Psi^*(x,t) \frac{\hbar}{i} \frac{\partial}{\partial x} \Psi(x,t) \mathrm{d}x \tag{6.58}$$

$$= \frac{\hbar}{i} \int_{-\infty}^{+\infty} \Psi^* \frac{\partial \Psi}{\partial x} \mathrm{d}x \tag{6.59}$$

Differentiating Eq. (6.59) with respect to time t, we shall get

$$\frac{\mathrm{d}\langle p_x \rangle}{\mathrm{d}t} = \frac{\hbar}{i} \int_{-\infty}^{+\infty} \left(\frac{\partial \Psi^*}{\partial t} \cdot \frac{\partial \Psi}{\partial x} + \Psi^* \frac{\partial^2 \Psi}{\partial x \cdot \partial t} \right) \mathrm{d}x \tag{6.60}$$

We know that the time-dependent Schrödinger equations for Ψ and Ψ^* are

$$i\hbar \frac{\partial \Psi}{\partial t} = -\frac{\hbar^2}{2m} \frac{\partial^2 \Psi}{\partial x^2} + V\Psi \tag{6.61}$$

$$\text{and } i\hbar \frac{\partial \Psi^*}{\partial t} = -\frac{\hbar^2}{2m} \frac{\partial^2 \Psi^*}{\partial x^2} + V \tag{6.62}$$

Differentiating Eq. (6.61) with respect to x, we shall get

$$i\hbar \frac{\partial^2 \Psi}{\partial x \cdot \partial t} = -\frac{\hbar^2}{2m} \frac{\partial^3 \Psi}{\partial x^3} + \frac{\partial (V\Psi)}{\partial x} \tag{6.63}$$

We may express Eq. (6.60) in the following form:

$$\frac{\mathrm{d}\langle x \rangle}{\mathrm{d}t} = \int_{-\infty}^{+\infty} \left(-i\hbar \frac{\partial \Psi^*}{\partial t} \cdot \frac{\partial \Psi}{\partial x} - \Psi^* i\hbar \frac{\partial^2 \Psi}{\partial x \cdot \partial t} \right) \mathrm{d}x$$

Substituting the expression for $-i\hbar \frac{\partial \Psi^*}{\partial t}$ and $\frac{\partial^2 \Psi}{\partial x \cdot \partial t}$ in the above equation, we get

$$\frac{\mathrm{d}\langle x\rangle}{\mathrm{d}t}=\int_{-\infty}^{+\infty}\left[\left(-\frac{\hbar^2}{2m}\frac{\partial^2\Psi^*}{\partial x^2}+V\Psi^*\right)\frac{\partial\Psi}{\partial x}-\Psi^*\int_{-\infty}^{+\infty}\left(-\frac{\hbar^2}{2m}\frac{\partial^3\Psi}{\partial x^3}+\frac{\partial(V\Psi)}{\partial x}\right)\right]\mathrm{d}x$$

$$=-\frac{\hbar^2}{2m}\int_{-\infty}^{+\infty}\left[\left(-\frac{\partial^2\Psi^*}{\partial x^2}\frac{\partial\Psi}{\partial x}+\Psi^*\frac{\partial^3\Psi}{\partial x^3}\right)\mathrm{d}x+\int_{-\infty}^{+\infty}V\Psi^*\frac{\partial\Psi}{\partial x}-\Psi^*\frac{\partial(V\Psi)}{\partial x}\right]\mathrm{d}x$$

$$=-\frac{\hbar^2}{2m}\int_{-\infty}^{+\infty}\left[\frac{\partial}{\partial x}\left(\frac{\partial\Psi}{\partial x}\frac{\partial\Psi^*}{\partial x}-\Psi^*\frac{\partial^2\Psi}{\partial x^2}\right)\right]\mathrm{d}x$$

$$=-\frac{\hbar^2}{2m}\left[\frac{\partial\Psi}{\partial x}\frac{\partial\Psi^*}{\partial x}-\Psi^*\frac{\partial^2\Psi}{\partial x^2}\right]_{-\infty}^{+\infty}-\int_{-\infty}^{+\infty}\Psi^*\frac{\partial V}{\partial x}\Psi\mathrm{d}x \tag{6.64}$$

As $x\to\infty$ or $-\infty$ and $\partial\Psi/\partial x\to 0$, the first term on the RHS of Eq. (6.64) will be zero. The second term represents the expectation value of the differential coefficient of the P.E, V with respect to x, i.e.,

$$\frac{\langle\partial V\rangle}{\partial x}=\int_{-\infty}^{+\infty}\Psi^*\frac{\partial V}{\partial x}\Psi\mathrm{d}x$$

$$\therefore\quad\frac{\mathrm{d}\langle p_x\rangle}{\partial t}=-\frac{\langle\partial V\rangle}{\partial x}$$

But $-\dfrac{\partial V}{\partial x}$ is the classical force F_x

$$\therefore\frac{\mathrm{d}\langle p_x\rangle}{\partial t}=\langle F_x\rangle\qquad\text{QED.}\tag{6.65}$$

6.6 Matrix Representation of Wave Function

Let us consider the complete orthonormal set of functions ϕ_1, ϕ_2, …, ϕ_n, which constitutes the basis of a co-ordinate in the Hilbert space. According to the expansion theorem, any function Ψ can be expanded in terms of these basis functions as follows:

$$|\Psi\rangle=\sum_i c_i|\phi_i\rangle \tag{6.66}$$

But *ci* may be expressed as

$$c_i=\langle\phi_i|\Psi\rangle \tag{6.67}$$

The coefficients c_1, c_2, …, c_n represent the components of the vector Ψ in the Hilbert space since values of ϕ_i^s are known; therefore, Ψ can be specified if values of ϕ_i^s are known. Hence, c_1, c_2, …, c_n can be said to form a representation of the wave function with respect to the basis ϕ_1, ϕ_2, …, ϕ_n. In matrix representation, we specify the vector Ψ by its components c_1, c_2, …, c_n and express it as column vector, i.e.,

$$\Psi=\begin{pmatrix}c_1\\c_2\\\vdots\\c_n\end{pmatrix} \tag{6.68}$$

This is called the matrix representation of the wave function. (Hilbert space: In the vector space, the unit vectors e_1, e_2, e_3, ... constitute the orthonormal basis. Or in other words, a space can be defined in which a set of functions $\phi_1(x)$, $\phi_2(x)$, ... constitutes the orthonormal unit vectors of the co-ordinate system. The corresponding infinite dimensional linear space is termed as the function space. In quantum mechanics, we often deal with complex functions, and the corresponding function space is known as the Hilbert space.)

6.7 Matrix Representation of Operator

Let us consider the Hamiltonian operator H, which alters the function ϕ into another function Ψ. Then we can write

$$|\Psi> = H|\phi> \tag{6.69}$$

Representing in terms of the basis function, we can write

$$\left|\Psi> = \sum_{j=1}^{n} c_j |\phi_j\rangle,\ c_j = <\right|\phi_j|\Psi\rangle \tag{6.70}$$

$$\text{and } \left|\phi> = \sum_{j=1}^{n} a_j |\phi_j\rangle,\ a_j = <\right|\phi_j|\phi\rangle \tag{6.71}$$

Substituting these expressions for Ψ and ϕ in Eq. (6.69), we get

$$\sum_{j=1}^{n} c_j |\phi_j\rangle = H\sum_{j=1}^{n} a_j |\phi_j\rangle = \sum_{j=1}^{n} H|\phi_j\rangle a_j$$

Multiplying from left by $\langle\phi_i|$, we can write

$$c_i = \sum_{j=1}^{n} \langle\phi_i|H\phi_j > a_j = \sum_{j=1}^{n} H_{ij} \cdot a_j,\ i = 1,2,\ldots,n \tag{6.72}$$

$$\text{where, } H_{ij} = \langle\phi_i|H\phi_j\rangle \tag{6.73}$$

Thus, the square array of numbers may be expressed as

$$\begin{bmatrix} H_{11} & H_{12} & \ldots & H_{1n} \\ H_{21} & H_{22} & \ldots & H_{21} \\ \vdots & \vdots & & \vdots \\ H_{n1} & H_{n2} & \ldots & H_{nn} \end{bmatrix}$$

constitutes the element of a matrix. This matrix is equivalent to the complete specification of the operator H. It is the matrix form of the operator H with respect to the basis ϕ_1, ϕ_2, ..., ϕ_n. The matrix elements are defined by Eq. (6.73). In terms of the matrix elements of operator H, Eq. (6.72) takes the following form:

$$\begin{bmatrix} c_1 \\ c_2 \\ \vdots \\ c_n \end{bmatrix} = \begin{bmatrix} H_{11} & H_{12} & \ldots & H_{1n} \\ H_{21} & H_{22} & \ldots & H_{21} \\ \vdots & \vdots & & \vdots \\ H_{n1} & H_{n2} & \ldots & H_{nn} \end{bmatrix} \begin{bmatrix} a_1 \\ a_2 \\ \vdots \\ a_n \end{bmatrix} \tag{6.74}$$

This is the matrix representation of the operator.

6.8 Properties of Matrix Elements

We are aware of the fact that the matrix elements obey the rule of matrix algebra for addition, subtraction, multiplication, etc. In quantum mechanics, we require two important properties of operators, which follow from the earlier definition of matrix elements.

Let A = a Hermitian operator.

Taking the help of Eq. (6.73) and $(\Psi_m, A\Psi_n) = (A\Psi_m, \Psi_n) = (\Psi_n, A\Psi_m)^*$, the matrix element can be expressed as

$$A_{ij} = \langle u_i | A | u_j \rangle = \langle Au_i | u_j >= \langle u_j | A | u_i \rangle^* = A_{ji}^* \tag{6.75}$$

This means that the diagonal elements (i, j) are real and the off-diagonal elements (i, j) and (j, i) on either side of the diagonal are complex conjugates. In other words, if ϕ_i^s constitutes a complete set of functions and A is a Hermitian operator, then the matrix representing the operator is Hermitian. If the functions u_i are Eigen functions of the Hermitian operator A itself, then

$$A_{ij} = \langle u_i | A | u_j \rangle = a_j \langle u_i | u_j > = a_j \delta_{ij} \tag{6.76}$$

That is, the final matrix is diagonal. This gives an important result that the matrix representation of an operator with respect to its own Eigen function becomes diagonal, and the matrix elements are the Eigen values of the said operator.

6.9 Matrix Form of the Schrödinger Equation

$$H\Psi = E\Psi \tag{6.77}$$

Let us consider a set of function $\phi_1, \phi_2, \ldots, \phi_n$ and express Ψ as a linear combination of ϕ's, that is

$$\Psi = \sum^n c_n \phi_n = \sum^n c_n |n\rangle \tag{6.78}$$

Operating H on Ψ, we get

$$H\Psi = \sum^n c_n H |n\rangle = E \sum^n c_n |n\rangle \tag{6.79}$$

Multiplying by bra $\langle m| = \phi_m$, we have

$$\sum^n \langle m | c_n H | n \rangle = E \sum^n c_m < m | n \rangle \tag{6.80}$$

$$\text{or } c_1 H_{11} + c_2 H_{22} + \ldots + Ec_n H_{mn}$$

$$= Ec_1 1 \mid 1 + Ec_2 2 \mid 2 + \ldots + Ec_n H_{mn}$$

$$\text{or } c_1 H_{11} + c_2 H_{22} + \ldots + 0 = Ec_1 + Ec_2 + \ldots + 0$$

$$\text{when } n = m, < m | n \rangle = 1$$

$$\text{and} < m | n \rangle = 0 \text{ when } m \neq n$$

$$\therefore c_m H_{mm} = Ec_m \tag{6.81}$$

Since $c_m H_{mm}$ commute,

$$H_{mm}c_m = Ec_m$$

There will be m separate matrix equations, i.e.,

$$H_{11}c_1 = Ec_1$$

$$H_{22}c_2 = Ec_2$$

$$H_{33}c_3 = Ec_3$$

$$\cdots$$

$$H_{nn}c_n = Ec_n$$

If the matrix is diagonalised, then

$$H = H_{mm} = \begin{bmatrix} H_{11} & H_{12} & \cdots & H_{1n} \\ H_{21} & H_{22} & \cdots & H_{2n} \\ \cdots & \cdots & \cdots & \cdots \\ \cdots & \cdots & \cdots & \cdots \\ H_{m1} & H_{m2} & \cdots & H_{mm} \end{bmatrix}$$

c_m is a column vector having $m = 1, 2, 3, \ldots, m$

$$\therefore c_m = \begin{bmatrix} c_1 \\ c_2 \\ \vdots \\ c_m \end{bmatrix}$$

Thus, the matrix form of the Schrödinger equation may be written as

$$\begin{bmatrix} H_{11} & H_{12} & \cdots & H_{1n} \\ H_{21} & H_{22} & \cdots & H_{2n} \\ \cdots & \cdots & \cdots & \cdots \\ \cdots & \cdots & \cdots & \cdots \\ H_{m1} & H_{m2} & \cdots & H_{mm} \end{bmatrix} \begin{bmatrix} c_1 \\ c_2 \\ \vdots \\ c_m \end{bmatrix} = E \begin{bmatrix} c_1 \\ c_2 \\ \vdots \\ c_m \end{bmatrix} \tag{6.82}$$

We may also write the time-dependent Schrödinger equation in matrix form.

6.9.1 Time-Dependent Schrödinger Equation in Matrix Form

The time-dependent Schrödinger equation is expressed as

$$i\hbar \frac{\mathrm{d}}{\mathrm{d}t} \Psi(\mathrm{r},t) = H\Psi(r,t) \tag{6.83}$$

This can also be written in terms of the basis vectors $|u_i\rangle$, that is,

$$\Psi(\mathrm{r},t) = \sum_{j=1} c_j(t) u_j(r) \tag{6.84}$$

Putting this value of Ψ in Eq. (6.83), we obtain

$$i\hbar\frac{\mathrm{d}}{\mathrm{d}t}\sum_{j=1} c_j(t)u_j(r) = H\sum_j c_j(t)u_j(r)$$

Multiplying by $\langle u_i|$ from left, we have $i\hbar\frac{\mathrm{d}}{\mathrm{d}t}\sum_j c_j(t)u_i \mid u_j(r) = \sum_j u_i \mid Hu_j c_j(t)$

$$\text{or } i\hbar\frac{\mathrm{d}c_i(t)}{\mathrm{d}t} = \sum_j H_{ij}c_j(t), \quad i = 1,2,3,\ldots \tag{6.85}$$

where H_{ij} = matrix elements of the Hamiltonian.

It is, thus, clear that the Schrödinger equation in matrix form yields a system of simultaneous differential equations for c_i(t), which is the time-dependent expansion coefficient.

BIBLIOGRAPHY

Atkins, P. and R. Friedman. 2007. *Molecular Quantum Mechanics*. New York: Oxford University Press

Flurry, R.L. Jr. 1983. *Quantum Chemistry*. Upper Saddle River, NJ: Prentice-Hall Inc.

Glasstone, S. 1973. *Theoretical Chemistry*. New Delhi: Affiliated East-West Press Pvt Ltd.

Johnson, C.S. Jr. and L.G. Pedersen. 1974. *Problems and Solutions in Quantum Chemistry and Physics*. London: Addison-Wesley Publishing Co.

Kaufman, E.D. 1966. *Advanced Concepts in Physical Chemistry*. New York: McGraw-Hill.

Veszprémi, T. and M. Fehér. 1999. *Quantum Chemistry*. New York: Plenum.

Solved Problems

Problem 1. Which of the following are Eigen functions for $\frac{\mathrm{d}}{\mathrm{d}x}$?

(a) x^2 (b) e^x (c) sin ax (d) $e^{-3.4\,x^2}$ (e) 37 (f) $\cos 4x + i\sin 4x$

Solution:

(a) $\frac{\mathrm{d}}{\mathrm{d}x}\left(x^2\right) = 2x \neq \text{constant time } x^2$

$\therefore$ It is not Eigen function for $\frac{\mathrm{d}}{\mathrm{d}x}$

(b) $\frac{\mathrm{d}}{\mathrm{d}x}e^x = 1\cdot e^x = \text{constant times } e^x$

$\therefore$ e^x is an Eigen function for $\mathrm{d}/\mathrm{d}x$.

(c) $\frac{\mathrm{d}}{\mathrm{d}x}(\sin ax) = a\cos ax \neq \text{constant times } \sin(ax)$

$\therefore$ It is not an Eigen function for $\mathrm{d}/\mathrm{d}x$.

(d) $\frac{\mathrm{d}}{\mathrm{d}x}\left(e^{-3.4x^2}\right) = 3.4 \times 2xe^{-3.4x^2}$

$$= 6.8x\ e^{-3.4x^2}$$

$$\neq \text{constant times } e^{-3.4x^2}$$

$\therefore$ It is not an Eigen function for $\mathrm{d}/\mathrm{d}x$.

(e) $\frac{d}{dx}(37) = 0 = 0 \times 37$

$= \text{constant times } 37$

$\therefore$ It is an Eigen function for d/dx.

(f) $(\cos 4x + i \sin 4x) = e^{i4x}$

$\therefore$ It is an Eigen function for d/dx. Answer.

Problem 2. Indicate which of the following functions are 'acceptable'? If one is not, give a reason. (a) $\Psi = \times$ (b) $\Psi = x^2$ (c) $\Psi = \sin x$ (d) $\Psi = e^{-x^2}$ (e) $\Psi = e^{-x}$

Solution:

(a) $\Psi = x$
It is not an acceptable function because it becomes infinite at $x = \pm\infty$.

(b) $\Psi = x^2$
It is not an acceptable function as it becomes infinite at $x = \pm\infty$.

(c) $\Psi = \sin x$
It is an acceptable function because it is single valued, continuous, and nowhere infinite.

(d) $\Psi = e^{-x^2}$
It is an acceptable function because it is single valued, continuous, and nowhere infinite.

(e) $\Psi = e^{-x}$
It is not an acceptable function because it becomes infinite at $x = -\infty$. Answer.

Problem 3. Show that if $\frac{1}{\pi}\sin(3.63x)$ is an Eigen function of the operator $\left[\left(-\frac{h^2}{8\pi^2 m}\right)\frac{d^2}{dx^2}\right]$, what is its Eigen value?

Solution: $\left\{\left(-\frac{h^2}{8\pi^2 m}\right)\frac{d^2}{d\pi^2}\right\}\frac{1}{\pi}\sin 3.63x$

$$= \left(-\frac{h^2}{8\pi^2 m}\right)\left(\frac{1}{\pi}\right)(3.63)\frac{d}{dx}\cos(3.23x)$$

$$= \left[(3.63)^2 \frac{h^2}{8\pi^2 m}\right]\cdot\left(\frac{1}{\pi}\sin(3.63x)\right)$$

$$= \text{constant time}\frac{1}{\pi}\sin(3.63x)$$

$$\therefore \text{Eigen value} = (3.63)^2 \frac{h^2}{8\pi^2 m}$$ Answer.

Problem 4. Show that the function $\Psi - 8\,e^{4x}$ is an Eigen function of the operator d/dx. What is the Eigen value?

Solution: $\frac{d}{dx}\left(8e^{4x}\right) = 8.4\ e^{4x} = 4.8e^{4x}$

Therefore, $8e^{4x}$ is an Eigen function of d/dx, and its Eigen value = 4. Answer.

Problem 5. Normalise the function $\cos\left(\frac{n\pi}{a}x\right)$ over the interval $-a < x < a$.

Solution: Let A = normalisation constant. Then, the normalised function will be

$$A\cos\left(\frac{n\pi x}{a}\right)$$

$$\therefore\ \Psi = A\cos\left(\frac{n\pi x}{a}\right) \text{ and } \Psi^* = A\cos\left(\frac{n\pi x}{a}\right)$$

$\therefore$ According to normalization condition,

$$\int_{-a}^{+a} \Psi\Psi^* \mathrm{d}x = 1$$

$$\therefore \int_{-a}^{+a} \Psi\Psi^* \mathrm{d}x = \int_{-a}^{+a} \left\{A\cos\left(\frac{n\pi x}{a}\right)\right\}\left\{A\cos\left(\frac{n\pi x}{a}\right)\right\}\mathrm{d}x = 1$$

$$\text{or } \int_{-a}^{+a} A^2\cos^2\frac{n\pi x}{a}\mathrm{d}x = 1$$

$$\text{or } \frac{1}{2}\int_{-a}^{+a} A^2\cos^2\frac{n\pi x}{a}\mathrm{d}x = 1$$

$$\text{or } \frac{A^2}{2}\int_{-a}^{+a}\left(1+\cos\frac{2n\pi x}{a}\right)\mathrm{d}x = 1$$

$$\text{or } \frac{A^2}{2}\int_{-a}^{+a}\mathrm{d}x + \int_{-a}^{+a}\cos\frac{2n\pi x}{a}\mathrm{d}x = 1$$

$$\text{or } \frac{A^2}{2}\left(2[x]_0^a + 2\left[\frac{2n\pi}{a}\sin\left(\frac{2n\pi x}{a}\right)\right]\right)_0^{+a} = 1$$

$$\text{or } \frac{A^2}{2}[2a+0] = 1$$

$$\text{or } A^2 = \frac{2}{2a} = \frac{1}{a}$$

$$\therefore A = \left(\frac{1}{a}\right)^{1/2} = a^{-1/2}$$

$\therefore$ Normalised function is $a^{-1/2}\cos(n\pi x/a)$ Answer.

Problem 6. Show that the functions $\sin(\pi x/a)$ and $\cos(\pi x/a)$ are orthogonal over the interval $0 < x < a$.

Solution: According to question, $\Psi = \sin(\pi x/a)$ and $\Psi^* = \cos(\pi x/a)$.

But for orthogonal condition,

$$\int_0^a \Psi\Psi^* \mathrm{d}x = 0$$

$$\text{or} \int_0^a \sin\left(\frac{\pi x}{a}\right)\cos\left(\frac{\pi x}{a}\right) d\,x$$

$$\text{or} \frac{1}{2}\int_0^a 2\sin \pi x/a \cos \pi x/a \,\mathrm{d}x$$

$$\text{or} \frac{1}{2}\int_0^a \sin 2\pi x/a \,\mathrm{d}x \quad [\because 2\sin\theta\cos\theta = \sin 2\theta]$$

$$\text{or} \frac{1}{2}\left[\frac{2\pi}{a}\cos\frac{2\pi x}{a}\right]_0^a = \frac{1}{2}\left[\frac{2\pi}{a}\cos 2\pi - \frac{2\pi}{a}\cos 0\right]_0^a$$

$$= \frac{1}{2}\times\frac{2\pi}{a}(\cos 2\pi - \cos 0)$$

$$= \frac{\pi}{a}(1-1) = \frac{\pi}{a}\times 0 = 0$$

$\therefore$ The two functions are orthogonal Answer.

Problem 7. Write the Hamiltonian operator for a free particle moving in one direction under the influence of zero potential energy.

Solution: We know that $H = \frac{1}{2}mv^2 + \text{PE}$.

But $\text{PE} = V = 0$

$$\therefore H = \frac{1}{2}mv^2 = \frac{1}{2m}m^2v_x^2 = \frac{p_x^2}{2m}$$

where p_x = linear momentum along x-axis

But for linear operator, the operator is $\frac{\hbar}{i}\cdot\frac{\mathrm{d}}{\mathrm{d}x}$

$$\therefore H = \frac{1}{2m}\left(\frac{\hbar}{i}\cdot\frac{\mathrm{d}}{\mathrm{d}x}\right)^2$$

$$= \frac{1}{2m}\left(\frac{\hbar}{i}\cdot\frac{\mathrm{d}}{\mathrm{d}x}\right)\left(\frac{\hbar}{i}\cdot\frac{\mathrm{d}}{\mathrm{d}x}\right)$$

$$= \frac{1}{2m}\frac{\hbar^2}{i^2}\cdot\frac{\mathrm{d}^2}{\mathrm{d}x^2}$$

$$= -\frac{\hbar^2}{2m}\cdot\frac{\mathrm{d}^2}{\mathrm{d}x^2}$$

Questions on Concepts

1. Prove that $\nabla^2\Psi = -\frac{4\pi^2}{\lambda^2}\cdot\Psi$
2. Derive time-independent Schrödinger's equation.
3. Derive time-dependent Schrödinger's equation.
4. Discuss the physical meaning of the wave function, Ψ.
5. (a) What do you mean by 'acceptable' wave function?
   (b) Which of the following wave functions are acceptable?
   a. $\tan x$
   b. $\cos x + \sin x$
   c. $\text{cosec}\, x$
   d. $\cos 2x + i \sin 2x$
   e. $\sin x$ [Answer (d) and (e) are acceptable wave functions].
6. (a) Explain normalisation of a wave function.
   (b) Show that e^{-ax} is an Eigen function of the operators $\mathrm{d}/\mathrm{d}x$ and $\mathrm{d}^2/\mathrm{d}x^2$ and find the corresponding Eigen values.
7. Show that $\sin nx$ and $\cos nx$ are both Eigen functions of the operator $\mathrm{d}^2/\mathrm{d}x^2$ but not of $\mathrm{d}/\mathrm{d}x$. What is the corresponding Eigen value in the former case?
8. Normalise the following:
   a. $\Psi = x + 4$ in the interval $-1 \le \pi \le 1$
   b. $\Psi = e^{-ix}$ in the interval $0 \le x \le 2\pi$
   c. $\Psi = \sin x + \sin 2x$ in the interval $0 \le x \le \pi$
9. (a) Explain orthogonality and also mention its proof.
   (b) Prove that the functions $\sin(\pi x/a)$ and $\cos(\pi x/a)$ are orthogonal over the interval $0 < x < a$.
10. Explain Eigen value and Eigen function. Show that the function $\Psi = xe^{-bx^2}$ is an Eigen function of the operator $\left(\mathrm{d}^2/\mathrm{d}x^{2-}4b^2x^2\right)$. Find its Eigen value.
11. Find out which of the following functions are Eigen functions of the operator d/dx?
    a. e^{-iax}
    b. $\cos ax$
    c. ax^2
    d. $\sin ax$
12. (a) Explain orthonormality.
    (b) What do you mean by degeneracy?
13. Show that the Laplacian operator $\nabla^2 = \frac{\partial^2}{\partial x^2} + \frac{\partial^2}{\partial y^2} + \frac{\partial^2}{\partial z^2}$ when transformed to spherical polar co-ordinates becomes $\nabla^2 = \frac{\partial^2}{\partial r^2} + \frac{2}{r}\frac{\partial}{\partial r} + \frac{1}{r^2}\frac{\partial^2}{\partial \theta^2} + \cos\theta/r^2\sin\theta\frac{\partial}{\partial\theta} + \frac{1}{r^2\sin^2\theta}\cdot\frac{\partial^2}{\partial\theta^2}$
14. Derive Ehrenfest's theorem or prove that

$$\text{(a)}\ \frac{\mathrm{d}\langle x\rangle}{\mathrm{d}t} = \frac{\langle p_x\rangle}{m}$$

$$\text{(b)}\ \frac{\mathrm{d}\langle x\rangle}{\mathrm{d}t} = \langle F_x\rangle$$

15. Write down the matrix form of a wave function.
16. How is an operator represented in the matrix form?
17. How will you achieve the matrix form of the Schrödinger equation?
18. Express the properties of matrix elements.

7

Playing with the Schrödinger Equation

We have become acquainted with the Schrödinger equation. Now we shall apply the Schrödinger equation in hypothetical problems. This essentially means that this equation will be applied to different simple systems. It is known that the electrons in atoms and molecules can execute three motions, namely, translational motion, vibrational motion, and rotational motion. But when the particle-like electron remains free or is under quantum mechanical constraints, it moves only linearly, i.e., in one direction. The example of such a system is the free particle in a box. In case of harmonic oscillator, the particle executes only vibrational motion, but the particle executes rotational motion only in case of rigid rotor or rotator.

7.1 Particle in a One-Dimensional Box

Let us consider a particle of mass m, which is constrained or restricted to motion in a one-dimensional potential box of length a. The particle is moving in a constant potential.

The potential (V) may be taken to be zero between $x = 0$ and $x = a$, i.e., inside the box, because its sole effect is to increase the energy by a constant amount, and outside the range $x = 0$ and $x = a$, the potential is infinity.

This situation is obtained by plotting a graph of potential energy (V) against distance x. It has been illustrated in Figure 7.1.

We shall apply the Schrödinger equation in one dimension to the free particle in the box. The Schrödinger equation in one dimension (along the x axis, say) may be expressed as

$$\frac{d^2\psi}{dx^2} + \frac{8\pi^2 m}{h^2}(E - V)\psi = 0$$

where ψ = wave function

E = energy along x axis

V = potential energy of the particle.

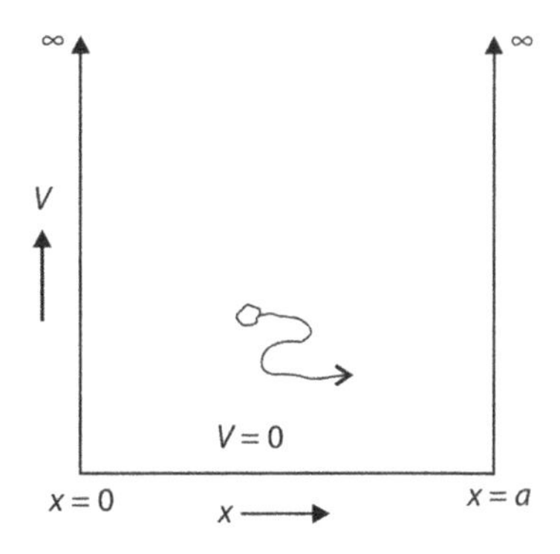

FIGURE 7.1 One-dimensional potential box of length ‘a’.

For a free particle, potential energy $V = 0$. Therefore, the Schrödinger equation takes the form

$$\frac{\mathrm{d}^2\Psi}{\mathrm{d}x^2} + \frac{8\pi^2 m}{h^2} E\psi = 0 \tag{7.2}$$

$$\text{or, } \frac{d^2\Psi}{dx^2} + k^2\psi = 0, \text{ where, } k^2 = \frac{8\pi^2 m}{h^2}$$

This is a second-order differential equation, whose general solution will be

$$\psi = A\sin kx + B\cos kx \tag{7.3}$$

where A and B are arbitrary constants. Applying the boundary condition of the problem, i.e., when $x = 0$, $\psi = 0$ and when $x = \text{a}$, $\psi = 0$ to Eq. (7.3), we may get an Eigen function of the system.

When $x = 0$, $\psi = 0$, putting these values in Eq. (7.3), we obtain

$$0 = A\sin 0 + B\cos 0 \text{ or } 0 = B \tag{7.4}$$

$$\text{Putting } B = 0 \text{ in Eq. (7.3), we have } \psi = A\sin kx \tag{7.5}$$

Further, we shall apply the other boundary condition, i.e., when $x = a$, $\psi = 0$ to Eq. (7.3), and we get

$$0 = A\sin ka \tag{7.6}$$

The LHS of the above equation may be written as sin $nx = 0$, where n = an integer. Therefore, Eq. (7.6) may be expressed as

$$A\sin kx = \sin n\pi = 0$$

But $A \neq 0$, and therefore,

$$\sin ka = \sin n\pi = 0$$

$$\text{or } ka = n\pi$$

$$\therefore\; k = \frac{n\pi}{a} \tag{7.7}$$

$$\therefore\; k^2 = \frac{n^2\pi^2}{a^2} \tag{7.8}$$

But $k^2 = \dfrac{8\pi^2 mE}{a^2}$

Putting this value of k^2 in Eq. (7.8), we get

$$\frac{8\pi^2 mE}{a^2} = \frac{n^2\pi^2}{a^2}$$

And the Eigen values will be

$$E_n = \frac{n^2h^2}{8m\,a^2} \tag{7.9}$$

where E_n = total energy and $n = 1, 2, 3, \ldots$

This is the energy of the particle in one-dimensional box, and the Eigen functions ψ_n will be obtained by putting $k = n\pi/a$ in Eq. (7.5) and it will be

$$\psi_n = A\sin\frac{n\pi}{a}x \tag{7.10}$$

It must be kept in mind that $k = n\pi/a$ is an essential condition for the solution of the wave equation to be acceptable, i.e., to be an Eigen function of the system.

Now we have to find out the value of the arbitrary constant A. For this purpose, normalisation condition is applied. The condition for normalisation is

$$\int \psi\psi^* \mathrm{d}\tau = 1 \tag{7.11}$$

where ψ^* = complex conjugate of ψ
$\mathrm{d}\tau$ = volume element.

Recall Eq. (7.5), which is

$$\psi = A\sin kx \text{ and } k = \frac{n\pi}{a}$$

$$\therefore\ \psi = A\sin\frac{n\pi}{a}x \tag{7.12}$$

In one dimension, the normalisation condition may be written as

$$\int \psi\psi^* \mathrm{d}x = 1$$

But the particle remains somewhere between the limits $x = 0$ and $x =$ a. Therefore, the above integral may be written as

$$\int_0^a \psi\psi^* \mathrm{d}x = 1$$

Since $\psi \approx \psi^*$, the above equation becomes

$$\int_0^a \psi^2 \mathrm{d}x = 1 \tag{7.13}$$

Putting the value of ψ in Eq. (7.13), we obtain

$$\int_0^a \left(A\sin\frac{n\pi x}{a}\right)^2 \mathrm{d}x = 1$$

$$\text{or } \int_0^a A^2\sin^2\frac{n\pi x}{a}\mathrm{d}x = 1$$

$$\text{or } \frac{A^2}{2}\int_0^a 2\sin^2\frac{n\pi x}{a}\mathrm{d}x = 1$$

$$\text{or } \frac{A^2}{2}\int_0^a \left(1-\cos\frac{2\,n\pi x}{a}\right)\mathrm{d}x = 1 \quad \left[\because\ 1-2\sin^2\theta = \cos 2\theta\right]$$

$$\text{or } \frac{A^2}{2}\left[\int_0^a \mathrm{d}x - \int_0^a \cos\frac{2\,n\pi x}{a}\right]\mathrm{d}x = 1$$

$$\text{or } \frac{A^2}{2}\left[x - \frac{a}{2n\pi}\sin\frac{2\,n\pi x}{a}\right]_0^a = 1$$

$$\text{or } \frac{A^2}{2}\left[a - \frac{a}{2n\pi}\sin 2\,n\pi\right] = 1$$

$$\text{or } \frac{A^2}{2}\cdot a = 1 \quad [\because\ \sin 2n\pi = 0]$$

$$\text{or } A^2 = \frac{2}{a}$$

$$\therefore\quad A = \left(\frac{2}{a}\right)^{1/2} \tag{7.14}$$

Therefore, the normalised Eigen function may be expressed as

$$\psi = \left(\frac{2}{a}\right)^{1/2}\sin\frac{n\pi x}{a} \quad \text{or in more general case}$$

$$\psi_n = \left(\frac{2}{a}\right)^{\frac{1}{2}}\sin\frac{\pi x}{a}x \tag{7.15}$$

and the probability of finding the particle at a particular value of $\times$ is expressed as

$$\psi_n^2 = \frac{2}{a}\sin^2\frac{n\pi}{a}x \tag{7.16}$$

7.1.1 Energy Level Diagram

The energy of a free particle in one-dimensional box is given in Eq. (7.9), which is

$$E_n = \frac{n^2h^2}{8ma^2}$$

From this, it is clear that the energy is quantised in units of $h^2/8ma^2$, the multiple being 1, 4, 9, 16, …, n^2. It is to be noted that the value of n cannot be zero; as a consequence of this, $E \neq 0$. Putting the value of $n = 1, 2, 3, 4, \ldots$, we shall get different Eigen values as follows:

n	E
1	$h^2/8ma^2$
2	$4k^2/8ma^2$
3	$9h^2/8ma^2$
4	$16h^2/8ma^2$
...	and so on ...

It is, thus, clear that $n = 1$ is the minimum value of n, and its corresponding energy is $h^2/8ma^2$, which is called the *zero point energy* (*ZPE*) of the system.

When energy vs. ψ and ψ^2 are plotted, the diagrams represented in Figure 7.2 obtained. The solid line (———) indicates energy vs. ψ, and the dotted line (– – – –) represents energy vs. ψ^2 graph.

It should be kept in mind that for the lowest energy state, half a wavelength of the wave function fits into the box, and there is no point inside the box, where the wave function amplitude is zero. Such a point in the diagram is known as a *node*.

It is clear from the figure that each Eigen function has $(n - 1)$ internal *nodes*. It is easy to understand from the figure that greater the number of nodes, greater is the kinetic energy from the energy equation $\left(E_n = n^2h^2/8ma^2\right)$. It is clear that the energy is inversely proportional to the mass of the particle and the square of the length of the box.

7.2 Particle in a Rectangular Three-Dimensional Box or Particle in a Three-Dimensional Box

We will consider the case of the three-dimensional box, with sides a, b, and c, as illustrated in Figure 7.3. An atom trapped in a crystal lattice is a suitable example of a particle thus confined. Naturally, the box for a real system will be less ideal as compared to the one we have selected here; some adjustment will, therefore, be required for a real system. Such confinement needs that the potential energy should

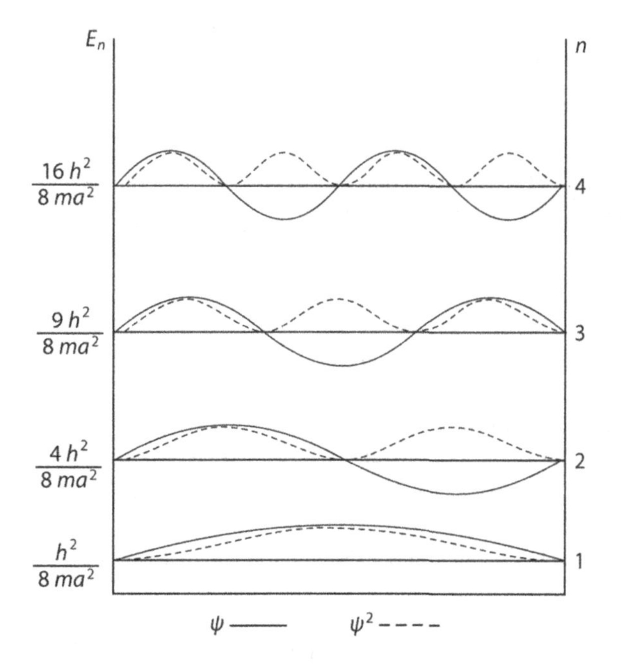

FIGURE 7.2 Energy level diagram with ψ and probability distribution ψ^2.

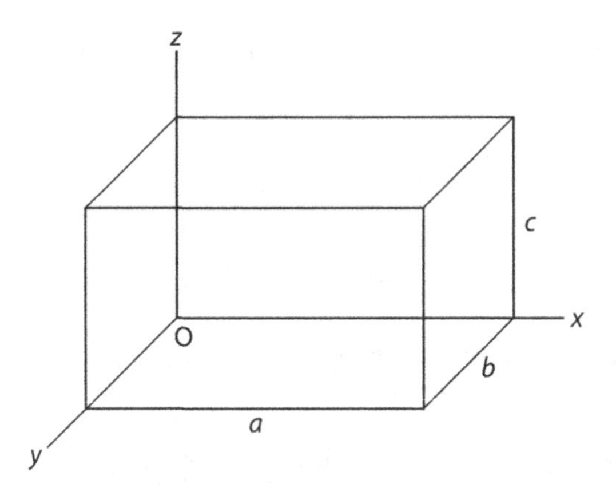

FIGURE 7.3 Particle in a three-dimensional box.

be infinity outside the box and finite inside the box. For the simplicity sake, the potential energy V is selected zero (i.e., $V = 0$) inside the box. The boundary conditions may be expressed as

$$V = \begin{cases} 0 \\ \infty \text{ for } 0 < x < a,\ 0 < y < b,\ 0 < z < c \end{cases}$$

where a, b, and c are the sides of the wall of the box.

Let the particle having mass m be inside the three-dimensional box. Keeping in view the above facts, the Schrödinger equation for the particle inside the box is given by

$$\frac{\partial^2 \psi}{\partial x^2} + \frac{\partial^2 \psi}{\partial y^2} + \frac{\partial^2 \psi}{\partial z^2} + \frac{8\pi^2 E}{h^2}\psi = 0 \tag{7.17}$$

We make assumption that the solutions of the above Schrödinger equation can be expressed in the product form of three functions as

$$\psi = X(x)Y(y)Z(z) \tag{7.18}$$

where X is a function of x only, Y is a function of y only, and Z is a function of z only. The importance of assuming this form of solution is that the above Schrödinger equation factorises into three equations, one in x, one in y, and one in z. This method of solving differential equations in more than one variable is called the separation of variables.

For simplicity sake, Eq. (7.18) may be written as

$$\psi = X \cdot Y \cdot Z. \tag{7.19}$$

in which the variables are not shown.

Since the functions Y and Z are independent of x, the differentiation of Eq. (7.19) with respect to x yields

$$\frac{\partial \psi}{\partial x} = YZ\frac{\mathrm{d}X}{\mathrm{d}x}$$ and further differention with respect to x gives

$$\frac{\partial^2 \psi}{\partial x^2} = YZ\frac{\mathrm{d}^2 X}{\mathrm{d}x^2} \tag{7.20}$$

Similarly, Eq. (7.19) is differentiated with respect to y, which gives

$\frac{\partial \psi}{\partial y} = XZ\frac{dY}{dy}$ and further differention with respect to y gives

$$\frac{\partial^2 \psi}{\partial y^2} = XZ\frac{d^2Y}{dy^2} \tag{7.21}$$

The double differentiation of Eq. (7.19) with respect to z yields

$$\frac{\partial^2 \psi}{\partial z^2} = XY\frac{d^2Z}{dz^2} \tag{7.22}$$

Substitution of Eqs. (7.20), (7.21), and (7.22) in Eq. (7.17) gives

$$YZ\frac{d^2X}{dx^2} + XZ\frac{d^2Y}{dy^2} + XY\frac{d^2Z}{dz^2} + \frac{8\pi^2 m}{h^2}EXYZ = 0$$

Dividing throughout by $\left(8\pi^2 m/h^2\right)XYZ$, we shall obtain

$$\frac{h^2}{8\pi^2 m}\left[\frac{1}{X}\frac{d^2X}{dx^2} + \frac{1}{Y}\frac{d^2Y}{dy^2} + \frac{1}{Z}\frac{d^2Z}{dz^2}\right] + E = 0 \tag{7.23}$$

Now we put the total energy as the sum of three energies along x, y, z axes, i.e.,

$$E = E_x + E_y + E_z \tag{7.24}$$

So that we can separate the variables in Eq. (7.23) to give

$$\frac{h^2}{8\pi^2 m}\frac{1}{X}\frac{d^2X}{dx^2} = -E_x \tag{7.25}$$

$$\frac{h^2}{8\pi^2 m}\frac{1}{Y}\frac{d^2Y}{dy^2} = -E_y \tag{7.26}$$

$$\frac{h^2}{8\pi^2 m}\frac{1}{Z}\frac{d^2Z}{dz^2} = -E_z \tag{7.27}$$

Now Eq. (7.25) may be expressed as

$$\frac{h^2}{8\pi^2 m}\cdot\frac{d^2X}{dx^2} = -E_x$$

$$\text{or} \quad \frac{d^2X}{dx^2} + \frac{8\pi^2 m}{h^2}E_x X = 0 \tag{7.28}$$

This equation is same as the equation for particle in a one-dimensional box.

Equation (7.28) is the second-order differential equation, the solution of which is given by

$$X = \left(\frac{2}{a}\right)^{1/2}\sin\frac{n_x \pi}{a}x \tag{7.29}$$

and $E_x = \frac{(n_x^2 h^2)}{8ma^2}$

Similarly, for Eqs. (7.26) and (7.27), we write the solutions as

$$Y = \left(\frac{2}{a}\right)^{1/2} \sin\frac{n_y \pi}{b} y \tag{7.30}$$

$$Z = \left(\frac{2}{a}\right)^{1/2} \sin\frac{n_z \pi}{c} z \tag{7.31}$$

Their energy will be written as

$$E_y = \frac{(n_y^2 h^2)}{8mb^2} \text{ and that of } E_z = \frac{(n_z^2 h^2)}{8mc^2}$$

By remembering $\psi = XYZ$ and by replacing the value of X, Y, and Z, we get

$$\text{or } \psi = \left(\frac{2}{a}\right)^{1/2} \sin\frac{n_x \pi}{a} x \cdot \left(\frac{2}{b}\right)^{1/2} \sin\frac{n_y \pi}{b} y \cdot \left(\frac{2}{c}\right)^{1/2} \sin\frac{n_z \pi}{c} z$$

$$\text{or } \psi = \left(\frac{2}{a}\right)^{1/2} \cdot \left(\frac{2}{b}\right)^{1/2} \cdot \left(\frac{2}{c}\right)^{1/2} \sin\frac{n_x \pi}{a} x \cdot \sin\frac{n_y \pi}{b} y \cdot \sin\frac{n_z \pi}{c} z$$

$$\text{or } \psi = \left(\frac{8}{abc}\right)^{1/2} \cdot \sin\frac{n_x \pi}{a} x \cdot \sin\frac{n_y \pi}{b} y \cdot \sin\frac{n_z \pi}{c} z$$

$$\text{or } \psi = \left(\frac{8}{v}\right)^{1/2} \cdot \sin\frac{n_x \pi}{a} x \cdot \sin\frac{n_y \pi}{b} y \cdot \sin\frac{n_z \pi}{c} z \tag{7.32}$$

where $a \cdot b \cdot c = v$.

Also recalling that

$$E = E_x + E_y + E_z$$

$$\therefore \ E = \frac{h^2}{8m}\left[\frac{n_x^2}{a^2} + \frac{n_y^2}{b^2} + \frac{n_z^2}{c^2}\right] \tag{7.33}$$

This is the energy of a particle in a three-dimensional box.

It is clear from Eq. (7.33) that there are thus three quantum numbers required to specify the energy of a particle. Let us suppose a particular case where $a = b = c$ (i.e., cubic box), and if this is put in Eq. (7.33), we shall get

$$E = \frac{h^2}{8ma^2}\left[n_x^2 + n_y^2 + n_z^2\right]$$

$$= \frac{h^2}{8ma^2} \Sigma n^2 \tag{7.34}$$

It is clear from Eq. (7.34) that the energy of the system depends on the sum of the squares of three quantum numbers.

7.2.1 Energy Levels for a Cubic Potential Box

We can find the energy levels by putting different values of *nx*, *ny*, and *nz* in Eq. (7.34).

Now we can construct the energy levels using Table 7.1. It is shown in Figure 7.4.

The following inferences are drawn from Eq. (7.34).

- Lesser the value of m of a particle, greater will be its energy, E.
- Lesser the size of the box, greater will be its energy, E.

Equation (7.34) also gives a clue that degeneracy will arise. Degeneracy means the number of equivalent energy states. It is shown in Table 7.1.

It is pertinent to note that for a minimum value of *nx*, *ny*, and *nz*, the energy will be minimum and its value $3h^2/8ma^2$. Thus, $3h^2/8ma^2$ is called the ZPE.

TABLE 7.1

Degenerate Energy Levels for a Particle in a Box

n_x	n_y	n_z	$\sum n^2$	Energy	Remarks
1	1	1	3	$3h^2/8ma^2$	Non-degenerate state
1	1	2	6		
1	2	1	6	$6h^2/8ma^2$	Triply degenerate state
2	1	1	6		
2	2	1	9		
2	1	2	9	$9h^2/8ma^2$	Triply degenerate state
1	2	2	9		
3	1	1	11		
1	3	1	11	$11h^2/8ma^2$	Triply degenerate state
1	1	3	11		
2	2	2	12	$12h^2/8ma^2$	Non-degenerate state

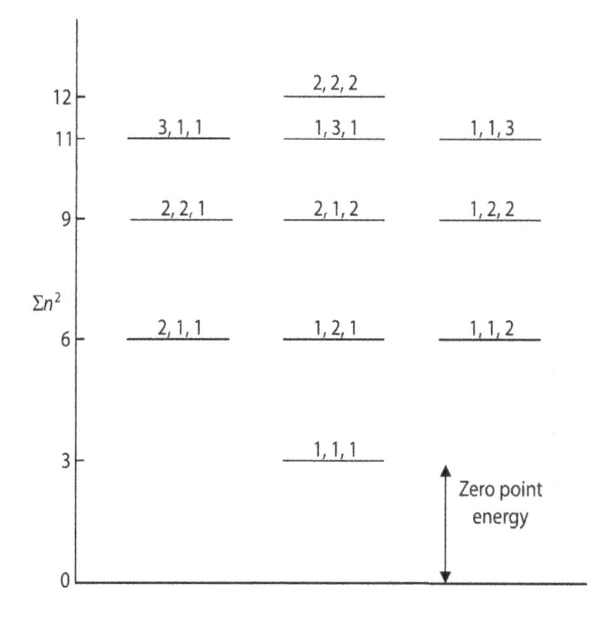

FIGURE 7.4 Degenerate energy levels in cubic box of length 'a'.

7.2.2 The Tunnel Effect or Tunnelling

The passage of particle (such as electrons) through a potential barrier is normally called the tunnelling or tunnel effect. In this effect, the particles leak through a narrow potential barrier, which constitutes a forbidden zone, provided the electrons are assumed to be classical particles. It is also a fact that there is a finite probability of electron/particle leaking/passing from one classically allowed zone to another, which arises as a consequence of quantum mechanics.

Actually, when the particle passes from one region to another through a barrier of finite thickness, some of it is reflected and some transmitted, and finally, when it reaches a third zone, it becomes wave-like, which may be seen in Figure 7.5. Thus, the tunnel effect is defined as follows.

> The passage of particle (such as electron) through a classically forbidden zone of a barrier having finite thickness, emerges from the remote side and becomes wave-like again is called the tunnelling or tunnel effect.

Now, we are going to study the tunnel effect in more detail. Let us suppose that the potential barrier has a finite thickness, as illustrated in Figure 7.5. Also, suppose that the mass of the particle leaking into the potential barrier has mass m.

The potential energy V has the following form:

- Zone 1 ($x < 0$): $V = 0$
- Zone 2 ($0 < x < a$): $V = V$ and $V > E$
- Zone 3 ($x > a$): $V = 0$

Under the above boundary condition, the Schrödinger equation takes the following forms in three different zones.

$$\text{Zone 1} \quad \frac{d^2\psi_1}{dx^2} + \frac{2mE}{\hbar^2}\psi_1 = 0 \tag{7.35}$$

$$\text{or,} \quad \frac{d^2\psi_1}{dx^2} + k_1^2\psi_1 = 0; \ k_1^2 = \frac{2mE}{\hbar^2}$$

$$\text{Zone 2} \quad \frac{d^2\psi_2}{dx^2} + \frac{2m}{\hbar^2}(V - E) = 0 \tag{7.36}$$

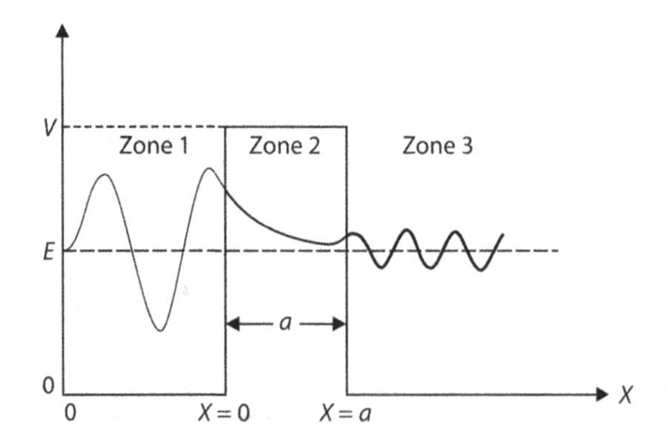

FIGURE 7.5 Particle confined behind a potential barrier of finite height and thickness.

$$\text{or,}\quad \frac{d^2\psi_2}{dx^2} + k_2^2\psi_2 = 0;\ k_2^2 = \frac{2mE}{\hbar^2}(V - E)$$

$$\text{Zone 3}\quad \frac{d^2\psi_3}{dx^2} + \frac{2m}{\hbar^2}\psi_3 = 0 \tag{7.37}$$

$$\text{or,}\quad \frac{d^2\psi_3}{dx^2} + k_2^2\psi_3 = 0;\ k_3^2 = \frac{2mE}{\hbar^2}$$

The solutions of the above differential equations may be expressed as

$$\text{Zone 1}\quad \psi_1 = C_1e^{ik_1x} + D_1e^{-ik_1x} \tag{7.38}$$

$$\text{Zone 2}\quad \psi_2 = C_2e^{ik_2x} + D_2e^{-ik_2x} \tag{7.38}$$

$$\text{Zone 3}\quad \psi_3 = C_3e^{ik_3x} + D_3e^{-ik_3x} \tag{7.39}$$

It has already been stated that in the second zone, $V > E$ and in the third zone, $E > V$, and same condition applies for the first zone, i.e., $E > V$, then K_1 and K_3 will both be real. Thus, ψ_1 and ψ_3 are oscillating functions in which C terms indicate particles moving from left to right, and D terms represent particles travelling from right to left.

Suppose the particles are incident on the barrier only from Zone 1, the situation is that the particles approach from left (the C_1 term), but most of the particles will be reflected back into Zone 1 (the D_1 term), whereas some particles may penetrate the barrier and continue moving to the right in Zone 3 (the C_3 term). Since no particles approach the barrier from right in Zone 3, then D_3 must be equal to zero.

Now the probability of finding a particle in Zone 3 in comparison to the probability of finding an incident particle in Zone 1 will be equal to $C_3C_3^*/C_1C_1^*$, which denotes the probability of penetration of the barrier. Further, we want to arrive at some notion of the factors that affect this probability, and for this, a relation between C_1 and C_3 must be established. This is done by considering the restrictions, which should be imposed on ψ_1, ψ_2, and ψ_3. These conditions are as follows:

$$\text{At } x = 0,\ \psi_1 = \psi_2 \text{ and } \frac{d\psi_1}{dx} = \frac{d\psi_2}{dx}$$

$$\text{At } x = a,\ \psi_2 = \psi_3 \text{ and } \frac{d\psi_2}{dx} = \frac{d\psi_3}{dx}$$

Now we shall apply this condition to Eqs. (7.38), (7.39), and (7.40), remembering that $D_3 = 0$.

For $\psi_1 = \psi_2$ at $x = 0$,

$$C_1 + D_1 = C_2 + D_2 \tag{7.41}$$

For $\frac{d\psi_1}{dx} = \frac{d\psi_2}{dx}$ at $x = 0$,

$$k_1(C_1 - D_1) = k_2(C_2 - D_2) \tag{7.42}$$

For $\psi_2 = \psi_3$ at $x = a$,

$$C_2e^{ik_2a} + D_2e^{-ik_2a} = C_3e^{ik_3a} \tag{7.43}$$

For $\frac{d\psi_2}{dx} = \frac{d\psi_3}{dx}$ at $x = a$,

$$k_2\left(C_2e^{ik_2a} - D_2e^{-ik_2a}\right) = k_3C_3e^{ik_3a} \tag{7.44}$$

From Eq. (7.41),

$$C_1 = C_2 + D_2 - D_1 \tag{7.45}$$

And from Eq. (7.42), $(C_1 - D_1) = \frac{k_2}{k_1}(C_2 - D_2)$

$$\text{or } D_1 = C_1 - \frac{k_2}{k_1}(C_2 - D_2) \tag{7.46}$$

Putting the value of D_1 in Eq. (7.45), we get

$$C_1 = C_2 + D_2 - C_1 + \frac{k_2}{k_1}(C_2 - D_2)$$

$$\text{or } 2C_1 = C_2 + D_2 + \frac{k_2}{k_1}(C_2 - D_2)$$

$$= C_2\left(1 + \frac{k_2}{k_1}\right) + D_2\left(1 - \frac{k_2}{k_1}\right)$$

$$\text{or } C_1 = \frac{C_2}{2}\left(1 + \frac{k_2}{k_1}\right) + \frac{D_2}{2}\left(1 - \frac{k_2}{k_1}\right) \tag{7.47}$$

From Eq. (7.43), we have

$$C_2e^{ik_2a} + D_2e^{-ik_2a} = C_3e^{ik_3a}$$

$$\text{or } C_2e^{ik_2a} = C_3e^{ik_3a} - D_2e^{-ik_2a}$$

$$\therefore C_2 = \frac{1}{e^{ik_2a}}\left(C_3e^{ik_3a} - D_2e^{-ik_2a}\right)$$

$$\text{or } C_2 = \left(C_3e^{ik_3a} - D_2e^{-ik_2a}\right)\cdot e^{ik_2a} \tag{7.48}$$

From Eq. (7.44), we have

$$k_2\left(C_2e^{ik_2a} - D_2e^{-ik_2a}\right) = k_3C_3e^{ik_3a}$$

$$\therefore \left(C_2e^{ik_2a} - D_2e^{ik_2a}\right) = \frac{k_3}{k_2}C_3e^{ik_3a}$$

$$\text{or } C_2 = \left(\frac{k_3}{k_2}C_3e^{ik_3a} + D_2e^{-ik_2a}\right)e^{-ik_2a} \tag{7.49}$$

Equating Eqs. (7.48) and (7.49), we obtain

$$\left(C_3e^{ik_3a} - D_2e^{-ik_2a}\right)\cdot e^{-ik_2a} = \left(\frac{k_3}{k_2}C_3e^{ik_3a} + D_2e^{-ik_2a}\right)e^{-ik_2a} \tag{7.50}$$

$$\text{or } C_3e^{ik_3a} - D_2e^{-ik_2a} = \frac{k_3}{k_2}C_3e^{ik_3a} + D_2e^{-ik_2a}$$

$$\text{or } C_3e^{ik_3a} - \frac{k_3}{k_2}C_3e^{ik_3a} = 2\ D_2e^{-ik_2a}$$

$$\text{or } C_3e^{ik_3a}\left(1 - k_3/k_2\right) = 2\ D_2e^{-ik_2a}$$

$$\therefore\ D_2 = \frac{1}{2}C_3\left(1 - k_3/k_2\right)e^{ik_3a}\cdot e^{-ik_2a} \tag{7.51}$$

Substituting the value of D_2 of Eq. (7.51) into Eq. (7.48), we get

$$C_2 = \left[C_3e^{ik_3a} - \left\{\frac{C_3}{2}\left(1 - \frac{k_3}{k_2}\right)e^{ik_3a}\cdot e^{ik_2a}\cdot e^{-ik_2a}\right\}\right]e^{-ik_2a}$$

$$\text{or } C_2 = C_3e^{ik_3a}\cdot e^{-ik_2a} - \left\{\frac{C_3}{2}\left(1 - \frac{k_3}{k_2}\right)e^{ik_3a}\cdot e^{ik_2a}\cdot e^{-ik_2a}\cdot e^{-ik_2a}\right\}$$

$$= C_3e^{ik_3a}\cdot e^{-ik_2a} - \left\{\frac{C_3}{2}\left(1 - \frac{k_3}{k_2}\right)e^{ik_3a}\cdot e^{ik_2a}\right\}$$

$$= C_3e^{ik_3a}\cdot e^{-ik_2a}\left[1 - \frac{1}{2}\left\{1 - \frac{k_3}{k_2}\right\}\right]$$

$$\text{or } C_2 = \left[1 - \frac{1}{2} + \frac{1}{2}\frac{k_3}{k_2}\right]\cdot C_3e^{ik_3a}\cdot e^{-ik_2a}$$

$$\text{or } C_2 = \left[\frac{1}{2} + \frac{1}{2}\frac{k_3}{k_2}\right]\cdot C_3e^{ik_3a}\cdot e^{-ik_2a}$$

$$\text{or } C_2 = \frac{1}{2}\left(1 + \frac{k_3}{k_2}\right)\cdot C_3e^{ik_3a}\cdot e^{-ik_2a} \tag{7.52}$$

Now we shall substitute the values of C_2 and D_2 from Eqs. (7.52) and (7.51) into Eq. (7.47), and we get

$$C_1 = \frac{C_2}{2}\left(1 + \frac{k_2}{k_1}\right) + \frac{D_2}{2}\left(1 - \frac{k_2}{k_1}\right)$$

$$= \frac{1}{2}\left(1 + \frac{k_2}{k_1}\right)\left[\frac{1}{2}\left(1 + \frac{k_3}{k_2}\right)C_3e^{ik_3a}\cdot e^{-ik_2a}\right] + \frac{1}{2}\left(1 - \frac{k_2}{k_1}\right)\left[\frac{1}{2}\left(1 - \frac{k_3}{k_2}\right)C_3e^{ik_3a}\cdot e^{-ik_2a}\right]$$

$$\text{or } C_1 = \frac{1}{4}C_3e^{ik_3a}\left[\left(1 + \frac{k_2}{k_1}\right)\left(1 + \frac{k_3}{k_2}\right)e^{-ik_2a} + \left(1 - \frac{k_2}{k_1}\right)\left(1 - \frac{k_3}{k_2}\right)e^{-ik_2a}\right]$$

$$= \frac{1}{4}C_3e^{ik_3a}\left[\left(1 + \frac{k_3}{k_1}\right)e^{-ik_2a} + \left(1 + \frac{k_3}{k_1}\right)e^{ik_2a} + \left(\frac{k_3}{k_2} + \frac{k_2}{k_1}\right)\left(e^{ik_2a} - e^{-ik_2a}\right)\right]$$

$$= \frac{1}{4}C_3e^{ik_3a}\left[\left(1 + \frac{k_3}{k_1}\right)(e^{-ik_2a} + e^{ik_2a}) - \left(\frac{k_3}{k_2} + \frac{k_2}{k_1}\right)\left(e^{ik_2a} - e^{-ik_2a}\right)\right]$$

$$= \frac{1}{4}C_3e^{ik_3a}\left[2\left(1 + \frac{k_3}{k_1}\right)\cos h(ik_2a) - 2\left(\frac{k_3}{k_2} + \frac{k_2}{k_1}\right)\sin h(ik_2a)\right]$$

$$\because \left[\cos hx = \frac{e^{ix} + e^{-ix}}{2} \text{ and } \sin hx = \frac{e^{ix} - e^{-ix}}{2}\right]$$

$$\text{or } C_1 = \frac{1}{2} C_3 e^{ik_3 a}\left[\left(1 + \frac{k_3}{k}\right)\cos h(ik_2 a) - \left(\frac{k_3}{k_2} + \frac{k_2}{k_1}\right)\sin h(ik_2 a)\right] \tag{7.53}$$

When $V_2 > E > V_{1/2}$, k_2 will be an imaginary quantity and ik_2 will be a real quantity.

Let us put $ik_2 = k_2'$ and substituting this in Eq. (7.53), we obtain

$$C_1 = \frac{1}{2} C_3 e^{ik_3 a}\left[\left(1 + \frac{k_3}{k_1}\right)\cos h(k_2' a) - \left(\frac{ik_3}{k_2'} + \frac{k_2'}{ik_1}\right)\sin h(k_2' a)\right]$$

$$\text{or } \frac{C_1}{C_3} = \frac{1}{2} e^{ik_3 a}\left[\left(1 + \frac{k_3}{k_1}\right)\cos h(k_2' a) - i\left(\frac{k_3}{k_2'} + \frac{k_2'}{k_1}\right)\sin h(k_2' a)\right] \tag{7.54}$$

Now we are interested in finding $C_1 C_1^* / C_3 C_3^*$, where C_1^* is the complex conjugate of C_1 and C_3^* is the complex conjugate of C_3.

Obviously, $C_1 C_1^* / C_3 C_3^*$ will be written as

$$\frac{C_1 C_1^*}{C_3 C_3^*} = \frac{1}{4}\left[\left(1 + \frac{k_3}{k_1}\right)^2 \cos h^2(k_2' a) - 2\left(\frac{k_3}{k_2'} - \frac{k_2'}{k_1}\right)^2 \sin h(k_2' a)\right]$$

$$\left[\because \ (a + ib)(a - ib) = a^2 + b^2\right] \tag{7.55}$$

The probability of transmission (T) of a particle through the barrier may be written as $C_3 C_3^* / C_1 C_1^*$, i.e., $T = C_3 C_3^* / C_1 C_1^*$, which is the reciprocal of the quantity of Eq. (7.55). It should be kept in mind that $T \neq 0$ unless $C_3 C_3^* / C_1 C_1^* = 0$. This situation is obtained when $\cos(k_2' a)$ and $\sin(k_2' a) = \infty$. It may be said that $C_1 C_1^* / C_3 C_3^*$ will be only infinite when $(k_2' a) \to \infty$. Since $k_2' a = \left[2m(V - E)/\hbar\right]^{1/2}$, the condition for zero probability of transmission through the barrier is $a\left[2m(V - E)\right]^{1/2} / \hbar = \infty$, and this condition will only be obtained when $V = \infty$ or $a = \infty$.

It means that the particle will penetrate to the potential barrier only and only if $E < V$ or the potential barrier is infinitely thick, and this effect is known as the *tunnel effect.*

Now let us suppose that $k_2' a \gg 1$. Under this condition, $\cos h(k_2' a) \approx \sin h\ (k_2' a) \approx e^{2k_2' a'}/2$. Putting this value in Eq. (7.55), we shall get

$$C_1 C_1^* / C_3 C_3^* = \frac{1}{4}\left[\left(1 + \frac{k_3}{k_1}\right)^2 \cdot \frac{1}{4} e^{2k_2' a} + \left(\frac{k_3}{k_2'} - \frac{k_2'}{k_1}\right)^2 \cdot \frac{1}{4} e^{2k_2' a}\right]$$

$$= \frac{e^{2k_2' a}}{16}\left[\left(1 + \frac{k_3}{k}\right)^2 + \left(\frac{k_3}{k_2'} - \frac{k_2'}{k_1}\right)^2\right]$$

Reciprocal of the above gives

$$\frac{C_3 C_3^*}{C_1 C_1^*} = T = \frac{16}{\left(1 + \frac{k_3}{k_1}\right)^2 + \left(\frac{k_3}{k_2'} - \frac{k_2'}{k_1}\right)^2} e^{-2k_2' a} \tag{7.56}$$

It is clear from the above equation that the probability of transmission decreases rapidly as the width of the barrier increases because of the presence of exponential term in Eq. (7.56). The exponential term $e^{-2k_2'a}$ is sometimes referred to as the *transparency factor.* The other important point to be noted is that T increases when transparency factor increases, i.e., as $-2k_2'a$ decreases, and the probability of transmission will also be greater when a, m, and $(V - E)$ will be smaller.

7.2.3 Importance of Tunnel Effect

The tunnel effect provides explanations for the following phenomena:

- The electrical breakdown of insulators.
- The switching action of a tunnel diode.
- The reverse breakdown of semi-conductor diode.
- The emission of α-particles from radio elements.
- In the explanation of oxidation-reduction reactions and in electrode reactions. In electrode reactions, the electrons must travel from one atom or molecule to another across a phase boundary.
- In chemical kinetics because the probability factor is related to the probability of reactants crossing the energy barrier.

7.2.4 Quantum Mechanical Explanation of Emission of α-Particles

It was initially suggested by G. Gammon in 1928 and independently by Gurney and Condon in 1929 regarding the explanation of empirical observations summarised in the Geiger–Nuttall law of radioactive disintegration. According to the theory of emission of α-particles from the radio elements, there is a probability of preformation of α-particles in the nucleus before its emission. The probability of preformation varies from 0.1 to 1, but the value approaches unity in case of even-Z and even-N nuclei. The potential energy of an α-particle nearby nucleus is illustrated in Figure 7.6.

It has been found by experiments that the α-particle is held by nucleus by very high binding forces. The kinetic energy of emitted α-particles has been found to be 4.2 MeV by experiment, whereas the potential energy has been found to be 8.78 MeV. These values have been obtained by carrying out experiments with U^{238} nucleus. The observation shows that the coulomb law is obeyed at least up to 8.78 MeV. The kinetic energy of α-particle is found to be too small to allow classical penetration. Here $E < V$, and there is a chance of tunnelling of α-particles through the potential barrier. It is because of the fact that a moving α-particle is regarded as a wave and it has small but finite probability of transmission through the potential barrier.

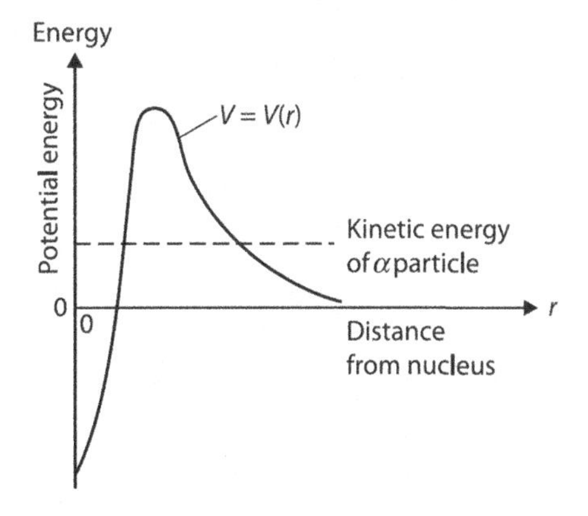

FIGURE 7.6 Potential energy diagram for nuclear binding forces holding an α-particle in the nucleus of an atom ($V(r)$ portion having positive slope indicates nuclear attraction, and portion with a negative slope denotes coulombic repulsion).

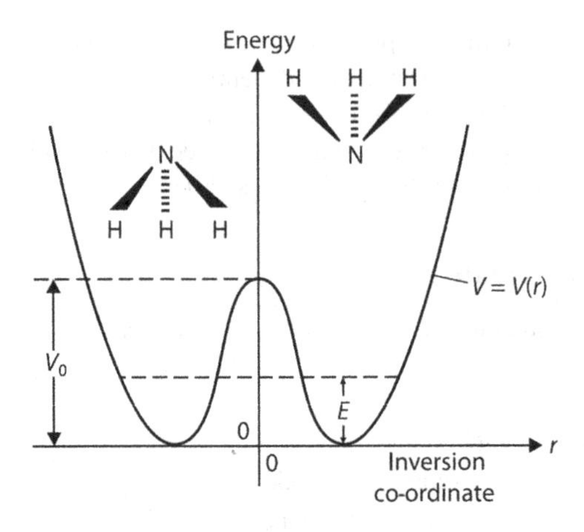

FIGURE 7.7 Umbrella inversion of ammonia.

One important point to be put forth is the umbrella inversion of pyramidal molecules, such as NH_3, PH_3, and AsH_3. The umbrella inversion takes place as a consequence of quantum mechanical tunnelling.

In such a case, proton of NH_3 and PH_3 or AsH_3 tunnel through the potential barrier keeping the atoms in pyramidal structure/shape. Potential energy vs. r diagram is illustrated in Figure 7.7. It is expected that in such cases, the transmission coefficient decreases as the mass of the heavy atom increases. There are some examples where the intermolecular electron transfer takes place via a quantum tunnelling mechanism.

7.3 Particle on a Ring

Let us suppose that a particle (electron) having mass m is restricted to move on the circular track of a ring on which the potential energy V is constant. The constant value of potential energy may be conveniently taken as zero, i.e., $V = 0$. Also suppose that C = circumference of the ring and r = radius of the ring (Figure 7.8).

Now, we shall apply the Schrödinger equation for the particle moving on a ring, i.e.,

$$\frac{d^2\psi}{dx^2} + \frac{8\pi^2 m}{h^2}(E - V)\psi = 0$$

But $V = 0$; therefore, the above equation takes the form

$$\frac{d^2\psi}{dx^2} + \frac{8\pi^2 m}{h^2}E\psi = 0 \tag{7.57}$$

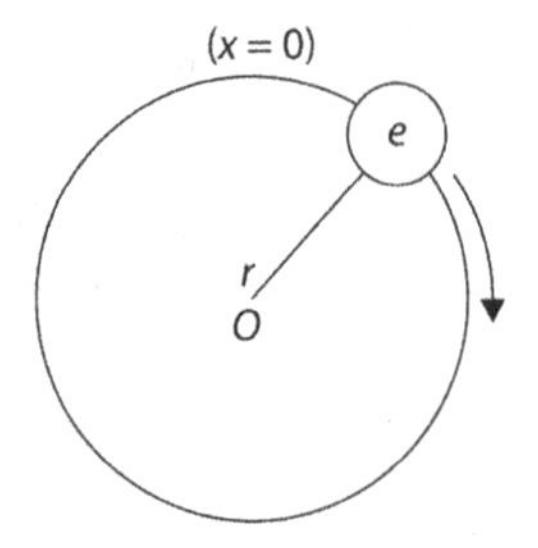

FIGURE 7.8 Particle moving on a ring.

An arbitrary point on the ring is selected as origin (i.e., $x = 0$), and the co-ordinate × varies along the circular track.

Since the wave function must be single-valued, we may write

$$\psi(x) = \psi(x + c), \text{ where } c = \text{circumference} \tag{7.58}$$

Equation (7.57) can be written in the form

$$\frac{d^2\psi}{dx^2} + K^2\psi = 0 \tag{7.59}$$

$$\text{where } K^2 = \frac{8\pi^2 mF}{h^2} \tag{7.60}$$

Equation (7.59) is a second-order differential equation, the general solution of which may be expressed as

$$\psi = A\sin Kx + B\cos Kx \tag{7.61}$$

where A and B are arbitrary constants.

The boundary conditions are here different from that of a one-dimensional box. Therefore, according to Eq. (7.58), we can write

$$\psi(0) = \psi(c) \tag{7.62}$$

Putting $x = 0$ in Eq. (7.61), we get

$$\psi(0) = A\sin Kx + B\cos Kx$$

$$\therefore \ \psi(0) = B = \psi(c) \tag{7.63}$$

$$\therefore \text{ Equation (7.61) takes the form}$$

$$B = A\sin Kc + B\cos Kc \tag{7.64}$$

Since the particle is moving on a ring with constant potential V, there is no discontinuity in V and consequently ψ and $d\psi/dx$ must be continuous.

Now since $\psi(0) = \psi(c)$

$$\therefore \ \left[\frac{d\psi}{dx}\right]_{x=0} = \left[\frac{d\psi}{dx}\right]_{x=c} \tag{7.65}$$

Differentiating Eq. (7.61) with respect to x, we get

$$\left[\frac{d\psi}{dx}\right] = AK\cos Kx - BK\sin Kx$$

$$\text{and} \left[\frac{d\psi}{dx}\right]_{x=0} = AK\cos K \times 0 - BK\sin K \times 0 = AK \tag{7.66}$$

$$\text{and} \ \left[\frac{d\psi}{dx}\right]_{x=c} = AK\cos Kc - BK\sin Kc \tag{7.67}$$

With the help of Eqs. (7.66) and (7.67), we can write

$$AK = AK\cos Kc - BK\sin Kc \tag{7.68}$$

Multiplying Eq. (7.64) by BK and Eq. (7.68) by A and then adding, we get

$$B^2K = ABK\sin Kc - B^2K\cos Kc$$

$$\underline{A^2K = A^2K\cos Kc - ABK\sin Kc}$$

$$\left(A^2 + B^2\right)k = A^2K\cos\ Kc + B^2K\cos Kc$$

$$\text{or } \left(A^2 + B^2\right)K = K\left(A^2 + B^2\right)\cos Kc$$

$$\text{or } \cos Kc = 1$$

$$\text{or } Kc = 2n\pi$$

$$\therefore\ Kc = \frac{2n\pi}{c},\ \text{ where, } n = 0,\ \pm 1,\ \pm 2,\ \pm 3 \tag{7.69}$$

since $\cos(-\theta) = \cos(\theta)$

Replacing the value of K in Eq. (7.61), we have

$$\psi = A\sin\frac{2n\pi}{c}x + B\cos\frac{2n\pi}{c}x \tag{7.70}$$

Using Eqs. (7.60) and (7.69), we can write

$$K^2 = \frac{4n^2\pi^2}{c^2} = \frac{8\pi^2 mE}{h^2}$$

$$\text{or } \frac{4n^2\pi^2}{c^2} = \frac{8\pi^2 mE}{h^2}$$

$$\text{or } E = \frac{n^2h^2}{2mc^2} \tag{7.71}$$

Putting $n = 0$, for which $E = 0$, Eq. (7.70) becomes $\psi(0) = B$

Therefore, there is a variation of ψ around the ring in the lowest state.

To find the value of A and B, normalisation is carried out, i.e., $\int_0^c \psi^2\, dx = 1.$

$$\text{or } \int_0^c \psi^2 dx = \int_0^c \left[A\sin\frac{2n\pi}{c}x + B\cos\frac{2n\pi}{c}x\right]^2 dx = 1$$

$$\text{or } \int_0^c \left[A^2\sin^2\frac{2n\pi}{c}x + B^2\cos^2\frac{2n\pi}{c}x + 2AB\sin\frac{2n\pi}{c}x\cdot\cos\frac{2n\pi}{c}x\right]dx = 1$$

$$\text{or } A^2\int_0^c \sin^2\frac{2n\pi}{c}x dx + \int_0^c B^2\cos^2\frac{2n\pi}{c}x + AB\int_0^c \sin\frac{4n\pi}{c}x dx = 1$$

The integration of the above yields

$$\left(A^2 + B^2\right)\frac{c}{2} = 1$$

$$A^2 + B^2 = \frac{2}{c}$$

$$\therefore \; A = \sqrt{\frac{2}{c}}\cos\alpha \;\text{ and }\; B = \sqrt{\frac{2}{c}}\sin\alpha \tag{7.72}$$

where α may have any value.

Therefore, the equation for wave function will be

$$\psi = \sqrt{\frac{2}{c}}\cos\,\alpha\,\sin\frac{2n\pi}{c}x + \sqrt{\frac{2}{c}}\sin\alpha\cos\frac{2n\pi}{c}x$$

$$\text{or } \psi = \sqrt{\frac{2}{c}}\sin\left(\frac{2n\pi}{c}x + a\right) \tag{7.73}$$

7.3.1 Particle on a Ring (Considering the Spherical Polar Co-Ordinates)

Let us consider a particle having mass m moving on the circumference c of a ring whose radius is r with a constant potential energy V. The value of V may be taken as zero for the simplicity sake.

For this present problem, it is simpler to transform the Cartesian co-ordinates x, y, and z by spherical polar co-ordinates. The relation between the two types of co-ordinates is shown in Figure 7.9. In the figure, the position of the particle may be defined by the length r, i.e., the distance of the particle from the origin together with angles θ and ϕ.

From Figure 7.9, we may write

$$x = r\sin\theta\cos\phi$$

$$y = r\sin\theta\sin\phi$$

$$z = r\cos\theta$$

With the help of these, the value of ∇^2 in terms of polar co-ordinates can be expressed as

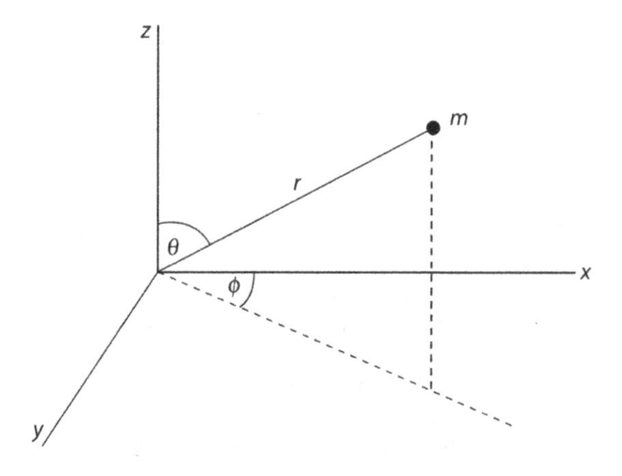

FIGURE 7.9 Cartesian and polar co-ordinates.

$$\nabla^2 = \frac{d^2}{dr^2} + 2\frac{1}{r}\frac{d}{dr} + \frac{1}{r^2 \sin\theta}\frac{d}{d\theta}\left(\sin\theta\frac{d}{d\theta}\right) + \frac{1}{r^2 \sin^2\theta}\cdot\frac{d^2}{d\phi^2} \tag{7.74a}$$

The present problem deals with the movement of the particle on the circular track of the ring, which may be supposed to lie in the xy plane in which $\theta = \pi/2$ and hence $\sin\theta = 1$. As it is a case of circular path having radius r, hence it will be constant. Since θ and r are constants in the present case, $d/d\theta$ and d/dr will be zero. Consequently, the first three terms of Eq. (7.74a) will disappear.

$$\nabla^2 = \frac{1}{r^2}\frac{d^2}{d\phi^2} \tag{7.74b}$$

Now, the Schrödinger equation is applied to the system, i.e.,

$$\nabla^2\psi + \frac{8\pi^2 m}{h^2}(E - V)\psi = 0 \tag{7.75}$$

$$\text{or } \nabla^2\psi + \frac{8\pi^2 mE}{h^2}\psi = 0 \quad [\therefore V = 0]$$

Replacing the value of ∇^2 from Eq. (7.74b), we shall get

$$\frac{1}{r^2}\frac{d^{2\psi}}{d\phi^2} + \frac{8\pi^2 mE}{h^2}\psi = 0 \tag{7.76}$$

$$\text{or } \frac{d^{2\psi}}{d\phi^2} + \frac{8\pi^2 mr^2 E}{h^2}\psi = 0$$

$$\text{or } \frac{d^{2\psi}}{d\phi^2} + \frac{8\pi^2 IE}{h^2}\psi = 0 \quad [\because mr^2 = I]$$

where I = moment of inertia

$$\text{or } \frac{d^{2\psi}}{d\phi^2} + M^2\psi = 0 \quad \left[\text{where } \frac{8\pi^2 IE}{h^2} = M^2\right] \tag{7.77}$$

$$\therefore E = M^2h^2/8\pi^2 I \quad \text{where } M = 0, \pm 1, \pm 2, \ldots$$

This is a second-order differential equation, the general solution of which will be given by

$$\psi = A\sin M\phi + B\cos M\phi \tag{7.78}$$

$$\text{and } \psi = C\,e^{iM\phi} + De^{-iM\phi} \tag{7.79}$$

Now we shall fix the boundary condition. For fixing the boundary condition, we have to think over the continuous motion of the particle. Since there is no barrier to the motion of the particle so long, it remains on the circular ring, and there is no need for the wave function to vanish at any point on the ring. It is known to us that the wave function must be single valued, so we can write

$$\Phi(2\pi + \phi) = \Phi(\phi) \tag{7.80}$$

In other words, we can say that when the particle remains on the ring, the variable is the angle ϕ, which can vary from 0 to 2π.

Now, we apply the normalisation condition, i.e.,

$$\int \psi\psi^* \mathrm{d}\tau = 1$$

In our case, we can write

$$\text{or} \quad \int_0^{2\pi} (A\sin M\phi + B\cos M\phi)^2 \, \mathrm{d}\phi = 1, \quad \text{Because } \psi\psi^* \sim \psi^2$$

$$\text{or} \quad \int_0^{2\pi} \left(A^2 \sin^2 M\phi + B^2 \cos^2 M\phi + 2AB\sin M\phi \cos M\phi\right) \mathrm{d}\phi = 1$$

$$\text{or} \quad \int_0^{2\pi} \left[\frac{A^2}{2}(1 - \cos 2M\phi) + \frac{B^2}{2}(1 + \cos 2M\phi) + AB\sin 2M\phi\right] \mathrm{d}\phi = 1$$

$$\text{or} \quad \int_0^{2\pi} \left[\frac{A^2}{2} - \frac{A^2}{2}\cos 2M\phi + \frac{B^2}{2} + \frac{B^2}{2}\cos 2M\phi + AB\sin 2M\phi\right] \mathrm{d}\phi = 1 \tag{7.81}$$

$$\text{or} \quad \int_0^{2\pi} \frac{A^2}{2} \mathrm{d}\phi - \int_0^{2\pi} \frac{A^2}{2} \cos 2M\phi \mathrm{d}\phi + \int_0^{2\pi} \frac{B^2}{2} \mathrm{d}\phi + \int_0^{2\pi} \frac{B^2}{2} \cos 2M\phi \mathrm{d}\phi + AB\int_0^{2\pi} \sin 2M\phi \mathrm{d}\phi = 1$$

The integration of sine and cosine functions between 0 and 2π gives zero value.

$$\text{or} \quad \left[\frac{A^2}{2}\phi + \frac{B^2}{2}\phi\right]_0^{2\pi} = 1$$

$$\text{or} \quad \frac{A^2}{2} \cdot 2\pi + \frac{B^2}{2} \cdot 2\pi = 1$$

$$\text{or} \quad A^2\pi + B^2\pi = 1 \tag{7.82}$$

Equation (7.82) is satisfied by the following relations:

$$A = \frac{1}{\sqrt{\pi}}\cos\alpha \text{ and } B = \frac{1}{\sqrt{\pi}}\sin\alpha$$

$$\therefore \quad \psi = \frac{1}{\sqrt{\pi}}\cos\alpha \sin M\phi + \frac{1}{\sqrt{\pi}}\sin\alpha \cos M\phi$$

$$\text{or} \quad \psi_s = \frac{1}{\sqrt{\pi}}\sin(M\phi + \alpha) \tag{7.83}$$

where s represents ψ in terms of a sine function.

Equation (7.82) is also satisfied by

$$A = -\frac{1}{\sqrt{\pi}}\sin\alpha \quad \text{and} \quad B = \frac{1}{\sqrt{\pi}}\cos\alpha$$

$$\therefore \quad \psi = -\frac{1}{\sqrt{\pi}}\sin\alpha \sin M\phi + \frac{1}{\sqrt{\pi}}\cos\alpha \cos M\phi$$

$$\therefore \quad \psi_c = \frac{1}{\sqrt{\pi}}\cos(M\phi + \alpha) \tag{7.84}$$

where c represents ψ in terms of cosine function.

Thus, it is clear that ψ_s and ψ_c are normalised but they are orthogonal also as

$$\int_0^{2\pi}\left[\frac{1}{\sqrt{\pi}}\sin(M_1\phi + \alpha)\right]\left[\frac{1}{\sqrt{\pi}}\cos(M_2\phi + \alpha)\right]\mathrm{d}\phi = 0,$$

where M_1 and M_2 are different values of the quantum number M, and α shifts the origin around the ring. Under the above condition, ψ_s and ψ_c are orthonormal.

It should be kept in mind that integration of Eq. (7.81) is not valid for $M = 0$ since the integration of the trigonometric terms amounts to coefficients of $1/2M$, which will be infinite if $M = 0$. For normalisation of the trigonometric solution to the wave equation for the particular value of $M = 0$, when $M = 0$ is put in equation $\psi = A\sin M\phi + B\cos M\phi$ to yield $\psi = B$, it can be normalised by writing

$$\int_0^{2\pi} B^2\mathrm{d}\phi = 1$$

$$\text{or} \quad \left[B^2\phi\right]_0^{2\pi} = 0$$

$$\text{or} \quad B^2 \cdot 2\pi = 1$$

$$\text{or} \quad B^2 = \frac{1}{2\pi}$$

$\therefore \; B = \left(\frac{1}{2\pi}\right)^{1/2}$ so that when $M = 0$, the ψ function has the constant value $\left(\frac{1}{2\pi}\right)^{1/2}$.

Our interest at present regarding the particle on a ring system is the angular momentum of the system. Let us now think over Eq. (7.79), which consists of C and D terms representing the motion in one direction only. For this, we restrict to the positive values of M, separating the two terms and express as

$$\psi_+ = Ce^{iM\phi} \tag{7.85}$$

$$\text{and} \quad \psi_- = De^{iM\phi} \tag{7.86}$$

Now we normalise the wave function ψ_+ by putting

$$\int_0^{2\pi} \psi_+\psi_+^*\mathrm{d}\phi = 1$$

$$\text{or} \quad \int_0^{2\pi}\left(Ce^{iM\phi}\right)\left(C^*e^{iM\phi}\right)\mathrm{d}\phi = 1$$

$$\text{or} \quad CC^*\int_0^{2\pi}\mathrm{d}\phi = 1$$

$$\text{or} \quad CC^*\left[\phi\right]_0^{2\pi} = 1$$

$$\text{or} \quad CC^* \cdot 2\pi = 1$$

$$\text{or } C^2 \cdot 2\pi = 1 \quad \left[\because\ CC^* \approx C^2\right]$$

$$\therefore \quad C = \left(\frac{1}{2\pi}\right)^{1/2}$$

Similarly, normalisation of Eq. (7.86) yields

$$D = \left(\frac{1}{2\pi}\right)^{1/2}$$

Putting the values of C and D in Eqs. (7.85) and (7.86), we get

$$\psi_+ = \left(\frac{1}{2\pi}\right)^{1/2} e^{iM\phi} \tag{7.87}$$

$$\psi_- = \left(\frac{1}{2\pi}\right)^{1/2} e^{-iM\phi} \tag{7.88}$$

These normalised solutions are orthogonal also as

$$\int_0^{2\pi} \left(\left(\frac{1}{2\pi}\right)^{1/2} e^{iM\phi}\right)\left(\left(\frac{1}{2\pi}\right)^{1/2} e^{-iM\phi}\right) d\phi = 0$$

If we shift the origin on the ring by an angle α, we can write

$$\psi_+ = \left(\frac{1}{\sqrt{2\pi}}\right)^{1/2} e^{i(M\phi+\alpha)} \quad \text{or}$$

$$\psi_+ = \left(\frac{1}{\sqrt{2\pi}}\right)^{1/2} e^{iM\phi} \quad \text{where } \alpha = 0 \text{ for simplicity sake.}$$

Now see Eq. (7.87) in which if $M = +1$ becomes similar to Eq. (7.88) in which $M = -1$. These could be combined as

$$\psi_\pm = \frac{1}{\sqrt{2\pi}} e^{\pm iM\phi} \tag{7.89}$$

Further, it would be possible to find out the angular momentum of the particle by operating $\psi_\pm$ with the angular momentum operator. It should be kept in mind that the ring on which the particle roams has been considered to lie in the XY plane so that the particle will have an angular momentum about the Z axis.

The operator for angular momentum about Z axis is given in polar co-ordinates by $(h/2\pi i)(d/d\phi)$, which operates on the function $\psi_\pm$ and yields

$$\frac{h}{2\pi i} \cdot \frac{d}{dx}(\psi_\pm) = \frac{h}{2\pi i} \cdot \frac{d}{dx}\left(\frac{1}{\sqrt{2\pi}} e^{iM\phi}\right)$$

$$= \frac{h}{2\pi i} \cdot \frac{d}{dx}\left(\frac{1}{\sqrt{2\pi}} e^{iM\phi}\right)$$

$$= \frac{Mh}{2\pi} \cdot \psi_\pm$$

$= L_z\psi_{\pm}$, where L_z is the angular momentum about the Z axis, which is quantised in the units of $h/2\pi$.

Now consider the equation

$$\psi_S = \frac{1}{\sqrt{\pi}}\sin(M\phi + \alpha)$$

and

$$\psi_s = \frac{1}{\sqrt{\pi}}\sin(M\phi + \alpha)$$

On putting $\alpha = 0$, we get

$$\psi_c = \frac{1}{\sqrt{\pi}}\sin M\phi$$

$$= \frac{1}{2i}\cdot\frac{1}{\sqrt{\pi}}\left(e^{iM\phi} - e^{-iM\phi}\right)$$

$$= \frac{1}{i\sqrt{2}}\left(\frac{1}{\sqrt{2\pi}}e^{iM\phi} - \frac{1}{\sqrt{2\pi}}e^{-iM\phi}\right)$$

$$\therefore\ \psi_s = \frac{1}{i\sqrt{2}}(\psi_+ + \psi_-) \tag{7.90}$$

Further considering the equation

$$\psi_c = \frac{1}{\sqrt{\pi}}\cos(M\phi + \alpha)$$

$$= \frac{1}{\sqrt{\pi}}\cos M\phi, \text{ where } \alpha = 0$$

$$= \frac{1}{2}\cdot\frac{1}{\sqrt{\pi}}\left(e^{iM\phi} + e^{-iM\phi}\right)$$

$$= \frac{1}{\sqrt{2}}\left(\frac{1}{\sqrt{2\pi}}e^{iM\phi} - \frac{1}{\sqrt{2\pi}}e^{-iM\phi}\right)$$

$$\therefore\ \psi_c = \frac{1}{\sqrt{2}}(\psi_+ + \psi_-) \tag{7.91}$$

In Eqs. (7.90) and (7.91), ψ_+ and ψ_- for a given value of M will be degenerate.

7.4 Particle on a Sphere

Let us consider a particle of mass m, which is free to roam anywhere on the surface of sphere having radius r. The radius r is constant, and the potential energy (V) of the particle confined on the surface of sphere is taken as zero as the particle is free to move on the surface, i.e., $V = 0$.

The Hamiltonian (H) for motion of particle in three-dimensional is given by

$$\hat{H} = -\frac{\hbar^2}{2m}\nabla^2 + V$$

$$= -\frac{\hbar^2}{2m}\nabla^2 \quad [\because V = 0]$$

But $\hat{H}\psi = E\psi$

$$\therefore\ -\frac{\hbar^2}{2m}\nabla^2\psi = E\psi \tag{7.92}$$

which is the *Schrödinger equation*. We have already discussed regarding $\nabla^2\psi$ in polar co-ordinates, which is expressed as

$$\nabla^2\psi = \frac{\mathrm{d}^2\psi}{\mathrm{d}r^2} + 2\frac{\mathrm{d}\psi}{r\mathrm{d}r} + \frac{1}{r^2\sin^2\theta}\frac{\mathrm{d}\psi}{\mathrm{d}\theta}\left(\sin\theta\frac{\mathrm{d}\psi}{\mathrm{d}\theta}\right) + \frac{1}{r^2\sin^2\theta}\frac{\mathrm{d}^2\psi}{\mathrm{d}\phi^2} \tag{7.93}$$

Putting these values in Eq. (7.92), we get

$$-\frac{\hbar^2}{2m}\left[\frac{\mathrm{d}^2\psi}{\mathrm{d}r^2} + \frac{2}{r}\frac{\mathrm{d}\psi}{\mathrm{d}r} + \frac{1}{r^2\sin^2\theta}\frac{\mathrm{d}\psi}{\mathrm{d}\theta}\left(\sin\theta\frac{\mathrm{d}\psi}{\mathrm{d}\theta}\right) + \frac{1}{r^2\sin^2\theta}\frac{\mathrm{d}^2\psi}{\mathrm{d}\phi^2}\right] = E\psi \tag{7.94}$$

We know that r is constant. If we assume $\theta = constant$, then the particle on a sphere becomes a particle on a ring problem. Also assume that θ is chosen such that $\sin\theta = 1$, i.e., $\theta = \pi/2$.

Taking into consideration the above facts, Eq. (7.94) reduces to

$$-\frac{\hbar^2}{2m}\cdot\frac{1}{r^2\sin^2\theta}\frac{\mathrm{d}^2\psi}{\mathrm{d}\phi^2} = E\psi$$

$$\text{or}\quad -\frac{\hbar^2}{2m}\cdot\frac{1}{r^2}\frac{\mathrm{d}^2\psi}{\mathrm{d}\phi^2} = E\psi$$

$$\text{or}\quad \frac{\mathrm{d}^2\psi}{\mathrm{d}\phi^2} = -2\frac{Emr^2}{\hbar^2}\psi = -\frac{2EI}{\hbar^2}\psi$$

$$\frac{\mathrm{d}^2\psi}{\mathrm{d}\phi^2} + m_l^2\psi = 0 \tag{7.95}$$

where $m_l^2 = \frac{2IE}{\hbar^2}$ $\quad\therefore E = \frac{m_l^2\hbar^2}{2I}$, I = moment of Inertia

$$\text{and } m_l = \pm\left(\frac{2IE}{\hbar^2}\right)^{\frac{1}{2}} = \pm\frac{(2IE)^{\frac{1}{2}}}{\hbar}$$

= a dimensional number positive (or negative) = 0, ±1, ±2, ±3, …

Equation (7.95) resembles the equation for a ring, and thus under certain circumstances, a sphere can be regarded as a stack of rings. The only difference is that in case of sphere, the particle can travel from ring to ring. This has been shown in Figure 7.10.

The normalised general solution of Eq. (7.95) is

$$\psi_{m_l}(\phi) = \frac{e^{im\phi}}{(2\pi)^{1/2}} \tag{7.96}$$

We now choose the acceptable solutions from among the general solutions by imposing the condition that the wave function ψ should be single valued. It means ψ must satisfy the cyclic boundary condition, i.e., $\psi(\phi) = \psi(\phi + 2\pi)$.

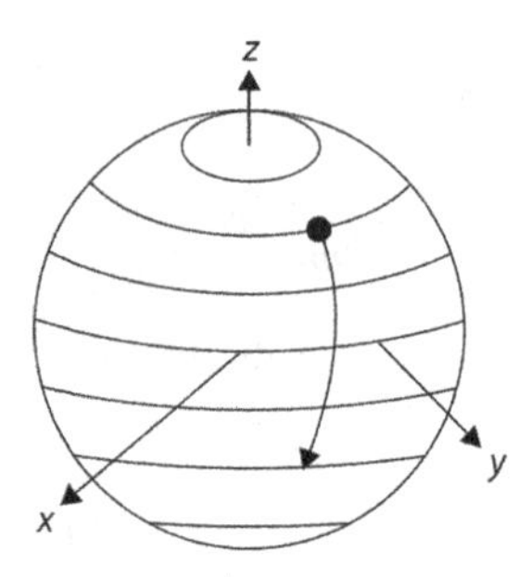

FIGURE 7.10 Movement of a particle from ring to ring, which is equivalent to the movement of the particle on the surface of the sphere.

Keeping this in mind, we can write

$$\psi_{m_l}(\phi+2\pi)=\frac{e^{im_l(\phi+2\pi)}}{\sqrt{2\pi}}=\frac{e^{im_l\phi}\cdot e^{2\pi im_l}}{\sqrt{2\pi}}$$
$$=\psi_{m_l}(\phi)e^{2\pi im_l} \tag{7.97}$$

We know that $e^{i\pi}=-1$, and therefore

$$\psi_{m_l}(\phi+2\pi)=(-1)^{2m_l}\,\psi(\phi) \tag{7.98}$$

Since we need $(-1)^{2m_l}=1$, $2m_l$ must be a positive or a negative even integer, including zero, and hence, $m_l=0,\pm1,\pm2,\pm3,\ldots$

7.4.1 The Legendre Polynomials

The Legendre polynomials are of significance in various quantum chemical problems. Actually, they are the basis for the wave functions for angular momentum and, hence, occur in problems dealing with spherical motion. The fact is actually that these polynomials describe the angular dependence of one electron system, which forms the basis of all the geometry of chemistry. It should be kept in mind that there are a number of methods for forming these polynomials, but the most convenient one is the solution of a differential equation.

Let us consider the following differential equation:

$$\left(1-x^2\right)\frac{\mathrm{d}y}{\mathrm{d}x}+2lxy=0 \quad \text{where, } l=\text{constant} \tag{7.99}$$

The above differential equation can be written in the following form on separating the variables:

$$\frac{\mathrm{d}y}{y}+\frac{2lx}{\left(1-x^2\right)}\mathrm{d}x=0 \tag{7.100}$$

On integration, we get

$$\int\frac{\mathrm{d}y}{y}+\int\frac{2lx}{\left(1-x^2\right)}\mathrm{d}x=\int 0$$

$$\text{or } \ln y+l\ln\left(1-x^2\right)=\text{constant}$$

$$\text{or } \ln\frac{y}{\left(1-x^2\right)^l} = \text{constant}$$

$$\text{or } \ln\frac{y}{\left(1-x^2\right)^l} = c$$

$$\therefore \; y = c\left(1-x^2\right)^l \quad \text{where, } c = \text{constant} \tag{7.101}$$

The lth derivative of Eq. (7.101) can be expressed as

$$\frac{\mathrm{d}^l y}{\mathrm{d}x^l} = C\frac{\mathrm{d}^l}{\mathrm{d}x^l}\left(1-x^2\right)^l \tag{7.102}$$

Differentiating Eq. (7.99) n times yields

$$\left(1-x^2\right)\frac{\mathrm{d}^{n+1}y}{\mathrm{d}x^{n+1}} + 2(l-n)\frac{\mathrm{d}^n y}{\mathrm{d}x^n} + n(2l-n+1)\frac{\mathrm{d}^{n-1}y}{\mathrm{d}x^{n-1}} = 0 \tag{7.103}$$

Considering $n = l + 1$ in special case, Eq. (7.103) becomes

$$\left(1-x^2\right)\frac{\mathrm{d}^{l+2}y}{\mathrm{d}x^{l+2}} - 2x\frac{\mathrm{d}^{l+1}y}{\mathrm{d}x^{l+1}} + l(l+1)\frac{\mathrm{d}^l y}{\mathrm{d}x^l} = 0 \tag{7.104}$$

Putting $\mathrm{d}^l y/\mathrm{d}x^l = z$, Eq. (7.104) takes the following form:

$$\left(1-x^2\right)\frac{\mathrm{d}^2 z}{\mathrm{d}x^2} - 2x\frac{\mathrm{d}z}{\mathrm{d}x} + l(l+1)z = 0 \tag{7.105}$$

$$\text{But } \frac{\mathrm{d}}{\mathrm{d}x}\left\{\left(1-x^2\right)\frac{\mathrm{d}z}{\mathrm{d}x}\right\} = \left(1-x^2\right)\frac{\mathrm{d}^2 z}{\mathrm{d}x^2} - 2x\frac{\mathrm{d}z}{\mathrm{d}x} \tag{7.106}$$

Therefore, Eq. (7.105) may be expressed as

$$\frac{\mathrm{d}}{\mathrm{d}x}\left\{\left(1-x^2\right)\frac{\mathrm{d}z}{\mathrm{d}x}\right\} + l(l+1)z = 0 \tag{7.107}$$

This is one of the forms of the *Legendre equation*. Now, we shall make use of Eq. (7.102), which follows that

$$z = c\frac{\mathrm{d}^l\left(1-x^2\right)^l}{\mathrm{d}x^l} \tag{7.108}$$

In special case, the arbitrary constant c may be given by

$$c = \frac{(-1)^l}{2^l l!} \tag{7.109}$$

The resulting solution is known as the *Legendre polynomial* of degree l, which is denoted by $P_l(x)$, and thus,

$$P_l(x) = \frac{(-1)^l}{2^l l!}\cdot\frac{\mathrm{d}^l\left(x^2-1\right)^l}{\mathrm{d}x^l} \tag{7.110}$$

This is called *Rodrigue's formula.* In this equation, the constant term (–1) has been combined with $(1-x^2)^l$ to yield $(x^2-1)^l$. Since the Legendre polynomial is a solution of Legendre Eq. (7.107), the latter may be expressed as

$$\frac{d}{dx}\left[(1-x^2)\frac{dP_l(x)}{dx}\right]+l(l+1)P_l(x)=0 \tag{7.111}$$

With the help of Eq. (7.110), the different *Legendre polynomials* can be evaluated, which have been mentioned below:

$$\text{When } l=0;\ P_0(x)=\frac{1}{2^0 0!}\cdot\frac{d^0}{dx^0}(x^2-1)^0=1$$

$$l=1;\ P_1(x)=\frac{1}{2^1 1!}\cdot\frac{d^1}{dx}(x^2-1)^1=\frac{1}{2}\cdot 2x=x$$

$$l=2;\ P_2(x)=\frac{1}{2^2 2!}\cdot\frac{d^2}{dx^2}(x^2-1)^2$$

$$=\frac{1}{4\cdot 2}\left[\frac{d^2}{dx^2}(x^4-2x^2+1)\right]$$

$$=\frac{1}{8}\left[12x^2-4\right]=\frac{1}{2}(3x^2-1)$$

$$l=3;\ P_3(x)=\frac{1}{2^3 3!}\cdot\frac{d^3}{dx^3}(x^2-1)^3$$

$$=\frac{1}{8\cdot 6}\frac{d^3}{dx^3}(x^6-3x^4+3x^2-1)$$

$$=\frac{1}{48}(120x^3-72x)=\frac{1}{2}(5x^3-3x)$$

$$\text{Similarly, for } l=4;\ P_4(x)=\frac{1}{8}(35x^2-30x^2+3)$$

It is observed that in case of each polynomial, the powers of x are either all odd or all even.

7.4.1.7 Normalisation of the Legendre Polynomial

The polynomials can be normalised by making use of the result

$$\int_{-1}^{+1} P_l(x)\cdot P_n(x)dx=\frac{2}{2l+1} \quad \text{for } l=n \tag{7.112}$$

and this system is considered in the interval $-1\le x\le 1$. Since $l=n$, the above equation may be expressed as

$$\int_{-1}^{+1} P_l(x)\cdot P_l(x)\mathrm{d}x = \int_{-1}^{+1} P_l^2(x)\cdot \mathrm{d}x$$

Suppose $l = 1$, then

$$\int_{-1}^{+1} P_1^2(x)\cdot \mathrm{d}x = \int_{-1}^{+1} x^2\cdot \mathrm{d}x = \left[\frac{x^3}{3}\right]_{-1}^{+1}$$

$$\left[\frac{1}{3}-\left(-\frac{1}{3}\right)\right] = \frac{2}{3} = \frac{2}{2l+1'} \quad (\text{when } \ l=1)$$

When $l = 2$, then

$$\int_{-1}^{+1} P_2(x)P_2(x)\cdot \mathrm{d}x = \int_{-1}^{+1} P_2^2(x)\mathrm{d}x$$

$$= \int_{-1}^{+1}\left\{\frac{1}{2}\left(3x^2-1\right)\right\}^2 \mathrm{d}x = \int_{-1}^{+1}\frac{1}{4}\left(9x^4-6x^2+1\right)\mathrm{d}x$$

$$= \frac{1}{4}\left[\int_{-1}^{+1} 9x^4\mathrm{d}x - \int_{-1}^{+1} 6x^2\mathrm{d}x + \int_{-1}^{+1}\mathrm{d}x\right]$$

$$= \frac{1}{4}\left[\frac{9x^5}{5}-\frac{6x^3}{3}+x\right]_{-1}^{+1}$$

$$= \frac{1}{4}\left[\left(\frac{9}{5}-2+1\right)-\left(-\frac{9}{5}-\frac{-6}{3}-1\right)\right]$$

$$= \frac{1}{4}\left[\left(\frac{9}{5}-1\right)-\left(-\frac{9}{5}+\frac{6}{3}-1\right)\right]$$

$$= \frac{1}{4}\left[\frac{4}{5}-\left(-\frac{9}{5}+1\right)\right]$$

$$= \frac{1}{4}\left[\frac{4}{5}+\frac{4}{5}\right] = \frac{2}{5} = \frac{2}{2l+1} \quad [\text{when } \ l=2]$$

which may be expressed in the general form as $2/2l+1\,[\text{when } l=l]$, and hence, the normalising factor will be $\sqrt{(2l+1)/2}$.

7.4.1.2 Orthogonality of the Legendre Polynomials

Further, we shall show that the Legendre polynomials form an orthogonal system in the boundary condition $-1 \le x \le 1$, that is

$$\int_{-1}^{+1} P_l(x)\cdot P_n(x)\mathrm{d}x = 0 \ \text{ for } \ l \ne n \tag{7.113}$$

Here, we shall show the orthogonal property of the Legendre polynomials by considering $P_0(x)$ and $P_1(x)$, i.e.,

$$\int_{-1}^{+1} P_l(x)\cdot P_n(x)\mathrm{d}x = \int_{-1}^{+1} 1\cdot x\cdot \mathrm{d}x \quad (\text{when } l \neq n)$$

$$\int_{-1}^{+1} x\cdot dx = \left[\frac{x^2}{2}\right]_{-1}^{+1}$$

$$= \left(\frac{1}{2}-\frac{1}{2}\right) = 0$$

Similarly, we can prove the orthogonal property of the Legendre polynomials by considering $P_1(x)$ and $P_2(x)$.

7.4.2 Associated Legendre Equation

If v is a solution of the Legendre equation

$$\left(1-x^2\right)\frac{\mathrm{d}^2y}{\mathrm{d}x^2} - 2x\frac{\mathrm{d}y}{\mathrm{d}x} + n(n+1)y = 0$$

then $\left(1-x^2\right)^{m/2}\dfrac{\mathrm{d}^m y}{\mathrm{d}x^m}$ is a solution of the associated Legendre equation.

Proof: Since v is a solution of the Legendre equation, we may write

$$\left(1-x^2\right)\frac{\mathrm{d}^2v}{\mathrm{d}x^2} - 2x\frac{\mathrm{d}v}{\mathrm{d}x} + n(n+1)v = 0 \tag{7.114}$$

Differentiating Eq. (7.114) m times with respect to x, by the Leibnitz theorem, we have

$$\left\{\left(1-x^2\right)\frac{\mathrm{d}^{m+2}v}{\mathrm{d}x^{m+2}} - m\cdot 2x\frac{\mathrm{d}^{m+1}v}{\mathrm{d}x^{m+1}} - \frac{m(m+1)}{2!}\cdot 2\cdot\frac{\mathrm{d}^m v}{\mathrm{d}x^m}\right\} - 2\left\{x\cdot\frac{\mathrm{d}^{m+1}v}{\mathrm{d}x^{m+1}} + m\cdot\frac{\mathrm{d}^m v}{\mathrm{d}x^m}\right\} + n(n+1)\frac{\mathrm{d}^m y}{\mathrm{d}x^m} = 0$$

$$\text{or}\quad \left(1-x^2\right)\frac{\mathrm{d}^{m+2}v}{\mathrm{d}x^{m+2}} - 2(m+1)x\frac{\mathrm{d}^{m+1}v}{\mathrm{d}x^{m+1}} + \left[n(n+1) - m(m+1)\right]\frac{\mathrm{d}^m v}{\mathrm{d}x^m} = 0$$

putting $v_1 = \dfrac{\mathrm{d}^m v}{\mathrm{d}x^m}$, we get

$$\left(1-x^2\right)\frac{\mathrm{d}^2v_1}{\mathrm{d}x^2} - 2(m+1)x\frac{\mathrm{d}v_1}{\mathrm{d}x} + \left[n(n+1) - m(m+1)\right]v_1 = 0 \tag{7.115}$$

Now, let $z = \left(1-x^2\right)^{m/2}\dfrac{\mathrm{d}^m y}{\mathrm{d}x^m} = \left(1-x^2\right)^{m/2} v_1$

So that $v_1 = z\left(1-x^2\right)^{-m/2}$

$$\text{or}\quad \frac{\mathrm{d}v_1}{\mathrm{d}x} = \left(1-x^2\right)^{-m/2}\frac{\mathrm{d}z}{\mathrm{d}x} + m\left(1-x^2\right)^{-(m/2)-1}\cdot xz$$

$$\text{and}\quad \frac{\mathrm{d}^2v_1}{\mathrm{d}x} = \left(1-x^2\right)^{-m/2}\frac{\mathrm{d}^2z}{\mathrm{d}x^2} + 2m\left(1-x^2\right)^{-(m/2)-1}x\frac{\mathrm{d}z}{\mathrm{d}x} + mz\left(1-x^2\right)^{-(m/2)-1} + m(m+2)x^2z\left(1-x^2\right)^{-(m/2)-2}$$

Putting this value in Eq. (7.115), we get

$$\left(1-x^2\right)\left\{\left(1-x^2\right)^{-m/2}\frac{\mathrm{d}^2z}{\mathrm{d}x^2}+2m\left(1-x^2\right)^{-(-m/2)-1}x\frac{\mathrm{d}z}{\mathrm{d}x}+mz\left(1-x^2\right)^{-(-m/2)-1}+m(m+2)x^2z\left(1-x^2\right)^{-(-m/2)-2}\right\}$$

$$-2(m+1)x\left\{\left(1-x^2\right)^{-m/2}\frac{\mathrm{d}z}{\mathrm{d}x}+mxz\left(1-x^2\right)^{-(m/2)-1}\right\}$$

$$+\{n(n+1)-m(m+1)\}\left(1-x^2\right)^{-m/2}z=0$$

$$\text{or } \left(1-x^2\right)^{-m/2}\left[\left(1-x^2\right)\frac{\mathrm{d}^2z}{\mathrm{d}x^2}-2x\frac{\mathrm{d}z}{\mathrm{d}x}+\left\{n(n+1)-\frac{m^2}{1-x^2}\right\}z\right]=0$$

$$\text{or } \left(1-x^2\right)\frac{d^2z}{dx^2}-2\frac{dz}{dx}+\left\{n(n+1)-\frac{m^2}{1-x^2}\right\}z=0$$

Hence, $z=\left(1-x^2\right)^{m/2}\dfrac{\mathrm{d}^my}{\mathrm{d}x^m}$ is the solution of the associated Legendre equation.

$$\left(1-x^2\right)\frac{\mathrm{d}^2y}{\mathrm{d}x^2}-2x\frac{\mathrm{d}y}{\mathrm{d}x}+\left\{n(n+1)-\frac{m^2}{1-x^2}\right\}=0 \tag{7.116}$$

or $\left(1-x^2\right)P_n''-2xP_n'+\{n(n+1)\}P_n=0$ when $m=0$ and $y=P_n$ Q.E.D.

7.4.3 Associated Legendre Functions

We have discussed above that $z=\left(1-x^2\right)^{m/2}\mathrm{d}^my/\mathrm{d}x^m$ is the regular solution of the associated Legendre equation, which may be relabelled as $P_n^m(x)$, i.e.,

$$z=P_n^m(x)=\left(1-x^2\right)^{m/2}\frac{\mathrm{d}^m}{\mathrm{d}x^m}P_n(x) \tag{7.117}$$

These are the associated Legendre functions. The reader should try to transform the associated Legendre equation in terms of spherical polar co-ordinates. The associate Legendre equation may be expressed in polar co-ordinates as

$$\frac{1}{\sin\theta}\cdot\frac{\mathrm{d}}{\mathrm{d}\theta}\left(\sin\theta\frac{\mathrm{d}z}{\mathrm{d}\theta}\right)+\left[n(n+1)-\frac{m^2}{\sin^2\theta}\right]z=0 \tag{7.118}$$

The Legendre functions can also be expressed as

$$P_l^m(x)=\left(1-x^2\right)^{m/2}\frac{\mathrm{d}^m}{\mathrm{d}x^m}P_l(x) \tag{7.119}$$

keeping in view the former notation.

The associated Legendre functions are orthogonal for different rank l but the same order m as per Eq. (7.119). The normalisation is provided by making use of such a factor that the normalised Legendre functions will be

$$\frac{2l+1(l-m)!}{(l-m)!}P_l^m(x) \tag{7.120}$$

TABLE 7.2
Associated Legendre Functions

$P_1^1(x) = (1-x^2)^{1/2} = \sin\theta$
$P_2^1(x) = 3x(1-x^2)^{\frac{1}{2}} = 3\cos\theta\sin\theta$
$P_2^2(x) = 3(1-x^2) = 3\sin^2\theta$
$P_3^1(x) = \frac{3}{2}(5x^2-1)(1-x^2)^{\frac{1}{2}} = \frac{3}{2}(5\cos^2\theta - 1)\sin\theta$
$P_3^2(x) = 15x(1-x^2) = 15\cos\theta\sin^2\theta$
$P_3^3(x) = 15(1-x^2)^{3/2} = 15\sin^3\theta$
$P_4^1(x) = \frac{5}{2}(7x^3-3x)(1-x^2)^{\frac{1}{2}} = 5/2(7\cos^2\theta - 3\cos\theta)\sin\theta$
$P_4^2(x) = \frac{15}{2}(7x^2-1)(1-x^2) = 15/2(\cos^2\theta - 1)\sin^2\theta$
$P_4^3(x) = 105x(1-x^2)^{3/2} = 105\cos\theta\sin^3\theta$
$P_4^4(x) = 105x(1-x^2)^2 = 105\sin^4\theta$

In addition to these, we may also develop a table of the associated Legendre function with the help of Eq. (7.117) (Table 7.2).

7.4.4 Spherical Harmonics

We should know that the angular part of the wave function is significant with regard to the formation of chemical bonds, which depends on the suitable overlap of the orbitals of the atoms concerned forming the bond. It should be kept in mind that the overlap involves the total wave function ψ but whether the overlap will take place or not will depend on the shape of the orbitals, and this is governed by the angular functions, which are known as *spherical harmonics*, which is generally denoted by $Y_l^m(\theta,\phi)$. The spherical harmonics are obtained by combining equations $\phi = 1/\sqrt{2\pi}e^{im\phi}$, where $m = 0, \pm1, \pm2, \ldots$

$$\text{and } \theta_l^m(\theta) = (-1)^m \sqrt{\frac{2l+l(l-m)!}{2(l+m)!}} P_l^m \cos\theta e^{im\phi}$$

After combining these two equations, we shall get

$$Y_l^m(\theta,\phi) = (-1)^m \sqrt{\frac{2l+l(l-m)!}{2(l+m)!}} P_l^m \cdot \cos\theta e^{im\phi} \tag{7.121}$$

when $l = 0, 1, 2, \ldots$ and $m = 0, \pm1, \pm2, \ldots$

The spherical harmonics are mutually orthogonal. The first few spherical harmonics have been given in Table 7.3.

The more usual method of representing the angular functions is by polar plot, as illustrated in Figure 7.11.

For a given value of θ, a line of length $\left[(\sqrt{3})/2(\sqrt{\pi})\right]\cos\theta$ is drawn from the origin making an angle θ with z axis. It is clear that $Y_{1,0}$ does not depend on θ, the line may orient with respect to the x axis and consequently a circle is generated, which is parallel to the xy plane. With the variation of θ from 0 to 2π, surfaces of two spheres will be generated whose centres lie on the z axis, which is shown in Figure 7.11. It is important to realise that the above plot has no physical importance, but it is merely a way of representing the angular function $Y_{1,0}$ by a plot.

TABLE 7.3

Spherical Harmonics

$Y_0^0(\theta,\phi) = \frac{1}{\sqrt{4\pi}}$
$Y_1^1(\theta,\phi) = -\sqrt{\frac{3}{8\pi}} \cdot \sin\theta e^{i\phi}$
$Y_1^0(\theta,\phi) = \sqrt{\frac{3}{4\pi}} \cdot \cos\theta$
$Y_1^{-1}(\theta,\phi) = +\sqrt{\frac{3}{8\pi}} \cdot \sin\theta e^{-i\phi}$
$Y_2^2(\theta,\phi) = \sqrt{\frac{5}{96\pi}} \cdot 3\ \sin^2\theta e^{2i\phi}$
$Y_2^1(\theta,\phi) = -\sqrt{\frac{5}{24\pi}} \cdot 3\ \sin\theta\cos\theta e^{i\phi}$
$Y_2^0(\theta,\phi) = \sqrt{\frac{5}{4\pi}}\left(\frac{3}{2}\cos^2\theta - \frac{1}{2}\right)$
$Y_2^{-1}(\theta,\phi) = \sqrt{\frac{5}{24\pi}} \cdot 3\ \sin\theta\cos\theta e^{-i\phi}$
$Y_2^{-2}(\theta,\phi) = \sqrt{\frac{5}{96\pi}} \cdot 3\ \sin\theta\cos\theta e^{-2i\phi}$

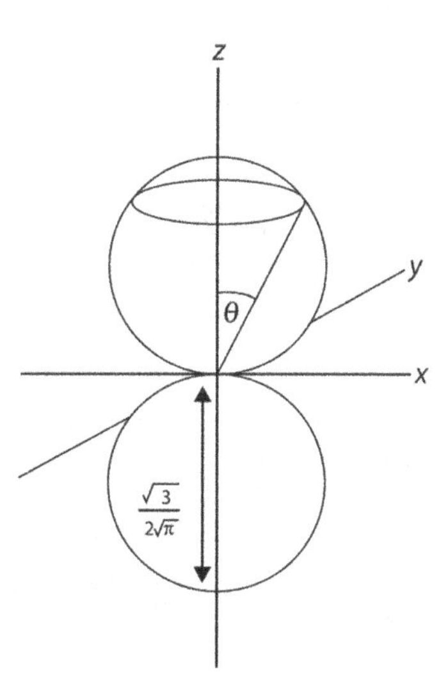

FIGURE 7.11 Polar plot of the $Y_{1,0}$ function for a given value of θ.

7.4.5 Particle on a Sphere

In Section 7.4, we have dealt with particle on a sphere, where a sphere was regarded as a stack of rings. We have learnt the Legendre polynomials, functions, and spherical harmonics. At this stage, we shall derive an equation for particle on sphere directly playing with Schrödinger equations.

Let the mass of the particle is m, which is free to move on the surface of sphere having radius r. V is the potential energy of the particle confined on the surface of sphere. V has been assumed to be zero be for simplicity sake, i.e., V = 0.

The Hamiltonian ($\hat{H}$) for motion of particle on the sphere is given by

$$\hat{H} = -\frac{\hbar^2}{2m}\nabla^2 + V$$

$$= -\frac{\hbar^2}{2m}\nabla^2 \quad [\because V = 0]$$

But $\hat{H}\psi = E\psi$

$$\therefore \quad \frac{\hbar^2}{2m}\nabla^2 = E\psi$$

which is the Schrödinger equation. $\nabla^2\psi$ in polar co-ordinates is expressed as

$$\nabla^2\psi = \frac{d^2\psi}{dr^2} + \frac{2}{r}\frac{d\psi}{dr} + \frac{1}{r^2\sin^2\theta}\frac{d}{d\theta}\left(\sin\theta\frac{d\psi}{d\theta}\right) + \frac{1}{r^2\sin^2\theta}\frac{d^2\psi}{d\phi^2}$$

Putting these values of $\nabla^2\psi$ in the above equation, we get

$$= -\frac{\hbar^2}{2m}\left[\frac{d^2\psi}{dr^2} + \frac{2}{r}\frac{d\psi}{dr} + \frac{1}{r^2\sin^2\theta}\frac{d}{d\theta}\left(\sin\theta\frac{d\psi}{d\theta}\right) + \frac{1}{r^2\sin^2\theta}\frac{d^2\psi}{d\phi^2}\right] = E\psi$$

Since r is constant, $\frac{d\psi}{dr} = \frac{d^2\psi}{dr^2} = 0$

Putting the values of $\frac{d\psi}{dr} = \frac{d^2\psi}{dr^2} = 0$ in the above equation yields

$$= -\frac{\hbar^2}{2m}\left[\frac{1}{r^2\sin^2\theta}\frac{d}{d\theta}\left(\sin\theta\frac{d\psi}{d\theta}\right) + \frac{1}{r^2\sin^2\theta}\frac{d^2\psi}{d\phi^2}\right] = E\psi$$

On simplification, we obtain

$$\frac{d^2\psi}{d\theta^2} + \frac{\cos\theta}{\sin\theta}\cdot\frac{d\psi}{d\theta} + \frac{1}{\sin^2\theta}\frac{d^2\psi}{d\phi^2} + \frac{2mr^2}{\hbar^2}E\psi = 0 \tag{7.122}$$

where ψ is function of θ and ϕ.

The following Schrödinger Eq. (7.122) can be solved by the method of separating the variables. For this purpose, let

$$\psi(\theta,\phi) = P(\theta)\cdot F(\phi) \tag{7.123}$$

That is, $\psi(\theta,\phi)$ may be expressed as a product of two functions. It should be kept in mind that P is a function of θ only and F is a function of ϕ only.

Differentiating Eq. (7.123) with respect to θ, we have

$$\frac{d\psi}{d\phi} = F(\phi)\cdot\frac{dp}{d\theta} \tag{7.124}$$

$$\text{and} \quad \frac{d^2\psi}{d\phi^2} = F(\phi)\cdot\frac{dp}{d\theta} \tag{7.125}$$

But differentiation with respect to ϕ yields

$$\frac{d\psi}{d\phi} = P(\theta).\frac{dF}{d\phi} \tag{7.126}$$

$$\text{and} \quad \frac{d^2\psi}{d\phi^2} = P(\theta)\cdot\frac{d^2F}{d\phi^2} \tag{7.127}$$

On substituting these values in Eq. (7.122), we have

$$F\frac{d^2P}{d\theta^2}\frac{\cos\theta}{\sin\theta}\cdot F\frac{dp}{d\theta}+\frac{1}{\sin^2\theta}\cdot P\frac{d^2F}{d\phi^2}+\frac{2mr^2}{\hbar^2}EPF=0$$

Multiplying throughout by $\sin^2\theta/PF$, we shall get

$$\frac{\sin^2\theta}{P}\cdot\frac{d^2P}{d\theta^2}+\cos\theta\cdot\sin\theta\frac{1}{P}\frac{dp}{d\theta}+\frac{1}{F}\frac{d^2F}{d\phi^2}+\frac{2mr^2}{\hbar^2}E=0$$

where $P = P(\theta)$ and $F = F(\theta)$

$$\frac{\sin^2\theta}{P}\cdot\frac{d^2P}{d\theta^2}+\cos\theta\cdot\sin\theta\frac{1}{P}\frac{dp}{d\theta}+\frac{2mr^2}{\hbar^2}\sin^2\theta E=-\frac{1}{F}\frac{d^2F}{d\phi^2} \tag{7.128}$$

It is clear that the LHS of Eq. (7.128) depends on θ only and the RHS depends on ϕ only. Therefore, both sides of the equation may be assumed to be a constant.

$$\text{Suppose} -\frac{1}{F}\frac{d^2F}{d\phi^2}=M^2$$

$$\text{and} \quad \frac{d^2F}{d\phi^2}+M^2F=0 \tag{7.129}$$

Again the LHS of Eq. (7.128) is also equal to M^2, i.e.,

$$\frac{\sin^2\theta}{P}\frac{d^2P}{d\theta^2}+\cos\theta\cdot\sin\theta\frac{1}{P}\frac{dp}{d\theta}+\frac{2mr^2}{\hbar^2}\sin^2\theta E=M^2 \tag{7.130}$$

Multiplying Eq. (7.130) by $\frac{P}{\sin^2\theta}$, we get

$$\frac{d^2P}{d\theta^2}+\frac{\cos\theta}{\sin\theta}\cdot\frac{dp}{d\theta}+\beta\cdot P=\frac{M^2P}{\sin^2\theta} \tag{7.131}$$

$$\text{where } \beta=\frac{2mr^2}{\hbar^2}E=\frac{2IE}{\hbar^2},\ I=\text{moment of inertia}$$

Equation (7.131) is called the *Legendre differential equation.* When $M = 0$, Eq. (7.131) transforms to

$$\frac{d^2P}{d\theta^2}+\frac{\cos\theta}{\sin\theta}\cdot\frac{dp}{d\theta}+\beta\cdot P=0 \tag{7.132}$$

The solution of this Legendre differential equation exists as the Legendre polynomials, which is given by

$$P_l(x)=\frac{1}{2^l\cdot l!}\frac{d^l}{dx^l}\left(x^2-1\right)^l \tag{7.133}$$

where $x = \cos\theta$ and l = an integer including zero.

From Eq. (7.133), we can find different Legendre polynomials.

Further, we shall solve Eq. (7.129). It is a second-order differential equation.

$$\text{i.e.,} \quad \frac{d^2F}{d\phi^2} + M^2F = 0$$

This equation is similar to the equation for particle on a ring we have already discussed. The solution of the above equation has been dealt under Section 7.3, and it has been found to be

$$F_+ = \frac{1}{\sqrt{\pi}}\sin M\phi$$

$$F- = \frac{1}{\sqrt{\pi}}\cos M\phi \tag{7.134}$$

$$F\pm = \frac{1}{\sqrt{2\pi}}e^{\pm iM\phi}$$

where M = a quantum number having values 0, ±1, ±2, …

Thus, we solved the Schrödinger equation for a particle on sphere.

7.5 Rigid Rotors

A rigid body rotating about a fixed axis is known as a rigid rotor. A rigid rotor involves two point masses m_1 and m_2 connected together by a rigid, weightless link of length r. The theory of rigid rotor is particularly significant when we consider the rotation of a diatomic molecule about an axis passing through the centre of gravity and perpendicular to its own axis. It has been illustrated in Figure 7.12.

In Figure 7.12,

r_1 = distance of the point mass m_1 from C, i.e., centre of mass

r_2 = distance of the point mass m_2 from C

$r = r_1 + r_2$

The moment of inertia (I) of the particles about a fixed axis is given by

$$I = m_1r_1^2 + m_2r_2^2 = \sum mr^2$$

From the definition of centre of mass $m_1r_1 = m_2r_2$

$$\text{or} \quad \frac{r_1}{1/m_1} = \frac{r_2}{1/m_2} = \frac{r_1 + r_2}{1/m_1 + 1/m_2} = \frac{r}{1/m_1 + 1/m_2}$$

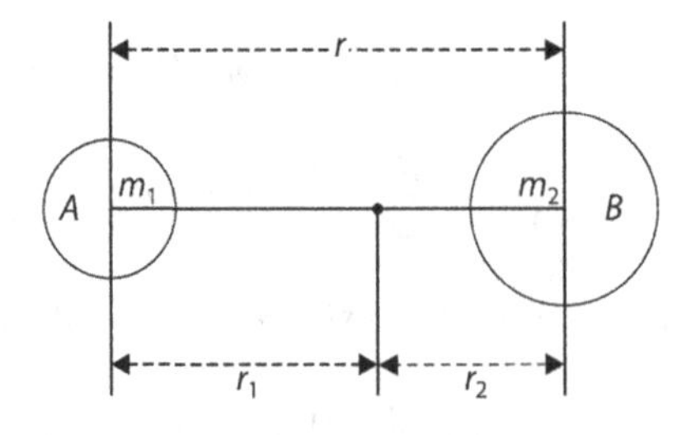

FIGURE 7.12 Rigid rotor.

$$\therefore\ m_1r_1 = m_2r_2 = \left(\frac{m_1}{m_1+m_2}\right)\cdot r$$

$$\therefore\ r_1 = \left(\frac{m_2}{m_1+m_2}\right)\cdot r \text{ and } r_2 = \left(\frac{m_1}{m_1+m_2}\right)\cdot r$$

Putting these values in equation $I = m_1r_1^2 + m_2r_2^2$, we get

$$I = m_1\left(\frac{m_2}{m_1+m_2}\right)^2 r^2 + m_2\left(\frac{m_1}{m_1+m_2}\right)^2 r^2$$

$$= \frac{m_1}{(m_1+m_2)^2}m_2 + \frac{m_1m_2\ r^2}{(m_1+m_2)^2}m_1$$

$$= \frac{m_1m_2\ r^2}{(m_1+m_2)^2}[m_2+m_1] = \frac{m_1m_2\ r^2}{(m_1+m_2)^2}[m_1+m_2]$$

$$= \frac{m_1m_2}{m_1+m_2}\ r^2$$

$$I = \mu r^2 \text{ where, } \mu = \text{reduced mass} = \frac{m_1m_2}{m_1+m_2}$$

Now, we are going to apply the Schrödinger equation to a rigid rotor. The time-independent Schrödinger equation is

$$\frac{d^2\psi}{dx^2} + \frac{d^2\psi}{dy^2} + \frac{d^2\psi}{dz^2} + \frac{2\mu}{\hbar^2}(E-V)\psi = 0$$

For free rotation, the potential $V(r) = 0$ and the Schrödinger equation becomes

$$\frac{d^2\psi}{dx^2} + \frac{d^2\psi}{dy^2} + \frac{d^2\psi}{dz^2} + \frac{2\mu}{\hbar^2}E\psi = 0$$

To solve this equation, it will be convenient to transform it from Cartesian co-ordinates to polar co-ordinates. The Schrödinger equation in polar co-ordinates is

$$\frac{1}{r^2}\frac{d}{dr}\left(r^2\frac{d\psi}{dr}\right) + \frac{1}{r^2\sin\theta}\cdot\frac{d}{d\theta}\left(\sin\theta\cdot\frac{d\psi}{d\theta}\right) + \frac{1}{r^2\sin\theta}\frac{d^2\psi}{d\phi^2} + \frac{2\mu}{\hbar^2}E\psi = 0 \tag{7.135}$$

Since r is constant, Eq. (7.135) reduces to

$$\frac{1}{r^2\sin\theta}\cdot\frac{d}{d\theta}\left(\sin\theta\cdot\frac{d\psi}{d\theta}\right) + \frac{1}{r^2\sin\theta}\frac{d^2\psi}{d\phi^2} + \frac{2\mu}{\hbar^2}E\psi = 0$$

$$\text{or}\quad \frac{1}{\sin\theta}\cdot\frac{d}{d\theta}\left(\sin\theta\cdot\frac{d\psi}{d\theta}\right) + \frac{1}{\sin^2\theta}\frac{d^2\psi}{d\phi^2} + \frac{2\mu E\psi}{\hbar^2} = 0$$

$$\text{or}\quad \frac{1}{\sin\theta}\cdot\frac{d}{d\theta}\left(\sin\theta\cdot\frac{d\psi}{d\theta}\right) + \frac{1}{\sin^2\theta}\frac{d^2\psi}{d\phi^2} + \frac{2IE}{\hbar^2} = 0 \tag{7.136}$$

Equation (7.136) is a differential equation consisting of two independent variables, θ and ϕ. Obviously, the wave function will depend only on the angles $\theta(0 \le \theta \le \pi)$ and $\phi(0 \le \phi \le 2\pi)$. It is now assumed that $\psi(\theta,\phi)$ is a product of two functions, i.e.,

$$\psi(\theta,\phi) = T(\theta)\cdot F(\phi) \tag{7.137}$$

where $T(\theta)$ is a function of θ only and $F(\theta)$ is a function of ϕ only. For simplicity, Eq. (7.137) may be expressed as

$$\psi = T \cdot F \tag{7.138}$$

Since F is independent of θ, differentiation of Eq. (7.138) with respect to θ yields

$$\frac{d\psi}{d\phi} = F \cdot \frac{dT}{d\theta} \tag{7.139}$$

Similarly, since T is independent of ϕ, the differentiation of Eq. (7.138) with respect to ϕ gives

$$\frac{d\psi}{d\phi} = T \cdot \frac{dF}{d\phi} \tag{7.140}$$

and further differentiation with respect to ϕ yields

$$\frac{d^2\psi}{d\phi^2} = T \cdot \frac{d^2F}{d\phi^2} \tag{7.141}$$

Putting these values in Eq. (7.136), we get

$$\frac{1}{\sin\theta}\frac{d}{d\theta}\left(\sin\theta \cdot F\frac{dT}{d\theta}\right) + \frac{T}{\sin^2\theta}\frac{d^2F}{d\phi^2} + \frac{21E}{\hbar^2}\psi = 0$$

$$\text{or} \quad \frac{1}{\sin\theta}\frac{d}{d\theta}\left(\sin\theta \cdot F\frac{dT}{d\theta}\right) + \frac{T}{\sin^2\theta}\frac{d^2F}{d\phi^2} + \frac{21E}{\hbar^2}TF = 0$$

$$\text{or} \quad \frac{F}{\sin\theta}\frac{d}{d\theta}\left(\sin\theta \cdot \frac{dT}{d\theta}\right) + \frac{T}{\sin^2\theta}\frac{d^2F}{d\phi^2} + \frac{21E}{\hbar^2}TF = 0$$

Multiplying this equation by $\frac{\sin^2\theta}{TF}$, we shall obtain

$$\frac{\sin\theta}{T}\frac{d}{d\theta}\left(\sin\theta\frac{dT}{d\theta}\right) + \frac{1}{F}\frac{d^2F}{d\phi^2} + \frac{21E}{\hbar^2}\sin^2\theta = 0$$

$$\text{or} \quad \frac{\sin\theta}{T}\frac{d}{d\theta}\left(\sin\theta\frac{dT}{d\theta}\right) + \frac{21E}{\hbar^2}\sin^2\theta = -\frac{1}{F}\frac{d^2F}{d\phi^2} \tag{7.142}$$

It is clear that the LHS of Eq. (7.142) consists of only one variable θ and the RHS comprises only one variable ϕ. Since the equation must hold true for all values of θ and ϕ, each side must be a constant (say m^2). Hence, Eq. (7.142) may be separately written as

$$-\frac{1}{F}\frac{d^2F}{d\phi^2} = m^2 \tag{7.143}$$

$$\text{and } \frac{\sin\theta}{T}\frac{d}{d\theta}\left(\sin\theta\frac{dT}{d\theta}\right)+\frac{21E}{\hbar^2}\sin^2\theta = m^2 \tag{7.144}$$

Equation (7.143) is called F equation, and Eq. (7.144) is called T equation. We shall solve these equations one by one.

7.5.1 *F* Equation

The F equation is $-\frac{1}{F}\frac{d^2F}{d\phi^2} = m^2$

$$\text{or } \frac{d^2F}{d\phi^2} + m^2F = 0 \tag{7.145}$$

$$\text{The solution of this equation is } F = A\ e^{\pm im\phi} \tag{7.146}$$

This is an acceptable wave function, provided m is an integer. Since F is required to be a single valued function, we can write

$$F(\phi) = F(\phi + 2\pi)$$

$$\text{or, } e^{\pm im\phi} = F\ e^{\pm im\phi}(\phi + 2\pi),$$

which requires that $e^{2\pi mi} = 1$. It will only be true if m is an integer, i.e., $m = 0, \pm1, \pm2, \pm3, \ldots$, which is called the magnetic quantum number. To normalise the wave function, we use the condition

$$\int_0^{2\pi} FF^* d\phi = 1$$

$$\text{or } \int_0^{2\pi} (Ae^{im\phi})(Ae^{-im\phi})d\phi = 1$$

$$\text{or } A^2\int_0^{2\pi} e^{im\phi}\cdot e^{-im\phi}\ d\phi = 1$$

$$\text{or } A^2\int_0^{2\pi}[d\phi] = 1$$

$$\text{or } A^2\int_0^{2\pi}[\phi]_0^{2\tau} = 1$$

$$\text{or } A^2\cdot 2\pi = 1$$

$$\therefore\ A = \frac{1}{2\pi}$$

$$\therefore\ A = \sqrt{\frac{1}{2\pi}} \tag{7.147}$$

Hence, the normalised wave function can be written as

$$F = \frac{1}{\sqrt{2\pi}} e^{im\phi};\ m = 0,\ \pm 1,\ \pm 2$$

This solution consists of the quantity i, which is a complex quantity. Sometimes, for certain purposes, it is required to have the solution in real form. We have already found the solution of the differential Eq. (7.145) in case of particle on a ring. We have also found the normalisation constant, which is $\frac{1}{\sqrt{\pi}}$, and the normalised wave functions have been found to be

$$\psi_c = \frac{1}{\sqrt{\pi}} \cos m\phi \ \text{ and } \ \psi_s = \frac{1}{\sqrt{\pi}} \sin m\phi \tag{7.148}$$

where c and s represent cosine and sine, respectively.

7.5.2 *T* Equation

The T equation is $\frac{\sin\theta}{T} \cdot \frac{\mathrm{d}}{\mathrm{d}\theta}\left(\sin\theta \frac{\mathrm{d}T}{\mathrm{d}\theta}\right) + \frac{2IE}{\hbar^2} \sin^2\theta = m^2$

Putting $\frac{2IE}{\hbar^2} = \beta$, the above equation takes the form

$$\frac{\sin\theta}{T} \cdot \frac{\mathrm{d}}{\mathrm{d}\theta}\left(\sin\theta \frac{\mathrm{d}T}{\mathrm{d}\theta}\right) + \beta \sin^2\theta - m^2 = 0 \tag{7.149}$$

Equation (7.149) is multiplied by $\frac{T}{\sin^2\theta}$ to get

$$\frac{1}{\sin\theta} \cdot \frac{\mathrm{d}}{\mathrm{d}\theta}\left(\sin\theta \frac{\mathrm{d}T}{\mathrm{d}\theta}\right) + \left(\beta - \frac{m^2}{\sin^2\theta}\right) T = 0$$

$$\text{or } \frac{1}{\sin\theta} \sin\theta \frac{\mathrm{d}^2 T}{\mathrm{d}\theta} + \cos\theta \frac{\mathrm{d}}{\mathrm{d}\theta} + \left(\beta - \frac{m^2}{\sin^2\theta}\right) T = 0 \tag{7.150}$$

Now, putting $q = \cos\theta$ to change the variable in the equation, we can write

$$\frac{\mathrm{d}T}{\mathrm{d}\theta} = \frac{\mathrm{d}T}{\mathrm{d}q} \cdot \frac{\mathrm{d}q}{\mathrm{d}\theta} = \frac{\mathrm{d}T}{\mathrm{d}\theta}(-\sin\theta) \tag{7.151}$$

$$\text{Further, } \frac{\mathrm{d}^2 T}{\mathrm{d}\theta^2} = \frac{\mathrm{d}}{\mathrm{d}\theta}\left(\frac{\mathrm{d}T}{\mathrm{d}\theta}\right) \tag{7.152}$$

Replacing from Eq. (7.151), we obtain

$$\frac{\mathrm{d}^2 T}{\mathrm{d}\theta^2} = \frac{\mathrm{d}}{\mathrm{d}\theta}\left(-\sin\theta \frac{\mathrm{d}T}{\mathrm{d}q}\right) \text{ and differentiating } (\sin\theta) \cdot \left(\frac{\mathrm{d}T}{\mathrm{d}q}\right) \text{ as a product, we get}$$

$$\frac{\mathrm{d}^2 T}{\mathrm{d}\theta^2} = -\sin\theta \frac{\mathrm{d}}{\mathrm{d}\theta}\left(\frac{\mathrm{d}T}{\mathrm{d}q}\right) + \frac{\mathrm{d}}{\mathrm{d}\theta} + \frac{\mathrm{d}T}{\mathrm{d}q}(-\cos\theta) \tag{7.153}$$

We know that

$$\frac{d}{d\theta} = \frac{d}{dq} \cdot \frac{dq}{d\theta}$$

For the first term of Eq. (7.153), the equation may be expressed as

$$\frac{d^2T}{d\theta^2} = -\sin\theta\left(\frac{d^2T}{dq^2} \cdot \frac{dq}{d\theta}\right) + \frac{dT}{dq}(-\cos\theta); \quad \text{Since } q = \cos\theta \tag{7.154}$$

$$\therefore \quad \frac{dq}{d\theta} = -\sin\theta$$

Putting this value in Eq. (7.153), we get

$$\frac{d^2T}{d\theta^2} = \sin^2\theta \frac{d^2T}{dq^2} - \cos\theta \frac{dT}{dq} \tag{7.155}$$

Now the values of $dT/d\theta$ and $d^2T/d\theta^2$ from Eqs. (7.151) and (7.155), respectively, are substituted in Eq. (7.150) to obtain

$$\frac{1}{\sin\theta}\left[\sin\theta\left(\sin^2\theta \cdot \frac{d^2T}{dq^2} - \cos\theta \frac{dT}{dq}\right) - \sin\theta\cos\theta \frac{dT}{dq}\right] + \left(\beta - \frac{m^2}{\sin^2\theta}\right)T = 0$$

$$\text{or} \left(\sin^2\theta \cdot \frac{d^2T}{dq^2} - 2\cos\theta \frac{dT}{dq}\right)\left(\beta - \frac{m^2}{\sin^2\theta}\right)T = 0 \tag{7.156}$$

Remembering $q = \cos\theta$ and therefore $\sin\theta = (1 - q^2)$, the above equation reduces to

$$\left(1 - q^2\right)\frac{d^2T}{dq^2} - 2q\frac{dT}{dq} + \left(\beta - \frac{m^2}{1 - q^2}\right)T = 0 \tag{7.157}$$

If now we put $\beta = l(l+1)$, Eq. (7.157) becomes

$$\left(1 - q^2\right)\frac{d^2T}{dq^2} - 2q\frac{dT}{dq} + \left(l(l+1) - \frac{m^2}{1 - q^2}\right)T = 0 \tag{7.158}$$

Equation (7.158) is a famous equation known as the associated Legendre equation. The solutions of Eq. (7.158) are the Legendre polynomials $p_l^m(q)$ of degree l and order m, which is expressed in the following form:

$$p_l^m(q) = \left(1 - q^2\right)^{m/2} \cdot \frac{d^m p_l(q)}{dq^m} \tag{7.159}$$

when $m \not> l$, here m indicates the number of differentiation operations.

It should be kept in mind that the solution of F equation illustrates that m may have positive as well as negative values, but we cannot differentiate a function a negative number of times; therefore, the function $T(\theta)$ may denoted by

$$T(\theta) = P_i^{|m|}(q) \tag{7.160}$$

The quantum number P has only positive values as it has arisen from the *associated Legendre polynomials*. In case we consider the rotational spectra of diatomic molecules, the quantum number l is replaced by J and $T(\theta)$ is represented by

$$T(\theta) = P_i^{|m|}(q) \tag{7.161}$$

where $J = 0,\ 1,\ 2,\ 3$
$m = -J, -(J-1) \ldots -1, 0, +1 \ldots \mp J.$

After the normalisation of Eq. (7.161), we shall get

$$T(\theta) = \left[\frac{(2J+1(J-|m|)!)}{2(J+|m|)!} P_J^{|m|}(\cos\theta) \right] \tag{7.162}$$

7.5.3 Energy Levels

We have seen that constant $\beta = l(l+1)$, but β is also related to energy as

$$\beta = \frac{21E}{\hbar^2}$$

$$\therefore \quad \frac{21E}{\hbar^2} = \beta = l(l+1) = J(j+1)$$

$$\text{or } \ E = \frac{\hbar^2}{2I} J(J+1) \tag{7.163}$$

This equation will give the Eigen values of the energy of the rotor, and J is known as the rotational quantum number.

We can also write the energy levels of rigid rotor by using the following equation:

$$\frac{2IE}{\hbar^2} = m, \text{ where } m = 0, \pm 1, \pm 2$$

$$\text{or } E = \frac{m\hbar^2}{2I}$$

Thus, by putting the values of m, n^2, and I, the energy levels of the rotor can be evaluated.

7.6 Hermite Polynomials

The Hermite polynomials, $H_n(x)$, may be defied by the generating function.

$$\sum_{n=0}^{\infty} H_n(x) \frac{t^n}{n!} = e^{2tx-t^2}. \tag{7.165}$$

Expanding the function on the RHS by using Taylor's theorem, Eq. (7.165) yields

$$\sum_{n=0}^{\infty} H_n(x) \frac{t^n}{n!} = \sum_{n=0}^{\infty} \left[\frac{d^n}{dt^n} e^{2tx-t^2} \right]_{t=0} \frac{t^n}{n!} \tag{7.166}$$

Now, after equating the coefficients of t^n in Eq. (7.166), we shall get

$$H_n(x) = \left[\frac{d^n}{dt^n} \cdot e^{2tx - t^2}\right]_{t=0}$$

$$= \left[\frac{d^n}{dt^n} \cdot e^{x^2 - x^2 + 2tx - t^2}\right]_{t=0}$$

$$= e^{x^2}\left[\frac{d^n}{dt^n} \cdot e^{-x^2 + 2tx - t^2}\right]_{t=0}$$

$$= e^{x^2}\left[\frac{d^n}{dt^n} \cdot e^{-(x-t)^2}\right]_{t=0}$$

It is important to note that $\left[\frac{d^n}{dt^n}.e^{-(x-t)^2}\right] = (-1)^n \frac{d^n}{dx^n} e^{-(x-t)^2}$

Using this, the above equation

$$e^{x^2}\left[\frac{d^n}{dt^n} e^{-(x-t)^2}\right]_{t=0} \quad \text{may be rewritten as}$$

$$e^{x^2}\left[(-1)^n \frac{d^n}{dx^n} e^{-(x-t)^2}\right]_{t=0}$$

$$= (-1)^n e^{x^2} \frac{d^n}{dx^n} e^{-x^2}$$

$$\therefore \; H_n(x) = (-1)^n e^{x^2} \frac{d^n}{dx^n} e^{-x^2} \tag{7.167}$$

This is Rodrigue's formula for finding out the Hermite polynomials.

Now we shall find out some Hermite polynomials using Rodrigue's formula by putting different values $n = 0, 1, 2, 3$

$$H_0(x) = (-1)^0 e^{x^2} \frac{d^0}{dx^0} e^{-x^2}$$

$$= 1 \cdot e^{x^2} \cdot e^{-x^2} = 1$$

$$H_1(x) = (-1)^1 e^{x^2} \frac{d}{dx} e^{-x^2}$$

$$= (-1)^1 e^{x^2} (-2x)^{e^{-x^2}} = 2x$$

$$H_2(x) = (-1)^2 e^{x^2} \frac{d^2}{dx^2} e^{-x^2} = 1 \cdot e^{x^2} \frac{d}{dx}\left[(-2x) \cdot e^{-x^2}\right]$$

$$= e^{x^2}\left[-2e^{-x^2} + (-2x)(-2x)e^{-x^2}\right]$$

$$= e^{x^2}\left[-2e^{-x^2} + 4x^2 e^{-x^2}\right]$$

$$= e^{x^2} \cdot e^{-x^2}\left[-2 + 4x^2\right] = \left(4x^2 - 2\right)$$

Similarly, we can find $H_3(x), H_4(x), H_5(x), H_6(x)$, etc., with the help of Rodrigue's formula. A few values of Hermite polynomials have been given below:

$$H_3(x) = 8x^3 - 12x$$

$$H_4(x) = 16x^4 - 48x^2 + 12$$

$$H_5(x) = 32x^5 - 160x^3 + 120x$$

$$H_6(x) = 64x^6 - 480x^4 + 720x^2 - 120$$

7.6.1 Orthogonal Properties of Hermite Polynomials

The Hermite polynomials exhibit orthogonal properties, i.e.,

$$A\int_{-\infty}^{+\infty} e^{-x^2} H_n(x) H_m(x) \mathrm{d}x = 0 \text{ if } m \neq n$$

$$\text{and } B\int_{-\infty}^{+\infty} e^{-x^2} H_n(x) H_m(x) \mathrm{d}x = \text{const. if } m = n$$

$$A\int_{-\infty}^{+\infty} e^{-x^2} H_n(x) H_m(x) \mathrm{d}x = 0 \text{ if } m \neq n$$

We shall make use of the values of Hermite polynomials to prove

$$\int_{-\infty}^{+\infty} e^{-x^2} H_o(x) H_1(x) \mathrm{d}x$$

$$= \int_{-\infty}^{+\infty} e^{-x^2} (1)(2x) \mathrm{d}x$$

$$= \int_{-\infty}^{+\infty} e^{-x^2} \cdot 2x \, \mathrm{d}x = 2\int_{-\infty}^{+\infty} xe^{-x^2} \mathrm{d}x$$

$$= 2\left[x\int e^{-x^2}\mathrm{d}x - \int \mathrm{d}(x)\int e^{-x^2}\mathrm{d}x\right]_{-\infty}^{+\infty}$$

$$= 2\left[x \cdot \pi - \int \pi \mathrm{d}x\right]_{-\infty}^{+\infty} \quad \left[\because \int e^{-x^2}\mathrm{d}x = \pi\right]$$

$$= 2\pi[x - x]_{-\infty}^{+\infty} = 2\pi \cdot 0 = 0$$

Similarly, we can prove that

$$\int_{-\infty}^{+\infty} e^{-x^2} H_1(x) \cdot H_2(x) \mathrm{d}x = 0$$

Thus, $\int_{-\infty}^{+\infty} e^{-x^2} H_n(x) \cdot H_m(x) \mathrm{d}x = 0$ Proved

a. To prove that

$$\int_{-\infty}^{+\infty} e^{-x^2} H_n(x) \cdot H_m(x) \mathrm{d}x = \pi = \text{constant} \quad \text{if} \quad m = n$$

When $m = n$, then the above equation may be written as

$$\int_{-\infty}^{+\infty} e^{-x^2} H_n(x) \cdot H_m(x) \mathrm{d}x = \text{constant}$$

$$\text{or} \quad \int_{-\infty}^{+\infty} e^{-x^2} H_0(x) \cdot H_0(x) \mathrm{d}x$$

$$\text{or} \quad \int_{-\infty}^{+\infty} e^{-x^2} (1)(1) \cdot \mathrm{d}x = \int_{-\infty}^{+\infty} e^{-x^2} \mathrm{d}x \quad [\because \; H_0(x) = 1]$$

This is a standard integral whose value is π.

$$\therefore \quad \int_{-\infty}^{+\infty} e^{-x^2} \mathrm{d}x = \pi \text{ constant}$$

Similarly, we can prove that

$$\int_{-\infty}^{\pm\infty} e^{-x^2} H_1(x) \cdot H_1 dx = \text{constant}$$

$$\text{or} \quad \int_{-\infty}^{+\infty} e^{-x^2} (2x)(2x) \mathrm{d}x = \int_{-\infty}^{+\infty} e^{-x^2} \cdot 4x^2 \mathrm{d}x$$

$$\text{or} \quad 4 \int_{-\infty}^{+\infty} x^2 e^{-x^2} \mathrm{d}x$$

This is also a standard integral whose value is found to be $30\pi^{1/2}$ = constant.

Thus, $\int_{-\infty}^{+\infty} e^{-x^2} H_n(n) \cdot H_m(x) \mathrm{d}x = \text{constant} \quad \text{if} \quad m = n$ Q.E.D

We shall take the help of Hermite polynomials while discussing harmonic oscillator.

7.7 Simple Harmonic Oscillator

7.7.1 Classical Treatment

We know that the model of simple harmonic oscillator (SHO) is used to perceive the meaning of vibrations of a diatomic molecule to understand the vibrations of polyatomic molecules and to have the knowledge of the theory of solids to understand the vibrations in the solids.

Simple harmonic oscillation is a topic of considerable significance in quantum mechanics; it describes the motion of a particle, which travels in a potential proportional to the square of the displacement.

Let us consider a particle of mass m executing simple harmonic motion along the x axis; the equilibrium position of the particle lies at $x = 0$. The restoring force F is proportional to the displacement but acts in the opposite direction and hence we may write

$$F = -Kx \tag{7.168}$$

where K is the proportionality constant known as force constant.

The potential energy of any particle is related to the force as

$$\frac{\mathrm{d}V}{\mathrm{d}x} = -F \tag{7.169}$$

$$\text{or } \frac{\mathrm{d}V}{\mathrm{d}x} = -F = -(-Kx) = Kx$$

$$\text{or } \mathrm{d}V = Kx\mathrm{d}x$$

On integration, we get

$$\int \mathrm{d}V = K\int x\mathrm{d}x$$

$$\text{or } V = K\frac{x^2}{2} + c \tag{7.170}$$

where V = potential energy
c = integration constant.

When potential energy vs. x is plotted, a graph is obtained, as shown in Figure 7.13.

It is clear from the graph that at $x = 0$, $V = 0$.

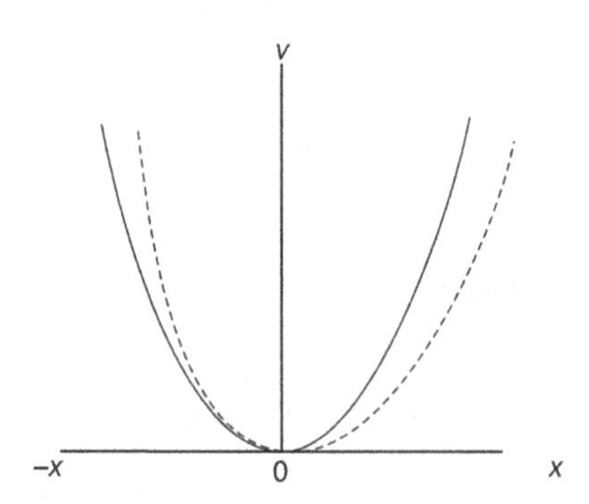

FIGURE 7.13 Potential energy of a linear harmonic oscillator.

Putting these values in Eq. (7.170), i.e., $V = \frac{1}{2}Kx^2 + C$, we get

$$0 = \frac{1}{2}K \times 0 + C \quad \therefore C = 0$$

$\therefore$ Equation (7.170) becomes

$$V = \frac{1}{2}Kx^2 \tag{7.171}$$

We have assumed above that the particle will execute simple harmonic motion and constitute a linear harmonic oscillator. For such oscillators,

$$m\frac{d^2x}{dt^2} = -Kx$$

$$\text{or } \frac{d^2x}{dt^2} = -\frac{K}{m}x \tag{7.172}$$

But in case of simple harmonic motion,

$$x = A \sin \omega t \tag{7.173}$$

$$\text{or } \quad x = A\sin 2\pi\upsilon \quad [\text{where } \upsilon = \text{frequency of vibration}]$$

$$\frac{d^2x}{dT} = (2\pi\upsilon)A\cos 2\pi\upsilon t$$

$$\frac{d^2x}{dt^2} = -(2\pi\upsilon)(2\pi\upsilon)A\cos 2\pi\upsilon t$$

$$= -\left(4\pi^2\upsilon^2\right) A\sin 2\pi\upsilon t$$

Equating Eqs. (7.172) and (7.174), we get

$$-k/m \; x = -4\pi^2\upsilon^2 x$$

$$\text{or } k/m = 4\pi^2\upsilon^2$$

$$k = 4\pi^2\upsilon^2 m \tag{7.175}$$

But we have obtained

$$\text{that } V = \frac{1}{2}Kx^2$$

$$\text{or, } V = \frac{1}{2}4\pi^2\upsilon^2 mx^2$$

$$V = 2\pi^2\upsilon^2 mx^2 \tag{7.176}$$

7.7.2 Quantum Mechanical Treatment

Now we shall put the value of potential energy V in one-dimensional Schrödinger equation:

$$\frac{d^2\psi}{dx^2} + \frac{8\pi^2 m}{h^2}(E - V)\psi = 0 \text{ to get}$$

$$\frac{d^2\psi}{dx^2} + \frac{8\pi^2 m}{h^2}\left(E - 2\pi^2\upsilon^2 m x^2\right)\psi = 0$$

$$\frac{d^2\psi}{dx^2} + \left(\frac{8\pi^2 mE}{h^2} - \frac{16\pi^4\upsilon^2 m^2}{h^2}\cdot x^2\right)\psi = 0$$

$$\text{or } \frac{d^2\psi}{dx^2} + \left(a - b^2x^2\right)\psi = 0 \tag{7.177}$$

$$\text{where } a = 8\pi^2 mE/h^2$$

$$\text{and } b^2 = \frac{16\pi^4\upsilon^2 m^2}{h^2}$$

$$\therefore\ b = \frac{4\pi^2\upsilon m}{h^2}$$

Equation (7.177) can be solved by the following two methods:

- Asymptotic solution
- Series solution

We shall discuss the methods one by one.

7.7.2.7 Asymptotic Solution

Let us suppose that Eq. (7.177) is valid for all values of x. If x is very large, then a can be neglected, and the solution obtained this way is known as *asymptotic solution*. Under the above condition, Eq. (7.177) takes the form

$$\frac{d^2\psi}{dx^2} - b^2x^2\psi = 0 \tag{7.178}$$

We then make x small and express ψ as a power series in x. The solutions obtained by the two methods are multiplied together to give the general solution.

Further, let e^{mx} be the solution of Eq. (7.178), then

$$\psi = e^{mx}$$

Differentiating with respect to x, we get $\frac{d\psi}{dx} = me^{mx}$ and $\frac{d^2\psi}{dx^2} = m^2e^{mx}$.

Putting the value of @@ in Eq. (7.178), we obtain

$$m^2e^{mx} - b^2x^2e^{mx} = 0$$

$$\text{or } e^{mx}\left(m^2 - b^2x^2\right) = 0$$

$$\text{But } e^{mx} \neq 0$$

$$\therefore\ m^2 = b^2x^2 = 0$$

$$\text{or } m^2 = b^2x^2 \quad \therefore m = bx$$

$$\text{Hence the solution becomes } \psi = e^{bx^2} \tag{7.179}$$

But on closer scrutiny, this is found not to be a solution of Eq. (7.178) because if we substitute this value of ψ in the said equation, it is not satisfied.

Let us then suppose that $\psi = e^{\rho x}$ is the true solution, then

$$\frac{d\psi}{dx} = 2x\rho e^{\rho x^2} \text{and}$$

$$\frac{d^2\psi}{dx^2} = 2\rho e^{\rho x^2} + (2x\rho)\rho \cdot e^{\rho x^2} \cdot (2_x)$$

$$= 2\rho e^{\rho x^2} + 4\rho^2 x^2 e^{\rho x^2} = e^{\rho x^2}\left[4x^2\rho^2 + 2\rho\right]$$

When $x \to \infty$, we can neglect the second term in comparison to the first and putting the remaining value in Eq. (7.178), we get

$$4x^2\rho^2 . e^{\rho x^2} - b^2x^2 e^{\rho x^2} = 0$$

$$\text{or } e^{\rho x^2}\left[4x^2\rho^2 - b^2x^2\right] = 0$$

$$\text{or } 4x^2\rho^2 - b^2x^2 = 0 \quad \left[\because\ e^{\rho x^2} \neq 0\right]$$

$$\text{or } 4\rho^2 = b^2 \quad \therefore\ p^2 = b^2/4 \quad \therefore\ \rho = \pm b/2$$

$$\therefore\ \Psi = e^{-bx^2/2}$$

But $\Psi = e^{-bx^2/2}$ is not the acceptable solution since it leads to $\Psi = \infty$ when x is large, while in actual case Ψ should be finite; therefore, the other one is the required solution.

$$\Psi = e^{-bx^2/2} \tag{7.180}$$

7.7.2.2 Series Solution

$$\text{Let } \Psi = a_0 + a_1x + a_2x^2 + \ldots + a_nx^n$$

$$= \sum a_n x^n = f(x) \tag{7.181}$$

Then the general solution of linear harmonic oscillator will be expressed as

$$\Psi = e^{-bx^2/2} \cdot f(x) \tag{7.182}$$

$f(x)$ is a power series in x, and this type of power series is called Hermite polynomials. Therefore, Eq. (7.182) may be written as

$$\Psi = e^{-bx^2/2} \cdot H_n(x) \tag{7.183}$$

where $H_n(x) =$ Hermite polynomials

$$\text{Now let } p = x\sqrt{b} \quad \therefore \frac{dp}{dx} = \sqrt{b}$$

$$\therefore \frac{d^2\psi}{dx^2} = \frac{d}{dx}\left(\frac{d\psi}{dx}\right) = \frac{d}{dx}\left(\frac{d\psi}{dp}\cdot\frac{dp}{dx}\right)$$

$$= \frac{d}{dp}\left(\frac{d\psi}{dp}\cdot\frac{dp}{dx}\cdot\frac{dp}{dx}\right)$$

$$= \frac{d^2\psi}{dp^2}\left(\frac{dp}{dx}\right)^2 = \frac{d^2\psi}{dp^2}\cdot b \tag{7.184}$$

Replacing this value in Eq. (7.177), i.e.,

$$\frac{d^2\psi}{dx^2} + \left(a - b^2x^2\right)\psi = 0 \text{ where, } a = 8\pi^2 mE/h^2 \text{ and } b = 4\pi^2 \upsilon m/h$$

We get

$$b\frac{d^2\psi}{dp^2} + \left(a - b^2x^2\right)\psi = 0$$

$$\text{or } b\frac{d^2\psi}{dp^2} + \left(a - b\cdot bx^2\right)\psi = 0$$

$$\text{or } b\frac{d^2\psi}{dp^2} + \left(a - bp^2\right)\psi = 0$$

$$\text{or } b\frac{d^2\psi}{dp^2} + \left(a/b - p^2\right)\psi = 0 \tag{7.185}$$

Since ψ should be finite and single valued when $p = 0$, then Eq. (7.185) becomes

$$\frac{d^2\psi}{dp^2} + \frac{a}{b}\psi = 0 \text{ when } P \to 0 \tag{7.186}$$

Now the general solution is

$$\psi = e^{-bx^2/2} \cdot H_n(x)$$

$$\text{or } \psi = e^{-p^2/2} \cdot H_n(x)$$

When $n = 0$, ψ becomes

$$\psi = e^{-p^2/2} H_0(x) = e^{-p^2/2} \cdot 1 \quad (\text{since } H_0(x) = 1)$$

$$= e^{-p^2/2}$$

$$\therefore \quad \frac{d\psi}{dp} = \left(\frac{-2p}{2}\right) \cdot e^{-p^2/2} = -pe^{-p^2/2}$$

$$\therefore \quad \frac{d^2\psi}{dp^2} = (-p)e^{-p^2/2}\left(\frac{-2p}{2}\right) + e^{-p^2/2}(-1)$$

$$= p^2 e^{-p^2/2} - e^{-\frac{p^2}{2}} = e^{-p^2/2}\left(p^2 - 1\right)$$

Putting this value in equation $d^2\psi/dp^2 + a/b^{\psi} = 0$,

$$\text{we shall get} = e^{-p^2/2}\left(p^2 - 1 + a/b\right) = 0$$

$$\text{But } e^{-p^2/2} \neq 0 \quad \therefore \left(p^2 - 1 + a/b\right) = 0$$

$$\therefore \quad \frac{a}{b} - 1 = 0 \text{ when } p \to 0$$

$$\therefore \quad \frac{a}{b} = 1 \text{ when } p = 0 \tag{7.187}$$

Now when $H_1(x) = 2x$, then $\psi = e^{-p^2/2} \cdot 2x = 2\, e^{-p^2/2} \cdot p/\sqrt{b} \quad \left[\because\ x = \frac{p}{\sqrt{b}}\right]$

$$= \frac{2}{\sqrt{b}}\left(p \cdot e^{-p^2/2}\right)$$

$$\therefore \quad \frac{d\psi}{dp} = \frac{2}{\sqrt{b}}\left[p\, e^{-p^2/2}(-p) + e^{-p^2/2} \cdot 1\right]$$

$$= \frac{2}{\sqrt{b}}\left[-p^2 e^{-p^2/2} + e^{-p^2/2}\right]$$

$$\therefore \quad \frac{d\psi}{dp} = \frac{2}{\sqrt{b}}\left[\left(-p^2\right)e^{-p^2/2} \cdot (-p) + e^{-p^2/2} \cdot (-2p) + e^{-p^2/2} \cdot (-p)\right]$$

$$= \frac{2}{\sqrt{b}}\left[-p^3 e^{-p^2/2} + 3p\, e^{-p^2/2}\right]$$

$$= \frac{2e^{-p^2/2}}{\sqrt{b}}\left[p^3 - 3p\right] \tag{7.188}$$

Now putting this value in the equation

$$\frac{d^2\psi}{dp^2} + a/b\psi = 0, \text{ we obtain}$$

$$\frac{2e^{-p^2/2}}{\sqrt{b}}\left[p^3-3p\right]+\left[\frac{a}{b}\cdot e^{-p^2/2}\cdot 2x\right]=0$$

$$=\frac{2e^{-p^2/2}}{\sqrt{b}}\left[p^3-3p\right]+\left[\frac{a}{b}\cdot e^{-p^2/2}\cdot 2\frac{p}{\sqrt{b}}\right]=0$$

$$=\frac{2}{\sqrt{b}}e^{-p^2/2}\left[p^3-3p+\frac{a}{b}\cdot p\right]=0$$

$$=\frac{2}{\sqrt{b}}e^{-p^2/2}\cdot p\left[p^2-p^3+a/b\right]=0$$

$$\text{But } p\cdot e^{-p^2/2}\neq 0$$

$$\therefore\; p^2-3\;a/b=0$$

$$\text{or } \frac{a}{b}=3 \text{ when } p=0 \tag{7.189}$$

Similarly, we can prove that $a/b \rightarrow 5,7,\ldots$, which may be expressed as general term as

$$\frac{a}{b}=(2n+1)$$

$$\text{But } a/b=\frac{8\pi^2 mE}{h^2}\bigg/\frac{4\pi^2\upsilon m}{h}=2E/h\upsilon$$

$$\therefore\; (2n+1)=2E/h\upsilon$$

$$\therefore\; 2E=(2n+1)/h\upsilon$$

$$\therefore\; E=\left(\frac{2n+1}{2}\right)h\upsilon=\left(n+\frac{1}{2}\right)h\upsilon \tag{7.190}$$

where n = vibrational quantum number = 0, 1, 2, 3, …

When $n=0$, $E=E_0$

$$\therefore E_0=(0+1/2)h\upsilon_0=\frac{1}{2}h\upsilon_0$$

$$\therefore\; E_0=\frac{1}{2}h\upsilon_0 \tag{7.191}$$

which is called ZPE.

Equation (7.190) is the equation for energy of harmonic oscillator.

The energy levels of linear harmonic oscillator are shown in Figure 7.14.

It is clear from the graph that for the lowest energy level (n = 0), there is a maximum probability of oscillator being at the equilibrium position. Further examination of the figure reveals that as the quantum number increases, the probability of finding the linear harmonic oscillator becomes greatest towards the extremities of its movement.

It is also pertinent to note that the energy of linear harmonic oscillator depends on the value of n.

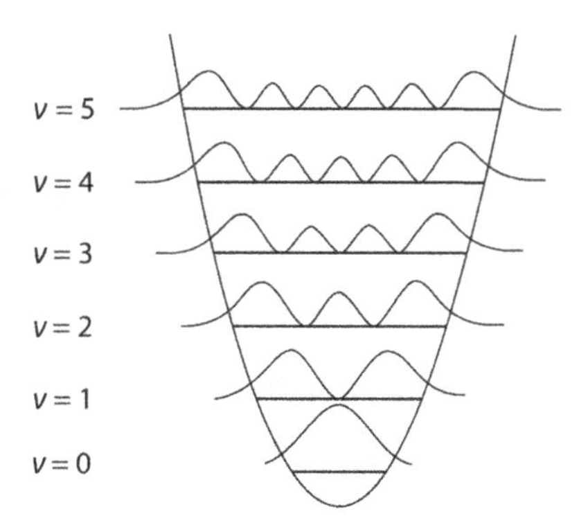

FIGURE 7.14 Energy levels of a linear harmonic oscillator.

7.7.3 Wave Function of Linear Harmonic Oscillator

We have found that the wave function of linear harmonic oscillator may be expressed as

$$\psi = e^{-p^2/2} H_n(x) \quad \text{where,} \quad p = x\sqrt{b}$$

$$\text{or} = e^{-bx^2/2} H_n(x) \tag{7.192}$$

Thus, the value of wave function of linear harmonic oscillator will depend on the value of Hermite polynomial. The value of $H_n(x)$ changes with the value of n. We have already given different values of Hermite polynomials by putting $n = 0, 1, 2, \ldots$.

Alternatively: The series solution can also be represented as follows. For this let

$$\psi = a_0 + a_1 x + a_2 x^2 + a_3 x^3 \ldots$$

$$\sum a_n x^n = f(x) \tag{7.193}$$

Then, the general solution of linear harmonic oscillator is represented by

$$\psi = e^{-bx^2/2} \sum a_n x^n = e^{-bx^2/2} \cdot f(x) \tag{7.194}$$

$$\text{Let} \quad x\sqrt{b} = y$$

$$\therefore \quad y^2 = bx^2$$

$$\therefore \quad \psi = e^{-y^2/2}$$

$$\frac{d\psi}{dy} = -ye^{-y^2/2} \text{ and}$$

$$\begin{aligned} \frac{d^2\psi}{dy^2} &= (-y)e^{-y^2/2}(-y) + e^{-y^2/2}(-1) \\ &= e^{-y^2/2}\left[y^2 - 1\right] = e^{-y^2/2}\left[bx^2 - 1\right] \end{aligned} \tag{7.195}$$

$$\text{again} \quad \psi = e^{-bx^2/2}$$

$$\therefore \quad \frac{d\psi}{dx} = (-b/2)(2x)e^{-bx^2/2} = (-bx)e^{-bx^2/2}$$

$$\therefore \quad \frac{d^2\psi}{dx^2} = -be^{-bx^2/2} + (-bx)\left(\frac{-2x}{2}\right)b \cdot e^{-bx^2/2}$$

$$= -be^{-bx^2/2} + b^2x^2e^{-bx^2/2} \tag{7.196}$$

$$= b\left(bx^2 - 1\right)e^{-bx^2/2}$$

Therefore, from Eqs. (7.195) and (7.196), we can write

$$b\frac{d^2\psi}{dy^2} = \frac{d^2\psi}{dx^2} \tag{7.197}$$

Therefore, Eq. (7.177) takes the form

$$b\frac{d^2\psi}{dy^2} + \left(a - by^2\right)\psi = 0 \tag{7.198}$$

$$\text{But } \psi = e^{-bx^2/2} \cdot f(x) \tag{7.199}$$

$$\text{or } \psi = e^{-y^2/2} \cdot \phi\left(\frac{y}{\sqrt{b}}\right)$$

$$\text{or} \quad \psi = e^{-y^2/2} \cdot f(y), \quad \text{where, } f(y) = \phi\left(\frac{y}{\sqrt{b}}\right)$$

Now let us express $f(y)$ in power series as $f(y) = a_0 + a_1y + a_2y^2 + \ldots$ We have written above that

$$\psi = e^{-y^2/2} \cdot f(y) = e^{-y^2/2} \cdot f \quad \left(\text{where } f(y) = f\right)$$

Differentiating this equation with respect to y, we shall get

$$\frac{d\psi}{dy} = e^{-y^2/2}f' + (-y)fe^{-y^2/2} = e^{-y^2/2}(f' - yf)$$

$$\text{and } \frac{d^2\psi}{dy^2} = e^{-y^2/2}\left[f'' - 2yf' + \left(y^2 - 1\right)f\right]$$

Putting the value of $d^2\psi/dy^2$ in the Eq. (7.177), we get

$$be^{-y^2/2}\left[f'' - 2yf' + \left(y^2 - 1\right)f\right] + \left(a - by^2\right)e^{-y^2/2} \cdot f = 0$$

$$\text{or } \left[y'' - 2yf'' + \left(y^2 - 1\right)f\right] + \left(\frac{a}{b} - y^2\right)f = 0$$

$$\text{But } \quad f = a_0 + a_1y + a_2y^2 + a_3y^3 + \ldots$$

On differentiating this equation, we get

$$f' = a_1 + 2a_2 y + 3a_3 y^2 + \ldots$$

$$\text{and} \quad f'' = 2a_2 + 6a_3 y + \ldots$$

Substituting the values in the above equation, we get

$$2a_2 + 6a_3 y + 4.3a_4 y^2 + \ldots + (n+1)(n+2)a_{n+2} y^n$$

$$-2y\left(a_1 + 2a_2 y + 3a_3 y^2 + \ldots\right)$$

$$+\left(a_0 + a_1 y + a_2 y^2 + \ldots\right)(a/b - 1) = 0$$

Now, the coefficient of individual power of y must be zero in order that the whole may be equal to zero.

$$\therefore \quad 2a_2 + (a/b - 1)a_0 = 0 \quad \text{for} \quad y^0$$

$$6a_3 - 2a_1 + (a/b - 1)a_1 = 0 \quad \text{for} \quad y^1$$

$$\text{and} \quad (n+1)(n+2)a_{n+2} - 2na_n + (a/b - 1)a_n = for \;\; y''$$

In general, we have

$$\frac{2n - (a/b - 1)}{(n+1)(n+2)} = \frac{a_{n+2}}{a_n} \tag{7.200}$$

This is called the *recursion formula*. If n is very large, then

$$\lim_{n \to \infty} \frac{a_{n+2}}{a_n} = \frac{2}{n} \approx 0$$

Therefore, from Eq. (7.200), we can write that

$$2n - (a/b - 1) = 0$$

$$\text{or} \quad 2n - a/b + 1 = 0$$

$$\text{or } a/b = (2n+1) \tag{7.201}$$

But we know that

$$a = 8\pi^2 mE/h^2 \text{ and } b = \frac{4\pi^2 vm}{h}$$

$$\therefore \; a/b = \frac{2E}{hv} \tag{7.202}$$

From Eqs. (7.201) and (7.202), we can write

$$\frac{2E}{hv} = (2n+1)$$

$$\text{or } \frac{E}{hv} = \left(\frac{2n+1}{2}\right) = \left(n + \frac{1}{2}\right)$$

$$\therefore \quad E = \left(n + \frac{1}{2}\right)hv \tag{7.203}$$

where n is either zero or any other integer.

The result implies that the Schrödinger equation for linear harmonic oscillator can have physically acceptable solutions only for a certain discrete value of energy. When $n = 0$, Eq. (7.203) takes the form

$$E_0 = \left(0 + \frac{1}{2}\right)hv_0 = \frac{1}{2}hv_0$$

$$\therefore \quad E_0 = \frac{1}{2}hv_0 \tag{7.204}$$

This is the lowest possible energy for linear harmonic oscillator, which is called the ZPE.

BIBLIOGRAPHY

Anderson, J.M. 1968. *Introduction to Quantum Chemistry*. New York: W.A. Benjamin Inc.
Flurry, R.L. Jr. 1983. *Quantum Chemistry*. Upper Saddle River, NJ: Prentice-Hall Inc.
Liboff, R.L. 1998. *Introductory Quantum Mechanics*, 3rd ed. New York: Addison-Wesley Publishing Co.
McQuarrie, D.A. and J.D. Simon. 2006. *Physical Chemistry*. New Delhi: Viva Books Pvt Ltd.
Peacock, T.E. 1968. *Foundations of Quantum Chemistry*. London: Wiley.
Pilar, F.L. 1990. *Elementary Quantum Chemistry*. New York: McGraw-Hill.
Rastogi, R.P. 1986. *An Introduction to Quantum Mechanics of Chemical Systems*. New Delhi: Oxford and IBH Publishing Co.

Solved Problems

Problem 1. Consider an electron in a one-dimensional box of length 258 pm.

a. What will be the ZPE of the system?
b. What electronic speed classically corresponds to the above-mentioned ZPE? Compare it to the speed of light.

Solution: (a) Given that a = 258 pm = 258×10^{-12}m; $m = 9.11 \times 10^{-31}$kg; n = 1; ZPE = E_0 =?; ZPE = $E_0 = n^2h^2/8ma^2$

$$= \frac{1^2\left(6.626\times10^{-34}\,\text{Js}\right)^2}{8\left(9.11\times10^{31}\text{kg}\right)\left(258\times10^{-12}\text{m}\right)^2}$$

$$= \frac{(6.626)^2\left(10^{-68}\right)}{8\times9.11\times10^{-31}\times(258)^2\times10^{-24}}$$

$$= \frac{43.90\times10^{-68}}{72.88\times66564\times10^{-31}\times10^{-24}}$$

$$= \frac{43.90 \times 10^{-68}}{48.512 \times 10^{-50}} = 0.90 \times 10^{-18}\,\text{J}$$

$$= 9.0 \times 10^{-19}\,\text{J}$$ Answer.

b. Since potential energy $V = 0$ in the box, E is all kinetic energy and it is equal to

$$E = mv^2/2$$

$$\therefore\ v^2 = \frac{2E}{m}$$

$$\text{or}\quad v = \left(\frac{2E}{m}\right)^{1/2}$$

$$= \left[\frac{2\left(9.0 \times 10^{-19} J\right)}{9.11 \times 10^{-31}}\right]^{1/2}$$

$$= \left[\frac{18 \times 10^{-19} J}{9.11 \times 10^{-31}}\right]^{1/2} = \left[1.976 \times 10^{12}\right]^{1/2}$$

$$= 1.40 \times 10^{6}\ \text{ms}^{-1}$$

We know that the velocity of light is $2.998 \times 10^8\ \text{ms}^{-1} = c$

$$\therefore\ \frac{v}{c} = \frac{1.40 \times 10^{6}\ \text{ms}^{-1}}{2.998 \times 10^{8}\ \text{ms}^{-1}} = 0.4669 \approx 0.467$$

or it is about 0.5% of the speed of light Answer.

Problem 2. Find the lowest kinetic energy of an electron in a three-dimensional box of dimensions 0.1×10^{-15}m, 1.5×0^{-15}m and 2×10^{-15}m.

Solution: Given that $a = 0.1 \times 10^{-15}$m; $b = 1.5 \times 10^{-15}$m; $c = 2 \times 10^{-15}$m

$$\text{For lowest KE,}\quad n_x = n_y = n_z = 1$$

$$m = 9.11 \times 10^{-31}\text{kg}$$

We know that

$$E = \frac{h^2}{8m}\left[\frac{n_x^2}{a^2} + \frac{n_y^2}{b^2} + \frac{n_z^2}{c^2}\right]$$

$$= \frac{h^2}{8m}\left[\frac{1}{a^2} + \frac{1}{b^2} + \frac{1}{c^2}\right]$$

$$= \frac{\left(6.627 \times 10^{-34}\right)^2}{8 \times 9.11 \times 10^{-31}}\left[\frac{1}{\left(0.1 \times 10^{-15}\right)^2} + \frac{1}{\left(1.5 \times 10^{-15}\right)^2} + \frac{1}{\left(2 \times 10^{-15}\right)^2}\right]$$

$$= \frac{43.917 \times 10^{-68}}{72.88 \times 10^{-31}} \left[\frac{1}{(0.1 \times 10^{-30})} + \frac{1}{(2.25 \times 10^{-30})} + \frac{1}{(4.10^{-30})} \right]$$

$$= 0.6025 \frac{\times 10^{-34} \times 1}{10^{-30}} \left(\frac{1}{0.001} + \frac{1}{2.25} + \frac{1}{4} \right)$$

$$= 0.6025 \times 10^{-7} (100 + 0.444 + 0.250)$$

$$= 0.6025 + 100.7 \times 10^{-7} = 60.67 \times 10^{-7}$$

$\therefore \; E = 6.067 \times 10^{-8}\,\text{J}$ Answer.

Problem 3. Determine the degeneracy of the energy level $17h^2/8ma^2$ of a particle in a cubical box.

Solution: The energy of a particle in box is given by

$$E = \frac{h^2}{8ma^2}\left(n_x^2 + n_y^2 + n_z^2\right) = \frac{17h^2}{8ma^2}$$

It is clear from this that

$$n_x^2 + n_y^2 + n_z^2 = 17$$

Possible arrangements for the sum of squared terms will be 17 as follows:

n_x	n_y	n_n
2	2	3
2	3	2
3	2	2

All three sets of quantum numbers give the same energy; therefore, the degree of degeneracy is 3 or we may say that the energy level is three-fold degenerate. Answer.

Problem 4. Calculate the percentage change in a given energy for a particle in a cubical box when edge of the cube is decreased by 10%.

Solution: We know that $E = \frac{h^2}{8ma^2}\left(n_x^2 + n_y^2 + n_z^2\right)$

Therefore, the ground state energy of a particle in a cubical box given by

$$E_{111} = \frac{3h^2}{8m^2}$$

where a = edge length of the cube. The edge has been decreased by 10%.

$$\therefore a' = a - \frac{10a}{100} = a - \frac{a}{10} = \frac{9a}{10}$$

$$\therefore E_{111}^1 = \frac{3h^2}{8ma^2} = \frac{3h^2}{8m(9a/10)^2}$$

$$= \frac{3h^2}{8ma^2} \times \frac{100}{81}$$

$$\therefore \quad E_{111}^{1} - E = \frac{3h^2}{8ma^2}\left(\frac{100}{81} - 1\right) = \frac{3h^2}{8ma^2} \times \frac{19}{81}$$

∴ % change in the given energy will be

$$\frac{3h^2}{8ma^2} \times \frac{19}{81} \times 100 = \frac{1900}{81} = 23.45\%$$

Answer.

Problem 5. A particle is moving in a one-dimensional box (of infinite height) of width 25 A. Calculate the probability of finding the particle within an interval of 5 A at the centre of the box when it is in its state of least energy.

Solution: We know that

$$\psi(x) = \left(\frac{2}{a}\right)^{1/2} \sin\frac{n\pi x}{a}$$

According to the question, when the particle is in the least energy state, n will be equal to 1. Hence, the above equation takes the form

$$\psi(x) = \left(\frac{2}{a}\right)^{1/2} \sin\frac{\pi x}{a}$$

At the centre of the box, $x = a/2$. The probability of finding the particle in the unit interval at the centre of the box is given by

$$[\psi(x)]^2 = \left[\left(\frac{2}{a}\right)^{1/2} \sin\frac{\pi(a/2)}{a}\right]^2 = \frac{2}{a}\sin^2 \pi/2 = 2/a$$

The probability P in the interval Δx is given by

$$P = [\psi(x)]^2 \,\Delta x = \frac{2}{a} \cdot \Delta x$$

In this case, $a = 25$ Å $= 25 \times 10^{-10}$m and

$$\therefore \quad \Delta x = 5\text{Å} = 5 \times 10^{-10}\,\text{m}$$

$$P = \frac{2 \times 5 \times 10^{-10}}{25 \times 10^{-10}} = 0.4$$

Answer.

Problem 6. For a particle in a one-dimensional box, prove that the average value of momentum along the x axis is zero.

Solution: The average value of momentum is given by

$$\bar{P}x = \left(\frac{2}{a}\right)\int_0^a \sin\left(\frac{n\pi x}{a}\right)\left(\frac{h}{2\pi i}\frac{d}{x}\right)\sin\left(\frac{n\pi x}{a}\right)dx$$

$$= \left(\frac{2}{a}\right)\left(\frac{h}{2\pi i}\right)\int_0^a \sin\frac{n\pi x}{a}\cos\left(\frac{n\pi x}{a}\right)dx$$

$$= \left(\frac{2}{a}\right)\frac{h}{4\pi i}\frac{1}{2}\int_0^a 2\sin\frac{n\pi x}{a}\cdot\cos\left(\frac{n\pi x}{a}\right)dx$$

$$= \left(\frac{2}{a}\right)\left(\frac{h}{4\pi i}\right)\int_0^a \sin\frac{2n\pi x}{a}dx$$

$$= \left(\frac{2}{a}\right)\left(\frac{h}{4\pi i}\right)\left(\frac{2n\pi}{a}\right)\int_0^a \sin\frac{2n\pi x}{a}dx$$

$$= \frac{nh}{2ia^2}\left[\cos\frac{2n\pi x}{a}\right]_0^a\cdot\frac{2n\pi}{a}$$

$$= \frac{n^2h\pi}{ia^3}[1-1]=0$$

Proved.

Problem 7. Using the approximate ground state wave function $\psi = \left(xa - x^2\right)$ for a particle in a one-dimensional box ($0 \le x \le a$), show that the approximate energy is $E = \frac{10}{\pi}\cdot\frac{\hbar^2}{8\pi^2 m}$

Solution: The wave function $\psi = \left(xa - x^2\right)$ and the boundary condition for one-dimensional box are $\psi = 0$ at $x = 0$ and $\psi = 0$ at $x = a$. Thus, ψ satisfies the boundary condition.

The approximate energy $E = \dfrac{\int_o^a \psi H\psi^* dx}{\int_0^a \psi\psi^* dx}$

now, $$\int_0^a \psi H\psi^* dx = \int_0^a \left(xa - x^2\right)\left(-\frac{h^2}{8\pi^2 m}\cdot\frac{d^2}{dx^2}\right)\times x(a-x)dx$$

$$= \int_0^a \left(xa - x^2\right)\left(-\frac{h^2}{8\pi^2 m}\right)\frac{d^2}{dx^2}\times\left(xa - x^2\right)dx$$

$$= \int_0^a \left(xa - x^2\right)\left(-\frac{h^2}{8\pi^2 m}\right)(-2)dx$$

$$= \int_0^a \frac{2h^2}{8\pi^2 m}\left(xa - x^2\right)dx$$

$$= \frac{2h^2}{8\pi^2 m}\left[\int_0^a xadx - \int_0^a x^2 dx\right]$$

$$= \frac{h^2}{4\pi^2 m}\left[\frac{x^2 a}{2} - \frac{x^3}{3}\right]_0^a = \frac{h^2}{4\pi^2 m}\left[\frac{a^3}{2} - \frac{a^3}{3}\right]$$

$$= \frac{a^3}{6}\left(\frac{h^2}{4\pi^2 m}\right) = \frac{a^2 h^2}{24\pi^2 m}$$

$$\text{further, } \int_0^a \psi\psi^* \mathrm{d}x = \int_0^a \left(xa - x^2\right)\left(xa - x^2\right)\mathrm{d}x$$

$$= \int_0^a \left(x^2a^2 + x^4 - 2ax^2\right)\mathrm{d}x$$

$$= \left(\frac{a^5}{3} + \frac{a^5}{5} - a^5/2\right)$$

$$= a^5\left[\frac{1}{3} + \frac{1}{5} - \frac{1}{2}\right]$$

$$= a^5\left[\frac{10 + 6 - 15}{30}\right] = \frac{a^5}{30}$$

$$\therefore\ E = \frac{h^2a^3}{24\pi^2 m}\Big/\frac{a^5}{30} = \frac{30h^2a^3}{24\pi^2 ma} = \frac{10}{\pi^2}\frac{h^2}{8ma^2}$$

Problem 8. (a) Normalise $\psi_1(x) = A_1 e^{-\alpha x^2}$ and $\psi_2(x) = A_2 e^{-\alpha x^2}$ over the interval $-\infty < x < \infty$, are the function orthogonal?

(b) Are the functions orthogonal in the interval $0 < x < \infty$.

Solution: (a) The condition of normalisation is $\int_{-\infty}^{+\infty} \psi_1(x)\ \psi_1(x)\mathrm{d}x = 1$

Putting the values of $\psi_1(x)$ in the above equation, we shall get

$$\int_{-\infty}^{+\infty} A_1 e^{-ax^2} A_1 e^{-ax^2} \mathrm{d}x = 1$$

$$\text{or } A_1^2 \int_{-\infty}^{+\infty} e^{-2ax^2} \mathrm{d}x = 1$$

$$\text{or } A_1^2 \left(\frac{\pi}{2\alpha}\right)^{1/2} = 1$$

$$A_1 = \left(\frac{2\alpha}{\pi}\right)^{1/4}$$

$$\text{Again } \int_{-\infty}^{\infty} \psi_2(x)\psi_2(x)\mathrm{d}x = 1$$

$$\text{or } \int_{-\infty}^{\infty} xe^{-\alpha x^2} xe^{-\alpha x^2} \mathrm{d}x = 1$$

$$\text{or } A_2^2 \int_{-\infty}^{\infty} x^2 e^{2\alpha x^2} \mathrm{d}x = 1$$

$$\text{or} \quad \frac{A_2^2}{4\alpha}\left(\frac{\pi}{2\alpha}\right)^{1/2} = 1$$

$$\therefore \; A_2^2 = \frac{4\alpha}{(\pi/2\alpha)^{1/2}}$$

$$\therefore \; A_2 = \frac{(4\alpha)^{1/2}}{(\pi/2\alpha)^{1/4}}$$

Now we shall test the orthogonality within the range $-\infty < x < \infty$ as follows:

$$\int_{-0}^{\infty} \psi_1(x)\psi_2(x)\mathrm{d}x = A_1A_2 \int_{-0}^{\infty} xe^{-2\alpha x^2}\mathrm{d}x$$

This integrand $xe^{-2\alpha x^2}$ changes sign as we move from x to $-x$. Therefore, the integral must vanish between the symmetrical ranges. Answer.

b. Now we are going to test the orthogonality over the range $-0 < x < \infty$.

$$\int_{0}^{\infty} \psi_1(x)\psi_2(x)\mathrm{d}x = A_1A_2 \int_{0}^{\infty} xe^{-2\alpha x^2}\mathrm{d}x = A_1A_2\left(\frac{1}{4\alpha}\right)$$

But this cannot be zero. So, these functions are not orthogonal within the range given. Answer.

Problem 9. A simple harmonic oscillator having two bodies of mass m_1 and m_2 is separated by a distance, which may change by a constant x, for a displacement x, the restoring force is equal to $-kx$. Express the Hamiltonian for this system.

Solution: According to the question, the separation of mass m_1 and m_2 can be shown as given below. The potential energy is given by

$$V = -\int_0^x -kx\mathrm{d}x = \int_0^x kx\mathrm{d}x = \frac{kx^2}{2}$$

The Hamiltonian for this system may be written as

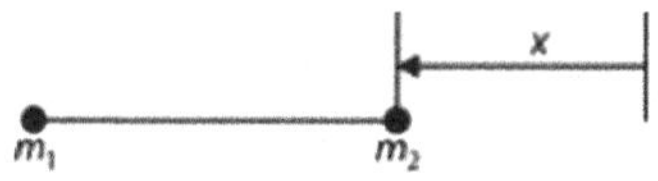

$$H = \frac{-\hbar}{2}\sum_{l=1}^{N}\frac{1}{m_i}\nabla_i^2 + V$$

In the present problem,

$$H = \frac{-\hbar}{2\mu}\frac{\mathrm{d}^2}{\mathrm{d}x^2} + \frac{kx^2}{2}$$

where μ = reduced mass
$= m_1m_2/m_1 + m_2$ Answer.

Problem 10. Show explicitly that $\psi_0(x)$ and $\psi_1(x)$ for the harmonic oscillator are orthogonal.

Solution: We know that

$$\psi_0(x) = \left(\frac{\alpha}{\pi}\right)^{1/4} e^{-\alpha x^2/2} \text{ and } \psi_1(x)\left(\frac{4\alpha^3}{\pi}\right)^{1/4} xe^{-\alpha x^2/2}$$

$$\therefore \int_{-\infty}^{\infty} \psi_0(x)\cdot\psi_1 \mathrm{d}x = \int_{-\infty}^{\infty} \left(\frac{\alpha}{\pi}\right)^{1/4} e^{-\alpha x^2/2} \left(\frac{4\alpha^3}{\pi}\right)^{1/4} xe^{-\alpha x^2/2} \mathrm{d}x$$

$$= \left(\frac{\alpha}{\pi}\right)^{1/4} \left(\frac{4\alpha^3}{\pi}\right)^{1/4} \int_{-\infty}^{\infty} xe^{-\alpha x^2/2} \cdot xe^{-\alpha x^2/2} \mathrm{d}x$$

$$= \left(\frac{4\alpha^4}{\pi^2}\right)^{1/4} \int_{-\infty}^{\infty} xe^{-\alpha x^2/2} \mathrm{d}x$$

$$\text{But } \int_{-\infty}^{\infty} xe^{-\alpha x^2/2} \mathrm{d}x = 0$$

$$\therefore \int_{-\infty}^{\infty} \psi_0(x)\psi_1(x) \mathrm{d}x = 0$$

Hence, $\psi_0(x)$ and $\psi_1(x)$ are orthogonal because the integral is an odd function of x. Proved.

Problem 11. Work out the energy Eigen value of a rigid rotor whose axis of rotation is along the z axis.

Solution: According to the question, since the axis of rotation is along the z axis, the motion of the rotor will be always in the xy plane. Consequently, θ will be

$$90°, \text{ and the equation} -\frac{\hbar^2}{2\mu}\left[\frac{1}{r^2 \sin\theta}\frac{d}{d\theta}\left(\sin\theta\frac{\mathrm{d}}{\mathrm{d}\theta}\right) + \frac{1}{r^2 sin^2\theta}\frac{\mathrm{d}^2}{\mathrm{d}\phi^2}\right]\psi = E\psi(\theta,\phi)$$

will take the following form:

$$\frac{d^2\psi}{d\phi^2} = \frac{-2IE\psi(\phi)}{\hbar^2}$$

If we write $2IE/\hbar^2 = m^2$, then the solution of the above equation will be

$$\psi(\phi) = e^{im\phi}; m = 0, \pm1, \pm2, \ldots$$

Then the energy Eigen value will be $E_m = h^2\psi/2I\,m^2$, where $m = 0, \pm1, \pm2, \ldots$. All levels except the one corresponding to $m = 0$ are two-fold degenerate. Answer.

Questions on Concepts

1. Derive an expression for energy of a particle in a one-dimensional bow within the limits of $0 < x < a$.
2. Suppose $\psi = A \sin kx$ for a free particle in a box, where $k = n\pi/a$. Find the value of A.
3. The energy of a free particle in one-dimensional is given by $E_n = n^2h^2/8ma^2$, where the terms have their usual significance. Find the ZPE for the free particle in a box and construct the energy level diagram.
4. Derive an expression for energy of a free particle in a three-dimensional box or a rectangular box.
5. Find the degenerate energy levels for a cubic potential box.
6. Find the energies of the six lowest energy levels of a free particle in a cubical box. Which of the levels are degenerate.
   [Hint: 3, 6, 9, 11, 12 and 14. $n^2\hbar^2/2ma^2$, first and fifth levels are non-degenerate; second, third, and fourth are three-fold degenerate.]
7. Determine the degree of degeneracy of the energy level $38\pi^2\hbar^2/2ma^2$ of a particle of a cubical potential box.
8. Illustrate that the wave functions for the one-dimensional particle in a box are orthogonal.
9. Discuss the degeneracies possible in a three-dimensional cubical box.
10. What do you mean by the tunnel effect? Prove that

$$\frac{C_3C_3^*}{C_1C_1^*} = \left[\frac{16}{\left(1+\frac{k_3}{k_1}\right)^2 + \left(\frac{k_3}{k_1}, -\frac{k_2^1}{k_1}\right)}\right] e^{-2k_2^1 a}$$

    where the terms are related to the tunnel effect.
11. What do you mean by transparency factor? What is the importance of tunnel effect?
12. a. Discuss the quantum mechanical explanation of emission of α particles.
    b. Explain the umbrella inversion of pyramidal molecules.
13. Find an expression for energy of a particle on a ring.

    or

    Prove that $E = n^2h^2/2mc^2$, where the terms have their usual significance.
14. Prove that the wave function of a particle on a ring is

$$\psi = \sqrt{\frac{2}{c}} \sin\left(\frac{2n\pi x}{c} + a\right)$$

15. When the wave function is

$$\psi = A \sin\frac{2n\pi}{c}x + B\cos\frac{2n\pi x}{c},$$ find the value of A and B.

16. Find the value of ψ function for a particle on a ring considering the spherical polar co-ordinate.
17. Suppose $\psi = A \sin M\Phi + B \cos M\Phi$ for a particle on a ring.

    Prove that $A = \frac{1}{\sqrt{\pi}}\cos\alpha$ and $B = \frac{1}{\sqrt{\pi}}\sin\alpha$

18. Confirm that the wave functions for a particle on a sphere may be written as $\psi(\theta,\phi) = \theta(\theta)\cdot\phi(\phi)$ by the method of separation of variables and find the value of θ.
19. Show that the Schrödinger equation for a particle free to rotate in three dimensions does actually separate into equations for the variation with of θ and ϕ.
20. Show that $E = m^2\hbar^2/2I$ for a particle on sphere.
21. Show under what conditions, the polar equation for a particle on a sphere becomes a particle on a ring? After getting the equation for a particle on a sphere equivalent to particle on a ring, solve it.
22. Show that the Legendre polynomials of degree l are expressed as

$$P_l(x) = \frac{1}{2^l l!}\frac{\mathrm{d}^l\left(x^2-1\right)^l}{\mathrm{d}x^l}$$

or

Derive Rodrigue's formula.

23. a. Discuss the normalisation of the Legendre polynomial.
    b. Show that the Legendre polynomials are orthogonal.
24. Show that $z = \left(1-x^2\right)^{m/2}\frac{\mathrm{d}^m y}{\mathrm{d}x^m}$

    Is the solution of the associated Legendre equation

$$\left(1-x^2\right)\frac{\mathrm{d}^2 y}{\mathrm{d}x^2} - 2x\frac{\mathrm{d}y}{\mathrm{d}x} + \left\{n(n+1) - \frac{m^2}{\left(1-x^2\right)}\right\}\cdot y = 0$$

25. How will you show that the normalised Legendre functions are

$$\frac{2l+1(l-m)!}{2(l+m)} + p_l^m(x)$$

26. Discuss spherical harmonics.
27. Find the F and T equations for a rigid rotor and solve them.
28. Prove that for a rigid rotor, the moment of inertia $I = \mu r^2$, where the terms have their usual significance.
29. Derive Rodrigue's formula for finding out the Hermite polynomials.
30. Discuss the orthogonal properties of Hermite polynomials.
31. Prove that

    a. $\int_{-\infty}^{+\infty} e^{-x^2} H_n(x)\cdot H_m(x)\mathrm{d}x = 0$ if $m \neq n$

    b. $\int_{-\infty}^{+\infty} e^{-x^2} H_n(x)\cdot H_m(x)\mathrm{d}x = \text{constant}$ if $m = n$

32. a. What do you mean by the SHO?
    b. Prove classically that the potential energy of an SHO is $V = 2\pi^2 v^2 m x^2$ when the terms have their usual significance.

33. Solve the equation for SHO

$$\frac{d^2y}{dx^2} + \left(a - b^2x^2\right)\psi = 0 \text{ asymptotically.}$$

34. Prove that $E = \left(n + 1/2\right)h\nu$ for an SHO by series solution.
35. Derive Recursion formula for an SHO.
36. a. Solve the Schrödinger equation for the linear harmonic Schrödinger and obtain its energy levels.
    b. Compare the energy values obtained quantum mechanically with those obtained classically and based on old quantum theory.
37. What is zero point energy of an SHO? Account for its occurrence.
38. Explain why there is a finite probability that the displacement of an LHO exceeds the classical value.
39. Show that the average value of the kinetic and potential energies of an SHO is separately equal to half of the total energy.

Numerical Problems

1. For a particle of mass 9×10^{-28} g confined to one-dimensional box 100 *A* long, calculate the number of energy levels lying between 9 and 10 eV. [Ans. $(n_1 - n_2) = 3$]
2. Show that when $n_x = x_y = n_2 = 0$ for a particle in a box, we have an infinitely large wavelength or uncertainty in position. [Ans. $-\lambda = \infty$ and $\Delta x = \infty$]
3. What is the degeneracy of the level for which
   a. $E = \dfrac{14h^2}{8ma^2}$ and
   b. $E = \dfrac{27h^2}{8ma^2}$ and what will be the values of n_x, n_y, and n_z. [Ans. (a) 6 (b) 3]
4. Find the height of the potential barrier for α particles emitted from Radon ${}_{86}R_n^{222}$ assuming that the effective nuclear radius is given by $r_0 = 1.5 \times 10^{-15}$ $A^{1/3}$m. [Ans. $V_0 = 26.62$ MeV]
5. Find the lowest energy of an electron in a cubic box having each side 1 A. [Ans. 18.03×10^{-18} J]
6. Using the ground state wave function for a particle in a one-dimensional box $0 \le x \le a$, prove that the approximate energy is

$$E = \frac{10}{\pi^2} \cdot \frac{14h^2}{8ma^2}$$

7. Calculate the ground state energy of a hydrogen atom electron assumed to be in a three-dimensional cubical box of length 0.1 nm if the ground state energy of the electron in a one-dimensional box of length 0.3 nm is 4 eV. Comment on the result. [Ans. $E_{111} = 27 \times 4$ eV $= 108$ eV]
8. Calculate the ZPE of a mass 1.68×10^{-27}kg executing SHO (force constant $k = 10$ Nm^{-1}). [Ans. 0.405×10^{-20} J]
9. Using the more general equation $x(t) = c\sin\left(wt + \phi\right)$, show that the total energy of harmonic oscillator is $E = 1/2Kc^2$.
10. Verify that $\psi(x) = e^{-x^2/2}$ is an asymptotic solution of equation

$$\frac{d^2y}{dx^2} + \left(\frac{\lambda}{\alpha} - x^2\right)\psi = 0$$

8

Hydrogen Atom

The hydrogen atom is the simplest atom, which is a system of two particles, namely, a proton and an electron. In this system, a single electron moves in the field of a nucleus of unit positive charge. This system is bound by electrostatic force of attraction. In other words, we can say that the electron of the hydrogen atom revolves around the nucleus in the field of the force of the nucleus. Since hydrogen is the simplest atom, it forms the basis for the theoretical treatment of more complex atomic systems. The Schrödinger equation for hydrogen atom can be solved in closed form. When the next atom in the order of increasing complexity is considered, for example, helium, which consists of two electrons moving in the field of the nucleus with charge two, the Schrödinger equation can no longer be solved in closed form. For the solution of such systems, the approximate methods are employed to get the wave function.

8.1 The Hydrogen Atom (Simple Solution of the Schrödinger Equation)

As already stated that the hydrogen atom is one electron system with a charge '$-e$' rotating around a central proton (p) with charge '$+e$'. The potential energy of the system is expressed as

$$V = \frac{(-e)(+e)}{r^2} \cdot r = -\frac{e^2}{r} \tag{8.1}$$

Putting the value of V in the Schrödinger equation, we have

$$\nabla^2\psi + \frac{8\pi^2 m}{h^2}\left(E + e^2/r\right)\psi = 0 \tag{8.2}$$

To solve the Schrödinger equation for hydrogen atom, let us consider

$$r = \left(x^2 + y^2 + z^2\right)^{1/2} \tag{8.3}$$

where r is the distance of electron from the nucleus.

Differentiating Eq. (8.3) with respect to x, we shall get

$$\frac{\partial r}{\partial x} = \frac{1}{2} \cdot \frac{2x}{\left(x^2 + y^2 + z^2\right)^{\frac{1}{2}}} = \frac{x}{r}$$

and $\dfrac{\partial \psi}{\partial x} = \dfrac{\partial \psi}{\partial r} \cdot \dfrac{\partial r}{\partial x} = \dfrac{x}{r} \cdot \dfrac{\partial \psi}{\psi r}$

Differentiating the aforementioned equation again with respect to x, we obtain

$$\frac{\partial^2\psi}{\partial x^2} = \frac{1}{r}\frac{\partial\psi}{\partial r} + \frac{x}{r}\cdot\frac{\partial^2\psi}{\partial r^2}\cdot\frac{\partial r}{\partial x} - \frac{x}{r^2}\cdot\frac{\partial\psi}{\partial r}\cdot\frac{\partial r}{\partial x}$$

$$= \frac{1}{r}\frac{\partial\psi}{\partial r} + \frac{x}{r}\cdot\frac{\partial^2\psi}{\partial r^2}\cdot\frac{x}{r} - \frac{x}{r^2}\cdot\frac{x}{r}\frac{\partial\psi}{\partial r}$$

$$= \frac{1}{r}\frac{\partial\psi}{\partial r} + \frac{x^2}{r^2}\cdot\frac{\partial^2\psi}{\partial r^2} - \frac{x^2}{r^3}\cdot\frac{\partial\psi}{\partial r}$$

Similarly, $\frac{\partial^2\psi}{\partial y^2} = \frac{1}{r}\frac{\partial\psi}{\partial r} + \frac{y^2}{r^2}\cdot\frac{\partial^2\psi}{\partial r^2} - \frac{y^2}{r^3}\cdot\frac{\partial\psi}{\partial r}$

and $\frac{\partial^2\psi}{\partial z^2} = \frac{1}{r}\frac{\partial\psi}{\partial r} + \frac{z^2}{r^2}\cdot\frac{\partial^2\psi}{\partial r^2} - \frac{z^2}{r^3}\cdot\frac{\partial\psi}{\partial r}$

On adding, $\frac{\partial^2\psi}{\partial x^2} + \frac{\partial^2\psi}{\partial y^2} + \frac{\partial^2\psi}{\partial z^2} = \frac{3}{r}\frac{\partial\psi}{\partial r} + \frac{(x^2+y^2+x^2)}{r^2}\cdot\frac{\partial^2\psi}{\partial r^2} - \frac{(x^2+y^2+x^2)}{r^3}\cdot\frac{\partial\psi}{\partial r}$

$$\text{or } \nabla^2\psi = \frac{3}{r}\frac{\partial\psi}{\partial r} + \frac{r^2}{r^2}\cdot\frac{\partial^2\psi}{\partial r^2} - \frac{r^3}{r^3}\cdot\frac{\partial\psi}{\partial r}$$

$$\text{or } \nabla^2\psi = \frac{3}{r}\frac{\partial\psi}{\partial r} + \frac{r^2}{r^2}\cdot\frac{\partial^2\psi}{\partial r^2} - \frac{1}{r}\cdot\frac{\partial\psi}{\partial r}$$

$$\text{or } \nabla^2\psi = \frac{2}{r}\frac{\partial\psi}{\partial r} + \frac{\partial^2\psi}{\partial r^2} \tag{8.4}$$

Putting the value of or $\nabla^2\Psi$ in Eq. (8.2), we shall get

$$\frac{\partial^2\psi}{\partial r^2} + \frac{2}{r}\cdot\frac{\partial\psi}{\partial r} + \frac{8\pi^2 m}{h^2}\left(E + \frac{e^2}{r}\right)\psi = 0 \tag{8.5}$$

The simplest solution of Eq. (8.5) may be given by

$$\Psi = e^{-ar} \tag{8.6}$$

Differentiating this equation with respect to r twice, we shall get

$$\frac{\partial\psi}{\partial r} = -ae^{-ar}$$

$$\text{and } \frac{\partial^2\psi}{\partial r} = (-a)(-a)e^{-ar} = a^2e^{-ar} \tag{8.7}$$

Putting these values in Eq. (8.5), we have

$$a^2e^{-ar} - \frac{2}{1}ae^{-ar} + \frac{8\pi^2 m}{h^2}\left(E + \frac{e^2}{r}\right)e^{-ar} = 0$$

or $a^2 - \frac{2}{r}a + \frac{8\pi^2 m}{h^2}\left(E + e^2/r\right) = 0$

since $e^{-ar} \neq 0$

$$\text{or} \left(a^2 + \frac{8\pi^2 mE}{h^2}\right) - \frac{2}{r}\left(\frac{4\pi^2 me^2}{h^2} - a\right) = 0 \tag{8.8}$$

Since Eq. (8.8) is true for any value of r, the sum of the first two terms, which is independent of r, and the third term involving r should be separately zero because the equation should also be satisfied when $r \to \infty$.

Thus, $\left(\frac{4\pi^2 me^2}{h^2} - a\right) = 0$

$$\therefore a = \frac{4\pi^2 me^2}{h^2}$$

$$\therefore a^2 = \frac{16\pi^4 m^2 e^4}{h^4} \tag{8.9}$$

Again $a^2 = \frac{8\pi^2 mE}{h^2} = 0$

$$\therefore a^2 = \frac{8\pi^2 mE}{h^2} = 0 \tag{8.10}$$

We can equate Eqs. (8.9) and (8.10) and shall get

$$\frac{16\pi^4 m^2 e^4}{h^4} = -\frac{8\pi^2 mE}{h^2}$$

$$\therefore E = -\frac{2\pi^2 me^4}{h^2} \tag{8.11}$$

This is the value of energy of hydrogen atom. It is clear that E is negative, which indicates that the electron is bound.

Now we can express the volume of a shell with radius r equal to $4\pi r^2 dr$, and the probability P of finding the electron in the shell would be $4\pi r^2 \psi^2 \approx 4\pi r^2 e^{-2ar}$. From this, we can find the radius of the shell where the probability of finding the electron is maximum.

Since $P = 4\pi r^2 e^{-2ar}$,

$$\therefore \frac{\mathrm{d}p}{\mathrm{d}r} = 4\pi\left[2re^{-2ar} + r^2(-2a)e^{-2ar}\right]$$

At maxima, $\frac{\mathrm{d}p}{\mathrm{d}r} = 0$

$$\therefore 4\pi e^{-2ar}\left[2r_{\max} - 2ar_{\max}^2\right] = 0$$

But $4\pi e^{-2ar} \neq 0$

$$\therefore 2r_{\max} - 2ar_{\max}^2 = 0$$

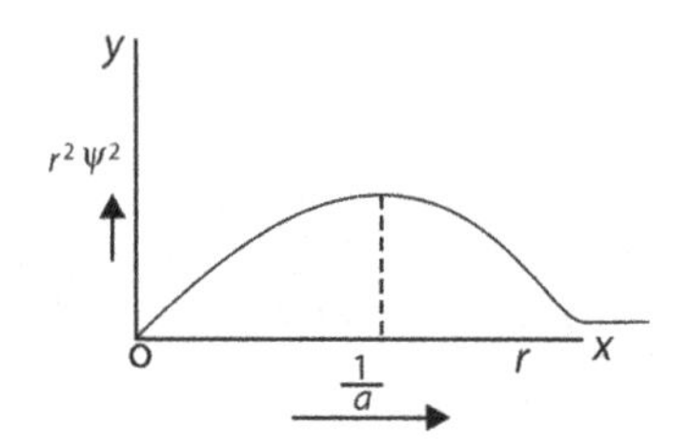

FIGURE 8.1 Probability vs. r graph.

$$\text{or } r_{max}\left(2-2ar_{max}\right)=0$$

$$\text{or } r_{max} \neq 0$$

$$\therefore 2-2ar_{max}=0$$

$$\text{or } 2ar_{max}=2$$

$$\text{or } ar_{max}=1$$

$$\therefore r_{max}=\frac{1}{a} \tag{8.12}$$

where $1/a$ = the Bohr radius.

If we plot $r^2\psi^2$, i.e., probability vs. r, the graph shown in Figure 8.1 is obtained.

It is clear from the graph that the probability of finding the electron will be maximum at $r_{max}=\frac{1}{a}$.

8.2 Generalised Solution of the Schrödinger Equation for Hydrogen Atom/Hydrogen-Like Species

The hydrogen atom is one electron system, and the electron revolves around the nucleus consisting of a proton.

Let $+Ze$ = charge of nucleus
$-e$ = charge on the electron
Z = number of proton = atomic number
r = the distance of the electron from the nucleus
m = mass of the electron
M = mass of the proton present in the nucleus

Since electron is negatively charged and nucleus is positively charged, there will be coulombic attraction between the two.

$$\therefore \text{The force of attraction} = \frac{(+Ze)(-e)}{4\pi\epsilon_0 r^2} = -\frac{Ze^2}{4\pi\epsilon_0 r^2}$$

$$\therefore \text{Potential, } V = -\frac{Ze^2}{4\pi\epsilon_0 r^2}\cdot r - \frac{Ze^2}{4\pi\epsilon_0 r}$$

$$\text{or} \quad V = -\frac{Ze^2}{4\pi\epsilon_0 r} \tag{8.13}$$

where ϵ_0 = permittivity of free space.

Substituting the value of V in the Schrödinger equation, we get

$$\nabla^2\psi + \frac{8\pi^2 m}{h^2}\left(E + \frac{Ze^2}{4\pi\epsilon_0 r}\right)\psi = 0 \tag{8.14}$$

It is more correct to replace m, the mass of the electron, by the reduced mass, i.e.,

$$\frac{mM}{m+M} = \mu$$

In case of hydrogen atom, $\mu = \dfrac{1840m^2}{m + 1840m}$

$$\text{or} \quad \mu = \frac{1840m^2}{m(1+1840)} = \frac{1840m^2}{1841m} = \frac{1840m}{1841} \approx m$$

But we shall use μ in place of m for a general case. Therefore, Eq. (8.14) can be expressed as

$$\nabla^2\psi + \frac{2\mu}{\hbar^2}\left(E + \frac{Ze^2}{4\pi\epsilon_0 r}\right)\psi = 0 \tag{8.15}$$

Expressing the Laplacian operator in polar coordinates, Eq. (8.15) changes into the following:

$$\frac{1}{r^2}\cdot\frac{\mathrm{d}}{\mathrm{d}r}\left(r^2\cdot\frac{\mathrm{d}\psi}{\mathrm{d}r}\right) + \frac{1}{r^2\sin\theta}\cdot\frac{\mathrm{d}}{\mathrm{d}\theta}\left(\sin\theta\cdot\frac{\mathrm{d}\psi}{\mathrm{d}\theta}\right) + \frac{1}{r^2\sin^2\theta}\cdot\frac{\mathrm{d}^2\psi}{\mathrm{d}\phi^2} + \frac{2\mu}{\hbar^2}\left(E + \frac{Ze^2}{4\pi\epsilon_0 r}\right)\psi = 0 \tag{8.16}$$

The process for solving Eq. (8.16) will be the same as used in the case of rigid rotor problem, but in this case, r is a variable so that ψ must be a function of three variables r, θ and ϕ and may be expressed as the product of three functions, i.e.,

$$\psi(r,\theta,\phi) = R(r)\cdot T(\theta)\cdot F(\phi) \tag{8.17}$$

where $R(r)$ is a function of r only, T is a function of θ only, and F is a function of ϕ only.

For the sake of simplicity, Eq. (8.17) may be expressed as

$$\Psi = RTF \tag{8.18}$$

Substituting the value of Ψ in Eq. (8.16), we obtain

$$\frac{1}{r^2}\cdot\frac{\mathrm{d}}{\mathrm{d}r}\left(r^2\cdot\frac{\mathrm{d}(RTF)}{\mathrm{d}r}\right) + \frac{1}{r^2\sin\theta}\cdot\frac{\mathrm{d}}{\mathrm{d}\theta}\left(\sin\theta\cdot\frac{\mathrm{d}(RTF)}{\mathrm{d}\theta}\right) + \frac{1}{r^2\sin^2\theta}\cdot\frac{\mathrm{d}^2(RTF)}{\mathrm{d}\phi^2} + \frac{2\mu}{\hbar^2}\left(E + \frac{Ze^2}{4\pi\epsilon_0 r}\right)RTF = 0$$

$$\text{or} \quad \frac{TF}{r^2}\cdot\frac{\mathrm{d}}{\mathrm{d}r}\left(r^2\cdot\frac{\mathrm{d}R}{\mathrm{d}r}\right) + \frac{RF}{r^2\sin\theta}\cdot\frac{\mathrm{d}}{\mathrm{d}\theta}\left(\sin\theta\cdot\frac{\mathrm{d}T}{\mathrm{d}\theta}\right) + \frac{RT}{r^2\sin^2\theta}\cdot\frac{\mathrm{d}^2F}{\mathrm{d}\phi^2} + \frac{2\mu}{\hbar^2}\left(E + \frac{Ze^2}{4\pi\epsilon_0 r}\right)RTF = 0 \tag{8.19}$$

Dividing Eq. (8.19) by RTF throughout, we get

where RTF represents R-equation, T equation and F equation respectively.

$$\frac{1}{r^2R}\cdot\frac{\mathrm{d}}{\mathrm{d}r}\left(r^2\cdot\frac{\mathrm{d}R}{\mathrm{d}r}\right) + \frac{1}{r^2T\sin\theta}\cdot\frac{\mathrm{d}}{\mathrm{d}\theta}\left(\sin\theta\cdot\frac{\mathrm{d}T}{\mathrm{d}\theta}\right) + \frac{1}{r^2\sin^2\theta}\cdot\frac{1}{F}\frac{\mathrm{d}^2F}{\mathrm{d}\phi^2} + \frac{2\mu}{\hbar^2}\left(E + \frac{Ze^2}{4\pi\epsilon_0 r}\right) = 0 \tag{8.20}$$

Now, multiplying Eq. (8.20) by $r^2 \sin^2 \theta$, we obtain

$$\frac{\sin\theta}{R}\cdot\frac{\mathrm{d}}{\mathrm{d}r}\left(r^2\cdot\frac{\mathrm{d}R}{\mathrm{d}r}\right)+\frac{\sin\theta}{\theta}\cdot\frac{\mathrm{d}}{\mathrm{d}\theta}\left(\sin\theta\cdot\frac{\mathrm{d}T}{\mathrm{d}\theta}\right)+\frac{1}{F}\cdot\frac{\mathrm{d}^2F}{\mathrm{d}\phi^2}+\frac{2\mu}{\hbar^2}\left(E+\frac{Ze^2}{4\pi\epsilon_0 r}\right)r^2\sin^2\theta=0 \tag{8.21}$$

On rearranging Eq. (8.21), we get

$$\frac{\sin^2\theta}{R}\cdot\frac{\mathrm{d}}{\mathrm{d}r}\left(r^2\cdot\frac{\mathrm{d}R}{\mathrm{d}r}\right)+\frac{\sin\theta}{T}\cdot\frac{\mathrm{d}}{\mathrm{d}\theta}\left(\sin\theta\cdot\frac{\mathrm{d}T}{\mathrm{d}\theta}\right)+\frac{2\mu}{\hbar^2}\left(E+\frac{Ze^2}{4\pi\epsilon_0 r}\right)r^2\sin^2\theta=-\frac{1}{F}\cdot\frac{\mathrm{d}^2F}{\mathrm{d}\phi^2} \tag{8.22}$$

The LHS of Eq. (8.22) is a function of r and θ, but the RHS is a function of ϕ only. This equality is only possible when each side of Eq. (8.22) will be equal to the same constant m^2, where m is called the *magnetic quantum number*. Therefore, the RHS of Eq. (8.22) becomes

$$-\frac{1}{F}\cdot\frac{\mathrm{d}^2F}{\mathrm{d}\phi^2}=m^2$$

$$\text{or}\ \frac{\mathrm{d}^2F}{\mathrm{d}\phi^2}=-m^2F$$

$$\text{or}\ \frac{\mathrm{d}^2F}{\mathrm{d}\phi^2}=m^2F=0 \tag{8.23}$$

This is a second-order differential equation and is called azimuthal wave equation or *F equation* or ϕ *equation*. We shall solve this equation later on.

The LHS of Eq. (8.22) may now be written as

$$\frac{\sin^2\theta}{R}\cdot\frac{\mathrm{d}}{\mathrm{d}r}\left(r^2\cdot\frac{\mathrm{d}R}{\mathrm{d}r}\right)+\frac{\sin\theta}{T}\cdot\frac{\mathrm{d}}{\mathrm{d}\theta}\left(\sin\theta\cdot\frac{\mathrm{d}T}{\mathrm{d}\theta}\right)+\frac{2\mu}{\hbar^2}\left(E+\frac{Ze^2}{4\pi\epsilon_0 r}\right)r^2\sin^2\theta=m^2 \tag{8.24}$$

Dividing this equation by $\sin^2 \theta$, we shall get

$$\frac{1}{R}\cdot\frac{\mathrm{d}}{\mathrm{d}r}\left(r^2\cdot\frac{\mathrm{d}R}{\mathrm{d}r}\right)+\frac{1}{T\sin\theta}\cdot\frac{\mathrm{d}}{\mathrm{d}\theta}\left(\sin\theta\cdot\frac{\mathrm{d}T}{\mathrm{d}\theta}\right)+\frac{2\mu r^2}{\hbar^2}\left(E+\frac{Ze^2}{4\pi\epsilon_0 r}\right)=\frac{m^2}{\sin^2\theta}$$

Now reversing the order of position or transposing the second term on the left side to the RHS, we obtain

$$\frac{1}{R}\cdot\frac{\mathrm{d}}{\mathrm{d}r}\left(r^2\cdot\frac{\mathrm{d}R}{\mathrm{d}r}\right)+\frac{2\mu r^2}{\hbar^2}\left(E+\frac{Ze^2}{4\pi\epsilon_0 r}\right)=\frac{m^2}{\sin^2\theta}-\frac{1}{T\sin\theta}\cdot\frac{\mathrm{d}}{\mathrm{d}\theta}\left(\sin\theta\cdot\frac{\mathrm{d}T}{\mathrm{d}\theta}\right) \tag{8.25}$$

The LHS of this equation is a function of r only, whereas the RHS is a function of θ only, and therefore, both sides of Eq. (8.25) must be equal to the same constant. Then, the LHS of Eq. (8.25) takes the form

$$\frac{1}{R}\cdot\frac{\mathrm{d}}{\mathrm{d}r}\left(r^2\cdot\frac{\mathrm{d}R}{\mathrm{d}r}\right)+\frac{2\mu r^2}{\hbar^2}\left(E+\frac{Ze^2}{4\pi\epsilon_0 r}\right)=\beta$$

Multiplying both sides by R/r^2, and on rearrangement, we get

$$\frac{1}{R}\cdot\frac{\mathrm{d}}{\mathrm{d}r}\left(r^2\cdot\frac{\mathrm{d}R}{\mathrm{d}r}\right)+\left[\frac{2\mu}{\hbar^2}\left(E+\frac{Ze^2}{4\pi\epsilon_0 r}\right)-\frac{\beta}{r^2}\right]R=0 \tag{8.26}$$

This is known as radial wave equation or R equation.

The RHS of Eq. (8.25) may also be equated to β, and then, we can write as

$$\frac{m^2}{\sin^2\theta} - \frac{1}{T\sin\theta}\cdot\frac{d}{d\theta}\left(\sin\theta\cdot\frac{dT}{d\theta}\right) = \beta$$

Multiplying both sides by T, we get

$$\left[\beta - \frac{m^2}{\sin^2\theta}\right]T + \frac{1}{\sin\theta}\cdot\frac{d}{d\theta}\left(\sin\theta\cdot\frac{dT}{d\theta}\right) = 0 \tag{8.27}$$

This equation is called the T equation or θ equation or polar wave equation.

Our next task is to solve Eq. (8.23), i.e., the *F equation*, Eq. (8.26), i.e., the *R equation*, and Eq. (8.27), i.e., the *T equation*.

8.3 Solution of the *F* Equation

The F equation is

$$\frac{d^2F}{d\phi^2} = -m^2F$$

$$\text{or } \frac{d^2F}{d\phi^2} + m^2F = 0$$

The solution of this equation is

$$F(\phi) = ce^{\pm im\phi} \tag{8.28}$$

where m represents the magnetic quantum number. Now since $F(\phi)$ must be single valued, it must have the same value for $\phi = 0$ and $\phi = 2\pi$, that is, after a complete rotation of the system. When $\phi = 0$, then Eq. (8.28) becomes

$$F(\phi) = ce^{\pm im\times 0} = c \tag{8.29}$$

When $\phi = 2\pi$, then

$$F(\phi) = ce^{\pm i2\pi m} \tag{8.30}$$

From Eqs. (8.29) and (8.30), we have

$$c = ce^{\pm i2\pi m}$$

$$\therefore e^{\pm i2\pi m} = 1 \tag{8.31}$$

$$\text{or } \cos 2\pi m \pm i\sin 2\pi m = 1$$

$$\therefore \cos 2\pi m + i\sin 2\pi m = 1 \tag{8.32}$$

$$\text{and } \cos 2\pi m - i\sin 2\pi m = 1 \tag{8.33}$$

Adding Eqs. (8.32) and (8.33) gives

$$2\cos 2\pi m = 2$$

$$\therefore \cos 2\pi m = 1 \tag{8.34}$$

$$\therefore 2\pi m = 0, \pm 2\pi, \pm 4\pi, \pm 6\pi, \ldots$$

$$\therefore m = 0, \pm 1, \pm 2, \pm 3 + \ldots$$

The positive and negative values of m are due to the fact that the rotation may occur either in the positive or in the negative direction.

Now we have to find the value of c in Eq. (8.28). This will be achieved by applying normalisation confined to the region $0 \le \phi \le 2\pi$. Hence, the normalisation should be carried out over the region between $\phi = 0$ and $\phi = 2\pi$.

Carrying out the normalisation of Eq. (8.28), we get

$$\int_0^{2\pi} FF^* \mathrm{d}\tau = 1$$

$$\text{or} \int_0^{2\pi} Ce^{+im\phi} Ce^{-im\phi} \mathrm{d}\phi = 1$$

$$\text{or } C^2 \int_0^{2\pi} e^{+im\phi} \cdot e^{-im\phi} \mathrm{d}\phi = 1$$

$$\text{or } C^2 \int_0^{2\pi} \mathrm{d}\phi = 1$$

$$\text{or } C^2 \cdot 2\pi = 1$$

$$\text{or } C^2 = \frac{1}{2\pi}$$

$$\therefore C = \frac{1}{\sqrt{2\pi}} \tag{8.35}$$

Therefore, the normalised solution can be expressed as

$$F(\phi) = \frac{1}{\sqrt{2\pi}} \cdot e^{im\phi} \tag{8.36}$$

where $m = 0, \pm 1, \pm 2, \pm 3, \ldots$

We know that the solution of F equation can also be written as

$$F(\phi) = (A \sin m\phi + B \cos m\phi) \tag{8.37}$$

And on normalisation of Eq. (8.37), we obtain

$$\int_0^{2\pi} \left(A^2 \sin^2 m\phi + B \cos m\phi\right)^2 \mathrm{d}\phi = 1$$

TABLE 8.1

Examples of $F(\phi)$

$\|m\|$	Complex Form	Real Form
0	$F_0 = \frac{1}{\sqrt{2\pi}}$	$F_0 = \frac{1}{\sqrt{2\pi}}$
1	$F_1 = \frac{1}{\sqrt{2\pi}} e^{i\phi}$	$F_{1\cos} = \frac{1}{\sqrt{2\pi}} \cos\phi$
	$F_{-1} = \frac{1}{\sqrt{2\pi}} e^{-i\phi}$	$F_{1\sin} = \frac{1}{\sqrt{\pi}} \sin\phi$
2	$F_2 = \frac{1}{\sqrt{2\pi}} e^{-i2\phi}$	$F_{2\cos} = \frac{1}{\sqrt{\pi}} \cos 2\phi$
	$F_{-2} = \frac{1}{\sqrt{2\pi}} e^{-i2\phi}$	$F_{2\sin} = \frac{1}{\sqrt{\pi}} \sin 2\phi$

$$\text{or} \int_0^{2\pi} \left(A^2 \sin^2 m\phi + B^2 \cos^2 m\phi + 2AB \sin m\phi \cdot \cos m\phi\right) d\phi = 1$$

$$\text{or} \int_0^{2\pi} \left[\frac{1}{2} A^2 (1 - \cos 2m\phi) + AB \sin 2m\phi + \frac{1}{2} B^2 (1 + \cos 2m\phi)\right] d\phi = 1$$

$$\text{or} \int_0^{2\pi} \left[\frac{A^2}{2} - \frac{A^2}{2} \cos 2m\phi + AB \sin 2m\phi + \frac{B^2}{2} + \frac{B^2}{2} \cos 2m\phi\right] d\phi = 1 \tag{8.38}$$

Since integration of the sine and cosine functions between 0 and 2π limits yields zero value, Eq. (8.38) reduces to

$$A^2\pi + B^2\pi = 1 \tag{8.39}$$

We can write $A = 1/\sqrt{\pi} \cos m\phi$ and $B = 1/\sqrt{\pi} \sin m\phi$, which satisfy Eq. (8.39).

Now we are going to give some of the normalised $F(\phi)$ in Table 8.1.

Thus, we see that the values of F_0 for complex form and for real form are the same for $m = 0$.

8.4 Solution of the *T* Equation or the Polar Wave Equation

The polar wave equation or the T equation is given by

$$\frac{1}{\sin\theta} \cdot \frac{d}{d\theta}\left(\sin\theta \cdot \frac{dT}{d\theta}\right) + \left(\beta - \frac{m^2}{\sin^2\theta}\right) T = 0$$

Let us set $z = \cos\theta$, then $(1 - z^2) = \sin^2\theta$

Also assume that $T(\theta) = P(z)$, or simply, $T = P$, then

$$\frac{dT}{d\theta} = \frac{dP}{dz} \cdot \frac{dz}{d\theta} = -\sin\theta \frac{dp}{dz} \quad \left[\therefore \frac{dp}{d\theta} = -\sin\theta\right]$$

$$\frac{d}{d\theta} = \frac{dz}{d\theta} \cdot \frac{d}{dz} = -\sin\theta \frac{dp}{dz} \tag{8.40}$$

Using Eq. (8.40), the aforementioned T equation is transformed into the following forms:

$$\frac{1}{\sin\theta}\left(-\sin\theta\frac{\mathrm{d}}{\mathrm{d}z}\right)\left[\sin\theta\left(-\sin\theta\frac{\mathrm{d}P}{\mathrm{d}z}\right)\right]+\left(\beta-\frac{m^2}{\sin^2\theta}\right)P=0 \tag{8.41}$$

$$\text{or } -\frac{\mathrm{d}}{\mathrm{d}z}\left(-\sin^2\theta\frac{\mathrm{d}P}{\mathrm{d}z}\right)+\left(\beta-\frac{m^2}{\sin^2\theta}\right)P=0 \tag{8.42}$$

$$\text{or } -\frac{\mathrm{d}}{\mathrm{d}z}\left[\left(1-z^2\right)\frac{\mathrm{d}P}{\mathrm{d}z}\right]+\left(\beta-\frac{m^2}{1-z^2}\right)P=0 \tag{8.43}$$

$$\text{or } \left(1-z^2\right)\frac{\mathrm{d}^2P}{\mathrm{d}z^2}-2z\frac{\mathrm{d}P}{\mathrm{d}z}+\left(\beta-\frac{m^2}{1-z^2}\right)P=0 \tag{8.44}$$

Equation (8.44) is of the form of differential equation, which is satisfied by the *associated Legendre functions*. When $\beta = l(l+1)$, Eq. (8.41) changes to

$$\left(1-z^2\right)\frac{\mathrm{d}^2P}{\mathrm{d}z^2}-2z\frac{\mathrm{d}P}{\mathrm{d}z}+\left(l(l+1)-\frac{m^2}{1-z^2}\right)P=0 \tag{8.45}$$

The associated Legendre functions of degree 'l' and order $|m|$, where, $l = 0, 1, 2, 3, \ldots$ and $|m| = 0, 1, 2, \ldots l$ are defined by Legendre polynomials, i.e.,

$$P_l^{|m|}(z)=\left(1-z^2\right)^{|m|/2}\cdot\frac{\mathrm{d}^{|m|}}{\mathrm{d}z^{|m|}}\cdot P_l(z) \tag{8.46}$$

and they satisfy the following differential equation:

$$\left(1-z^2\right)\frac{\mathrm{d}^2P_l^{|m|}(z)}{\mathrm{d}z^2}-2z\frac{\mathrm{d}P_l^{|m|}}{\mathrm{d}z}+\left[l(l+1)-\frac{m^2}{1-z^2}\right]P_l^{|m|}(z)=0 \tag{8.47}$$

On comparing Eq. (8.47) with Eqs. (8.41) to (8.44), which is our transformed T equation, we find that they are of the same form. We require here only to identify $P(z)$ with $P_l^{|m|}(z)$ and β with $l(l+1)$. Finally, we can write

$$T(\theta)=N_\theta P_l^{|m|}(z) \tag{8.48}$$

where N_θ = normalisation constant.

It is now necessary to normalise $T(\theta)$. We have already assumed that $z = \cos\theta$, and this is a variable; hence, the electron will be confined between $z = -1$ and $z = +1$. Therefore, the value of the integral $\int_{-1}^{+1} P_l^{|m|}(z)P_l^{|m|}(z)\mathrm{d}z$

$$=\frac{2}{2l+1}\cdot\frac{(l+|m|)!}{(l-|m|)!}\delta_{ll'} \tag{8.49}$$

has been found out. $\delta_{ll'}$ is called the *Kronecker delta*. The value $\delta_{ll'} = 0$ when $l \neq l'$ and $\delta_{ll'} = 1$ when $l = l'$. It should be noted that the change from $z = -1$ to $z = +1$ corresponds to the variation of θ from 0 to π.

TABLE 8.2

Values of the Legendre Functions $P_l^{|m|}$

$P_0^0(\cos\theta) = 1$
$P_1^0(\cos\theta) = \cos\theta$
$P_2^0(\cos\theta) = \frac{1}{2}(3\cos^2\theta - 1)$
$P_3^0(\cos\theta) = \frac{1}{2}(5\cos^3\theta - 3\cos\theta)$
$P_1^1(\cos\theta) = \sin\theta$
$P_2^1(\cos\theta) = 3\sin\theta\cos\theta$
$P_3^1(\cos\theta) = \frac{3}{2}\sin\theta(5\cos^2\theta - 1)$
... and so on

Therefore, on applying normalisation condition, we may write

$$\int_0^{\pi} T(\theta_{ml}^*) \cdot T(\theta_{ml}^*) \sin\theta \, d\theta = 1 \tag{8.50}$$

which leads to

$$N_\theta^2 \int_{-1}^{+1} P_l^{|m|}(z) \cdot P_l^{|m|}(z) \, dz = 1 \tag{8.51}$$

In terms of the associated Legendre functions, now by substituting the value of integral from Eq. (8.49), we get

$$N_\theta^2 = \frac{(2l+1)(l-|m|)!}{2(l+|m|)!} \tag{8.52}$$

$$\therefore N_\theta = \left[\frac{(2l+1)(l-|m|)!}{2(l+|m|)!}\right]^{1/2}$$

Therefore, the function $T(\theta)$ or simply T is expressed as

$$T(\theta) \text{ or } T = \left[\frac{(2l+1)(l-|m|)!}{2(l+|m|)!}\right]^{1/2} \cdot P_l^{|m|}(\cos\theta) \tag{8.53}$$

Thus, the function $T(\theta)$ is obtained.

The values of the Legendre functions for some values of 1 and $|m|$ are mentioned in Table 8.2 using Eq. (8.53).

8.5 The Laguerre Differential Equation

The Laguerre differential equation is represented by

$$x\frac{d^2y}{dx^2} + (1-x)\frac{dy}{dx} + \lambda y = 0 \tag{8.54}$$

where λ = constant.

Equation (8.54) may be solved by series integration.

Let us suppose that

$$y = \sum_{r=0}^{\infty} a_r x^{k+r} \tag{8.55}$$

is the solution of Eq. (8.54); differentiating Eq. (8.55), we get

$$\frac{dy}{dx} = \sum_{r=0}^{\infty} a_r (k+r) x^{k+r-1}$$

And further differentiation results

$$\frac{dy}{dx^2} = \sum_{r=0}^{\infty} a_r (k+r)(k+r-1) x^{k+r-1}$$

Now we shall substitute the value of dy/dx and d^{2}y/dx^2 in Eq. (8.54), and it takes the following form:

$$\sum_{r=0}^{\infty} a_r (k+r)(k+r-1) x^{k+r-1} + (1-x)(k+r) x^{k+r-1} + \lambda x^{k+r} = 0$$

$$\text{or} \sum_{r=0}^{\infty} a_r \left[(k+r)^2 x^{k+r-1} - (k+r-\lambda) x^{k+r} \right] = 0 \tag{8.56}$$

Since Eq. (8.56) is an identity, we can equate to zero the coefficient of different powers of x.

When the coefficient of the lowest power of x, i.e., of x^{k-1} is equated to zero, we obtain $a_0 k^2 = 0$.

But $a_0 \neq 0 \therefore k = 0$

Now equating the coefficient of x^{k+r} to zero, we get

$$a_{r+1}(k+r+1)^2 - a_r(k+r-\lambda) = 0$$

$\therefore$ Putting $k = 0$ in this equation, we get

$$a_{r+1} = \frac{r-\lambda}{(r+1)^2} \cdot ar \tag{8.57}$$

Putting $r = 0, 1, 2, \ldots$ in Eq. (8.57), we have

$$a_1 = \frac{-\lambda}{1} a_0 = (-1)\lambda a_0$$

$$a_2 = \frac{1-\lambda}{2^2} \cdot a_2 = (-1)^2 \frac{\lambda(\lambda-1)}{(2!)} a_0$$

$$a_3 = \frac{2-\lambda}{3^2} \cdot a_2 = \frac{(-1)^3 \lambda(\lambda-1)(\lambda-2)}{(3!)^2} a_0 \text{ and so on}$$

$$\therefore a_r = (-1)^r \lambda(\lambda-1) \frac{(\lambda-r+1)}{(r!)^2} a_0$$

Therefore, from Eq. (8.55), we may write

$$y = \sum_{r=0}^{\infty} a_r x^r = a_0 + a_1 a + a_2 x^2 + + a_r x^r +$$

$$\text{or } y = a_0 \left[1 - \lambda x + \frac{\lambda(\lambda-1)}{(2!)^2} x^2 + + (-1)^r \frac{\lambda(\lambda-1)(\lambda-2)(\lambda-r+1)x^r}{(r!)^2} + \right] \tag{8.58}$$

If $\lambda = n$, then

$$y = a_0 \left[1 - \frac{n}{1^2} \cdot x + \frac{n(n-1)}{(2!)^2} \cdot x^2 + + \frac{(-1)^r n(n-1)(n-r+1)}{(r!)^2} \cdot x^r \right]$$

(The highest power of x being n, i.e., x^n.)

$$\text{or } y = a_0 \sum_{r=0}^{\infty} (-1)^r \frac{n!}{(n-r)!(r!)^2} \cdot x^r \tag{8.59}$$

This is the solution of the Laguerre equation $x \frac{d^2 y}{dx^2} + (1-x)\frac{dy}{dx} + \lambda y = 0$.

8.5.1 Laguerre Polynomials

Equation (8.59) is the solution of the Laguerre equation. But when $a_0 = 1$, it is called the Laguerre polynomial of the order n and generally represented by $L_n(x)$.

∴ The Laguerre polynomial is expressed by

$$L_n(x) = \sum_{r=0}^{\infty} (-1)^r \frac{n!}{(n-r)!(r!)^2} \cdot x^r \tag{8.60}$$

Some authors define the Laguerre polynomial by considering $a_0 = n!$

∴ Equation (8.60) becomes

$$L_n(x) = n! \sum_{r=0}^{\infty} (-1)^r \frac{n!}{(n-r)!(r!)^2} \cdot x^r$$

$$\text{or } L_n(x) = \sum_{r=0}^{\infty} (-1)^r \frac{(n!)^2}{(n-r)!(r!)^2} \cdot x^r \tag{8.61}$$

Furthermore, we are going to express the other form of the Laguerre polynomial, i.e., the *Rodrigues formula.*

8.5.2 The Rodrigues Formula for the Laguerre Polynomials

The Rodrigues formula for the Laguerre polynomials is given by

$$L_n(x) = \frac{e^x}{n!} \frac{d^n}{dx^n} \left(n^n e^{-x} \right)$$

We prove this formula as follows:

Proof: The RHS of this formula is $\frac{e^x}{n!}\frac{d^n}{dx^n}\left(n^n e^{-x}\right)$.

Using the Leibnitz theorem, we can write

$$= \frac{e^x}{n!}\left[x^n(-1)^n e^{-x} + n \cdot nx^{n-1}(-1)^{n-1} e^{-x} + \frac{n(n-1)}{2} \cdot n(n-1)x^{n-2}(-1)^{n-2} e^{-x} + \ldots + n!e^{-x}\right]$$

$$= \frac{e^x}{n!} \cdot e^{-x}\left[(-1)^n x^n + (-1)^{n-1} \cdot \frac{n.n!}{(n-1)!} x^{n-1} + \ldots + n!\right]$$

$$= \sum_{r=0}^{\infty}(-1)^r \frac{n!}{(r!)^2 (n-r)!} \cdot x^r = L_n(x)$$

If we make $L_n(x) = \sum_{r=0}^{\infty}(-1)^r \frac{(n!)^2}{(n-r)!(r!)^2} \cdot x^r$, then

$$L_n(x) = \frac{e^x}{n!} \cdot \frac{d^n}{dx^n}\left(x^n e^{-x}\right) \tag{8.62}$$

Using Eq. (8.62), we are now able to find the first few Laguerre polynomials by putting $n = 0, 1, 2, 3, \ldots$, which has been illustrated in Table 8.3.

TABLE 8.3

Few Laguerre Polynomials

$$L_0(x) = \frac{e^x}{0!} \cdot \frac{d^0}{dx^0}\left(x^0 e^{-x}\right) = 1$$

$$L_1(x) = \frac{e^x}{1!} \cdot \frac{d}{dx}\left(xe^{-x}\right) = e^x\left(e^{-x} - xe^{-x}\right) = (1 - x)$$

$$L_2(x) = \frac{e^x}{2!} \cdot \frac{d^2}{dx^2}\left(x^2 e^{-x}\right) = \frac{e^x}{2!}\frac{d}{dx}\left(2xe^{-x} - x^2 e^{-x}\right)$$

$$= \frac{e^x}{2!}\left(2e^{-x} - 2xe^{-x} - 2xe^{-x} + x^2 e^{-x}\right)$$

$$= \frac{e^x}{2!} \cdot e^x\left(2 - 2x - 2x - x^2\right)$$

$$= \frac{1}{2!}\left(x^2 - 4x + 2\right)$$

Similarly, $L_3(x) = \frac{1}{3!}\left(-x^3 + 9x^2 - 18x + 6\right)$

$$L_4(x) = \frac{1}{4!}\left(x^4 + 16x^3 - 72x^2 - 96x + 24\right)$$

$$L_5(x) = \frac{1}{5!}\left(-x^5 + 25x^4 - 200x^3 + 600x^2 - 600x + 24\right)$$

and $L_6(x) = \frac{1}{6!}\left(-x^6 - 36x^5 + 450x^4 - 2400x^3 + 5400x^2 - 4320x + 720\right)$

and so on

8.5.3 The Laguerre Associated Equation and Its Solution

The Laguerre associated equation is represented by

$$x\frac{d^2y}{dx^2}+(k+1-x)\frac{dy}{dx}+ny=0 \tag{8.63}$$

If z is the solution of the Laguerre equation of the order $n + k$, then d^kz/dx^k will satisfy the Laguerre associated equation.

This fact is proved as follows.

Proof: The Laguerre associated and the Laguerre equations of order $n + k$ are given by Eqs. (8.63) and (8.64), respectively.

$$x\frac{d^2y}{dx^2}+(1-x)\frac{dy}{dx}+(n+k)y=0 \tag{8.64}$$

Since z is the solution of Eq. (8.64), we can write

$$x\frac{d^2z}{dx^2}+(1-x)\frac{dz}{dx}+(n+k)z=0 \tag{8.65}$$

Differentiating Eq. (8.65) k times, we get

$$\frac{d^k}{dx^k}+\left(x\frac{d^2z}{dx^2}\right)+\frac{d^k}{dx^k}\left[(1-x)\frac{dz}{dx}\right]+(n+k)\frac{d^kz}{dx^k}=0 \tag{8.66}$$

Applying the Leibnitz theorem to Eq. (8.66) reduces it to

$$\frac{d^{k+2}z}{dx^{k+2}}x+{}^kC_1\frac{d^{k+1}z}{dx^{k+1}}(1)+\frac{d^{k+1}z}{dx^{k+1}}\cdot(1-x)+{}^kC_1\frac{d^kz}{dx^k}\cdot(-1)+(n+k)\frac{d^kz}{dx^k}=0$$

$$\text{or } x\frac{d^{k+2}z}{dx^{k+2}}+(k+1-x)\frac{d^{k+1}z}{dx^{k+1}}+n\frac{d^kz}{dx^k}=0 \quad \therefore {}^kC_1=k$$

$$\text{or } x\frac{d^2}{dx^2}\left(\frac{d^kz}{dx^k}\right)+(k+1-x)\frac{d}{dx}\left(\frac{d^kz}{dx^k}\right)+n\frac{d^kz}{dx^k}=0 \tag{8.67}$$

In view of Eq. (8.63), Eq. (8.67) shows that d^kz/dx^k is a solution of Eq. (8.63).

Because the Laguerre polynomials satisfy the corresponding Laguerre equation, therefore, $L_{n+k}(x)$ satisfies Eq. (8.64). It is clear that as proved above, we see that $d^kz/dx^k\ L_{n+k}(x)$ is a solution of the Laguerre associated equation (Eq. 8.63). Obviously, $(-1)^k\left(d^kz/dx^k\right)L_{n+k}(x)$ is also the solution of Eq. (8.63) as $(-1)^k$ is constant. We now define this expression as the associated Laguerre polynomial.

8.5.4 Associated Laguerre Polynomials

The associated Laguerre polynomials of degree n and order k are defined by the following mathematical expression:

$$L_n^k(x)=(-1)^k\frac{d^k}{dx^k}L_{n+k}(x) \tag{8.68}$$

where $L_n^k(x)$ = associated Laguerre polynomials of degree n and order k and $L_{n+k}(x)$ = Laguerre polynomial of degree $n + k$.

Now we shall show that

$$L_n^k(x)=\sum_{r=0}^{n}(-1)^r\cdot\frac{(n+k)!}{(n-r)!(k+r)!r!}\cdot x^r$$

We have defined that

$$L_n^k(x)=(-1)^k\frac{\mathrm{d}^k}{\mathrm{d}x^k}L_{n+k}(x)$$

$$\text{or } L_n^k(x)=(-1)^k\frac{\mathrm{d}^k}{\mathrm{d}x^k}\sum_{r=0}^{n+k}(-1)^r\frac{(n+k)!}{(n+k-r)!(r!)^2}\cdot x^r$$

$$\left(\text{This has been written by using the definition } L_n(x)=\sum_{r=0}^{n+k}(-1)^r\frac{n!}{(n-r)!(r!)^2}\cdot x^r.\right)$$

$$\text{or } L_n^k(x)=(-1)^k\sum_{r=0}^{n+k}(-1)^r\frac{(n+k)!}{(n+k-r)!(r!)^2}\cdot\frac{\mathrm{d}^k}{\mathrm{d}x^k}x^r$$

$$=(-1)^k\left[\sum_{r=0}^{n-1}(-1)^r\frac{(n+k)!}{(n+k-r)!(r!)^2}\cdot\frac{\mathrm{d}^k}{\mathrm{d}x^k}x^r+\sum_{r=k}^{n+1}(-1)^r\frac{(n+k)!}{(n+k-r)!(r!)^2}\cdot\frac{\mathrm{d}^k}{\mathrm{d}x^k}x^r\right]$$

$$=(-1)^k\left[0+\sum_{r=k}^{n+k}(-1)^r\frac{(n+k)!}{(n+k-r)!(r!)^2}\times\frac{r!}{(r-k)!}\cdot x^{r-k}\right]$$

$$\left(\therefore\quad \frac{dk}{dx^k}\cdot x^r=\begin{cases}0 & \text{if } r<k\\ \dfrac{r!}{(r-k)!}\cdot x^{r-k} & \text{if } r>k\end{cases}\right)$$

$$=(-1)^k\sum_{s=0}^{n}(-1)^{k+r}\frac{(n+k)!}{(n+k-s-k)(s+k)!s!}\cdot x^s$$

Here, we have considered s as a new variable of summation such that $r=s+k$, i.e., $s=r-k$ so that when $r=k$, $s=0$ and when $r=k+n$, $s=n$.

Now, we may write

$$L_n^k(x)=(-1)^{2k}\sum_{s=0}^{n}(-1)^s\frac{(n+k)!}{(n-s)!(s+k)!s!}\cdot x^s$$

$$=\sum_{r=0}^{n}(-1)^r\frac{(n+k)!}{(n-r)!(s+k)!r!}\cdot x^r$$

$$=\sum_{r=0}^{n}(-1)^r\frac{(n+k)!}{(n-r)!(r+k)!r!}\cdot x^r \qquad \text{Proved. (8.69)}$$

8.5.5 The Rodrigues Formula for the Associated Laguerre Polynomials

Rodrigues gave the following formula for the associated Laguerre polynomials:

$$L_n^k(x) = \frac{e^x \cdot x^{-k}}{n!} \cdot \frac{d^n}{dx^n}\left(e^{-x} \cdot x^{n+k}\right) \tag{8.70}$$

We are going to prove it in the following manner.

Proof: We shall prove Eq. (8.70) by considering the RHS, i.e.,

$$\frac{e^x \cdot x^{-k}}{n!} \cdot \frac{d^n}{dx^n}\left(e^{-x} \cdot x^{n+k}\right)$$

Applying the Leibnitz theorem to Eq. (8.70), we write

$$\frac{e^x \cdot x^{-k}}{n!} \sum_{r=o}^{n} {}^nC_r D^{n-r} x^{n+k} \cdot D^r e^{-x}$$

$$= \frac{e^x \cdot x^{-k}}{n!} \sum_{r=0}^{n} {}^nC_r \cdot \frac{(n+k)!}{\{n+k-(n-r)\}!} \cdot x^{n+k-(n-r)} \cdot (-1)^r e^{-x}$$

$$\left(\because D^n x^m = \frac{m!}{(m-n)!} x^{m-n} \quad \text{and} \quad D^n e^{ax} = a^n e^{ax}\right)$$

$$= \sum_{r=0}^{n} \frac{e^x x^{-k}}{n!} \cdot \frac{n!}{r!(n-r)!} \cdot \frac{(n+k)!}{(k+r)!} x^{k+r} \cdot (-1)^r e^{-x}$$

$$= \sum_{r=0}^{n} (-1)^r \cdot \frac{(n+k)!}{(n-r)!(k+r)!r!} \cdot x^r$$

$$= L_n^k(x) = \text{L.H.S}$$

Hence, it has been proved.

Using the Rodrigues formula for the associated Laguerre polynomials, i.e.,

$$L_n^k(x) = \frac{e^x \cdot x^{-k}}{n!} \cdot \frac{d^n}{dx^n}\left(e^{-x} \cdot x^{n+k}\right),$$

we can find some associated Laguerre polynomials. Calculated values of some associated Laguerre polynomials are given in Table 8.4.

TABLE 8.4

Some Associated Laguerre Polynomials

$L_0^0 = 1$	$L_0^2 = 2$
$L_1^0 = -x + 1$	$L_1^2 = -6x + 18$
$2!L_2^1 = x^2 - 4x + 2$	$2!L_2^2 = 12x^2 - 96x + 144$
$L_0^1 = 1$	$L_0^3 = 6$
$L_1^1 = -2x + 4$	$L_1^3 = -24x + 96$
$2!L_2^1 = 3x^2 - 18x + 18$	$2!L_2^3 = 6x^2 - 600x + 1200$

8.6 Solution of the Radial Equation

The radial equation or the R equation is

$$\frac{1}{r^2}\cdot\frac{\mathrm{d}}{\mathrm{d}r}\left(r^2\frac{\mathrm{d}R}{\mathrm{d}r}\right)+\left[\frac{2\mu}{\hbar^2}\left(E+\frac{ze^2}{4\pi\epsilon_0 r}\right)-\frac{\beta}{r^2}\right]R=0$$

Substituting $\beta = l(l+1)$, the radial equation takes the following form:

$$\frac{1}{r^2}\cdot\frac{\mathrm{d}}{\mathrm{d}r}\left(r^2\frac{\mathrm{d}R}{\mathrm{d}r}\right)+\left[\frac{2\mu}{\hbar^2}\left(E+\frac{ze^2}{4\pi\epsilon_0 r}\right)-\frac{l(l+1)}{r^2}\right]R=0$$

where l is known as the orbital angular momentum quantum number.

For the found state, the energy E remains negative, and hence, the radial equation becomes

$$\frac{1}{r^2}\cdot\frac{\mathrm{d}}{\mathrm{d}r}\left(r^2\frac{\mathrm{d}R}{\mathrm{d}r}\right)+\left[\frac{2\mu(-E)}{\hbar^2}+\frac{2\mu ze^2}{4\pi\epsilon_0\hbar^2 r}-\frac{l(l+1)}{r^2}\right]R=0 \tag{8.71}$$

At this stage, to solve the radial equation, we shall change the variable for mathematical convenience. For this, let us suppose that

$$\rho = 2\alpha r \tag{8.72}$$

where α = constant and its value is so chosen that the resulting equation should look simpler.

If a function s is differentiable with respect to r, then we can write

$$\frac{\mathrm{d}s}{\mathrm{d}r}=\frac{\mathrm{d}s}{\mathrm{d}\rho}\cdot\frac{\mathrm{d}\rho}{\mathrm{d}r}=2\alpha\frac{\mathrm{d}s}{\mathrm{d}\rho}\qquad\left[\begin{array}{c}\therefore \rho=2\alpha r\\ \therefore \dfrac{\mathrm{d}\rho}{\mathrm{d}r}=2\alpha\end{array}\right]$$

Since $\dfrac{\mathrm{d}\rho}{\mathrm{d}r}=2\alpha$

$$\therefore \frac{\mathrm{d}}{\mathrm{d}r}=2\alpha\frac{\mathrm{d}}{\mathrm{d}\rho} \tag{8.73}$$

$$\text{and}\quad \frac{\mathrm{d}^2}{\mathrm{d}r^2}=4\alpha^2\frac{\mathrm{d}^2}{\mathrm{d}\rho^2} \tag{8.74}$$

With the help of these values, we may transform the first term of Eq. (8.71) as

$$\frac{1}{r^2}\cdot\frac{\mathrm{d}}{\mathrm{d}r}\left(r^2\frac{\mathrm{d}R}{\mathrm{d}r}\right)=\frac{1}{r^2}\left(r^2\frac{\mathrm{d}^2R}{\mathrm{d}r^2}+2r\frac{\mathrm{d}R}{\mathrm{d}r}\right)$$

$$=\left(\frac{\mathrm{d}^2R}{\mathrm{d}r^2}+\frac{2}{r}\frac{\mathrm{d}R}{\mathrm{d}r}\right)$$

$$=4\alpha^2\frac{\mathrm{d}^2R}{\mathrm{d}\rho^2}+\frac{2}{(\rho/2\alpha)}\cdot 2\alpha\cdot\frac{\mathrm{d}R}{\mathrm{d}\rho}$$

$$=4\alpha^2\left[\frac{\mathrm{d}^2R}{\mathrm{d}\rho^2}+\frac{2}{\rho}\cdot\frac{\mathrm{d}R}{\mathrm{d}\rho}\right] \tag{8.75}$$

Substituting this value in Eq. (8.71), we obtain

$$4\alpha^2\left[\frac{d^2R}{d\rho^2}+\frac{2}{\rho}\cdot\frac{dR}{d\rho}\right]+\left[\frac{2\mu(-E)}{\hbar^2}-\frac{2\mu ze^2}{4\pi\epsilon_0\hbar^2 r}-\frac{l(l+1)}{r^2}\right]R=0$$

Dividing throughout by $4\alpha^2$, we have

$$\frac{d^2R}{d\rho^2}+\frac{2}{\rho}\cdot\frac{dR}{d\rho}+\left[\frac{2\mu(-E)}{\hbar^2\cdot 4\alpha^2}-\frac{2\mu ze^2}{4\pi\epsilon_0\hbar^2 r}\cdot\frac{1}{4\alpha^2}-\frac{l(l+1)}{4\alpha^2\cdot r^2}\right]R=0$$

$$\text{or } \frac{d^2R}{d\rho^2}+\frac{2}{\rho}\cdot\frac{dR}{d\rho}+\left[\frac{2\mu(-E)}{\hbar^2\cdot 4\alpha^2}-\frac{2\mu ze^2}{4\pi\epsilon_0\hbar^2\cdot 2\alpha\rho}-\frac{l(l+1)}{\rho^2}\right]R=0$$

$$\text{or } \frac{d^2R}{d\rho^2}+\frac{2}{\rho}\cdot\frac{dR}{d\rho}+\left[-\frac{1}{4}+\frac{n}{\rho}-\frac{l(l+1)}{\rho^2}\right]R=0 \tag{8.76}$$

when $-\frac{2\mu E}{\hbar^2}=\alpha^2$ and $\frac{2\mu ze^2}{4\pi\epsilon_0\hbar^2\cdot 2\alpha}=n=\frac{\mu ze^2}{4\pi\epsilon_0\hbar^2\alpha}$

Now we have to solve Eq. (8.76). For this, we shall find the asymptotic solution pertaining to the large p limit, where the R equation is simplified. Equation (8.76) reduces to the following form when $p\to\infty$:

$$\frac{d^2R}{d\rho^2}-\frac{R}{4}=0 \tag{8.77}$$

when $\rho\to\infty$, $\frac{2}{\rho}\frac{dR}{d\rho}\to 0$, $\frac{n}{\rho}\to 0$, and $\frac{l(l+1)}{\rho^2}\to 0$

Therefore, the solutions of Eq. (8.77) are given by

$$R(\rho)=e^{\rho/2} \text{ and } R(\rho)=e^{-\rho/2} \tag{8.78}$$

Since p varies from zero to ∞, the former of these solutions will increase as p increases, and thus, it will lead to an unacceptable wave function. The other solution, i.e., $R(\rho)=e^{-\rho/2}$ will be satisfactory and acceptable since it decreases to zero as p, and hence, *r*, distance of electron from nucleus, increases to infinity.

Keeping in mind the asymptotic solution, a possible solution to Eq. (8.76) may be expressed as

$$R(\rho)=e^{\rho/2}F(\rho)$$

where $F(\rho)$ is another function depending on ρ. From various considerations, it seems that $F(\rho)$ may be split into two factors ρ^l and $G(\rho)$, which may be expressed as $F(\rho)=\rho'\cdot G(\rho)$,

wherein l has the same significance as in Eq. (8.76).

Thus, $R(\rho)$ may be written as

$$R(\rho)=e^{-\rho/2}\cdot\rho^l G(\rho)$$

$$\text{or } R(\rho)=G(\rho)\cdot\rho^l\cdot e^{-\rho/2} \tag{8.79}$$

For simplicity sake, we shall write $R(\rho)=R$ and $G(\rho)=G$ only.

Differentiating this equation with respect to ρ, we have

$$\frac{dR}{d\rho}=\rho^{l}\cdot e^{-\rho/2}\cdot\frac{dG}{d\rho}+G\left[e^{-\rho/2}\cdot l\rho^{l-1}+\rho^{l}\cdot e^{-\rho/2}\cdot\left(\frac{-1}{2}\right)\right]$$

$$=\rho^{l}\cdot e^{-\rho/2}\cdot\frac{dG}{d\rho}+G\rho^{l}e^{-\rho/2}\left(\frac{l}{\rho}-\frac{1}{2}\right) \quad (8.80)$$

$$=\rho^{l}\cdot e^{-\rho/2}\cdot\frac{dG}{d\rho}+R\left(\frac{l}{\rho}-\frac{1}{2}\right) \quad (8.81)$$

Further differentiation will result in

$$\frac{d^2R}{d\rho^2}=\rho^{l}\cdot e^{-\rho/2}\cdot\frac{d^2G}{d\rho^2}+\frac{dG}{d\rho}\left[e^{-\rho/2}\cdot l\cdot\rho^{l-1}+\rho^{l}\cdot e^{-\rho/2}\cdot\left(\frac{-1}{2}\right)\right]$$
$$+\left(\frac{l}{\rho}-\frac{1}{2}\right)\frac{dR}{d\rho}+Rl\left(\frac{-1}{\rho^2}\right)$$

Now, we shall substitute the value of $dR/d\rho$ from Eq. (8.81) and get

$$\frac{d^2R}{d\rho^2}=\rho^{l}\cdot e^{-\rho/2}.\frac{d^2G}{d\rho^2}+\frac{dG}{d\rho}\left[e^{-\frac{\rho}{2}}\cdot l\cdot\rho^{l-1}-\frac{1}{2}\rho^{l}\cdot e^{-\frac{\rho}{2}}\right]+\left(\frac{l}{\rho}-\frac{1}{2}\right)$$
$$\times\left[\rho^{l}e^{-\rho/2}\frac{dG}{d\rho}+R\left(\frac{l}{\rho}-\frac{1}{2}\right)\right]-\frac{Rl}{\rho^2}$$

$$\text{or}\ \frac{d^2R}{d\rho^2}=\rho^{l}\cdot e^{-\rho/2}\cdot\frac{d^2G}{d\rho^2}+\frac{dG}{d\rho}e^{-\rho/2}\cdot\rho^{l}\left(\frac{l}{\rho}-\frac{1}{2}\right)+\left(\frac{l}{\rho}-\frac{1}{2}\right)\times\rho^{l}e^{-\rho/2}\cdot\frac{dG}{d\rho}+R\left(\frac{l}{\rho}-\frac{1}{2}\right)^2-\frac{Rl}{\rho^2}$$

$$\text{or}\ \frac{d^2R}{d\rho^2}=\rho^{l}\cdot e^{-\rho/2}\cdot\frac{d^2G}{d\rho^2}+\rho^{l}\cdot e^{-\rho/2}\cdot 2\left(\frac{l}{\rho}-\frac{1}{2}\right)\frac{dG}{d\rho}+\left[\left(\frac{l}{\rho}-\frac{1}{2}\right)^2-\frac{1}{\rho^2}\right]G\rho^{l}e^{-\rho/2} \quad (8.82)$$

Substituting Eqs (8.82), (8.80), and (8.79) in Eq. (8.76) and cancelling the common factor $\rho^{l}e^{-\rho/2}$, we obtain

$$\text{or}\left[\frac{d^2G}{d\rho^2}+2\left(\frac{l}{\rho}-\frac{1}{2}\right)\frac{dG}{d\rho}+\left\{\left(\frac{1}{\rho}-\frac{1}{2}\right)^2-\frac{1}{\rho^2}\right\}G\right]+\frac{2}{\rho}\left[\frac{dG}{d\rho}+G\left(\frac{l}{\rho}-\frac{1}{2}\right)\right]$$
$$+\left[\frac{-1}{4}+\frac{n}{\rho}-\frac{l(l+1)}{\rho^2}\right]G=0 \quad (8.83)$$

The last term of the LHS may be simplified as given in the following:

$$\left(\frac{l^2}{\rho^2}-\frac{l}{\rho}+\frac{1}{4}-\frac{l}{\rho^2}+\frac{2l}{\rho^2}-\frac{1}{\rho}-\frac{1}{4}+\frac{n}{\rho}-\frac{l^2}{\rho^2}-\frac{l}{\rho^2}\right)G=\left(\frac{n-l-1}{\rho}\right)G \quad (8.84)$$

The second term is given by

$$2\left(\frac{l}{\rho}-\frac{1}{2}\right)+\frac{2}{\rho}=2\left(\frac{l}{\rho}-\frac{1}{2}+\frac{1}{\rho}\right)=2\left(\frac{2l-\rho+2}{2\rho}\right) \tag{8.85}$$

Considering the values in Eqs. (8.84) and (8.85), Eq. (8.83) reduces to

$$\frac{\mathrm{d}^2G}{\mathrm{d}\rho}+\left(\frac{2l-\rho+2}{\rho}\right)\cdot\frac{\mathrm{d}G}{\mathrm{d}\rho}+\frac{(n-l-1)}{\rho}\cdot G=0$$

$$\text{or } \rho\cdot\frac{\mathrm{d}^2G}{\mathrm{d}\rho^2}+(2l-\rho+2)\cdot\frac{\mathrm{d}G}{\mathrm{d}\rho}+(n-l-1)\cdot G=0 \tag{8.86}$$

The differential equation, for which the associated Laguerre polynomial L_k^p is a solution, can be represented as

$$\rho\cdot\frac{d^2L_k^p(\rho)}{\mathrm{d}\rho^2}+(p+1-\rho)\frac{\mathrm{d}}{\mathrm{d}\rho}L_k^p(\rho)+(k-p)L_k^p(\rho)=0 \tag{8.87}$$

Now comparison of Eq. (8.86) with (8.87) shows that $G(\rho)$ may be identified with the Laguerre polynomial $L_k^p(\rho)$, with

$$p=2l+1 \tag{8.88}$$

$$\text{and } k=n+l \tag{8.89}$$

Thus, the polynomial multiplied by a constant factor will be the required solution of Eq. (8.86), which may be expressed as

$$G(\rho)=CL_{n+1}^{2l+1}(\rho) \tag{8.90}$$

where C is a constant, which may be equal to the normalisation factor. Therefore, the complete expression for $R(\rho)$ function may be written with the help of Eq. (8.79), which is

$$R(\rho)=G(\rho)\cdot\rho^l\cdot e^{-\rho/2}$$

Putting the value of $G(\rho)$ from Eq. (8.90), the complete expression for the function $R(\rho)$ may be given by

$$R(\rho)=Ce^{-\rho/2}\rho^lL_{n+1}^{2l+1}(\rho) \tag{8.91}$$

This is obviously an acceptable function for the associated Laguerre polynomial.

Before proceeding further, we find the value of energy (E_n) and the Bohr radius 'a_0'.

We have assumed before that

$$\alpha^2=\frac{-2\mu E}{\hbar^2} \tag{8.92}$$

$$\text{and } n=\frac{\mu ze^2}{4\pi\epsilon_0\hbar^2\alpha}$$

$$\therefore \alpha=\frac{\mu ze^2}{4\pi\epsilon_0\hbar^2 n}$$

$$\therefore \alpha^2 = \frac{\mu^2 z^2 e^4}{16\pi^2 \in_0^2 \hbar^4 n^2} \tag{8.93}$$

From Eqs. (8.92) and (8.93), we can write that

$$\frac{-2\mu E}{\hbar^2} = \frac{\mu^2 z^2 e^4}{16\pi^2 \in_0^2 \hbar^4 n^2}$$

$$\text{or } \frac{-2E}{\hbar^2} = \frac{\mu\, z^2 e^4}{16\pi^2 \in_0^2 \hbar^4} \cdot \frac{1}{n^2}$$

$$\text{or } E = \left(\frac{-\mu z^2 e^4}{32\pi^2 \in_0^2 \hbar^2} \right) \cdot \frac{1}{n^2} \tag{8.94}$$

where E = energy of hydrogen atom
and n = principal quantum number.

We have obtained before that

$$\alpha = \frac{\mu z e^2}{4\pi^2 \in_0 \hbar^2 n}$$

$$\text{or } \frac{z}{\alpha} = \frac{4\pi \in_0 \hbar^2}{\mu e^2} \cdot \frac{1}{n} \tag{8.95}$$

when $n = 1$, z/α is known as the Bohr radius, which is represented by a_0.

$$\therefore a_0 = \left(\frac{4\pi \in_0 \hbar^2}{\mu e^2} \right) = \text{Bohr radius/radius of the first Bohr orbit.}$$

From Eq. (8.95), we have $\frac{z}{\alpha} = a_0 \cdot \frac{1}{n}$

$$\therefore \alpha = \frac{z}{\alpha_0 n} \text{ and } p = 2\alpha r = \left(\frac{2z}{a_0 n} \right) \cdot |r \tag{8.96}$$

The value of a_0 has been found to be 0.53 A.

In Eq. (8.94), we have introduced the principal quantum member, In Eqs. (8.88) and (8.89), we have used, $p = 2l + 1$ and $k = n + l$.

It is also a fact that k must always be greater than or equal to p if the associated polynomial is to be different from zero. By making use of this fact, it is possible to establish the permitted values of n and l, where l is the azimuthal quantum number. Using the aforementioned equations, we can write that

$$(n + l) \geq (2l + 1)$$

$$\therefore l \leq (n - 1) \tag{8.97}$$

The possible values of n are, thus, 1, 2, 3, and hence, the values of l will be 0, 1, 2, $(n - 1)$.

While dealing with the rigid rotor problem, we have mentioned that the possible values of m (i.e., magnetic quantum number) will be 0, ±1, ±2, …, ±I. It should be noted that when there will be no perturbing field, $(2l + 1)$ values will be of same energy, but in the magnetic field, the energy levels are separated and there will be possibilities of $(2l + 1)$ orientations.

Thus, it is clear that the quantum mechanics has provided us three quantum member, namely, *n*, *l*, and *m*.

8.6.1 Normalisation of the Radial Wave Function

We have derived the value of radial wave function in terms of *p*, which is equal to

$$Ce^{-\rho/2} \cdot \rho^l \cdot L_{n+l}^{2l+1}(\rho).$$

$$\text{i.e.,} \quad R(\rho) = Ce^{-\rho/2} \cdot \rho^l L_{n+l}^{2l+1}(\rho)$$

Since ρ is dependent on *r*, we may write

$$R(p) = R(r) \tag{8.98}$$

Then the aforementioned equation may be written as

$$R(r) = Ce^{-\rho/2} \cdot \rho^l \cdot L_{n+l}^{2l+1}(\rho) \tag{8.99}$$

We have also found out the value of *p* as mentioned in Eq. (8.96), i.e.,

$\rho = \dfrac{2zr}{a_0 n}$, where the terms have their usual significance.

The normalisation constant '*c*' [in Eq. (8.99)] can be determined from

$$\int_0^\infty R^2(r) r^2 \mathrm{d}r = 1 \tag{8.100}$$

The factor r^2 has been inserted to convert the length d*r* into volume element. From the aforementioned equation

$\rho = \dfrac{2zr}{a_0 n'}$, we have

$$r = \frac{a_0 n}{2z} \cdot \rho \quad \therefore \mathrm{d}r = \frac{n a_0}{2z} \mathrm{d}\rho$$

$$\text{But} \int_0^\infty \rho^2 \left| L_{n+l}^{2l+1}(\rho) \right|^2 \mathrm{d}\rho = \frac{2n\{(n+l)!\}^3}{(n-l-1)!} \tag{8.101}$$

$$\text{And} \int_0^\infty R^2(r) r^2 \mathrm{d}r = c^2 \cdot \frac{1}{\alpha^3} \int_0^\infty \rho^2 \left| L_{n+l}^{2l+1} \right|^2 \mathrm{d}\rho = 1$$

Putting the value of the LHS of Eq. (8.101), we get

$$c^2 \cdot \frac{1}{\alpha^3} \cdot \frac{2n\{(n+l)!\}^3}{(n-l-1)!} = 1$$

$$\text{or } c^2 \cdot \frac{1}{(2z/na_0)^3} \cdot \frac{2n\{(n+l)!\}^3}{(n-l-1)!} = 1 \quad \left[\therefore \alpha = \frac{2z}{na_0}\right]$$

$$\text{or } c^2 = \left(\frac{2z}{na_0}\right)^3 \frac{(n+l-1)!}{2n\{(n+l)!\}^3}$$

$$\therefore c = \pm\left[\left(\frac{2z}{na_0}\right)^3 \frac{(n-l-1)!}{2n\{(n+l)!\}^3}\right]^{1/2} \tag{8.102}$$

Thus, the value of normalisation constant is determined, and the normalised wave function will be expressed as

$$R(r) \text{ or } R_{n,l}(r) = -\left[\left(\frac{2z}{na_0}\right)^3 \cdot \frac{(n-l-1)!}{2n\{(n+l)!\}^3}\right]^{1/2} \cdot e^{-\rho/2} \cdot \rho^l L_{n+l}^{2l+1}(\rho)$$

$$\text{or } R_{n,l}(r) = -\frac{\sqrt{4(n-l-1)!}}{n^4\left[(n+l)!\right]^3} \cdot \left(\frac{a}{a_0}\right)^{3/2} \cdot \rho^l e^{-\rho/2} \cdot L_{n+l}^{2l+1}(\rho) \tag{8.103}$$

$$\text{or } R_{n,l}(r) = -\frac{\sqrt{4(n-l-1)!}}{n^4\left[(n+l)!\right]^3} \cdot \left(\frac{z}{a_0}\right)^3 \cdot e^{\frac{-zr}{na_0}} \cdot \left(\frac{2zr}{na_0}\right)^l \cdot L_{n+l}^{2l+1}\left(\frac{2zr}{na_0}\right) \tag{8.104}$$

where $\rho = \dfrac{2zr}{na_0'}$.

The minus sign in Eq. (8.103) is selected so as to make R_1,0 positive.

The subscript n, l has been added in the LHS of Eq. (8.103) to indicate that the function consists of the quantum member n and l.

Now we are going to calculate the radial wave function with the help of Eq. (8.103).

1. For $n = 1$, (*k shell*), $l = 0$ (*Is orbital*)

$$R_{1,0} = -\sqrt{\frac{4}{1^4} \cdot \frac{0!}{(1^!)^3}} \cdot \left(\frac{z}{a_0}\right)^{3/2} \cdot \rho^0 \cdot e^{-\rho/2} L_1^1(\rho)$$

$$= -2\left(\frac{z}{a_0}\right)^{3/2} \cdot e^{-\rho/2} L_1^1(\rho)$$

We know that

$$L_1(\rho) = -\rho + 1 \quad \therefore \quad L_1^1 = \frac{d}{d\rho}(-\rho + 1) = -1$$

$$\therefore R_{1,0} = (-2)\left(\frac{z}{a_0}\right)^{3/2} \cdot e^{-\rho/2}(-1)$$

$$= 2\left(\frac{z}{a_0}\right)^{3/2} \cdot e^{-\rho/2}$$

2. For $n = 2$ (L shell)
   a. $l = 0$ (2s orbital)

$$R_{2,0}(r) = -\sqrt{\frac{4}{16}\cdot\frac{1!}{(2!)^3}}\left(\frac{z}{a_0}\right)^{3/2}\cdot\rho^0\cdot e^{-\rho/2}L_2^1(\rho)$$

$$= -\sqrt{\frac{4\times 1}{16\times 8}}\left(\frac{z}{a_0}\right)^{3/2}\cdot e^{-\rho/2}L_2^1(\rho)$$

$$= -\sqrt{\frac{1}{16\times 2}}\left(\frac{z}{a_0}\right)^{3/2}\cdot e^{-\rho/2}L_2^1(\rho)$$

$$= -\frac{1}{4\sqrt{2}}\left(\frac{z}{a_0}\right)^{3/2}\cdot e^{-\rho/2}L_2^1(\rho)$$

But $L_2(\rho) = \rho^2 - 4\rho + 2$

$$\therefore L_2^1(\rho) = \frac{\mathrm{d}}{\mathrm{d}_\rho}\left(\rho^2 - 4\rho + 2\right) = 2\rho - 4$$

$$\therefore R_{2,0}(r) = -\frac{1}{4\sqrt{2}}\left(\frac{z}{a_0}\right)^{3/2}\cdot e^{-\rho/2}\cdot(2\rho - 4)$$

$$= (-)(-2)\frac{1}{4\sqrt{2}}\left(\frac{z}{a_0}\right)^{3/2}\cdot e^{-\rho/2}\cdot(2-\rho) = \frac{1}{2\sqrt{2}}\left(\frac{z}{a_0}\right)^{3/2}\cdot e^{-\rho/2}\cdot(2-\rho)$$

b. $l = 1$ (2p orbital)

$$R_{2,1}(r) = -\sqrt{\frac{4}{2^4}\cdot\frac{0!}{(3!)^3}}\left(\frac{z}{a_0}\right)^{3/2} .e^{-\rho/2}\rho^0 L_3^3(\rho)$$

$$= -\sqrt{\frac{4\times 1}{16\times 216}}\left(\frac{z}{a_0}\right)^{3/2} .e^{-\rho/2}\rho^1 L_3^3(\rho)$$

$$= -\frac{1}{2.6\sqrt{6}}\left(\frac{z}{a_0}\right)^{3/2} .e^{-\rho/2}\rho^1 L_3^3(\rho)$$

But $L_3(\rho) = -\rho^3 + 9\rho^2 - 18\rho + 6$

$$\therefore L_3^3(\rho) = \frac{\mathrm{d}^3}{\mathrm{d}\rho^3}\left(-\rho^3 + 9\rho^2 - 18\rho + 6\right) = -6$$

$$R_{2,1}(r) = -\frac{1}{2\times 6\sqrt{6}}\left(\frac{z}{a_0}\right)^{3/2}\cdot\rho(-6)$$

$$= -\frac{(-1)(-6)}{2\times 6\sqrt{6}}\left(\frac{z}{a_0}\right)^{3/2}\cdot\rho$$

$$= \frac{1}{2\sqrt{6}}\left(\frac{z}{a_0}\right)^{3/2}\cdot\rho$$

3. For $n = 3$ (M shell)
   a. $l = 0$ (3s orbital)

$$\therefore R_{3,0}(r) = -\sqrt{\frac{4}{3^4} \cdot \frac{2!}{(3!)^3}} \left(\frac{z}{a_0}\right)^{3/2} \cdot e^{-\rho/2} \rho^0 L_3^1(\rho)$$

$$= -\sqrt{\frac{4}{3^4} \cdot \frac{2!}{216}} \left(\frac{z}{a_0}\right)^{3/2} \cdot e^{-\rho/2} L_3^1(\rho)$$

But $L_3(\rho) = -\rho^3 + 9\rho^2 - 18\rho + 6$

$$L_3^1(\rho) = \frac{\mathrm{d}}{\mathrm{d}_\rho}\left(-\rho^3 + 9\rho^2 - 18\rho + 6\right)$$

$$= \left(-3\rho^2 + 18\rho - 18\right) = -3\left(\rho^2 - 6\rho + 6\right)$$

$$\therefore R_{3,0}(R) = -\frac{2}{3^2}\sqrt{\frac{1}{216}} \left(\frac{z}{a_0}\right)^{3/2} \cdot e^{-\rho/2}\left(\rho^2 - 6\rho + 6\right)(-3)$$

$$= -\frac{(-2)(-3)}{3^2} \times \frac{1}{6\sqrt{6}} \left(\frac{z}{a_0}\right)^{3/2} \cdot e^{-\rho/2}\left(\rho^2 - 6\rho + 6\right)$$

$$= \frac{1}{9\sqrt{6}} \left(\frac{z}{a_0}\right)^{3/2} \cdot e^{-\rho/2}\left(\rho^2 - 6\rho + 6\right)$$

   b. For $l = 1$ (3p orbital)

$$\therefore R_{3,1}(r) = -\sqrt{\frac{4}{3^4} \cdot \frac{1!}{(4!)^3}} \left(\frac{z}{a_0}\right)^{3/2} \cdot e^{-\rho/2} \rho^1 \left(L_4^3(\rho)\right)$$

$$= -\frac{2 \times 1}{3^2 \times 24 \times 2\sqrt{6}} \cdot \left(\frac{z}{a_0}\right)^{3/2} \cdot e^{-\rho/2} \rho L_4^3(\rho)$$

But $L_4 = \rho^4 - 16\rho^3 + 72\rho^2 - 96\rho + 24$

$$\therefore L_4^3 = \frac{\mathrm{d}^3}{\mathrm{d}\rho^3}\left(\rho^4 - 16\rho^3 + 72\rho^2 - 96\rho + 24\right)$$

$$= \left(24\rho - 96\right) = -24\left(4 - \rho\right)$$

$$\therefore R_{3,1}(r) = \frac{(-2)(-24)}{3^2 \times 24 \times 2\sqrt{6}} \cdot \left(\frac{z}{a_0}\right)^{3/2} \cdot e^{-\rho/2} \rho\left(4 - \rho\right)$$

$$= \frac{1}{9\sqrt{6}} \cdot \left(\frac{a}{a_0}\right)^{3/2} \cdot e^{-\rho/2} \rho\left(4 - \rho\right)$$

c. For $l = 2$ (3d orbital)

$$\therefore R_{3,2}(r) = -\sqrt{\frac{4}{3^4}\cdot\frac{0!}{(5!)^3}}\cdot\left(\frac{z}{a_0}\right)^{3/2}\cdot e^{-\rho/2}\rho^2\cdot L_5^5(\rho)$$

$$= -\sqrt{\frac{4\times 1}{3^4\times(5!)^3}}\cdot\left(\frac{z}{a_0}\right)^{3/2}\cdot e^{-\rho/2}\rho^2 L_5^5(\rho)$$

$$= -\frac{2\times 1}{3^2\times 5!\sqrt{5!}}\cdot\left(\frac{z}{a_0}\right)^{3/2}\cdot e^{-\rho/2}\rho^2 L_5^5(\rho)$$

$$= -\frac{2}{5!\times 3^2\times(5!)^{1/2}}\cdot\left(\frac{z}{a_0}\right)^{3/2}\cdot e^{-\rho/2}\rho^2 L_5^5(\rho)$$

But we know that

$$L_5(\rho) = -\rho^5 + 25\rho^4 - 200\rho^3 + 600\rho^2 - 600\rho + 120$$

$$\therefore L_5^5(\rho) = \frac{d^5}{d\rho^5}\left(-\rho^5 + 25\rho^4 - 200\rho^3 + 600\rho^2 - 600\rho + 120\right) = -5!$$

$$\therefore \frac{(-2)}{9\times 5!\sqrt{5!}}\cdot\left(\frac{z}{a_0}\right)^{3/2}.e^{-\rho/2}\rho^2(-5!)$$

$$= \frac{(-2)(-5!)}{9\times 5!\sqrt{5!}}\cdot\left(\frac{z}{a_0}\right)^{3/2}.e^{-\rho/2}\rho^2$$

$$= \frac{2}{9\sqrt{120}}\cdot\left(\frac{z}{a_0}\right)^{3/2}.e^{-\rho/2}\rho^2$$

$$= \frac{1}{9\sqrt{30}}\cdot\left(\frac{z}{a_0}\right)^{3/2}.e^{-\rho/2}\rho^2$$

Similarly, we can calculate for $n = 4$ (indicates N shell), i.e., for $l = 0$, $l = 1$, $l = 2$, and $l = 3$; some radial wave functions have been given in Table 8.5.

8.6.2 Complete Wave Function for the H Atom

The complete wave function for hydrogen atom may be written as

$$\psi_{lmm}(r,\theta,\phi) = R_{nl}(r)\cdot T_{lm}(\theta)\cdot F_m(\phi) \tag{8.105}$$

where $n = 1, 2, 3, 4$; $l = 0, 1, 2, 3$ $(n - 1)$; and $m = 0, \pm1, \pm2, \pm3$

We have already determined the expressions for $R_{nl}(r)$, $T_{lm}(\theta)$, and $F_m(\phi)$, which are given by

$$R_{nl}(r) = -\sqrt{\frac{4(n-l-1)!}{n^4[(n+l)!]^3}}\cdot\left(\frac{z}{a_0}\right)^{3/2}\cdot\rho^l\cdot e^{-\rho/2}\cdot L_{n+l}^{2l+1}(\rho)$$

where $\rho = \dfrac{2zr}{na_0}$ and for $H, \rho = \dfrac{2r}{a_0}$.

TABLE 8.5

Radial Wave Functions

$R_{1,0}(r)$	$2\left(\frac{z}{a_0}\right)^{3/2} \cdot e^{-\rho/2}$
$R_{2,0}(r)$	$\frac{1}{2\sqrt{2}}\left(\frac{z}{a_0}\right)^{3/2} \cdot e^{-\rho/2} \cdot (2-\rho)$
$R_{2,1}(r)$	$\frac{1}{2\sqrt{6}}\left(\frac{z}{a_0}\right)^{3/2} \rho \cdot e^{-\rho/2}$
$R_{3,0}(r)$	$\frac{1}{9\sqrt{6}}\left(\frac{z}{a_0}\right)^{3/2} \cdot e^{-\rho/2}\left(\rho^2-6\rho+6\right)$
$R_{3,1}(r)$	$\frac{1}{9\sqrt{6}}\left(\frac{z}{a_0}\right)^{3/2} \cdot e^{-\rho/2}\rho(4-\rho)$
$R_{3,2}(r)$	$\frac{1}{9\sqrt{30}}\left(\frac{z}{a_0}\right)^{3/2} \cdot e^{-\rho/2} \cdot \rho^2$
$R_{4,0}(r)$	$\frac{1}{96}\left(\frac{z}{a_0}\right)^{3/2} \cdot e^{-\rho/2}\left(24-36\rho+12\rho^2-\rho^3\right)$
$R_{4,1}(r)$	$\frac{1}{32\sqrt{15}}\left(\frac{z}{a_0}\right)^{3/2} \cdot e^{-\rho/2}\rho\left(\rho^2-10\rho+20\right)$
$R_{4,2}(r)$	$\frac{1}{96\sqrt{5}}\left(\frac{z}{a_0}\right)^{3/2} \cdot e^{-\rho/2} \cdot \rho^2(6-\rho)$
$R_{4,3}(r)$	$\frac{1}{96\sqrt{35}}\left(\frac{z}{a_0}\right)^{3/2} \cdot e^{-\rho/2} \cdot \rho^3$

Because $z = 1$ and $n = 1$

$$T_{l,m}(\theta) = \left[\frac{(2l+1)(l-|m|)!}{2(l+|m|)!}\right]^{1/2} \cdot P_l^{|m|}(\cos\theta)$$

$$\text{and } F_m(\phi) = \frac{1}{\sqrt{2\pi}} \cdot e^{\pm im\phi}$$

$$\text{or } F_m(\phi) = \frac{1}{\sqrt{2\pi}} \cdot (\cos m\phi \pm i \sin m\phi)$$

In finding out the complete wave function for the H atom, we shall take only real form. Using the aforementioned three formulae, the complete wave functions for the H atom and H-like atoms may be written for different states as shown in Table 8.6.

$$\rho = \frac{2zr}{na_0} \quad (\text{for } H, z = 1 \text{ and } n = 1)$$

If it is required to convert the functions in terms of r, one should put $\rho = 2zr/na_0$ in the functions described in Table 8.6.

TABLE 8.6

Complete Wave Functions of Hydrogen/Hydrogen-Like Atoms

n	l	M	ψ	Functions
1	0	0	1s	$\frac{1}{\sqrt{\pi}}\left(\frac{z}{a_0}\right)^{3/2}\cdot e^{-\rho/2}$
2	0	0	2s	$\frac{1}{4\sqrt{2\pi}}\left(\frac{z}{a_0}\right)^{3/2}\left(2-\frac{\rho}{2}\right)\cdot e^{-\rho/4}$
2	1	0	$2p_z$	$\frac{1}{4\sqrt{2\pi}}\left(\frac{z}{a_0}\right)^{3/2}\frac{\rho}{2}\cdot e^{-\rho/4}\cos\theta$
2	1	+1	$2p_x$	$\frac{1}{4\sqrt{2\pi}}\left(\frac{z}{a_0}\right)^{3/2}\frac{\rho}{2}\cdot e^{-\rho/4}\sin\theta\cos\phi$
2	1	−1	$2p_y$	$\frac{1}{4\sqrt{2\pi}}\left(\frac{z}{a_0}\right)^{3/2}\frac{\rho}{2}\cdot e^{-\rho/4}\sin\theta\cos\phi$
3	0	0	3s	$\frac{1}{81\sqrt{3\pi}}\left(\frac{z}{a_0}\right)^{3/2}\left(27-9\rho+\frac{\rho^2}{2}\right)\cdot e^{-\rho/6}$
3	1	0	$3p_z$	$\frac{\sqrt{2}}{81\sqrt{\pi}}\left(\frac{z}{a_0}\right)^{3/2}\left(6-\frac{\rho}{2}\right)\frac{\rho}{2}\cdot e^{-\rho/6}\cos\theta$
3	1	+1	$3p_x$	$\frac{\sqrt{2}}{81\sqrt{\pi}}\left(\frac{z}{a_0}\right)^{3/2}\left(6-\frac{\rho}{2}\right)\frac{\rho}{2}\cdot e^{-\rho/6}\sin\theta\cos\theta$
3	1	−1	$3p_y$	$\frac{\sqrt{2}}{81\sqrt{\pi}}\left(\frac{z}{a_0}\right)^{3/2}\left(6-\frac{\rho}{2}\right)\frac{\rho}{2}\cdot e^{-\rho/6}\sin\theta\cos\phi$
3	2	0	$3d_z^2$	$\frac{1}{81\sqrt{6\pi}}\left(\frac{z}{a_0}\right)^{3/2}\frac{\rho^2}{4}\cdot e^{-\rho/6}\left(3\cos^2\theta-1\right)$
3	2	+1	$3d_{xz}$	$\frac{\sqrt{2}}{81\sqrt{\pi}}\left(\frac{z}{a_0}\right)^{3/2}\frac{\rho^2}{4}\cdot e^{-\rho/6}\sin\theta\cos\theta\cos\phi$
3	2	−1	$3d_{yz}$	$\frac{\sqrt{2}}{81\sqrt{\pi}}\left(\frac{z}{a_0}\right)^{3/2}\frac{\rho^2}{4}\cdot e^{-\rho/6}\sin\theta\cos\theta\sin\phi$
3	2	+2	$3d_{x^2-y^2}$	$\frac{1}{81\sqrt{2\pi}}\left(\frac{z}{a_0}\right)^{3/2}\frac{\rho^2}{4}\cdot e^{-\rho/6}\sin^2\theta\cos 2\phi$
3	2	2	$3d_{xy}$	$\frac{1}{81\sqrt{2\pi}}\left(\frac{z}{a_0}\right)^{3/2}\frac{\rho^2}{4}\cdot e^{-\rho/6}\sin^2\theta\sin 2\phi$

8.6.3 Hydrogenic Atomic Orbital

We shall start with the lowest state in hydrogen and reach the higher quantum number and energy states. For the lowest energy level or shell $n = l$, $l = 0$, $m = 0$, and the energy Eigen function represented by $\psi_{100}(r,\theta,\phi) = 1/\sqrt{\pi}\,(z/a_0)^{3/2}\cdot e^{-\rho/2}$, which may be expressed in terms of r and a.u. as given in the following:

$$\psi_{100} = \psi(r) = \psi_{1s} = \frac{1}{\sqrt{\pi}}.e^{-r} \tag{8.106}$$

The orbital is denoted by ψ_{1s} In Eq. (8.106), the symbol ψ_{1s} designates the wave function of the electron in the $n = 1$ orbital. The aforementioned function ψ_{100} or ψ_{1s} is independent of θ and ϕ as the equation does not contain the term.

The value of the wave function is a function of the radial distance r from the centre of the atom. $\psi(r)$ function is for the 1*s* orbital of hydrogen. Therefore, the probability density for the electron at a given point will be expressed as

$$p(r) = \left(\frac{1}{\sqrt{\pi}} e^{-r}\right)^2 = \psi^2(r)$$

$$\text{or } \psi^2(r) = \frac{1}{\pi} \cdot e^{-2r} \tag{8.107}$$

$$= \text{probability density } p(r)$$

When $\psi(r)$ in a.u. for 1*s* orbital is plotted against *r* (in a.u.), the graph shown in Figure 8.2 is obtained.

But when $p(r) = \psi^2(r) = 1/\pi e^{-2r}$ (in a.u.) is plotted against *r* (in a.u.), the radial plot shown in Figure 8.3 is obtained.

It is clear that for s orbitals, the probability of finding an electron at the nucleus is finite, i.e., nonzero, whereas, for all other orbitals, the value of $\psi^2(r) = p(r)$ at the nucleus of the atom is zero. The electron density of 1*s* electron has been shown in Figure 8.4.

The density of the dots is actually the pictorial representation of the probability density.

In this case, the contour is a circle, which represents the cross section of the spherical orbital of 1*s*.

When the principal quantum number is two, i.e., $n = 2$, then l has two values, i.e., $l = 0$ and 1. The orbital with $l = 0$ represents 2*s* orbital, and the mathematical expression for ψ_{200} or ψ_{2s} is given by

$$\psi_{200} = \psi_{2s} = \frac{1}{4\sqrt{2\pi}}\left(\frac{z}{a_0}\right)^{3/2} \cdot \left(2 - \frac{\rho}{2}\right) \cdot e^{-\rho/4},$$

which is expressed in terms of *r* analytically as

$$\psi_{200} = \psi_{2s} = \frac{1}{4\sqrt{2\pi}} . e^{-r/2} \tag{8.108}$$

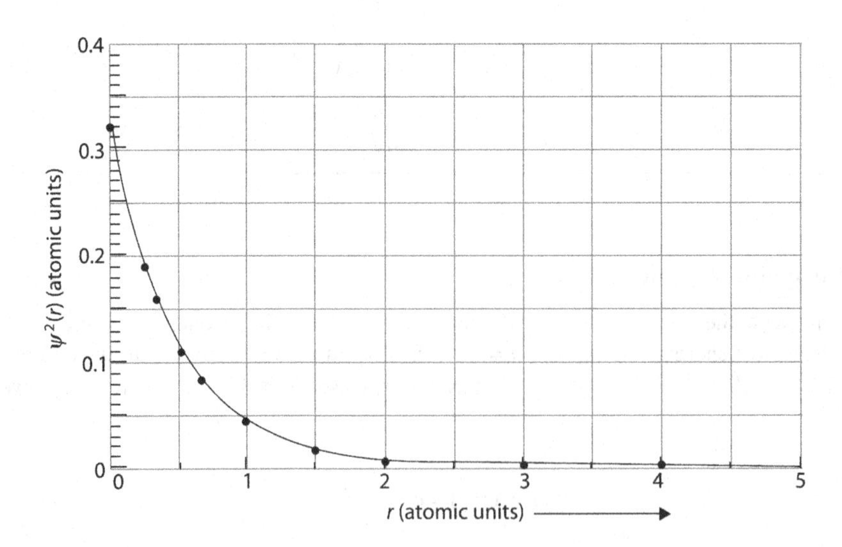

FIGURE 8.2 Plot of $\psi(1s) = \frac{1}{\sqrt{\pi}} e^{-r}$ against *r*.

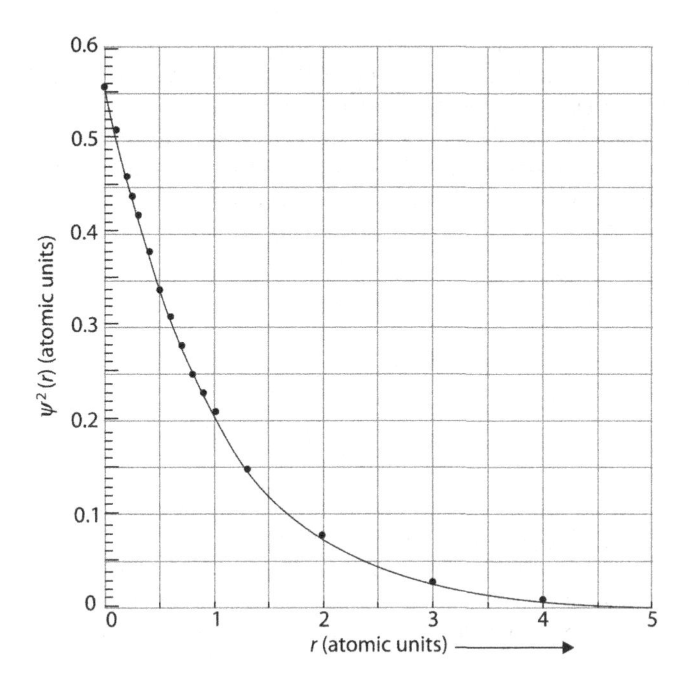

FIGURE 8.3 Plot of $p(r)$ vs. r.

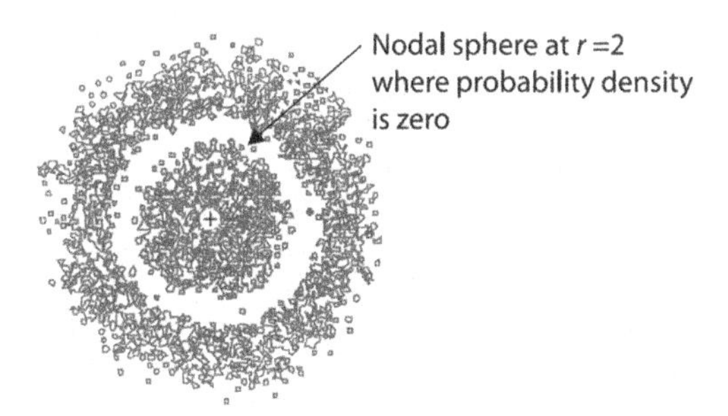

FIGURE 8.4 Electron density of 1*s* electron.

The main differences between the 2*s* and 1*s* orbitals for hydrogen atom are that 2*s* orbital is larger than 1*s* orbital, and it is clear from Eq. (8.108) that for $r = 2$, the 2*s* wave function is zero, i.e., $\psi_{2s} = 0$.

A surface on which a wave function is zero is known as a *node* (or nodal surface). Thus, it is clear that 2*s* orbital of H atom consists of a *nodal sphere* with a radius of 2 a.u. as illustrated in Figure 8.5.

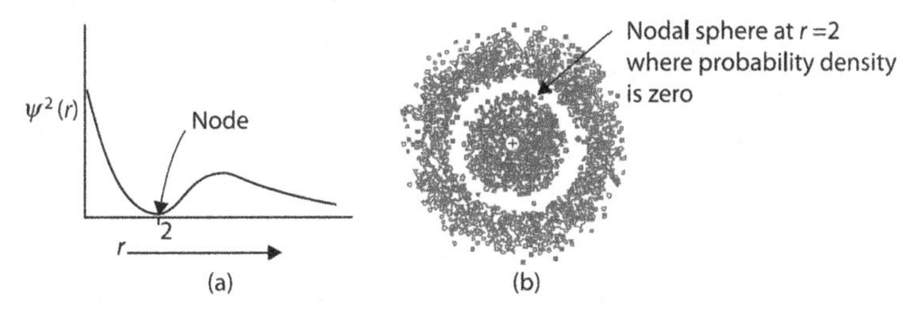

FIGURE 8.5 (a) The graph of $\psi^2(r)$ vs. r for 2*s* orbital, and (b) a cross section through the probability function drawn in 3D and the probability density is denoted by stippling.

It is important to keep in mind that all s orbitals are spherically symmetrical. It is also important to note that orbitals having $l = 0$ and different values of n differ only with regard to the effective volume and the number of nodes. The shape of s orbitals is shown in Figure 8.6.

In the second shell with $n = 2$, an orbital with $l = 1$ is encountered. When $l = 1$, three orbitals are obtained, since magnetic quantum number $m = -1$, 0, and +1, and these three orbitals have directional properties. The wave function $\psi_{210} = \psi_{2p_z}$ is mathematically represented by

$$\psi_{210} = \psi_{2p_z} = \frac{1}{4\sqrt{2\pi}}\left(\frac{z}{a_0}\right)^{3/2} \cdot \frac{\rho}{2} \cdot e^{-\rho/4} \cos\theta$$

and in terms of polar coordinates and a.u., it is expressed as

$$\psi_{210} = \psi_{2p_z} = \frac{1}{4\sqrt{2\pi}} \cdot (\cos\theta) \cdot r \cdot e^{-r/2} \tag{8.109}$$

It is pertinent to note that the $2p_z$ orbital has regions of greatest concentration or probability along the z-axis. The electron density of $2p_z$ orbital is illustrated in Figure 8.7.

When we put $\theta = 90°$ in Eq. (8.109), ψ_{2p_z} becomes zero, i.e., ψ_{210} becomes zero. Examining the aforementioned representation, it is said that the probability of finding the electron in the xy plane is zero. It is also inferred that this nodal plane having the atomic nucleus is a characteristic of all p orbitals.

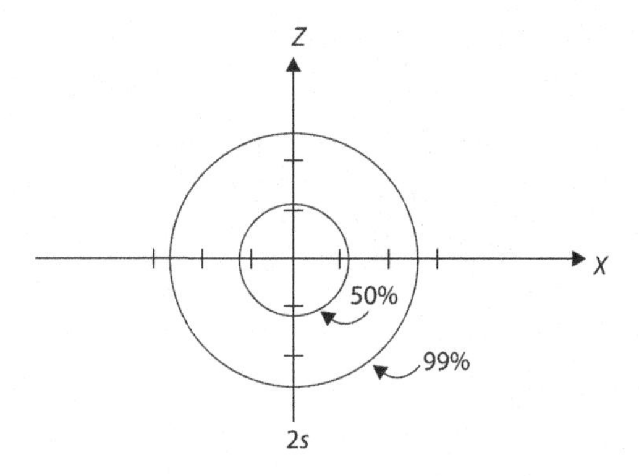

FIGURE 8.6 Shape of an s orbital.

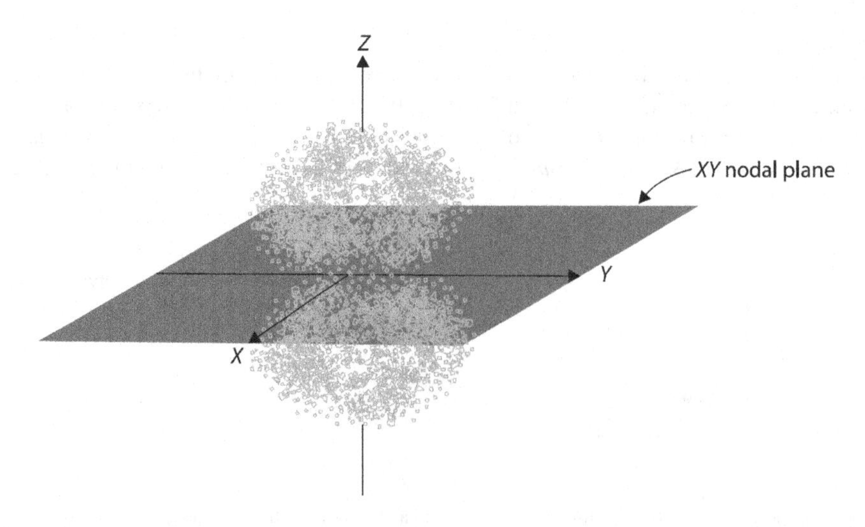

FIGURE 8.7 Representation of electron.density of $2p_z$ orbital.

The $2p_z$ orbital can also be represented with a contour diagram or a spatial notation. The contour diagram is obtained by plotting lines of constant probability density, which is $|\psi^2|$

$$\text{or } \psi^2, \text{ which will be equal to } p(r) = \psi^2_{2p_z} = \frac{r^2 \cos^2\theta}{32\pi} e^{-r} \tag{8.110}$$

The contour diagram of the $2p_z$ orbital is shown in Figure 8.8, and it represents lines of constant ψ^2 in the yz plane, which have been so selected in 3D that they enclose 50% or 99% of the total probability density. The $2p_z$ orbital is symmetrical around the z-axis.

It should be kept in mind that the 99% probability shell portrayed as a surface. The plus (+) and minus (−) on the two lobes indicate the relative signs of ψ, which should not be confused with the electric charge. Mind that there is no probability of finding the electron on xy plane. Such a surface is called a nodal surface, which should not be planar. The finally obtained spatial notation of $2p_Z$ orbital is illustrated in Figure 8.9.

Now we shall analyse by putting $\theta = 0$ and $\theta = 90°$, again for $\theta = 90°$ and $\theta = 180°$. The value of $2p_z$ wave function will be positive for positive z values and negative for negative z values. From $\theta = 0$ to $90°$, the value of z will be positive and from $\theta = 90°$ to $180°$, the value of z will be negative, which is indicative of the fact that the signs of the two lobes of $2p_z$ orbital in their spatial representation depend on the

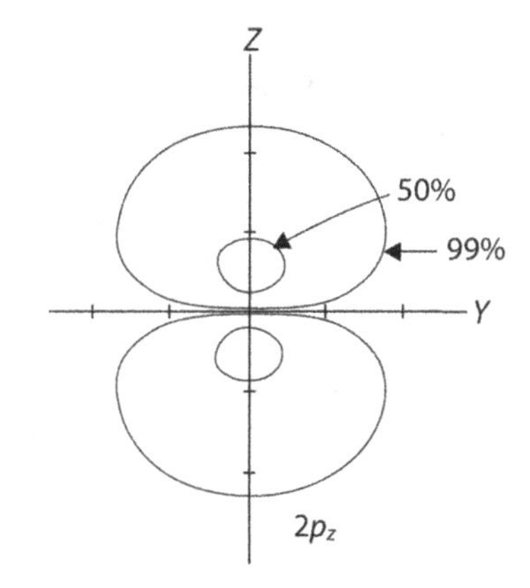

FIGURE 8.8 Contour of $2p_z$.

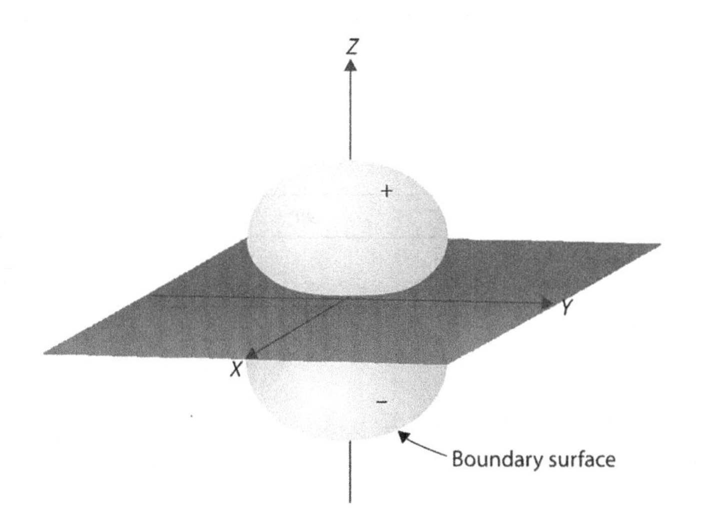

FIGURE 8.9 Spatial notation of $2p_z$ orbital.

positive and negative values of z. It should be kept in mind that all p functions change sign when inverted at the atomic nucleus and are known as antisymmetric. In contrast, s orbital functions are symmetrical due to the fact that the inversion does not produce a change in the algebraic sign.

The equation for the wave function ψ_{211} or ψ_{2p_x} is

$$\frac{1}{4\sqrt{2}\pi}\left(\frac{z}{a_0}\right)^{3/2} \cdot \frac{\rho}{2} \cdot e^{-\rho/4} \sin\theta \cos\phi$$

which may be expressed in terms of r as

$$\psi_{2p_x} = \frac{1}{4\sqrt{2}\pi} \cdot r \cdot e^{-r/2} \sin\theta \cos\phi \tag{8.111}$$

and the wave function $\psi_{2,1,-1}$, i.e., ψ_{2p_y} may be expressed in terms of r as

$$\psi_{2p_y} = \frac{1}{4\sqrt{\pi}} \cdot r \cdot e^{-r/2} \cdot \sin\theta \sin\phi \tag{8.112}$$

From Eq. (8.111), it is clear that ψ_{2p_x} consists of $\sin\theta \cdot \cos\phi$ when $\phi = 0$, $\cos\phi = 1$. Consequently, the value of $\sin\theta$ will be positive in the X-direction as θ moves from 0° to π. The trigonometrical values of $\sin\theta$ from 0° to π have been given in Table 8.7.

The polar plot of $\sin\theta$ with its values in the +X-direction will give another circle for $\phi = \pi$ or 180° (i.e., along –X-direction). As ϕ moves from 90° to 360°, the negatives of $\cos\phi$ are obtained which are illustrated in Table 8.8.

The polar plot of $\cos\phi$ with its values for different angles from $\pi/2$ to 2π results a circle in the –X-direction. It must be kept in mind that as ϕ increases, the value of $\sin\theta\cos\phi$ for any value of θ decreases due to the $\cos\phi$ factor. Thus the complete polar plot of $2p_x$ function contains two spheres, one along +X-direction another along –X-direction.

Similarly, for $2p_y$ orbital, we can show by polar plot that it consists of two spheres: one along +Y direction and another along –Y direction, considering the $2p_y$ wave function and that is

$$\psi_{2,1,-1} = \psi_{2p_y} = \frac{1}{4\sqrt{2}\pi}\left(\frac{z}{a_0}\right)^{3/2} \cdot r \cdot e^{-r/2} \sin\theta \cdot \sin\phi \tag{8.113}$$

The $2p_y$ orbital will have XZ plane as the nodal plane. The complete set of $2p_x$, $2p_y$, and $2p_z$ orbitals is shown in Figure 8.10. It should be noted that these three equivalent p orbitals differ only in their spatial orientations.

TABLE 8.7

Trigonometrical Values of $\sin\theta$ from 0 to π

$\theta \rightarrow$	0°	30°	60°	90°	120°	150°	180°
$\sin\theta$	0	$\frac{1}{2}$	$\frac{\sqrt{3}}{2}$	1	$\frac{\sqrt{3}}{2}$	$\frac{1}{2}$	0

TABLE 8.8

Negatives of $\cos\phi$ as ϕ Moves from 90° to 360°

$\theta \rightarrow$	90°	120°	150°	180°	210°	270°	360°
$\cos\phi \rightarrow$	0	$\frac{-1}{2}$	$\frac{-\sqrt{3}}{2}$	–1	$\frac{-\sqrt{3}}{2}$	$\frac{-1}{2}$	0

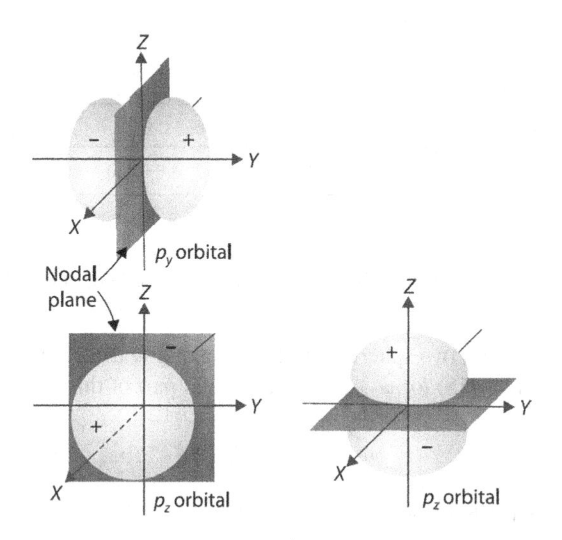

FIGURE 8.10 Boundary surfaces for p_x, p_y, and p_z orbitals.

When the value of principal quantum number $n = 3$, l can have three values, i.e., 0, 1, and 2. As for 3*s* orbital, $l = 0$, it will be similar to the *s* orbitals previously described. It is a general result that the number of nodes will be equal to $(n - 1)$, which indicates that, for 3*s* orbital, there will be two nodes.

When $l = 1$, there will be three 3*p* orbitals. When the magnetic quantum number for 2*p* and 3*p* orbitals will remain same, they will have the same angular dependence, although the boundary contours of 3*p* orbitals are more complicated with respect to 2*p* orbitals due to the presence of an additional node. However, the outer part of 3*p* orbital looks similar to the 2*p* orbital.

When $n = 3$ and $l = 2$, the $l = 2$ value will give five possible values of the magnetic quantum number, and thus, it will result in 5*d* orbitals. The wave functions for 5d orbitals are mentioned in the following:

$$\psi_{3d_z^2} = \frac{1}{81\sqrt{6\pi}} \cdot \left(\frac{z}{a_0}\right)^{3/2} \cdot \frac{\rho^2}{4} \cdot e^{-\rho/6}\left(3\cos^2\theta - 1\right)$$

$$= \frac{1}{81\sqrt{6\pi}} \cdot r^2 \cdot e^{-r/3}\left(3\cos^2\theta - 1\right) \text{ in a.u.} \tag{8.114}$$

$$\text{Similarly } \psi_{3\text{dxz}} = \frac{\sqrt{2}}{81\sqrt{\pi}} \cdot r^2 \cdot e^{-r/3} \sin\theta\cos\theta\cos\phi \text{ in a.u.} \tag{8.115}$$

$$\psi_{3\text{dyz}} = \frac{\sqrt{2}}{81\sqrt{\pi}} \cdot r^2 \cdot e^{-r/3} \sin\theta\cos\theta\cos\phi \text{ in a.u.} \tag{8.116}$$

$$\psi_{3\text{d}x^2-y^2} = \frac{1}{81\sqrt{2\pi}} \cdot r^2 \cdot e^{-r/3} \sin^2\theta\cos 2\phi \text{ in a.u.} \tag{8.117}$$

$$\text{and } \psi_{3\text{dxy}} = \frac{1}{81\sqrt{2\pi}} \cdot r^2 \cdot e^{-r/3} \sin^2\theta\sin 2\phi \text{ in a.u.} \tag{8.118}$$

From Eq. (8.114), it is clear that the angular part of $\psi_{3d_z^2}$ consists of $(3\cos^2\theta - 1)$. This gives a clue towards understanding the fact that $\psi_{3d_z^2}$ will have the maximum value 2 at $\theta = 0°$ and 180°. The value 2 will be obtained by putting $\theta = 0°$ or 180° in $(3\cos^2\theta - 1)$.

But $\psi_{3d_z^2}$ will have zero value when $3\cos^2\theta = 1$ or $\cos^2\theta = 1/3$ or $\cos\theta = 1/\sqrt{3}$, which gives $\theta = 54.74°$ and 125.26°. It means that $\psi_{3d_z^2}$ orbital has a nodal plane. By putting $\theta = 90°$ in $3\cos^2\theta - 1$, we get

TABLE 8.9

Values of $\psi_{3d_z^2}$

$\theta \rightarrow$	0°	54.74°	90°	125.26°	180°
$\psi_{3d_z^2} = \left(3\cos^2\theta - 1\right) \rightarrow$	2	0	−1	0	2

$\psi_{3d_z^2} = -1$ in the xy plane. It means that in the xy plane, $\psi_{3d_z^2}$ is negative. The aforementioned values of $\psi_{3d_z^2}$ may be summarised in Table 8.9.

Thus, the polar plot will contain two spherical lobes: one along $+z$ direction and one along z direction. Now, we shall consider Eq. (8.115) for $\psi_{3d_{xz}}$, and the angular part of this can be represented as

$$\psi_{3d_{xz}} = \sin\theta\cos\theta \cdot \cos\phi$$

It can be inferred from the equation $\psi_{3d_{xz}} = \sin\theta\cos\theta.\cos\phi$ that

$$\psi_{3d_{xz}} = 0 \text{ at } \theta = 0°, 90° \text{ and } 180° \text{ and for } \phi = 90° \text{ and } 270°$$

$$\psi_{3d_{xz}} = \frac{1}{2} \text{ at } \theta = 45°, \phi = 0° \text{ and } \theta = 135°, \phi = 180°$$

$$= \text{maximum positive value.}$$

$$\psi_{3d_{xz}} = \frac{-1}{2} \text{ at } \theta = 45°, \phi = 180° \text{ and } \theta = 135°, \phi = 0°$$

$$= \text{maximum negative value.}$$

The polar plot will consist of spheres and the lines where the planes yz and xz cut the spheres form the nodal lines. The function $\psi_{3d_{yz}}$ consists of sin θ cos θ sin ϕ as the angular part. We find different values by putting different θ and ϕ values in equation $\psi_{3d_{yz}} = \sin\theta\cos\theta\sin\phi$, and by polar plot, we arrive at a function similar to $\psi_{3d_{xz}}$ function. The only difference is that the nodal lines lie in the xy and xz planes.

From Eq. (8.117), it is clear that the angular part of the equation is $\psi_{3d_{x^2-y^2}} = \sin^2\theta\cos 2\phi$. We find that

$$\text{at } \theta = 0 \text{ and } \theta = 180° \quad \psi_{3d_{x^2-y^2}} = 0$$

$$\text{at } \phi = 45° \text{ and } \phi = 135° \quad \psi_{3d_{x^2-y^2}} = 0$$

It means that the polar plot will result in spheres, and these spheres will cut each other and form two nodal lines.

Similarly, the value of $\psi_{3d_{xy}} = \sin^2\theta\sin^2 2\phi$ will be zero at $\theta = 0°$ and 180° and at $\phi = 90°$ and 270°, and therefore, it has nodal lines in the xz and yz planes. The five $3d$ orbitals of hydrogen are shown in Figure 8.11.

It is to be noted that the $3d_{z^2}$ orbital with magnetic quantum number $m = 0$ has a different shape than the other d orbitals d_{z^2}, which is symmetric with respect to z-axis. Furthermore, it is to be noticed that the signs labelling the various lobes indicate that the $3d$ orbital wave function remains either positive or negative in the region. The d orbital functions are found to be symmetrical since inversion at the origin does not change the algebraic sign.

8.6.4 Radial Wave Function

The radial portion of the hydrogenic wave function is given in Table 8.5. It is to be noted that this function depends on the quantum numbers n and l, even though the energy depends only on n. Since the radial

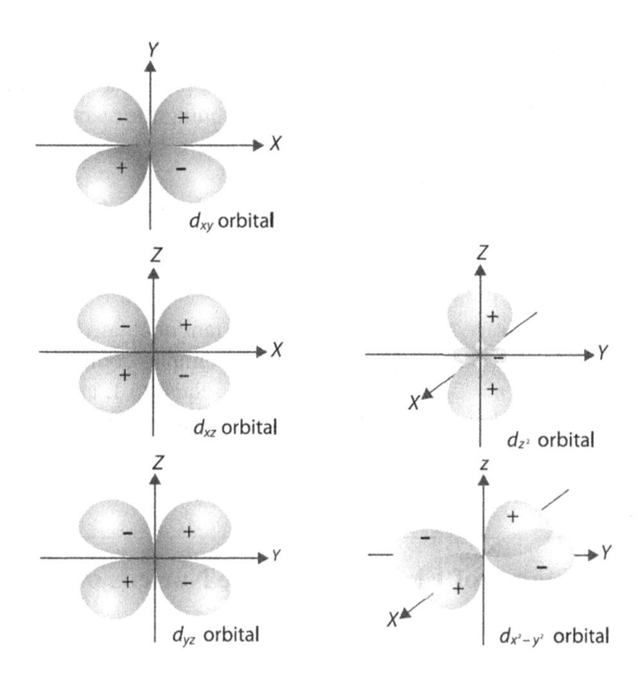

FIGURE 8.11 Five 3*d* orbitals.

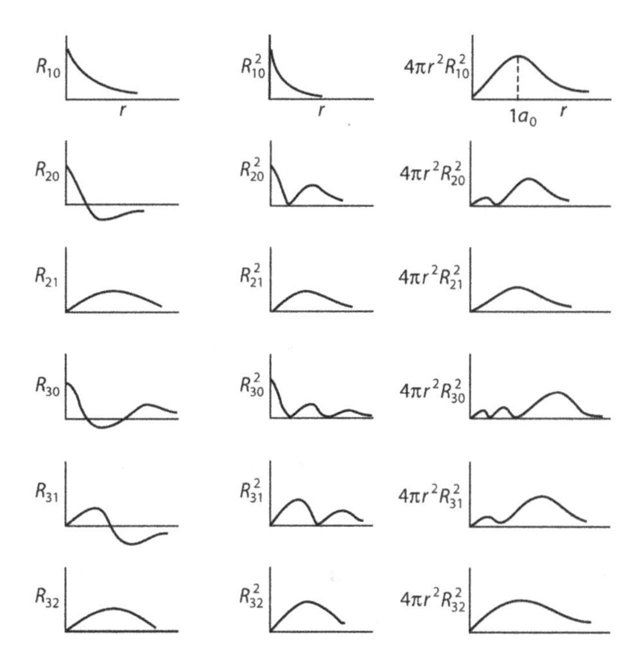

FIGURE 8.12 $R_{nl}(r)$, $R_{nl}^2(r)$, and $4\pi r^2 R_{nl}^2(r)$ vs. *r* graph.

portion of the hydrogenic wave function contains *p*, which is equal to $2zr/na_0$, we have plotted $R_n l\ (r)$ vs. *r*, $R_{nl}^2(r)$ vs. *n*, and $4\pi r^2 R_{nl}^2(r)$ vs. *r*. Several plots have been illustrated in Figure 8.12.

The first two, i.e., R_{nl} vs. *r* and R_{nl}^2 vs. *r* have already been defined. Only difference is that we have plotted ψ vs. *r* and ψ^2 vs. *r*. The third, i.e., $4\pi r^2 R_{nl}^2(r)$ vs. *r* graph is new, which gives the probability of finding the electron in a spherical shell of thickness r at a radius r (where $4\pi r^2$ = surface area of sphere of radius r). $R_{nl}^2(r)$ is the electron density function, while $4\pi r^2 R_{nl}^2(r)$ is known as the radial distribution function. Actually, it gives the idea of overall probability of finding the electron at r or between r and $r + dr$.

Now we shall point towards the *radial nodes*.

When the value of principal quantum number is n, then the total number of radial nodes will be $(n - 1)$. At the point of node, there is zero probability of finding the electron.

Another important point to be noted is that for a z value of 1 and an n of 1, the maximum in the radial distribution function comes at one Bohr radius.

The nodal characteristics of a wave function have an important qualitative significance. For a given function, the more nodes there are, the higher energy. For example, comparing the s functions, the one with $n = 1$ has no nodes. For $n = 2$, the number of node is equal to 1, and for $n = 3$, it is 2, and so on. It should be kept in mind that both the number of nodes and the energies increase with increasing value of n.

8.7 Most Probable Distance of Electron from the Nucleus of H Atom

The ψ_{1s} function of H atom is given by

$$\psi_{1s} = \frac{1}{\sqrt{\pi}}\left(\frac{1}{a_0}\right)^{3/2} e^{-r/a_0} = Ne^{-r/a_0}$$

$$\therefore\ \psi_{1s}^2 = N^2 e^{-2r/a_0}$$

For the most probable condition,

$$\frac{dP}{dr} = 0$$

$$\therefore\ P(r) = \psi_{1s}^2 . r^2 = r^2 N^2 e^{-2r/a_0}$$

$$\therefore\ \frac{dP(r)}{dr} = N^2\left(2r - \frac{2r^2}{a_0}\right)e^{-2r/a_0} = 0$$

$$\text{or } 2r - \frac{2r^2}{a_0} = 0 \text{ or } r = a_0 = 0.529\,\text{Å} = \text{Bohr radius}$$

8.7.1 Average Distance of Electron from the Nucleus of H Atom

The average distance $(\bar{r})$ can be obtained from the expectation value postulate of quantum mechanics. We know that

$$(\bar{r}) = \frac{\int \psi_{1s}^* r \psi_{1s} d\tau}{\int \psi_{1s}^2}$$

where $d\tau$ = volume element in polar form = $r^2 dr \sin\theta\, d\theta\, d\phi$

$$\text{or } \bar{r} = \frac{\int_0^\infty Ne^{-r} \cdot r \cdot Ne^{-r} r^2 \mathrm{d}r \int_0^\pi \sin\theta \mathrm{d}\theta \int_0^{2\pi} \mathrm{d}\phi}{\int_0^\infty \left(Ne^{-r}\right)^2 r^2 \mathrm{d}r \int_0^\pi \sin\theta \mathrm{d}\theta \int_0^{2\pi} \mathrm{d}\phi} \quad \left[\therefore \psi_s = Ne^{-r} \text{ in a.u.}\right]$$

$$= \frac{\int_0^\infty r^3 e^{-2r} \mathrm{d}r \cdot 4\pi}{\int_0^\infty r^2 e^{-2r} \mathrm{d}r \cdot 4\pi}$$

$$= \frac{\int_0^\infty r^3 e^{-2r} \mathrm{d}r}{\int_0^\infty r^2 e^{-2r} \mathrm{d}r} \left[\therefore \int_0^\infty r^n e^{-ar} \mathrm{d}r = \frac{n}{a^{n+1}}\right]$$

$$= \frac{3!/2^4}{2!/2^3} = \frac{3}{2} \text{a.u.} = \frac{3}{2} a_0 = \frac{3}{2} \times 0.529 \text{ Å}$$

$= 0.7935$ Å = average distance of electron from the nucleus of H atom

BIBLIOGRAPHY

Merzbacher, E. 1970. *Quantum Mechanics*. New York: Wiley.
Messiah, A. 1961. *Quantum Mechanics*, vol. I. Amsterdam: North-Holland Publishing Company.
Pauling, L. and E.B. Wilson. 1935. *Introduction to Quantum Mechanics*. New York: McGraw-Hill Book Co. Inc.
Pilar, F.L. 1990. *Elementary Quantum Chemistry*. New York: McGraw-Hill Book Co. IncMcGraw-Hill.
Slater, J.C. 1960. *Quantum Theory of Atomic Structure*, vol. I. New York: McGraw-Hill Book Co. Inc.

Solved Problems

Problem 1. Show that the ψ_{1s} function for hydrogen atom is orthogonal to ψ_{2s} functions.

Solution: We know that ψ_{1s} for Hydrogen $= \frac{1}{\sqrt{\pi}}\left(\frac{1}{a_0}\right)^{3/2} e^{-r/a_0}$

And xp_{2s} for hydrogen $= \frac{1}{4\sqrt{2\pi}}\left(\frac{1}{a_0}\right)^{3/2}\left(2-\frac{r}{a_0}\right)e^{-r/2a_0}$

$$\therefore \psi_{1s}\psi_{2s} = \left[\frac{1}{\sqrt{\pi}}\left(\frac{1}{a_0}\right)^{3/2} \cdot e^{-r/a_0}\right]\left[\frac{1}{4\sqrt{2\pi}}\left(\frac{1}{a_0}\right)^{3/2}\left(2-\frac{r}{a_0}\right)e^{-r/2a_0}\right]$$

$$= \frac{1}{4\sqrt{2\pi}} \cdot \left(\frac{1}{a_0}\right)^3\left(2-\frac{r}{a_0}\right)e^{-3r/2a_0}$$

$$\text{or } \psi_{1s} \cdot \psi_{2s}\mathrm{d}\tau = \frac{1}{4\sqrt{2\pi}}\left(\frac{1}{a_0}\right)^3\left(2-\frac{r}{a_0}\right)e^{-3r/2a_0}\mathrm{d}\tau$$

$$\text{or } \psi_{1s} \cdot \psi_{2s}\mathrm{d}\tau = \frac{1}{4\sqrt{2\pi}}\left(\frac{1}{a_0}\right)^3\left(2-\frac{r}{a_0}\right)e^{-3r/2a_0} \cdot 4\pi r^2 \mathrm{d}r \quad \left[\therefore \mathrm{d}\tau = 4\pi r^2 \mathrm{d}r\right]$$

$$= \frac{1}{\sqrt{2}}\left(\frac{1}{a_0}\right)^3 r^2\left(2-\frac{r}{a_0}\right)e^{-3r/2a_0} \mathrm{d}r$$

The condition of orthogonality is

$$\int_0^\infty \psi_{1s} \cdot \psi_{2s} \, d\tau = 0$$

$$\therefore \quad \frac{1}{\sqrt{2}}\left(\frac{1}{a_0}\right)^3 \int_0^\infty r^2\left(2 - \frac{r}{a_0}\right) e^{-3r/2a_0} dr$$

$$= \frac{1}{\sqrt{2}}\left(\frac{1}{a_0}\right)^3 \left[\int_0^\infty 2r^2 e^{-3r/2a_0} dr - \int_0^\infty \frac{r^3}{a_0} e^{-3r/2a_0} dr\right]$$

$$= \frac{1}{\sqrt{2}}\left(\frac{1}{a_0}\right)^3 \left[\frac{2 \times L2!}{(3/2)^{2+1}} - \frac{3!}{(3/2)^{3+1}}\right] \left[\therefore \quad \int_0^\infty x^n e^{-ax} dx = \frac{n!}{a^{n+1}}\right]$$

$$= \frac{1}{\sqrt{2}}\left(\frac{1}{a_0}\right)^3 \left[\frac{2 \times 2}{27/8} - \frac{2 \times 3 \times 2^4}{3^4}\right]$$

$$= \frac{1}{\sqrt{2}}\left(\frac{1}{a_0}\right)^3 \left[\frac{32}{27} - \frac{32}{27}\right] = 0$$

$$\therefore \quad \int_0^\infty \psi_{1s}\psi_{2s} d\tau = 0$$

This ψ_{1s} for hydrogen atom is orthogonal to ψ_{2s}. Answer.

Problem 2. Suppose an electron in hydrogen atom resides in 2 p_z orbital, what will be the relative probability density that it will have at $\theta = \pi/2$ and $\theta = \pi/4$?

Solution: The wave function ψ_{2p_z} is represented by

$$\psi_{2p_z} = \frac{1}{4\sqrt{2\pi}}\left(\frac{1}{a_0}\right)^{3/2} \cdot \frac{r}{a_0} e^{-r/2a_0} \cos\theta$$

And the probability density will be $\psi^2_{2p_z}$

$$\psi^2_{2p_z} = \frac{1}{32\pi}\left(\frac{1}{a_0}\right)^5 \cdot r^2 e^{-r/a_0} \cos^2\theta$$

$$\text{When } \theta = \pi/2, \ \psi^2_{2p_z} = \frac{1}{32\pi}\left(\frac{1}{a_0}\right)^5 \cdot r^2 e^{-r/a_0} \cos^2 \pi/2$$

$$= 0 \quad \left[\therefore \cos\frac{\pi}{2} = 0\right]$$

$$\text{When } \theta = \pi/4, \ \psi^2_{2p_z} = \frac{1}{32\pi}\left(\frac{1}{a_0}\right)^5 \cdot r^2 e^{-r/a_0} \cos^2 \pi/4$$

$$\text{or } \psi_{2p_z}^2 = \frac{1}{32\pi}\left(\frac{1}{a_0}\right)^5 \cdot r^2 e^{-r/a_0}(1/2)$$

$$\text{or } \psi_{2p_z}^2 = \frac{1}{64\pi}\left(\frac{1}{a_0}\right)^5 \cdot r^2 e^{-r/a_0}$$ Answer.

Problem 3. Show that the hydrogen wave functions for $l = 0$ have maximum value along Cartesian axes.

Solution: We know that $l = 1$ represents p orbital.

$$\therefore \psi_{2p_z} = \frac{1}{4\sqrt{2\pi}}\left(\frac{1}{a_0}\right)^{3/2} \cdot \frac{r}{a_0} e^{-r/2a_0} \cos\theta$$

Applying the condition for maxima, i.e., $d\psi/d\theta = 0$, we shall differentiate the aforementioned equation with respect to θ. On differentiation, we get

$$\frac{d\psi_{2p_z}}{d\theta} = \frac{1}{4\sqrt{2\pi}}\left(\frac{1}{a_0}\right)^{3/2} . \frac{r}{a_0} e^{-r/2a_0}(-\sin\theta) = 0$$

Hence, θ is either 0° or π because sin θ value will be zero for $\theta = 0°$ and also for $\theta = \pi$ or 180°. It clearly indicates that the maxima lies along the z-axis.

Similarly, $\psi_{2p_x} = \frac{1}{4\sqrt{2\pi}}\left(\frac{1}{a_0}\right)^{3/2} \cdot \frac{r}{a_0} e^{-r/2a_0} \cdot \sin\theta \cdot \cos\phi$

At maxima,

$$\therefore \frac{d\psi_{2p_x}}{d\theta} = 0,$$

$$\therefore \frac{d\psi_{2p_x}}{d\theta} = \frac{1}{4\sqrt{2\pi}}\left(\frac{1}{a_0}\right)^{3/2} \cdot \frac{r}{a_0} e^{-r/2a_0} \frac{d}{d\theta}(\sin\theta)\cos\phi = 0$$

$$= \frac{1}{4\sqrt{2\pi}}\left(\frac{1}{a_0}\right)^{3/2} \cdot \frac{r}{a_0} \cdot e^{-r/2a_0} \cos\theta\cos\phi = 0$$

This will be zero when $\theta = \pi/2$ or $3\pi/2$. Hence, maxima will be along the x-axis.

$$\text{And for } \psi_{2p_y} = \frac{1}{4\sqrt{2\pi}}\left(\frac{1}{a_0}\right)^{3/2} \cdot \frac{r}{a_0} e^{-r/2a_0} \cdot \sin\theta \cdot \sin\phi$$

∴ At maxima

$$\frac{d\psi_{2p_x}}{d\theta} = 0 = \frac{1}{4\sqrt{2\pi}}\left(\frac{1}{a_0}\right)^{3/2} \cdot \frac{r}{a_0} \cdot e^{-r/2a_0} \cdot \frac{d}{d\theta}(\sin\theta)\sin\phi$$

$$= \frac{1}{4\sqrt{2\pi}}\left(\frac{1}{a_0}\right)^{3/2} \cdot \frac{r}{a_0} \cdot e^{-r/2a_0} \cdot \cos\theta \cdot \sin\phi = 0$$

This will be zero when $\theta = \pi/2$; hence, the maxima will lie along Y-axis. Answer.

Problem 4. Show that the angular wave functions for l = 2, m = 0 and l = 1, $m = 1, \psi_i(\theta \cdot \phi) = \left(\frac{15}{16\pi}\right)^{1/2}(3\cos^2\theta - 1)$ and $\psi_j(\theta, \phi) = \left(\frac{3}{4\pi}\right)^{1/2} \sin\theta\cos\phi$ are orthogonal.

Solution: For solving this, we have to show that $\int_{\text{all space}} \psi_i \psi_j \, d\tau = 0$.

Putting the value of ψ_i and ψ_j, we get

$$\int_{\theta=0}^{\pi}\int_{\phi=0}^{\pi}\left(\frac{45}{64\pi^2}\right)^{1/2}(3\cos^2\theta - 1)\sin\theta\cos\phi d\theta d\phi$$

$$= \left(\frac{45}{64\pi^2}\right)^{1/2}\int_0^{\pi}(3\cos^2\theta - 1)\sin\theta d\theta \cdot \int_0^{\pi}\cos\phi d\phi = 0$$

As the value of second integral will be zero, obviously the first integral. Thus, $\psi_i(\theta, \phi)$ and $\psi_j(\theta, \phi)$ will be orthogonal. Answer.

Problem 5. Consider $\Psi_{100} = A\ e^{-\rho/2}$. Normalise this function considering $\int_0^{\infty}\psi_{100}^2 d\tau = 1$, where $d\tau$ = $r^2\sin\theta d\theta d\phi dr$ = volume element. Find the value of A.

Solution: Given that $\psi_{100} = Ae^{-\rho/2}$, where $\rho = \frac{2zr}{a_0}$

$$\therefore \quad \psi_{100}^2 = A^2e^{-\rho}$$

$$\therefore \quad \psi_{100}^2 d\tau = A^2e^{-\rho}.d\tau = A^2e^{-2zr/a_0}d\tau$$

$$\text{or} \quad \int_0^{\infty}\psi_{100}^2 r^2\sin\theta d\theta d\phi dr = 1$$

$$\text{or} \quad \int_0^{\infty}A^2e^{-2zr/a_0}r^2\sin\theta.d\theta d\phi dr = 1$$

$$\text{or} \quad A^2\int_0^{\infty}\int_0^{\pi}\int_0^{2\pi}e^{-2zr/a_0}r^2\sin\theta d\theta d\phi dr = 1$$

$$\text{or} \quad 2.2\pi.A^2\int_0^{\infty}e^{-2zr/a_0}r^2 dr = 1$$

$$\text{or} \quad 4\pi A^2\frac{2!}{\left(\frac{2z}{a_0}\right)^3} = \frac{4\pi A^2 \times 2}{8\left(\frac{z}{a_0}\right)^3} = 1$$

$$\text{or} \quad A^2 = \frac{1}{\pi}\left(\frac{z}{a_0}\right)^3$$

$$\therefore A = \frac{1}{\sqrt{\pi}}\left(\frac{z}{a_0}\right)^{3/2} \qquad \text{Answer.}$$

Problem 6. Considering the ground state of hydrogen atom, find the expectation value of the radius vector r of the electron.

Solution: We know that $\psi_{100} = \frac{1}{\sqrt{\pi}}\left(\frac{1}{a_0}\right)^{3/2} \cdot e^{-r/a_0}$

Therefore, the expectation value

$$\bar{r} = \int \psi_{100}^* \cdot r\psi_{100} \mathrm{d}\tau$$

$$= \frac{1}{\pi a_0^3} \int_0^\infty r^3 e^{-2r/a_0} \mathrm{d}r \int_0^\pi \int_0^{2\pi} \sin\theta \mathrm{d}\theta \mathrm{d}\phi$$

From the standard table of integral, we know that $\int_0^\pi \int_0^{2\pi} \sin\theta \mathrm{d}\theta \mathrm{d}\phi = 4\pi.$

$$\therefore \bar{r} = \frac{4\pi}{\pi a_0^3} \int_0^\infty r^3 e^{-2r/a_0} \mathrm{d}r$$

$$= \frac{4\pi}{\pi a_0^3} \cdot \frac{3!}{\left(\frac{2}{a_0}\right)^4} \qquad \left[\therefore x^n \cdot e^{-ax} \mathrm{d}x = \frac{n!}{a^{n+1}}\right]$$

$$= \frac{4\pi}{a_0^3} \cdot \frac{1 \cdot 2 \cdot 3}{2^4 / a_0^4} = \frac{4}{a_0^3} \cdot \frac{a_0^4 \cdot 1 \cdot 2 \cdot 3}{2^4}$$

$$= \frac{4 \times 1 \times 2 \times 3}{2^4} \cdot a_0$$

$$= \frac{3}{2} a_0, \text{ where } a_0 = \text{Bohr radius}$$

$$= \text{the expectation value of } r \qquad \text{Answer.}$$

Problem 7. Find the most probable distance of the electron of the hydrogen atom in its $2p$ state. What will be the radial probability at this distance?

Solution: We know that the radial probability density $P_{nl}(r) = r^2 |R_{nl}|^2$ where

$$R_{21} = \left(\frac{1}{2a_0}\right)^{3/2} \cdot \frac{1}{a_0\sqrt{3}} r \cdot e^{-r/2a_0}$$

$$\therefore P_{21}(r) = r^2 R_{21}^2 = r^2 \left\{\left(\frac{1}{2a_0}\right)^{3/2} \cdot \left(\frac{1}{a_0\sqrt{3}} \cdot re^{-r/2a_0}\right)\right\}^2$$

$$\text{or } P_{21}(r) = r^2 \cdot \frac{1}{8a_0^3 3a_0^2} \cdot r^2 e^{-r/a_0}$$

$$\text{or } P_{21}(r) = \frac{1}{24a_0^5} \cdot r^4 e^{-r/a_0}$$

$$P_{21} \text{ will be maximum when } \frac{dP_{21}}{dr} = 0$$

$$\text{or } \frac{dP_{21}}{dr} = \frac{1}{24a_0^5}\left(4r^3 - \frac{r^4}{a_0}\right)e^{-r/a_0} = 0$$

$$\text{or } 4r^3 - \frac{r^4}{a_0} = 0 \text{ or } 4r^3 = \frac{r^4}{a_0}$$

or $r = 4a_0$. Thus, the most probable distance is four times the Bohr radius.

$$\therefore \text{The radial probability } P_{nl}(r) = r^2|R_{nl}|^2 = (4a_0)^2\left\{\left(\frac{1}{2a_0}\right)^{3/2}\frac{1}{a_0\sqrt{3}}r \cdot e^{-r/2a_0}\right\}^2$$

$$= 16a_0 \cdot \frac{1}{8a_0^3} \cdot \frac{1}{3a_0^2} \cdot 16a_0^2 \cdot e^{-4a_0/a_0} = \frac{32}{3a_0}e^{-4} \qquad \text{Answer.}$$

Problem 8. Prove that the 1*s*, 2*p*, and 3*d* orbitals of a hydrogen-like atom show a single maximum in the radial probability curves. Find the values at which these maxima take place.

Solution: We know that the radial probability density $P_{nl} = r^2 |R_{nl}|^2$ where

$$R_{10} = R_{1s} = ce^{-zr/a_0}$$

$$R_{21} = R_{2p} = cre^{-zr/2a_0} \text{ and}$$

$$R_{32} = R_{3d} = ce^{-zr/3a_0}$$

The condition of maxima is $\dfrac{dP_{nl}}{dr} = 0$

$$\therefore \frac{dP_{10}}{dr} = c\left(2r - \frac{2zr^2}{a_0}\right)e^{-2zr/a_0} = 0$$

On solving this, $r = a_0/z$

Again $\dfrac{dP_{21}}{dr} = 0 = c\left(4r^3 - \dfrac{zr^4}{a_0}\right)e^{-2zr/a_0} = 0$

And on solving, we shall get $r = 4a_0/z$.

Similarly $dP_{32}/dr = 0$ will give $r = 9a_0/z$.

In general, $r_{max} = \dfrac{n^2 a_0}{z}$ Answer.

Problem 9. The wave function for the electron in the ground state of H atom is $\psi_{1s} = 1/\sqrt{\pi}\,(1/a_0)^{3/2}\,e^{-r/a_0}$, where a_0 = Bohr radius. Find the probability that the electron will be found somewhere between 0.9 and 1.1. What will be the probability that this electron will remain beyond $2a_0$?

Solution: Given that $\psi_{1s} = \dfrac{1}{\sqrt{\pi}}\left(\dfrac{1}{a_0}\right)^{3/2} \cdot e^{-r/a_0}$

$\therefore$ The probability of finding the electron between 0.9 and 1.1 will be equal to

$$\int_{0.9a_0}^{1.1a_0} \psi_{1s}^2 \cdot 4\pi r^2 \mathrm{d}r$$

$$\text{or } P = \int_{0.9a_0}^{1.1a_0} \left[\left\{\left(\frac{1}{\sqrt{\pi}}\right)\cdot\left(\frac{1}{a_0}\right)^{3/2} \cdot e^{-r/a_0}\right\}\right]^2 \cdot 4\pi r^2 \mathrm{d}r$$

$$= \frac{4}{a_0^3}\int_{0.9a_0}^{1.1a_0} r^2 e^{-2r/a_0} \mathrm{d}r$$

Putting $x = \dfrac{2r}{a_0} \therefore \mathrm{d}x = \dfrac{2\mathrm{d}r}{a_0}$ and $r = \dfrac{x}{2}a_0$ and $\mathrm{d}r = \dfrac{a_0}{2}\cdot \mathrm{d}x$

$$\therefore P = \frac{4}{a_0^3}\int_{1.8}^{2.2}\left(\frac{x}{2}a_0\right)^2 e^{-x}\left(\frac{a_0}{2}\right)\mathrm{d}x$$

$$= \frac{4}{a_0^3}\int_{1.8}^{2.2}\frac{x^2}{4}\cdot a_0^2 \cdot \frac{a_0}{2} e^{-x}\mathrm{d}x$$

$$= \frac{1}{2}\int_{1.8}^{2.2} x^2 e^{-x}\mathrm{d}x = \frac{1}{2}\left[e^{-x}\left(-x^2 - 2x - 2\right)\right]_{1.8}^{2.2}$$

$$= -\frac{1}{2}\left[e^{-x}\left(x^2 + 2x + 2\right)\right]_{1.8}^{2.2}$$

$$= \frac{1}{2}e^{-x}\left[x^2 + 2x + 2\right]_{2.2}^{1.8}$$

$$= \frac{1}{2}\left[e^{-1.8}\left(1.8^2 + 2\times 1.8 + 2\right) - e^{-2.2}\left(2.2^2 + 2\times 2.2 + 2\right)\right]$$

$$= \frac{1}{2}[0.1653 \times 8.84 - 0.1108 \times 11.24]$$

$$= \frac{1}{2}[1.460 - 1.245] = \frac{0.2146}{2} = 0.1073 \qquad \text{Answer.}$$

Now the probability that the electron will be found beyond $2a_0$

$$P' = \frac{4}{a_0^3}\int_{2a_0}^{\infty} e^{-2r/a_0} r^2 \mathrm{d}r$$

$$= \frac{1}{2}\left[e^{-x}\left(x^2 + 2x + 2\right)\right]_{\infty}^{4} \text{ as carried out above}$$

$$= \frac{1}{2}\left[e^{-4}\left(4^2 + 2\times 4 + 2\right)\right]$$

$$= \frac{1}{2}\left[e^{-4}(16 + 8 + 2)\right] = \frac{1}{2}\left[e^{-4}\times 26\right] = 13e^{-4} = 0.238 \qquad \text{Answer.}$$

Problem 10. At time $t = 0$, the wave function for H atom is given by $\psi(r,0)=1/\sqrt{10}\times\left(2\psi_{100}+\psi_{210}+\sqrt{2}\ \psi_{111}+\sqrt{3}\psi_{2,1,-1}\right)$, where the subscripts are values of the quantum number n, l, and m.

What will be the expectation value for the energy of the system?

Solution: The expectation value of the energy of the system $\langle E\rangle = \langle\psi|H|\psi\rangle$
According to the question, we may write

$$\langle E\rangle = \frac{1}{10}\left(2\psi_{100}+\psi_{210}+\sqrt{2}\ \psi_{111}+\sqrt{3}\psi_{2,1,-1}\right)|H|\left(2\psi_{100}+\psi_{210}+\sqrt{2}\ \psi_{111}+\sqrt{3}\psi_{2,1,-1}\right)$$

$$= \frac{1}{10}\left\langle\left(2\psi_{100}+\psi_{210}+\sqrt{2}\ \psi_{111}+\sqrt{3}\psi_{2,1,-1}\right)\middle|\left(2E_1\psi_{100}+E_2\psi_{210}+\sqrt{2}\ E_2\psi_{211}+\sqrt{3}\ E_2\psi_{2,1,-1}\right)\right\rangle$$

$$= \frac{1}{10}(2E_1+E_2+2E_2+3E_2)$$

$$= \frac{1}{10}(4E_1+6E_2)$$

Since $E_1 = -13.8\text{eV}$ and $E_2 = -3.4\text{eV}$

$$\therefore \langle E\rangle = \frac{1}{10}(-13.8\times 4-6\times 3.4)\text{eV}$$

$$= \frac{1}{10}(-55.2-20.4)\text{eV}$$

$$= -7.56\text{eV}$$

Problem 11. If ψ_1 and ψ_2 are wave functions for a degenerate energy state E, prove that any linear combination $c_1\psi_1+c_2\psi_2$ will also be a wave function.

Solution: According to the question, ψ_1 and ψ_2 are wave functions for energy state E, and then we can write

$$\frac{\hbar^2}{8\pi^2 m}\frac{d^2\psi_1}{dx^2}+(E-V)\psi_1=0 \qquad \text{(i)}$$

$$\text{and}\quad \frac{\hbar^2}{8\pi^2 m}\frac{d^2\psi_2}{dx^2}+(E-V)\psi_2=0 \qquad \text{(ii)}$$

Now multiplying Eqs. (i) and (ii) by constants c_1 and c_2, respectively, and then adding, we shall get

$$\frac{\hbar^2}{8\pi^2 m}\cdot\frac{d^2}{dx^2}\left(c_1\psi_1+c_2\psi_2\right)+(E-V)\left(c_1\psi_1+c_2\psi_2\right)=0$$

From this, it is concluded that $(c_1\psi_1+c_2\psi_2)$ is also a wave function of energy state. Answer.

Problem 12. Normalise the wave function $\psi = re^{-r/2}\cos\theta$.

Solution: We know that the condition for normalisation is $\int_0^\infty \psi\psi^*\,d\tau = 1$

Let A be the normalisation factor, then we can write that

$$\int_0^\infty\left(Are^{-r/2}\cos\theta\right)\left(Are^{-r/2}\cos\theta\right)d\tau = 1$$

Putting the values of dτ, we have

$$\text{or } A^2\left[\int_0^\infty r^2 e^{-r} r^2 \mathrm{d}r \int_0^\pi \cos^2\theta \cdot \sin\theta \mathrm{d}\theta \int_0^{2\pi} \mathrm{d}\phi\right] = 1$$

$$\text{or } A^2\left[\int_0^\infty r^4 e^{-r} \mathrm{d}r \cdot \int_0^\pi \cos^2\theta \cdot \sin\theta \mathrm{d}\theta \int_0^{2\pi} \mathrm{d}\phi\right] = 1$$

Now we shall write the values of integrals, which are

$$A^2\left[\frac{4!}{(1)^5}\cdot\left(\frac{-\cos^3\theta}{3}\right)\right]_0^\pi \cdot 2\pi = 1 \left[\therefore \int_0^\infty x^n e^{-ax}\mathrm{d}x = \frac{n!}{a^{n+1}}\right]$$

$$\text{or } A^2\left[24\cdot(2/3)\cdot 2\pi\right] = 1$$

$$\text{or } A^2\left[\frac{48}{3}\cdot 2\pi\right] = 1$$

$$\text{or } A^2 \cdot 32\pi = 1$$

$$\text{or } A^2 = \frac{1}{32\pi}$$

$$\therefore A = \frac{1}{\sqrt{32\pi}}$$ Answer.

Problem 13. Find the average distance of the electron from the nucleus of H atom in the 2*s* state.

Solution: We can write

$$\overline{r_{2s}} = \frac{1}{32\pi a_0^3}\int_0^\infty \left(2 - \frac{r}{a_0}\right)^2 \left(e^{-r/a_0}\right) r \cdot r \, \mathrm{d}r \int_0^\pi \sin\theta \, \mathrm{d}\theta \cdot \int_0^{2\pi} \mathrm{d}\phi$$

$$= \frac{1}{8a_0^3}\int_0^\infty \left(4r^3 - \frac{4r^4}{a_0} + \frac{r^5}{a_0^2}\right)^2 e^{-r/a_0} \, \mathrm{d}r$$

$$= \frac{1}{8a_0^3}\left[\frac{4.3!}{(1/a_0)^4} - \frac{4.4!}{a_0(1/a_0)^5} + \frac{5!}{a_0^2(1/a_0)^6}\right]$$

$$= \frac{1}{8a_0^3}\left[4\times1\times2\times3a_0^4 - 4\times1\times2\times3\times4a_0^4 + 1\times2\times3\times4\times5a_0^4\right]$$

$$= \frac{1\times a_0^4}{8a_0^3}[24 - 96 + 120]$$

$$= a_0[3 - 12 + 15] = 6a_0$$ Answer.

Problem 14. Show that $1s$ and $2s$ orbitals for H atom are orthogonal.

Solution: We know that

$$\psi_{1s} = \left(\frac{1}{\pi a_0^3}\right)^{1/2} e^{-r/a_0} = \frac{1}{\sqrt{\pi}} e^{-r} \left(\text{in a.u.}\right)$$

$$\psi_{2s} = \frac{1}{4\sqrt{2\pi}}(2-r)e^{-r/2}(\text{in a.u.})$$

$$\therefore \int \psi_{1s}\psi_{2s} d\tau = \frac{4\pi}{4\sqrt{2\pi}} \int_0^\infty e^{-r} \cdot r^2(2-r)e^{-r/2} dr \quad \left(\because d\tau = 4\pi \int r^2 dr\right)$$

$$= \frac{1}{\sqrt{2}}\left[2\int_0^8 r^2 e^{-3r/2} dr - \int_0^\infty r^3 e^{-3r/2} dr\right]$$

We can evaluate the integral by standard table as

$$\int_0^\infty r^2 e^{-3r/2} dr = \frac{2!}{\left(\frac{3}{2}\right)^3} = \frac{16}{27}$$

$$\int_0^\infty r^3 e^{-3r/2} dr = \frac{3!}{\left(\frac{3}{2}\right)^4} = \frac{32}{27}$$

$$\therefore \int \psi_{1s}\psi_{2s} d\tau = \frac{1}{\sqrt{2}}\left[\frac{2\times 16}{27} - \frac{32}{27}\right]$$

$$= \frac{1}{\sqrt{2}}\left[\frac{32}{27} - \frac{32}{27}\right]$$

$$= \frac{1}{\sqrt{2}} \times 0 = 0$$

Therefore, $1s$ and $2s$ orbitals of H atom are orthogonal. Proved.

Problem 15. Show that the spherical harmonics for $3d_z^2$ and $3d_{x^2-y^2}$ orbitals are orthogonal to each other.
Solution: Given that

$$3d_z^2 = \psi(\theta,\phi) = \left(\frac{5}{16\pi}\right)^{1/2} \left(3\cos^2\theta - 1\right)$$

$$3d_{x^2-y^2} = \psi(\theta,\phi) = \left(\frac{15}{16\pi}\right)^{1/2} \sin^2\theta \cos 2\phi$$

Now according to the question,

$$\int_0^{2\pi}\left(\frac{5}{16\pi}\right)^{1/2}\left(3\cos^2\theta-1\right)\left(\frac{15}{16\pi}\right)^{1/2}\sin^2\theta\cos 2\phi\cdot d\phi\int_0^{\pi}\sin\theta d\theta$$

$$=\frac{5\sqrt{3}}{16\pi}\int_0^{\pi}\sin^3\theta\left(3\cos^2\theta-1\right)d\theta\int_0^{2\pi}\cos 2\phi d\phi$$

$$=\frac{5\sqrt{3}}{16\pi}\left[3\int_0^{\pi}\sin^3\theta\cos^2\theta d\theta-\int_0^{\pi}\sin^3\theta d\theta\right]\int_0^{2\pi}\cos 2\phi d\phi$$

$$=\frac{5\sqrt{3}}{16\pi}\left[3\left(\frac{4}{15}\right)-\frac{4}{3}\right]\times 0=0$$

Hence, $3d_z^2$ and $3d_{x^2-y^2}$ orbitals are orthogonal. Answer.

Questions on Concepts

1. Find the simple solution of hydrogen atom.
2. Let the radial equation of hydrogen atom be expressed as

$$\frac{\partial^2\psi}{\partial r^2}+\frac{2}{r}\frac{\partial\psi}{\partial r}+\frac{8\pi^2 m}{h^2}\left(E+e^2/r\right)\psi=0$$

   Suppose the solution of this equation be $\psi = e^{-ar}$, then find the value of E of hydrogen atom.
3. Arrive at the generalised solution of the Schrödinger equation for hydrogen atom.
4. Find the solution of F equation, i.e., $d^2F/3\phi^2 = -m^2F$ and also show that the constant $c = 1/\sqrt{2}\pi$.
5. How will you find the solution of T equation for hydrogen atom?
6. What are the Laguerre polynomials? Derive the Rodrigues formula of the Laguerre polynomials.
7. Derive Legendre polynomials and associated Legendre polynomials.
8. Show that $R(\rho)ce^{-\rho/2}\cdot\rho^2 L_{n+1}^{2l+1}(\rho)$, where the terms have their usual significance.
9. How will you normalise the radial wave function?
10. Calculate the expectation value of potential energy V of the electron in the $1s$ state of hydrogen atom.
11. Evaluate the most probable distance of the electron hydrogen atom in its $3d$ state.
12. What will be the probability of finding the 1s electron of the hydrogen atom at a distance $a_0/2$ from the nucleus?
13. Calculate (a) the most probable distance and (b) the average distance of electron from the nucleus of hydrogen atom. [Ans. (a) $r = a_0$, (b) $3/2a_0$]
14. Discuss the use of Legendre functions in solving the θ equation of H atom.
15. Explain the probability distribution curve? What do you mean by radial distribution function?
16. Write down the radial wave functions for the first three orbitals of H atom and discuss their importance.
17. Draw the angular part of hydrogen like s, p, d, and f orbitals showing only two dimensions of the three-dimensional function on the basis of the following angular wave functions:

$$s \text{ orbital}, l = 0, m = 0, \theta\phi = \left(\frac{1}{4\pi}\right)^{1/2}$$

$$p_z \text{ orbital}, l = 1, m = 0, \theta\phi = \left(\frac{3}{4\pi}\right)^{1/2} \cos\theta$$

$$d_z^2 \text{ orbital}, l = 2, m = 0, \theta\phi = \left(\frac{5}{16\pi}\right)^{1/2} \left(3\cos^2\theta - 1\right)$$

18. The wave function for the electron in the ground state of hydrogen atom is $\psi_{1s} = \left(\pi a_0^3\right)^{-1/2} e^{-r/a_0}$. What is the probability of finding the electron somewhere inside a small sphere of radius 1.0×10^{-12} m centred on the nucleus? [Ans. 9.0×10^{-6}]
19. Show that for H atom, 1*s* and 2*s* orbitals are orthogonal.
20. What are nodes? State the number of radial and angular nodes in a hydrogen-like atom.
21. Calculate the probability that the electron in the ground state of hydrogen atom is between (a) $0.9a_0$ and $1.1a_0$ and (b) beyond $2a_0$. [Ans. (a) 0.108, (b) 0.238]
22. Calculate the probability that the electron in the ground state of hydrogen atom is found between (a) 0 and a_0 and (b) $0.99a_0$ and $1.01a_0$. [Ans. (a) 0.32, (b) 1.1×10^{-2}]
23. An electron in the hydrogen atom resides in $2p_z$ orbital. What will be the relative probability density that it will have at $\theta = \pi/2$ and $\theta = \pi/4$?
24. Normalise the wave function $\psi = re^{-r/2} \cos\theta$ and find the normalisation constant *N*. [Ans. $N = 1/\sqrt{32\pi}$]
25. Calculate the average distance of the electron from the nucleus. [Ans. $6a_0$]
26. Find the average distance of the electron in the 2*pz* orbital from the nucleus of *H* atom. [Ans. $5a_0$]
27. It is given that for hydrogen atom $\hat{H} = -1/2\nabla^2 - 1/r$ and $\psi_{1s} = 1/\sqrt{\pi}e^{-r}$ in a.u. Calculate the ground-state energy of hydrogen atom in eV. [Ans. –13.6 eV]

[Hint: According to the question, $\hat{H} = -\frac{\nabla^2}{2} - \frac{1}{r}$ and $\psi_{1s} = \frac{1}{\sqrt{\pi}} e^{-r}$]

We have to find the ground-state energy, i.e., $\bar{E}_{1s}$. We know that $\bar{E}_{1s} = \int_0^\infty \psi_{1s}\hat{H}\psi_{1s}^* d\tau$.

Putting the values of ψ_{1s}, $\hat{H}$ and ψ_{1s}^*, and $d\tau$, we shall get

$$\bar{E}_{1s} = \frac{1}{\pi}\int_0^\infty e^{-r}\left[-\frac{1}{2r^2}\frac{d}{dr}\left(r^2\frac{d}{dr}\right) - \frac{1}{r}\right]e^{-r}r^2 dr \cdot \int_0^\pi \sin\theta d\theta \int_0^{2\pi} d\phi$$

$$= -\frac{4\pi}{2\pi}\int_0^\infty r^2 e^{-2r} dr \quad \left[\text{where } \int_0^{2\pi} d\phi = 2\pi \text{ and} \int_0^\pi \sin\theta d\theta = -2\right]$$

$$= -2\int_0^\infty r^2 e^{-2r} dr$$

$$= -2.\frac{2!}{(2)^3} \quad \left[\therefore \int_0^\infty x^n e^{-ax} dx = \frac{n!}{(a)^{n+1}}\right]$$

$$= \frac{-2\times 2}{8} = -\frac{1}{2}\text{a.u.}$$

$$= -2.18\times 10^{-18}\,\text{J} = -13.6\text{eV}$$

9

Approximate Methods

The hydrogenic system, i.e., hydrogen atom or any one-electron-containing-an-ion system, is the only chemical system for which an exact solution of the Schrödinger equation is known. In case of the problem related to many-electron atomic systems and molecules, the equations obtained are found to be too complex to be solved exactly. In other words, we can say that apart from a few simple cases, the solution of the Schrödinger equation is impossible in closed form. In many-electron systems, the electrons interact with each other, and the solutions of the equation obtained become difficult due to many potential energy terms.

Under the aforementioned circumstances, approximate methods or theories have been applied to get the solution. These approximate methods or theories are as follows:

- Perturbation theory or method
- Variation theorem or method

We shall acquire knowledge about each theory or method one by one.

9.1 Perturbation Theory/Method for Nondegenerate States

If a chemical system is subjected to an electric or a magnetic field, the system becomes disturbed or perturbed. Under this situation, the system under consideration differs from an exactly solvable system by only a small disturbance, enabling an approximation to be carried out by expanding in powers of a smallness parameter.

Let H = perturbed Hamiltonian,

H^0 = zero-order Hamiltonian without disturbance or perturbation of the system or zero-order unperturbed Hamiltonian whose solution is known, and

$(H - H^0)$ = change/difference between perturbed and unperturbed Hamiltonian. H differs from H^0 by only a small amount.

Then, the effect of $(H - H^0)$ can be taken into account by utilising the perturbation theory. Furthermore, let us consider that H^0 has a spectrum of given values:

$E_0^0, E_1^0, E_2^0, ------ E_n^0$ having a set of Eigen functions $\Psi_0^0, \Psi_1^0, \Psi_2^0, ------ \Psi_n^0$.

Then, we can express the Schrödinger equation as

$$H^0\Psi_i^0 = E_i^0\Psi_i^0 \tag{9.1}$$

It has been assumed that H is slightly different from H^0 due to perturbation. Then, H can be expressed as a polynomial in a perturbation parameter as

$$H = H^0 + \lambda H' + \lambda^2 H'' + \lambda^3 H''' + \tag{9.2}$$

where H' = first-order perturbation correction,

H'' = second-order perturbation correction,

H''' = third-order perturbation correction and so on (the magnitude of perturbation is found to be inversely proportional to order), and

λ = perturbation parameter which varies from 0 to 1.

It is to be noted that the given function in Eq. (9.1) will constitute a complex orthogonal set. In case of small perturbation, the true energy E_k of Kth level will be very close to E_k^0, which is the true energy under unperturbed condition. We can also expand E_k in power series in the perturbation parameter having first term E_k^0.

$$E_k = E_k^0 + \lambda E_k'' + \lambda^2 E_k'' + \lambda^3 E_k''' + \cdots \tag{9.3}$$

Similarly, $\psi_k'^0$ constitutes a complete set, and Ψ_k is very close to ψ_k^0. Hence, we can express the perturbed wave function for K[th] level in power series of λ as given in the following:

$$\psi_k = \psi_k^0 + \lambda\psi_k' + \lambda^2\psi_k'' + \lambda^3\psi_k'' + \cdots \tag{9.4}$$

Under small perturbation, Eqs. (9.1), (9.3), and (9.4) will fastly converge.

The Schrödinger equation for the perturbed system may be expressed as

$$H\psi_k = E_k\psi_k \tag{9.5}$$

Now putting the values of H, E, and Ψ from their series expansion in Eq. (9.5), we get

$$\begin{aligned}&\left(H^0 + \lambda H' + \lambda^2 H'' + \text{-----}\right)\left(\psi_k^0 + \lambda\psi_k' + \lambda^2\psi_k'' + \text{------}\right)\\ &= \left(E_k^0 + \lambda E_k' + \lambda^2 E_k'' + \text{------}\right)\left(\psi_k^0 + \lambda\psi_k' + \lambda^2\psi_k'' + \text{------}\right)\end{aligned}$$

Here, we have considered the perturbation parameter up to λ^2 for simplicity.

On multiplication, the aforementioned equation takes the following form:

$$H^0\psi_k^0 + \lambda H^0\psi_k' + \lambda^2 H^0\psi_k'' + \lambda H'\psi_k^0 + \lambda^2 H'\psi_k' + \lambda^3 H'\psi_k'' + \lambda^2 H''\psi_k^0 + \lambda^3 H''\psi_k' + \lambda^4 H''\psi_k'' \text{-----}$$

$$E_k^0\psi_k^0 + \lambda E_k^0\psi_k' + \lambda^2 E_k^0\psi_k'' + \lambda E_k'\psi_k^0 + \lambda^2 E_k'\psi_k' + \lambda^3 E_k'\psi_k'' + \lambda^2 E_k''\psi_k^0 + \lambda^3 E_k''\psi_k' + \lambda^4 E_k''\psi_k'' \text{----}$$

Or

$$\begin{aligned}&\left(H^0\psi_k^0 - E_k^0\psi_k^0\right) + \lambda\left(H^0\psi_k' + H'\psi_k^0 - E_k^0\psi' - E_k'\psi_k^0\right)\\ &+ \lambda^2\left(H^0\psi_k'' + H'\psi_k' + H''\psi_k^0 - E_k^0\psi_k'' - E_k'\psi_k' - E_k''\psi_k^0\right) + \text{----} = 0\end{aligned} \tag{9.6}$$

This has been done by collecting the coefficients of like powers of λ. Since λ is arbitrary, the coefficient of each power will be zero, as given in the following:

$$H^0\psi_k^0 - E_k^0\psi_k^0 = 0 \tag{9.7}$$

$$H^0\psi_k' + H'\psi_k^0 = E_k^0\psi_k' + E_k'\psi_k^0 \tag{9.8}$$

$$H^0\psi_k'' + H'\psi_k' + H''\psi_k^0 = E_k^0\psi_k'' + E_k'\psi_k' + E_k''\psi_k^0 \tag{9.9}$$

Equations (9.7–9.9) represent zero order, first order, and second order, respectively.

At this stage, we consider the first-order perturbation in which correction to energy and correction to wave function will be dealt with.

9.1.1 First-Order Perturbation

9.1.1.1 Correction to Energy

Let us consider Eq. (9.8), which is

$$H^0\psi_k' + H'\psi_k^0 - E_k^0\psi_k' - E_k'\psi_k^0 = 0$$

Here, we assume that the first-order correction to the wave function can be expressed as a linear function of the wave function of the unperturbed problem, namely

$$\psi'_k = \sum_l a_l \psi_l^0 \tag{9.10}$$

Multiplying the LHS and the RHS by H^0 and applying $H^0\psi_l^0 = E_l^0\psi_l^0$, we have

$$H^0\psi'_k = \sum_l a_l H^0\psi_l^0 = \sum_l a_l E_l^0\psi_l^0 \tag{9.11}$$

Equation (9.8) can be written as

$$\left(H^0 - E_k^0\right)\psi'_k = (E'_k - H')\psi_k^0$$

Replacing the value of ψ'_k in the aforementioned equation, we get

$$\left(H^0 - E_k^0\right)\left(\sum_l a_l\psi_l^0\right) = (E'_k - H')\psi_k^0$$

$$\text{or} \quad H^0\sum_l a_l\psi_l^0 - E_k^0\sum_l a_l\psi_l^0 = (E'_k - H')\psi_k^0$$

$$\text{or,} \quad \sum_l a_l H^0\psi_l^0 - E_k^0\sum_l a_l\psi_l^0 = (E'_k - H')\psi_k^0$$

$$\text{or,} \quad \sum_l a_l H^0\psi_l^0 - E_k^0\sum_l a_l\psi_l^0 = (E'_k - H')\psi_k^0 \tag{9.12}$$

Now multiplying the LHS and the RHS by $\overset{0}{\psi}{}_K^*$ and then on integration, we get

$$\int \overset{0}{\psi}{}_K^* \sum_l a_l\left(E_l^0 - E_k^0\right)\psi_l^0 \mathrm{d}\tau = \int \overset{0}{\psi}{}_K^*(E'_k - H')\psi_k^0\mathrm{d}\tau\overset{0}{\psi}{}_K^* \tag{9.13}$$

$$\sum_l a_l\left(E_l^0 - E_k^0\right)\int \overset{0}{\psi}{}_K^*\psi_1^0\mathrm{d}\tau = E'_K\int \overset{0}{\psi}{}_K^*\psi_1^0\mathrm{d}\tau - \int \overset{0}{\psi}{}_K^* H'\psi_k^0\mathrm{d}\tau \tag{9.14}$$

Since the functions ψ_l^0 are all orthogonal, the LHS will be zero and the aforementioned equation may be expressed as

$$0 = E'_k - \int \overset{0}{\psi}{}_K^* H'\psi_1^0\mathrm{d}\tau \qquad \left[\because \int \overset{0}{\psi}{}_K^*\psi_k^0\mathrm{d}\tau = 1\right] \tag{9.15}$$

$$E'_k = \int \overset{0}{\psi}{}_K^* H'\psi_k^0\mathrm{d}\tau \tag{9.16}$$

This is the first-order correction to energy, which may also be represented as

$$E_k' = E^0 + H_{kk}' \tag{9.17}$$

$$\text{when } \left[\overset{0}{\psi}{}_K^* H' \psi_k^0 \mathrm{d}\tau = H_{kk}' \right]$$

9.1.1.2 Correction to Wave Function

It is now necessary to find the first-order correction to wave function on considering Eq. (9.10). Equation (9.10) is $\psi_k' = \sum_l a_l \psi_l'$ in which the coefficient a_1 has to be determined.

To find a_l, Eq. (9.12) has to be considered, which is

$$\sum_l a_l \left(E_l^0 - E_k^0\right)\psi_1^0 = \left(E_k' - H'\right)\psi_k^0$$

This equation is multiplied by $\overset{0}{\psi}{}_l^*$, and then, integration is carried out. Simplification and rearrangement of terms yield

$$a_l\left(E_l^0 - E_k^0\right)\int \overset{0}{\psi}{}_l^* \psi_l^0 \mathrm{d}\tau = E_k' \int \overset{0}{\psi}{}_l^* \psi_k^0 \mathrm{d}\tau - \int \overset{0}{\psi}{}_l^* H' \psi_k^0 \mathrm{d}\tau$$

$$\text{or, } a_l\left(E_l^0 - E_k^0\right)\times 1 = E_k' \times 0 - \int \overset{0}{\psi}{}_l^* H' \psi_k^0 \mathrm{d}\tau$$

$$\text{or,} a_l\left(E_l^0 - E_k^0\right) = -\int \overset{0}{\psi}{}_l^* H' \psi_k^0 \mathrm{d}\tau$$

$$= H_{lk}' \tag{9.18}$$

$$\text{or } a_l = -\frac{H_{lk}'}{\left(E_l^0 - E_k^0\right)} \tag{9.19}$$

where $l \neq k$.

$$\because \psi_k' = \sum_l a_l \psi_l^0$$

$$\therefore \psi_k' = -\sum_l \frac{H_{lk}'}{\left(E_l^0 - E_k^0\right)}\psi_l^0 = \sum_l \frac{H_{lk}'}{\left(E_l^0 - E_k^0\right)}\psi_l^0$$

Therefore, the first-order perturbed wave function will be

$$\psi_k' = \sum_l \frac{H_{lk}'}{\left(E_l^0 - E_k^0\right)}\psi_l^0 \tag{9.20}$$

9.1.2 Second-Order Perturbation

9.1.2.1 Correction to Energy

The second-order perturbation correction to the energy and the wave functions are found by a straightforward process using Eq. (9.9) and by applying same methods as we have used in case of first-order perturbation corrections.

Equation (9.9) is as follows:

$$H^0\psi_k'' + H'\psi_k' + H''\psi_k^0 - E_k^0\psi_k'' - E_k'\psi_k' - E_k''\psi_k^0 = 0$$

For getting the value of energy correction, let us assume that the second-order correction to the wave functions can be expressed as a linear function of the wave functions of the unperturbed condition, namely

$$\psi_k'' = \sum_l b_l\psi_l^0 \tag{9.21}$$

and then, we can write

$$H^0\psi_k'' = \sum_l b_l H^0\psi_l^0 = \sum_l b_l E_l^0\psi_l^0 \tag{9.22}$$

Putting the values of ψ_k'' and $H^0\psi_k''$ in the aforementioned equation for second-order perturbation, we get

$$\sum_l b_l H^0\psi_l^0 - E_k^0\sum_l b_l\psi_l^0 + H'\psi_k' - H''\psi_k^0 - E_k''\psi_k^0 = 0$$

$$b_l\left(E_l^0 - E_k^0\right)\sum_l \psi_l^0 = (E_k'' - H'')\psi_k^0 + (E_k' - H')\psi_k' \tag{9.23}$$

Multiplying each term by $\overset{0}{\psi_l^*}$ and then integrating, we obtain

$$b_l\left(E_l^0 - E_k^0\right)\int\sum_l \psi_l^0\overset{0}{\psi_l^*}\,\mathrm{d}\tau = \int\left(E_k'' - H''\right)\psi_k^0\overset{0}{\psi_k^*}\,\mathrm{d}\tau + \int\left(E_k' - H'\right)\psi_k'\overset{0}{\psi_k^*}\,\mathrm{d}\tau \tag{9.24}$$

Since $\sum_l \psi_l^0\overset{0}{\psi_l^*}$ will be zero due to orthogonality, the LHS of Eq. (9.24) will be zero.

$$\therefore\ b_l\left(E_l^0 - E_k^0\right)\times 0 = \int E_k''\psi_k^0\overset{0}{\psi_k^*}\,\mathrm{d}\tau - \int H''\psi_k^0\overset{0}{\psi_k^*}\,\mathrm{d}\tau + \int E_k'\psi_k'\overset{0}{\psi_k^*}\,\mathrm{d}\tau - \int H'\psi_k'\overset{0}{\psi_k^*}\,\mathrm{d}\tau$$

$$\text{or, } 0 = E_k''\int\psi_k^0\overset{0}{\psi_k^*}\,\mathrm{d}\tau - \int\psi_k^0 H''\overset{0}{\psi_k^*}\,\mathrm{d}\tau + E_k'\int\psi_k'\overset{0}{\psi_k^*}\,\mathrm{d}\tau - \int H'\psi_k'\overset{0}{\psi_k^*}\,\mathrm{d}\tau$$

$$\text{or}\quad 0 = E_k'' - \int\psi_k^0 H''\overset{0}{\psi_k^*}\,\mathrm{d}\tau - \int\psi_k' H'\overset{0}{\psi_k^*}\,\mathrm{d}\tau$$

$$\left[\because \int\psi_k^0\overset{0}{\psi_k^*}\,\mathrm{d}\tau\right] = 1 \text{ and} \left[\because \int\psi_k'\overset{0}{\psi_k^*}\,\mathrm{d}\tau\right] = 0$$

$$\text{or}\quad E_k'' = \int\psi_k^0 H''\overset{0}{\psi_k^*}\,\mathrm{d}\tau + \int\psi_k' H'\overset{0}{\psi_k^*}\,\mathrm{d}\tau$$

$$\text{or}\quad E_k'' = H_{kk}'' + \int\psi_k' H'\overset{0}{\psi_k^*}\,\mathrm{d}\tau$$

$$= H_{kk}'' + \sum_l \left(\frac{H_{lk}'}{\left(E_k^0 - E_l^0\right)} \right) \int \psi_k^* H' \psi_l^0 \, d\tau$$

$$= H_{kk}'' + \sum_l \left(\frac{H_{lk}'}{\left(E_k^0 - E_l^0\right)} \right) H_{lk}' \left[\therefore \int \psi_i^* H' \psi_l^0 \, d\tau = H_{ki}' \right]$$

$$or\ E_k'' = \int \sum_{l \neq l} \left(\frac{H_{lk}'}{\left(E_k^0 - E_l^0\right)} \right) H_{lk}' + H_{kk}'' \tag{9.25}$$

Equation (9.25) is the fundamental result of second-order correction to energy (E_k'').

9.1.2.2 Second-Order Correction to Wave Functions

Now our next task is to find the expression for second-order correction to wave function. We perform this task in the similar manner as we have performed in the case of first-order correction to wave function. We have already assumed that

$$\Psi_k'' = \sum_l b_l \Psi_l^0$$

which implies determining the coefficient b_l. To find b_l, again consider the equation

$$H^0 \Psi_k'' + H' \Psi_k' + H'' \Psi_k^0 - E_k^0 \Psi_k'' - E_k' \Psi_k' - E_k'' \Psi_k^0 = 0$$

Using equation $\Psi_k'' = \sum_l b_l \Psi_l^0$

we can write

$$H^0 \Psi_k'' = \sum_l b_l H^0 \Psi_l^0 = \sum_l b_l E_l^0 \Psi_l^0 \tag{9.26}$$

$$\text{and } E_k^0 \Psi_k'' = \sum_l b_l E_l^0 \Psi_l^0 \tag{9.27}$$

Substituting Eqs. (9.26) and (9.27) in the aforementioned equation, we obtain

$$\sum_l b_l E_l^0 \Psi_l^0 + H' \Psi_k' + H'' \Psi_k^0 - \sum_l E_k^o b_l \Psi_l^0 - E_k' \Psi_k' - E_k'' \Psi_k^0 = 0$$

$$b_l \left(E_l^0 - E_k^0\right) \sum_l \Psi_l^0 + H' \sum_l a_l \Psi_l^0 + H'' \Psi_k^0 - E_k' \sum a_l \Psi_l^0 - E_k'' \Psi_k^0 = 0$$

Multiplying this equation by $\overset{0}{\psi_l^*}$ throughout, we get

$$b_l \left(E_l^0 - E_k^0\right) \sum_l \Psi_l^0 \overset{0}{\psi_l^*} + H' \sum_l a_l \Psi_l^0 \overset{0}{\psi_l^*} + H'' \Psi_k^0 \overset{0}{\psi_l^*} - E_k' \sum a_l \Psi_l^0 \overset{0}{\psi_l^*} - E_k'' \Psi_k^0 \overset{0}{\psi_l^*} = 0$$

Integrating this equation, we find

$$b_l \left(E_l^0 - E_k^0\right) \int \sum_l \Psi_l^0 \overset{0}{\psi_l^*} \, d\tau + H' \int \sum a_l \Psi_l^0 \overset{0}{\psi_k^*} \, d\tau + H'' \int \Psi_k^0 \overset{0}{\psi_k^*} \, d\tau$$

$$- E_k' \int \sum_l a_l \Psi_l^0 \overset{0}{\psi_k^*} \, d\tau - E_k'' \int \Psi_k^0 \overset{0}{\psi_k^*} \, d\tau = 0 \tag{9.28}$$

$$\text{or}\, b_l\left(E_l^0 - E_k^0\right) + \sum_l \Psi_l H' \overset{0}{\psi_l^*}\, d\tau - H'' \times 0 - E_k' \sum_l a_l \int \Psi_l^0 \overset{0}{\psi_l^*}\, d\tau - E_k'' \int \Psi_k^0 \overset{0}{\psi_l^*}\, d\tau = 0$$

$$\text{or}\, b_l\left(E_l^0 - E_k^0\right) + \sum_l a_l \cdot H_{ll}'' - E_k' \sum a_l - E_{k\times 0}'' \times 0$$

$$\text{or}\, b_l\left(E_l^0 - E_k^0\right) = E_k' \sum a_l + \sum a_l H_{ll}' = 0 \tag{9.29}$$

We know that $a_l = \dfrac{-H_{lk}'}{\left(E_l^0 - E_k^0\right)}$

and $E_k' = H_{kk}'$

Putting these values of a_l and E_k' in Eq. (9.29), we get

$$b_l\left(E_l^0 - E_k^0\right) = \sum_l -\frac{H_{lk}'}{\left(E_l^0 - E_k^0\right)} \cdot H_{kk}' + \sum_l -\frac{H_{lk}'}{\left(E_l^0 - E_k^0\right)} H_{ll}'$$

$$= \sum_l \frac{H_{lk}' \cdot H_{kk}'}{\left(E_k^0 - E_l^0\right)} - \sum_l \frac{H_{lk}' \cdot H_{ll}'}{\left(E_l^0 - E_k^0\right)}$$

$$\text{or}\, b_l = \left[\sum_l \frac{H_{lk}' \cdot H_{kk}'}{\left(E_k^0 - E_l^0\right)\left(E_l^0 - E_k^0\right)} - \sum_l \frac{H_{lk}' \cdot H_{ll}'}{\left(E_l^0 - E_k^0\right)}\right] \tag{9.30}$$

Putting the values of b_l, from Eq. (9.30), we can write

$$\Psi_k'' = \sum \left[\sum_l \frac{H_{lk}' \cdot H_{kk}'}{\left(E_k^0 - E_l^0\right)\left(E_l^0 - E_k^0\right)} - \sum_l \frac{H_{lk}' \cdot H_{ll}'}{\left(E_l^0 - E_k^0\right)^2}\right] \Psi_l^0 \tag{9.31}$$

Equation (9.31) represents the second-order correction to wave function.

9.2 Bra–ket Notation or Dirac's Notation

Bra–ket notation has been introduced by Dirac; that is why it is also called Dirac's notation. This notation is frequently used in the literature due to its economical form. It should be kept in mind that the state of a system can be denoted by a vector, which is the state vector in the vector space. *Bra* and *ket* are also called *bra* vector and *ket* vector: Now we give some knowledge about *bra* and *ket.*

A ket or ket vector is similar to the wave function for a state. Dirac gave the symbol | > to indicate a group of ket, and the symbol | *m* > represents the ket vector or *ket*, which corresponds to the state m of the system. In case of superposition of two states, it is described by a linear combination of the corresponding *kets.*

A *bra* or *bra vector* is similar to the complex conjugate of the wave function for a state. The symbol < | represents a group of bras, and when we write the symbol <*n*|, it indicates the *bra* or *bra vector*, which corresponds to the state *n* of the system. In short, we can say that *bra* and *ket* are represented by < | and | >, respectively, and with regard to *m* and *n* states, they are denoted by < *m*| and |*n* >, respectively.

According to Dirac, the scalar product of a bra and a ket corresponds to the integral of the product of complex conjugate of the wave function for one state and the wave function for another state, which is represented by < *n* | *m* >. It should be noted that the words 'bra' and 'ket' have been derived from the word 'bracket' by deleting the letter 'c'. The aforementioned facts will be more clear by considering some examples.

Suppose ϕ^* and Ψ are two wave functions and ϕ^* is the complex conjugate of Ψ. Then, $\int\phi^* \psi \, d\tau$ can be represented in bra–ket shorthand notation as $< \phi^* | \psi >$.

$$\text{It means } \underset{\text{bra}}{\underline{<\phi^*}} \; \underset{\text{ket}}{\underline{|\psi >}} = \int \phi^* \psi \, d\tau \tag{9.32}$$

The advantage of bra–ket notation is that many relations in quantum mechanics may be expressed more succinctly compared with using integral representations.

Similarly, if the integral product of the wave functions is represented by $\phi_m^* \, \phi n \, d\tau$, then its Dirac's notation will be represented by

$$\phi_m | \phi_n > \equiv < m | n > \tag{9.33}$$

If an operator ($\hat{A}$) lies between the wave functions, then the integral products are represented by

$$\int \phi_m^* \hat{A} \phi_n \; d\tau$$

then its Dirac's notation will be

$$< \phi_m \left| \hat{A} \right| \phi_n > \equiv < m \left| \hat{A} \right| n > \equiv A_{mn} \tag{9.34}$$

$$\text{and} \int \phi_m^* H \phi_n d\tau \equiv \langle m | H | n \rangle \equiv H_{mm} \tag{9.35}$$

But for usual notation $\left[\int \phi_m^* \phi_n d\tau \right]^* = \int \phi_n^* \phi_m d\tau$, Dirac's notation will be represented by $< \phi_m | \phi_n >^* = < \phi_n | \phi_m >$ or simply

$$< m | n >^* = \langle n | m \rangle$$

We have learnt about Hermitian operator earlier. So for Hermitian operator A, we can write

$$\int \phi_m^* A \phi_n d\tau = \int \phi_n \left(A \phi_m \right)^* d\tau = \left[\int \phi_m^* A \phi_n d\tau \right]^* \tag{9.36}$$

and this can be simply represented in Dirac's notation as given in the following:

$$\langle m | A | n \rangle = \langle n | A | m \rangle^* \tag{9.37}$$

It should be kept in mind that any function Ψ can be expressed as a sum of a complete set of orthonormal function ϕ as given in the following:

$$\psi = \sum_m C_m \phi_m$$

and its Dirac's notation will be

$$| \psi > = \sum_m C_m | \phi_m > = \sum_m C_m | m > \tag{9.38}$$

$$\text{and } \int \phi_n^* \Psi d = \sum_m C_m \int \phi_n^* \phi_m d\tau = C_n$$

where $C_n = \int \phi_n^* \psi d\tau$

which is denoted in Dirac's notation as

$$\langle n|\Psi\rangle = \sum_m C_m \langle n|m\rangle = C_n \quad (9.39)$$

and $C_n =< n|\Psi >$

$$\text{but } \psi = \sum_m C_m \phi_m = \sum_m \int \phi_m^* \psi d\tau \phi_m \quad (9.40)$$

will be expressed in Dirac's notation as

$$|\Psi >= \sum_m \langle m|\Psi\rangle |m >= \sum_m |m >< m|\Psi > \quad (9.41)$$

At this stage, we need to understand about Hermitian operators and Hermitian matrix.

Suppose that operators A for which

$$\int \phi^* \hat{A} \psi d\tau = \int \psi \left(\hat{A}\phi\right)^* d\tau \quad (9.42)$$

for arbitrary function Ψ and ϕ are called Hermitian operators. It is also known that Hermitian operators always have real Eigen values and their Eigen functions having different Eigen values are orthogonal. Therefore, the Hermitian operators may be considered to represent as observable dynamical quantities.

There is a close relation between Hermitian operators and Hermitian matrix.

Suppose that there is a set of general functions ϕm, where $m = 1, 2, 3, 4,\ldots$.

We may define a matrix of complex numbers as given in the following:

$$\int \phi_m^* \hat{A} \phi d\tau = A_{mn} \quad (9.43)$$

Equation (9.42) implies that $A_{mn} = A_{mn}^*$, which is nothing but the definition of Hermitian matrix. In Dirac's notation, the matrix element A_{mn} is $< \phi_m|\hat{A}\phi_n >$. However, there is a symmetry implied by Eq. (9.42) in the manner A acts. It may be regarded as acting forward on ϕ_n or a complex conjugate, backwards on ϕ_m. It implies that $< \phi_m\hat{A}|\phi_n > = <\phi_m|\hat{A}\phi_n>$. So we usually write both equivalent expressions in symmetric form as $<\phi_m|\hat{A}\phi_n>$.

9.2.1 Expression for First-Order Correction to Energy for Nondegenerate State Using Dirac's Notation

Let us consider Eq. (9.8), which is

$$H^0\Psi_k' + H'\Psi_k^0 - E_k^0\Psi_k' - E_k'\Psi_k^0 = 0$$

$$\text{or } H^0\Psi_k' + H'\Psi_k^0 - E_k^0\Psi_k' + E_k\Psi_k^0$$

Multiplying the aforementioned equation by $< \Psi_k^0 |$, we get

$$\langle \Psi_k^0 | H^0 \Psi_k' \rangle + \langle \Psi_k^0 | H' \Psi_k^0 \rangle = E_k^0 \langle \Psi_k^0 | \Psi_k' \rangle + E_k' \langle \Psi_k^0 | \Psi_k^0 \rangle \tag{9.44}$$

Since H^0 is a Hermitian,

$$\langle \Psi_k^0 | H^0 \Psi_k' \rangle = \langle H^0 \Psi_k^0 | \Psi_k' \rangle = \langle E_k^0 \langle \Psi_k^0 | \Psi_k' \rangle = E_k^0 \langle \Psi_k^0 | \Psi_k' \rangle$$

Putting this value in Eq. (9.44), i.e.,

$$\langle \Psi_k^0 | H^0 \Psi_k' \rangle = E_k^0 \langle \Psi_k^0 | \Psi_k^0 \rangle, \text{ we got}$$

$$E_k^0 \langle \Psi_k^0 | \Psi_k' \rangle + \langle \Psi_k^0 | H' \Psi_k' \rangle = E_k^0 \langle \Psi_k^0 | \Psi_k' \rangle + E_k' \langle \Psi_k^0 | \Psi_k^0 \rangle$$

$$\text{or} \quad \langle \Psi_k^0 | H' \Psi_k^0 \rangle = E_k' \langle \Psi_k^0 | \Psi_k^0 \rangle \tag{9.45}$$

But $\langle \Psi_k^0 | \Psi_k^0 | \rangle = 1$, Eq. (9.45) takes the form

$$E_k' = \langle \Psi_k^0 | H' | \Psi_k^0 \rangle \tag{9.46}$$

This may also be written as

$$E_k' = \langle k | H' | k \rangle \tag{9.47}$$

which is often termed as matrix elements, and Eqs. (9.46) and (9.47) are known as the first-order correction to the energy.

9.2.2 First-Order Correction to Wave Function for Nondegenerate State Using Dirac's Notation

Again, let us consider Eq. (9.8), i.e.,

$$H^0 \Psi_k' + H' \Psi_k^0 = E_k^0 \Psi_k' + E_k' \Psi_k^0$$

$$\text{or} \quad H' \Psi_k^0 + H^0 \Psi_k' = E_k' \Psi_k^0 + E_k^0 \Psi_k'$$

for first-order correction to wave function. To get this, let us assume that the first-order correction to the wave function can be expressed as a linear function of the unperturbed problem, i.e.,

$$\Psi_k' = \sum_l a_l \Psi_l^0 \tag{9.48}$$

Substituting Eq. (9.48) in Eq. (9.8) and multiplying it by $\langle \Psi_l^0 |$, we get

$$\langle \Psi_k^0 | H' | \Psi_l^0 \rangle + H^0 \sum a_l \Psi_l^0 \langle \Psi_l^0 | = E_k' \langle \Psi_k^0 | \Psi_l^0 \rangle + \sum a_l E_k^0 \langle \Psi_l^0 | \Psi_k^0 \rangle$$

$$\text{or,} \quad \langle \Psi_k^0 | H' | \Psi_l^0 \rangle + \sum a_l H^0 \Psi_l^0 \langle \Psi_l^0 | = E_k' \langle \Psi_k^0 | \Psi_l^0 \rangle + E_k^0 \cdot \sum a_l \Psi_l^0 \langle \Psi_l^0 |$$

$$\text{or,} \quad \langle \Psi_k^0 | H' | \Psi_l^0 \rangle + \sum a_l E_l^0 \langle \Psi_l^0 | \Psi_l^0 \rangle = E_k' \langle \Psi_k^0 | \Psi_l^0 \rangle + \sum a_l \; E_k^0 \langle \Psi_l^0 | \Psi_k^0 \rangle$$

$$\text{or,}\quad \langle\Psi_k^0|H'|\Psi_l^0\rangle+\sum a_l E_l^0\langle\Psi_l^0|\Psi_l^0\rangle = E_k'\langle\Psi_l^0|\Psi_k^0\rangle+\sum E_k^0\ a_l\langle\Psi_l^0|\Psi_k^0\rangle$$

$$\text{or,}\quad \langle\Psi_k^0|H'|\Psi_l^0\rangle+\sum a_l E_l^0 = \sum a_l E_k^0$$

$$\text{or,} a_l = \left(E_l^0 - E_k^0\right) = -\langle\Psi_k^0|H'|\Psi_l^0\rangle$$

$$\text{or,} a_l = \frac{-\Psi_k^0|H'|\Psi_l^0}{\left(E_l^0 - E_k^0\right)}$$

$$\therefore \psi_k' = \frac{\psi_k^0 \mid H' \mid \psi_l^0}{\left(E_k^0 - E_l^0\right)} = \frac{\psi_l^0 \mid H' \mid \psi_k^0}{\left(E_k^0 - E_l^0\right)} \tag{9.49}$$

Equation (9.49) is the first-order correction to the wave function.

9.2.3 Second-Order Correction to the Energy Using Dirac's Notation

The second-order perturbation equation is

$H^0\Psi_k''+H'\Psi_k'+H''\Psi_k^0 = E_k^0\Psi_k''+E_k'\Psi_k'+E_k''\Psi_k^0$ in which H″ in which H'' may be neglected and the rest of the equation will be

$$H^0\Psi_k''+H'\Psi_k' = E_k^0\Psi_k''+E_k'\Psi_k'+E_k''\Psi_k^0 \tag{9.50}$$

Multiplying Eq. (9.50) by $\langle\Psi_k^0,$ we get

$$\langle\Psi_k^0|H^0\Psi_k''\rangle+\langle\Psi_k^0|H'\Psi_k'\rangle = E_k^0\langle\Psi_k^0|\Psi_k''\rangle+E_k'\langle\Psi_k^0|\Psi_k'\rangle+E_k''\langle\Psi_k^0|\Psi_k^0\rangle \tag{9.51}$$

Here, we take the advantage of Hermiticity of H^0, and we can write

$$\langle\Psi_k^0|\Psi_k''\rangle = \langle H^0\Psi_k^0|\Psi_k''\rangle = E_k^0\langle\Psi_k^0|\Psi_k''\rangle \tag{9.52}$$

Substituting $E_k^0\langle\Psi_k^0|\Psi_k''\rangle$ in place of the first term of Eq. (9.51), we find that the first term of the LHS cancels the first term of the RHS. In Eq. (9.51), $E_k''\langle\Psi_k^0|\Psi_k^0\rangle = E_k''$ because $\langle\Psi_k^0|\Psi_k^0\rangle = 1$. Therefore, Eq. (9.51) reduces to

$$\langle\Psi_k^0|\Psi_k''\rangle = E_k'\langle\Psi_k^0|\Psi_k'\rangle+E_k''$$

$$\therefore\ E_k'' = \langle\Psi_k^0|H'|\Psi_k'\rangle - E_k'\langle\Psi_k^0|\Psi_k'\rangle \tag{9.53}$$

$$\text{But,}\ \langle\Psi_k^0|\Psi_k' >= \Psi a_l < \Psi_k^0|\Psi_k^0\rangle = 0$$

$$\therefore\ E_k'' = \langle\Psi_k^0|H'|\Psi_k'\rangle - E_k'\times 0$$

$$\text{or,}\quad E_k'' = \langle\Psi_k^0|H'|\Psi_k'\rangle \tag{9.54}$$

and, $\Psi_k' = \Sigma\dfrac{\langle\psi_l^0|H'|\psi_k^0\rangle}{\left(E_k^0 - E_l^0\right)}\psi_l^0$ from Eq. (9.49). Substituting this value in Eq. (9.54), we get

$$\text{or} \quad E_k'' = \langle \Psi_k^0 | H' | \Psi_k' \Sigma \frac{\langle \psi_l^0 | H' | \psi_k^0 \rangle}{\left(E_k^0 - E_l^0\right)} \psi_l^0$$

$$\text{or} \quad E_k'' = \Sigma \frac{\left(\langle \psi_k^0 | H' | \psi_l^0 \rangle\right)\left(\langle \psi_l^0 | H' | \psi_k^0 \rangle\right)}{\left(E_k^0 - E_l^0\right)}$$

$$\text{or,} \quad E_k'' = \frac{\Sigma | < \Psi_k^0 | H' | \Psi_{ki}^0 >}{\left(E_k^0 - E_1^0\right)} \tag{9.55}$$

This is the fundamental equation for second-order correction to energy using Dirac's notation.

9.2.4 Alternatively: Second-Order Correction to the Energy Using Dirac's Notation

The second-order perturbation equation is written as

$$H^0 \psi_n'' + H' \psi_n' + H'' \psi_n^0 = E_n^0 \psi_n'' + E_n' \psi_n' + E_n'' \psi_n^0$$

On rearrangement, we may write

$$(H^0 + E_n^0)\psi_n'' + (H' + E_n')\psi_n' + (H'' + E_n'')\psi_n^0 = 0 \tag{9.56}$$

$$\text{Since } \psi_n'' = \Sigma g_m \psi_n^0 \text{ and } \psi_n' = \Sigma c_m \psi_m^0$$

we can write Eq. (9.56) on substitution of the value of ψ_n'' and ψ_n' as

$$\Sigma g_m \left(H^0 + E_n^0\right)\psi_m^0 + \Sigma c_m \left(H' + E_n'\right)\psi_m^0 + \left(H'' + E_n''\right)\psi_n^0 = 0 \tag{9.57}$$

Now writing the wave function in terms of ket, we get

$$\Sigma g_m \left(H^0 + E_n^0\right)\psi_m^0 \rangle \Sigma (H' + E_n') c_m |\psi_m^0\rangle + (H'' + E_n'')\psi_n^0 = 0$$

Taking the scalar product with bra $\langle \psi_n^0 |$, the aforementioned equation takes the form

$$\Sigma g_m \left(E_m^0 + E_n^0\right)\langle \psi_n^0 | \psi_m^0 \rangle \Sigma (H' - E_n') c_m \langle \psi_n^0 | \psi_m^0 \rangle + (H'' + E_n'')\langle \psi_n^0 | \psi_n^0 = 0 \tag{9.58}$$

$$\text{where } H^0 \psi_m^0 = E^0 \psi_m^0$$

The first term of the LHS of Eq. (9.58) will be zero as $\langle \psi_n^0 | \psi_m^0 = 0$ due to orthogonality. Therefore, Eq. (9.58) will be

$$\Sigma (H' - E_n') c_m \langle \psi_n^0 | \psi_m^0 \rangle + (H'' - E_n'') c_m \langle \psi_n^0 | \psi_n^0 \rangle = 0$$

$$\text{or } \Sigma c_m \langle \psi_n^0 | H' | \psi_m^0 \rangle - 0 + H_{nn}'' - E_n'' = 0$$

$$\text{or } E_n'' = H_{nn}'' \Sigma c_m \langle \psi_n^0 | H' | \psi_m^0 \rangle \tag{9.59}$$

Putting the value of C_m in Eq. (9.59), we get

$$\text{or} \quad E_n'' = H_{nn}'' + \frac{\langle \psi_m^0 | H' | \psi_n^0 \rangle \langle \psi_n^0 | H' | \psi_m^0 \rangle}{\left(E_m^0 - E_n^0\right)}$$

$$\text{or} \quad E_n'' = H_{nn}'' + \frac{\left| \langle \psi_m^0 | H' | \psi_n^0 \rangle \right|^2}{\left(E_m^0 - E_n^0\right)} \tag{9.60}$$

$$E_n'' = H_{nn}'' + \frac{\left| H'_{mn} \right|^2}{\left(E_m^0 - E_n^0\right)} \tag{9.61}$$

Equations (9.60) and (9.61) represent the second-order correction to the energy.

9.2.5 Second-Order Correction to Wave Function Using Dirac's Notation

We use Eq. (9.57) for finding out the second-order correction to wave function, i.e.,

$$\Sigma g_m \left(H^0 - E_n^0\right) \psi_m^0 + \Sigma \left(H' + E_n'\right) c_m \psi_m^0 + \left(H'' + E_n''\right) \psi_n^0 = 0$$

$$\text{or } \Sigma g_m \left(H^0 - E_n^0\right) \psi_m^0 + \Sigma \left(H' + E_n'\right) c_m \, \psi_m^0 \rangle + \left(H'' - E_n''\right) \psi_n^0 = 0$$

Writing the wave function in terms of ket, we get

$$\Sigma g_m \left(E_m^0 - E_n^0\right) | \psi_m^0 + \Sigma \left(H' - E_n'\right) c_m \, | \, \psi_m^0 \rangle + \left(H'' - E_n''\right) \psi_n^0 = 0$$

or simply,

$$\Sigma g_m \left(E_m^0 - E_n^0\right) | m \rangle + \Sigma \left(H' - E_n'\right) c_m | m \rangle + \left(H'' - E_n''\right) | \, n = 0 \tag{9.62}$$

Now scalar product with bra $\langle m|$ gives

$$\Sigma g_m \left(E_m^0 - E_n^0\right) \langle m | m + \Sigma \langle m \left(H' - E_n'\right) c_m | m \rangle + \left(H'' - E_n''\right) \langle m | n \rangle = 0$$

$$\text{or } g_m \left(E_m^0 - E_n^0\right) + \Sigma \langle m \left(H' - E_n'\right) c_m | m \geq 0$$

$$\text{or } g_m \left(E_m^0 - E_n^0\right) + \Sigma c_m \langle m | H' | m \rangle - c_m \langle m | E_n' | m \rangle = 0$$

$$\therefore g_m \left(E_m^0 - E_n^0\right) + \Sigma c_m \left[E_n' - < m | H' | m \right]$$

$$\because g_m = \frac{\Sigma c_m \left[E_n' - \langle m | H' | m \rangle \right]}{\left(E_m^0 - E_n^0\right)} \tag{9.63}$$

$$\therefore \psi_n'' = \Sigma g_m \psi_m^0 = \frac{\Sigma c_m \left[E_n' - \langle m | H' | m \rangle \right]}{\left(E_m^0 - E_n^0\right)} \psi_m^0 \tag{9.64}$$

Equation (9.64) represents the second-order correction to wave function.

9.3 Perturbation Theory: A Degenerate Case

The energy level has several independent wave functions, which satisfy the wave equation. The Schrödinger wave equations for the unperturbed and the perturbed states are represented by

$$H^0\psi^0 = E^0\psi^0 \tag{9.65}$$

$$H\psi = E\psi \tag{9.66}$$

It is clear that a large number of solutions are possible for a given energy states represented by $H^0\psi^0 = E^0\psi^0$, i.e. $\psi^0_{m1}, \psi^0_{m2}, \psi^0_{m3}, ----\psi^0_{mn}$ are degenerate states of energy level E^0_m.

Defining a new state by taking into consideration the linear combination of initially chosen wave functions, we can write

$$\chi'_{ml} = \sum_{m'l'} A_{m'l'}, \text{ where } l = 1,2,3,------\alpha \text{ For } \alpha - \text{fold degeneracy}$$

Now let us write the perturbed Hamiltonian wave function and energy Eigen values in power series.

$$H = H_0 + \lambda' H' + \lambda'' H'' + \cdots \tag{9.67}$$

$$\chi_{ml} = \chi^0_{ml} + \lambda'\chi'_{ml} + \lambda''\chi''_{ml} + \tag{9.68}$$

$$\text{and } E_{ml} = E^0_{ml} + \lambda' E'_{ml} + \lambda'' E''_{ml} + \tag{9.69}$$

The Schrödinger equation for perturbed state having the wave function ψ_{ml} can be expressed as

$$H\Psi_{ml} = E_{ml}\psi_{ml} \text{ or } H\psi_{ml}\psi_{ml} = 0 \tag{9.70}$$

Substituting the values of H, Ψ_{ml}, and E_{ml} in Eq. (9.70), we obtain

$$\left(H^0 + \lambda' H' + \lambda'' H'' + \cdots\right)\left(\chi^0_{ml} + \lambda'\chi'_{ml} + \lambda''\chi''_{ml} + \cdots\right)$$

$$-\left(E^0_{ml} + \lambda' E'_{ml} + \lambda'' E''_{ml}\right)\left(\chi^0_{ml} + \lambda'\chi'_{ml} + \lambda''\chi''_{ml} + \cdots\right) = 0) \tag{9.71}$$

From Eq. (9.71), we can express a set of equations for the coefficient of different powers of λ which are as follows:

$$\lambda^0 : H^0\chi^0_{ml} - E^0_{ml}\chi^0_{ml} = 0 \tag{9.72}$$

$$\lambda' : \left(H^0 - E^0_{ml}\right)\chi'_{ml} + \left(H' - E'_{ml}\right)\chi^0_{ml} = 0 \tag{9.73}$$

$$\lambda'' : \left(H^0 - E^0_{ml}\right)\chi''_{ml} + \left(H' - E'_{ml}\right)\chi'_{ml} + \left(H'' - E''_{ml}\right)\chi^0_{ml} = 0 \tag{9.74}$$

Now, we derive equation for first-order correction to energy and wave function in degenerate case.

9.3.1 First-Order Correction to Energy

The first-order Eq. (9.73) is

$$\left(H^0 - E^0_{ml}\right)\chi'_{ml} + \left(H' - E'_{ml}\right)\chi^0_{ml} = 0$$

putting $\chi'_{ml} = \sum_{m'l'} A_{m'l'}\psi^0_{m'l'}$. In this equation, we get

$$\sum_{m'l'} A_{m'l'}\left(H^0 - E^0_{ml}\right)\psi^0_{m'l'} + \sum_{l'=1}^{\infty}\left(H' - E'_{ml}\right)\chi^0_{ml} = 0$$

$$\text{or} \sum_{m'l'} A_{m'l'}\left(E^0_{m'l'} - E^0_{ml}\right)\psi^0_{m'l'} + \sum_{l'=1}^{\infty}\left(H' - E'_{ml}\right)\chi^0_{ml} = 0 \tag{9.75}$$

Taking scalar product with ψ^*_{mj} and integrating over space, we get

$$\sum_{m'l'} A_{m'l'}\left(E^0_{m'l'} - E^0_{ml}\right)\int^0 \psi^*_{mj}\psi^0_{m'l'}\mathrm{d}\tau + \sum_{l'=1}^{\infty}\int^0 \psi^*_{mj}H'\chi^0_{m},\mathrm{d}\tau - \sum_{l'=1}^{\infty}\int^0 \psi^*_{mj}E'_{ml}\chi^0_{m}\mathrm{d}\tau = 0 \tag{9.76}$$

The first term of the LHS of Eq. (9.75) will be zero, as ψ^*_{mj} and E^0_{ml} are orthogonal for $m \neq n$ and the energy term $\left(E^0_{m'l'} - E^0_{ml}\right)$ will be zero for $m'l' = ml$.

Equation (9.76) then takes the form

$$\sum_{l'=1}^{\infty}\int^0 \psi^*_{mj}H'\chi^0_{m},\mathrm{d}\tau - \sum_{l'=1}^{\infty}\int^0 \psi^*_{mj}E'_{ml}\chi^0_{m}\mathrm{d}\tau = 0. \tag{9.77}$$

Let us put $\int^0 \Psi^*_{mj}H'\chi^0_{ml} = H'_{jl'}$

and $\int^0 \Psi^*_{mj}\chi^0_{ml}\mathrm{d}\tau = \Delta_{jl'}$

And introducing these symbols in Eq. (9.77), we get

$$\sum_{l'=1}^{\infty}\left[H'_{jl'} - \Delta_{jl'}E'_{ml}\right] = 0,\ j = 1,2,3,\ldots\alpha \tag{9.78}$$

Equation (9.77) is a homogeneous linear simultaneous equation, which can be solved by obtaining secular equation for nonzero linear solutions, the determinant for which must be zero, and it will take the following form:

$$\begin{vmatrix} H'_{11} - \Delta_{11}E'_{ml} & H'_{12} - \Delta_{12}E'_{ml} & ----- & H'_{1\alpha} - \Delta_{1\alpha}E'_{ml} \\ H'_{21} - \Delta_{21}E'_{ml} & H'_{22} - \Delta_{22}E'_{ml} & ----- & H'_{2\alpha} - \Delta_{2\alpha}E'_{ml} \\ --- & --- & --- & --- \\ H'_{\alpha1} - \Delta_{\alpha1}E'_{ml} & H'_{\alpha1} - \Delta_{\alpha1}E'_{ml} & ----- & H'_{\alpha\alpha} - \Delta_{\alpha\alpha}E'_{ml} \end{vmatrix} = 0$$

Applying the condition that

$\Delta_{jl'} = \begin{cases} 1, \text{ if } j = l' \\ 0, \text{ if } j \neq l' \end{cases}$, the determinant is transformed into simpler form as

$$\begin{vmatrix} H'_{11} - E'_{ml} & H'_{12} - E'_{ml} & ----- & H'_{1\alpha} - E'_{ml} \\ H'_{21} - E'_{ml} & H'_{22} - E'_{ml} & ----- & H'_{2\alpha} - E'_{ml} \\ --- & --- & --- & --- \\ H'_{\alpha1} - E'_{ml} & H'_{\alpha1} - E'_{ml} & ----- & H'_{\alpha\alpha} - E'_{ml} \end{vmatrix} = 0$$

By diagonalising the secular determinant, we can write the aforementioned equation as

$$\begin{vmatrix} H'_{11} - E'_{ml} & 0\ldots & \ldots\ldots & 0 \\ 0 & H'_{22} - E'_{ml} & \ldots\ldots & 0 \\ - - - & - - - & - - - & - - - \\ 0 & 0 & \ldots\ldots & H'_{\alpha\alpha} - E'_{ml} \end{vmatrix} = 0$$

This will give the energy values in the form of Hamiltonians, i.e.,

$$E'_{ml} = H'_{11}, H'_{22}\ , H'_{33} \ldots, H'_{\infty\infty}$$

The states will be clear on actual application.

9.3.2 First-Order Correction to Wave Function

For this purpose, we utilise Eq. (9.75), which is

$$\sum_{m'l'} A'_{m'l'}\left(E^0_{m'l'} - E^0_{ml}\right)\psi^0_{m'l'} + \sum_{l'=1}^{\infty}\left(H' - E'_{ml}\right)\int \overset{0}{\psi}{}^*_{m'l'}\, x^0_{ml}\mathrm{d}\tau = 0$$

Now considering the scalar product with $\overset{0}{\psi}{}^*_{m'l'}$, and integrating, we get

$$\sum_{m'l'} A'_{m'l'}\left(E^0_{m'l'} - E^0_{ml}\right)\int \overset{0}{\psi}{}^*_{m'l'}\overset{0}{\psi}{}^*_{m'l'}\ \mathrm{d}\tau + \sum_{l'=1}^{\infty}\left(H' - E'_{ml}\right)\int \overset{0}{\psi}{}^*_{m'l'}\, x^0_{ml}\mathrm{d}\tau = 0 \qquad (9.79)$$

But $\int \overset{0}{\psi}{}^*_{m'l'}\overset{0}{\psi}{}^*_{m'l'}\,\mathrm{d}\tau = 1$ is due to normalisation condition. So Eq. (9.79) takes the form

$$\sum_{m'l'} A'_{m'l'}\left(E^0_{m'l'} - E^0_{ml}\right) + \sum_{l'=1}^{\infty} \overset{0}{\psi}{}^*_{m'l'}\left(H' - E'_{ml}\right)x^0_{ml}\mathrm{d}\tau = 0$$

$$\therefore \quad A'_{m'l'} = \frac{-\sum_{l'=1}^{\infty} \overset{0}{\psi}{}^*_{m'l'}\left(H' - E'_{ml}\right)x^0_{ml}\mathrm{d}\tau}{\left(E^0_{m'l'} - E^0_{ml}\right)}$$

$$\text{But}\, x'_{ml} = A'_{m'l'}\,\overset{0}{\psi}{}^*_{m'l'} \qquad (9.80)$$

Putting the value of $A'_{m'l'}$ in Eq. (9.80), we get

$$X'_{ml} = \frac{-\sum_{l'=1}^{\infty} \overset{0}{\psi}{}^*_{m'l'}\left(H' - E'_{ml}\right)x^0_{ml}\mathrm{d}\tau}{\left(E^0_{m'l'} - E^0_{ml}\right)}\psi^0_{m'l'} \qquad (9.81)$$

Equation (9.81) represents the first-order correction to wave function in degenerate case.

9.3.3 Alternative Way to Handle Degenerate Perturbation Theory: Twofold Degeneracy

Let us consider the following fundamental equation:

$$H^0\psi_a^0 = E^0\psi_{a'}^0$$

$$H^0\psi_b^0 = E^0\psi_{b'}^0$$

$$<\psi_a^0|\psi_a^0> = 0 \tag{9.82}$$

in which ψ_a^0 and ψ_a^0 are both normalised.

We can represent the wave function ψ^0 as a linear combination of the two states ψ_a^0 and ψ_b^0:

$$\text{i.e.}\, \psi^0 = \alpha\psi_a^0 + \beta\psi_{b}^0 \tag{9.83}$$

And it is still an Eigen state of H^0, having the same Eigen value E_0.

$$H^0\psi^0 = E^0\psi^0 \tag{9.84}$$

Actually, we want to solve the Schrödinger equation, i.e.,

$$H_\psi = E\psi \tag{9.85}$$

$$\left.\begin{aligned} \text{Having}\quad & H = H^0 + \lambda^1 H^1, \\ & \psi = \psi^0 + \lambda^1\psi^1 + \lambda^2\psi^2 + \ldots\ldots \\ \text{and}\quad & E = E^0 + \lambda^1 E^1 + \lambda^2 E^2 + \ldots\ldots \end{aligned}\right\} \tag{9.86}$$

Substituting these values in $H\psi = E\psi$, we get

$$\left(H^0 + \lambda^1 H^1 \ldots\right)\left(\Psi^0 + \lambda^1\psi^1 \ldots\right) = \left(E^0 + \lambda^1 E^1\right)\left(\Psi^0 + \lambda^1\psi^1 \ldots\right)$$

$$or,\quad H^0\Psi^0 + H^0\Psi^1\lambda^1 + H^1\Psi^0\lambda^1 + \lambda^2 H^1\Psi^1 = E^0\Psi^0 + E^0\Psi^1\lambda^1 + E^1\Psi^1\lambda^1 + E^1\Psi^1\lambda^2 + \ldots$$

Collecting the like powers of λ, we get

$$H^0\Psi^0 + H^0\Psi^1\lambda^1 + H^1\Psi^0\lambda^1 = E^0\Psi^0 + E^0\Psi^1\lambda^1 + E^1\Psi^0\lambda^1$$

But $H^0\Psi^0 = E^0\Psi^0$; therefore, the first terms of each side cancel and the remaining equation will be

$$H^0\Psi^1\lambda^1 + H^1\Psi^0\lambda^1 = E^0\Psi^1\lambda^1 + E^1\Psi^0\lambda^1$$

$$\text{or}\left(H^0\Psi^1 + H^1\Psi^0\right)\lambda^1 = \left(E^0\Psi^1 + E^1\Psi^0\right)\lambda^1$$

$$\therefore\qquad H^0\Psi^1 + H^1\Psi^0 = E^0\Psi^1 + E^1\Psi^0 \tag{9.87}$$

Multiplying the aforementioned equation with ψ_a^0, we can write

$$<\psi_a^0|H^0\psi^1> + <\psi_a^0|H^1\psi^0> = E^0 <\psi_a^0|\psi^1> + E^1 <\psi_a^0|\psi^0>$$

Sin H^0 is Hermitian, therefore

$$\left\langle\psi_a|H^0\psi^1\right\rangle = E^0\left\langle\psi_a^0|\psi^1\right\rangle$$

and the first term on the left cancels the first term on the right. The rest of the equation will be

$$<\psi_a^0 \left| H' \psi^1 \right> = E' \left\langle \psi_a^0 \middle| \psi^1 \right\rangle \tag{9.88}$$

Putting the value of $\psi_a^0 = \alpha\psi_a^0 + \beta\psi_b^0$, we get

$$\left\langle \psi_a^0 \middle| H^1 \left(\alpha\psi_a^0 + \beta\psi_b^0\right) \right\rangle = E^1 \psi_a^0 \left| \left(\alpha\psi_a^0 + \beta\psi_b^0\right) \right>$$

$$= \alpha < \psi_a^0 \left| E^1 \right| \psi_a^0 > + \beta < \psi_a^0 \left| E^1 \right| \psi_b^0 >$$

$$\text{or,} \quad \alpha < \psi_a^0 \left| H^1 \right| \psi_a^0 > + \beta < \psi_a^0 \left| H^1 \right| \psi_b^0 > + \alpha < \psi_a^0 \left| E^1 \right| \psi_a^0 > + \beta < \psi_a^0 \left| E^1 \right| \psi_a^0 >$$

$$\text{or,} \quad \alpha < \psi_a^0 \left| H^1 \right| \psi_a^0 > + \beta \left\langle \psi_a^0 \middle| H^1 \middle| \psi_b^0 \right> = \alpha E^1 + \beta E^1 < \psi_a^0 \left| \psi_b^0 \right\rangle$$

$$= \alpha E^1 + \beta E^1 \times 0 \left[\because \left\langle \psi_a^0 \middle| \psi_b^0 \right\rangle = 0 \right]$$

$$= \alpha E^1$$

We can write the aforementioned equation in compact form as

$$\alpha\omega_{aa} + \beta\omega_{ab} = \alpha E^1 \tag{9.89}$$

$$\text{Where,} \quad W_{ij} = \psi_i^0 | H^1 | \psi_j, (i \cdot j = a \cdot b)$$

and w's are known matrix elements of H^1 with respect to the unperturbed wave functions ψ_a^0 and ψ_b^0.
Similarly, multiplying Eq. (9.87) with ψ_b^0, we arrive at

$$\alpha\omega_{ba} + \beta\omega_{bb} = \beta E^1 \tag{9.90}$$

From Eq. (9.89), we may write

$$\beta\omega_{ab} = \left(\alpha E^1 - \alpha\omega_{aa}\right) = \alpha\left(E^1 - \omega_{aa}\right)$$

$$\therefore \quad \beta = \frac{\left(\alpha E^1 - \omega_{aa}\right)}{\omega_{ab}}$$

By putting the value of β in Eq. (9.90), we obtain

$$\alpha\omega_{ba} + \frac{\alpha\left(E^1 - \omega_{aa}\right)}{\omega_{ab}} \cdot \omega_{bb} = \frac{\alpha\left(E^1 - \omega_{aa}\right)}{\omega_{ab}} \cdot E^1$$

$$\text{or}\, \alpha\omega_{ba} \cdot \omega_{ab} + \alpha\left(E^1 - \omega_{aa}\right)\omega_{bb} = \alpha\left(E^1 - \omega_{aa}\right)E^1$$

$$\alpha\left(E^1\right)^2 - \alpha\omega_{aa}E^1 - \alpha E^1\omega_{bb} + \alpha\omega_{aa}\omega_{bb} - \alpha\omega_{ba}\omega_{ab} = 0$$

$$\alpha\left[\left(E^1\right)^2 - E^1\left(\omega_{aa} + \omega_{bb}\right) + \left(\omega_{aa}\omega_{bb} - \omega_{ba}\omega_{ab}\right)\right] = 0$$

$$\alpha \neq 0,$$

$$\therefore \quad \left[\left(E'\right)^2 - E'\left(\omega_{aa} + \omega_{bb}\right) + \left(\omega_{aa}\omega_{bb} - \omega_{ba}\omega_{ab}\right)\right] = 0 \tag{9.91}$$

This is a quadratic equation of the form $ax^2 + bx + c = 0$, and the solution of Eq. (9.91) will be

$$E'\pm = \frac{\left(W_{aa} + W_{bb}\right) \pm \sqrt{\left(W_{aa} + W_{bb}\right)^2 - 4\left(W_{aa} \cdot W_{bb} - W_{ba} \cdot W_{ab}\right)}}{2}$$

$$= \frac{\left(W_{aa} + W_{bb}\right) \pm \sqrt{\left(W_{aa} - W_{bb}\right)^2 - 4\left(W_{aa} \cdot W_{bb} - W_{ba} \cdot W_{ab}\right) + 4\left(W_{aa} \cdot W_{bb}\right)}}{2}$$

$$= \frac{1}{2}\left[\left(W_{aa} + W_{bb}\right) \pm \sqrt{\left(W_{aa} - W_{bb}\right)^2 + 4\left(W_{aa} \cdot W_{bb}\right)}\right]$$

$$\therefore \; E^1\pm = \frac{1}{2}\left[\left(W_{aa} + W_{bb}\right) \pm \sqrt{\left(W_{aa} - W_{bb}\right)^2 + 4\left|W_{ab}\right|^2}\right] \tag{9.92}$$

Thus, we got two roots of the two perturbed energies.

9.4 Application of Perturbation Theory

The application of perturbation theory has been used in connection with the energy change when a system is disturbed/perturbed. The disturbance in the system may be caused by replacing the atom by another atom in a molecule or the system may be disturbed by electric polarizability, etc., From the amount of perturbation, one can calculate the energy. Here, we are going to illustrate the foregoing results using well-known examples.

9.4.1 Anharmonic Oscillator

A simple harmonic oscillator (SHO) is perturbed by an additional term bx^4 in V(x), i.e., potential energy, where b is a constant. Now we calculate the first-order correction to the ground-state energy.

For this, let us assume the oscillator whose Hamiltonian is given by

$$H = \frac{-h^2}{8\pi^2 m} \cdot \frac{d^2}{dx^2} + \frac{1}{2}kx^2 + bx^4 \tag{9.93}$$

where $H^0 = \frac{-h^2}{8\pi^2 m} \cdot \frac{d^2}{dx^2} + \frac{1}{2}kx^2$

and $H' = bx^4$ is a perturbed Hamiltonian.

It is to be noted that the first three Eigen functions of the harmonic oscillator can be expressed in terms of the appropriate Hermite polynomials, which are

$$H_0 = 1$$

$$H_1 = 2y$$

$$H_2 = 4y^2 - 2, \quad \text{where, } y = \sqrt{\beta x} \text{ and } y^2 = \beta x^2$$

With the help these, we can have the first three normalised wave functions as follows:

$$\psi_0 = \left(\sqrt{\frac{\beta}{\Pi}}\right)^{1/2} \cdot e^{-y^2/2} = \left(\sqrt{\frac{\beta}{\Pi}}\right)^{1/2} \cdot e^{-\beta x^2/2}$$

$$\psi_1 = \left(\sqrt{\frac{\beta}{4\Pi}}\right)^{1/2} 2y \cdot e^{-y^2/2} = \left(\sqrt{\frac{\beta}{4\Pi}}\right)^{1/2} \cdot 2\sqrt{\beta x} \cdot e^{-\beta x^2/2}$$

$$\psi_2 = \left(\sqrt{\frac{\beta}{4\Pi}}\right)^{1/2} \cdot \left(4\beta x^2 - 2\right) \cdot e^{-\beta x^2/2}$$

Now we find the effect of the perturbation bx^4 on the energy of the ground state for which it is necessary to evaluate the integral of the type $\int_{-\infty}^{+\infty} \psi_i H' \psi_i \mathrm{d}\tau$. In our case, $H' = bx^4$ and $\psi_i = \psi_j = \psi_0$.

Now putting the value of H' and $\psi_0'^s$ in the said integral, we can obtain the first-order perturbed energy as follows:

$$E' = \int_{-\infty}^{+\infty} \psi_0 H' \psi_0 \mathrm{d}\tau$$

$$\text{or,} \quad E' = \int_{-\infty}^{+\infty} \left(\sqrt{\frac{\beta}{\Pi}}\right)^{1/2} e^{-\beta x^2/2} \cdot bx^4 \cdot \left(\sqrt{\frac{\beta}{\Pi}}\right)^{1/2} \cdot e^{-\beta x^2/2} \mathrm{d}x$$

$$\text{or,} \quad E' = \int_{-\infty}^{+\infty} \left(\sqrt{\frac{\beta}{\Pi}}\right)^{1/2} \left(\sqrt{\frac{\beta}{\Pi}}\right)^{1/2} \cdot b \int_{-\infty}^{+\infty} x^4 \left(e^{-\beta x^2/2}\right)^2 \mathrm{d}x$$

$$\text{or,} \quad E' = b \cdot \sqrt{\frac{\beta}{\Pi}} \int_{-\infty}^{+\infty} x^4 \cdot e^{-\beta x^2} \mathrm{d}x$$

$$= 2b \sqrt{\frac{\beta}{\Pi}} \int_{-\infty}^{+\infty} x^4 \left(e^{-\beta x^2}\right)^2 \mathrm{d}x \tag{9.94}$$

Now, we see that the integral in Eq. (9.94) is of the form

$\int_0^\infty x^4 e^{-\alpha x^2} \mathrm{d}x$ which is equal to

$\frac{1.3------(2n-1)}{2^n + 1} \cdot \sqrt{\frac{\Pi}{a^{2n+1}}}$. Putting $n = 2$ in the standard integral, we get the value of $\int_0^\infty x^4 e^{-\beta x^2} \mathrm{d}x$

$$= \frac{1.3}{2^{2+1}} \sqrt{\frac{\pi}{\beta^5}} = \frac{2}{2^3} \sqrt{\frac{\pi}{\beta^5}}$$

$$=\frac{3}{8}\sqrt{\frac{\pi}{\beta^5}} \tag{9.95}$$

Putting the value of the integral in Eq. (9.94), we get the value of E', i.e., perturbed energy as

$$E'=\left(2b\sqrt{2\beta/\pi}\right)\left(3/8\sqrt{\pi/\beta^5}\right)$$

$$\text{or,}\quad E'=\left(2b\sqrt{\beta/\pi\cdot\pi/\beta^5}\right)(3/8)$$

$$\text{or,}\quad E'=\frac{2\times 3b}{8}\sqrt{1/\beta^4}=\frac{2\times 3b}{8\beta^2}=\frac{3b}{4\beta^2}$$

Thus, the value of perturbed energy in the general state will be

$$E'=\frac{3b}{4\beta^2} \tag{9.96}$$

9.4.2 Electronic Polarisability of Hydrogen Atom

Let us suppose that an electric field E is applied to a hydrogen atom in its ground state. Since the ground state of hydrogen atom is spherically symmetric, we may select E to lie along the z axis, and hence, $H' = -eEz$, where the terms have their usual significance.

This $H' = -eEz$ will be equal to $-\mu.E$ and will be also equal to $er\,E\cos\theta$.

$$\text{it means } H'=-\mu E=erE\cos\theta=-eEZ \tag{9.97}$$

where r = position vector of the electron,

θ = angle between the position vector and z axis, and

μ = electric dipole moment = $-er$.

This interacts with E, i.e., electric field, which provides an additional potential energy to the electron. Since this additional energy term is very less, it will amount to perturbation on

$$H^0=\frac{-h^2}{2\mu}\nabla^2-\frac{ke^2}{r};\text{where } k=\frac{1}{4\Pi_{\varepsilon 0}} \tag{9.98}$$

The ground-state hydrogen atom is nondegenerate, and the first-order energy correction has already been evaluated, which is written as

$$E_1^1=\int_0^\infty \psi_0^* H'\psi_0 d\tau=\psi_0^*\left|eEr\cos\theta\right|\psi_0> \tag{9.99}$$

$$\text{but}\quad \psi_0=\frac{1}{\left(\Pi a_0^3\right)}\cdot e^{-2r/a_0}$$

$$\therefore\quad \psi_0^*\psi_0\approx\psi_0^2=\frac{1}{\left(\Pi a_0^3\right)}\cdot e^{-2r/a_0}$$

We also know that

$$d\tau = r^2 \sin\theta dr d\theta d\phi \tag{9.100}$$

In Eq. (9.99), the volume element

$$d\tau = \sin\theta d\theta. \tag{9.101}$$

We have to find the value of θ part of the integral, i.e.,

$$\int_0^{\pi} \cos\theta \cdot \sin\theta d\theta = -\int_0^{\pi} \cos\theta \cdot d(\cos\theta)$$

Putting $\cos\theta = x$, the aforementioned equation changes to

$$-\int_0^{\pi} \cos\theta \cdot d(\cos\theta) = -\int_0^{\pi} x dx = -\left[\frac{x^2}{2}\right]_0^{\pi}$$

Putting $x = \cos\theta$, we get

$$-\left[\frac{x^2}{2}\right]_0^{\pi} = -\left[\frac{\cos^2\theta}{2}\right]_0^{\pi} = -\left[\frac{\cos^2\pi - \cos^2 0}{2}\right]$$

$$= \left[\frac{1-1}{2}\right] = 0$$

It means the integral $\int_0^{\pi} \cos\theta \cdot \sin\theta d\theta = 0$

$$\therefore \quad E_1^1 = 0 \tag{9.102}$$

Hence, the ground state of hydrogen will not exhibit a first-order *Stark effect.* (When an atom is placed or put in a uniform electric field, the energy levels are shifted. As a consequence of shifting of energy levels, spectral lines split. This effect is called the stark effect.)

Since the first-order energy becomes zero, i.e., $E^1 = 0$, we calculate the energy shift to second order. The second-order correction to energy is expressed as

$$E_1^2 = \left\langle \Psi_1^0 \middle| H^1 \middle| \psi_1^1 \right\rangle \tag{9.103}$$

The calculation can be easily done by Dalgarno and Lewis method by assuming that

$$F\psi_1^0 = \psi_1^1 \cdot \tag{9.104}$$

The method depends on finding out the operator F.

Putting the value of ψ_1^1 in Eq. (9.103), we get

$$E_1^2 = \left\langle \psi_1^0 \middle| H^1 \middle| F\psi_1^0 \right\rangle = \left\langle \psi_1^0 \middle| H^1 F \middle| \psi_1^0 \right\rangle \tag{9.105}$$

$$\text{and} \quad H^1\psi_1^0 + H^0 F\psi_1^0 = E_1^1\psi_1^0 + FE_1^0\psi_1^0 \tag{9.106}$$

This is obtained by putting $F\psi_1^0 = \psi_1^1$
in equation $H^1\psi_1^0 + H^0\psi_1^1 = E_1^1\psi_1^0 + E_1^0\psi_1^1$ ·
Since $E_1^1 = 0$ (from Eq. (9.102)) and $E_1^0\Psi_1^0 = H^0\Psi_1^0$, Eq. (106) reduces to

$$\left(FH^0 - H^0F\right)\psi_1^0 = H'\psi_1^0 \tag{9.107}$$

Putting the value of H^0 in Eq. (9.107) from Eq. (9.98), we get

$$\left[F\left(\frac{-h^2}{2\mu}\nabla^2 - \frac{ke^2}{r}\right) + \left(\frac{-h^2}{2\mu}\nabla^2 - \frac{ke^2}{r}\right)F\right]\psi_1^0 = H'\psi_1^0$$

$$or\frac{-h^2}{2\mu}\cdot F\nabla^2\psi_1^0 - \frac{Fke^2}{r}\psi_1^0 + \frac{h^2}{\mu}\cdot\nabla^2\left(F\psi_1^0\right) + \frac{ke^2}{r}F\psi_1^0 = H'\psi_1^0$$

$$\text{or}\frac{-h^2}{2\mu}\left[\nabla^2\left(F\psi_1^0\right) - F\nabla^2\psi_1^0\right] = H'\psi_1^0 \tag{9.108}$$

Now, we find the value of $\nabla^2\left(F\psi_1^0\right)$ for replacing it in Eq. (9.108):

$$\nabla^2\left(F\psi_1^0\right) = \frac{\mathrm{d}}{\mathrm{d}x}\left[\frac{\mathrm{d}}{\mathrm{d}x}\left(F\psi_1^0\right)\right]$$

$$= \frac{\mathrm{d}}{\mathrm{d}x}\left[\frac{\mathrm{d}F}{\mathrm{d}x}\cdot\psi_1^0 + F\frac{\mathrm{d}\psi_1^0}{\mathrm{d}x}\right]$$

$$= \frac{\mathrm{d}^2F}{\mathrm{d}x^2}\psi_1^0 + \frac{\mathrm{d}F}{\mathrm{d}x}\cdot\frac{\mathrm{d}\psi_1^0}{\mathrm{d}x} + \frac{\mathrm{d}F}{\mathrm{d}x}\cdot\frac{\mathrm{d}\psi_1^0}{\mathrm{d}x} + F\frac{\mathrm{d}^2\psi_1^0}{\mathrm{d}x^2}$$

$$= \nabla^2F\psi_1^0 + 2\nabla F\nabla\cdot\psi_1^0 + F\nabla^2\psi_1^0$$

After replacing this value in Eq. (9.108), we get

$$\frac{\hbar^2}{2\mu}\left[\nabla^2F\psi_1^0 + \nabla^2F\psi_1^0 - 2\nabla F\cdot\nabla\psi_1^0 - F\nabla^2\psi_1^0\right] = H'\psi_1^0$$

$$\text{or}\frac{\hbar^2}{2\mu}\left[\psi_1^0\nabla^2F + 2\nabla\psi_1^0\cdot\nabla F\right] = H^1\psi_1^0 \tag{9.109}$$

We know that

$$\psi_1^0 = \frac{1}{\left(\pi a_0^3\right)^{1/2}}\cdot e^{-r/a_0},$$

So putting this value in Eq. (9.109), we get

$$\frac{\hbar^2}{2\mu}\left[\frac{1}{\left(\pi a_0^3\right)^{1/2}}\cdot e^{-r/a_0}\nabla^2F + 2\nabla\frac{1}{\left(\pi a_0^3\right)^{1/2}}\cdot e^{-r/a_0}\cdot\nabla F\right] = H'\frac{1}{\left(\pi a_0^3\right)^{1/2}}\cdot e^{-r/a_0}$$

$$\text{or,}\ \frac{\hbar^2}{2\mu}\left[\frac{1}{\left(\pi a_0^3\right)^{1/2}}\cdot e^{-r/a_0}\nabla^2 F+\frac{2}{\left(\pi a_0^3\right)^{1/2}}\cdot e^{-r/a_0}\cdot\nabla F\right]=\frac{H'}{\left(\pi a_0^3\right)^{1/2}}\cdot e^{-r/a_0}$$

$$\text{or,}\ \frac{\hbar^2}{2\mu}\left[\frac{1}{\left(\pi a_0^3\right)^{1/2}}\cdot e^{-\frac{r}{a_0}}\nabla^2 F-\frac{2}{a_0}\cdot\frac{1}{\left(\pi a_0^3\right)^{1/2}}\cdot e^{-r/a_0}\cdot\leq\nabla F\right]=\frac{H'}{\left(\pi a_0^3\right)^{1/2}}\cdot e^{-r/a_0}$$

$$\frac{\hbar^2}{2\mu}\left[\nabla^2 F-\frac{2}{a_0}\nabla F\right]\cdot=H'$$

$$\nabla^2 F-\frac{2}{a_0}\cdot\frac{\mathrm{d}F}{\mathrm{d}r}=\frac{2\mu}{\hbar^2}\cdot eEr\cos\theta. \tag{9.110}$$

The angular part of the RHS of Eq. (9.110) comes from cos θ, which is Legendre polynomial P_1. Therefore, F may be considered to be of the following form:

$$F=f[r]P_1 \tag{9.111}$$

where P_1 = Legendre polynomial.

Substituting the value of F in Eq. (9.110) results in an equation which is satisfied by $f(r)$ and leads to

$$F=\frac{\mu}{\hbar^2}e\cdot Ea_0\left(\frac{r}{2}+a_0\right)r\cos\theta. \tag{9.112}$$

Now, we calculate the second-order correction in energy, i.e.,

$$E_1^2=\left\langle\psi_1^0\middle|H^1F\middle|\psi_1^0\right.$$

$$=\int_0^\infty \begin{matrix}0\\ \psi_1^*\end{matrix}\ \psi_1^0\left(H^1F\right)\mathrm{d}\tau$$

$$=\frac{-\mu}{\hbar^2}e^2E^2a_0\cdot\frac{1}{\pi a_0^3}\int_0^{\pi}\int_0^{2\pi}\cos^2\theta\sin\theta\mathrm{d}\theta\mathrm{d}\phi$$

$$=\int_0^\infty\left(\frac{r^5}{2}+a_0r^4\right)e^{-2r/a_0}\mathrm{d}r$$

After putting the value of the first part of integrals, we can write

$$E_1^2=\frac{-4\mu}{3\hbar^2}\frac{e^2E^2}{a_0^2}\cdot\int_0^\infty\left(\frac{r^5}{2}+a_0r^4\right)e^{-2r/a_0}\mathrm{d}r \tag{9.113}$$

Now we solve the integral, i.e.,

$$\int_0^\infty\left(\frac{r^5}{2}+a_0r^4\right)e^{-2r/a_0}\mathrm{d}r$$

which can be written as

$$= \int_0^\infty \left(\frac{r^5}{2} e^{-2r/a_0} \mathrm{d}r \right) + \int_0^\infty \left(a_0 r^4 \right) e^{-2r/a_0} \mathrm{d}r$$

$$= \frac{1}{2} \cdot \frac{5!}{\left(\frac{2}{a_0} \right)^6} + \frac{a_0 \cdot 4!}{\left(\frac{2}{a_0} \right)^5} \quad \left[\because \int_0^\infty x^n e^{-ax} \mathrm{d}x = \frac{n!}{a^{n+1}} \right]$$

$$= \frac{1}{2} \cdot \frac{5! \cdot a_0^6}{2^6} + \frac{4! a_0 \cdot a_0^5}{2^5}$$

$$= a_0^6 \left[\frac{5!}{2^7} + \frac{4!}{2^5} \right] = \frac{27}{16} a_0^6$$

Putting the value of the integral, we get

$$E_1^2 = \frac{-4\mu}{3\hbar^2} \frac{e^2 E^2}{a_0^2} \cdot \frac{27}{16} a_0^6$$

$$E_1^2 = \frac{-9\mu}{4\hbar^2} e^2 E^2 a_0^4 \tag{9.114}$$

Putting the value of $a_0 = \dfrac{h^2}{k\mu e^2}$, in Eq. (9.114), we get

$$E_1^2 = \frac{-9}{4k} \mathrm{a}_0^3 \, E^2 \tag{9.115}$$

This is the value of second-order correction to energy of the hydrogen atom.

9.4.3 Helium Atom

The helium atom consists of two electrons in orbit around a nucleus having charge $+ze$. The charge $+ze$ may be assumed to remain at the origin, and there are two electrons with radius vector r_1 and r_2, as represented in Figure 9.1.

The Hamiltonian of helium system is expressed in the mathematical form after neglecting the nuclear motion as

$$H = \left(-\frac{\hbar^2}{2m} \nabla_1^2 - \frac{Ze^2}{4\pi\varepsilon_0 r_1} \right) + \left(\frac{-\hbar^2}{2m} \nabla_2^2 - \frac{Ze^2}{4\pi\varepsilon_0 r_2} \right) + \frac{Ze^2}{4\pi\varepsilon_0 r_{12}} \tag{9.116}$$

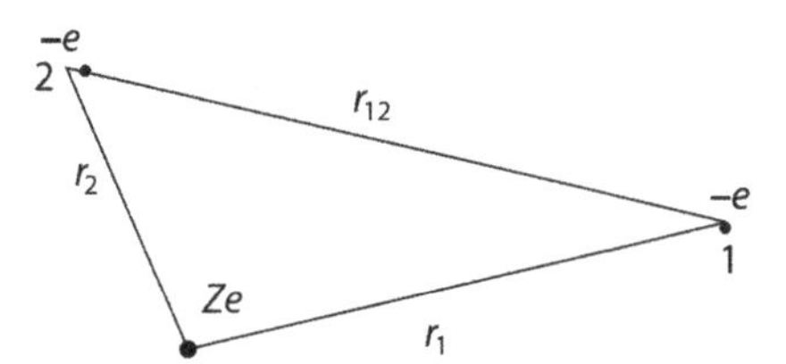

FIGURE 9.1 He atom.

where the subscripts 1 and 2 refer to two electrons. The potential energy expression in the first and second brackets refers to the interaction between nucleus and the electrons 1 and 2, respectively. The last term refers to the interaction energy (repulsion between two electrons) between the electrons, and ∇_1 and ∇_2 represent the coordinates of electrons 1 and 2, respectively.

The first and the second brackets terms may be written as

$$H_1 = \left(\frac{-\hbar^2}{2m} \nabla_1^2 - \frac{Ze^2}{4\pi\varepsilon_0 r_1} \right)$$

$$H_2 = \left(\frac{-\hbar^2}{2m} \nabla_2^2 - \frac{Ze^2}{4\pi\varepsilon_0 r_2} \right)$$

Equation (9.116) may be rewritten as

$$H = H_1 + H_2 + \frac{Ze^2}{4\pi\varepsilon_0 r_{12}} \tag{9.117}$$

where H_1 and H_2 are unperturbed Hamiltonian and $Ze^2/4\pi\varepsilon_0 r_{12}$ represents the perturbed Hamiltonian.

It is to be noted that the Schrödinger equation corresponding to H_1 and H_2 are exactly solvable and the solutions are actually the hydrogenic wave functions, i.e.,

$$\left(\frac{-\hbar^2}{2m} \nabla_1^2 - \frac{Ze^2}{4\pi\varepsilon_0 r_1} \right) \psi_1^0 = E_1^0 \psi_1^0 \tag{9.118}$$

$$\text{where} \quad E_1^0 = \frac{-Z^2 me^4}{(4\pi\varepsilon_0)^2 2\hbar^2}; \psi_1^0 = \frac{z^3}{\pi a_0^3}^{1/2} \cdot e^{\frac{-z_1}{a_0}} \tag{9.119}$$

$$\text{and} \quad E_2^0 \psi_2^0 = \left(-\frac{\hbar^2}{2m} \nabla_2^2 - \frac{Ze^2}{4\pi\varepsilon_0 r_2} \right) \psi_2^0 \tag{9.120}$$

$$\text{where} \quad E_2^0 = \frac{-Z^2 me^4}{(4\pi\varepsilon_0)^2 2\hbar^2}; \psi_2^0 = \left(\frac{z^3}{(\pi a_0)^3} \right) e^{\frac{-z_2}{40}} \tag{9.121}$$

Here, the value of $a_0 = \frac{4\pi\varepsilon_0 \hbar^2}{me^2}$.

The wave function here will be represented by the product of ψ_1^0 and ψ_2^0, i.e.,

$$\psi^0 = \psi_1^0 \psi_2^0$$

The energy Eigen value corresponding to H_1 and H_2 will be

$$E_1^0 + E_2^0 = -2Z^2 W_H \tag{9.122}$$

$$W_H = \frac{me^4}{(4\pi\varepsilon_0)^2 2\hbar^2} = 13.6\text{eV} \tag{9.123}$$

Therefore, the first-order correction to the ground-state energy will be expressed as

$$E_1 = \int\int \psi_1^{0*} \psi_2^{0*} \frac{ze^2}{4\pi\varepsilon_0 r_{12}} \psi_1^0 \psi_2^0 d\tau_1 d\tau_2 \tag{9.124}$$

Putting the values of ψ_1^0 and ψ_2^0, we obtain

$$E_1 = \frac{z^6 e^2}{4\pi\varepsilon_0 \pi^2 a_0^6} \int\int \frac{1}{r_{12}} \cdot e^{\frac{-2z}{a_0}(r_1+r_2)} d\tau_1 d\tau_2 \tag{9.125}$$

where $d\tau_1 = r_1^2 \sin\theta_1 dr_1 d\theta_1 d\phi_1$ and $d\tau_2 = r_2^2 \sin\theta_2 dr_2 d\theta_2 d\phi_2$.

The integration of Eq. (9.125) may be carried out by expanding $1/r_{12}$ in terms of Legendre polynomial. The expansion may be expressed as

$$\frac{1}{r_{12}} = \sum_{l=0}^{\infty} \sum_{m=-1}^{l} \frac{4\pi}{2l+1} \frac{r_<^l}{r_>^{l+1}} Y_{lm}^*(\theta_1\phi_1) Y_{lm}'(\theta_2\phi_2) \tag{9.126}$$

where $r_<$ = Smaller of r_1 and r_2 and

$r_>$ = Larger of r_1 and r_2.

Putting the value of $1/r_{12}$ in Eq. (9.125) and multiplying by $4\pi Y_{oo} Y_{oo}^*$ which equals 1, we obtain

$$E_1 = \frac{16z^6 e^2}{4\pi\varepsilon_0 a_0^6} \sum_{l=0}^{\infty} \sum_{m=-1}^{l} \frac{1}{2l+1} \int_0^{\infty}\int_0^{\infty} e^{\frac{-2z_1}{4_0}} e^{\frac{-2m_2}{\alpha_0}} \cdot \frac{r_<'}{r_>^{l+1}} \cdot r_1^2 r_2^2 dr_1 dr_2$$

$$\times \int_0^{2\pi}\int_0^{\pi} Y_{lm}^*(\theta_1\phi_1) Y_{00}(\theta_1\phi_1) \sin\theta_1 d\phi_1 d\theta_1.$$

$$\int_0^{2\pi}\int_0^{\pi} Y_{ao}^*(\theta_2\phi_2) Y_{lm}(\theta_2\phi_2) \sin\theta_2 d\phi_2 d\theta_2 \tag{9.127}$$

Since the spherical harmonics are normalised, the angular part will yield $\delta_{l0}\delta_{mo}\delta_{l0}\delta_{mo}$. All the terms in series will be zero except $l = m = 0$ due to the presence of Kronecker deltas. Then, we can write

$$E_1 = \frac{16z^6 e^2}{4\pi\varepsilon_0 a_0^6} \int_0^{\infty}\int_0^{\infty} e^{\frac{-2u_1}{a_0}} \cdot e^{\frac{-2\pi_2}{a_0}} \cdot \frac{1}{r_>} \cdot r_1^2 r_2^2 dr_1 dr_2 \tag{9.128}$$

Here, first we integrate over r_1 and then over r_2.

But $r_> = r_2$ in the range $0 \le r_1 \le r_2$ and $r_> = r_1$ in the range $r_2 \le r_1 \le \infty$.

Then, the value of E_1 will be written as

$$E_1 = \frac{16z^6 e^2}{4\pi\varepsilon_0 a_0^6} \int_0^{\infty} e^{\frac{-2\pi_2}{a_0}} \cdot r_2^2 \left(\int_0^{r_2} e^{\frac{-2\pi_1}{2\sigma_0}} \cdot r_1^2 dr_1 \right)$$

$$dr_2 + \frac{16z^6 e^2}{4\pi\varepsilon_0 a_0^6} \int_0^{\infty} e^{\frac{-2\pi_2}{4_0}} \cdot r_2^2 \left(\int_{r_2}^{\infty} e^{\frac{-2m_1}{a_0}} \cdot r_1 dr_1 \right) dr_2 \tag{9.129}$$

Now, with the help of the values of the standard integrals

1. $\int_0^\infty x^n e^{-ax} dx = \frac{n!}{a^{n+1}}, n \geq 0, a > 0$

2. $\int x e^{ax} dx = \frac{ax-1}{a^2} e^{ax}$ and

3. $\int x^2 e^{ax} dx = \left(\frac{x^2}{a} - \frac{2x}{a^2} + \frac{2}{a^3} \right) e^{ax}$

we can get the values of integrals in Eq. (9.129) when integrated over r_1 and r_2 as

$$E_1 = \frac{5}{4} Z \cdot W_H. \tag{9.130}$$

This equation gives the values of E_1 for first-order perturbation energy in the ground state, and therefore, the total energy in the ground state will be

$$E = \left(-2z^2 W_H + \frac{5}{4} z \cdot W_H \right)$$

$$\text{or}, E = -\left(2z^2 - \frac{5}{4} z. \right) W_H \tag{9.131}$$

Thus, we see that the perturbation theory is a powerful tool for obtaining approximate solutions to wave equation having no exact solutions, provided they differ only by a small amount from a known solutions.

9.4.4 Alternatively: The Helium Atom

The helium atom consists of two electrons in the 1s orbit with the nucleus having charge $+ze$. Nucleus is supposed to be in motionless, and the Hamiltonian of He atom may be expressed as

$$H = -\frac{\hbar^2}{2m} \nabla_1^2 - \frac{ze^2}{r_1} - \frac{\hbar^2}{2m} \nabla_2^2 - \frac{ze^2}{r_2} + \frac{e^2}{r_{12}} \tag{9.132}$$

where r_1 and r_2 = position vectors of the two electrons relative to the nucleus, $r_{12} = |r_1 - r_2|$

$$Z = 2$$

The unperturbed Hamiltonian H^0 of the system can be written as

$$H^0(1,2) = H^0(1) + H^0(2)$$

$$= \left[\frac{-\hbar^2}{2m} \nabla_1^2 - \frac{ze^2}{r_1} \right] + \left[\frac{-\hbar^2}{2m} \nabla_2^2 - \frac{ze^2}{r_2} \right] \tag{9.133}$$

The Eigen equation can be expressed as

$$H^0(1,2) \cdot \psi^0(1,2) = E^0 \psi^0(1,2) \tag{9.134}$$

$$\text{with} \, \psi^0(1,2) = \psi_n(1) \cdot \psi_{n^1}(2) \tag{9.135}$$

$$H^0(1) \cdot \psi_n(1) = E_n \psi_n(1) \tag{9.136}$$

$$H^0(2)\cdot\psi_{n'}(2) = E_{n'}\psi_{n'}(2). \tag{9.137}$$

$$\text{and } E^0 = E_n + E_{n'} \tag{9.138}$$

Since the two electrons of He belong to the ground state, we should write $n = n'$, and hence, from the theory of hydrogen atom, we can safely write

$$\psi_n(1) = \left(\frac{z^3}{\pi a_0^3}\right)^{1/2} \cdot e^{-Zr_1/a_0} \tag{9.139}$$

$$\psi_n(2) = \left(\frac{z^3}{\pi a_0^3}\right)^{1/2} \cdot e^{-Zr_2/a_0} \tag{9.140}$$

$$E_n = E_{n'} = E_{100} = \left(\frac{-Z^2\cdot e^2}{2a_0^3}\right) \tag{9.141}$$

$$\text{where } a_0 = \frac{\hbar^2}{me^2}$$

100 in E_{100} indicates that for each electron $n = 1$, $l = 0$, and $m = 0$.

It is very important to note that though the two electrons are identical, there is no exchange of degeneracy in the ground state of wave function $\psi^0(1, 2)$ as the wave function does not alter because of the exchange of $1 \leftrightarrow 2$. This is why it may be said that $\psi^0(1, 2)$ is a nondegenerate wave function of $H^0(1, 2)$.

Now, $H'(1,2) = \frac{e^2}{r_{12}}$ = interaction energy between the two electrons.

$$E^1 = \langle \Psi^0 | H' | \psi^0 \rangle$$

$$= \int \psi^*(1,2) H' \psi^0(1,2) d\tau_1 d\tau_2$$

$$= \left(\frac{z^3}{\pi a_0^3}\right)^{1/2} \int e^{\left[-\frac{2}{ao}(r_1+r_2)\right]} \frac{e^2}{r12} \cdot e^{\left[-\frac{z}{a0}(r1+r2)\right]} d\tau_1 \cdot d\tau_2 \tag{9.142}$$

$$\text{where } \begin{array}{l} d\tau_1 = r_1^2 dr_1 \sin\theta_1 d\theta_1 d\phi_1 \\ d\tau_2 = r_2^2 dr_2 \sin\theta_2 d\theta_2 d\phi_2 \end{array} \tag{9.143}$$

The integral in the Eq. (9.142) can be found out by keeping in mind that

$$\frac{1}{r_{12}} = \frac{1}{|r_1 - r_2|} = \frac{1}{\left[r_1^2 + r_2^2 - r_1 r_2 \cos\theta\right]^{1/2}}$$

where θ = angle between the position vectors r_1 and r_2.

$$\text{Hence, } \frac{1}{r_{12}} = \frac{1}{r_2}\left[\sum_l \left(\frac{r_1}{r_2}\right)^l \cdot p_l(\cos\theta)\right], r_2 > r_1$$

$$= \frac{1}{r_1}\left[\sum_l \left(\frac{r_2}{r_1}\right)^l \cdot p_l(\cos\theta)\right], r_1 > r_2$$

But from the addition theorem related to spherical harmonics, $P_l\cos(\theta)$ may be expressed as

$$P_l(\cos\theta) = \frac{4\pi}{2l+1}\sum_{m=-l}^{m=l} Y_{lm}^*(\theta_1\phi_1)Y_{lm}(\theta_2,\phi_2) \tag{9.144}$$

The theorem helps us to perform the angle integrals in Eq. (9.142) easily keeping in view the fact that the spherical harmonics is orthonormal.

$$E^1 = (4\pi^2)[\int_0^\infty \left[1/r_1 e^{-2Z/a_0(r_1+r_2)} \cdot r_1^2 dr_1 + \int_0^{r2} \frac{1}{r_2} e^{-2Z/a_0(r_1+r_2)} \cdot r_1^2 dr_1\right] r_2^2 dr_2$$

$$E^1 = \frac{5}{8}\frac{Ze^2}{a_0} \tag{9.145}$$

This is the value of energy correction.

9.5 Variation Theorem/Method

9.5.1 Variation Method

The Schrödinger equation is

$$H\psi = E\psi$$

where ψ = an arbitrary normalised wave function

$$\text{or}\,\psi^* \; H\psi = \psi^* \; E\psi$$

On integration, we write

$$\int \psi^* H\psi d\tau = \int \psi^* E\psi d\tau$$

$$= E\int \psi^* \psi d\tau$$

$$E = \frac{\int \Psi^* H\Psi d\tau}{\int \Psi^* \Psi d\tau} = \int \psi^* H\psi d\tau \tag{9.146}$$

Since $\int\psi^*\psi\, d\tau = 1$, Ψ is normalised.

The method of variation actually consists in evaluating the integral $\int\psi^* H\psi d\tau$ having trial wave function ψ, which is dependent upon a number of parameters, and these parameters are varied until the energy expressed by $E = \int\psi^* H\psi d\tau$ is minimum and then ψ will be the correct wave function for the Schrödinger equation. Now, we prove the variation theorem.

9.5.2 Variation Theorem

If ϕ be the trial wave function, H be the Hamiltonian operator, and E_0 represents the minimum energy of the system, then $\int\phi^* H\phi d\tau \geq E_0$, and ϕ is normalised.

Proof: Let ϕ = a trial wave function.

E_0 = minimum energy of the system.

We can write

$$\int \phi^* (H - E_0)\phi d\tau = \int \phi^* H\phi d\tau - \int \phi^* E_0\phi d\tau$$

$$= \int \phi^* H\phi d\tau - E_0 \int \phi^* \phi d\tau$$

$$= \int \phi^* H\phi d\tau - E_0 \left[\because \int \phi^* \phi d\tau = 1\right]$$

We have to prove that $(\int\phi^* H\phi d\tau - E_0) \geq 0$

$$\text{or,} \int \phi^* H\phi d\tau \geq E_0 \tag{9.147}$$

Now expanding the variation function ϕ as a linear series in terms of a normalised and orthogonal set of function in ψ, we can write

$$\phi = \sum a_k \psi_k \tag{9.148}$$

where the term a_k is the coefficients of the term ψ_k.

Furthermore, substituting $\phi = \sum a_k \psi$ in $\int \phi^* (H - E_0)\phi d\tau$, we have

$$\int \sum a_k^* \psi_k^* (H - E_0) \sum a_k \psi_k d\tau$$

$$= \sum a_k^* a_k \left(\int \psi_k^* H \psi_k d\tau - E_0 \int \Psi_k^* \psi_k d\tau \right)$$

$$= \sum a_k^* a_k \left[\int \psi_k^* H \psi_k d\tau - E_0 \right] \left[\because \int \psi^* \psi d\tau = 1 \right]$$

Let us represent $\int \psi_k^* H \psi_k d\tau = E_k$.

$$\therefore \int \phi^* (H - E_0)\phi d\tau = a_k^* a_k (E_k - E_0) \tag{9.149}$$

Now ψ_0 is the function for the ground state and ψ_1 ψ_2 ... are the functions corresponding to higher energy levels.

Since $a_k^* a_k$ is always positive, $(E_k - E_0)$ will be positive, where ψ_k refers to higher energy state. Consequently, $\int\phi^*(H - E_0)\phi d\tau$ = positive when $E_k = E_0$, i.e., minimum; then

$$\int \phi^* (H - E_0)\phi d\tau = 0$$

and hence, $\int \phi^* (H - E_0)\phi d\tau \geq 0$

$$\text{or} \int \phi^* H\phi d\tau \geq E \tag{9.150}$$

It means that the integral $\int\phi^* H\phi \, d\tau$ can never be less than the true minimum energy of the system.

Alternatively

Let us consider the integral $\int\phi^* H\phi \, \phi \, d\tau$ where

H = the Hamiltonian for the system and

ϕ = the sum function, which consists of a variable parameter.

It is then necessary to show that the integral

$$E = \int \phi^* H\phi d\tau \tag{9.151}$$

is always an upper limit to the lowest energy level E_0 of the system or we have to prove that

$$E_k \geq E_0 \tag{9.152}$$

It is to be noted that the variation function ϕ is fully unrestricted and its choice may be arbitrary, but more closely it should be chosen according to the physical reality, which it denotes, then more closely E will approach $E_{0'}$ i.e., $E \to E_0$. If by chance, the variation function selected happened to be the true ground-state function, then

$$\int \phi^* H\phi d\tau = \int \phi_0^* H\psi_0 d\tau = E_0 \int \psi_0^* \psi_0 d\tau = E_0 \tag{9.153}$$

Now ϕ can be expended as a linear series in the complete set, which makes the Eigen function of H, we have

$$\phi = \sum_i a_i \psi_i \tag{9.154}$$

$$\text{where} \sum_n a_n^* a_n = 1 \quad \text{and} \quad a_m^* a_n \int \psi_m^* \psi_n d\tau = 0$$

Putting the value of Eq. (9.154) in Eq. (9.151), we get

$$E = \sum_{nn'} a_n^* a_{n'} \int \psi_n^* H \psi_{n'} d\tau = \sum_n a_n^* a_n E_n \tag{9.155}$$

Since ψ_i is the true Eigen function of H and

$$H\psi_n = E\psi_n \tag{9.156}$$

Subtracting E_0 from both sides, we get

$$E - E_0 = \sum_n a_n^* a_n (E_n - E_0) \tag{9.157}$$

Since $E_n \geq E_0$ for all values of n and $a_n^* a_n$ is always positive, the RHS of Eq. (9.157) is always greater than zero, except in case where variation function is true ground-state wave function. Thus, we have

$$\sum_n a_n^* a_n (E - E_0) \geq 0$$

which clearly implies that $E - E_0 > 0$, and hence, $E > E_0$, so that E is always an upper bound to E_0.

Alternatively

Let us guess a normalised trial wave function ψ_t for the ground state and compute the expectation value of the Hamiltonian.

$$\langle \psi_t | H | \psi_t \rangle = \int \psi_t^* H \psi_t \mathrm{d}\tau \tag{9.158}$$

We are going to prove that this value can never be less than the true ground-state energy.

For this, we expand ψ_t in terms of the actual Eigen functions ψ_n of H, i.e.,

$$\psi_t = \sum_n a_n \psi_n \tag{9.159}$$

Where normalisation of ψ_t requires

$$\sum_n |a_n|^2 = 1 \tag{9.160}$$

Using Eq. (9.159) in Eq. (9.158), we can write

$$< H \geq \int \left(\sum_n a_n^* a_n \right) H \left(\sum_m a_m \psi_m \right) \mathrm{d}\tau$$

$$= \sum_n \sum_m a_n^* a_m \int \psi_n^* H \psi_m \mathrm{d}\tau$$

$$= \sum_n \sum_m a_n^* a_m E_m \delta_{nm}$$

$$= \sum_n |a_n|^2 E_n \tag{9.161}$$

As $E_n > E_0$ (where E_0 = ground-state energy), we can surely say from Eq. (9.161) that

$$\langle H \rangle \geq \sum_n |a_n|^2 E_0 = E_0 \sum_n |a_n|^2 \tag{9.162}$$

So using Eq. (9.160), we get from Eq. (9.162)

$$< \psi_t | H | \psi_t \geq E_0 \tag{9.163}$$

The strategy is thus to make a plausible guess at ψ_t and include some adjustable parameter.

Alternatively

Let us consider the Schrödinger equation describing some system

$$H\psi_i = E_i\psi_i \tag{9.164}$$

where i is a label symbolising the state of the system. There exists a complete set of Eigen value and the corresponding set of Eigen functions.

The ψ_i that are solutions to Eq. (9.164) are linear independent. That is,

$$\int \psi_i^* \psi_j \mathrm{d}\tau = \delta_{ij} \tag{9.165}$$

where i and j denote the states, the integration is over all space, and δ_{ij} is the Kronecker delta function. δ_{ij} becomes unity when $i = j$ and zero when $i \neq J$. The function forms a complete orthonormal function space. A complete function is known as Hilbert space. Any other arbitrary function, say u, in the same space can be formed as a linear combination of these:

$$u = \sum_i a_i\psi_i \tag{9.166}$$

Normalisation of u requires that

$$\sum_i a_i^* a_i = 1 \tag{9.167}$$

Let now consider the *expectation value* of Hamiltonian (<H>) with respect to u:

$$\langle H \rangle = \int u^* H u \mathrm{d}\tau$$

$$= \sum_i \sum_j \int a_i^* \psi_i^* H a_j \psi_j \mathrm{d}\tau$$

$$= \sum_i \sum_j a_i^* a_j \int \psi_i^* H \psi_j \mathrm{d}\tau \tag{9.168}$$

But because of Eq. (9.164), $H\psi_j = E_j\,\psi_j$, giving

$$\langle H \rangle = \sum_i \sum_j a_i^* a_j E_j \int \psi_i^* \psi_j \mathrm{d}\tau$$

$$= \sum_i \sum_i \int a_i^* a_j E_j \delta_{ij}$$

$$= \sum_i a_i^* a_i E_i \tag{9.169}$$

Now let us subtract the ground-state energy, E_0 from both the sides of Eq. (9.169). We get

$$\langle H \rangle - E_0 = \sum_i a_i^* a_i (E_i - E_0) \tag{9.170}$$

Since $a_i^* a_i$ is positive and $(E_i - E_0)$ is also positive,

$$\langle H \rangle - E_0 \geq 0 \tag{9.171}$$

This is a very important result, which gives clear-cut idea that the expectation value of the given Hamiltonian with respect to any arbitrary normalised function, $\int u^* H u \, d\tau$, always lies above the true energy of the ground state.

The variation theory is perhaps the most useful approximation method in computational chemistry. It has a severe limitation. In the general case, it will give only the lowest energy state, of a given spin and symmetry of the system under consideration.

9.5.3 Computation of Energy Eigen Value and Wave Function by Variation Method

The ground-state Schrödinger equation is:

$H^0\psi^0 = E^0\psi^0$ and the Schrödinger equation in the variable state is written as

$$H\psi = E\psi$$

With these two equations, we can say that

$$H^0 \rightarrow H$$

$$E^0 \rightarrow E$$

$$\psi^0 \rightarrow \psi$$

The changes are due to the occurrence of several potential terms.

Now the expectation energy value can be written as

$$\langle \psi | H | \psi \rangle = E \langle \psi | \psi \rangle$$

$$\therefore \bar{E} = \frac{\langle \psi | H | \psi \rangle}{\langle \psi | \psi \rangle} \tag{9.172}$$

The problem is how to compute the expectation energy value from the aforementioned equation.

It will be computed by taking into consideration the minimum energy value and normalised function.

Now, we use the fact that the wave function can be expressed as a liner combination of a set of orthonormal Eigen functions ϕ_1, ϕ_2 ... of the same system having Eigen values E_1, E_2 ..., i.e.,

$$\psi = a_1\phi_1 + a_2\phi_2 + \cdots \tag{9.173}$$

where a_1 and a_2 are the coefficients included for normalising the ψ function. Thus, we can write

$$\bar{E} = \frac{\langle \psi | H | \psi >}{\langle \psi | \psi \rangle}$$

or in the form of integral, it can be expressed as

$$\bar{E} = \frac{\int \psi H \psi \, d\tau}{\int \psi \psi \, d\tau}$$

Putting the value of Ψ, we have

$$\bar{E} = \frac{\int (a_1\phi_1 + a_2\phi_2) H (a_1\phi_1 + a_2\phi_2) \mathrm{d}\tau}{\int (a_1\phi_1 + a_2\phi_2)(a_1\phi_1 + a_2\phi_2) \mathrm{d}\tau} \quad (9.174)$$

$$\text{or, } \bar{E} = \frac{\int (a_1^2\phi_1 H\phi_1 \mathrm{d}\tau) + \int (a_2^2\phi_2 H\phi_2 \mathrm{d}\tau) + \int (2a_1a_2\phi_1 H\phi_2 \mathrm{d}\tau)}{\int (a_1^2\phi_1^2 \mathrm{d}\tau) + \int a_2^2\phi_2^2 \mathrm{d}\tau + \int 2a_1a_2\phi_1\phi_2 \mathrm{d}\tau}$$

$$\text{or, } \bar{E} = \frac{a_1^2 \int (\phi_1 H\phi_1 \mathrm{d}\tau + a_2^2 (\phi_2 H\phi_2 \mathrm{d}\tau) + 2a_1a_2 \int (\phi_1 H\phi_2 \mathrm{d}\tau)}{a_1^2 \int (\phi_1^2 \mathrm{d}\tau) + a_2^2 \int \phi_2^2 \mathrm{d}\tau + 2a_1a_2 \int \phi_1\phi_2 \mathrm{d}\tau}$$

$$\text{or } \bar{E} = \frac{a_1^2 H_{11} + a_2^2 H_{22} + 2a_1a_2 H_{12}}{a_1^2 S_{11} + a_2^2 S_{22} + 2a_1a_2 S_{12}} \quad (9.175)$$

$$\text{where } H_{11} = \int \phi_1 H\phi_1 \mathrm{d}\tau, \quad H_{12} = \int \phi_1 H\phi_2 \mathrm{d}\tau$$

$$H_{22} = \int \phi_2 H\phi_2 \mathrm{d}\tau, H_{21} = \int \phi_2 H\phi_1 \mathrm{d}\tau$$

$$s_{11} = \int \phi_1\phi_1 \mathrm{d}\tau, s_{22} = \int \phi_2\phi_2 \mathrm{d}\tau$$

$$S_{12} = \int \phi_1\phi_2 \mathrm{d}\tau, S_{21} = \int \phi_2\phi_1 \mathrm{d}\tau$$

But $H_{12} = H_{21}$ because of Hermitian nature.

Equation (9.175) can be written as

$$\bar{E}\left(a_1^2 S_{11} + a_2^2 S_{22} + 2a_1a_2 S_{12}\right) = \left(a_1^2 H_{11} + a_2^2 H_{22} + 2a_1a_2 H_{12}\right) \quad (9.176)$$

Now, we apply the following condition to get the minimum energy:

$$\frac{\mathrm{d}\bar{E}}{da_1} = 0 \text{ and } \frac{\mathrm{d}\bar{E}}{da_2} = 0$$

Differentiating Eq. (9.176) with respect to a_1, we get

$$\frac{d\bar{E}}{\mathrm{d}a_1}\left(a_1^2 S_{11} + a_2^2 S_{22} + 2a_1a_2 S_{12}\right) + \bar{E}(2a_1 S_{11} + 0 + 2a_2 S_{12}) = 2a_1 H_{11} + 0 + 2a_2 H_{12}$$

$$\text{or } 0 + \bar{E}(2a_1 s_{11} + 2a_2 s_{12}) = 2a_1 H_{11} + 2a_2 H_{12}$$

$$\text{or } \bar{E}(2a_1 s_{11} + 2a_2 s_{12}) = (2a_1 H_{11} + 2a_2 H_{12})$$

$$\text{or } 2a_1\left[H_{11} - \bar{E}s_{11}\right] + 2a_2\left[H_{12} - \bar{E}s_{12}\right] = 0$$

$$\text{or } a_1\left[H_{11} - \bar{E}s_{11}\right] + a_2\left[H_{12} - \bar{E}s_{12}\right] = 0 \tag{9.177}$$

Furthermore, differentiating Eq. (9.176) with respect to a_2, we get

$$\frac{\mathrm{d}\bar{E}}{\mathrm{d}a_2}\left(a_1^2 s_{11} + a_2^2 s_{22} + 2a_1 a_2 s_{12}\right) = \frac{\mathrm{d}\bar{E}}{\mathrm{d}a_2}\left(a_1^2 H_{11} + a_2^2 H_{22} + 2a_1 a_2 H_{12}\right)$$
$$+ \ \bar{E}\left(0 + 2a_2 s_{22} + 2a_1 s_{12}\right)$$

$$\text{or } 0 + \bar{E}\left(2a_2 s_{22} + 2a_1 s_{12}\right) = \left(0 + 2a_2 H_{22} + 2a_1 H_{22}\right)$$

$$\text{or } \bar{E}\left(2a_1 s_{22} + 2a_1 s_{12}\right) = \left(2a_2 H_{22} + 2a_1 H_{12}\right)$$

$$\text{or } 2a_2\left[H_{22} - \bar{E}s_{22}\right] + 2a_1\left[H_{12} - \bar{E}s_{12}\right] = 0$$

$$\text{or } a_2\left[H_{22} - \bar{E}s_{22}\right] + a_1\left[H_{12} - \bar{E}s_{12}\right] = 0 \tag{9.178}$$

From Eqs (9.177) and (9.178), we can write the secular determinant as

$$\begin{vmatrix} H_{11} - \bar{E}_{s_{11}} & H_{12} - \bar{E}s_{12} \\ H_{12} - \bar{E}s_{12} & H_{22} - \bar{E}s_{22} \end{vmatrix} = 0 \tag{9.179}$$

From this, we can compute the value of energy in terms of Hamiltonian on diagonalisation as

$$\begin{vmatrix} H_{11} - \bar{E}_{s_{11}} & 0 \\ 0 & H_{22} - \bar{E}s_{22} \end{vmatrix} = 0$$

$$\bar{E} = \frac{H_{11}}{S_{11}} \ or \ \frac{H_{22}}{S_{22}}$$

$$\text{or}, \bar{E} = \frac{\int \phi_1 H \phi_1 \mathrm{d}\tau}{\int \phi_1 \phi_2 \mathrm{d}\tau} \ \text{or} \ \frac{\int \phi_2 H \phi_2 \mathrm{d}\tau}{\int \phi_2 \phi_2 \mathrm{d}\tau} \tag{9.180}$$

Thus, the energy is computed.

9.5.4 Computation of Wave Function

From Eq. (9.173), we can write $\psi = a_1\phi_1 + a_2\phi_2$, and its complex conjugate will be expressed as

$$\psi^* = a_1\phi_1^* + a_2\phi_2^*$$

Now applying normalisation condition, we have

$$\int \psi^* \psi \mathrm{d}\tau = 1$$

Putting the value of ψ and ψ^* in the aforementioned equation, we have

$$\int \left(a_1\phi_1^* + a_2\phi_2^*\right)\left(a_1\phi_1 + a_2\phi_2\right)\mathrm{d}\tau = 1$$

$$\text{or,} \int a_1^2\phi_1^*\phi_1\mathrm{d}\tau + \int a_1a_2\phi_1^*\phi_2\mathrm{d}\tau + \int a_1a_2\phi_2^*\phi_1\mathrm{d}\tau + \int a_2^2\phi_2^*\phi_2\mathrm{d}\tau = 1$$

$$\text{or,}\, a_1^2 + 0 + 0 + a_2^2 = 1$$

$$\text{or,}\, a_1^2 + a_2^2 = 1$$

$$\text{or,} \sum a_i^2 = 1 \quad (9.181)$$

Thus, knowing the value of a_i, ψ, wave function can be computed.

9.6 Application of Variation Principle/Method

9.6.1 Estimation of Energy of the Ground State of the Simple Harmonic Oscillator Using the Trial Function Ae^{-ax^2}

The Schrödinger wave equation for this case is $H\psi = E\psi$

$$\text{or}\left[-\frac{h^2}{2m}\cdot\frac{\mathrm{d}^2}{\mathrm{d}x^2} + \frac{k}{2}x^2\right]\psi = E\psi \quad (9.182)$$

Now, let

$$\psi = Ae^{-\alpha x^2} \quad (9.183)$$

where c = normalising constant.

On normalising the function, we may write

$$\int_{-\infty}^{+\infty} \psi^*\psi dx = \int_{-\infty}^{+\infty} \left(Ae^{-\alpha x^2}\right)^* \left(Ae^{-\alpha x^2}\right)\mathrm{d}x = 1$$

$$= A^2 \int_{-\infty}^{+\infty} e^{-2\alpha x^2}\mathrm{d}x = 1$$

$$\text{or,}\, A^2\sqrt{\frac{\pi}{2\alpha}} = 1 \quad \left[\because \int_{-\alpha}^{+\alpha} e^{-2\alpha x^2} = \sqrt{\frac{\pi}{2\alpha}}\right]$$

$$\text{or,}\, A^2 = \frac{1}{\sqrt{\frac{\pi}{2\alpha}}} = \left(\frac{2\alpha}{\pi}\right)^{1/2}$$

$$A = \left(\frac{2\alpha}{\pi}\right)^{1/4}$$

$$\psi = \left(\frac{2\alpha}{\pi}\right)^{1/4} \cdot e^{-\alpha x^2} \tag{9.184}$$

and $H\psi = \left[-\frac{\hbar^2}{2m}\cdot\frac{d^2}{dx^2} + \frac{kx^2}{2}\right] Ae^{-\alpha x^2}$

We have

$$\frac{d}{dx}\left(Ae^{-\alpha x^2}\right) = Ae^{-\alpha x^2}(-2\alpha x)$$

$$= -Ae^{-\alpha x^2}(2\alpha x)$$

$$= -2\alpha x\psi$$

So

$$\frac{d^2}{dx^2}\left(Ae^{-\alpha x^2}\right) = \frac{d}{dx}(-2\alpha x\psi) - 2\alpha\psi$$

$$= -2\alpha x(-2\alpha x\psi) - 2\alpha\psi$$

$$= \left(4\alpha^2 x^2\psi - 2\alpha\psi\right)$$

$$H\psi = \left[-\frac{\hbar^2}{2m}\left(4\alpha^2 x^2\psi - 2\alpha\psi\right) + \frac{kx^2}{2}\psi\right]$$

$$= \left(-\frac{\hbar^2}{2m}\cdot 4\alpha^2 + \frac{k}{2}\right)x^2\psi + \frac{\hbar^2}{2m}\cdot 2\alpha\psi$$

$$= \left(-\frac{2\hbar^2\alpha^2}{m} + \frac{k}{2}\right)x^2\psi + \frac{\hbar^2}{m}\cdot\alpha\psi \tag{9.185}$$

$$\text{But } E = \frac{\langle\psi|H|\psi\rangle}{\langle\psi|\psi\rangle} = \langle\psi|H|\psi\rangle \quad \left[\because \langle\psi|\psi = 1\rangle\right]$$

$$E = \langle\psi|H|\psi\rangle$$

$$= \int_{-\infty}^{+\infty}\psi^* H\psi\, dx$$

$$= \int_{-\infty}^{+\infty} \psi^* \left(-\frac{2\hbar^2\alpha^2}{m} + \frac{k}{2} \right) x^2 \psi \, dx + \int_{-\infty}^{+\infty} \frac{\hbar^2}{m} \alpha \psi^* \psi \, dx \text{(on putting the value of } H\psi \text{)}$$

$$= \left(-\frac{2\hbar^2\alpha^2}{m} + \frac{k}{2} \right) \int_{-\infty}^{+\infty} \psi^* x^2 \psi \, dx + \int_{-\infty}^{+\infty} \frac{\hbar^2}{m} \alpha \psi^* \psi \, dx$$

$$= \left(-\frac{2\hbar^2\alpha^2}{m} + \frac{k}{2} \right) (A^2) \int_{-\infty}^{+\infty} x^2 e^{-2\alpha x^2} dx + \frac{\hbar^2}{m} \alpha \int_{-\infty}^{+\infty} \psi^* \psi \, dx$$

$$= \left(-\frac{2\hbar^2\alpha^2}{m} + \frac{k}{2} \right) (A^2) \int_{-\infty}^{+\infty} x^2 e^{-2\alpha x^2} dx + \frac{\hbar^2}{m} \alpha \qquad \left[\because \int_{-\infty}^{\infty} \psi^* \psi \, dx = 1 \right]$$

$$= \left(-\frac{2\hbar^2\alpha^2}{m} + \frac{k}{2} \right) \left(\frac{2\alpha}{\pi} \right)^{1/2} \int_{-\infty}^{+\infty} x^2 e^{-2\alpha x^2} dx + \frac{\hbar^2}{m} \alpha$$

$$= \left(-\frac{2\hbar^2\alpha^2}{m} + \frac{k}{2} \right) \left(\frac{2\alpha}{\pi} \right)^{1/2} \cdot 2 \int_{0}^{\infty} x^2 e^{-2\alpha x^2} dx + \frac{\hbar^2}{m} \alpha$$

$$= \left(-\frac{2\hbar^2\alpha^2}{m} + \frac{k}{2} \right) \left(\frac{2\alpha}{\pi} \right)^{1/2} \cdot 2 \times \frac{1}{4} \sqrt{\frac{\pi}{(2\alpha)^3}} + \frac{\hbar^2}{m} \alpha \qquad \left[\int_{0}^{\alpha} x^2 e^{-2\alpha x^2} dx = \frac{1}{4} \sqrt{\frac{\pi}{(2\alpha)^3}} \right]$$

$$= \left(-\frac{2\hbar^2\alpha^2}{m} + \frac{k}{2} \right) \left(\frac{2\alpha}{\pi} \right)^{1/2} \cdot \frac{1}{2} \sqrt{\frac{\pi}{8\alpha^3}} + \frac{\hbar^2}{m} \alpha$$

$$= \left(-\frac{2\hbar^2\alpha^2}{m} + \frac{k}{2} \right) \left(\frac{2\alpha}{\pi} \right)^{1/2} \cdot \frac{1}{2 \cdot 2\alpha} \sqrt{\frac{\pi}{2\alpha}} + \frac{\hbar^2}{m} \alpha$$

$$= \left(-\frac{2\hbar^2\alpha^2}{m} + \frac{k}{2} \right) \cdot \frac{1}{4\alpha} + \frac{\hbar^2}{m} \alpha$$

$$= -\frac{2\hbar^2\alpha^2}{m} \cdot \frac{1}{4\alpha} + \frac{k}{2} \cdot \frac{1}{4\alpha} + \frac{\hbar^2}{m} \alpha$$

$$= -\frac{\hbar^2\alpha}{2m} + \frac{k}{8\alpha} + \frac{\hbar^2}{m} \alpha$$

$$= \left(-\frac{\hbar^2\alpha}{2m} + \frac{\hbar^2}{m} \alpha \right) + \frac{k}{8\alpha}$$

$$= \frac{\hbar^2 \alpha}{2m} + \frac{k}{8\alpha} \tag{9.186}$$

Now, we minimise the energy E with respect to α, i.e., $\mathrm{d}E/\mathrm{d}\alpha = 0$.

From Eq. (9.186), we obtain

$$\frac{\mathrm{d}E}{\mathrm{d}\alpha} = \frac{\hbar^2}{2m} - \frac{k}{8\alpha^2} = 0$$

$$\text{or,} \quad \frac{k}{8\alpha^2} = \frac{\hbar^2}{2m}$$

$$\text{or,} \quad \frac{k}{4\alpha^2} = \frac{\hbar^2}{m}$$

$$\text{or,} \quad 4\alpha^2 = \frac{km}{\hbar^2} \quad \text{or,} \quad \alpha^2 = \frac{km}{4\hbar^2}$$

$$\text{i.e.,} \quad \alpha = \sqrt{\frac{km}{4\hbar^2}} \tag{9.187}$$

Putting the value of α in Eq. (9.186), we get

$$E = \frac{\hbar^2}{2m} \cdot \sqrt{\frac{km}{4\hbar^2}} + \frac{k}{8\left(\sqrt{km/4\hbar^2}\right)}$$

$$= \frac{\hbar^2}{2m} \cdot \frac{\sqrt{km}}{2\hbar} + \frac{k}{8\left(\sqrt{km/4\hbar^2}\right)}$$

$$= \frac{\hbar}{4} \cdot \sqrt{\frac{km}{m^2}} + \frac{k\sqrt{4\hbar^2}}{8\sqrt{km}} = \frac{\hbar}{4} \cdot \sqrt{\frac{k}{m}} + \frac{k \cdot 2\hbar}{8\sqrt{km}}$$

$$= \frac{\hbar}{4} \cdot \sqrt{\frac{k}{m}} + \frac{\hbar}{4}\sqrt{\frac{k^2}{km}} = \frac{\hbar}{4} \cdot \sqrt{\frac{k}{m}} + \frac{\hbar}{4}\sqrt{\frac{k}{m}}$$

$$= 2 \cdot \frac{\hbar}{4} \cdot \sqrt{\frac{k}{m}} = \frac{\hbar}{2} \cdot \sqrt{\frac{k}{m}} = \frac{1}{2} \cdot h\nu_0 \tag{9.188}$$

$$\text{where } \nu_0 = \hbar \cdot \sqrt{\frac{k}{m}}$$

Thus, the energy of SHO will be $E = \frac{1}{2} h\nu_0$.

9.6.2 Ground State of Helium Atom

Helium atom consists of two electrons moving in the field of nucleus having charge two. Here, in this case, there is not only interaction between nucleus and the electrons but also interaction between the two electrons.

The electronic interaction will be considered and included by using variation theory and including a parameter called nuclear charge which is optimised with respect to energy. The nuclear charge z' will be taken into consideration here.

Let us assume a variation function for the electronic wave function of the He atom, which is a simple product of two 1s hydrogen-like orbitals in which the variation parameters will be z'. The variation function will then be expressed as

$$\phi = \phi_1 \cdot \phi_2 = \left(\frac{z'}{a_0}\right)^3 e^{\frac{z' r_1}{a_0}} \cdot e^{-\frac{z' r_2}{a_0}} \tag{9.189}$$

where z' = effective nuclear charge,
r_1 and r_2 = the distance between electrons 1 and 2 from the nucleus,
ϕ_1 = the function of electron 1, and
ϕ_2 = the function of electron 2.

Now we write the Hamiltonian for system as

$$H = -\frac{\hbar^2}{2m}\left(\nabla_1^2 + \nabla_2^2\right) - ze^2\left(\frac{1}{r_1} + \frac{1}{r_2}\right) + \frac{1}{r_{12}} \tag{9.190}$$

Here, z = correct nuclear charge to describe the interaction between the electrons with nucleus

Since ϕ_1 and ϕ_2 are hydrogen-like wave functions having nuclear charge $z'e$, they are Eigen functions of the corresponding hydrogen-like Hamiltonian H_1 or H_2, viz.,

$$H_1\phi_1 = \left(-\frac{\hbar^2}{2m}\nabla_1^2 - \frac{\hbar^2}{2m}\nabla_2^2\right)\phi_1 = -z'E_H\phi_1 \tag{9.191}$$

where $-E_H$ = energy of 1s orbital of hydrogen

A similar equation also holds true for ϕ_2.

From Eq. (9.191), we can substitute for the kinetic energy operator in Eq. (9.190) and find for the variation integral $E = \int \phi H \phi \, d\tau$.

$$E = -2z'E_H + (z' - z)e^2 \int_0^\infty \phi^* \left(\frac{1}{r_1} + \frac{1}{r_2}\right)\phi \, d\tau + \int_0^\infty \frac{\phi^* \phi}{r_{12}} d\tau \tag{9.192}$$

Upon substitution of Eq. (9.189), the first integral of Eq. 9.192 becomes

$$e^2 \int_0^\infty \phi^* \left(\frac{1}{r_1} + \frac{1}{r_2}\right)\phi \, d\tau = e^2 \int_0^\infty \left(\frac{\phi_1^2}{r_1} + \frac{\phi_2^2}{r_2}\right) d\tau$$

$$= 2e^2 \int_0^\infty \left(\frac{\phi_1^2}{r_1}\right) d\tau \tag{9.193}$$

The last integral in Eq. (9.193) is simply found out and gives $z'e^2/\alpha_0$ or $2z'E_H$. The total contribution will be $4z'E_H$. The second integral in Eq. (9.192) is not so easy to evaluate and consists of an expansion of r_{12}. It is found to have the value $5/4z'E_H$.

Now inserting the values of the integrals in Eq. (9.192), the variation integral has the value

$$E=\left\{-2z'^2+\frac{5}{4}z'+4z'(z'-z)\right\}E_H \tag{9.194}$$

For optimising, we differentiate E with respect to z' and put equal to zero, i.e., $dE/dz' = 0$. This will yield

$$(-4z'+5/4+8z'-4z)E_H=0 \tag{9.195}$$

$$\text{or,}\quad (5/4+4z'-4z)E_H=0$$

$$\text{or,}\quad \frac{5}{4}+4(z'-z)=0$$

$$\text{or,}\quad 4(z'-z)=-\frac{5}{4}$$

$$\text{or,}\quad (z'-z)=-\frac{5}{16}$$

$$\therefore z'=\left(z-\frac{5}{16}\right) \tag{9.196}$$

By putting the value of z' in Eq. (9.192), we obtain

$$E=\ -2\left(z-\frac{5}{6}\right)E_H \tag{9.197}$$

$\frac{5}{6}$ in Eq. (9.196) is called the screening constant. It indicates the screening of the nucleus by the other electron.

9.6.3 Ground State of Hydrogen Atom

Here, we find out the value of energy of hydrogen atom by the application of variation theory. For this, we choose a trial function

$$\psi = Ae^{-ar^2} \tag{9.198}$$

where A = normalisation constant

We now choose to normalise the aforementioned function so that

$$\int_0^\infty \psi\psi^* d\tau = 1$$

$$\text{or}\quad \int_0^\infty (Ae^{-ar^2})\left(Ae^{-ar^2}\right)d\tau = 1$$

$$\text{or} \quad \int_0^\infty A^2 e^{-2ar^2} r^2 dr = 1 \quad [\because d\tau = r^2 dr] \tag{9.199}$$

$$\text{or} \quad A^2 \int_0^\infty e^{-2ar^2} r^2 dr = A^2 \int_0^\infty r^2 e^{-2ar^2} dr = 1$$

$$\text{or} \quad A^2 \frac{1}{4}\sqrt{\frac{\pi}{(2a)^3}} = 1 \qquad \left[\because \int x^2 e^{-ax^2} dx = \frac{1}{4}\sqrt{\frac{\pi}{a^3}}\right]$$

$$\text{or} \quad A^2 . \frac{1}{4}\sqrt{\frac{\pi}{8a^3}} = 1$$

$$\text{or} \quad A^2 = \frac{4}{\sqrt{\frac{\pi}{8a^3}}} = 4\sqrt{\frac{8a^3}{\pi}} = 8\sqrt{\frac{2a^3}{\pi}} \tag{9.200}$$

Now the integral must be evaluated as

$$E = \int_0^\infty \psi^* \left[\frac{h^2}{2m}\nabla^2 - \frac{e^2}{(4\pi\varepsilon_0)r}\right] \psi r^2 dr$$

Since ψ does not depend on θ and ϕ, only the r-dependent part of ∇^2 will be taken into consideration, i.e.,

$$E = A^2 \int_0^\infty e^{-ar^2} \left\{-\frac{h^2}{2m}\left[\frac{1}{r^2}\frac{d}{dr}\left(r^2 \frac{d}{dr}\right)\right] - \frac{e^2}{(4\pi\varepsilon_0)r}\right\} \tag{9.201}$$

On differentiation, we get

$$E = A^2 \int_0^\infty e^{-ar^2} \left\{-\frac{h^2}{2m}(-6ae^{-ar2} + 4a^2 r^2 e^{-ar^2} - \frac{e^2}{(4\pi\varepsilon_0)r}\right\} e^{-ar^2} r^2 dr$$

$$= \frac{3A^2 \cdot ah^2}{m}\int_0^\infty e^{-2ar^2} r^2 dr - \frac{2A^2 \cdot ah^2}{m}\int_0^\infty e^{-2ar^2} r^4 dr - \frac{A^2 e^2}{4\pi\varepsilon_0}\int_0^\infty e^{-ar^2} r dr$$

Now, the values of the standard integrals are

$$A^2 \int_0^\infty e^{-2ar^2} r^4 dr = 8\sqrt{\frac{2a^3}{\pi}}\left(\frac{3}{32}a\sqrt{\frac{\pi}{2a^3}}\right) = \frac{3}{4a} \tag{9.202}$$

$$A^2 \int_0^\infty e^{-2ar^2} r dr = \frac{1}{4a}\left(8\sqrt{\frac{2a^3}{\pi}}\right) = \frac{2}{a}\sqrt{\frac{2a^3}{\pi}} \tag{9.203}$$

$$\therefore \quad E = \frac{3ah^2}{m} - \frac{2a^2h^2}{m}\left(\frac{3}{4a}\right) - \frac{e^2}{(4\pi\varepsilon_0)}\left(\frac{2}{a}\sqrt{\frac{2a^3}{\pi}}\right)$$

$$= +\frac{3ah^2}{m} - \frac{2e^2}{4\pi\varepsilon_0}\sqrt{\frac{2a}{\pi}} \tag{9.204}$$

But for a minimum E, we require that

$$\frac{dE}{da} = \frac{3h^2}{2m} = \frac{e^2}{4\pi\varepsilon_0}\sqrt{\frac{2}{\pi a}} = 0$$

And therefore,

$$a = \frac{8}{9}\left[\frac{m^2e^4}{(4\pi\varepsilon_0)^2\ \pi h^4}\right] \tag{9.205}$$

Putting this value for 'a' in Eq. (9.204), we get

$$E = -\frac{8}{3\pi}\left[\frac{me^4}{2h^2(4\pi\varepsilon_0)^2}\right] = -0.85\left[\frac{me^4}{2h^2(4\pi\varepsilon_0)^2}\right] \tag{9.206}$$

This gives the energy of hydrogen atom with the help of the variation method of approximation.

BIBLIOGRAPHY

Atkins, P. and R. Friedman. 2007. *Molecular Quantum Mechanics*. New York: Oxford University Press

Eyring, H., J. Walter and G.E. Kimball. 1944. *Quantum Chemistry*. New York: Wiley

Lowe, J.P. and K.A. Peterson. 2006. *Quantum Chemistry*, 3rd ed. London: Academic Press

Margenau, H. and G.M. Murphy. 1943. *The Mathematics of Physics and Chemistry*. Princeton, NJ: D. Van Nostrand

McQuarrie, D.A. 2007. *Quantum Chemistry*. New Delhi: Viva Books Pvt. Ltd.

Pauling, L. and E.B. Wilson. 1935. *Introduction to Quantum Mechanics*. New York: McGraw-Hill Book Co. Inc.

Schiff, L.I. 1955. *Quantum Mechanics*. New York: McGraw-Hill

Solved Problems

Based on perturbation theory

Problem 1. Use the first-order perturbation to estimate the energy of a particle in a one-dimensional box from $x = 0$ to $x = a$. With a slanted bottom, such that $\frac{V_0 x}{a}, 0 \le x \le a$.

Solution: According to the question,

$$\psi_n^0 = \left(\frac{2}{a}\right)^{1/2} \quad \sin\frac{n\pi x}{a}$$

$$E_n^0 = \frac{n^2h^2}{8ma^2}$$

The perturbation Hamiltonian $H' = \frac{x}{a}V_0$.
V_0 = height of the potential at $x = a$
The Hamiltonian H for perturbed system = $H^0 + H'$
where H' = Hamiltonian due to perturbation

$$\therefore \quad E + E' = ?$$

$$\therefore \quad \text{Perturbed energy} = E'$$

$$\therefore \quad E' = \frac{\psi_n^0|H'|\psi_n^0}{\psi_n^0 \mid \psi_n^0} = \psi_n^0|H'|\psi_n^0 \quad = \psi_n^0\left|\frac{V_0 x}{a}\right|\psi_n^0$$

$$= \int_0^a \Psi_n^* \cdot \frac{V_0 x}{a} \cdot a \cdot \psi_n^0 \mathrm{d}x \overset{0}{} \quad = \int_0^a \left(\frac{2}{a}\right)^{1/2} \sin\frac{n\pi x}{a} \cdot \frac{V_0 x}{a} \cdot \int_0^a \left(\frac{2}{a}\right)^{1/2} \sin\frac{n\pi x}{a} \cdot \mathrm{d}x$$

$$= \int_0^a \left(\frac{2}{a}\right) \cdot \frac{V_0}{a} \cdot \sin^2\frac{n\pi x}{a} \cdot \mathrm{d}x = \left(\frac{2}{a}\right) \cdot \left(\frac{V_0}{2a}\right) \int_0^a x \cdot 2\sin^2\frac{n\pi x}{a} \cdot \mathrm{d}x = \frac{V_0}{a^2}\left[\int_0^a x\left(1 - \cos\frac{2n\pi x}{a}\right)\mathrm{d}x\right]$$

$$= \frac{V_0}{a^2}\left[\int_0^a x \cdot \mathrm{d}x - \int_0^a x \cdot \cos\frac{2n\pi x}{a}\mathrm{d}x\right] = \frac{V_0}{a^2}\left\{\left[\frac{x^2}{2}\right]_0^a - \int_0^q x \cdot \cos\frac{2n\pi x}{a}\mathrm{d}x\right\}$$

$$= \frac{V_0}{a^2}\left[\frac{a^2}{2} - \int_0^a x \cdot \cos\frac{2n\pi x}{a}\mathrm{d}x\right]$$

Now let

$$I = \int_0^a x \cdot \cos\frac{2n\pi x}{a}\mathrm{d}x$$

$$= x \cdot \int_0^a \cos\frac{2n\pi x}{a} - \int_0^a \left\{1 \cdot \int \cos\frac{2n\pi x}{a}\right\}\mathrm{d}x$$

This has been done according to integration by parts, i.e.,

$$\int (u \cdot v)\mathrm{d}x = u\int v\mathrm{d}x - \int\left\{\frac{\mathrm{d}u}{\mathrm{d}x}\int v.\mathrm{d}x\right\}\mathrm{d}x$$

Or $$I = x\left[\frac{\frac{\sin 2n\,\pi x}{a}}{\frac{2n\,\pi}{a}}\right]_0^a - \left[\int_0^a \frac{\frac{\sin 2n\,\pi x}{a}}{\frac{2n\,\pi}{a}}\right]\mathrm{d}x$$

$$= \frac{ax}{2n\pi} \cdot \left[\frac{\sin 2n\,\pi x}{a}\right]_0^a - \frac{ax}{2n\pi} \cdot \int_0^a \frac{\sin 2n\,\pi x}{a}\mathrm{d}x = \frac{ax}{2n\pi} \cdot [\sin 2n\,\pi x] - \frac{a^2}{4n^2\pi^2} \cdot \left[\cos\frac{2n\,\pi x}{a}\right]_0^a$$

$$= \frac{ax}{2n\pi} \cdot [\sin 2n\,\pi x] + \frac{a^2}{4n^2\pi^2} \cdot [\cos 2n\,\pi - \cos 0]_0^a = \frac{ax}{2n\pi} \cdot (\sin 2n\,\pi x) + \frac{a^2}{4n^2\pi^2}(1 - 1)$$

$$= 0 + 0 = 0$$

$$\therefore \quad I = 0$$

$$\text{or} \quad E' = \frac{V_0}{a^2}\left[\frac{a^2}{2} - 0\right] = \frac{V_0 a^2}{2a^2} = \frac{V_0}{2}$$

$$\therefore \quad \text{Total energy} = E_n^0 + E'$$

$$= \frac{n^2h^2}{8ma^2} + \frac{V_0}{2}$$

It is, thus, clear that in case of first-order perturbation, the energy is raised by a constant amount $V_0/2$, as illustrated in the aforementioned figure. Answer.

Problem 2. Apply first-order perturbation theory to the ground state of a system with the potential energy:

$$V(x) = \begin{cases} \infty \text{if } x < -a \text{ or } a < x \\ k\dfrac{x^2}{2} \text{if } -a < x < a \end{cases}$$

Solution: Here, we consider H^0 to be the particle in a box Hamiltonian with $-a \le x \le a$ not $0 \le x \le a$.

$$H' = \frac{kx^2}{2}$$

Here,

$$\psi_n^0 = \sqrt{\frac{2}{2a}}\sin\left(\frac{\pi(x+a)}{2a}\right);$$

$$E_n^0 = \frac{h^2}{8m(2a)^2} = \frac{h^2}{32ma^2}$$

$$\therefore \quad E' = \psi_n^0 | H' | \psi_n^0$$

$$= \int_{-a}^{+a} \sqrt{\frac{2}{2a}}\sin\left(\frac{\pi(x+a)}{2a}\right) \cdot \frac{kx^2}{2} \cdot \sqrt{\frac{2}{2a}}\sin\left(\frac{\pi(x+a)}{2a}\right) dx$$

$$= \left(\sqrt{\frac{2}{2a}}\right)\left(\sqrt{\frac{2a}{2a}}\right)\int_{-a}^{+a} \sin^2\left(\frac{\pi(x+a)}{2a}\right) \cdot \frac{kx^2}{2} dx$$

$$= \frac{1}{a}\int_{-a}^{+a} \frac{kx^2}{2} \cdot \sin^2\left(\frac{\pi(x+a)}{2a}\right) \cdot dx$$

Now, from trigonometric identity, we may write

$$\sin\left(\frac{\pi x}{2a} + \frac{\pi}{2}\right) = \sin\frac{\pi x}{2a} \cdot \cos\frac{\pi}{2} + \cos\frac{\pi x}{2a} \cdot \sin\frac{\pi}{2}$$

$$= 0 + \cos\frac{\pi x}{2a} \cdot 1 = \cos\frac{\pi x}{2a}$$

$\therefore$ The aforementioned equation may be expressed as

$$\frac{k}{2a}\int_{-a}^{+a} x^2 \cdot \sin^2\left(\frac{\pi(x+a)}{2a}\right)\cdot dx = \frac{k}{2a}\int_{-a}^{+a} x^2 \cdot \cos^2\left(\frac{\pi x}{2a}\right)\cdot dx$$

$$= \frac{k}{2a}\left(\frac{2a}{\pi}\right)^3 \cdot \int_{-\pi/2}^{+\pi/2} y^2 \cdot \cos^2 y \cdot dy$$

where $\frac{\pi x}{2a} = y \therefore \frac{\pi dx}{2a} = dy$ and $\therefore = \left(\frac{2ay}{\pi}\right)^2 = \frac{4a^2}{\pi^2}\cdot y^2$

$$\therefore \quad x^2 dx = \left(\frac{4a^2}{\pi^2}\cdot\right)\left(\frac{2a}{\pi}\right)dy = \frac{8a^3}{\pi^3}\cdot dy$$

$$= \left(\frac{2a}{\pi}\right)^3$$

a

$\therefore$ The aforementioned equation becomes

$$E' = \frac{8ka^2}{\pi^3}\int_0^{\pi/2} y^2 \cos^2 y \cdot dy$$

$$= \frac{8ka^2}{\pi^3}\left[\int_0^{\pi/2} y^2 dy - \int_0^{\pi/2} y^2 \sin^2 y \cdot dy\right]$$

Here, we have converted the integral from $-\frac{\pi}{2}$ to $+\frac{\pi}{2}$ to twice the integral from 0 to $\frac{\pi}{2}$.

The first integral $\int y^2 \; dy = \frac{y^3}{3}$ and the value of the standard integral $\int y^2 \sin^2 y dy$

$$= \frac{y^3}{6} - \left(\frac{y^2}{4} - \frac{1}{8}\right)\sin 2y - \frac{y\cos 2y}{4}$$

$\therefore$ Putting the values of the two integrals in the aforementioned equation, we get

$$E' = \frac{8ka^2}{\pi^3}\left[\frac{y^3}{3} - \left\{\frac{y^3}{6} - \left(\frac{y^2}{4} - \frac{1}{8}\right)\sin 2\, y - \frac{y\cos 2\, y}{4}\right\}\right]_0^{\pi/2}$$

$$= \frac{8ka^2}{\pi^3}\left[\frac{y^3}{3} - \frac{y^3}{6} + \left(\frac{y^2}{4} - \frac{1}{8}\right)\sin 2\, y + \frac{y\cos 2\, y}{4}\right]_0^{\pi/2}$$

$$= \frac{8ka^2}{\pi^3}\left[\frac{y^3}{6} + \left(\frac{y^2}{4} - \frac{1}{8}\right)\sin 2\, y + \frac{y\cos 2\, y}{4}\right]_0^{\pi/2}$$

$$= \frac{8ka^2}{\pi^3}\left[\frac{(\pi/2)^3}{6} + 0 - \frac{\pi}{8}\right] = \frac{8ka^2}{\pi^3}\left[\frac{\pi^3}{48} - \frac{\pi}{8}\right] = \left[\frac{ka^2}{6} - \frac{ka^2}{\pi^2}\right] = ka^2\left[\frac{1}{6} - \frac{1}{\pi^2}\right]$$

$$= ka^2\left[\frac{1}{6} - \frac{1}{9.86}\right] \approx ka^2\left[\frac{1}{6} - \frac{1}{10}\right] \approx \frac{ka^2}{15}$$

$$\therefore \quad E = E_n^0 + E' = \frac{h^2}{32ma^2} + \frac{ka^2}{15} = \frac{h^2}{32ma^2} + (0.066)ka^2 \qquad \text{Answer.}$$

Problem 3. Find the first-order correction to the ground-state energy of an anharmonic oscillator of mass '*m*', and angular frequency *w* is subjected to a potential $V(x) = 1/2mw^2x^2 + bx^4$
where b = a parameter independent of *x*. The ground-state wave function is

$$\psi_0^0 = \left(\frac{mw}{\pi h}\right)^{1/4} e^{-mwx^2/2h}$$

Solution: Given that $H' = bx^4$ and $\psi_0^0 = \left(\frac{mw}{\pi h}\right)^{1/4} e^{\frac{-mwx^2}{2h}}$

We have to find E_0^1.

The first-order correction to energy is given by

$$E_0^1 = \psi_0 |H'| \psi_0$$

$$= \int_{-\infty}^{+\infty} \left(\frac{mw}{\pi h}\right)^{1/4} e^{\frac{-mwx^2}{2h}} \cdot bx^4 \cdot \left(\frac{mw}{\pi h}\right)^{1/4} e^{\frac{-mwx^2}{2h}} dx$$

$$= \left(\frac{mw}{\pi h}\right)^{1/2} \cdot b \int_{-\infty}^{+\infty} x^4\, e^{\frac{-mwx^2}{h}} dx = \left(\frac{mw}{\pi h}\right)^{1/2} \cdot 2b \int_0^{\infty} x^4\, e^{\frac{-mwx^2}{h}} dx$$

Using the standard integral value, i.e.,

$$\int_0^{\infty} x^4\, e^{-ax^2} dx = \frac{3\sqrt{\pi}}{8} \cdot \frac{1}{a^{5/2}}$$

We can write by putting the value of integral as

$$E_0^1 = \left(\frac{mw}{\pi h}\right)^{1/2} \cdot 2b \cdot \frac{3\sqrt{\pi}}{8} \cdot \frac{1}{a^{5/2}}$$

$$= b\left(\frac{mw}{\pi h}\right)^{1/2} \cdot \frac{3\sqrt{\pi}}{4} \cdot \left(\frac{1}{\left(\frac{mw}{\pi h}\right)^{5/2}}\right)$$

$$= b\left(\frac{mw}{\pi h}\right)^{1/2} \cdot \frac{3\sqrt{\pi}}{4} \cdot \left(\frac{h}{mw}\right)^{5/2}$$

$$= \frac{3bh^2}{4m^2w^2}$$

After simplification,

$$E_0^1 = \frac{3bh^2}{4m^2w^2} \qquad \text{Answer.}$$

Problem 4. A particle is trapped in an infinite square well of bottom a, and its unperturbed wave function is $\psi_n^0 = \left(\frac{2}{a}\right)^{1/2} \sin\frac{n\pi x}{a}$. If the system is perturbed by raising the floor of the well by a constant quantity V_0, then calculate the first- and second-order corrections to the energy of nth state.

Solution: Given that $H' = V_0$ (constant)

The first-order correction to the energy of nth state is given by

$$E_n' = \psi_n^0 | H' | \psi_n^0 = \psi_n^0 | V_0 | \psi_n^0$$

$$= V_0 \psi_n^0 \mid \psi_n^0 = V_0 \left[\because\ \psi_n^0 \mid \psi_n^0\right] = 1$$

Thus, the corrected energy levels are raised by V_0.

Now, we find out the second-order correction. We know that

$$E_n'' = \sum_m{}' \frac{\left|\psi_m^0 | H' | \psi_n^0\right|^2}{E_n^0 - E_m^0}$$

$$= \sum_m{}' V_0^2 \left|\psi_m^0 \mid \psi_n^0\right|^2$$

$$= V_0^2 \sum \left|\psi_m^0 \mid \psi_n^0\right|^2$$

$$= V_0^2 x 0 = 0 \left[\because\ \psi_n^0 \mid \psi_n^0 = 0\right]$$

$$\therefore E_n'' = 0$$

Hence, the second-order correction to energy amounts to zero. Answer.

Problem 5. Due to electronic difference, the p-electron in a bond between two different atoms such as in $C = N^-$ ion does not behave exactly similar to a particle in a flat box but like a particle in a box, which has a slightly higher potential energy on one side than the other, considering a perturbation of $H' = Kx$ for the ground state ψ_1 of a particle in a box system.

a. Sketch the perturbed system.
b. Assuming that the only correction to the real ground-state wave function is the second particle in a box wave function ψ_2, find out the coefficient a_2 and calculate the first-order corrected wave function.

Solution:

a. The system will look like following figure in which the sloped line represents the real bottom of the box.

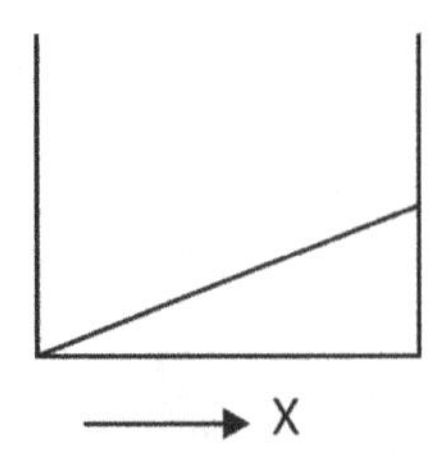

b. In order to get the value of a_2, we are required to evaluate the following expression:

$$a_2 = \frac{\int_0^a \psi_2^* \cdot kx \cdot \psi_1 \, dx}{E_1 - E_2}$$

where ψ_2^* = wave function of Particle in a box and
ψ_1 = wave function of Particle in a box.

The values of ψ_2^* and ψ_1 are known, and the corresponding energies are also known substituting the values of

$$\psi_2^* = \left(\frac{2}{a}\right)^{1/2} \sin\frac{2n\pi}{a}$$

$$\psi_1 = \left(\frac{2}{a}\right)^{1/2} \sin\frac{n\pi}{a}$$

$$E_1 = \frac{h^2}{8ma^2} \text{ and } E_2 = \frac{2^2 h^2}{8ma^2} \quad \left[\because \; E = \frac{n^2 h^2}{8ma^2}\right]$$

In the aforementioned expression for a_2

$$a_2 = \frac{\int_0^a \left(\frac{2}{a}\right)^{1/2} \sin\frac{2\pi x}{a} \cdot kx \cdot \left(\frac{2}{a}\right)^{\frac{1}{2}} \sin\frac{\pi x}{a} dx}{\frac{h^2}{8ma^2} - \frac{2^2 h^2}{8ma^2}} = \frac{\left(\frac{2k}{a}\right)\int_0^a x \cdot \sin\frac{2\pi x}{a} \cdot \sin\frac{\pi x}{a} dx}{\frac{-3h^2}{8ma^2}}$$

or

$$a_2 = \frac{\left(\frac{2k}{a}\right)\frac{1}{2}\int_0^a 2x \cdot \sin\frac{2\pi x}{a} \cdot \sin\frac{\pi x}{a} dx}{\frac{-3h^2}{8ma^2}}$$

$$= \frac{\left(\frac{k}{a}\right)\int_0^a \left[x \cdot 2\sin\frac{2\pi x}{a} \cdot \sin\frac{\pi x}{a} dx\right]}{\frac{-3h^2}{8ma^2}}$$

a

$$= \frac{\left(\frac{k}{a}\right)\int_0^a x\left[\cos\frac{\pi x}{a} - \cos\frac{3\pi x}{a}\right] dx}{\frac{-3h^2}{8ma^2}}$$

$$\left[\because \; 2\sin ax \cdot \sin bx = \cos(a-b) - \cos(a+b)\right]$$

$$= \frac{\left(\frac{k}{a}\right)\left[\int_0^a x\cos\frac{\pi x}{a} dx - \int_0^a x\cos\frac{3\pi x}{a} dx\right]}{\frac{-3h^2}{8ma^2}}$$

Putting the values of the standard integrals, we can write

$$a_2 = \frac{\frac{k}{a}\left[\frac{a^2}{\pi^2}\cos\frac{\pi x}{a} + \frac{ax}{\pi}\sin\frac{\pi x}{a} - \frac{a^2}{9\pi^2}\cos\frac{3\pi x}{a} - \frac{ax}{3\pi}\sin\frac{3\pi x}{a}\right]_0^a}{\frac{-3h^2}{8ma^2}}$$

$$= \frac{\frac{k}{a}\left[-\frac{a^2}{\pi^2} + 0 + \frac{a^2}{9\pi^2} - 0 - \frac{a^2}{\pi^2} - 0 + \frac{a^2}{9\pi^2} + 0\right]}{\frac{-3h^2}{8ma^2}} = \frac{\frac{k}{a}\left[-\frac{2a^2}{9\pi^2} - \frac{2a^2}{\pi^2}\right]}{\frac{-3h^2}{8ma^2}} = \frac{\frac{k}{a}\cdot\frac{2a^2}{\pi^2}\left[\frac{1}{9} - \frac{1}{1}\right]}{\frac{-3h^2}{8ma^2}}$$

$$= \frac{-\frac{2a^2k}{\pi^2 a}\cdot\frac{8}{9}}{\frac{-3h^2}{8ma^2}} = \frac{128kma^3}{27\pi^2 2h^2}$$

$$\therefore \quad a_2 = \frac{128kma^3}{27\pi^2 2h^2}$$

The approximate wave function may be expressed as

$$\psi_1,\text{ real } \approx \psi_1,\text{ PIAB+} = \frac{128kma^3}{27\pi^2 2h^2}\cdot\psi_2,\text{ PIAB}$$

Putting the values for a cyanide species $m = m_e$, $k = 1\times10^{-7}kg^m/s^2$

$$a \approx 1.15\text{Å}\left(1.15\times10^{-10}m\right),$$

We can evaluate the aforementioned expression and obtain

$$\psi_1,\text{ real } \approx \psi_1\text{ PIAB} + 0.1516\cdot\psi_2\text{ PIAB}$$

where PIAB = Particle in a box. Answer.

Problem 6. Suppose that an electron is present in a potential box having the length a. When an electric field ε is turned on in the × direction, the electron experiences a force of magnitude $-e\varepsilon$, and the potential function has added to it the term $+e\varepsilon x$. The potential then has the form shown in the figure shown in the following:

a. What is the lowest allowed energy in a first-order approximation for the electron? It may be assumed that $e\varepsilon a << $ ground-state energy in the absence of the electric field.
b. Use first-order perturbation theory to obtain an approximation to the ground-state wave function and evaluate the first term in correction.

Solution:

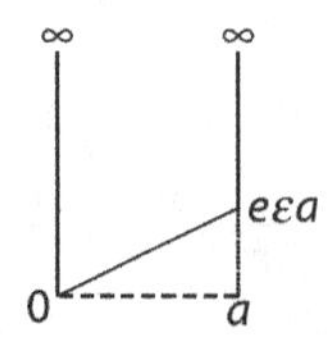

a. For the aforementioned considered system, let $H = H^0 + H'$

where $H^0 = \dfrac{-h^2}{2m} \cdot \dfrac{d^2}{dx^2}$ and $H' = e\varepsilon x$.

We know that $H_0\psi_0 = E_0\psi_0$ and $E_n = E_n^0 + E_n'$, where $E_n^0 = \dfrac{n^2h^2}{8ma^2}$

$$\text{and} \quad \psi_n^0 = \left(\frac{2}{a}\right)^{1/2} \sin\frac{n\pi x}{a}$$

$$\therefore \quad E_n' = \int_0^a \left(\psi_n^0\right)^* H'\psi_n^0 dx$$

$$= \int_0^a \left(\psi_n^0\right)^* \cdot e\varepsilon x\psi_n^0 dx$$

$$= e\varepsilon x \text{ where } x = \int_0^a \left(\frac{2}{a}\right)^{1/2} \sin\frac{n\pi x}{a} \cdot x\left(\frac{2}{a}\right)^{1/2} \sin\frac{n\pi x}{a} dx$$

$$= \int_0^a \left(\frac{2}{a}\right) x \sin^2\frac{n\pi x}{a} dx$$

$$\therefore \quad E_n' = e\varepsilon \int_0^a \left(\frac{2}{a}\right) x \sin^2\frac{n\pi x}{a} dx$$

$$= \left(\frac{2e\varepsilon}{a}\right)\int_0^a x\sin^2\frac{n\pi x}{a} dx = \frac{2e\varepsilon}{a}\int_0^a x\sin^2\frac{n\pi x}{a} dx$$

$$= \frac{2e\varepsilon}{a} \cdot \frac{a^2}{4} \quad = \left[\because \int_0^a x\sin^2\frac{n\pi x}{a} dx = \frac{a^2}{4}\right]$$

$$= \frac{e\varepsilon a}{2}$$

Therefore, $E_1 = E_1^0 + E_1^1 = \dfrac{h^2}{8ma^2} + \dfrac{e\varepsilon a}{2}$ Answer.

b. We know that the wave function from first-order perturbation theory is expressed as

$$\psi_1 = \psi_1^0 + \sum_{k \neq 1} \frac{k|H'|1}{E_l^0 - E_k^0}\psi_k^0$$

$$\therefore \quad \text{Correction term} = \frac{2|H'|1}{E_1^0 - E_2^0} = \frac{\int_0^a \left(\psi_2^0\right)^* e\ \varepsilon x\left(\psi_1^0\right)dx}{\left(\dfrac{h^2}{8ma^2} - \dfrac{4h^2}{8ma^2}\right)}$$

where $E_1^0 = \dfrac{1^2 \cdot h^2}{8ma^2}$ and $E_2^0 = \dfrac{2^2 \cdot h^2}{8ma^2}$

$$= \frac{\int_0^a \left(\psi_2^0\right)^* e\ \varepsilon x\left(\psi_0^1\right)\mathrm{d}x}{\dfrac{-3h^2}{8ma^2}}$$

Now, $\int_0^a \left(\psi_2^0\right)^* e\ \varepsilon x\left(\psi_0^1\right)\mathrm{d}x = \dfrac{2e\ \varepsilon}{a}\int_0^a \sin\dfrac{2\pi x}{a}\cdot x\sin\dfrac{\pi x}{a}\mathrm{d}x$

Let us put $y = \dfrac{\pi x}{a}$ such that

$$x = \frac{ay}{\pi}$$

$$\text{or}, \mathrm{d}x = \frac{a}{\pi}\mathrm{d}y$$

Putting the value of x and dx in the aforementioned integral, we get

$$\int_0^a \left(\psi_2^0\right)^* e\varepsilon x\left(\psi_1^0\right)\mathrm{d}x$$

$$= \frac{2e\ \varepsilon}{a}\int_0^a \sin\frac{2\pi x}{a}\cdot x\sin\frac{\pi x}{a}\mathrm{d}x$$

$$= \frac{2e\ \varepsilon}{a}\int_0^a \sin 2y\cdot\frac{ay}{\pi}\cdot\sin y\cdot\frac{a}{\pi}\mathrm{d}y$$

$$= \left(\frac{2e\ \varepsilon}{a}\right)\left(\frac{a^2}{\pi^2}\right)\int_0^\pi \sin 2y\cdot y\cdot\sin y\mathrm{d}y$$

$$= \left(\frac{2e\ \varepsilon}{a}\right)\left(\frac{a^2}{\pi^2}\right)\int_0^\pi y\cdot 2\sin y\cos y\cdot\sin y\mathrm{d}y$$

$$= \left(\frac{2e\ \varepsilon}{a}\right)\left(\frac{a^2}{\pi^2}\right)\int_0^\pi 2y\sin^2 y\cos y\mathrm{d}y$$

$$= \left(\frac{2e\ \varepsilon}{a}\right)\left(\frac{a^2}{\pi^2}\right)\int_0^\pi y\sin^2 y\cos y\mathrm{d}y$$

$$= \left(\frac{2e\ \varepsilon}{a}\right)\left(\frac{2a^2}{\pi^2}\right)\left[\int_0^\pi \left(y\cos y\mathrm{d}y - y\cos^3 y\mathrm{d}y\right)\right] \quad \text{since } \left[\sin^2 y = 1-\cos^2 y\right]$$

Let $I_1 = \int_0^{\pi} y \cos y dy$ and $I_2 = \int_0^{\pi} \cos^3 y dy$

Now, we solve the integrals one by one.

$$I_1 = \int_0^{\pi} y \cos y dy$$

$$= \int_0^{\pi} \left[y \cdot \sin y - \left(\int 1 \cdot \int \cos y dy \right) dy \right]$$

$$= \left[\left(y \cdot \sin y + \cos y \right) \right]_0^{\pi}$$

$$= \left(0 + \cos \pi - 0 - \cos y \right) = (-1 - 1) = -2$$

$$I_2 = \int_0^{\pi} y \cos y^3 dy'$$

$$= \int_0^{\pi} y \left(\frac{1}{4} \cos 3y + \frac{3}{4} \cos y \right) dy \quad \left[\because \cos 3y = 4 \cos^3 y - 3 \cos y \right]$$

$$= \frac{1}{4} \left[y \cdot \int \cos 3y \ dy - \int \left(\int \cos 3y \ dy \right) dy \right]_0^{\pi}$$

$$= \frac{1}{4} \left[y \frac{\sin 3y}{3} + \frac{\cos 3y}{9} \right]_0^{\pi}$$

$$= \frac{1}{4} \left[0 + \frac{\cos 3\pi}{9} - 0 - \frac{\cos 0}{9} \right] = \frac{1}{4} \left[\frac{-1}{9} - \frac{1}{9} \right] = \frac{1}{4} \left[\frac{-2}{9} \right]$$

Now putting the values of I1 and I2 in the aforementioned equation, we obtain the correction term

$$= \frac{\left(\frac{2e\varepsilon}{a} \right) \left(\frac{2a^2}{\pi^2} \right) (-2) \left(\frac{1}{4} \times \frac{(-2)}{9} \right)}{\frac{-3h^2}{8ma^2}} = \frac{\frac{-2e \ \varepsilon}{a} \cdot \frac{2a^2}{\pi^2} \cdot \frac{1}{9}}{\frac{3h^2}{8ma^2}}$$

$$= \left(\frac{-2e \ \varepsilon a}{\pi^2} \right) \left(\frac{1}{9} \right) \left(\frac{8ma^2}{3h^2} \right)$$

$$= \left(\frac{-2e \ \varepsilon a}{3\pi^2} \right) \frac{8}{9} \left(\frac{ma^2}{h^2} \right)$$

$$= -0.889 \left(\frac{2e \ \varepsilon a}{3\pi^2} \right) \left(\frac{ma^2}{h^2} \right) \text{ and finall}$$

$$\psi_1 \cong \psi_1^0 + (-0.889) \left(\frac{2}{3\pi^2} \right) (e \cdot \varepsilon a) \left(\frac{ma^2}{h^2} \right) \psi_2^0$$

$$\text{or,}\quad \psi_1 \cong \psi_1^0 + \left(\frac{1.778}{3\pi^2}\right)(e\cdot\varepsilon a)\left(\frac{ma^2}{h^2}\right)\psi_2^0 \qquad \text{Answer.}$$

Problem 7. A rotor with a moment of inertia I and electric dipole moment μ executes rotational motion in a plane. Find the first- and second-order corrections to the energy levels when the rotor is acted on by an electric field ε in the plane of rotation. The energy Eigen values and Eigen functions of a plane of rotor are

$$E_m = \frac{\hbar^2 m^2}{2I}, m = 0, \pm1, \pm2$$

And normalised Eigen functions are

$$\psi(\phi) = \frac{1}{\sqrt{2}} e^{im\phi}, m = 0, \pm1, \pm2$$

Solution: Given that $E_m = \dfrac{h^2 m^2}{2I}$ and $\psi(\phi) = \dfrac{1}{\sqrt{2\pi}} e^{im\phi}$

The perturbation = H'

$$= -\mu\varepsilon\cos\phi = \frac{-\mu\varepsilon}{2}\left(e^{i\phi} + e^{-i\phi}\right)$$

We also have $\langle n/H'/n\rangle$

$$= \int_0^{2\pi} \frac{1}{\sqrt{2\pi}} e^{-im\phi}\left(-\mu\varepsilon\cos\phi\right) \times \frac{1}{\sqrt{2\pi}} e^{im\phi}\,d\phi$$

$$= \frac{-1}{2\pi}\int_0^{2\pi} e^{-im\phi}\cdot e^{+im\phi}\cdot \mu\varepsilon\phi\,d\phi$$

$$= \frac{-\mu\varepsilon}{2\pi}\int_0^{2\pi} \cos\phi\ d\phi$$

$$\therefore\quad E_n' = n/H'/n = \frac{-\mu\varepsilon}{2\pi}\int_0^{2\pi}\cos\phi\ d\phi$$

$$= \frac{-\mu\varepsilon}{2\pi}\left[\sin\phi\right]_0^{2\pi} = 0$$

$$\text{And } E_n'' = \sum_m' \frac{|n/H'/m|^2}{E_n^0 - E_m^0}$$

$$\text{where}\quad |n/H'/m| = \frac{-\mu\varepsilon}{4\pi}\int_0^{2\pi} e^{-in\phi}\left(e^{i\phi} + e^{-i\phi}\right)e^{im\phi}\,d\phi$$

$$= \frac{-\mu\varepsilon}{4\pi}\left[\int_0^{2\pi} e^{-(m+1-n)\phi}\,d\phi + \int_0^{2\pi} e^{-(m+1-n)\phi}\,d\phi\right]$$

The integrals will be finite when $m = n - 1$ and when $m = n + 1$

$$\therefore \quad E_n'' = \left(\frac{\mu\varepsilon}{4\pi}\right)^2 \left(\frac{4\pi^2}{E_n^0 - E_{n-1}} + \frac{4\pi^2}{E_n^0 - E_{n+1}}\right)$$

$$= \left(\frac{\mu\varepsilon}{4\pi}\right)^2 \cdot \frac{4\pi^2 \cdot 2I}{\hbar^2}\left(\frac{1}{2n-1} - \frac{1}{2n+1}\right)$$

$$= \frac{\mu^2\varepsilon^2}{16\pi^2} \cdot \frac{8\pi^2 I}{\hbar^2}\left(\frac{2n+1-2n+1}{4n^2-1}\right)$$

$$= \frac{\mu^2\varepsilon^2 I}{\hbar^2\left(4n^2-1\right)}$$

Answer.

Problem 8. A rigid rotor is perturbed in a plane, and the perturbation is $H' = V_0/2\ (3 \cos 2\phi - 1)$, where V_0 = constant. Estimate the ground-state energy up to second order in the perturbation.
Solution: The energy Eigen value and Eigen functions are represented by

$$E_m = \frac{m^{2h}}{2I}, m = 0, \pm 1, \pm 2$$

$$\text{and } \psi_m(\phi) = \frac{1}{\sqrt{2\pi}} e^{im\phi}$$

In this case, all levels are double degenerate except the ground state.
∴ The first-order correction to the ground-state energy may be given by

$$E_0' = \psi / H' / \psi = \psi \left| \frac{V_0}{2}\left(3\cos^2\phi - 1\right) \right| \psi$$

$$= \psi \left| \frac{3V_0}{2}\left(\cos^2\phi\right) \right| \psi - \psi \left| \frac{3V_0}{2} \right| \psi$$

$$= \frac{3V_0}{4} - \frac{V_0}{2} = \frac{V_0}{4} = \text{first-order correction to energy}$$

$$\text{And} \qquad E_0'' = \sum_m \frac{\left|\langle 0/H'/m\rangle\right|^2}{E_0^0 - E_m^0}$$

$$\text{But} \quad \langle 0 / H' / m\rangle = \frac{V_0}{2}\int_0^{2\pi} \frac{1}{\sqrt{2\pi}}\left(3\cos^2\phi - 1\right) \cdot \frac{1}{\sqrt{2\pi}} e^{im\phi} d\phi$$

$$= \frac{3V_0}{4\pi}\int_0^{2\pi} \cos^2\phi e^{im\phi} d\phi - \frac{V_0}{4\pi}\int_0^{2\pi} e^{im\phi} d\phi$$

Using $\cos^2\phi = \left(\frac{1+\cos 2\phi}{2}\right)$ and the second integral will vanish

$$\therefore \quad \langle 0/H'/m \rangle = \frac{3V_0}{8\pi} \int_0^{2\pi} (1+\cos 2\phi) e^{im\phi} \mathrm{d}\phi$$

$$= \frac{3V_0}{8\pi} \int_0^{2\pi} \cos 2\phi e^{im\phi} \mathrm{d}\phi$$

because the other integrals will vanish.

Expressing cos 2ϕ in exponential form, we obtain

$$\langle 0/H'/m \rangle = \frac{3V_0}{8\pi} \int_0^{2\pi} \left(e^{i2\phi} + e^{-i2\phi}\right) e^{im\phi} \mathrm{d}\phi$$

$$= \frac{3V_0}{16\pi} \int_0^{2\pi} e^{i(m+2)\phi} \mathrm{d}\phi + \frac{3V_0}{16\pi} \int_0^{2\pi} e^{i(m-2)\phi} \mathrm{d}\phi$$

The value of the first integral will be finite if $m = -2$ and second integral will be finite if $m = +2$, and their values will be equal to $3V_0/8$.

$$E_{\pm 2} = \frac{2\hbar^2}{I}, E_0 = 0$$

Hence,

$$E_0^0 - E_2^0 = E_0^0 - E_{0-2}^0 = \frac{-2\hbar^2}{I}$$

$$\text{and } E_0'' = \frac{\left(\frac{3V_0}{8}\right)^2}{-2\hbar^2/I} + \frac{\left(\frac{3V_0}{8}\right)^2}{-2\hbar^2/I} = \frac{-9V_0^2 I}{64\hbar^2}$$

Answer.

Problem 9. An SHO has reduced mass μ and force constant k. The SHO is subjected to the quartic perturbation $H' = ax^4$. Obtain an expression for the first-order correction to the nth Eigen value of this oscillator, and also obtain the first nonvanishing term in the first-order correction for the ground-state wave function.

Solution: Given that the quartric perturbation $H' = ax^4$ and the Hamiltonian

$$H = H^0 + H' = \left(\frac{-h^2}{2\mu} \frac{\mathrm{d}^2}{\mathrm{d}x^2} + \frac{1}{2}kx^2 \right) + ax^4$$

$$E_n^0 = (n + /1_2) hv$$

$$\psi_n^0(x) = \left(\frac{\infty}{2^n \cdot n!} \frac{1}{\sqrt{\pi}} \right)^{1/2} H_n(\xi) e^{-\xi^2/2}$$

Let us put $\propto x = \xi$

$$\therefore \quad \infty \, \mathrm{d}x = \mathrm{d}\xi \quad \therefore \quad \mathrm{d}x = \mathrm{d}\xi / \alpha$$

$$\text{But } E_n' = n/ax^4/n$$

$$= \int_{-\infty}^{+\infty} (\psi_n^0)^* |ax^4| \psi_n^0 \mathrm{d}x$$

$$= \int_{-\infty}^{+\infty} \left(\frac{\propto}{2^n \cdot n!\sqrt{\pi}}\right)^{1/2} H_n e^{-\xi^2/2} \cdot ax^4 \cdot \left(\frac{\propto}{2^n \cdot n!\sqrt{\pi}}\right)^{1/2} H_n e^{-\xi^2/2} \mathrm{d}\xi$$

$$= \int_{-\infty}^{+\infty} \left(\frac{\propto}{2^n \cdot n!\sqrt{\pi}}\right) \frac{a}{\propto^5} \cdot H_n^2 \cdot \xi^4 \cdot e^{-\xi^2} \mathrm{d}\xi \; [\because \alpha x = \xi]$$

$$= \left(\frac{\propto}{2^n \cdot n!\sqrt{\pi}}\right)\left(\frac{a}{\propto^5}\right) \int_{-\infty}^{+\infty} H_n^2 \cdot \xi^4 \cdot e^{-\xi^2} \mathrm{d}\xi \qquad \text{(i)}$$

From Recursion formula, we can write

$$\xi Hn = \frac{H_{n+1}}{2} + nH_{n-1} \qquad \text{(ii)}$$

$$\text{or, } \xi H_n = \frac{\xi H_{n+1}}{2} + \xi n H_{n-1} \qquad \text{(iii)}$$

When $n \rightarrow n + 1$, Eq. (ii) takes the form

$$\xi H_{n-1} = \frac{H_{n+2}}{2} + n + 1H_n \qquad \text{(iv)}$$

When $n \rightarrow n - 1$, Eq. (ii) becomes

$$\xi H_{n-1} = \frac{H_n}{2} + (n-1)H_{n-2} \qquad \text{(v)}$$

Substituting these values in Eq. (iii), we get

$$\xi^2 H_n = \frac{1}{2}\left[\frac{H_{n+2}}{2} + (n+1)H_n\right] + n\left[\frac{H_n}{2} + (n+1)H_{n-2}\right]$$

$$\text{or } \xi^2 H_n = \frac{H_{n+2}}{4} + \frac{(n+1)H_n}{2} + \frac{nH_n}{2} + (n-1)H_{n-2}$$

$$\text{or } \xi^2 H_n = \frac{H_{n+2}}{4} + \frac{(2n+1)H_n}{2} + n(n-1)H_{n-2} \qquad \text{(vi)}$$

Squaring Eq. (vi), we obtain

$$\xi^4 H_n^2 = \frac{H_{n+2}^2}{16} + \left(\frac{2n+1}{2}\right)^2 H_n^2 + \{n(n-1)\}^2 H_{n-2}^2 \text{ Product terms} \qquad \text{(vii)}$$

From expressions (i) and (vii), we have

$$E_n^1 = \left(\frac{\propto}{2^n \cdot n!\sqrt{\pi}}\right)\left(\frac{a}{\propto^5}\right)\int_{-\infty}^{+\infty}\left[\frac{H_{n+2}^2}{16} + \left(\frac{2n+1}{2}\right)^2 \frac{H_n^2}{4} + \{n(n-1)\}^2 H_{n-2}^2 + \text{product terms}\right] e^{-\xi^2} d\xi$$

$$= \left(\frac{\propto}{2^n \cdot n!\sqrt{\pi}}\right)\left(\frac{a}{\propto^5}\right)\left[\int_{-\infty}^{+\infty}\frac{H_{n+2}^2}{16} e^{-\xi^2} d\xi + \left(\frac{2n+1}{2}\right)^2 H_n^2 e^{-\xi^2}\ d\xi + \{n(n-1)\}^2 H_{n-2}^2 - e^{-\xi^2} d\xi\right]$$

$+0.$ Product terms

$$= \left(\frac{\propto}{2^n \cdot n!\sqrt{\pi}}\right)\left(\frac{a}{\propto^5}\right)\left[\int_{-\infty}^{+\infty}\frac{H_{n+2}^2}{16} e^{-\xi^2} d\xi + \left(\frac{2n+1}{2}\right)^2 H_n^2 e^{-\xi^2}\ d\xi + \{n(n-1)\}^2 H_{n-2}^2 - e^{-\xi^2} d\xi\right] \quad \text{(viii)}$$

under orthonormalisation condition,

$$\text{But} \int_{-\infty}^{+\infty} H_{n+2}^2\ e^{-\xi^2} d\xi = 2^{n-2}(n-2)!\sqrt{\pi}$$

$$\int_{-\infty}^{+\infty} H_n^2\ e^{-\xi^2} d\xi = 2^n n!\sqrt{\pi}$$

$$\text{and} \int_{-\infty}^{+\infty} H_{n-2}^2 - e^{\xi^2} d\xi = 2^{n-2}(n-2)!\sqrt{\pi}$$

Substituting these values in Eq. (viii), we have

$$E_n^1 = \left(\frac{\propto}{2^n \cdot n!\sqrt{\pi}}\right)\left(\frac{a}{\propto^5}\right)\left[\frac{2^{n+2}\cdot(n+2)!\sqrt{\pi}}{16} + \left(\frac{2n+1}{2}\right)^2 \cdot 2^n \cdot n!\sqrt{\pi} + \{n(n-1)\}^2 \cdot 2^n \cdot n!\sqrt{\pi}\, 2^{n-2}(n-2)!\sqrt{\pi}\right]$$

$$E_n^1 = \left(\frac{\propto}{2^n \cdot n!\sqrt{\pi}}\right)\left(\frac{a}{\propto^5}\right)2^n \cdot n!\sqrt{\pi}\left[\frac{2^2\cdot(n+1)(n+2)}{16} + \left(\frac{2n+1}{2}\right)^2 \frac{\{n(n-1)\}^2 2^{-2}}{n(n-1)}\right]$$

$$\text{or}\quad E_n^1 = \frac{a}{\propto^4}\left[\frac{(n+1)(n+2)}{4} + \left(\frac{2n+1}{4}\right)^2 + \frac{n(n-1)}{4}\right]$$

$$\text{or}\quad E_n^1 = \frac{a}{4\propto^4}\left[n^2 + 3n + 2 + 4n^2 + 4n + 1 + n^2 - n\right]$$

$$\text{or}\quad E_n^1 = \frac{a}{4\propto^4}\left[6n^2 + 6n + 3\right]$$

$$\text{or}\quad E_n^1 = \frac{3a}{4\propto^4}\left[2n^2 + 2n + 1\right]$$

$$= \frac{3a}{2\propto^4}\left[n^2 + n + 1/2\right]$$

This is the expression for first-order energy correction.

Wave function: The ground-state Eigen function corrected to the first order will be

$$\psi_n^1 = \psi_n^0 + \sum_{k \neq n} \frac{k|H'|n}{\left[E_n^0 - E_k^0\right]} \psi_k^0$$
$$= \psi_n^0 + \frac{1|H'|0}{E_0^0 - E_1^0} \psi_1^0 + \psi_n^0 + \frac{2|H'|0}{E_0^0 - E_2^0} \psi_2^0 + \frac{3|H'|0}{E_0^0 - E_3^0} \psi_3^0 + \ldots\ldots \tag{ix}$$

where the second and fourth terms on the right side of the aforementioned expression vanish because the integrals are odd function of x, but the third term will not vanish.

$$\therefore \psi_n^0 + \frac{2|H'|0}{E_0^0 - E_2^0} \psi_2^0 \tag{x}$$

We know that $E_n^0 = (n + 1/2)h\upsilon$

$$\therefore E_0^0 = \frac{2\upsilon}{2}$$
$$E_1^0 = 3/2\ h\upsilon$$
$$E_2^0 = 5/2\ h\upsilon$$
$$E_3^0 = 7/2\ h\upsilon$$

$$\therefore E_0^0 - E_1^0 = -h\upsilon$$
$$E_0^0 - E_2^0 = -2h\upsilon$$

$$\text{and } E_0^0 - E_3^0 = -3h\upsilon$$

It is also known that

$$\psi_n^0 = \left(\frac{\propto}{2^n \cdot n!\sqrt{\pi}}\right) H_n e^{-\xi^2/2}$$

$$\psi_0^0 = \left(\frac{\propto}{\sqrt{\pi}}\right) H_0 e^{-\xi^2/2} = \frac{\propto}{\sqrt{\pi}} e^{-\xi^2/2}$$

$$\psi_1^0 = \left(\frac{\propto}{2\sqrt{\pi}}\right) H_1 e^{-\xi^2/2} \left[\text{where } H_0,\ H_1 \text{ and } H_2 \text{ are values of Hermite polynomials}\right]$$

$$\psi_2^0 = \left(\frac{\propto}{2^2 2!\sqrt{\pi}}\right) H_2 e^{-\xi^2/2}$$

These values, when put, will give the first-order correction to wave function. Answer.

Based on Variation Theory

Problem 10. Estimate the variational energy of a particle in a hard one-dimensional box of length a, using the variation function $\psi(x) = Ax(a - x)$.

Solution: First of all, we normalise the wave function as given in the following.

$$\int_0^a \psi^*(x)\psi(x)\mathrm{d}x = 1$$

$$\text{or } \int_0^a Ax(a-x)\cdot Ax(a-z)\ \mathrm{d}x = 1$$

$$\text{or } A^2\int_0^a x^2(a-x)^2\ \mathrm{d}x = 1$$

$$\text{or } A^2\int_0^a x^2\left(a^2+x^2-2ax\right)\ \mathrm{d}x = 1$$

$$\text{or } A^2\int_0^a \left(a^2x^2+x^4-2ax^3\right)\ \mathrm{d}x = 1$$

$$\text{or } A^2\left[\int_0^a a^2x^2\ \mathrm{d}x + \int_0^a x^4\ \mathrm{d}x - \int_0^a 2a\ x^3\ \mathrm{d}x\right] = 1$$

$$\text{or } A^2\left[\frac{a^2x^3}{3}+\frac{x^5}{5}-2a\frac{x^4}{4}\right]_0^a = 1$$

$$\text{or } A^2\left[\frac{a^5}{3}+\frac{a^5}{5}-\frac{2a^5}{4}\right] = 1$$

$$\text{or } A^2a^5\left[\frac{1}{3}+\frac{1}{5}-\frac{1}{2}\right] = 1$$

$$\text{or } A^2a^5\left[\frac{10+6-15}{30}\right] = 1$$

$$\text{or } \frac{A^2a^5}{30} = 1$$

$$\text{or } A^2 = \frac{30}{a^5} \quad \text{or, } A = \left(\frac{30}{a^5}\right)^{1/2}$$

According to the question, there are no variable parameters, and hence, the variational energy will not be minimised.

$$\therefore E = -A^2 \frac{\hbar^2}{2m} \int_0^a x(a-z) \frac{d^2}{dx^2} [x(a-z)dx] dx$$

$$= -A^2 \frac{\hbar^2}{2m} \int_0^a x(a-x) \frac{d}{dx} \left[\frac{d}{dx} (ax - x^2) \right] dx$$

$$= -A^2 \frac{\hbar^2}{2m} \int_0^a x(a-x) \frac{d}{dx} [a - 2x] dx$$

$$= -A^2 \frac{\hbar^2}{2m} \int_0^a x(a-x)(-2) dx$$

$$= \frac{A^2 \hbar^2}{m} \int_0^a (ax - x^2)\, dx$$

$$= \frac{A^2 \hbar^2}{m} \left[\int_0^a ax\, dx - \int_0^a x^2\, dx \right]$$

$$= \frac{A^2 \hbar^2}{m} \left[\frac{ax^2}{2} - \frac{x^3}{3} \right]_0^a$$

$$= \frac{A^2 \hbar^2}{m} \left[\frac{a^3}{2} - \frac{a^3}{3} \right] = \frac{A^2 \hbar^2}{m} \cdot \frac{a^3}{6}$$

$$= \frac{30}{a^5} \frac{\hbar^5}{m} \cdot \frac{a^3}{6} = \frac{5\hbar^2}{ma^2} = \frac{5\left(\frac{h}{2\pi} \right)^2}{ma^2}$$

$$\frac{5h^2}{4\pi^2 ma^2} = \frac{5}{4\pi^2} \cdot \frac{h^2}{ma^2} = 0.12665 \frac{h^2}{ma^2}$$

$$\text{whereas } E_0 = \frac{h^2}{8ma^2} = \frac{1}{8} \cdot \frac{h^2}{ma^2} = 0.125 \frac{h^2}{ma^2}$$

$\therefore$ The variational energy is $\frac{0.00165}{0.125} \times 100\%$ higher, i.e., 1.32% higher than the true ground-state energy. Answer.

Problem 11. Calculate the ground-state energy for one-dimensional harmonic oscillator.

$$H = -\frac{h^2}{2m} \cdot \frac{d^2}{dx^2} + \frac{1}{2} mw^2 x^2.$$

Considering the trial wave function $\psi = Ae^{-bx^2}$, where b is a constant and A will be found out by normalisation.

Solution: First of all, A will be determined by normalisation.

$$\int_{-\infty}^{+\infty}\psi^*\psi\,dx = \int_{-\infty}^{+\infty}\left(Ae^{-dx^2}\right)\left(Ae^{-dx^2}\right)\,dx = 1$$

$$\text{or}\quad A^2\int_{-\infty}^{+\infty}e^{-2bx^2}\,dx = 1$$

$$\text{or}\quad A^2\left(\frac{\pi}{2b}\right)^{1/2} = 1 \qquad \because \int_{-\infty}^{+\infty}e^{-2bx^2}\,dx = \left(\frac{\pi}{2b}\right)^{1/2}$$

$$\text{or}\quad A^2 = \frac{1}{\left(\frac{\pi}{2b}\right)^{1/2}} = \left(\frac{2b}{\pi}\right)^{1/2}$$

$$\text{or}\quad A = \left(\frac{2b}{\pi}\right)^{1/4}$$

Now, $\langle H\rangle = \langle T\rangle + \langle V\rangle$

$$\text{But } T = \frac{-\hbar^2}{2m}|A|^2\int_{-\infty}^{+\infty}e^{-bx^2}\frac{d^2}{dx^2}\left(e^{-bx^2}\right)\,dx$$

We have to find out the values of

$$\frac{d^2}{dx^2}\left(e^{-bx^2}\right) = \frac{d}{dx}\left[\frac{d}{dx}e^{-bx^2}\right]$$

$$= \frac{d}{dx}\left[e^{-bx^2}(-2bx)\right] = (-2b)\frac{d}{dx}\left(xe^{-bx^2}\right)$$

$$= (-2b)\left[1\cdot e^{-bx^2} + x\cdot e^{-bx^2}(-2bx)\right]$$

$$= (-2b)\left[e^{-bx^2} - 2bx^2e^{-bx^2}\right]$$

$$= \left[-2be^{-bx^2} + 4b^2x^2e^{-bx^2}\right]$$

Putting this values in the equation of energy, i.e.,

$$T = \frac{-\hbar^2}{2m}|A|^2\int_{-\infty}^{+\infty}e^{-bx^2}\left[-2be^{-bx^2} + 4b^2x^2e^{-bx^2}\right]dx$$

$$= \frac{-\hbar^2}{2m}|A|^2\left[-2b\int_{-\infty}^{\infty}e^{-2bx^2}dx + 4b^2\int_{-\infty}^{\infty}x^2e^{-2bx^2}dx\right]$$

$$= \frac{-\hbar^2}{2m}|A|^2\left[(-2b)\left(\frac{\pi}{2b}\right)^{1/2} + 4b^2\left(\frac{\sqrt{\pi}}{2}\cdot\frac{1}{(2b)^{3/2}}\right)\right]$$

$$= \frac{-\hbar^2}{2m}|A|^2\left[(-2b)\left(\frac{\pi}{2b}\right)^{1/2} + \left(\frac{4b^2}{2\times 2b}.\sqrt{\pi/2b}\right)\right]$$

$$= \frac{-\hbar^2}{2m}\left[\left(\frac{2b}{\pi}\right)^{1/2} - (2b)\left(\frac{\pi}{2b}\right)^{1/2} + b\left(\frac{\pi}{2b}\right)^{1/2}\cdot\left(\frac{2b}{\pi}\right)^{1/2}\right]$$

$$= \frac{-\hbar^2}{2m}[(-2b)+b] = \frac{-\hbar^2}{2m}b$$

$$\text{and } V = \frac{1}{2}mw^2|A|^2\int_{-\infty}^{\infty} e^{-2bx^2}\cdot x^2\mathrm{d}x$$

$$= \frac{1}{2}mw^2|A|^2\int_{-\infty}^{\infty} x^2 e^{-2bx^2}\mathrm{d}x$$

$$= \frac{1}{2}mw^2\left(\frac{2b}{\pi}\right)^{1/2}\left(\frac{\sqrt{2}}{2}\cdot\frac{1}{(2b)^{3/2}}\right)$$

$$= \frac{1}{2}mw^2\left(\frac{2b}{\pi}\right)^{1/2}\left(\frac{1}{4b}\cdot\frac{\sqrt{\pi}}{2b}\right)$$

$$= \frac{1}{2}mw^2\frac{1}{4b} = \frac{mw^2}{8b}$$

$$\therefore H = \frac{\hbar^2}{2m}b + \frac{mw^2}{8b}$$

Now we minimise $\phi H\rangle$ to get the tightest bound.

$$\therefore \frac{\mathrm{d}H}{\mathrm{d}b} = \frac{\hbar^2}{2m} - \frac{mw^2}{8b^2} = 0$$

$$\text{or } \frac{\hbar^2}{2m} = \frac{mw^2}{8b^2}$$

$$\text{or } 8b^2 = \frac{2m^2w^2}{\hbar^2} \text{ or, } b^2 = \frac{2m^2w^2}{8\hbar^2} = \frac{m^2w^2}{4\hbar^2}$$

$$\therefore b = \frac{mw}{2\hbar}$$

Putting the values of b in the aforementioned $\langle H\rangle$ equation, we have

$$H_{\min} = \frac{\hbar^2}{2m}\left(\frac{mw}{2\hbar}\right) + \frac{mw^2}{8\frac{mw}{2\hbar}}$$

$$= \frac{\hbar^2}{4m\hbar}.mw + \frac{2\hbar mw^2}{8mw}$$

$$= \frac{\hbar w}{4} + \frac{hw}{4} = \frac{1}{2}hw$$

Problem 12. The Schrödinger equation of a particle confined to the $+x$ axis direction is given by $-\frac{\hbar^2}{2m}\cdot\frac{d^2\psi}{dx^2} = mgx\psi = E\psi$ having $\psi(0) = (0), \to 0\ as x \to \infty$ and E is the energy Eigen value. Use the trial function xe^{-ax} and find the best value of a parameter.

Solution: Given that $H = -\frac{-\hbar^2}{2m}\cdot\frac{d^2}{dx^2} + mgx$ and $\psi = xe^{-ax}$

With the help of the second equation, we obtain

$$\psi \mid \psi = \int_0^\infty \left(xe^{-ax}\right)\left(xe^{-ax}\right) dx$$

$$= \int_0^\infty x^2 e^{-2ax} dx$$

$$= \frac{1}{4a^3}\left[\therefore \int_0^\infty x^n e^{-ax} dx = \frac{n!}{a^{n+1}}\right]$$

$$\text{and } \psi\left|\frac{-\hbar^2}{2m}\cdot\frac{d^2}{dx^2}\right|\psi = \frac{\hbar^2}{m}a\int_0^\infty xe^{-2ax}dx - \frac{\hbar^2}{2m}a^2\int_0^\infty x^2e^{-2ax} dx$$

$$= \frac{\hbar^2}{4am} - \frac{\hbar^2}{8am} = \frac{\hbar^2}{8am}$$

$$\therefore \psi|\max|\psi = mg\int_0^\infty x^3 e^{-2ax} dx \quad [\text{Heare } n = 3]$$

$$= mg\frac{n!}{(2a)^{n+1}} = \frac{(mg)1.2.3}{(2a)^{3+1}} = \frac{2.3mg}{2^4.a^4}$$

$$= \frac{3mg}{8a^4}$$

$$\therefore \frac{\psi|H|\psi}{\psi \mid \psi} = \frac{\left[\hbar^2/8am\right] + \left[3mg/8a^4\right]}{\frac{1}{4a^3}}$$

$$= \frac{\hbar^2a^2}{2m} + \frac{3}{2}\frac{mg}{a}$$

Minimising $\langle H\rangle$ with respect to a, we obtain

$$\frac{dH_{\min}}{da} = \frac{\hbar^2}{2m}\cdot 2a - \frac{3}{2}\frac{mg}{a^2} = 0$$

$$\text{or} \quad \frac{\hbar^2}{2m} \cdot 2a = \frac{3}{2}\frac{mg}{a^2}$$

$$\text{or} \quad 4a^3 = \frac{(2m)(3mg)}{\hbar^2} = \frac{6m^2 g}{\hbar^2}$$

$$\text{or} \quad a^3 = \frac{6m^2 g}{4\hbar^2} = \frac{3}{2}\frac{m^2 g}{\hbar^2}$$

$$\therefore a = \left(\frac{3m^2 g}{2\hbar^2}\right)^{1/3}$$

= Best value of a for keeping H minimum Answer

Problem 13. Estimate the ground-state energy of particle in one-dimensional box of infinite height, equal to a, using the trial wave function $\psi = A(a^2 - x^2)x\alpha$, where α is the variational parameter.

Solution: The given trial wave function is well behaved and satisfies the boundary conditions, $\psi(0) = 0$ and $\psi(a) = 0$. We assume $a = 1$ for simplicity sake and then normalise.

The normalisation condition is

$$\psi \mid \psi = \int_0^1 \psi^* \, \psi \, d\tau = 1$$

Since the box is one-dimensional, $d\tau = dx$ and $0 \leq x \leq 1$.

$$\therefore \int_0^1 \psi^* \psi \, dx = 1$$

$$\therefore \int_0^1 A^2 (1 - x^2)^2 x^{2\alpha} dx = 1$$

$$\text{or} \quad A^2 \left[\int_0^1 x^{2\alpha} \, dx + \int_0^1 x^{2\alpha+4} \, dx - \int_0^1 x^{2\alpha+2} \, dx \right] = 1$$

$$\text{or} \quad A^2 \left[\frac{1}{2\alpha+1} + \frac{1}{2\alpha+5} + \frac{2}{2\alpha+3} \right] = 1$$

On simplification, we can write

$$A = \left[\frac{1}{8}(2\alpha+1)(2\alpha+3)(2\alpha+5) \right]^{1/2}$$

$$\therefore \psi = \left[\frac{(2\alpha+1)(2\alpha+3)(2\alpha+5)}{8} \right]^{1/2} \cdot (1 - x^2) x^{\alpha}$$

but $H = -\dfrac{\hbar^2}{2m} \cdot \dfrac{d^2}{dx^2}$ [for $V(x) = 0$ inside the box]

$$\text{And}\, E \equiv H = \psi|H|\psi = \int_0^1 \psi^* H\psi \mathrm{d}x$$

$$= -\frac{\hbar^2}{2m} \cdot \frac{(2\alpha+1)(2\alpha+3)(2\alpha+5)}{8} \int_0^1 (1-x^2)x^\alpha \cdot \frac{\mathrm{d}^2}{\mathrm{d}x^2}\left[(1-x^2)x^\alpha\right]\mathrm{d}x$$

This is an easy integral, which will yield

$$E = \frac{\hbar^2}{4m}\left(\frac{2\alpha+5}{2\alpha+1}\right)(2\alpha^2+2\alpha^2+2\alpha-1)$$

$$= \frac{\hbar^2}{4m}\left[\frac{4\alpha^3+14\alpha^2+8\alpha-5}{(2\alpha-1)}\right]$$

In atomic c units, $\hbar = m = 1$ we can write

$$E = \frac{4\alpha^3+14\alpha^2+8\alpha-5}{4(2\alpha-1)} \quad \text{in } a.u$$

Now, we minimise it on differentiation with respect to α and put it equal to zero, i.e.,

$$\left(\frac{\mathrm{d}E}{\mathrm{d}\alpha}\right)_{\propto 0} = 0, \quad \text{we get}$$

$$8\alpha_0^3 + 8\alpha_0^2 - 14\alpha_0 + 1 = 0$$

The roots of this equation may be found out numerically. One root of this equation for our purpose will be $\propto_0 = 0.862$.

Substituting the value of $\propto_0$ in the aforementioned energy equation, we obtain

$$E_1 = 5.13\ a.u \qquad \text{Answer.}$$

Problem 14. The function Ae^{-cr^2} is utilised as an approximation to the 1s wave function for H atom. Estimate the value of A from the normalisation condition, and apply the variation method to obtain the best value of the parameter 'c'. What is the minimum error that is made when the energy is estimated with this function?

Solution: Given that $\psi = Ae^{-cr^2}$; C =?; Minimum error =?

Now we normalise the function $\psi = Ae^{-cr^2}$ from the normalisation condition so that

$$\int_0^\infty \psi^*\psi \mathrm{d}r = 1$$

$$\text{or} \quad \int_0^\infty \left(Ae^{-cr^2}\right)\left(Ae^{-cr^2}\right) \mathrm{d}r = 1$$

$$\text{or} \quad A^2\int_0^\infty e^{-2cr^2} r^2 \mathrm{d}x = 1 \qquad \text{(i)}$$

$$\text{or}\quad A^2\int_0^\infty e^{-2cr^2}r^2\mathrm{d}r = \frac{A^2}{4}\sqrt{\frac{\pi}{8c^3}} = 1 \qquad \left[\because \int_0^\infty x^2e^{-ax^2}\cdot\mathrm{d}x = \frac{1}{4}\sqrt{\frac{\pi}{a^3}}\right]$$

$$\text{or}\quad \frac{A^2}{4}\sqrt{\frac{\pi}{4\times 2c^3}} = 1$$

$$\text{or}\quad \frac{A^2}{4\times 2}\sqrt{\frac{\pi}{2c^3}} = 1 \qquad \text{or,}\ \frac{A^2}{8}\sqrt{\frac{\pi}{2c^3}} = 1$$

$$\text{or}\quad A^2 = 8\sqrt{\frac{2c^3}{\pi}} \tag{ii}$$

In Eq. (i), we might have used $4\pi r^2\mathrm{d}r$ as the volume element, but we used $r^2\mathrm{d}r$ as volume element for simplicity sake.

Now,

$$E = \int_0^\infty \psi^*\left[\frac{-\hbar^2}{2m}\nabla^2 - \frac{e^2}{4\pi\varepsilon_0}\right]\psi r^2\mathrm{d}r$$

$$= A^2e^{-cr^2}\left\{\frac{-\hbar^2}{2m}\left[\frac{1}{r^2}\frac{\mathrm{d}}{\mathrm{d}r}\left(r^2\frac{\mathrm{d}}{\mathrm{d}r}\right)\right] - \frac{e^2}{(4\pi\varepsilon_0)r}\right\}e^{-cr^2}r^2\ \mathrm{d}r \tag{iii}$$

After carrying out differentiation, we obtain

$$E = A^2\int_0^\infty e^{-cr^2}\left[\frac{-\hbar^2}{2m}\left(-6ce^{-cr^2} + 4c^2r^2e^{-cr^2}\right) - \frac{e^2}{(4\pi\varepsilon_0)r}\right]e^{-cr^2}r^2\ \mathrm{d}r$$

$$= \frac{3A^2c\hbar^2}{m}\int_0^\infty e^{-2cr^2}r^2\ \mathrm{d}r - \frac{2A^2c^2\hbar^2}{m}\int_0^\infty e^{-2cr^2}r^4\ \mathrm{d}r - \frac{A^2e^2}{(4\pi\varepsilon_0)}\int_0^\infty e^{-2cr^2}r\mathrm{d}r$$

We use the value of standard integrals and write down

$$A^2\int_0^\infty e^{-2cr^2}r^4\ \mathrm{d}r = 8\sqrt{\frac{2c^3}{\pi}}\left(\frac{3}{32}c\sqrt{\frac{\pi}{2c^3}}\right) = \frac{3}{4c}$$

$$A^2\int_0^\infty e^{-2cr^2}r\mathrm{d}r = \frac{1}{4c}\left(8\sqrt{\frac{2c^3}{\pi}}\right) = \frac{2}{c}\sqrt{\frac{2c^3}{\pi}}$$

Putting these in the aforementioned equation, we obtain

$$E = \frac{3c\hbar^2}{m} - \frac{2c^2\hbar^2}{m}\left(\frac{3}{4c}\right) - \frac{e^2}{(4\pi\varepsilon_0)}\left(\frac{2}{c}\sqrt{\frac{2c^3}{\pi}}\right)$$

$$= \frac{+3c\hbar^2}{2m} - \frac{2e^2}{(4\pi\varepsilon_0)}\sqrt{\frac{2c^3}{\pi}} \tag{iv}$$

To get the minimum values of energy, we differentiate E with respect to c and get

$$\frac{dE}{dc} = \frac{3\hbar^2}{2m} - \frac{e^2}{(4\pi\varepsilon_0)}\sqrt{\frac{2}{\pi c}} = 0$$

$$\text{or} \quad \frac{3\hbar^2}{2m} = \frac{e^2}{(4\pi\varepsilon_0)}\frac{2}{\pi c}$$

Squaring both the sides, we have

$$\frac{9\hbar^2}{4m^2} = \frac{e^2}{(4\pi\varepsilon_0)^2}\frac{2}{\pi c}$$

$$\therefore c = \frac{8}{9}\left[\frac{e^4 m^2}{(4\pi\varepsilon_0)^2 \pi\hbar^2}\right]$$

Substituting the value of c in Eq. (iv), we get

$$E = \frac{-8}{3\pi}\left[\frac{me^4}{2\hbar^2(4\pi\varepsilon_0)^2}\right] = -0.848\left[\frac{me^4}{2\hbar^2(4\pi\varepsilon_0)^2}\right]$$

But the ground-state energy of the hydrogen atom is equal to

$$-\left[\frac{me^4}{2\hbar^2(4\pi\varepsilon_0)^2}\right]$$

Hence, the variation function will yield an error of about 15%. Answer.

Problem 15. Consider a particle in a one-dimensional box and apply variation method to the particle. Let $V = 0$ for $-1 \leq x \leq +1$ and $V = \infty$ otherwise. Use $f_1 = (1 - x^2)$ and $f_2 = (1 - x^4)$ to build up the trail function $\psi = c_1 f_1 + c_2 f_2$.

a. Estimate the energy with the function and compare with the exact solution.
b. Find the 'best' values of c_1 and c_2.

Solution: Given that

$$f_1 = (1 - x^2), f_2 = (1 - x^4)$$

$$\psi = c_1 f_1 + c_2 f_2, \quad V = 0 \quad \text{for} -1 \leq x \leq +1$$

Applying the variational principle,

$$E = \frac{\int_{-1}^{+1} \psi^* H\psi dx}{\int \psi^* \psi dx} = \frac{\psi|H|\psi}{\psi \mid \psi} \tag{i}$$

$$\text{or} \quad E\psi \mid \psi = \psi|H|\psi$$

$$\text{or} \quad E(c_1 f_1 + c_2 f_2)(c_1 f_1 + c_2 f_2) = (c_1 f_1^* + c_2 f_2^*)H(c_1 f_1 + c_2 f_2) \tag{ii}$$

$$\text{or} \quad E\left[c_1^2 \int f_1^2 dx + 2c_1 c_2 \int f_1 f_2 dx + c_2^2 \int f_2^2 dx\right] = c_1^2 \int f_1^* H f_1 dx + 2c_1 c_2 \int f_1 H f_2 dx + c_2^2 \int f_2^* H f_2 dx$$

Differentiating E, with respect to c_1 we get

$$\frac{\partial E}{\partial c_1}\left[c_1^2 f_1 \mid f_2 + 2c_1c_2 f_1 \mid f_2 + c_2^2 f_2 \mid f_2\right] + E\left[2c_1 f_1 \mid f_2 + 2c_2 f_2 \mid f_2 + 0\right]$$

$$= 2c_1 f_1 |H| f_1 + 2c_2 f_1 |H| f_2 + 0 \qquad \text{(iii)}$$

Again, differentiating 'E' with respect to c_2, we obtain

$$\frac{\partial E}{\partial c_2}\left[c_1^2 f_1 \mid f_2 + 2c_1c_2 f_1 \mid f_2 + c_2^2 f_2 \mid f_2\right] + E\left[0 + 2c_1 f_1 \mid f_2 + 2c_2 f_2 \mid f_2\right]$$

$$= 0 + 2c_1 f_1 |H| f_2 + 2c_2 f_2 |H| f_2 \qquad \text{(iv)}$$

Putting $\frac{\partial E}{\partial c_1} = \frac{\partial E}{\partial c_2} = 0$, Eqs. (iii) and (iv) can be written as

$$= 2c_1 f_1 |H| f_1 - 2c_1 E f_1 \mid f_1 + 2c_2 f_1 |H| f_2 - 2c_2 E f_1 \mid f_2 = 0 \qquad \text{(v)}$$

$$\text{and } 2c_1 f_1 |H| f_2 - 2c_1 E f_1 \mid f_2 + 2c_2 f_2 |H| f_2 - 2c_2 E f_2 \mid f_2 = 0 \qquad \text{(vi)}$$

For possible solution, the determinant of coefficients c_1 and c_2 should be zero. The secular determinant with the help of Eqs. (v) and (vi) can be written as

$$\begin{vmatrix} f_1|H|f_1 - Ef_1 \mid f_1, f_1|H|f_2, -Ef_1 \mid f_2 \\ f_1|H|f_2 - Ef_1 \mid f_2, f_2|H|f_2, -Ef_2 \mid f_2 \end{vmatrix} = 0$$

$$\text{or} \begin{vmatrix} H_{11} - ES_{11}, H_{12} - ES_{12} \\ H_{21} - ES_{21}, H_{22} - ES_{22} \end{vmatrix} = 0$$

where

$$H_{11} = f_1|H|f_1$$

$$H_{22} = f_2|H|f_2$$

$$H_{12} = f_1|H|f_2$$

$$H_{21} = f_2|H|f_1$$

$$S_{11} = f_1 \mid f_1$$

$$S_{22} = f_2 \mid f_2$$

$$S_{12} = f_1 \mid f_2$$

$$S_{21} = f_2 \mid f_1$$

$$H_{11} = \int_{-1}^{+1} \left(1 - x^2\right) \left[\frac{-\hbar^2}{2m} \cdot \frac{d^2}{dx^2}\right] \left(1 - x^2\right) dx$$

$$= \int_{-1}^{+1} \left(1 - x^2\right) \frac{\hbar^2}{m} dx$$

$$= \frac{\hbar^2}{m}\left[x - \frac{x^3}{3}\right]_{-1}^{+1} = \frac{2\hbar^2}{m}\left[1 - \frac{1}{3}\right] = \frac{4\hbar^2}{3m}$$

$$H_{22} = \int_{-1}^{+1} \left(1 - x^4\right) \left[\frac{-\hbar^2}{2m} \cdot \frac{d^2}{dx^2}\right] \left(1 - x^4\right) dx$$

$$= \frac{6\hbar^2}{m} \int_{-1}^{+1} \left(1 - x^4\right) x^2 dx = \frac{6\hbar^2}{m} \int_{-1}^{+1} x^2 dx - \int_{-1}^{+1} x^6 dx$$

$$= \frac{6\hbar^2}{m}\left[\left[\frac{x^3}{3}\right]_{-1}^{+1} - \left[\frac{x^7}{7}\right]_{-1}^{+1}\right]$$

$$= \frac{12\hbar^2}{m}\left[\frac{1}{3} - \frac{1}{7}\right] = \frac{12\hbar^2}{m}\left[\frac{7-3}{21}\right] = \frac{48\hbar^2}{21m} = \frac{16\hbar^2}{7m}$$

$$H_{12} = \int_{-1}^{+1} \left(1 - x^2\right) \left[\frac{-\hbar^2}{2m} \cdot \frac{d^2}{dx^2}\right] \left(1 - x^4\right) dx$$

$$= \int_{-1}^{+} \left(1 - x^2\right) 12x^2 \frac{\hbar^2}{2m} dx$$

$$= \frac{6\hbar^2}{m} \int_{-1}^{+1} \left(x^2 - x^4\right) = \frac{6\hbar^2}{m}\left[\frac{x^3}{3} - \frac{x^5}{5}\right]_{-1}^{+1}$$

$$= \frac{12\hbar^2}{m}\left[\frac{1}{3} - \frac{1}{5}\right] = \frac{12\hbar^2}{m} \frac{2}{15} = \frac{8\hbar^2}{5m} = H_{21}$$

$$H_{21} = \int_{-1}^{+1} \left(1 - x^4\right) \left[\frac{-\hbar^2}{2m} \cdot \leq \frac{d^2}{dx^2}\right] \left(1 - x^2\right) dx$$

$$= \int_{-1}^{+} \left(1 - x^4\right) \frac{\hbar^2}{m} dx$$

$$= \frac{\hbar^2}{m} \int_{-1}^{+1} \left(1 - x^4\right) dx = \frac{\hbar^2}{m}\left[x - \frac{x^5}{5}\right]_{-1}^{+1}$$

$$= \frac{2\hbar^2}{m}\left[1 - \frac{1}{5}\right] = \frac{8\hbar^2}{5m}$$

$$S_{11} = \int_{-1}^{+1} (1-x^2)^2 = \int_{-1}^{+1} (1-2x^2+x^4)\,dx$$

$$= 2\left[x - \frac{2x^3}{3} + \frac{x^5}{5}\right]_0^1 = 2\left[1 - \frac{2}{3} + \frac{1}{5}\right]_0^1 = \frac{16}{15}$$

$$S_{22} = \int_{-1}^{+1} (1-x^4)^2 = \int_{-1}^{+1} (1-2x^4+x^8)\,dx$$

$$= 2\left[x - \frac{2x^5}{5} + \frac{x^9}{9}\right]_0^1 = 2\left[1 - \frac{2}{5} + \frac{1}{9}\right] = 2\left[\frac{45-18+5}{45}\right] = \frac{64}{45}$$

$$S_{12} = S_{21} = \int_{-1}^{+1} (1-x^2)(1-x^4)\,dx$$

$$= 2\int_{-1}^{+1} (1-x^2-x^4+x^6)\,dx$$

$$= 2\left[x - \frac{x^3}{3} - \frac{x^5}{5} + \frac{x^7}{7}\right]_0^1 = 2\left[1 - \frac{1}{3} - \frac{1}{5} + \frac{1}{7}\right] = \frac{2\times 64}{105}$$

$$= \frac{128}{105}$$

Substituting these values in the secular determinant, we obtain

$$\begin{vmatrix} \frac{4\hbar^2}{3m} - E\cdot\frac{16}{15}, & \frac{8\hbar^2}{5m} - E\cdot\frac{128}{105} \\ \frac{8\hbar^2}{5m} - E\cdot\frac{128}{105}, & \frac{16\hbar^2}{7m} - E\cdot\frac{128}{105} \end{vmatrix} = 0$$

$$\text{or, } \frac{\hbar^2}{m}\begin{vmatrix} \left(\frac{4}{3} - \frac{16Em}{15\hbar^2}\right), & \left(\frac{8}{5} - \frac{128Em}{105\hbar^2}\right) \\ \left(\frac{8}{5} - \frac{128Em}{105\hbar^2}\right), & \left(\frac{16}{7} - \frac{128Em}{105\hbar^2}\right) \end{vmatrix} = 0$$

Putting $\frac{Em}{\hbar^2} = w$, we have

$$\begin{vmatrix} \left(\frac{4}{3} - \frac{16}{15}w\right), & \left(\frac{8}{5} - \frac{128}{105}w\right) \\ \left(\frac{8}{5} - \frac{128}{105}w\right), & \left(\frac{16}{7} - \frac{128}{105}w\right) \end{vmatrix} = 0$$

$$\text{or} \begin{vmatrix} ((1.33-1.066w),(1.6-1.22w) \\ (1.60-1.22w),(2.285)-1.422w \end{vmatrix} = 0$$

$$\text{or } (1.33-1.066w)(2.285-1.422w)-(1.60-122w)^2 = 0$$

$$\text{or} \quad \begin{array}{c} (1.33\times 2.285-1.33\times 1.422w-1.066w\times 2.285+1.066\times 1.422w)^2 \\ -\left[(1.60)^2-2\times 1.60\times 1.22w+(1.22)^2 w^2 = 0\right] \end{array}$$

$$\text{or } 3.039-1.891w-2.435w+1.515w^2-2.56+3.904w+1.488w^2 = 0$$

$$\text{or } 0.479-4.326w+3.904w+1.515w^2-1.488w^2 = 0$$

$$\text{or } 0.479-0.422w+0.027w^2 = 0$$

$$\text{or } 0.027w^2-0.422w+0.479 = 0$$

$$\text{or } w^2-15.63w+17.74 = 0$$

Solving this quadratic equation, the roots will be

$$w = 1.23 \text{ and } 14.4$$

Thus, Eigen values have the upper bounds

$$w_1 = 1.23\frac{\hbar^2}{m} \text{ and } w_3 = 14.4\frac{\hbar^2}{m}$$

The exact Eigen values have been found to be

$$E_1 = 1.23\frac{\hbar^2}{m} \text{ and } E_2 = 11.10\frac{\hbar^2}{m}$$

c. The Eigen factors can be found out with the help of the equation

$$\begin{vmatrix} \left(\frac{4}{3}-\frac{16}{15}w\right),\left(\frac{8}{5}-\frac{128}{105}w\right) \\ \left(\frac{8}{5}-\frac{128}{105}w\right),\left(\frac{16}{7}-\frac{64}{45}w\right) \end{vmatrix} = 0$$

$$\left(\frac{4}{5}-\frac{16}{15}w\right)c_1+\left(\frac{8}{5}-\frac{64}{45}w\right)c_2 = 0$$

$$\text{or } (1.3333-1.0666+1.23)c_1+(1.6-1.2191\times 1.23)c_2 = 0$$

$$\text{or } (1.3333-1.3119)c_1+(1.6-1.4994)c_2 = 0$$

$$\text{or } 0.0214c_1 + 0.1006c_2 = 0$$

$$\text{or } 0.1006c_2 = -0.0214c_1$$

$$\therefore \quad C_2 = \frac{-0.0214}{0.1006}c_1 = -0.2127C_1$$

Now, the value of c_1 can be determined by the normalisation condition, i.e.,

$$\int \psi^* \psi \, dx = 1 = c_1^2\left[S_{11} - 2(0.2127)S_{12} + 0.0449S_{22}\right]$$

$$= c_1^2\left[1.0666 - 2(0.2127)1.219 + 0.0449 \times 1.2222\right]$$

$$= c_1^2\left[1.0666 - 0.5185 + 0.05487\right]$$

$$= c_1^2(0.60297)$$

$$\therefore \quad c_1^2 = \frac{1}{0.60297} = 1.6584 \qquad \therefore \quad c_1 = \sqrt{1.6584} \qquad \therefore c_1 = 1.2877$$

$$\therefore c_2 = -02127C_1 = -0.2127 \times 1.2877 = -0.273$$

Answer.

Questions on Concepts

1. What do you mean by perturbation theory? Find first-order perturbation correction to energy.
2. How will you find first-order perturbation correction to wave function?
3. Find second-order perturbation correction to energy.
4. How will you arrive at the second-order correction to wave function?
5. Explain bra–ket notation?
6. Find the expression for first-order correction to energy for nondegenerate state using Dirac's notation.
7. Find first-order perturbation correction to wave function for nondegenerate state with the help of Dirac's notation.
8. Prove that $\psi_k' = \sum \frac{\langle \psi_l^0 | H' | \psi_k^0 \rangle}{\left(E_k^0 - E_l^0\right)} \Psi_l^0.$
9. Prove that $E_k'' = \sum \frac{\left|\langle \psi_k^0 | H' | \psi_l^0 \rangle\right|^2}{\left(E_k^0 - E_l^0\right)}.$
10. Prove that $E_n'' = H_{mn}'' + \frac{|H_{mm}'|^2}{E_m^0 - E_n^0}.$
11. Prove that $\psi_k = \sum \frac{cm\left[E_n'm|H'|m\right]}{\left(E_m^0 - E_n^0\right)} \psi_m^0.$
12. What do you mean by degenerate state? Find expression for first-order perturbation correction to energy for degenerate state.
13. Obtain an expression for first-order correction to wave function for a degenerate state.
14. Apply perturbation theory to find an expression for the a harmonic oscillator.

15. Find the electric polarisibility of hydrogen atom by perturbation method.
16. Prove that $E_1^2 = \frac{-q}{4k} a_0^3 E^2$, where the terms have their usual significance.
17. Apply perturbation theory to the He atom and find $E_1 = \frac{5}{4} ZW_H$
18. The second-order correction to the energy of the ground state is always negative, why?
19. What is the criterion of validity of perturbation theory?
20. What is variation principle?
    Prove that $\int \phi^* H \phi d\tau \geq E_0$ or, $\int \phi^* H \phi d\tau \geq 0$
21. How will you compute the energy Eigen value by variation method?
22. Compute the wave function by variation method.
23. Apply the variational principle in the estimation of energy of the ground state of the simple harmonic oscillator using the trial wave function $\psi = Ae^{-\alpha x^2}$.
24. Apply variation method to the ground state of He atom in calculating its energy, i.e.,

$$E = -2(Z - 5/16)E_H$$

25. Apply variation principle to the ground state of H atom to find its energy, i.e.,

$$E = -0.85\left[me^4/2\hbar^2(4\pi E_0)^2\right]$$

26. A particle exists in a one-dimensional square well potential of which is modified by the addition of a term c_1x/a where c_1 is a constant and 'a' is the length of the well. Calculate the energy of the system with the aid of perturbation theory when c_1 is small.

$$\left[\text{Ans. } E_n = \frac{n^2h^2}{8ma^2} + \frac{c_1}{2}\right]$$

27. Suppose that for a real system, a real wave function is a linear combination of two orthonormal basis function where the energy integrals are as follows:

$$H_{11} = -15 \qquad H_{22} = -4 \qquad \text{and} \qquad H_{12} = H_{21} = -1$$

Estimate the energy of the system and find the coefficients of expansion.

$$\psi_a = c_1\psi_1 + c_2\psi_2$$

$$\left[\text{Ans. } E_1 = -15.09,\ E_2 = -3.91,\ C_1 = 0.996, \text{ and } C_2 = 0.0896\right]$$

28. An anharmonic oscillator has the potential function $V = \frac{1}{2}kx^2 + cx^4$, where c is anharmonicity constant. Find the energy correction to the ground state of the anharmonic oscillator in terms of c.

$$\left[\text{Ans. } E_{\text{perturb}} \frac{3c}{4\alpha^2}\right]$$

29. A linear harmonic oscillator is perturbed by an electric field of strength ε. If the oscillating mass has the charge $-e$, the perturbing Hamiltonian becomes $H' = +e\varepsilon x$. Find the perturbation correction to the energy through second order.

$$\left[\text{Ans. } E_n = E_n^0 + E_n' + E_n'' = (n + 1/2)h\nu - \frac{e^2\varepsilon^2}{8\pi^2\mu\nu^2}\right]$$

30. A simple harmonic oscillator having mass m_0 and angular frequency w is perturbed by an extra potential bx^3. Calculate the second-order correction to the ground-state energy of the oscillator.

$$\left[\text{Ans. } E_0'' = \frac{-11b^2\hbar^2}{8m_0^3w^4}\right]$$

31. A simple harmonic oscillator of mass 'm' and having angular frequency w is perturbed by the aid of $1/2bx^2$ as potential. Find the first- and second-order corrections to the ground-state energy.

$$\left[\text{Ans. } E' = \frac{b\hbar}{4mw}; E''_{=} = \hbar b^2 / 16m^2w^3\right]$$

32. A one-dimensional simple harmonic oscillator is subjected to a perturbation $H' = -eFx$, where F = electric field applied in the $+x$ direction. Show that $E'_n = 0$, $E''_n = -e^2F^2 2k$.
33. Calculate the variation energy of a particle in a one-dimensional box of length 'a' with trial function $t\psi(x) = Ax^3(a^3 - x^3)$. Estimate the percentage error.

$$\left[\text{Ans. } E_{\min} = 0.1976\frac{\hbar^2}{ma^2}, \text{ too high by about } 58\%\right]$$

34. Obtain the variation energy of a harmonic oscillator with the trial function $\psi(x) = A/(b^2 + x^2)$ where b = a variable parameter. Minimise the energy and find the percent error from the correct ground-state energy.
35. Use the variation to calculate an upper bound to the ground-state energy of a particle in a box of length 'a'. Mention the percent $\psi = A\sin^2(\pi x/a)$ error from the correct energy.

$$\left[\text{Ans. } E_{\min} = \hbar^2/6ma^2, 33.3\%\right]$$

36. An oscillating particle having mass 'm' has the potential function $\psi = cx^4$ where c is a constant. Obtain the formula for the variational energy with the trial function $\psi = Ae^{-bx^2}$ where 'b' is variational parameter.

$$\left[\text{Ans. } E_{\min} = (0.660)\left(\frac{c\hbar^4}{m^2}\right)^{1/3}\right]$$

37. Apply the variation method to the ground state of the hydrogen atom using the trial function.

$$\psi = \{A(1 - r/b)\} \text{ if } 0 < r < b$$
$$0 \quad \text{if } b < r$$

Compare your result with the correct energy, $-2.1787 \times 10^{-18} J$.

$$\left[\text{Ans. } E_{\min} = -1.3624 \times 10^{-18} J\right]$$

38. Use the trial function $\psi(x) = e^{-\alpha x^2}$, where α is a variable parameter to calculate the ground-state energy of a one-dimensional simple harmonic oscillator.

$$\left[\text{Ans. } E = \hbar w/2\right]$$

39. Calculate, by variation method, the energy of the first excited state of a linear harmonic oscillator using the trial function $\psi = Nxe^{-\lambda x^2}$ where λ = variable parameter.

$$\left[\text{Ans. } E = 3\,\hbar w/2\right]$$

40. A particle of mass 'm' is moving in one-dimensional box defined by the potential $V = 0$, $0 \leq x \leq$ a and $V = \infty$ otherwise. Find the ground-state energy with the trial function $\psi(x) = Ax(a - x)$, $0 \leq x \leq a$.

$$\left[\text{Ans.} = 10\,\hbar^2/2ma^2\right]$$

10

Diatomic Molecules

We have already discussed the quantum mechanics of a system having one electron and one nuclear charge, i.e., one electron system consisting of a nucleus with one charge only. We know that chemistry is concerned with the collection of atoms called molecules. One of the serious flaws of Bohr's theory is that it failed to predict chemical bonding. Modern quantum theory became successful in predicting the condition of chemical bonding and that of molecules. For some simple molecules, the existing quantum chemical results are at least as good as the experimental findings. For larger molecules, however, approximations are needed. Most of the rest study will be centred around the applicability of approximate methods to do the qualitative forecasts about the electronic nature of molecules.

In this chapter, we shall deal with the molecular orbital theory (MOT) and valence bond (VB) approach to the formation of diatomic molecules. It is to be noted that whether we deal with the MOT approach or VB approach, we have to take measures related to the electrons and nuclei (proton is heavier than electrons). Since the nuclei are heavier than the electrons, the nuclei move very slowly in comparison with the electrons. This idea was first put forth by Born and Oppenheìmer in 1927. Their idea opined that the vibrational and rotational motions of a molecule are separable from the electronic motions.

Before dealing with the MOT and VB approach, we shall deal with the Born–Oppenheìmer approximation in the light of wave mechanics.

10.1 Born–Oppenheìmer Approximation

The Born–Oppenheìmer approximation is an adiabatic approximation in which the motion of the atomic nuclei is considered to be much slower than the motion of the electrons. When we want to estimate the motion of electrons, the nuclei can be considered to be in the fixed position.

In the Born–Oppenheìmer approximation, the molecular potential energy varies with the internuclear distance. Formulation of Hamiltonian may be done in two ways:

1. By keeping the internuclear distance constant (R $\rightarrow$ 0 for large nucleus–nucleus potential energy)
2. By considering the motion of the nucleus. R should be variable for
   i. Diatomic molecules will be considered as harmonic oscillator and rigid rotator
   ii. The molecule will be considered as an anharmonic oscillator and nonrigid rotator, which will be a general case to study hyperfine structure in band spectra

It is important to note that the Born–Oppenheìmer approximation is very reliable in the case of ground electronic states but less reliable for excited states.

Now, we are going to formulate the Born–Oppenheìmer approximation for that the molecular wave function can be expressed as a product, i.e.,

$$\psi = \Phi(r_i,\ R_p)\chi_i(R_p)\cdot\rho(\alpha,\beta,\gamma) \tag{10.1}$$

where Φ = electronic wave function, which is a function of electronic coordinate r_i and nuclear coordinate R_p,

χ = the wave function for nuclear vibration, which is a function of nuclear coordinates, and

ρ = the wave function for rotation of the whole molecule, which is a function of three angles α, β, and γ.

The aforementioned equation specifies the orientation of the molecule with regard to space fixed axis. Actually, this assumption of simple product form results in the Born–Oppenheìmer approximation.

The wave function should be multiplied by a function, which should describe the translation of the molecule as a whole. In many places, we can neglect this function, as it leads to wave functions for a particle in a box problem and takes only the motion of the particles about the centre of mass of the considered molecule.

Now, we are going to divide the Hamiltonian into three parts in the following manner:

$$H = T_N + T_e + V \tag{10.2}$$

$$\text{where } T_N = \sum_p \frac{1}{2M_p}\nabla_p^2 = \text{nuclear kinetic energy in atomic units } (\hbar = 1) \tag{10.3}$$

$$M_p = \textit{m}\text{ass of the } p^{\text{th}}\text{nucleus}$$

$$T_e = \sum_i \frac{1}{2M_p}\nabla_p^2 = \text{electronic energy expressed considsering atomic units} \tag{10.4}$$

$$M = \text{mass of an electron, which is 1 in atomic units}$$

$$\text{and } V = \sum_{p<q}\frac{Z_pZ_q}{R_{pq}} + \sum_{i<j}\frac{1}{r_{ij}} - \sum_{i,p}\frac{Z_p}{r_{ip}} \tag{10.5}$$

$$\text{where } \sum_{p<q}\frac{Z_pZ_q}{R_{pq}} = \text{nucleus} - \text{nucleus interation}$$

$$\sum_{i<j}\frac{1}{r_{ij}} = \text{electron} - \text{electron interaction}$$

$$\sum_{i,p}\frac{Z_p}{r_{ip}} = \text{nucleus} - \text{electron interaction}$$

$$V = \text{potential energy of the molecule}$$

$$Z_p = \text{charge on nucleus } p$$

The electronic Hamiltonian may now be given by

$$H_e = T_e + V \tag{10.6}$$

and hence, the electronic Schrödinger equation may be written as

$$\left[T_e + V\left(r_i, R_p\right)\right]\Phi_a\left(r_i, R_p\right) = E_e\left(R_p\right)\Phi_a\left(r_i, R_p\right) \tag{10.7}$$

Since both V and Φ are explicit functions of nuclear coordinates, E_e will also be a function of the nuclear coordinates. It is clear that for any fixed nuclear position, the set of electronic wave functions will be orthonormal, and thus, we can write

$$\int \Phi_a^*(r_i, R_p)\Phi_b(r_i, R_p)\mathrm{d}T_e = 0 \text{ for } a \neq b \text{ and}$$

$$= 1 \text{ for } a = b \tag{10.8}$$

It may be assumed that we are able to solve Eq. (10.7) and go on taking into account complete Schrödinger equation:

$$H_\psi = E_\psi \tag{10.9}$$

Putting the value of Eqs. (10.1) and (10.2) in Eq. (10.9), we obtain the ath electronic state.

$$(T_N + T_e + V)(\Phi_a \chi_\rho) = E(\Phi_a \chi_\rho) \tag{10.10}$$

It should be kept in mind that *x, p, E*, and E_e are considered to apply to the ath electronic state. Multiplying from left by Φ and integrating over electronic coordinates, we obtain

$$\left(\int \Phi_a^* T_N \Phi_a \,\mathrm{d}\tau_e\right)(\chi_\rho) + \left(\int \Phi_a^* H_e \Phi_a \,\mathrm{d}\tau_e\right)(\chi_\rho) = E\left(\int \Phi_a^* \Phi_a \,\mathrm{d}\tau_e\right)(\chi_\rho) \tag{10.11}$$

Now, we shall make use of Eqs. (10.7) and (10.8) to get

$$\left(\int \Phi_a^* T_N \Phi_a \,\mathrm{d}\tau_e\right)(\chi_\rho) + E_e(R_\rho)(\chi_\rho) = E(\chi_\rho)$$

Furthermore, we may write

$$\left(\int \Phi_a^* T_N \Phi_a\right)(\chi_\rho) = (\chi_\rho)\left(\int \Phi_a^* T_N \Phi_a \,\mathrm{d}\tau\right)(\chi_\rho) + T_N(\chi_\rho) \tag{10.12}$$

Since TN consists of the derivative with respect to the nuclear coordinates, it operates on all the wave functions to its right. Using Eq. (10.12), we get

$$T_N(\chi_\rho) + (\chi_\rho)\left[E_e + \left(\int \Phi_a^* \tau_N \Phi_a \,\mathrm{d}\tau\right)\right] = E(\chi_\rho) \tag{10.13}$$

In this equation, $\int \Phi_a^* \tau_N \Phi_a \tau_e$ is an additional term, which may be expressed as

$$H' = \int \Phi_a^* T_N \Phi_a \,\mathrm{d}\tau_e \tag{10.14}$$

This term is the effective potential energy, which determines the motion of the nucleus, and H' can be expressed as

$$H'_{aa} = \int \Phi_a^*(r_i, R_p)\left(\sum_p \frac{1}{2M_p}\nabla_p^2\right)\Phi_a(r_i, R_p)\mathrm{d}\tau_e \tag{10.15}$$

It is to be noted that the Born–Oppenheimer approximation is valid if H' and the similar terms that connect two different electronic states such that

$$H'_{ab} = \int \Phi_a^* \tau_N \Phi_a \mathrm{d}\tau \tag{10.16}$$

are small. We can further say that H'_{aa} and H'_{ab} will be small if electronic states do not vary fast with a change in R_p. It is to be noted that x will be nonzero for a small range of R_p for most stable molecules about the equilibrium position. Also to be noted that in this range, ϕ may be taken as independent of R_p so that H' and H'' tend to zero. Under this situation, the Schrödinger equation for the nuclei may be expressed as

$$T_N(\chi_P) + E_e(\chi_P) = E(\chi_P) \tag{10.17}$$

This equation represents both internal motion of the nuclei and rotation. It should be kept in mind that the Eigen value E in the equation is the total energy of the molecule within the Born–Oppenheimer approximation.

10.2 Hydrogen Molecule Ion

The simplest possible molecular system would need two nuclei (in order to be a molecule) and one electron (to provide bonding). In other words, we can say that the simplest molecular system is the one that consists of only one electron moving in the field of two hydrogen nuclei. The @@ ion is a system, which contains two hydrogen nuclei '*a*' and '*b*' and an electron '*e*' moving between the two attracting centres '*a*' and '*b*', i.e., internuclear distance is '*R*'. This situation has been illustrated in Figure 10.1. The main characteristics of such a system are as follows:

- There is attraction between electron and nuclei.
- There is repulsion between the nuclei '*a*' and '*b*'.

Accordingly, the Hamiltonian for H_2^+ ion may be expressed as

$$H = -\frac{\hbar^2}{2m}\nabla_p^2 - \frac{e^2}{r_a} - \frac{e^2}{r_b} + \frac{e^2}{R} \tag{10.18}$$

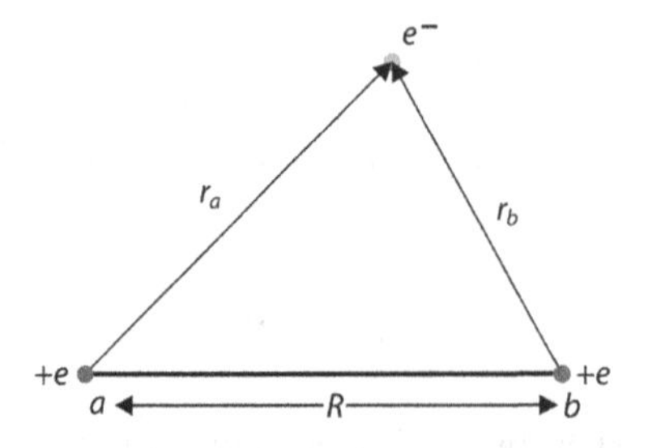

FIGURE 10.1 The hydrogen molecule ion.

where r_a and r_b = distance between the electron and the two nuclei, respectively

$+e$ = nuclear charge

R = internuclear distance

We have to multiply the r term by $1/4\pi\epsilon_0$, but we have left it for simplicity.

In the case of large value of R, the electron will be associated with either a or b, and it would behave just like hydrogen atom.

ψ_a = Eigen function when the electron is associated with a.

ψ_b = Eigen function when the electron is associated with b.

The wave function ψ of H_2^+ ion will be determined by the linear combination of ψ_a and ψ_b and expressed as

$$\psi = c_1\psi_a + c_2\psi_b \tag{10.19}$$

where c_1 and c_2 = coefficients, which are adjustable parameters.

We know that

$$H\psi = E\psi \tag{10.20}$$

Multiplying both sides by ψ, we get

$$\psi H\psi = E\psi^2$$

On integrating over the entire space $d\tau$, we get

$$\int \psi H\psi d\tau = \int E\psi^2 d\tau$$

$$\text{or} \quad E = \frac{\int \psi H\psi d\tau}{\int \psi^2 d\tau}$$

Now putting the value of ψ in the aforementioned equation, we obtain

$$E = \frac{\int (c_1\psi_a + c_2\psi_b)H((c_1\psi_a + c_2\psi_b)d\tau)}{(c_1\psi_a + c_2\psi_b)^2 d\tau}$$

$$= \frac{c_1^2\int \psi_a H\psi_a d\tau + c_1c_2\int \psi_a H\psi_b d\tau + c_1c_2\int \psi_b H\psi_a d\tau + c_2^2\int \psi_b H\psi_b d\tau}{(c_1\psi_a + c_2\psi_b)^2 d\tau} \tag{10.21}$$

$$= \frac{c_1^2 H_{aa} + c_2^2 H_{bb} + 2c_1c_2H_{ab}}{c_1^2 + c_2^2 + 2c_1c_2S}\text{[on normalisation]} \tag{10.22}$$

where $\int \psi_a H\psi_a d\tau = H_{aa} = E_0 =$ energy of H atom with nucleus 'a' and

$\int \psi_b H\psi_b d\tau = H_{bb} = E_0 =$ energy of H atom with nucleus 'b'

$$\therefore \quad H_{aa} = H_{bb} = E_0.$$

This is because of the fact that hydrogen atoms are the same whether formed by the electron in association with nucleus a or b.

$\int \psi_a \psi_b d\tau = S$ = overlap integral, which indicates the simultaneous probability of wave functions ψ_a and ψ_b in the same region

$\int \psi_a \psi_b d\tau = H_{ab}$ = resonance integral = K

$\int \psi_a \psi_b d\tau = H_{ba}$

But $H_{ab} = H_{ba}$

To find the value of E, we shall take the help of the variational principle. The value of E will be minimum for values of C_1 and C_2 and equated to zero, i.e.,

$$\frac{dE}{dc_1} = 0 \text{ and } \frac{dE}{dc_2} = 0 \text{ when } \frac{dE}{dc_1} = 0$$

Then Eq. (10.22) can be written as

$$0 = \frac{(2c_1H_{aa} + 2c_2H_{ab})(c_1^2 + c_2^2 + 2c_1c_2S) - (2c_1 + 2c_2S) \times (c_1^2H_{aa} + c_2^2H_{bb} + 2c_1c_2H_{ab})}{(c_1^2 + c_2^2 + 2c_1c_2S)^2}$$

$$\text{or } \frac{(2c_1H_{aa} + 2c_2H_{ab})(c_1^2 + c_2^2 + 2c_1c_2S)}{(c_1^2 + c_2^2 + 2c_1c_2S)^2} - \frac{(2c_1 + 2c_2S) \times (c_1^2H_{aa} + c_2^2H_{bb} + 2c_1c_2H_{ab})}{(c_1^2 + c_2^2 + 2c_1c_2S)^2} = 0$$

$$\text{or} \quad \frac{(2c_1H_{aa} + 2c_2H_{ab})}{(c_1^2 + c_2^2 + 2c_1c_2S)} - \frac{(2c_1 + 2c_2S)E}{(c_1^2 + c_2^2 + 2c_1c_2S)} = 0 \left[\because \frac{(c_1^2H_{aa} + c_2^2H_{bb} + 2c_1c_2H_{ab})}{(c_1^2 + c_2^2 + 2c_1c_2)} \right] = E$$

$$\text{or} \quad \frac{(2c_1H_{aa} + 2c_2H_{ab}) - (2c_1 + 2c_2S)E}{(c_1^2 + c_2^2 + 2c_1c_2S)} = 0$$

$$\text{or} \quad (2c_1H_{aa} + 2c_2H_{ab}) - (2c_1 + 2c_2S)E = 0$$

$$\text{or} \quad 2(c_1H_{aa} + c_2H_{ab}) - 2(c_1 + c_2S)E = 0$$

$$\text{or} \quad (c_1H_{aa} + c_2H_{ab}) - (c_1 + c_2S)E = 0$$

$$\text{or} \quad c_1H_{aa} + c_2H_{ab} - c_1E - c_2SE = 0$$

$$\text{or} \quad c_1(H_{aa} - E) + c_2(H_{ab} - SE) = 0 \tag{10.23}$$

$$\text{similiarly, when} \frac{dE}{dc_2} = 0 \text{ then } c_1(H_{ab} - SE) + c_2(H_{bb} - E) = 0 \tag{10.24}$$

Equations (10.23) and (10.24) are called *secular equations.* When these equations are solved, we get E in term of $E_0 = H_{aa} = H_{bb}$.

The secular determinant from Eqs. (10.23) and (10.24) can be expressed as

$$\begin{vmatrix} (H_{aa} - E), & (H_{ab} - SE) \\ (H_{ab} - SE), & (H_{bb} - E) \end{vmatrix} = 0 \quad [\because H_{aa} = H_{bb}] \tag{10.25}$$

or $(H_{aa} - E)^2 = (H_{ab} - \mathrm{SE})^2$

$$\therefore \quad (H_{aa} - E) = \pm(H_{ab} - \mathrm{SE}) \tag{10.26}$$

I. Taking the positive value, we can write

$$H_{aa} - E = H_{ab} - \text{SE}$$

$$or \quad H_{aa} - H_{ab} = E - \text{SE} = E(1 - S)$$

$$\therefore \quad E(1 - S) = H_{aa} - H_{ab}$$

$$\therefore \quad E = \frac{H_{aa} - H_{ab}}{(1 - S)} = E_1 \text{ (say)}$$

$$\therefore \quad E_1 = \frac{(E_0 - H_{ab})}{(1 - S)} \quad \because H_{aa} = E_0 \tag{10.27}$$

II. Taking the negative value, we can write

$$H_{aa} - E = -(H_{ab} - \text{SE})$$

$$\text{or} \quad H_{aa} - E = -H_{ab} + \text{SE}$$

$$\text{or} \quad H_{aa} + H_{ab} = E + \text{SE}$$

$$\text{or} \quad (E_0 + H_{ab}) = E(1 + S)$$

$$\therefore \quad E = \frac{(E_0 + H_{ab})}{(1 + S)} = E_2 \text{ (say)}$$

$$\therefore \quad E_2 = \frac{(E_0 + H_{ab})}{(1 + S)} \tag{10.28}$$

Equations (10.27) and (10.28) may be written in short as

$$E_{\pm} = \frac{(E_0 \pm H_{ab})}{(1 \pm S)} \tag{10.29}$$

In approximate works, S is sometimes neglected, and then, Eq. (10.29) takes the form

$$E_{\pm} = (E_0 \pm H_{ab}) \tag{10.30}$$

Equations (10.29) and (10.30) exhibit how an original pair of degenerate levels of energy E_0 is split into two, one lying higher and the other existing lower than before. On the basis of Eqs. (10.27) and (10.28), we can safely say that energy is split up to two levels E_1 and E_2.

Furthermore, since $E_0(E_H)$ is negative and H_{ab} is also negative, the numerator of the RHS of Eq. (10.27) will be greater than $E_0 = (H_{aa})$ due to the fact that RHS $= -\dfrac{E_0 + H_{ab}}{1 - S}$.

Because $s \approx 0.5$, $E_1 = \dfrac{-E_0 + H_{ab}}{0.5}$, which is still higher than $(-E_0 + H_{ab})$, which itself is greater than E_0.

Thus, we can say that $E_1 > E_0$.

When $S = 0.5$ is put in Eq. (10.28), we may write

$$E_2 = \frac{-E_0 + H_{ab}}{1.5}, \text{ which will be less than } E_0 \text{ i.e., } E_2 < E_0.$$

The aforementioned facts can be represented as

$$\begin{array}{l} \underline{\hspace{4cm}} E_1 \\ \underline{\hspace{4cm}} E_0 \\ \underline{\hspace{4cm}} E_2 \end{array}$$

Thus, due to the formation of the molecules, the energy levels of H_2^+ split up into two energy levels E_1 and E_2 or E_+ and E_-, in which $E_1 > E_0$ and $E_2 < E_0$.

It is known to us that each energy state can accommodate two electrons with opposite spin, and if the electrons in the molecule/molecular ion can be accommodated in the lower energy state, then the molecular form will be stabler than the atomic form. In the case of H_2^+ ion, there is only one electron, which can be perfectly accommodated in the lower energy level, and hence H_2^+ ion will be more stable than H atom.

The Schrödinger equation for hydrogen atom is given by

$$\nabla^2\Psi + \frac{2m}{\hbar^2}(E - V)\Psi = 0$$

When expressed in atomic unit, it takes the form

$$\nabla^2\Psi + 2(E - V)\Psi = 0 \tag{10.31}$$

Now, we shall go back to one dimension because of the spherical symmetry, and the aforementioned equation is represented by

$$\frac{1}{r^2}\frac{\mathrm{d}}{\mathrm{d}r}\left(r^2\frac{\mathrm{d}\Psi}{\mathrm{d}r}\right) + 2(E - V)\Psi = 0 \tag{10.32}$$

$$\nabla^2\Psi + 2E\Psi - 2V\Psi = 0 \tag{10.33}$$

$$\text{or} \quad E\Psi = \left(\frac{-\nabla^2}{2} + V\right)\Psi \qquad \therefore H = \left(\frac{-\nabla^2}{2} + V\right) \tag{10.34}$$

But V for H_2^+ ion is expressed as

$$V = -\frac{1}{r_a} - \frac{1}{r_b} + \frac{1}{R}$$

$$\therefore \quad \int \Psi_a H \Psi_a \mathrm{d}\tau = \int \Psi_a \left(\frac{-\nabla^2}{2} + V\right)\Psi_a \mathrm{d}\tau$$

$$= \int \Psi_a \left(\frac{-\nabla^2}{2} - \frac{1}{r_a} - \frac{1}{r_b} + \frac{1}{R}\right)\Psi_a \mathrm{d}\tau$$

$$= \int \Psi_a \left(\frac{-\nabla^2}{2} - \frac{1}{r_a}\right)\Psi_a \mathrm{d}\tau - \int \frac{\Psi_a^2 \mathrm{d}\tau}{r_b} + \int \frac{\Psi_a^2 \mathrm{d}\tau}{R}$$

$$= E_0 - \int \frac{\Psi_a^2 \mathrm{d}\tau}{r_b} + \int \frac{\Psi_a^2 \mathrm{d}\tau}{R} \tag{10.35}$$

where $\int \Psi_a \left(\frac{-\nabla^2}{2} + V \right) \Psi_a d\tau$ = energy of hydrogen atom with nucleus '*a*'

$-\int \frac{\Psi_a^2 d\tau}{r_b}$ = energy of interaction of charge density associated with orbital Y_a with the nucleus '*b*'.

= coulombic force = J

R = constant distance between the nuclei '*a*' and '*b*'

and $\int \Psi_a^2 d\tau = 1$

Thus,

$$H_{aa} = \int \Psi_a H \Psi_a d\tau = E_0 + J + \frac{1}{R} \tag{10.36}$$

$$\text{similarly,} \quad H_{bb} = \int \Psi_b H \Psi_b d\tau = E_0 + J + \frac{1}{R} \tag{10.37}$$

$$\int \Psi_a H \Psi_b d\tau = \int \Psi_a \left(\frac{-\nabla^2}{2} - \frac{1}{r_a} - \frac{1}{r_b} + \frac{1}{R} \right) \Psi_b d\tau$$

$$= \int \Psi_a \left(\frac{-\nabla^2}{2} - \frac{1}{r_a} \right) \Psi_b d\tau - \int \frac{\Psi_a \Psi_b d\tau}{r_b} + \frac{1}{R} \int \Psi_a \Psi_b d\tau$$

Now, we shall multiply the aforementioned equation by @@. The first term takes the form

$$\iint \Psi_a \left(\frac{-\nabla^2}{2} - \frac{1}{r_a} \right) \Psi_b \Psi_a^2 d\tau = \int \Psi_a \left(\frac{-\nabla^2}{2} - \frac{1}{r_a} \right) \Psi_b d\tau \cdot \int \Psi_a \Psi_b d\tau$$

$$= E_0 \cdot S \qquad \left[\because \int \Psi_a \left(\frac{-\nabla^2}{2} - \frac{1}{r_a} \right) \Psi_b d\tau = E_0 \right]$$

$$\text{and} \quad \int \Psi_a H \Psi_b d\tau = E_0 \cdot S - \int \frac{\Psi_a \Psi_b d\tau}{r_b} + \frac{1}{R} \cdot S \qquad \left[\because \int \Psi_a \Psi_b d\tau = S \right]$$

$$= E_0 \cdot S + K + \frac{S}{R} \tag{10.38}$$

where $\Psi_a \Psi_b d\tau$ = simultaneous probability of two Eigen functions Ψ_a and Ψ_b in the same region $d\tau$.

$-\int \frac{\Psi_a \Psi_b d\tau}{r_b} = K$ = energy of interaction of the electron with nucleus '*b*' = resonance energy

$$\text{similarly,} \quad \int \Psi_b H \Psi_b d\tau = E_0 \cdot S + K + \frac{S}{R} \tag{10.39}$$

Therefore, $E_1 = \dfrac{H_{aa} - H_{ab}}{1-S}$ [from Eq. (10.27)]

$$= \frac{E_0 + J + \dfrac{1}{R} - E_0 S - K - \dfrac{S}{R}}{1-S}$$

$$= \frac{(E_0 - E_0 S)}{(1-S)} + \frac{\dfrac{1}{R}(1-S)}{(1-S)} + \frac{J-K}{1-S}$$

$$= E_0 + \frac{1}{R} + \frac{J-K}{1-S} \tag{10.40}$$

Similarly, the value of E_2 may be given by

$$E_2 = E_0 + \frac{1}{R} + \frac{J-K}{1-S} \tag{10.41}$$

where E_0 = atomic energy

$\dfrac{1}{R}$ = coulombic contribution which is very less

K = resonance energy = exchange energy

Hence, whole of the binding energy of H_2^+ ion is because of resonance energy K, considering S negligible. It should be kept in mind that K arises because of the oscillation of the electron between two nuclei.

Equation (10.41) is indicative of the fact that the energy of H_2^+ is less than that of a system having a separate atom of hydrogen and a proton by the value of $J + K/1 + S$, since the energy is negative. The energy with the positive sign indicates the bond energy.

It remains now to evaluate the overlap integral S and the two interaction integrals H_{aa} and H_{ab} or E_{aa} and E_{ab}. These integrals are coulomb integral (J) and the resonance integral (K). These are evaluated by transforming the variables to elliptical coordinates. These coordinates are utilised for problems consisting of two centres A and B, and R is the distance between A and B. The lines AP and BP define a plane, and the line obtained by the intersection of this plane with XY plane defines the angle Φ. The point P is defined by the coordinates μ, v, and Φ, and the point P specifies the distance r_a and r_b along the line AP and BP, respectively, and the angle Φ.

Elliptical coordinates μ and v are then defined as follows:

$$\mu = \frac{r_a + r_b}{R} \quad \text{and} \quad v = \frac{r_a - r_b}{R} \quad \text{and the third coordinate is the angle } \Phi. \tag{10.42}$$

Try to understand that keeping μ constant defines an ellipsoid of revolution with the points A and B as foci. Surfaces of constant v represent the paraboloids of revolution about the z axis. These surfaces have been illustrated in Figure 10.2.

The equations that express x, y, and z in terms of μ, v, and Φ are as follows:

$$\left.\begin{aligned} x &= \frac{R}{2}\left(\mu^2 - 1\right)\left(1 - v^2\right)\cos\Phi \\ y &= \frac{R}{2}\left(\mu^2 - 1\right)\left(1 - v^2\right)\sin\Phi \\ z &= \frac{R}{2}\mu v \end{aligned}\right\} \tag{10.43}$$

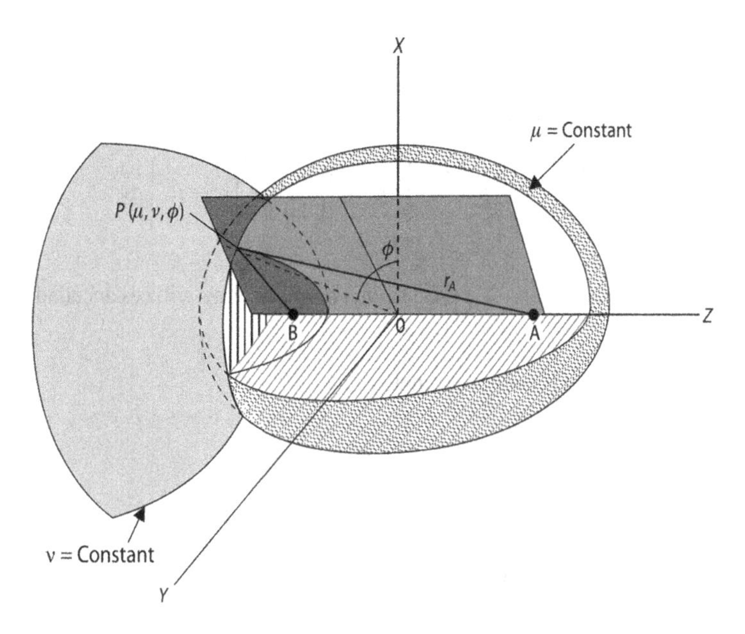

FIGURE 10.2 Confocal ellipsoidal coordinates. (*Note*: μ = the ellipsoids of revolution about z axis; ν = surfaces of paraboloids of revolution and the intersection of these two surfaces is a circle Φ = representing a point $P(\mu, \nu, \phi)$ on the circle.)

In problems of quantum mechanics, we are often required to solve the integrals over all space. The differential volume is $d\tau$, which must be known for the elliptical coordinate system, and the limits of integration are mentioned, which include all spaces as

$$\left.\begin{aligned} d\tau = \frac{R^3}{8}\left(\mu^2 - \nu^2\right)d\mu d\nu d\Phi \;\; 1 \le \mu \le \infty \\ - \le \nu \le +1 \\ 0 \le \Phi \le 2\pi \end{aligned}\right\} \tag{10.44}$$

We shall derive the value of volume element by arc length method. From Eq. (10.43), we may write

$$dx = \frac{R}{2}\left[\left(1-\nu^2\right)^{\frac{1}{2}} \cdot \frac{1}{2}\left(1-\mu^2\right)^{\frac{1}{2}} \cdot 2\mu\cos\Phi d\mu - \frac{1}{2}\left(\mu^2-1\right)^{\frac{1}{2}} \cdot \left(1-\nu^2\right)^{\frac{1}{2}} 2\nu\cos\Phi d\nu - \left(\mu^2-1\right)^{\frac{1}{2}} \cdot \left(1-\nu^2\right)^{\frac{1}{2}} \sin\Phi d\Phi\right]$$

$$= \frac{R}{2}\left[\left(\frac{1-\nu^2}{1-\mu^2}\right)^{\frac{1}{2}} \mu d\,\mu\cos\Phi - \left(\frac{\mu^2-1}{1-\nu^2}\right)^{\frac{1}{2}} \nu d\nu\cos\Phi - \left(\mu^2-1\right)^{\frac{1}{2}}\left(1-\nu^2\right)^{\frac{1}{2}}\sin\Phi d\Phi\right] \tag{10.45}$$

$$dy = \frac{R}{2}\left[\left(\frac{1-\nu^2}{\mu^2-1}\right)^{\frac{1}{2}} \mu d\,\mu\cos\Phi - \left(\frac{\mu^2-1}{1-\nu^2}\right)^{\frac{1}{2}} \nu d\nu\cos\Phi - \left(\mu^2-1\right)^{\frac{1}{2}}\left(1-\nu^2\right)^{\frac{1}{2}}\sin\Phi d\Phi\right] \tag{10.46}$$

$$dz = \frac{R}{2}\left[\nu d\mu + \mu d\nu\right] \tag{10.47}$$

Now $dx^2 + dy^2 + dz^2 = ds^2$ = arc length

$$\therefore \quad ds^2 = dx^2 + dy^2 + dz^2$$

$$= \frac{R^2}{4}\left[\left\{\left(\frac{1-v^2}{\mu^2-1}\right)\mu^2 + v^2\right\}d\mu^2 + \left\{\left(\frac{\mu^2-1}{1-v^2}\right)v^2 + \mu^2\right\}dv^2\right.$$

$$\left. +\left(\mu^2-1\right)\left(1-v^2\right)\, d\Phi^2 + \text{other products terms will be cancelled}\right] \tag{10.48}$$

But according to orthogonal curvilinear coordinates

$$ds^2 = h_\mu^2\, d\mu^2 + h_v^2 dv^2 + h_\Phi^2\, d\Phi^2 \tag{10.49}$$

where h's are scalars.

On comparison of Eqs. (10.49) with (10.48), we may conclude that

$$h_\mu = \frac{R}{2}\left\{\left(\frac{1-v^2}{\mu^2-1}\right)\mu^2 + v^2\right\}^{1/2} = \frac{R}{2}\left\{\frac{\left(1-v^2\right)\mu^2 + v^2\left(\mu^2-1\right)}{\mu^2-1}\right\}^{1/2}$$

$$h_v = \frac{R}{2}\left\{\left(\frac{\mu^2-1}{1-v^2}\right)v^2 + \mu^2\right\}^{1/2} = \frac{R}{2}\left\{\frac{\left(\mu^2-1\right)v^2 + \left(1-v^2\right)}{\left(1-v^2\right)}\right\}^{1/2}$$

$$h_\Phi = \frac{R}{2}\left[\left(\mu^2-1\right)\left(1-v^2\right)\right]^{1/2}$$

$\therefore$ volume element $d\tau = h_\mu \cdot h_v \cdot h_\Phi d\mu \cdot dv \cdot d\Phi$

$$\therefore \quad h_\mu \cdot h_v \cdot h_\Phi = \frac{R^3}{2}\left\{\frac{\left(1-v^2\right)\mu^2 + v^2\left(\mu^2-1\right)}{\mu^2-1} \cdot \frac{\left(\mu^2-1\right)v^2 + \mu^2\left(1-v^2\right)}{\left(1-v^2\right)} \cdot \left(\mu^2-1\right)\left(1-v^2\right)\right\}^{1/2}$$

$$= \frac{R^3}{2}\left\{\left(\mu^2 - v^2\mu^2 + v^2\mu^2 - v^2\right)\left(\mu^2 v^2 - v^2 + \mu^2 - \mu^2 v^2\right)\right\}^{1/2}$$

$$= \frac{R^3}{2}\left\{\left(\mu^2 - v^2\right)^2\right\}^{\frac{1}{2}}$$

$$= \frac{R^3}{2}\left(\mu^2 - v^2\right) \tag{10.50}$$

$$\therefore \quad d\tau = h_\mu \cdot h_v \cdot h_\Phi d\mu \cdot dv \cdot d\Phi$$

$$\therefore \quad d\tau = \frac{R^3}{2}\left(\mu^2 - v^2\right) d\mu \cdot dv \cdot d\Phi \tag{10.51}$$

Now, we shall use the value of the volume element over all space ($d\tau$) in evaluating the overlap integral, Couloumb integral, and resonance integral ahead.

[NB: Volume element is defined as $\vec{a} \cdot \vec{b} \times \vec{c}$ or $\vec{ds_1} \cdot \vec{ds_2} \times \vec{ds_3}$ or [abc] or $[ds_1 \cdot ds_2 ds_2]$ in triple vector scalar product.]

10.2.1 Evaluation of Overlap Integral

$$S = \int \Psi_a \Psi_b \mathrm{d}\tau$$

The wave functions are $\Psi_a = \frac{1}{\sqrt{\Pi}} e^{-r_a}$ and $\Psi_b = \frac{1}{\sqrt{\Pi}} e^{-r_b}$. Putting these values of Ψ_a and Ψ_b in the aforementioned equation, we get

$$S = \int \frac{1}{\sqrt{\pi}} e^{-r_a} \cdot \frac{1}{\sqrt{\Pi}} e^{-r_b} \mathrm{d}\tau = \frac{1}{\pi} \int e^{-(r_a + r_b)} \mathrm{d}\tau$$

Transforming into elliptical coordinates keeping in mind that $(r_a + r_b) = \mu R$ (according to Eq. (10.42))

$$\therefore \quad S = \frac{R^3}{8\pi} \int_{\mu=1}^{\infty} \int_{v=-1}^{+1} \int_{\phi=0}^{2\pi} e^{-\mu R} \left(\mu^2 - v^2\right) \mathrm{d}\mu \cdot \mathrm{d}v \cdot \mathrm{d}\phi$$

$$= \frac{R^3}{8\pi} \int_1^{\infty} e^{-\mu R} \mathrm{d}\mu \int_{-1}^{1} \left(\mu^2 - v^2\right) \mathrm{d}v \int_0^{2\pi} \mathrm{d}\phi$$

$$= \frac{R^3}{8\pi} 2\pi \int_1^{\infty} e^{-\mu R} \mathrm{d}\mu \mathrm{d}v \int_{-1}^{1} \left(\mu^2 - v^2\right) \mathrm{d}v \mathrm{d}\mu \qquad \left[\because \int_0^{2\pi} \mathrm{d}\phi = 2\pi\right]$$

$$= \frac{R^3}{4} \int_1^{\infty} e^{-\mu R} \mu^2 \mathrm{d}\mu - \frac{R^3}{4} 2/3 \int_1^{\infty} e^{-\mu R} \mathrm{d}\mu$$

$$= \frac{R^3}{4} \int_1^{\infty} \mu^2 e^{-\mu R} \mathrm{d}\mu - \frac{R^3}{6} \int_1^{\infty} e^{-\mu R} \mathrm{d}\mu$$

$$\left[\therefore \quad \int_{-1}^{+1} \mathrm{d}v = 2 \text{ and } \int_{-1}^{+1} v^2 \mathrm{d}v = 2/3\right] \tag{10.52}$$

Putting the value of standard integrals in Eq. (10.52), we shall get

$$S = \frac{R^3}{4} \cdot \frac{1}{R^3} e^{-R} \left[R^2 + 2R + 2\right] - \frac{R^3}{6} \cdot \frac{1}{R} e^{-R}$$

$$\text{Since } \int_1^{\infty} x^n e^{-ax} \mathrm{d}x = \frac{1}{x^{n+1}} e^{-a} \left[x^n + nx^{n-1} + n(n-1)x^{n-2} + \cdots\right]$$

$$\therefore \quad S = \frac{e^{-R}}{2} \left[R^2 + 2R + 2\right] - \frac{R^2}{6} e^{-R}$$

$$\text{or} \quad S = e^{-R} \left[\frac{R^2}{2} + R + 1 - \frac{R^2}{6}\right]$$

$$\text{or} \quad S = e^{-R}\left[\frac{R^2}{2} - \frac{R^2}{6} + R + 1\right]$$

$$\therefore \quad S = e^{-R}\left[\frac{R^2}{3} + R + 1\right] \tag{10.53}$$

This is the value of overlap integral.

10.2.2 Evaluation of the Coulomb Integral

We know that

$$J = \int \Psi_a^2 \, d\tau = \frac{1}{\Pi}\int \frac{e^{-2r_a}}{r_b} d\tau$$

$$\text{But} \quad \mu = \frac{r_a + r_b}{R}, \; v = \frac{r_a - r_b}{R}$$

$\therefore 2r_a = (\mu + v)R$
and $r_b = (\mu - v)^R/2$

$$\therefore \quad J = \frac{R^3}{8\pi}\int_1^\infty \int_{-1}^{+1} \int_0^{2\pi} \frac{e^{-(\mu + v)R}}{\frac{R}{2(\mu - v)}}(\mu^2 - v^2)\,d\mu dv d\Phi$$

$$= \frac{R^2}{2}\int_1^\infty \int_{-1}^{+1} e^{-(\mu + v)R} \cdot (\mu + v)\,d\mu dv$$

$$= \frac{R^2}{2}\left[\int_1^\infty \int_{-1}^{+1} e^{-(\mu + v)R} \; \mu d\mu dv + \int_1^\infty e^{-(\mu + v)R} d\mu v dv\right]$$

$$= \frac{R^2}{2}\left[\int_1^\infty e^{-\mu R}\mu d\mu \cdot \int_{-1}^{+1} e^{-\mu R} dv + \int_1^\infty e^{-\mu R} d\mu \cdot \int_{-1}^{+1} e^{-vR} v dv\right]$$

$$\text{Now,} \int_{-1}^{+1} e^{-vR}\,dv = \left[e^{-vR} \cdot \left(-\frac{1}{R}\right)\right]_{-1}^{+1} = \left[e^{-R} \times \left(-\frac{1}{R}\right)\right] - \left[e^{R}\left(-\frac{1}{R}\right)\right] = \frac{e^R - e^{-R}}{R}$$

$$\text{and} \quad \int_{-1}^{+1} e^{-vR} v\,dv = \left[v \cdot \left(\frac{-1}{R}\right)e^{-vR} - \int \left(\frac{-1}{R}\right)e^{-vR} dv\right]_{-1}^{+1}$$

$$= \left[\left(-\frac{1}{R}\right)e^{-R}\right] - \left[\left(-\frac{1}{R}\right)e^{R}\right] - \frac{1}{R^2}\left[e^{-R} - e^{R}\right]$$

$$= \frac{e^R - e^{-R}}{R^2} - \frac{e^R + e^{-R}}{R}$$

$$\therefore J=\frac{R^2}{2}\left[\frac{1}{R^2}\cdot e^{-R}(R+1)\times\frac{e^R-e^{-R}}{R}\right]+\frac{R^2}{2}\left[\frac{1}{R}e^{-R}\right]\left[\frac{e^R-e^{-R}}{R^2}-\frac{e^R+e^{-R}}{R}\right]$$

$$=\frac{1}{2R}(R+1)\left(1-e^{-2R}\right)+\frac{1}{2R}\left(1-e^{-2R}\right)-\frac{1}{2}\left(1+e^{-2R}\right)$$

$$=\frac{1}{2R}\left(1-e^{-2R}\right)(R+2)-\frac{1}{2}\left(1+e^{-2R}\right)$$

$$=\left(\frac{R}{2R}+\frac{1}{R}\right)\left(1-e^{-2R}\right)-\frac{1}{2}-\frac{1}{2}e^{-2R}$$

$$=\frac{1}{R}-\frac{1}{R}e^{-2R}+\frac{1}{2}-\frac{1}{2}e^{-2R}-\frac{1}{2}-\frac{1}{2}e^{-2R}$$

$$=\frac{1}{R}-\frac{1}{R}e^{-2R}-e^{-2R}=\frac{1}{R}\left[1-e^{-2R}+\frac{1}{R}\cdot\mathrm{Re}^{-2R}\right]$$

$$\therefore\quad J=\frac{1}{R}\left(1-e^{-2R}-\mathrm{Re}^{-2R}\right)=\frac{1}{R}\left[1-e^{-2R}(1+R)\right]\tag{10.54}$$

This is the value of the coulomb integral.

10.2.3 Evaluation of Resonance Integral or Exchange Integral

We know that the resonance integral

$$K=\int\frac{\Psi_a\Psi_b}{r_b}=\frac{1}{\pi}\int\frac{e^{-(r_a-r_b)}}{r_b}$$

$$\text{Putting } \mu=\frac{(r_a+r_b)}{R}\text{ and } v=\frac{(r_a-r_b)}{R}$$

$\therefore r_b=(\mu-v)R/2$

Now transforming the aforementioned equations into elliptical coordinates, we obtain

$$K=\frac{R^3}{8\pi}\int_1^{\infty}\int_{-1}^{+1}\int_0^{2\pi}\frac{e^{-\mu R}}{R/2(\mu-v)}\left(\mu^2-v^2\right)\mathrm{d}\mu\mathrm{d}v\mathrm{d}\Phi$$

$$=\frac{R^3}{8\pi}\cdot 2\pi\cdot 2/R\int_1^{\infty}\int_{-1}^{1}e^{-\mu R}(\mu+v)\mathrm{d}\mu\mathrm{d}v$$

$$=\frac{R^2}{2}\int_1^{\infty}\int_{-1}^{1}e^{-\mu R}\mu\mathrm{d}\mu\mathrm{d}v+\frac{R^2}{2}\int_1^{\infty}\int_{-1}^{1}e^{-\mu R}\mathrm{d}\mu\mathrm{d}v$$

$$=\frac{R^2}{2}\times 2\cdot\frac{1}{R^2}e^{-R}(R+1)+\frac{R^2}{2}\times 0\int_1^{\infty}e^{-\mu R}\mathrm{d}v$$

$$=e^{-R}(R+1)\tag{10.55}$$

This is the value of the resonance integral.

When R is large, it can be observed that $S \to 0$, H_{aa} approaches E_0 and H_{ab} approaches zero. Thus, it is concluded that both E_1 and E_2 tend to become E_0, i.e., the energy of the 1s orbital of hydrogen.

10.3 Evaluation of Ψ and Ψ² (Probability)

On putting the value of E in the secular Eqs. (10.23) and (10.24), we shall find that

$$c_1 = c_2 \left(\text{for symmetric solution}\right) \tag{10.56}$$

$$\text{and } c_1 = -c_2 \left(\text{for symmetric solution}\right) \tag{10.57}$$

Thus, we can express the two molecular orbital (MO) wave functions as

$$\Psi_S = c_1 \left(\Psi_a + \Psi_b\right) \tag{10.58}$$

$$\text{and } \Psi_A = c_2 \left(\Psi_a - \Psi_b\right) \tag{10.59}$$

In order to evaluate c_1 and c_2, we shall make use of the normalisation condition, i.e.,

$$\int \Psi_S^2 d\tau = 1$$

$$\text{or } \int c_1^2 \left(\Psi_a + \Psi_b\right)^2 d\tau = 1$$

$$\text{or } \int c_1^2 \left(\Psi_a^2 + \Psi_b^2 + 2\Psi_a\Psi_b\right) d\tau = 1$$

$$\text{or } \int c_1^2 \Psi_a^2 d\tau + c_1^2 \int \Psi_b^2 d\tau + 2c_1^2 \int \Psi_a\Psi_b d\tau = 1$$

$$\text{or } c_1^2 + c_1^2 + 2c_1^2 S_{ab} = 1$$

$$\text{or } 2c_1^2 \left(1 + S_{ab}\right) = 1$$

$$c_1^2 = \frac{1}{2(1+S_{ab})} \qquad \therefore c_1 = \pm\frac{1}{\sqrt{2(1+S_{ab})}} \tag{10.60}$$

$$\text{and similarly, } c_2 = \pm\frac{1}{\sqrt{2(1+S_{ab})}}$$

$$\therefore \quad \Psi_S = \frac{1}{\sqrt{2(1+S_{ab})}}\left(\Psi_a + \Psi_b\right) = \frac{1}{\sqrt{(2+2S_{ab})}}\left(\Psi_a + \Psi_b\right) \tag{10.61}$$

$$\text{Similarly, } \Psi_A = \frac{1}{\sqrt{2(1-S_{ab})}}\left(\Psi_a - \Psi_b\right) = \frac{1}{\sqrt{(2-2S_{ab})}}\left(\Psi_a - \Psi_b\right) \tag{10.62}$$

$$\therefore \quad \text{Probability, } \Psi_S^2 = \left\{ \frac{1}{\sqrt{(2+2S_{ab})}} (\Psi_a - \Psi_b) \right\}^2$$

$$= \frac{1}{(2+2S_{ab})} \left(\Psi_a^2 + 2\Psi_a \Psi_b + \Psi_b^2 \right)$$

$$= \frac{1}{2(1+S_{ab})} \left(\Psi_a^2 + 2\Psi_a \Psi_b + \Psi_b^2 \right)$$

When S_{ab} <<< 1, then Sab may be neglected and the aforementioned equation becomes

$$P_1 = \Psi_S^2 = \frac{1}{2} \left(\Psi_a^2 + 2\Psi_a \Psi_b + \Psi_b^2 \right) = \frac{1}{2} (\Psi_a - \Psi_b)^2 \tag{10.63}$$

$$\text{Similarly, } P_2 = \Psi_A^2 = \left\{ \frac{1}{\sqrt{2(1-S_{ab})}} (\Psi_a - \Psi_b) \right\}^2$$

$$= \frac{1}{2(1-S_{ab})} \left(\Psi_a^2 - 2\Psi_a \Psi_b + \Psi_b^2 \right)$$

But when S_{ab} <<< 1, then the aforementioned equation takes the form

$$\Psi_A^2 = \frac{1}{2} \left(\Psi_a^2 - 2\Psi_a \Psi_b + \Psi_b^2 \right)^2$$

$$= 1/2(\Psi_a - \Psi_b)^2 \tag{10.64}$$

We see that

$$P_1 (\text{probability}) = \frac{1}{(2+2S_{ab})} \left(\Psi_a^2 + 2\Psi_a \Psi_b + \Psi_b^2 \right)$$

$$\text{and } P_2 (\text{probability}) = \frac{1}{(2-2S_{ab})} \left(\Psi_a^2 + \Psi_b^2 - 2\Psi_a \Psi_b \right)$$

If we plot a graph between the electron density/probability and the internuclear distance, curves are obtained as depicted in Figure 10.3.

It is obvious from the figure that P_1 has the value $4/(_{2+2S})$ in the midway between the nuclei. It actually rises to a maximum in the proximity of the nuclei, and after that, it decreases to zero on the side remote from the internuclear space.

For Ψ_2 orbital, $P_2 \rightarrow 0$ midway between the nuclei. Electron density plots reveal that $P1$ has a substantial electron density between the nuclei, whereas P_2 is electronically deficient in the internuclear space. The lower energy state indicates a stable molecular entity and is the bonding state, whereas the higher energy state is the antibonding state.

Here, it can be easily observed that Ψ_1 state leads to a stable molecule, whereas the higher state is energetically unstable with respect to the isolated hydrogen atom. This has been illustrated in Figure 10.4.

In the plot, the depth of the minimum in the energy vs. r curve is called the dissociation energy (D_e), and this is the energy, which is able to break the bond. When D_e is larger, the molecule becomes more stable, and low value of D_e obviously indicates that the molecule is not very stable.

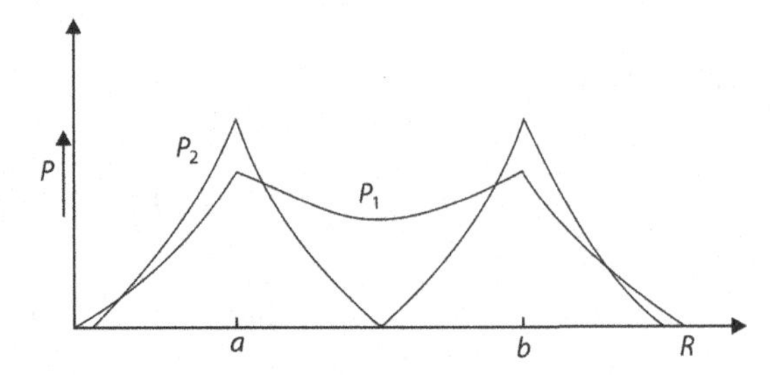

FIGURE 10.3 Electron density/probability of finding electron for H_2^+ orbitals.

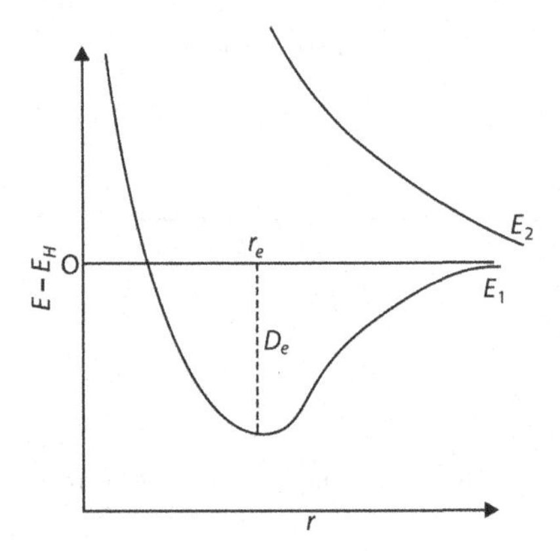

FIGURE 10.4 Potential energy vs. r curve for H_2^+.

In the case of H_2^+ ion, D_e is found to be 1.76 eV and r_e = 1.32 Å. Experimental value of D_e = 2.791 eV at r_e = 1.06Å.

10.4 Hydrogen Molecule (Spin Independent)

The hydrogen molecule (H_2) consists of two positively charged nuclei '*a*' and '*b*' and two electrons 1 and 2, as illustrated in Figure 10.5. In other words, we can say that hydrogen molecule with two protons and two electrons plays an important role in the theory of molecular electronic structure.

For the hydrogen molecule, we have no exact solution in closed form, but luckily, it is possible with the aid of fairly simple process to construct approximate wave functions, which consist of some of the fundamental characteristics that one can expect for an exact solution.

Various distances of the molecule are shown in the figure. It should be kept in mind that when the two atoms remain far apart, their mutual repulsion energy will be zero, i.e., V = 0. The system will then look as two separate H atoms, but if the atoms remain in their normal state, the appropriate wave function

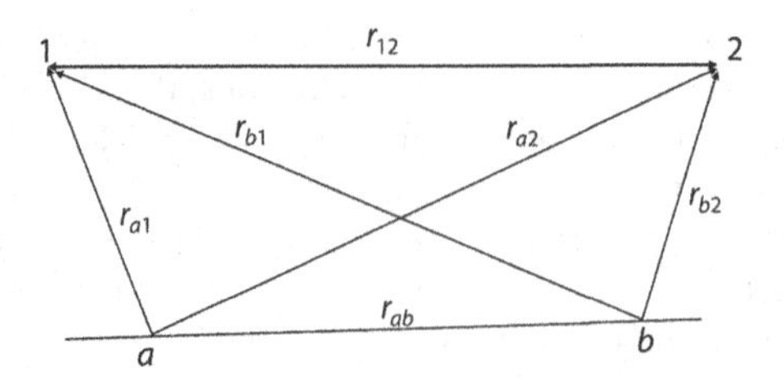

FIGURE 10.5 Hydrogen molecule.

(excluding spin) will be just like hydrogenic 1s wave function $\left[\Psi_{100} = \frac{1}{\sqrt{11}}\left(\frac{1}{a_0}\right)^{\frac{3}{2}} \cdot e^{-\frac{r}{a_0}}\right]$. These orbitals may be illustrated by $\Psi_a(1)$ and $\Psi_b(2)$, which clearly indicates that electron (1) is associated with nucleus (a) and electron (2) with nucleus (b).

$\Psi_a^2(1)$ and $\Psi_b^2(2)$ = probability of electron (**1**) being in the field of nucleus (a) and electron (2) in the field of nucleus (b):

$$\Psi_I^2 = \Psi_a^2(1)\cdot\Psi_b^2(2)$$

$$\text{or} \quad \Psi_I = \Psi_a(1)\cdot\Psi_b(2) \tag{10.65}$$

Similarly, the simultaneous probability of electron (2) on the nucleus (a) and electron (1) on (b) is given as

$$\Psi_{II}^2 = \Psi_a^2(2)\cdot\Psi_b^2(1)$$

$$\text{or} \quad \Psi_{II} = \Psi_a(2)\cdot\Psi_b(1) \tag{10.66}$$

When two atoms of hydrogen are brought together to form a molecule, a simple wave function would be the product of two 1s atomic wave functions, as given in Eqs. (10.65) and (10.66). Since we cannot distinguish between electron (1) and electron (2), the wave function $\Psi_a(2)\Psi_b(1)$ is as much acceptable as $\Psi_a(1)\Psi_b(2)$. For this, we must assume that the complete wave function should of the type

$$\Psi = C_1\Psi_I + C_2\Psi_{II} \tag{10.67}$$

This concept was put forward by *Heitler* and *London.*

In a hydrogen molecule, there are two electrons; therefore, two Laplacian operators will be represented by ∇_1^2 and ∇_2^2, and the Schrödinger equation may be expressed as

$$\left(\nabla_1^2 + \nabla_2^2\right)\Psi + \frac{\hbar^2}{2m}(E - V)\Psi = 0 \tag{10.68}$$

Expressing E in terms of $h\, e^2/a_0$ and r in terms of a_0, the Schrödinger equation may be represented as

$$\left(\nabla_1^2 + \nabla_2^2\right)\Psi + 2(E - V)\Psi = 0$$

$$\text{or} \quad -\left(\frac{\nabla_1^2}{2} + \frac{\nabla_2^2}{2}\right)\Psi + V\Psi = E\Psi = H\Psi \tag{10.69}$$

$$\text{or} \quad H = \left(-\frac{\nabla_1^2}{2} - \frac{\nabla_2^2}{2} + V\right)$$

But V for H_2 molecule is expressed by

$$V = \left(-\frac{1}{r_a(1)} - \frac{1}{r_a(2)} - \frac{1}{r_b(1)} - \frac{1}{r_b(2)} + \frac{1}{r_{ab}} + \frac{1}{r_{12}}\right) \tag{10.70}$$

It should be kept in mind that r terms must be multiplied but this has not been left for simplicity.

$$\text{or} \quad H = -\frac{\nabla_1^2}{2} - \frac{\nabla_2^2}{2} + \left(-\frac{1}{r_a(1)} - \frac{1}{r_a(2)} - \frac{1}{r_b(1)} - \frac{1}{r_b(2)} + \frac{1}{r_{ab}} + \frac{1}{r_{12}} \right)$$

Now we can write for E as

$$E = \frac{\int \Psi H \Psi \mathrm{d}\tau}{\int \Psi^2 \mathrm{d}\tau}$$

Putting the value of Ψ from Eq. (10.67), we obtain

$$E = \frac{\iint (c_1\Psi_I + c_2\Psi_{II}) H (c_1\Psi_I + c_2\Psi_{II}) \mathrm{d}\tau_1 \mathrm{d}\tau_2}{\iint (c_1\Psi_I + c_2\Psi_{II})^2 \mathrm{d}\tau_1 \mathrm{d}\tau_2}$$

$$\text{or} \quad \left[\frac{c_1^2 \iint \Psi_I H \Psi_I \mathrm{d}\tau_1 \mathrm{d}\tau_2 + c_2^2 \iint \Psi_{II} H \Psi_{II} \mathrm{d}\tau_1 \mathrm{d}\tau_2 + c_1 c_2 \iint \Psi_I H \Psi_{II} \mathrm{d}\tau_1 \mathrm{d}\tau_2 + c_1 c_2 \iint \Psi_{II} H \Psi_I \mathrm{d}\tau_1 \mathrm{d}\tau_2}{c_1^2 \iint \Psi_I^2 \mathrm{d}\tau_1 \mathrm{d}\tau_2 + c_2^2 \iint \Psi_{II}^2 \mathrm{d}\tau_1 \mathrm{d}\tau_2 + 2c_1 c_2 \iint \Psi_I \Psi_{II} \mathrm{d}\tau_2} \right]$$

$$= \frac{c_1^2 H_{11} + c_2^2 H_{22} + 2c_1^2 H_{12}}{c_1^2 + c_2^2 + 2c_1 c_2 S}, \quad \text{where,} \quad S = \iint \Psi_I \Psi_{II} \mathrm{d}\tau_1 \mathrm{d}\tau_2 \tag{10.71}$$

and 'S' is called *overlap* or *orthogonality* integral.

Applying variation method for getting minimum, for which condition is

$$\frac{\mathrm{d}E}{\mathrm{d}c_1} = \frac{\mathrm{d}E}{\mathrm{d}c_2} = 0$$

Therefore,

$$\frac{\mathrm{d}E}{\mathrm{d}c_1} = \frac{(2c_1 H_{11} + 2c_2 H_{12})(c_1^2 + c_2^2 + 2c_1 c_2 S) - (2c_1 + 2\ c_2 S)(c_1^2 H_{11} + c_2^2 H_{22} + 2c_1 c_2 H_{12})}{(c_1^2 + c_2^2 + 2c_1 c_2 S)^2}$$

Multiplying the aforementioned equation by $\dfrac{c_1^2 + c_2^2 + 2c_1 c_2 S}{2}$, we obtain

$$(c_1 H_{11} + c_2 H_{12}) - \frac{(c_1 + c_2 S)(c_1^2 H_{11} + c_2^2 H_{22} + 2c_1 c_2 H_{12})}{(c_1^2 + c_2^2 + 2c_1 c_2 S)} = 0$$

$$\text{or} \quad (c_1 H_{11} + c_2 H_{12}) - (c_1 + c_2 S) E = 0 \quad [\text{from Eq. (10.23)}] \tag{10.72}$$

Similarly, $\frac{dE}{dc_2} = 0$ and we can write

$$\frac{dE}{dc_2} = (c_2 H_{22} + c_1 H_{21}) - (c_2 + c_1 S)E = 0 \quad (10.73)$$

On rearranging Eqs. (10.72) and (10.73), we may write

$$\left.\begin{array}{l} c_1(H_{11} - E) + c_2(H_{12} - SE) = 0 \\ c_1(H_{21} - SE) + c_2(H_{22} - E) = 0 \end{array}\right\} \quad (10.74)$$

This is the secular equation, and the corresponding secular determinant may be expressed as

$$\begin{vmatrix} H_{11} - E & H_{12} - SE \\ H_{21} - SE & H_{22} - E \end{vmatrix} = 0 \quad (10.75)$$

The terms H_{11} and H_{22} are coulombic integral, i.e.,

$$\begin{aligned} H_{11} &= \iint \Psi_1 H \Psi_1 \, d\tau_1 \, d\tau_2 \\ &= \iint \Psi_a(1)\Psi_b(2) H \Psi_a(1)\Psi_b(2) d\tau_1 \, d\tau_2 \\ &= \iint \Psi_a(1)\Psi_b(2)\left(-\frac{\nabla_1^2}{2} - \frac{\nabla_2^2}{2}\right) + \left(-\frac{1}{r_a(1)} - \frac{1}{r_a(2)} - \frac{1}{r_b(1)} - \frac{1}{r_b(2)} + \frac{1}{r_{ab}} + \frac{1}{r_{12}}\right)\Psi_a(1) H \Psi_b(2) d\tau_1 \, d\tau_2 \\ &= \iint \Psi_a(1)\Psi_b(2)\left(-\frac{\nabla_1^2}{2} - \frac{1}{r_a(1)}\right)\Psi_a(1)\Psi_b(2) d\tau_1 \, d\tau_2 \\ &\quad + \iint \Psi_a(1)\Psi_b(2)\left(-\frac{1}{r_a(2)} - \frac{1}{r_a(1)} + \frac{1}{r_{ab}} + \frac{1}{r_{12}}\right)\Psi_a(1)\Psi_b(2) d\tau_1 \, d\tau_2 \\ &\quad + \iint \Psi_a(1)\Psi_b(2)\left(-\frac{1}{r_a(2)} - \frac{1}{r_a(1)} + \frac{1}{r_{ab}} + \frac{1}{r_{12}}\right)\Psi_a(1)\Psi_b(2) d\tau_1 \, d\tau_2 \\ &= E_0 + E_0 + J \\ &= 2E_0 + J \end{aligned} \quad (10.76)$$

where E_0 = energy of an isolated H atom in its normal state.

Since both $\Psi_a(1)$ and $\Psi_b(2)$ are normalised, the first two terms = $2E_0$.

J = attraction between the charged clouds around (b) and the nucleus (a) and the attraction between the charged cloud around (a) and the nucleus (b).

Furthermore, it consists of also coulombic repulsion between two charged clouds of density $\Psi^2_{a(1)}$ and $\Psi^2_{b(1)}$ and the coulomb term between the two nuclei. The coulombic interaction is represented by J.

Similarly, it can be found that

$$H_{22} = 2E_0 + J \quad (10.77)$$

And the integrals

$$H_{12} = H_{21} = \iint \Psi_I H \Psi_{II}\, d\tau_1\, d\tau_2$$

$$= \iint \Psi_a(1)\Psi_b(2)\left(-\frac{\nabla_1^2}{2} - \frac{\nabla_2^2}{2} - \frac{1}{r_a(1)} - \frac{1}{r_a(2)} - \frac{1}{r_b(1)} - \frac{1}{r_b(2)} + \frac{1}{r_{ab}} + \frac{1}{r_{12}}\right)\Psi_b(1)\Psi_a(1)\, d\tau_1\, d\tau_2$$

$$= \iint \Psi_a(1)\Psi_b(2)\left(-\frac{\nabla_1^2}{2} - \frac{1}{r_a(1)}\right)\Psi_b(1)\Psi_a(1)\, d\tau_1\, d\tau_2$$

$$+ \iint \Psi_a(1)\Psi_b(2)\left(-\frac{\nabla_2^2}{2} - \frac{1}{r_b(2)}\right)\Psi_b(1)\Psi_a(1)\, d\tau_1\, d\tau_2$$

$$+ \iint \Psi_a(1)\Psi_b(2)\left(\left(\frac{1}{r_a(1)} - \frac{1}{r_a(2)} - \frac{1}{r_b(1)} - \frac{1}{r_b(2)} + \frac{1}{r_{ab}} + \frac{1}{r_{12}}\right)\right)\Psi_b(1)\Psi_a(1)\, d\tau_1\, d\tau_2 \quad (10.78)$$

The first term in Eq. (10.78) is $\iint \Psi_a(1)\Psi_b(2)\left(-\frac{\nabla_1^2}{2} - \frac{1}{r_a(1)}\right)\Psi_b(1)\Psi_a(1)\, d\tau_1\, d\tau_2$

Now, we shall replace $H = \left(-\frac{\nabla_1^2}{2} - \frac{1}{r_a(1)}\right) = E_0$, and the first term is written as

$$= E_0 \iint \Psi_a(1)\Psi_b(2)\Psi_a(2)\Psi_b(1)\, d\tau_1\, d\tau_2$$

$$= E_0 \iint \Psi_I \Psi_{II}\, d\tau_1\, d\tau_2$$

$$= E_0 \cdot S, \quad \text{where} \quad S = \iint \Psi_I \Psi_{II}\, d\tau_1\, d\tau_2 \quad (10.79)$$

The second term in Eq. (10.78) will be also be equal to $E_0 \cdot S$, i.e.,

$$\text{second term} = E_0 \cdot S \quad (10.80)$$

The third term arises due to the fact that electrons are exchanged between the two nuclei and symbolised by K.

$$\therefore \quad H_{12} = H_{21} = 2E_0 \cdot S + K \quad (10.81)$$

From determinant Eq. (10.75), we can write

$$(H_{11} - E)^2 = (H_{12} - SE)^2$$

$$H_{11} - E = \pm(H_{12} - SE) \quad (10.82)$$

i. Now if $H_{11} - \mathrm{E} = H_{12} - \mathrm{SE}$

$$\text{Then } E = \frac{H_{12} - H_{11}}{S - 1} = (2E_0S + K - 2E_0 - J)/(S - 1)$$

$$\text{or} \quad E_1 = 2E_0 + \frac{K - J}{S - 1} \tag{10.83}$$

where $E = E_1$ for condition (i)

ii. If $H_{11} - E = -H_{12} - \mathrm{SE}$,
Then $E\,(1 + S) = H_{11} + H_{12}$

$$\therefore \quad E_2 = \frac{H_{11} - H_{12}}{(1+S)} \quad \text{where } E = E_2$$

$$= \frac{2E_0S + K + 2E_0 + J}{(1+S)}$$

$$= 2E_0 + \frac{J + K}{1 + S} \tag{10.84}$$

In both Eqs. (10.83) and (10.84), the first term is the energy for H atoms, and the second term represents the change of energy due to interactions between two hydrogen atoms. It should be kept in mind that K, J, and S can be expressed as functions of r_{ab}, and $(E - 2E_0)$ is equal to the potential energy of the system.

If we are able to evaluate $(E - 2E_0)$ for various distances of separation of the two atoms, two types of energy values E_2 and E_1 will be found.

The plot of $(E - 2E_0)$ vs. distance r will give the following type of curves (Figure 10.6). It is clear from the graph that only in the case of Eigen function corresponding to E_2, the stable molecule is possible. Ψ_1 and Ψ_2 are antisymmetric and symmetric wave functions, respectively.

Calculation of c_1 and c_2

c_1 and c_2 can be evaluated from Eqs. (10.67) and (10.74), i.e.,

$$c_1(H_{11} - E) + c_2(H_{12} - SE) = 0$$

$$\text{or} \quad \frac{c_1}{c_2} = \frac{H_{12} - SE}{H_{11} - E} \pm 1$$

$$\text{or} \quad \frac{c_1}{c_2} = \pm 1 \text{ or } c_1 = \pm c_2 \tag{10.85}$$

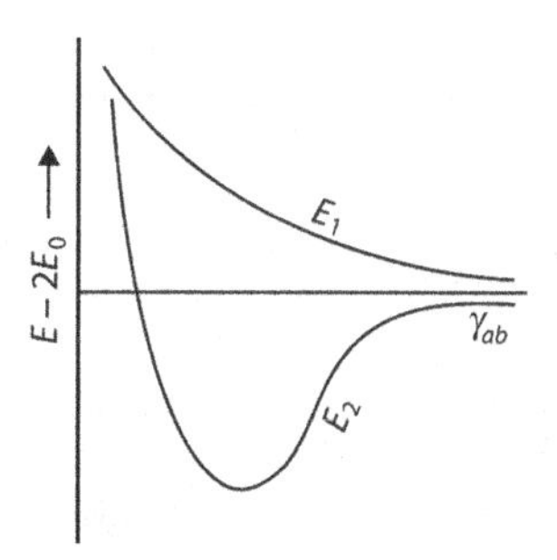

FIGURE 10.6 $(E - 2E_0)$ vs. r graph.

Then the symmetric Eigen function corresponding to E_2 is given by

$$\Psi_S = c_1\Psi_I + c_2\Psi_{II} = c_s(\Psi_I + \Psi_{II}), \text{ where } c_1 = c_2 = c_s \quad (10.86)$$

The Eigen function corresponding to E_1 is Ψ_A, and the value of

$$\Psi_A = c_1\Psi_I - c_2\Psi_{II} = c_A(\Psi_I + \Psi_{II}) \quad (10.87)$$

Now, we can find the value of cs and c_A with the help of normalisation condition. Since Ψs and Ψa are real,

$$\iint \Psi_S^2 \, d\tau_1 \, d\tau_2 = 1 \quad (10.88)$$

$$\text{and} \iint \Psi_A^2 \, d\tau_1 \, d\tau_2 = 1 \quad (10.89)$$

From Eq. (10.88), we can write

$$\iint \Psi_S^2 \, d\tau_1 \, d\tau_2 = c_s^2 \iint (\Psi_I + \Psi_{II})^2 \, d\tau_1 \, d\tau_2$$

$$= c_s^2 \left[\iint \Psi_I^2 \, d\tau_1 \, d\tau_2 + \iint \Psi_{II}^2 \, d\tau_1 \, d\tau_2 + 2\iint \Psi_I \Psi_{II} \, d\tau_1 \, d\tau_2 \right]$$

$$= c_s^2 [1 + 1 + 2S] = 1$$

$$\therefore \quad c_s = \frac{1}{\sqrt{2+2S}} \quad (10.90)$$

Similarly, we can show that

$$c_A = \frac{1}{\sqrt{2-2S}} \quad (10.91)$$

$$\therefore \quad \Psi_S = \frac{1}{\sqrt{2+2S}} \cdot (\Psi_I + \Psi_{II}) = \frac{1}{\sqrt{2+2S}} \left[\Psi_a(1) \cdot \Psi_b(2) + \Psi_a(2) \cdot \Psi_b(1) \right]$$

$$\text{and } \Psi_A = \frac{1}{\sqrt{2-2S}} \cdot (\Psi_I - \Psi_{II}) = \frac{1}{\sqrt{2-2S}} \left[\Psi_a(1) \cdot \Psi_b(2) - \Psi_a(2) \cdot \Psi_b(1) \right] \quad (10.92)$$

It is to be noted that if in Ψs, the coordinates of either the nuclei or the electrons are interchanged, the value of the Eigen functions remains unaltered. This Eigen function is called the *symmetric Eigen function*, which consists of the positional coordinates of both the electrons and the nuclei. On the other hand, if either of these pairs of coordinates in Ψ_A are interchanged, the value of Ψ_A changes its sign, and this is referred to as *antisymmetric*.

10.5 Linear Combination of Atomic Orbitals

When atomic orbitals (AOs) combine linearly, the combination is called the linear combination of atomic orbitals (LCAOs). As a result of the linear combination of AOs, MOs are formed. The LCAO approximation arises due to the fact that with an electron, which is very close or near to a nucleus, the potential energy is dominated by the interaction between the electron and the nucleus.

Thus, very nearer or close to the nucleus of an atom X in a molecule, the wave function of the molecule remains similar to the wave function of the atom X. It is to be noted that the LCAO approximation shows the enhancement in electron density associated with the chemical binding. The LCAO method takes into account the symmetry of the molecule utilising the symmetry-adapted linear combination (SALC). The SALCs are constructed by group theory adequate for the symmetry group of molecule. SALCs are utilised in the formation of MOs.

To understand clearly, let us assume that if $\Psi_a^{(1)}$ and $\Psi_b^{(2)}$ are two wave functions for two electrons 1 and 2 of the AO having similar energies in a molecule, then we can express the molecular wave function as follows:

$$\overline{\Psi} = a_1\Psi_a^{(1)} \pm a_2\Psi_b^{(1)} \tag{10.93}$$

where a_1 and a_2 are constants. Equation (10.93) has been used in the formation of MO from AOs. This process is known as LCAO. For forming MOs, the following conditions have to be satisfied:

i. $\Psi_a^{(1)}$ and $\Psi_b^{(1)}$ must be wave functions of orbital having similar energies.
ii. The AOs must have the similar symmetry with respect to the intermolecular axis.

We can also have another solution of $\overline{\Psi} =$ in the form of product of two wave functions such that

$$\overline{\Psi_2} = \Psi_a^{(1)} \cdot \Psi_b^{(2)} \tag{10.94}$$

which is based on the fact that the composite probability would be the product of the individual probabilities. It should be kept in mind that Ψ_2 would be a solution of the wave equation for two electrons having no interaction.

If Φ_1 and Φ_2 are the wave functions of two electrons 1 and 2 and they are normalised, then the MO can be expressed in the following way:

$$\psi^+ = c_1\Phi_1 + c_2\Phi_2 \tag{10.95}$$

$$\psi^- = c_1\Phi_1 - c_2\Phi_2 \tag{10.96}$$

The values of c_1 and c_2 are found out by applying (i) normalisation condition and (ii) orthogonality theorem. Now applying the normalisation condition for Ψ^*, we can express

$$\int \psi^+\psi^{-*}\,dr = \int (c_1\Phi_1 + c_2\Phi_2)(c_1\Phi_1^* + c_2\Phi_2^*)d\tau$$

$$= \int (c_1\Phi_1 \cdot c_1\Phi_1^*)d\tau + \int (c_2\Phi_2 \cdot c_2\Phi_2^*)d\tau + \int (c_1\Phi_1^* \cdot c_2\Phi_2)d\tau + \int (c_2\Phi_2^* \cdot c_1\Phi_1)d\tau = 1$$

$$= c_1^2\int \Phi_1\Phi_1^*\,d\tau + c_2^2\int \Phi_2\Phi_2^*\,d\tau - c_1c_2\int \Phi_1^*\Phi_2\,d\tau + c_2c_1\int \Phi_2^*\Phi_1\,d\tau = 1$$

$$= c_1^2 + c_2^2 + 0 + 0 = 1$$

$$\text{or} \quad \int \psi^+ \psi^{-*} \, d\tau = c_1^2 + c_2^2 = 1 \tag{10.97}$$

Now applying orthogonality theorem, we can write as

$$\int \psi^+ \psi^- \, d\tau = \int (c_1 \Phi_1 + c_2 \Phi_2)(c_1 \Phi_1 - c_2 \Phi_2) d\tau = 0$$

$$= \int c_1^2 \Phi_1^2 \, d\tau - \int c_2^2 \Phi_2^2 \, d\tau = 0$$

$$\text{or} \quad \int \psi^+ \psi^- \, d\tau = c_1^2 \int \Phi_1^2 \, d\tau - c_2^2 \int \Phi_2^2 \, d\tau = 0$$

$$= c_1^2 - c_2^2 = 0$$

$$\text{or} \quad c_1^2 = c_2^2$$

$$\therefore \quad c_1 = c_2 \tag{10.98}$$

Putting the value of $c_1 = c_2$ in Eq. (10.97), we get

$$c_1^2 - c_1^2 = 1$$

$$2c_1^2 = 1$$

$$\text{or} \quad c_1^2 = \frac{1}{2}$$

$$\therefore \quad c_1 = \pm \frac{1}{\sqrt{2}}$$

And, therefore, Eqs. (10.95) and (10.96) can be expressed jointly as

$$\psi^{\pm} = \frac{1}{\sqrt{2}} (\Phi_1 \pm \Phi_2) \tag{10.99}$$

Thus, the molecular wave function is found out.

10.6 Molecular Orbital Theory

We have already studied that in dealing with the diatomic molecule, our many electron wave functions had to be expressed as the product of one electron functions. The pertinent question is, if we are to form a molecular wave function as a linear combination of atomic wave functions, do we use a product of terms, which are themselves a linear combination of one electron wave functions, or do we use a linear combination of terms, each of which is a product of one electron terms? The safest answer is, both type. Actually, they are two different approximations to molecular wave functions. The first leads to go with MOT, whereas the second leads to VB approach.

On the basis of the given fact, we are in a position to write the total MO wave function for ground state of the H_2 molecule as the product of two one-electron MOs as

$$\Psi_{MO} = 1s\sigma(1) \cdot 1s\sigma(2) \tag{10.100}$$

If we put in the LCAO form for $1s\sigma$ MO, we obtain

$$\Psi_{\text{MO}} = A\left[1s_A(1)+1s_B(1)\right]\left[1s_A(2)+1s_B(2)\right] \tag{10.101}$$

where A = normalising constant.

The normalisation constant can be found as found previously in the following manner:

$$\begin{aligned}\Psi_{\text{MO}} \mid \Psi_{\text{MO}} = A^2 \langle 1s_A(1)1s_A(2)+1s_B(1)1s_B(2)+1s_A(1)1s_B(2)+1s_B(1)1s_A(2) \mid \\ +1s_A(1)1s_A(2)+1s_B(1)1s_B(2)+1s_A(1)1s_B(2)+1s_B(1)1s_A(2)\rangle = 1\end{aligned}$$

These can be expressed in short as used previously, i.e.,

$$\langle \Psi_{\text{MO}} \mid \Psi_{\text{MO}} \rangle = A^2\left(4+8S+4S^2\right) = 1 \tag{10.102}$$

$$\text{or} \quad 4A^2(1+S)^2 = 1$$

$$\text{or} \quad A^2 = \frac{1}{4(1+S)^2} = 1$$

$$\therefore \quad A^2 = \frac{1}{4(1+S)^2}$$

$$\therefore \quad A = \frac{1}{2(1+S)} \tag{10.103}$$

It should be kept in mind that you are already acquainted with S in this chapter. The MO can be estimated as

$$\begin{aligned} E_{\text{MO}} &= \langle \Psi_{\text{MO}} | H | \Psi_{\text{MO}} \rangle \\ &= 2E_0 + \frac{\langle aa \mid aa\rangle + \langle aa \mid bb\rangle + \langle ab \mid ab\rangle + \langle 4aa \mid ab\rangle}{2(1+S)^2} - \frac{\langle 2B \mid aa\rangle + \langle 2A \mid ab\rangle}{1+S} \end{aligned} \tag{10.104}$$

where the abbreviation represents the following:

$$\langle aa \mid aa\rangle = \left\langle 1s_A(1)1s_A(2) \left| \frac{1}{r_{12}} \right| 1s_A(1)1s_A(2) \right\rangle$$

$$\langle aa \mid bb\rangle = \left\langle 1s_A(1)1s_B(2) \left| \frac{1}{r_{12}} \right| 1s_A(1)1s_B(2) \right\rangle$$

$$\langle aa \mid ab\rangle = \left\langle 1s_A(1)1s_A(2) \left| \frac{1}{r_{12}} \right| 1s_A(1)1s_B(2) \right\rangle$$

$$\langle aa \mid ab\rangle = \left\langle 1s_A(1)1s_A(2) \left| \frac{1}{r_{12}} \right| 1s_A(1)1s_A(2) \right\rangle$$

$$\langle B \mid aa \rangle = \left\langle 1s_A(1) \middle| \frac{1}{r_{B1}} \middle| 1s_A(1) \right\rangle$$

$$\langle A \mid ab \rangle = \left\langle 1s_A(1) \middle| \frac{1}{r_{A1}} \middle| 1s_B(1) \right\rangle$$

All the given integrals have already been evaluated, so the symbols used should not be confused with.

If we calculate at different values of r_{AB} and hydrogen ls AO with nuclear charge = 1, a minimum is observed in the total energy at $r = 1.59$ Bohrs. But the total energy at $r = 1.59$ Bohrs has been found to be –1.0974 Hartree, and the energy of two isolated H atoms is found to be –1.0 Hartree. Therefore, the predicted value of dissociation energy will be 0.0974 Hartree.

Now a graph is plotted between the ground-state energy of H_2 molecule and atomic separation r in a.u. as shown in Figure 10.7.

It is found that the nature of both LCAO MO function and the function with ξ optimised are correct qualitatively near-equilibrium internuclear distance. However, both go to the wrong separated atom limit which is –0.75 Hartree rather than –1.0 Hartree. It may be assumed to be a common failure of simple MO calculations.

The cause of this failure can be achieved from the form of the molecular wave function, Ψ_{MO} of equations

$$\left. \begin{aligned} \Psi_1 &= \frac{1}{\sqrt{2(1+S)}}(1s_a + 1s_b) \\ \text{and} \quad \Psi_2 &= \frac{1}{\sqrt{2(1-S)}}(1s_a - 1s_b) \end{aligned} \right\} \tag{10.105}$$

and Ψ_{MO} may be expressed in the following form:

$$\Psi_{MO} = \Psi_1(1)\Psi_1(2)\frac{1}{\sqrt{2}}\left[\alpha(1)\beta(2) - \beta(1)\alpha(2)\right] \tag{10.106}$$

which is written in the form of the Slater determinant, i.e.,

$$\Psi_{MO} = \Psi_1(1)\Psi_1(2)\frac{1}{\sqrt{2}}\begin{vmatrix} \alpha(1) & \alpha(2) \\ \beta(1) & \beta(2) \end{vmatrix}$$

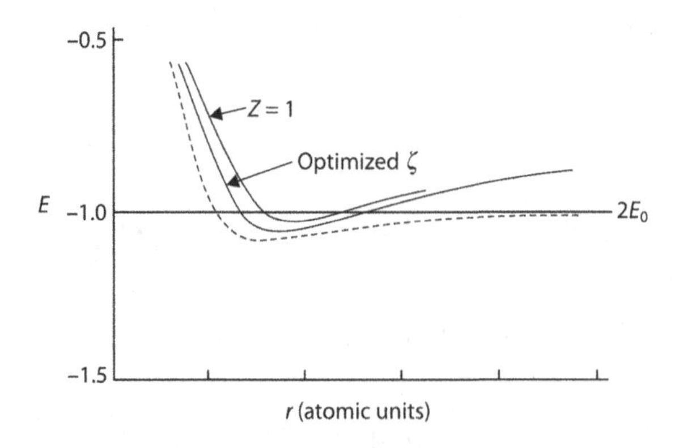

FIGURE 10.7 *E* vs. *r* graph.

after including the spin part of the wave function.

If we neglect the spin part of the wave function, we can express Ψ_{MO} after inserting the form of Ψ_1 as contained in Eq. (10.105) in the following form:

$$\Psi_{MO} = \frac{1}{\sqrt{2(1+S)}}\left[1s_a(1)1s_a(2) + 1s_b(1)1s_b(2) + 1s_a(1)1s_b(2) + 1s_b(1)1s_a(2)\right] \quad (10.107)$$

We can infer from first and second terms of Eq. (10.107) that both electrons are associated with the same portion, i.e., $H_a^- - H_b^+$ and $H_b^- - H_a^+$. It is well known to us that the ionisation potential of H atom is high and the electron affinity value is low, indicating that the reaction $2H \rightarrow H^+ + H^-$ should be highly endothermic and from this, it can be concluded that the chance of ionic form of H_2 molecule is very rare.

In Eq. (10.107), the third and the fourth terms show that the electrons are shared equally and, hence, the H_2 molecule will have covalent structure.

Let us define Ψ_{cov} and Ψ_{ion} as

$$\Psi_{cov} = \frac{1s_a(1)1s_a(2) + 1s_b(1)1s_b(2)}{\sqrt{2(1+S^2)}} \quad (10.108)$$

$$\text{and} \quad \Psi_{ion} = \frac{1s_a(1)1s_a(2) + 1s_b(1)1s_b(2)}{\sqrt{2(1+S^2)}} \quad (10.109)$$

Then, Ψ_{MO} can be written as

$$\Psi_{MO} = (\Psi_{cov} + \Psi_{ion}) \quad (10.110)$$

It is easily observed that the MO approach provides equal weight to the covalent and ionic structures. It should be kept in mind that this is highly unreasonable in the case of H_2 molecule, and hence, this is responsible for the poor agreement between the observed and the calculated results.

It is recalled that the MO wave function Ψ_{MO}, as written in Eq. (10.106), considers the spin correlation. This equation gives a clue to understand that an α electron and an electron may come nearer; hence, the relative probability of finding electron 1 in different regions of space does not depend on the position of the electron 2. If an electron is found at a point in space, the probability of the second electron to come near the first electron should be very less. This is because of the coulomb repulsion. It can be said in another way that the position of electrons in a many electron system should be correlated. A true wave function must consider this charge correlation of electrons. The Ψcov describes a very strong charge correlation since in this wave function, when one electron comes close to the nucleus '*a*', the other electron becomes associated with the nucleus '*b*'.

Ψ_{ion} ignores this correlation, and in the equal contribution limit of Ψ_{cov} and Ψ_{ion}, the charge correlation is disregarded completely.

10.7 Valence Bond Treatment of H_2 Molecule

Heitler and London (1927) first of all performed the quantum mechanical treatment of H_2 molecule. The VB method assumes the molecules as comprising atoms held together by localised bonds. The VB method considers molecules as made up of atomic cores, i.e., nuclei and inner shell electrons and bonding valence electrons. In the case of H_2, both the electrons are valence electrons.

Let us suppose that a system consists of two atoms a and b. Then, the total wave function will be the product of the wave functions of the constituents. Let 1sa and $1s_b$ represent the 1s orbital of two hydrogen atoms in their ground state. $1s_a(1)$ and $1s_b(2)$ are assumed to represent that the electron 1 is near the nucleus 'a' and electron 2 is near the nucleus 'b'. In this situation, the wave function of the system may be expressed as $1s_a(1)1s_b(2)$. It should be kept in mind that when the two H atoms will be closer to each other, it will be difficult to say that the electron 1 is near the nucleus 'a' and electron 2 is near the nucleus 'b'. It is, thus, important to note that we cannot distinguish the wave function $1s_a(1)1s_b(2)$ from $1s_a(2)1s_b(1)$ as the electrons are identical. The special feature of VB method of H_2 is to select the two wave functions as follows:

$$\Psi(1,2) = c_1\Phi_1 + c_2\Phi_2 \tag{10.111}$$

$$\left.\begin{aligned} \text{where } \Phi_1 &= 1s_a(1)1s_b(2) \\ \text{and } \Phi_2 &= 1s_a(2)1s_b(1) \end{aligned}\right\} \tag{10.112}$$

In H_2 molecule, both the atoms are hydrogen atoms; therefore, by symmetry, we can express

$$C_1^2 = C_2^2 \text{ or } C_1 = \pm C_2$$

Now, we can write the energy expression as

$$E = \frac{\langle \Psi(1,2)|H|\Psi(1,2)\rangle}{\langle \Psi(1,2)\Psi(1,2)\rangle} \tag{10.113}$$

After applying the variation principle, one can write the secular equation as

$$\left.\begin{aligned} C_1(H_{11} - E) + C_2(H_{12} - S^2E) = 0 \\ C_1(H_{21} - S^2E) + C_2(H_{22} - E) = 0 \end{aligned}\right\} \tag{10.114}$$

and the corresponding secular determinant will be simply written as

$$\begin{vmatrix} H_{11} - E & H_{12} - S^2E \\ H_{21} - S^2E & H_{22} - E \end{vmatrix} = 0 \tag{10.115}$$

$$\begin{aligned} \text{where } H_{11} &= \langle \Phi_1 |H| \Phi_1 \rangle \\ H_{22} &= \langle \Phi_2 |H| \Phi_2 \rangle \\ H_{12} = H_{21} &= \langle \Phi_1 |H| \Phi_2 \rangle \\ S^2 &= \langle \Phi_1 | \Phi_1 \rangle \end{aligned}$$

and the Hamiltonian H is given by

$$H = -\frac{\nabla_1^2}{2} - \frac{\nabla_2^2}{2} - \frac{1}{r_a(1)} - \frac{1}{r_b(1)} - \frac{1}{r_a(2)} - \frac{1}{r_b(2)} - \frac{1}{r_{12}} + \frac{1}{R} \tag{10.116}$$

This expression has already been dealt for H_2 molecule.

Since $H_{11} = H_{22}$ and $H_{12} = H_{21}$, the secular determinant leads to

$$(H_{11} - E)^2 = \left(H_{12} - S^2 E\right)^2$$

$$\text{or} \quad (H_{11} - E) = \pm\left(H_{12} - S^2 E\right)$$

If, $H_{11} - E = H_{12} - S^2E$, then

$$\left(E - S^2 E\right) = (H_{11} - H_{12})$$

$$\text{or} \quad E\left(1 - S^2\right) = (H_{11} - H_{12})$$

$$\therefore \quad E = \frac{(H_{11} - H_{12})}{\left(1 - S^2\right)}$$

The energy in this equation will be represented as

$$E_+ = \frac{(H_{11} - H_{12})}{\left(1 - S^2\right)} \tag{10.117}$$

and if $(H_{11} - E) = -(H_{12} - S^2E)$, then

$$H_{11} - E = -H_{12} - S^2 E$$

$$\text{or} \quad (H_{11} - H_{12}) = E + S^2 E$$

$$\text{or} \quad E\left(1 + S^2\right) = H_{11} + H_{12}$$

$$\therefore \quad E = \frac{H_{11} + H_{12}}{1 + S^2}$$

This E will be represented by E_-.

$$\therefore \quad E_- = \frac{H_{11} + H_{12}}{1 + S^2} \tag{10.118}$$

where E_+ and E_- represent the value of Hamiltonian of H_2 molecule corresponding to symmetric and antisymmetric wave functions defined as follows:

$$\Psi_+ = C_1(\Phi_1 + \Phi_2) \tag{10.119}$$

$$\Psi_- = C_1(\Phi_1 - \Phi_2) \tag{10.120}$$

The normalisation condition gives

$$\int \Psi_+^2 \, d\tau_1 \, d\tau_2 = 1$$

or $$\int C_1^2 (\Phi_1 + \Phi_2)^2 \, d\tau_1 \, d\tau_2 = 1$$

or $$C_1^2 \left[\int \Phi_1^2 \, d\tau_1 \, d\tau_2 + \int \Phi_2^2 \, d\tau_1 \, d\tau_2 + 2\int \Phi_1 \Phi_2 \, d\tau_1 \, d\tau_2 \right] = 1$$

or $$C_1^2 \left[\int 1s_a^2(1) d\tau_1 \int 1s_b^2(2) d\tau_2 + \int 1s_a^2(2) d\tau_2 \int 1s_b^2(1) d\tau_1 + 2\int 1s_a(1) 1s_b(1) d\tau_1 \right]$$

$$\int 1s_b(2) 1s_a(2) d\tau_2] = 1$$

or $$C_1^2 \left[2 + 2S^2 \right] = 1$$

$$\text{or} \quad C_1^2 = \frac{1}{2+2S^2} \quad \text{or} \quad C_1 = \frac{1}{\sqrt{2+2S^2}} \tag{10.121}$$

$$\therefore \quad \Psi_+ = \frac{1}{\sqrt{2+2S^2}} (\Phi_1 + \Phi_2) \tag{10.122}$$

Similarly, $\Psi-$ will be found if

$$C_1 = \frac{1}{\sqrt{2+2S^2}}$$

when Eq. (10.120) is taken into consideration, and it will be

$$\therefore \quad \Psi_- = \frac{1}{\sqrt{2+2S^2}} (\Phi_1 - \Phi_2) \tag{10.123}$$

where Ψ_+ and Ψ_- are normalised wave functions of H_2 molecule.

Now, we shall evaluate the integrals H_{11} and H_{12}. After splitting the electronic Hamiltonian on the basis of the following equations,

$$H = H' + 1/R, \text{ where } H' = H^C(1) + H^C(2) + \frac{1}{r_{12}} \tag{10.124}$$

$$\text{and} \quad H^C(1) = \frac{-\nabla_1^2}{2} - \frac{1}{r_{a1}} - \frac{1}{r_{b1}} \tag{10.125}$$

$$H^C(2) = \frac{-\nabla_2^2}{2} - \frac{1}{r_{a2}} - \frac{1}{r_{b2}}, \text{ we may write} \tag{10.126}$$

$$H_{11} = 1\left\langle s_a(1) 1s_b(2) \middle| \left(H^C(1) + H^C(2) + \frac{1}{r_{12}} \right) \middle| s_a(1) 1s_b(2) \right\rangle$$

$$= \left\langle 1s_a(1) \middle| H^C(1) \middle| 1s_a(1) \right\rangle + \left\langle 1s_b(2) \middle| H^C(2) \middle| 1s_b(2) \right\rangle + \langle aa \mid bb \rangle \tag{10.127}$$

pressed as

$$\langle aa \mid bb \rangle = \int\int 1s_a(1)1s_a(1)\left|\frac{1}{r_{12}}\right|1s_b(2)1s_b(2)d\tau_1\,d\tau_2 \tag{10.128}$$

$$\left\langle 1s_a(1)\middle|H^C(1)\middle|1s_a(1)\right\rangle = \left\langle 1s_a(1)\middle|-\frac{-\nabla_1^2}{2}-\frac{1}{r_{a1}}\middle|1s_a(1)\right\rangle - \left\langle 1s_a(1)\middle|\frac{1}{r_{b1}}\middle|1s_a(1)\right\rangle$$

$$\text{or} \quad H_{11} = E_H - P \tag{10.129}$$

$$\text{where } P = \left\langle 1s_a(1)\middle|\frac{1}{r_{b1}}\middle|1s_a(1)\right\rangle$$

The value of $\langle 1\ s_b(2)|H^c(2)|1s_b(2)\rangle$ will be the same as that found in Eq. (10.129), since $1s_a$ and $1s_b$ refer two H atoms. Therefore, one should write

$$H_{11} = 2E_H - 2P + \langle aa|bb\rangle \tag{10.130}$$

Similarly, we shall also find the value of H_{12} as follows:

$$H_{11} = \left\langle 1s_a(1)1s_b(2) \mid (H^C(1) + H^C(2) + \frac{1}{r_{12}}) \mid 1s_b(1)1s_b(2)\right\rangle$$

$$= \left\langle 1s_a(1)\middle|H^C(1)\middle|1s_a(1)\right\rangle S + \left\langle 1s_a(2) \mid H^C(2)1s_b(2)\right\rangle S + \langle ab / ab \rangle \tag{10.131}$$

where $\langle ab|ab\rangle$ is two-centre exchange integral, and it is defined as

$$\langle ab / ab\rangle = \int\int 1s_a(1)1s_b(1)\left|\frac{1}{r_{12}}\right|1s_a(2)1s_b(2)d\tau_1\,d\tau_2 \tag{10.132}$$

$$\left\langle 1s_a(1)\middle|H^C(1)\middle|1s_a(1)\right\rangle = \left\langle 1s_a(1)\middle|-\frac{-\nabla_1^2}{2}-\frac{1}{r_{b1}}\middle|1s_b(1)\right\rangle - \left\langle 1s_a(1)\middle|\frac{1}{r_{a1}}\middle|1s_b(1)\right\rangle$$

$$= E_0 S - Q$$

where

$$Q = 1s_a\left|\frac{1}{r_{12}}\right|1s_a$$

The expression for the second term of Eq. (10.131) will be same as that in Eq. (10.133) and hence

$$H_{12} = 2E_0 S^2 - 2SQ + \langle ab / ab \rangle \quad (10.134)$$

The value of $E_H = -13.6\,\text{eV}$, and P and Q are positive and their nature is that they increase with the decrease of R. For this, the following equations may be observed:

$$P = \frac{1}{R}\left[1 - (1+R)e^{-2R}\right]$$

$$Q = (1+R)e^{-R}$$

From the aforementioned equation for H_{11} and H_{12}, it is clear that they are negative quantities, which imply that $E_+ < E_-$. Finally, we can express the total ground-state energy, E_0 of H_2 molecule as

$$E_0(R) = E_+ + \frac{1}{R} = \frac{H_{11} + H_{11}}{1+S^2} + \frac{1}{R}$$

$$= 2E_H - 1\frac{P+QS}{1+S^2} + \frac{1}{1+S^2}\left[\langle aa|bb\rangle + \langle ab/ab\rangle\right] + \frac{1}{R} \quad (10.135)$$

It is clear from this equation that all the integrals in Eq. (10.135) decrease with the increase of R and it becomes zero at $R = \infty$ and hence $E_0(R) \rightarrow 2E_H$ when the atoms of H_2 molecule are infinitely separated. Thus, it is clear that the VB approach gives a suitable description of the molecular dissociation.

Heitler and London estimated the total energy $E_+ + \frac{1}{R}$ as a function of R and found the dissociation energy to be equal to 3.16 eV and equilibrium bond length to be equal to 0.87Å.

10.8 Configuration Interaction

We have to search a way to exclude the ionic terms from the MO function at large internuclear distance so that the MO energy might be improved. Let us suppose that a MO wave function is constructed from the product of two one-electron $1s\sigma^*$ orbital of H_2^+. From the LCAO construction, we may write

$$\Psi'_{\text{MO}} = \frac{1}{2(1-S)}\left[1s_a(1) - 1s_b(1)\right] - \left[1s_a(2) - 1s_b(2)\right]$$

$$= \frac{1}{2(1-S)}\left[1s_a(1)1s_a(2) + 1s_b(1)1s_b(2) - 1s_a(1)1s_b(2) - 1s_b(1)1s_a(2)\right] \quad (10.136)$$

It is observed that the covalent and ionic terms in this function appear with opposite sign. We know that

$$\Psi_{\text{MO}} = N\left[1s_a(1)1s_b(2) + 1s_b(1)1s_a(2) + 1s_a(1)1s_a(2) + 1s_b(1)1s_b(2)\right] \quad (10.137)$$

where N = normalisation constant.

If Eq. (10.136) is subtracted from Eq. (10.137), at large values of R, then the ionic terms will be excluded. The value of $N = \frac{1}{2(1-S)}$. On subtraction we shall get

$$\Psi_{\text{MO}} - \Psi'_{\text{MO}} = \frac{2}{2(1-S)}\left[1s_a(1)1s_b(2) + 1s_b(1)1s_a(2)\right] \quad (10.138)$$

But at $R \to \infty$, S = 0, Therefore, Eq. (10.138) becomes

$$\begin{aligned}\Psi_{\text{MO}} - \Psi'_{\text{MO}} &= \frac{2}{2(1-0)}\left[1s_a(1)1s_b(2) + 1s_b(1)1s_a(2)\right] \\ &= \left[1s_a(1)1s_b(2) + 1s_b(1)1s_a(2)\right]\end{aligned} \tag{10.139}$$

Thus, the ionic terms are removed or excluded.

Now, let us suppose that an improved wave function Ψ is a linear combination of Eqs. (10.136) and (10.137), which is expressed as

$$\Psi = c_1\Psi_{\text{MO}} + c_2\Psi'_{\text{MO}} \tag{10.140}$$

where the coefficients are treated as variational parameters.

In effect, we are mixing the $\left(1\sigma_g^+\right)^2$ and $\left(1\sigma_u^+\right)^2$ configurations, thereby performing a configuration interaction calculation.

The variational treatment is similar to the one we used for finding out the LACO function for H_2^+. It should be kept in mind that the letters *g* and *u* refer to the German words *gerade* (even) and *ungerade* (odd), respectively. It means that the two MOs may be even or odd with respect to certain symmetry operation.

The resulting secular determinant will take the form

$$\begin{vmatrix} \langle \Psi_{\text{MO}} | H | \Psi_{\text{MO}} \rangle - E & \langle \Psi_{\text{MO}} | H | \Psi'_{\text{MO}} \rangle \\ \langle \Psi_{\text{MO}} | H | \Psi'_{\text{MO}} \rangle & \langle \Psi'_{\text{MO}} | H | \Psi'_{\text{MO}} \rangle - E \end{vmatrix} = 0 \tag{10.141}$$

There is no overlap term in the off-diagonal elements because of the fact that the functions $1\sigma_g^+$ and $1\sigma_u^+$ are orthogonal.

If the configuration interaction calculation is carried out as a function of *R* and if a variationally found effective nuclear charge is included, then an equilibrium internuclear separation, i.e., r_e is found to be 1.45 Bohrs. At r_e = 1.45 Bohrs, the total energy is found to be –1.477 Hartree, and the dissociation energy is found to be 0.1477 Hartree. The optimum nuclear charge ζ at this separation has been found to be 1.193. The energy is found to be better than either the MO or the VB energy.

It is, thus, observed that MO treatment with configuration interaction produces better results with respect to the VB treatment. The VB treatment may also be improved if some way can be found out for adding some ionic contribution.

Let us construct a wave function of the following form:

$$\Psi = c\Psi_{\text{VB}} + c_2\Psi'_{\text{VB}} \tag{10.142}$$

$$\text{where } \Psi'_{\text{VB}} = \frac{1}{\sqrt{2\left(1+S^2\right)}}\left[1s_a(1)1s_a(2) + 1s_b(1)1s_b(2)\right] \tag{10.143}$$

Here, the two terms of Ψ'_{VB} correspond to putting both electrons on atom 'a' and both on 'b', respectively. Eq. (10.142) is again a configuration interaction wave function.

Ψ_{VB} corresponds to a covalent bond configuration, whereas Ψ_{VB} corresponds to the sum of two equally weighted ionic configurations. Again, the coefficients are found variationally. The resulting determinant may be expressed as

$$\begin{vmatrix} \langle \Psi_{\text{VB}} | H | \Psi_{\text{VB}} \rangle - E & \langle \Psi_{VB} | H | \Psi'_{\text{VB}} \rangle - \dfrac{2SE}{1+S^2} \\ \langle \Psi_{\text{VB}} | H | \Psi'_{\text{VB}} \rangle - \dfrac{2SE}{1+S^2} & \langle \Psi'_{\text{VB}} | H | \Psi'_{\text{VB}} \rangle - E \end{vmatrix} = 0 \tag{10.144}$$

At very large R, $S \to 0$, so the determinant changes to

$$\begin{vmatrix} \langle \Psi'_{VB}|H|\Psi_{VB}\rangle - E & \langle \Psi_{VB}|H|\Psi'_{VB}\rangle \\ \langle \Psi_{VB}|H|\Psi'_{VB}\rangle & \langle \Psi'_{VB}|H|\Psi'_{VB}\rangle - E \end{vmatrix} = 0$$

In other words, we can write

$$(H_{11} - E)(H_{22} - E) = H_{12} \cdot H_{21}$$

$$\text{or} \quad H_{11} - E = \pm H_{12}$$

$$\therefore \quad E = H_{11} \pm H_{12}$$

$$\text{Further,} \quad E_+ = H_{11} + H_{12} \text{ and } E_- = H_{11} - H_{12} \tag{10.145}$$

Thus, the energies can be found out.

10.9 Comparison of the Molecular Orbital and Valence Bond Theories

Here, we are going to compare the MO and the VB theories (VBTs) of the H_2 molecule in the ground state. The VB method begins with individual atoms and assumes interaction between them. Since the electrons are indistinguishable, the wave functions of the molecule will be linear combinations of two state functions $1s_a(1)1s_b(2)$ and $1s_a(2)1s_b(1)$, where $1s_a$ and $1s_b$ are the AOs centred on atoms 'a' and 'b', respectively. Therefore, the VB wave function can be written as

$$\Psi_{VB} = \left[C_1 1s_a(1)1s_b(2) + C_2 1s_a(2)1s_b(1)\right] \tag{10.146}$$

where c_1 and c_2 are coefficients.

These two terms indicate the covalent characteristics of the bond. This is termed as the simple VB wave function of Heitler and London.

The MO method, on the other hand, begins with the consideration that the molecule contains MOs, which are constructed by the linear combination of AOs. In the case of MOT, electrons revolve about the polycentric nuclei of the molecule. There is no individual existence of AOs. This means that the orbitals of linked atoms lose their individual existence.

In the case of H_2 molecule, MO wave functions with respect to electrons 1 and 2 may be expressed as

$$\Psi_1 = c_1 1s_a(1) + c_2 1s_b(1)$$

$$\Psi_2 = c_1 1s_a(2) + c_2 1s_b(2)$$

Therefore, complete ground-state molecular wave function for H_2 molecule will be represented by the product of Ψ_1 and Ψ_2 as given in the following:

$$\begin{aligned} \Psi_{MO} = \Psi_1\Psi_2 &= \left[c_1 1s_a(1) + c_2 1s_b\right]\left[c_1 1s_a(2) + c_2 1s_b(2)\right] \\ &= c_1c_2 1s_a(1)1s_b(2) + c_1c_2 1s_a(2)1s_b(1) + c_1^2 1s_a(1)1s_a(2) + c_2^2 1s_a(1)1s_b(2) \end{aligned} \tag{10.147}$$

where the first two terms represent the covalent nature of the wave function and the last two terms symbolise the structures in which both electrons lie on nucleus 'a' and 'b', thereby giving rise to the ionic structure of H_2 molecule, i.e., $H^+_a{\cdot}H^-_a$ or $H^-{\cdot}H^+$. The equation for Ψ_{VB} differs from the equation for Ψ_{MO} with regard to the absence of ionic terms $1s_a(1)1s_a(2)$ and $1s_b(1)1s_b(2)$. Such ionic terms are ignored in the VB approach. In the case of H_2 molecule, the results of the VB approach are found to be better than the MO approach, which gives a clue towards understanding that such terms are not significant. However, in the case of heteronuclear molecules, such ionic terms become significant, and these should be included in the molecular wave function (Ψ_{MO}).

Thus, we are in a position to state that VB approach underestimates the ionic terms, whereas MO approach gives emphasis on the ionic terms. Therefore, the modified VB wave function must include the ionic terms and may be written as

$$\Psi_{VB} = c_1 1s_a(1)1s_b(2) + c_2 1s_a(2)1s_b(1) + c_3 1s_a(1)s_a(2) + c_4 1s_b(1)1s_b(2) \quad (10.148)$$

It is to be noted that most of the molecular calculations are done using MO approach due to the fact that the complicated mathematics is involved in VB approach. Again, MO approach is easier for the description of the excited states in molecule, and the aromatic stability can be satisfactorily explained.

10.10 Symmetric and Antisymmetric Wave Functions

A wave function is said to be symmetric if the interchange of any pair of particles among its arguments leaves the wave function unaltered or unchanged.

A wave function is said to be antisymmetric if the interchange of any pair of particles changes the sign of Ψ or wave function.

We shall clarify these statements by taking an example.

Let us suppose that the probability of finding particle 1 is at the point r_1 and particle 2 is at the point r_2 at a given time. Then the probability may be expressed as

$$\left|\Psi[r_1(1), r_2(2)]\right|^2 dr_1 dr_2 \quad (10.149)$$

where Ψ = wave function describing the system.

Because the two particles are identical, they must be physically indistinguishable. Therefore, the aforementioned probability may also be written as

$$\left|\Psi[r_2(1), r_1(2)]\right|^2 dr_1 dr_2 \quad (10.150)$$

which means that the probability of finding particle 1 is at point r_2 and particle 2 at the point r_1 at a given time. It is clear from the aforementioned equations that they differ only in the exchange of coordinates of two identical, indistinguishable particles. In the absence of vector fields (e.g., a magnetic field), the wave function Ψ can always be selected as real; hence, we may write

$$\Psi^2[r_1(1), r_2(2)] = \Psi^2[r_2(1), r_1(2)] \quad (10.151)$$

So either

$$\Psi[r_1(1), r_2(2)] = \Psi[r_2(1), r_1(2)] \quad (10.152)$$

or

$$\Psi[r_1(1), r_2(2)] = -\Psi[r_2(1), r_1(2)] \quad (10.153)$$

Expression (10.152) states that the wave function Ψ is symmetric with respect to an interchange of the full coordinates (space and spin) of a pair of identical particles. Expression (10.153) states that the wave function Ψ is antisymmetric with respect to an interchange of the full coordinates (space and spin) of a pair of identical particles.

In quantum mechanics, the total wave function Ψ is assumed to be a simple product of individual one particle wave functions, i.e.,

$$\Psi[r_1(1), r_2(2) ------ r_n(N), t] = \Psi_1[r_1(1), t]\Psi_2[r_2(1), r_1(2), t]... \quad (10.154)$$

Neither condition (Eqs. 10.152 and 10.153) is met if we use the form of Eq. (10.154). Instead of this, we must use

$$\Psi[r_1(1), r_2(2)] = \frac{1}{\sqrt{2}}[\Psi_1\{r_1(1)\}\Psi_2\{r_2(2)\} + \Psi_2\{r_1(1)\}\Psi_1\{r_2(2)\}] \quad (10.155)$$

and $$\Psi[r_1(1), r_2(2)] = \frac{1}{\sqrt{2}}[\Psi_1\{r_1(1)\}\Psi_2\{r_2(2)\} - \Psi_2\{r_1(1)\}\Psi_1\{r_2(2)\}] \quad (10.156)$$

from Eqs. (10.152) and (10.153).

In short, we may express the aforementioned expression as

$$\Psi(r_1, r_2) = \frac{1}{\sqrt{2}}[\Psi_1(r_1)\Psi_2(r_2) + \Psi_2(r_1)\Psi_1(r_2)] \quad (10.157)$$

and $$\Psi(r_1, r_2) = \frac{1}{\sqrt{2}}[\Psi_1(r_1)\Psi_2(r_2) - \Psi_2(r_1)\Psi_1(r_2)] \quad (10.158)$$

from Eqs. (10.152) and (10.153), respectively. The factor $\frac{1}{\sqrt{2}}$ is for normalisation. The +(–) case is known as symmetric (antisymmetric).

In the antisymmetric case, we note that $\Psi \to 0$ where $\Psi_1 = \Psi_2$, i.e., when particles are in the same state. Hence, if the particles are described by an antisymmetric wave function, there will be zero probability that they will coincide, which means that they tend to keep away from each other.

Now we shall consider four important postulates.

1. All fundamental particles are denoted by wave functions, which are either symmetric or anti-symmetric with respect to the interchange of full coordinates (space and spin) of a pair of identical particles.
2. Particles never go from one symmetry type to another symmetry type.
3. All particles having half integral spin are denoted by antisymmetric wave function.
4. All particles having zero or integral spin are described by symmetric wave function.

Now we are going to show that the symmetry behaviour of a wave function does not alter in time. The Schrödinger wave equation for n identical particles is written as

$$i\hbar \frac{d}{dt}\Psi(1, 2, ----, n, t) = H(1, 2, ---, n, t) \quad (10.159)$$

where each of the numbers indicates all the coordinates (position and spin) of one of the particles. The Hamiltonian H is symmetrical in its arguments, because the identity of the particles means that they can be substituted for each other without altering H.

If Ψ_s is symmetric at a particular time t, then $H\Psi_s$ will be also symmetric, and according to Eq. (10.159), $d\Psi_S/dt$ is symmetric. Then the wave function at an infinitesimally later time dt will be given by $\frac{d\Psi}{dt}$ and will be also symmetric.

Such a step-by-step integration of the wave equation can, in principle, be continued for an arbitrary large time intervals, and Ψ_s is observed to remain symmetric always.

Similarly, if Ψ_A is antisymmetric at any time, $H\Psi_A$ and, hence, $d\Psi_A/dt$ will be antisymmetric and the integration of the wave equation exhibits that Ψ_A is always antisymmetric.

Keeping in view the aforementioned facts, we can safely say that the symmetry character of a wave function does not alter with time.

10.11 Pauli's Exclusion Principle

In the case of H_2 molecule, there are two electrons and both H atoms of the molecule remain in 1s state. Let us suppose that both the electrons have the same spin, i.e., either both are of α type or both are of β type. Then, we can write

$$\left.\begin{aligned} \Psi_I &= \Psi_{1s}\alpha(1)+\Psi_{1s}\beta(2) \\ \Psi_{II} &= \Psi_{1s}\alpha(2)+\Psi_{1s}\beta(1) \end{aligned}\right\} \tag{10.160}$$

This may be expressed in determinantal form as

$$\begin{vmatrix} a(1) & a(2) \\ \beta(1) & \beta(2) \end{vmatrix} = 0 \tag{10.161}$$

This determinant will be equal to zero due to the fact that $\alpha(1) \equiv \alpha(2)$ and $\beta(1) \equiv \beta(2)$. Since the determinant is zero, Ψ_I and Ψ_{II} must be zero, but such a state cannot exist. Hence, two electrons cannot have the same spin when all other quantum numbers are identical. Therefore, for wave functions to exist, two electrons cannot have the same spin. This is called *Pauli's exclusion principle.*

Thus, $\Psi_I = \Psi_{1s}\alpha(1) + \Psi_{1s}\beta(2)$

and $\Psi_{II} = \Psi_{1s}\alpha(2) + \Psi_{1s}\beta(1)$ in place of the written equation.

This is expressed in determinantal form as follows:

$$\begin{vmatrix} a(1) & \beta(2) \\ a(2) & \beta(1) \end{vmatrix} \neq 0 \text{ because two rows are not identical.}$$

Therefore, the wave function will be governed by Eq. (10.161) and can be written as

$$\Psi = \{a(1)\beta(2) - a(2)\beta(1)\} \tag{10.162}$$

which is antisymmetric in nature and is permissible by the Pauli's principle. Therefore, the total wave function by the Pauli's principle must be antisymmetric. The rigorous statement of the Pauli's principle is that *wave functions of electrons must be antisymmetric.*

With two electrons, four spin functions are possible, of which three are symmetric and one is antisymmetric, which are shown in the following:

$$\left.\begin{aligned} &a(1)a(2)+\beta(1)\beta(2) \\ &a(1)\beta(2)+a(2)\beta(1) \\ &\beta(1)\beta(2)+a(1)\beta(2) \end{aligned}\right\} \text{Symmetric}(S)$$

and $\quad a(1)\beta(2)-\beta(1)a(2)$ Antisymmetric(A)

Multiplying the two orbital wave functions Ψ_s and Ψ_A for an H_2 molecule by the spin Eigen functions, a complete set of eight Eigen functions will result. Among these, only the antisymmetric wave functions will be accepted. The product of two symmetric and two antisymmetric wave functions will be always symmetric, but the product of one symmetric and one antisymmetric wave function must be antisymmetric. Hence, we can write the following:

$$\Psi_S\left[a(1)\beta(2)-\beta(1)a(2)\right]$$
$$\Psi_A\left[a(1)a(2)+\beta(1)\beta(2)\right]$$
$$\Psi_A\left[\beta(1)\beta(2)+a(1)a(2)\right]$$
$$\Psi_A\left[a(1)\beta(2)+\beta(1)a(2)\right]$$

Thus, there is only one stable state of H_2 with symmetrical orbital wave function. On the other hand, there are three states of unstable or repulsive form with antisymmetric orbital Eigen function. The stable form of H_2 molecule is the singlet state, and unstable state is the triplet state. The three states constituting the triplet states have almost the same energy, but these differ in spin quantum number value.

10.12 Antisymmetric Wave Function and Slater Determinant

Let 1, 2, 3... $n = n$ number of electrons,

a, *b*, *c*... $n = n$ orbital wave function having single electron,

α or β = spin wave function associated with orbital wave function to complete the wave function of the electron, and

$a\alpha$, $b\beta$, $c\beta$ ..., na = available wave function when a general case is considered.

Now suppose that electron 1 occupies the orbital '*a*', whereas electron two occupies the orbital '*b*', electron three occupies the orbital '*c*', and so on. A possible wave function for the whole system will be obtained by multiplying together the single-electron wave function. Thus, Ψ (wave function) can be expressed as

$$\Psi = (a\alpha)_1\,(b\beta)_2\,(c\beta)_3 \ldots (n\alpha)_n \tag{10.163}$$

where the numerals indicate the electron that occupies each particular orbital. Since the energy of the system will not alter if the coordinates of any pair of electron (say 1 and 3) were interchanged; therefore, the satisfactory wave function would be represented by

$$\Psi = (a\alpha)_3\,(b\beta)_2\,(c\beta)_1 \ldots (n\alpha)_n \tag{10.164}$$

If other interchanges are carried out, then it will lead to equivalent wave function, and hence, it is possible to write a general expression for the wave function of the system as follows:

$$\Psi = p(a\alpha)_1\,(b\beta)_2\,(c\beta)_3 \ldots (n\alpha)_n \tag{10.165}$$

where p = permutation operator, which represents the operation of exchanging the coordinates of the orbitals by pair of electrons.

Up until now, no consideration has been given to the question whether these wave functions satisfy Pauli's principle or not. However, we have already seen that the complete wave function of a system of two or more electrons must be antisymmetric, which means an interchange in the coordinates of any pair of electrons must change the sign of the wave function. This condition will be satisfied if the linear combination of the function given in Eq. (10.165) is carried out, i.e.,

$$\Psi = \frac{1}{\sqrt{n!}} \sum \pm P (a\alpha)_1 (b\beta)_2 (c\beta)_3 \ldots (n\alpha)_n \tag{10.166}$$

where $\frac{1}{\sqrt{n!}}$ = normalisation constant.± = + or − sign is used before each term in the summation on the basis of the particular permutation obtained from the beginning one by an even or odd number, respectively, considering the exchanges in the coordinates of the pairs of electrons.

It is, thus, clear that each successive interchange of the coordinates of any pair of electrons will change the sign but not the magnitude of the wave equation described in Eq. (10.166). This type of approximate wave function of a system is called an antisymmetric wave function since the summation in Eq. (10.166) may also be expressed in the form of the determinant:

$$\Psi = \frac{1}{\sqrt{n!}} \begin{vmatrix} (a\alpha)_1 & (b\beta)_1 & (c\beta)_1 \ldots & (n\alpha)_1 \\ (a\alpha)_2 & (b\beta)_2 & (c\beta)_2 \ldots & (n\alpha)_1 \\ (a\alpha)_n & (b\beta)_n & (c\beta)_n \ldots & (n\alpha)_n \end{vmatrix} \tag{10.167}$$

in which the diagonal elements are as found in Eq. (10.163), and this type of determinant is known as the *Slater determinant.*

The wave functions represented in Eqs. (10.166) and (10.167) correspond to the particular arrangement of spin given in Eq. (10.163), i.e., the diagonal of the determinant. Since every one of the n electron may have spin α or β spin, there will be 2^n way for arranging the spin functions α or β among the n electrons. Consequently, there will be 2^n determinants like the aforementioned equation. Now it is safe to state that the overall wave functions that satisfy the Pauli's principle are often expressed by the Slater determinant.

The properties of the Slater determinant follow the same property as those that we have learnt in the case of determinant in mathematics.

The important properties are as follows:

- If any two rows or columns of a Slater determinant remain the same, the value of the determinant will be zero, e.g.,

$$\begin{vmatrix} 1s(1)\alpha(1) & 1s(1)\alpha(1) \\ 1s(2)\alpha(2) & 1s(2)\alpha(2) \end{vmatrix} = 0$$

Here, the two rows are same; therefore, the determinant vanishes.

- If two rows or columns of a Slater determinant are interchanged, the resulting determinant will have the same magnitude but negative of the original determinant, e.g.,

$$\begin{vmatrix} a(1) & b(1) & c(1) \\ a(2) & b(2) & c(2) \\ a(3) & b(3) & c(3) \end{vmatrix} = -\begin{bmatrix} a(2) & b(2) & c(2) \\ a(1) & b(1) & c(1) \\ a(3) & b(3) & c(3) \end{bmatrix}$$

It is clear that both the determinants will have the same magnitude, but the value of the second one will be with a negative sign.

Alternatively

Let us consider the two possible approximate wave functions of He, which are as follows:

$$\Psi_{\text{He},1} = \frac{1}{\sqrt{2}}\left[(1s_1a)(1s_2\beta)+(1s_1\beta)(1s_2a)\right] \tag{10.168}$$

and $$\Psi_{\text{He},1} = \frac{1}{\sqrt{2}}\left[(1s_1a)(1s_2\beta)-(1s_1\beta)(1s_2a)\right] \tag{10.169}$$

The question arises whether these two are antisymmetric? We shall answer this question after interchanging electrons 1 and 2 in the first wave function of Eq. (10.168). By doing so, we obtain

$$\Psi(2,1) = \frac{1}{\sqrt{2}}\left[(1s_2a)(1s_1\beta)+(1s_2\beta)(1s_1a)\right]$$

This will be identified as the original wave function $\Psi(1,2)$, which is only algebraically rearranged.

However, on exchanging electron in Eq. (10.169), we get

$$\Psi(2,1) = \frac{1}{\sqrt{2}}\left[(1s_2a)(1s_1\beta)-(1s_2\beta)(1s_1a)\right] \tag{10.170}$$

which can be expressed algebraically as $-\Psi(1,2)$. Hence, this wave function will be called antisymmetric with respect to the exchange of electrons, and consequently, it is a proper wave function for the spin orbitals of He. It means that Eq. (10.169), but not Eq. (10.164), represents the correct form for spin–orbital wave function of He in the ground state. Thus, the simple statement of Pauli's principle is that *wave functions of electrons must be antisymmetric with respect to the exchange of electrons.* This comes from the identification that Eq. (10.169) is the only acceptable wave function for He and can be expressed in the form of a matrix determinant.

For He, the determinant will be written as

$$\Psi_{\text{He}} = \frac{1}{\sqrt{2}}\begin{vmatrix} 1s_1\alpha & 1s_1\beta \\ 1s_2\alpha & 1s_2\beta \end{vmatrix} \tag{10.171}$$

Such notation of writing antisymmetrised wave function is called the *Slater determinant* where $\frac{1}{\sqrt{2}}$ is the normalisation factor. In Eq. (10.171), α and β represent spin and s represents state (space).

Let us consider the case of $_3Li^7$ whose electronic configuration is $1s^22s^1$, in which 1s electron can hold $\downarrow\uparrow$, whereas $2s\alpha$ can hold either $\downarrow$ or $\uparrow$.

Hence, in the case of lithium, the wave function can be expressed in the determinant form as

$$\Psi = \frac{1}{\sqrt{3!}}\begin{vmatrix} 1s_1\alpha & 1s_1\beta & 1s_1\alpha \\ 1s_2\alpha & 1s_2\beta & 1s_2\alpha \\ 1s_3\alpha & 1s_3\beta & 1s_3\alpha \end{vmatrix} \text{ or } \frac{1}{\sqrt{3!}}\begin{vmatrix} 1s_1\alpha & 1s_1\beta & 1s_1\beta \\ 1s_2\alpha & 1s_2\beta & 1s_2\beta \\ 1s_3\alpha & 1s_3\beta & 1s_3\beta \end{vmatrix} \tag{10.172}$$

It is to be noted that in both the cases, two columns of the determinant indicate the same spin orbital for two of the three electrons. For n electrons, including normalisation constant, the wave function can be expressed as

$$\Psi = \frac{1}{\sqrt{n!}}\begin{vmatrix} 1s_1\alpha & 1s_1\beta & \cdots & 1s_1\alpha \\ 1s_2\alpha & 1s_2\beta & \cdots & 1s_2\alpha \\ 1s_n\alpha & 1s_n\beta & \cdots & 1s_n\alpha \end{vmatrix}$$

This determinant is called the general form of the Slater determinant.

10.13 Bonding and Antibonding Orbitals

We may use the equation for the wave function of H_2 molecule as

$$\Psi(r_1) = N_{MO}\left[a(1)+b(1)\right] \tag{10.173}$$

to describe its ground state. a(l) and b(1) are AOs. By intuition, one can deduce from the symmetry of a diatomic molecule like H_2 that the two AOs should appear in a symmetric manner.

But nothing stops them from having coefficients with different signs. We may write Eq. (10.173) and its counterpart with a negative sign as follows:

$$\Psi_g(r_1) = \frac{1}{\sqrt{2(1+S)}}\left[a(1)+b(1)\right] \tag{10.174}$$

$$\Psi_M(r_1) = \frac{1}{\sqrt{2(1-S)}}\left[a(1)-b(1)\right] \tag{10.175}$$

where letters g and u represent the first letter of German words *gerade* (even) and *ungerade* (odd), respectively. $\Psi_g(r_1)$ represents the bonding orbital (BO), and $\Psi_u(r_1)$ represents the antibonding orbital.

The two MOs in the equations can be used to build up two-electron functions for different states of the molecule under consideration. If we limit ourselves to singlets, then the spatial function will necessarily be symmetric when r_1 and r_2 are exchanged.

For the ground state, we may write

$$\Phi_0(r_1,r_2) = \Psi_g(r_1)\Psi_g(r_2) \tag{10.176}$$

$$\Phi_0(r_1,r_2) = \Psi_M(r_1)\Psi_M(r_2) \tag{10.177}$$

which will be also symmetric and will correspond to a possible spatial component for a singlet.

Now, we are in a position to evaluate the total energy of H_2 molecule in a state, with spatial function described in Eq. (10.177), i.e.,

$$E_1(R) = \langle\Phi_1|H|\Phi_1\rangle \tag{10.178}$$

A triplet function for a two-electron system can also be expressed in the following form:

$$\Psi_t(x_1,x_2) = \Phi(r_1,r_2)\Phi_t(\xi_1,\xi_2) \tag{10.179}$$

but with the spin function selected as one of the following three triplet functions as given in the following:

$$\alpha_1\alpha_2 = \frac{1}{\sqrt{2}}\left[\alpha_1\beta_2+\beta_1 a_2\right], \beta_1\beta_2 \tag{10.180}$$

With the two orbitals in Eqs. (10.174) and (10.175), there is only one possibility for $\Phi(r_1, r_2)$ in Eq. (10.179), which is

$$\Phi(r_1,r_2) = \frac{1}{\sqrt{2}}\left[\Psi_g(r_1)\Psi_M(r_2)-\Psi_M(r_1)\Psi_g(r_2)\right] \tag{10.181}$$

The three states described can be denoted with the following notations:

$$(g\alpha, g\beta), (u\alpha, u\beta), (g\alpha, u\alpha) \tag{10.182}$$

The ground-state function indicates a state having a minimum in the potential curve for a given equilibrium distance R_e. When one of the electrons is promoted to the spin orbital $u\alpha$, we obtain a triplet state with a repulsive potential curve, which remains without minimum.

Suppose both the electrons are placed in u orbital having different spins, then we obtain an excited singlet state with a potential curve, which lies even higher. This is the background that we call Ψ_g as the BO and Ψ_u as the antibonding orbital.

10.14 Electron Density in Molecular Hydrogen

Let $\rho(1)$ = density function for electron 1 in a molecule and

$\rho(1)d\tau_1$ = density function for electron 1 within the volume element $d\tau_1$

= sum of all the probabilities with regard to electron 1 in $d\tau_1$.

As far as other electrons are concerned, they are found in all possible regions of space.

H_2 molecule under consideration is a two-electron system, and its ground state function $\Psi(1, 2)$ is given by

$$\Psi_0 = \frac{1}{\sqrt{2}}\left|\Psi_1(1)\overline{\Psi_1}(2)\right| \text{ according to MOT} \tag{10.183}$$

where $\overline{\Psi_1}(2) = \begin{vmatrix} \alpha(1) & \alpha(2) \\ \beta(1) & \beta(2) \end{vmatrix}$ and α and β represent the spins. If α represents $\uparrow$, then β will represent $\downarrow$.

$$\text{Therefore, } \rho(1)d\tau_1 = \left[\int \Psi_0^2(1,2)d\tau_2\right]d\tau_1 \tag{10.184}$$

where $\int \Psi_0^2(1,2)d\tau_2 = \rho(1)$

If we follow Eq. (10.183), we may write

$$\Psi_0(1,2) = \frac{1}{\sqrt{2}}\left[\Psi_1(1)\overline{\Psi_1}(2) - \Psi_1(2)\overline{\Psi_1}(1)\right]$$

$$\therefore \int \Psi_0^2(1,2)d\tau_2 \left(\frac{1}{\sqrt{2}}\right)^2 \int\left[\Psi_1^2(1)\overline{\Psi_1^2} + \Psi_1^2(1)\overline{\Psi_1^2}(1) - 2\Psi_1(1)\overline{\Psi_1}(2).\Psi_1(2)\overline{\Psi_1}(1)\right]d\tau_2$$

$$= \frac{1}{2}\left[\Psi_1^2\int\overline{\Psi_1^2}(2)d\tau_2 + \overline{\Psi_1^2}(1)\int\Psi_1^2(2)d\tau_2 - 2\Psi_1(1)\overline{\Psi_1}(1)\int\overline{\Psi_1}(2)\Psi_1(2)d\tau_2\right] \tag{10.185}$$

$$or, \ \Psi_0^2(1,2)d\tau_2 = \frac{1}{2}\left[\Psi_1^2(1) + \overline{\Psi_1^2}(1)\right] \tag{10.186}$$

where $\int\overline{\Psi_1^2}(2)d\tau_2 = 1, \int\Psi_1^2(2)d\tau_2 = 1$ and $\overline{\Psi_1}(2)\Psi_1(2)d\tau_2 = 0$ due to orthogonal property.

$$\therefore \ \rho(1) = \frac{1}{2}\left[\Psi_1^2(1) + \overline{\Psi_1^2}(1)\right] \tag{10.187}$$

But according to VBT,

$$\Psi_1 = \frac{1}{\sqrt{2(1+S)}}(1s_a + 1s_b$$

$$\therefore \quad \Psi_1^2(1) = \frac{1}{2(1+S)}\left[1s_a^2(1) + 1s_b^2(1) + 2\cdot 1s_a(1)\cdot 1s_b(1)\right]$$

$$\therefore \quad \rho(1) = \Psi_1^2(1) = \frac{\left[1s_a^2(1) + 1s_b^2(1) + 2\cdot 1s_a(1)\cdot 1s_b(1)\right]}{2(1+S)} \tag{10.188}$$

Similarly, $\Psi_1^2(1)$ will have the same value and will be represented by $\rho(2)$, i.e., $\rho(1) = \rho(2)$.

Since electrons 1 and 2 are indistinguishable, the total one electron density function will be expressed as

$$\rho = \rho(1) + \rho(2) = 2\rho(1) \tag{10.189}$$

$$\text{where } \rho(2) = \frac{\left[1s_a^2(2) + 1s_b^2(2) + 2\cdot 1s_a(1)\cdot 1s_b(2)\right]}{2(1+S)}$$

$$\therefore \quad \rho = 2\rho(1) = \frac{2}{2(1+S)}\left[1s_a^2 + 1s_b^2 + 2\cdot 1s_a\cdot 1s_b\right] \tag{10.190}$$

$$\therefore \quad \int \rho\, d\tau_1 = \left(\frac{1}{1+S}\right)\int 1S_a^2\, d\tau_1 + \int 1S_b^2\, d\tau_1 \int 1S_a\cdot 1S_b\, d\tau_1$$

$$= \left(\frac{1}{1+S}\right)[1+1+2S]$$

$$= \frac{1}{1+S} + \frac{1}{1+S} + \frac{2S}{1+S}$$

$$\therefore \quad \int \rho d\tau_1 = \frac{2}{1+S} + \frac{2S}{1+S} = \frac{2(1+S)}{(1+S)} = 2 \tag{10.191}$$

Thus, it is vivid that of the two electrons in H_2, $\frac{1}{1+S}$ electrons are associated with each H atom and $\frac{2s}{1+S}$ electron will remain present in the internuclear region of the bond, which has been named as electron population by *R.S. Mulliken.*

Now according to VBT, we substitute

$$\Psi_+ = \frac{\Phi_1 + \Phi_2}{\sqrt{2\left(1+S^2\right)}}$$

Then, the spinless density function will be expressed as

$$\rho(1) = \int \Psi_+^2 d\tau_2 = \frac{1}{2\left(1+S^2\right)}\left[\int \Phi_1^2 d\tau_2 + \int \Phi_2^2 d\tau_2 + 2\int \Phi_1\Phi_2 d\tau_2\right]$$

which is equivalent to

$$\rho(1) = \frac{1s_a^2(1) + 1s_b^2(1) + 2 \times 1s_a(1)1s_b(1)}{2(1+S^2)}$$

Hence, the total electron density will be given by

$$\rho = \rho(1) + \rho(2) = 2\rho(1)$$

$$= \frac{2}{2(1+S^2)}\left[1s_a^2 + 1s_b^2 + 2 \times 1s_a 1s_b\right]$$

$$\therefore \quad \rho = \frac{1}{(1+S^2)}\left[1s_a^2 + 1s_b^2 + 2 \times 1s_a 1s_b\right] \tag{10.192}$$

It is, thus, clear that according to VBT, $\frac{1}{(1+S^2)}$ electrons will remain associated with each H atom and $\frac{2s}{(1+S^2)}$ will be found in the internuclear region. It is concluded that with regard to the covalent bonding, both MOT and VBT agree to the same point.

Because $0 < s < 1$, the MOT arrives at a bigger build-up of electron density in the overlap space in comparison with the VBT.

10.15 Excited State of H_2 Molecule

Here, we shall examine how far the MOT is able to describe the excited electronic state of H_2 molecule. On the basis of LCAO, we obtained two orbitals Ψ_1 and Ψ_2, which have the following values:

$$\left.\begin{aligned} \Psi_1 &= \frac{1}{\sqrt{2(1+S)}}(1s_a + 1s_b) \\ \text{and} \quad \Psi_2 &= \frac{1}{\sqrt{2(1-S)}}(1s_a - 1s_b) \end{aligned}\right\} \tag{10.193}$$

In the ground state, both the electrons of H_2 molecule reside in the Ψ_1 orbital. It should be kept in mind that the first excited-state configuration above the ground state should be as a result of the promotion of an electron from the Ψ_1 orbital to the Ψ_2 orbital.

For $\Psi_1'\Psi_2'$, the following functions can be framed:

$$\left.\begin{aligned} \chi_1 &= \frac{1}{\sqrt{2}}\left[\Psi_1(1)\Psi_2(2) + \Psi_1(2)\Psi_2(1)\right] \\ \text{and} \quad \chi_1' &= \frac{1}{\sqrt{2}}\left[\Psi_1(1)\Psi_2(2) - \Psi_1(2)\Psi_2(1)\right] \end{aligned}\right\} \tag{10.194}$$

Now let us consider an operator p, which interchanges the coordinates. Then, after operation, we can write

$$\left.\begin{aligned} \hat{p}\chi_1 &= \frac{1}{\sqrt{2}}\left[\Psi_1(2)\Psi_2(1)+\Psi_1(1)\Psi_2(1)\right]=\chi_1 \\ \hat{p}\chi_1' &= \frac{1}{\sqrt{2}}\left[\Psi_1(1)\Psi_2(2)-\Psi_1(2)\Psi_2(1)\right]=\chi_1' \end{aligned}\right\} \tag{10.195}$$

It is, thus, clear that χ_1 is symmetric and χ_1 is antisymmetric with regard to the interchange of electrons.

It is known to us that for the two electrons, there will be four possible spin functions, and to get the complete wave functions, the spin functions should be multiplied with the aforementioned space functions.

The spin function will be expressed as

$$\left.\begin{aligned} \sigma_1 &= \alpha(1)\alpha(2) \\ \sigma_2 &= \beta(1)\beta(2) \\ \sigma_3 &= \alpha(1)\beta(2) \\ \sigma_4 &= \beta(1)\alpha(2) \end{aligned}\right\} \tag{10.196}$$

The Eigen values of these functions for the operator $\hat{S}_z$ can be written as $\widehat{S_z} = \widehat{S_{z1}} + \widehat{S_{z2}}$ and, respectively, have values +1, –1, and 0 in the units of $h/2\pi$. The equation for σ_3 and σ_4 are doubly degenerate and clearly represent two similar/equivalent spin states. When operated with p, it will result in the following:

$$\hat{p}\sigma_3 = \hat{p}\alpha(1)\beta(2) = \alpha(2)\beta(1) = \sigma_4$$

$$\hat{p}\sigma_4 = \hat{p}\beta(1)\alpha(2) = \alpha(1)\beta(2) = \sigma_3$$

$$\text{Therefore, } \hat{p}(\sigma_3+\sigma_4) = \sigma_3+\sigma_4$$

$$\hat{p}(\sigma_3-\sigma_4) = -(\sigma_3-\sigma_4)$$

It is, therefore, essential to consider the linear combination of σ_3 and σ_4 functions. The following four spin functions can be constructed for a two-electron system, i.e.,

$$\Phi_s = \frac{1}{\sqrt{2}}\left[\alpha(1)\beta(2)-\beta(1)\alpha(2)\right] \tag{10.197}$$

$$\Phi_T = \left\{\begin{array}{c} \alpha(1)\alpha(2) \\ \frac{1}{\sqrt{2}}\left[\beta(1)\alpha(2)+\beta(2)\alpha(1)\right] \\ \beta(1)\beta(2) \end{array}\right. \tag{10.198}$$

where $\frac{1}{\sqrt{2}}$ = normalisation factor of the spin functions,

Φ_s = spin function for the singlet state and has Eigen value –1 for the $\hat{p}$ operator and consequently Φ_s is antisymmetric, and

Φ_T = spin function for the triplet state, which has three components having Eigen value +1 for the operator $\hat{p}$ and hence symmetric.

Now according to the Pauli's exclusion principle, the total wave function, including electron spin, must be antisymmetric. Thus, the following combinations are obtained:

$$\left.\begin{aligned} {}^1\psi_1 &= \chi_1\Phi_s \\ {}^3\psi_1 &= \chi_1'\Phi_T \end{aligned}\right\} \tag{10.199}$$

where the superscripts 1 and 3 in Ψ_1 indicate the multiplicity of states.

After putting the value of χ_1 and Φ_s, we can write

$$\begin{aligned} {}^1\psi_1 &= \frac{1}{\sqrt{2}}\left[\Psi_1(1)\Psi_2(2)+\Psi_1(2)\Psi_2(1)\right]\frac{1}{\sqrt{2}}\left[\alpha(1)\beta(2)-\beta(1)\alpha(2)\right] \\ &= \frac{1}{2}\left[\Psi_1(1)\alpha(1)\Psi_2(2)\beta(2)-\Psi_1(1)\beta(1)\Psi_2(2)\alpha(2)+\Psi_1(2)\beta(2)\Psi_2(1)\alpha(1)-\Psi_1(2)\alpha(2)\Psi_2(1)\beta(1)\right] \\ &= \frac{1}{2}\left[\{\Psi_1(1)\Psi_2(2)-\Psi_1(2)\Psi_2(1)\}-\{\Psi_1(1)\Psi_2(2)-\Psi_1(2)\Psi_2(1)\}\right] \\ &= \frac{1}{2}\left[\left|\Psi_1(1)\Psi_2(2)\right|-\left|\Psi_1(1)\Psi_2(2)\right|\right] \end{aligned} \tag{10.200}$$

Similarly, the wave function for the triplet state may be expressed in the determinant form as

$$\frac{1}{\sqrt{2}}\left|\Psi_1(1)\Psi_2(2)\right|$$

$$\begin{aligned} \text{or}\quad {}^3\psi_1 &= \frac{1}{\sqrt{2}}\left[\left|\Psi_1(1)\Psi_2(2)\right|+\left|\Psi_1(1)\Psi_2(2)\right|\right] \\ &= \frac{1}{\sqrt{2}}\left|\Psi_1(1)\Psi_2(2)\right| \end{aligned} \tag{10.201}$$

Furthermore, we shall estimate the electronic energies of the excited states ${}^1\psi_1$ and ${}^3\psi_1$.

$$E_{\text{el}}\left({}^1\psi_1\right) = {}^1\psi_1\left|H_e'\right|{}^1\psi_1 \tag{10.202}$$

${}^1\psi_1$ is normalised.

$$\text{where } H_e' = H^c(1)+H^c(2)+\frac{1}{r_{12}}$$

On substituting for ${}^1\psi_1$, we shall get

$$E_{\text{el}}\left({}^1\psi_1\right) = \iint \frac{1}{4}\left[\left|\Psi_1(1)\overline{\Psi_2}(2)\right|-\overline{\Psi_1}(1)\Psi_2(2)\right]H_e'\left[\left|\Psi_1(1)\overline{\Psi_2}(2)\right|\right]-\left|\overline{\Psi_1}(1)\Psi_2(2)\right|\Big]d\tau_1\,d\tau_2 \tag{10.203}$$

It should be kept in mind that the integral is over the entire space.

We may represent the aforementioned integral into two integrals in the following manner:

$$I_1 = \iint \left[\left| \Psi_1(1)\overline{\Psi_2}(2) \right| \left| H_e' \right| \left| \Psi_1(1)\overline{\Psi_2}(2) \right| \right] d\tau_1 \, d\tau_2$$

$$= \iint \left[\left| \overline{\Psi_1}(1)\Psi_2(2) \right| \left| H_e' \right| \left| \overline{\Psi_1}(1)\Psi_2(2) \right| \right] d\tau_1 \, d\tau_2 \tag{10.204}$$

$$\text{and} \quad \iint \left[\left| \Psi_1(1)\overline{\Psi_2}(2) \right| \left| H_e' \right| \left| \overline{\Psi_1}(1)\Psi_2(2) \right| \right] d\tau_1 \, d\tau_2$$

$$= \iint \left[\left| \overline{\Psi_1}(1)\Psi_2(2) \right| \left| H_e' \right| \left| \Psi_1(1)\overline{\Psi_2}(2) \right| \right] d\tau_1 \, d\tau_2 \tag{10.205}$$

$$\therefore \quad E_{\text{el}}\left({}^1\overline{\Psi_1} \right) = \frac{1}{4}[2I_1 - 2I_2] = \frac{1}{2}[I_1 - I_2] \tag{10.206}$$

Expanding the determinants in I_1, we obtain

$$I_1 = \iint \left[\Psi_1(1)\overline{\Psi_2}(2) - \Psi_1(2)\overline{\Psi_2}(1) \right] \left| H_e' \right| \left[\Psi_1(1)\overline{\Psi_2}(2) - \Psi_1(2)\overline{\Psi_2}(1) \right] d\tau_1 \, d\tau_2 \tag{10.207}$$

Let $\Psi_1(1)\overline{\Psi_2}(2) = A$, $\Psi_1(2)\overline{\Psi_2}(1) = B$, the Eq. (10.207) the Eq. (10.207) take the form as –

$$I_1 = \langle A|H_e'|A\rangle + \langle B|H_e'|B\rangle - \langle A|H_e'|B\rangle - \langle B|H_e'|A\rangle$$

Since H' is Hermitian,

$$\langle A|H_e'|B\rangle = \langle B|H_e'|A\rangle$$

Putting this value, we get

$$I_1 = \langle A|H_e'|A\rangle + \langle B|H_e'|B\rangle - \langle 2A|H_e'|B \ \rangle \tag{10.208}$$

Substituting for H_e' in $\langle A|H_e'|A\rangle$, we can write

$$\langle A|H_e'|A\rangle = \left\langle \Psi_1(1)\Psi_2(2) \middle| \left(H^c(1) + H^c(2) + \frac{1}{r_{12}} \right) \middle| \Psi_1(1)\Psi_2(2) \right\rangle$$

$$= \langle \Psi_1(1)|H^c(1)|\Psi_1(1)\rangle \langle \overline{\Psi_2}(2) \mid \overline{\Psi_2}(2)\rangle + \langle \overline{\Psi_2}(2)|H^c(2)|\overline{\Psi_2}(2)\rangle \langle \Psi_1(1)\Psi_1(1)\rangle$$

$$+ \left\langle \Psi_1(1)\overline{\Psi_2}(2) \middle| \frac{1}{r_{12}} \middle| \Psi_1(1)\overline{\Psi_2}(2) \ \right\rangle \tag{10.209}$$

It is to be noted that the spin functions and orbital functions are normalised; therefore, Eq. (10.209) can be given by

$$\langle A|H_e'|A\rangle = E_1 + E_2 + J_{12} \tag{10.210}$$

where the first term of Eq. (10.209) represents E_1, the second term represents E_2, and the third term represents J_{12}. It should be kept in mind that J_{12} represents the repulsion between the electron 1 of the Ψ_1 orbital and electron 2 of the Ψ_2 orbital.

Similar procedure results in

$$\langle B|H_e'|B\rangle = E_1 + E_2 + J_{12} \tag{10.211}$$

whereas the integral

$$\begin{aligned}\langle B|H_e'|B\rangle &= \langle A|H_e'|B\rangle \\ &= \langle \Psi_1(1)|H^c(1)|\overline{\Psi_2}(1)\rangle\langle \overline{\Psi_2}(2)\,|\,\Psi_1(2)\rangle + \langle \overline{\Psi_2}(2)|H^c(2)|\Psi_1(2)\rangle\langle \Psi_1(1)\overline{\Psi_2}(1)\rangle \\ &\quad + \left\langle \Psi_1(1)\overline{\Psi_2}(2)\left|\frac{1}{r_{12}}\right|\Psi_1(2)\overline{\Psi_2}(1)\right\rangle = 0\end{aligned} \tag{10.212}$$

due to the orthogonality of the orbital and spin function.

$$\begin{aligned}\therefore\quad I_1 &= \text{sum of Eqs.}(10.210)\text{and } (10.211) \\ &= 2E_1 + 2E_2 + 2J_{12}\end{aligned} \tag{10.213}$$

Now, we shall find the value of I_2. I_2 can be written as

$$\begin{aligned}I_2 &= \iint \left[\Psi_1(1)\overline{\Psi_2}(2) - \Psi_1(2)\overline{\Psi_2}(1)\right]H_e'\left[\overline{\Psi_1}(1)\Psi_2(2) - \overline{\Psi_1}(2)\Psi_2(1)\right]d\tau_1\, d\tau_2 \\ &= \langle A|H_e'|C\rangle - \langle A|H_e'|D\rangle - \langle B|H_e'|C\rangle + \langle B|H_e'|D\rangle\end{aligned} \tag{10.214}$$

where A and E are defined before, but $C = \overline{\Psi_1}(1)\Psi_2(2)$ and $\mathrm{D} = \overline{\Psi_1}(2)\Psi_2(1)$.

The first and fourth terms of Eq. (10.214) become zero due to the orthogonality of the spin function, whereas the second and third terms will be equal and are given by

$$\begin{aligned}\langle A|H_e'|D\rangle &= \langle B|H_e'|C\rangle \\ &= \left\langle \Psi_1(1)\overline{\Psi_2}(2)\left|\frac{1}{r_{12}}\right|\overline{\Psi_1}(2)\Psi_2(1)\right\rangle\end{aligned} \tag{10.215}$$

Actually, this equation consists of the repulsion between two electrons of H_2, but electron 1 is associated with Ψ_1 and Ψ_2 both and similar is the situation with electron 2. It can be viewed that the electrons are exchanged between the two orbitals, and hence, this type of integral will be represented by K and is known as the exchange integral.

$$\therefore\quad I_2 = -2K_{12} \tag{10.216}$$

It should be noted that if one electron is associated with two orbitals having different spins, then the exchange integral will be zero. Thus, substituting the values of I_1 and I_2 from earlier in Eq. (10.206), we shall get

$$E_{\text{el}}\left({}^{1}\psi_{1}\right) = E_{1} + E_{2} + J_{12} + K_{12} \tag{10.217}$$

and $$E_{\text{el}}\left({}^{3}\psi_{1}\right) = E_{1} + E_{2} + J_{12} + K_{12} \tag{10.218}$$

$$\therefore \quad E_{\text{el}}\left({}^{1}\psi_{1}\right) - E_{el}\left({}^{3}\psi_{1}\right) = 2K_{12} \tag{10.219}$$

Since J and K integrals symbolise the electron repulsion, the triplet state will be more stable. This gives an idea that the states of highest spin multiplicity will have maximum stability, which is Hund's rule.

10.16 Electronic Transition in Hydrogen Molecule

When the hydrogen molecule will absorb electromagnetic radiation, it will be promoted from its ground state to an excited state. Suppose the energy difference between the ground state and the excited state is ΔE, then this will be equal to $h\nu$, i.e., $\Delta E = h\nu = hc/\lambda$, where the terms have their usual significance.

Thus, we can expect that the said molecule will absorb radiation of a particular wavelength (λ) corresponding to ΔE (energy difference between the ground state and the excited state). It should be kept in mind that such type of transition will depend on the square of the transition dipole moment integral, which is expressed/defined in a.u. as

$$\left\langle \Psi_{g} \left| \sum_{i} r_{i} \right| \Psi_{\text{ex}} \right\rangle$$

where r_i = position vector of the electron i in the molecule,

Ψ_g = ground-state wave function, and

Ψ_{ex} = excited-state wave function.

In the aforementioned definition, the summation in the integral is overall and the electron remains in the molecule.

Let us suppose the transition from the singlet ground state Ψ_0 to the excited triplet state ${}^{3}\Psi_1$ of the H_2 molecule. Then, the transition dipole moment is expressed as

$$M\left(\Psi_{0} \rightarrow {}^{3}\psi_{1}\right) = \left\langle \Psi_{g} \left| \sum_{i} r_{i} \right| {}^{3}\psi_{1} \right\rangle \text{on the basis of the above equation.}$$

where $M(\Psi_0 \rightarrow {}^{3}\psi_1)$ = transition dipole moment

This integral is over the spin–orbital space.

Try to recall the following equations:

$$\Psi_{0} = \Psi_{1}(1)\Psi_{1}(2)\frac{1}{\sqrt{2}}\left[\alpha(1)\beta(2) - \beta(1)\alpha(1)\right]$$

$$\Phi_{S} = \frac{1}{\sqrt{2}}\left[\alpha(1)\beta(2) - \beta(1)\alpha(2)\right]$$

$${}^{1}\psi_{1} = \chi_{1}\Phi_{S}$$

$${}^{3}\psi_{1} = \chi_{1}'\Phi_{T}$$

On separating the orbital functions and spin functions, using the aforementioned equation, we can write

$$\psi_0 = \chi_0 \Phi_S$$

and ${}^3\psi_1 = \chi_1' \Phi_T$ where $\chi_0 = \Psi_1(1)\Psi_1(2)$ and

$$\therefore \quad M\left(\Psi_0 \rightarrow {}^3\psi_1\right) = \langle \chi_0 | \Sigma r_i | \chi' \rangle_1 \langle \Phi_S \mid \Phi_T \rangle \tag{10.220}$$

In this equation, the two integrals indicate the spin and orbital coordinates of the electron.

Since Φ_s and Φ_T are orthogonal, $\langle \Phi_S | \Phi_T \rangle = 0$.

This suggests that the transition from the singlet to the triplet state is forbidden. If, however, the spin and the orbital functions cannot be separated because of the spin–orbit interaction, then M may not be zero.

Now, we shall try to compute the transition dipole moment of an optical transition in H_2 molecule from Ψ_0 to ${}^1\psi_1$. If the spin and the orbital functions be separated as before, then the equation can be written as

$$M\left(\Psi_0 \rightarrow {}^1\psi_1\right) = \left\langle \chi_0 \left| \sum_i r_i \right| \chi_1 \right\rangle \langle \Phi_S \mid \Phi_s \rangle \tag{10.221}$$

Since Φ is normalised, $\langle \Phi_S | \Phi_S \rangle = 1$.

$$M\left(\Psi_0 \rightarrow {}^1\psi_1\right) = \left\langle \chi_0 \left| \sum_i r_i \right| \chi_1 \right\rangle \tag{10.222}$$

Furthermore, putting the value of χ_0 and χ_1 in Eq. (10.222), we shall get

$$\begin{aligned} M\left(\Psi_0 \rightarrow {}^1\psi_1\right) &= \frac{1}{\sqrt{2}} \left\langle \Psi_1(1)\Psi_1(2) \left| \vec{r_1} + \vec{r_2} \right| \left[\Psi_1(1)\Psi_1(2) + \Psi_1(2)\Psi_2(1)\right] \right\rangle \\ &= \frac{1}{\sqrt{2}} \left\langle \Psi_1(1)\Psi_1(2) \left| \vec{r_1} \right| \left\{\Psi_1(1)\Psi_2(2) + \Psi_1(2)\Psi_2(1)\right\} \right\rangle \\ &\quad + \left\langle \Psi_1(1)\Psi_2(2) \left| \vec{r_2} \right| \left\{\Psi_1(1)\Psi_2(2) + \Psi_1(2)\Psi_2(1)\right\} \right\rangle \\ &= \frac{1}{\sqrt{2}} \Big[\left\langle \Psi_1(1) \left| \vec{r_1} \right| \Psi_1(1) \right\rangle \left\langle \Psi_1(2) \mid \Psi_2(2) \right\rangle + \left\langle \Psi_1(1) \left| \vec{r_1} \right| \Psi_2(1) \right\rangle \left\langle \Psi_1(2) \mid \Psi_1(2) \right\rangle \\ &\quad + \left\langle \Psi_1(2) \left| \vec{r_2} \right| \Psi_2(2) \right\rangle \left\langle \Psi_1(1) \mid \Psi_1(1) \right\rangle + \left\langle \Psi_1(2) \left| \vec{r_2} \right| \Psi_2(2) \right\rangle \left\langle \Psi_1(1) \mid \Psi_1(1) \right\rangle \Big] \end{aligned} \tag{10.223}$$

Due to orthogonality, the first and fourth terms will vanish because

$$\left\langle \Psi_1(2) \mid \Psi_2(2) \right\rangle = 0 \text{ and}$$

$$\left\langle \Psi_1(1) \mid \Psi_2(1) \right\rangle = 0$$

But $\langle \Psi_1(2) | \Psi_1(2) \rangle = 1$

and $\langle \Psi_1(1) | \Psi_1(1) \rangle = 1$

due to normalisation condition. Therefore, the remaining part of the second and third terms will be written as

$$M\left(\Psi_0 \rightarrow {}^1\psi_1\right) = \frac{1}{\sqrt{2}} \left[\left\langle \Psi_1(1) \left| \vec{r_1} \right| \Psi_2(1) \right\rangle + \left\langle \Psi_1(2) \mid \vec{r_2} \mid \Psi_2(2) \right\rangle \right] \tag{10.224}$$

Since the two electrons cannot be distinguished, the aforementioned equation may be expressed as

$$M\left(\Psi_0 \rightarrow {}^1\psi_1\right) = \frac{2\times 1}{\sqrt{2}}\left[\left\langle \Psi_1(1)\middle|\vec{r_1}\middle|\Psi_2(1)\right\rangle\right] = \sqrt{2}\left[\left\langle \Psi_1(1)\middle|\vec{r_1}\middle|\Psi_2(1)\right\rangle\right] \tag{10.225}$$

But we know that

$$\Psi_1 = \frac{(1s_a + 1s_b)}{\sqrt{2(1+S)}} \tag{10.226}$$

$$\Psi_2 = \frac{(1s_a - 1s_b)}{\sqrt{2(1-S)}} \tag{10.227}$$

Putting the value of Ψ_1 and Ψ_2 in Eq. (10.225), we shall get

$$\begin{aligned}
M\left(\Psi_0 \rightarrow {}^1\psi_1\right) &= \sqrt{2}\cdot\frac{1}{\sqrt{2(1+S)}}\cdot\frac{1}{\sqrt{2(1-S)}} \times \left\langle \{1s_a(1) + 1s_b(1)\}\middle|\vec{r_1}\middle|\{1s_a(1) - 1s_b(1)\}\right\rangle \\
&= \frac{\sqrt{2}\cdot}{\sqrt{4\left(1-S^2\right)}}\cdot\left[\left\langle 1s_a(1)\middle|\vec{r_1}\middle|1s_a(1)\right\rangle - \left\langle 1s_b(1)\middle|\vec{r_1}\middle|1s_b(1)\right\rangle\right] \\
&= \frac{1}{\sqrt{2\left(1-S^2\right)}}\cdot\left[\left\langle 1s_a(1)\middle|\vec{r_1}\middle|1s_a(1)\right\rangle - \left\langle 1s_b(1)\middle|\vec{r_1}\middle|1s_b(1)\right\rangle\right] \\
&= \frac{1}{\sqrt{2\left(1-S^2\right)}}\cdot\left[\left\langle 1s_a(1)\middle|\vec{r_1}\middle|1s_a(1)\right\rangle - \left\langle 1s_b(1)\middle|\vec{r_1}\middle|1s_b(1)\right\rangle\right]
\end{aligned} \tag{10.228}$$

It is clear from Eq. (10.228) that since the 1sa orbital is spherically symmetrical, the position of the electron will coincide with the position of the nucleus 'a'. Similarly, the position of nucleus 'b' as $1s_b$ is also spherically symmetrical. Thus, the bracketed portion of Eq. (10.228) may be written as equal to $\left[\vec{r_a} - \vec{r_b}\right]$, and on putting this value in the aforementioned equation, we arrive at

$$M\left(\Psi_0 \rightarrow {}^1\psi_1\right) = \frac{1}{\sqrt{2\left(1-S^2\right)}}\cdot\left[\vec{r_a} - \vec{r_b}\right] \tag{10.229}$$

$$M\left(\Psi_0 \rightarrow {}^1\psi_1\right) = \frac{1}{\sqrt{2\left(1-S^2\right)}}\cdot[R] \tag{10.230}$$

where $\vec{r_a}$ and $\vec{r_b}$ are position vectors of nuclei a and b, respectively, and R is the bond length of H_2 molecule.

10.17 Homopolar Diatomic or Homonuclear Diatomic Molecules

The development of MOT is due basically to Mulliken (1932), and its advancement has been made by other workers. We are aware of the fact that MOs are formed out of two AOs, one on each atom.

If the atoms are identical, then the bond between them will be *homopolar*, and in such a situation of diatomic molecule, the MO will be mathematically expressed as

$$\Psi = \frac{N}{\sqrt{2}}(\Phi_a \pm \Phi_b) \tag{10.231}$$

where N = normalisation factor

and Φ_a and Φ_b = participating AOs

The plus and minus combination in Eq. (10.231) will usually be *bonding* and *antibonding*, respectively. It should be kept in mind that at large distances, the energies of both the MOs will tend to be the energy of the AO representing Φ_A.

An MO, which concentrates electron density in the region between the nuclei, thereby decreasing the energy of the system, is known as *bonding MO*. On the other hand, an *antibonding MO* is characterised by largely decreased electron density between the nuclei.

The formation of bonding MO in H_2 (for example) takes place by adding two ls orbitals. Let us suppose that one hydrogen atom is represented by H_a and another by H_b, and their corresponding AOs are represented by $1s_a$ and $1s_b$, respectively. When $1s_a$ and $1s_b$ AOs having one electron each are combined together, their combinations are represented by $(1s_a + 1s_b)$ and $(1s_a - 1s_b)$, i.e., the two hydrogen atoms (H_a and H_b) combine to yield two types of MOs. One is called σ (sigma) MO, and the other is called σ^* (sigma star) MO.

The σ MO is the bonding MO, whereas σ^* MO is the antibonding MO. The schematic formation of bonding σ and antibonding σ^* MOs is illustrated in Figure 10.8.

In the first combination of the two 1s valence orbitals, electrons density increases in the region between the nuclei. The σ MO is symmetrical around the internuclear line.

In the second subtractive combination of the two 1s valence orbitals, electron density decreases in the region between the nuclei. A nodal plane is also formed in which there is zero probability of finding an electron. The σ^* MO is symmetrical around the internuclear line.

A plot of energies of σ and σ^* is shown in Figure 10.9, which is a function of the internuclear separation.

We have illustrated that from two 1s AOs having the same energy, we can construct two MOs. The bonding MO is lower energy than the original AOs, and the energy of the antibonding orbital is higher, which has been illustrated in Figure 10.10.

10.17.1 Molecules with *s* and *p* Valence Atomic Orbitals

Now we will discuss the diatomic molecules having 2*s* and 2*p* valence AOs and will illustrate how the MOs will be formed in such a situation, for example, in the case of N_2. We will follow the method of making linear (i.e., additive and subtractive) combinations of proper AOs. We are aware of the fact that 2*s* orbitals have less energy than 2*p* orbitals.

Let us first consider the 2*s* AOs. The two MOs deduced from 2*s* orbitals are similar to the H_2 MOs. The combination $(2s_a + 2s_b)$ results in σ BOs, whereas the combination $(2s_a - 2s_b)$ yields the relative high-energy-containing orbitals, which has a nodal plane, and the probability of finding the electron is zero.

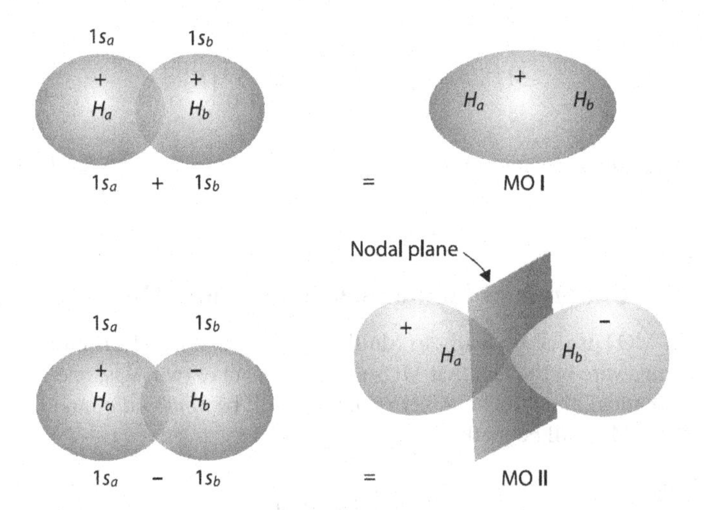

FIGURE 10.8 (a) Formation of sigma molecular orbital (MO) of hydrogen; (b) formation of sigma antibonding MO.

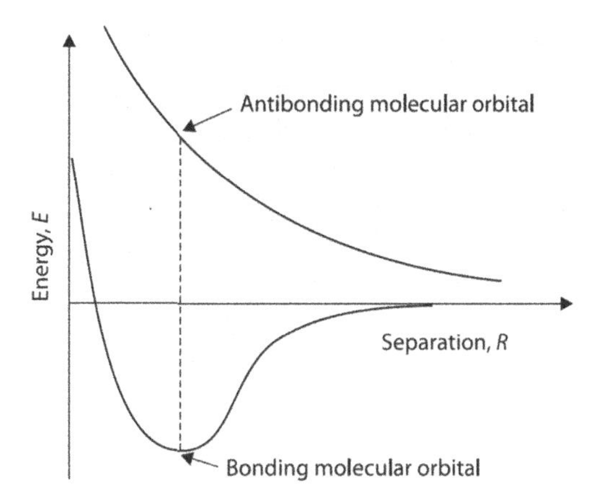

FIGURE 10.9 Energy vs. internuclear separation.

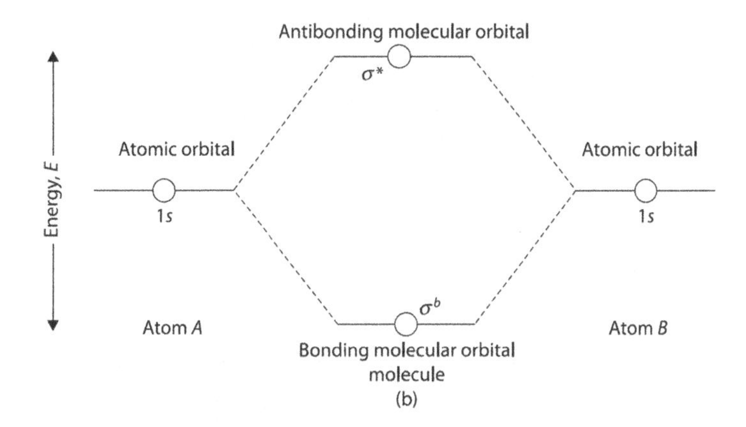

FIGURE 10.10 Position of σ and σ^* with respect to energy in hydrogen molecule.

Electron density is forced out of the bonding region, and thus, this orbital is termed as antibonding MO σ^*, which is illustrated in Figure 10.11.

The three $2p$ orbitals are directed along the coordinate axes X, Y, and Z. The line, which joins the nuclei in diatomic molecule, is normally represented by X axis. The two sets of axes, i.e., Y and Z are parallel, and the X axis is common to both the nuclei. In Figure 10.11, it has been illustrated that there are two different types of p orbital in a diatomic molecule. One p orbital of each atom lies along the internuclear axis and is known as p_x orbital. If we take the example of $2p_x$ orbitals, they are directed towards each other and overlap in the X direction. This is not the situation with the other two p orbitals in $2p_y$ and $2p_z$ because these do not overlap above and below it.

In Figure 10.11, plus and minus signs denote only the signs of the wave function, not the electrical charges. The AOs used in the top row are $2s$ orbitals, in the middle row, $2p_x$, and in the bottom row, it is $2p_y$ or the equivalent $2p_z$ orbitals.

The combination $(2s_a + 2s_b)$ represents the orbital symmetrical around X axis, and this is why it is σ MO. It is found that the electron density is larger in the overlap region, which clearly indicates that $2s_a + 2s_b$ is the BO σ_x or σ_x^b. The other combination is $2s_a - 2s_b$ which reduces the electron density in the overlap area. Since in $2s_a - 2s_b$, there is a nodal plane half-way between the nuclei, $2s_a - 2s_b$ combination will represent the antibonding MO σ_x^*. The σ_x^b and σ_x^* have been shown in Figure 10.11.

Now, we shall examine the MO formed from $2p_y$ and $2p_z$ orbitals. The combination $y_a + y_b$ is a somewhat new type of MO. Since p orbitals have a node at the nucleus, the MO $y_a + y_b$ has a nodal plane, which has the internuclear line. Thus, if we rotate the MO $y_a + y_b$ through 180°, it becomes $y_a - y_b$, as shown in Figure 10.12.

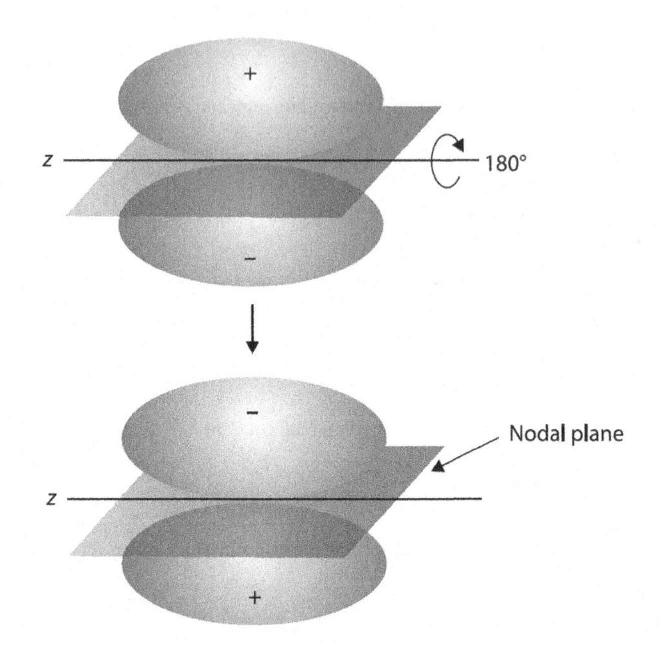

FIGURE 10.12 Overlap of two $2p_x$ orbitals.

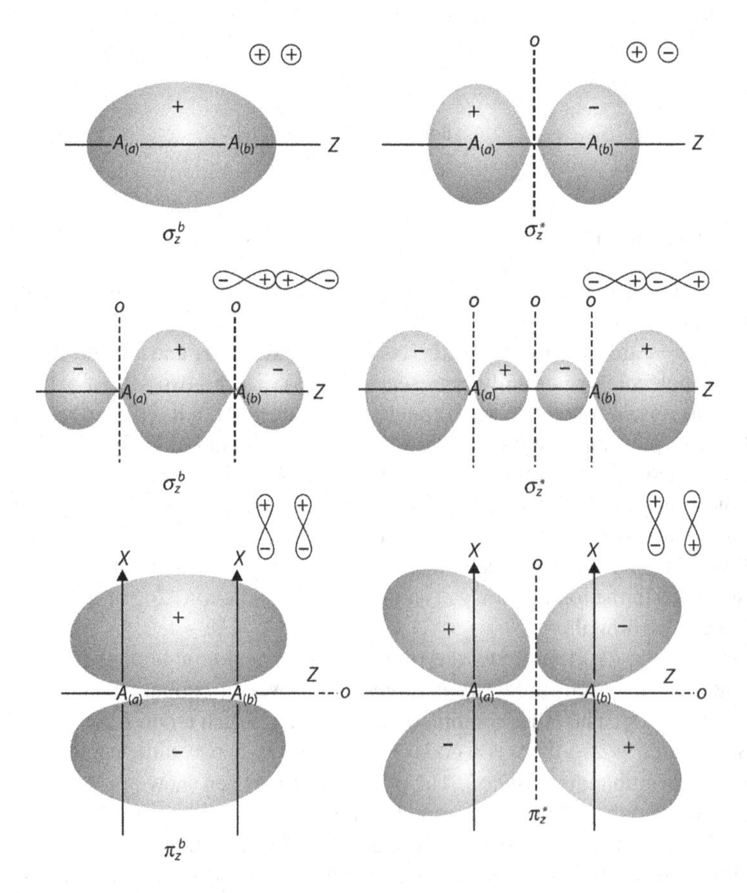

FIGURE 10.11 Six different kinds of molecular orbitals.

This type of MO is known as Pi or π orbital. Actually π MO is produced from the combinations of parallel p orbitals above and below the internuclear line. Since $y_a + y_b$ orbital concentrates electron density in the bonding region, it is called π bonding or π^b orbital. It will be better to represent this as π_y^b because it is formed by the combination of $2p_y$ orbitals. Similarly, when $z_a + z_b$ orbitals will combine, it will give π_z^b orbitals, i.e., π BO.

Now let us assume the combinations $y_a - y_b$ and $z_a - z_b$, which have a nodal plane consisting of the internuclear axis. They are antibonding orbitals (π^*) since they have reduced electron density between the nuclei. The π_y^b and π_y^* MOs are illustrated in Figure 10.12.

Finally, we can say that we began with eight valence orbitals (one 2s and three 2p orbitals on each atom) and constructed eight MOs (σ_s^b, σ_s^*, $\sigma_x^b, \sigma_x^*, \pi_y^b$, π_y^*, π_z^b, and π_z^*). The energy of these MOs is in the following order:

$$\sigma_s^b < \sigma_s^* < \pi_y^b = \pi_z^b < \sigma_x^b < \pi_y^* = \pi_z^* < \sigma_x^*$$

10.17.2 Electronic Configuration of Homonuclear Diatomic Molecules

The electronic configuration of the molecule is written by putting the electrons in different energy levels following Pauli's exclusion principle. Thus, a pair of electrons will be first occupied in the lowest MO level. In the case when the MOs are degenerate, Hund's rule will be followed. Now we are going to write down the electronic configurations of some homonuclear diatomic molecule.

- H_2: The number of electrons in H_2 molecule is two, which results from two H AOs. The MO electron configuration will be as $H_2[\sigma 1s^2]$.
- He_2: The overlap of two He 1s orbitals will result in σ1s and σ*1s MOs with two electrons in each. The electronic configuration of He_2 will be written as

$$\text{He}_2\left[\left(\sigma 1s^2\right)\left(\sigma^* 1s^2\right)\right]$$

In this case, the number of bonding electrons is 2 and that of antibonding electrons is also 2. Therefore, $= \frac{n_b - n_a}{2} = \frac{2-2}{2} = 0$. Thus, the bond order is zero. Therefore, He_2 will not exist.

Similarly, for other diatomics, the electronic configuration can be written as

- Li_2 : $KK\sigma 2s^2$, where $KK = \sigma 1s^2 \sigma^* 1s^2$
- Be_2 : $KK\sigma 2s^2 \sigma^* 2s^2$
- B_2 : $KK\sigma 2s^2 \sigma^* 2s^2 \sigma 2p_x^2$
- C_2 : $KK\sigma 2s^2 \sigma^* 2s^2 \sigma 2p_x^2 \pi 2p_y^1 \pi 2p_z^1$
- N_2 : $KK\sigma 2s^2 \sigma^* 2s^2 \sigma 2p_x^2 \pi 2p_y^2 \pi 2p_z^2$
- O_2 : $KK\sigma 2s^2 \sigma^* 2s^2 \sigma 2p_x^2 \pi 2p_y^2 \pi 2p_z^2 \pi^* 2p_z^1 \pi^* 2p_y^1$
- F_2 : $KK\sigma 2s^2 \sigma^* 2s^2 \sigma 2p_x^2 \pi 2p_y^2 \pi 2p_z^2 \pi^* 2p_z^2 \pi^* 2p_y^2$

The MO electronic configuration gives an insight into the nature of the bonds. It also gives a clue to understand the magnetic character of the molecule by counting the number of unpaired electrons. For example, N_2 is diamagnetic because it contains only paired electrons, whereas N_2^+ and O_2 are paramagnetic because of the fact that these two contain unpaired electrons.

10.18 Heteropolar Diatomic or Heteronuclear Diatomic Molecules

In the case of heteronuclear diatomic molecules also, like AB, we can apply the LCAO approximation. In AB, B has a larger electronegativity than A. The LCAO can be written as

$$\Psi = c_A\Phi_A + c_B\Phi_B \tag{10.232}$$

where Φ_A and Φ_B = AOs and

c_A and c_B = constants but so selected that the energy of the MO is minimum.

In the case of heteronuclear diatomics, $c_A \neq c_B$. The following conditions must be satisfied for the combinations to take place.

- Ψ_A and Ψ_B are orbitals having similar energy for the isolated atoms.
- The AOs must have the same symmetry with respect to the internuclear axis.
- Overlapping of AOs should be sufficient. It should be kept in mind that c_A and c_B described earlier measure the polarity between A and B when c_A and c_B will be unequal and c_A^2 / c_B^2 value should be very much different from unity. c_A^2 / c_B^2 is the measure of polarity of AB type of molecule. When the ratio of $c_A^2 / c_B^2 < 1$, the bond will possess certain ionic character.

  Now we shall consider the electronic configuration of some heteronuclear diatomics.
- HF: $\mathrm{H}\,(1s) + \mathrm{F}\left(1s^2 2s^2 2p^5\right) \rightarrow \mathrm{HF}\left[\sigma 1s^2 \sigma 2s^2 \sigma 2p_x^2 \pi 2p_y^2 \pi 2p_z^2\right]$
- NO: $\mathrm{N}\left(1s^2 2s^2 2p^2\right) + \mathrm{O}\left(1s^2 2s^2 2p^4\right) \rightarrow \mathrm{NO}\left[\mathrm{KK}\sigma 2s^2 \sigma^* 2s^2 \sigma 2p_x^2 \pi 2p_y^2 2p_z^2 \pi^* 2p_y^1\right]$
- BN: $\mathrm{B}\left(1s^2 2s^2 2p^1\right) + \mathrm{N}\left(1s^2 2s^2 2p^3\right) \rightarrow \mathrm{BN}\left[\mathrm{KK}\sigma 2s^2 \sigma^* 2s^2 \pi 2p_y^2 \pi 2p_z^1 \sigma 2p_x^1\right]$
- CO: $\mathrm{C}\left(1s^2 2s^2 2p^2\right) + \mathrm{O}\left(1s^2 2s^2 2p^4\right) \rightarrow \mathrm{CO}\left[\mathrm{KK}\sigma 2s^2 \sigma^* 2s^2 \pi 2p_y^2 \pi 2p_z^2 \sigma 2p_x^2\right]$

Similarly, we can write down the electronic configuration of other heteronuclear diatomics. It is pertinent to note that in the case of heteronuclear molecule, a bonding MO leans towards the more electronegative atom, whereas its antibonding partner leans towards the less electronegative atom.

BIBLIOGRAPHY

Bethe, H.A. and E.E. Salpeter. 1957. *Quantum Mechanics of One and Two Electron Atoms*. New York: Academic Press.

Dandel, R., R. Lefebvre and C. Moser. 1959. *Quantum Chemistry*. New York: Interscience Publishers Inc.

Dewar, M.J.S. 1969. *The Molecular Orbital Theory of Organic Chemistry*. New York: McGraw-Hill.

Jean, Y. and F. Voltron. 1993. *An Introduction to Molecular Orbitals*. New York: Oxford University Press.

Levine, I.N. 2000. *Quantum Chemistry*, 4th ed. New Jersey: Prentice Hall.

Pauling, L. and E.B. Wilson. 1935. *Introduction to Quantum Mechanics*. New York: McGraw-Hill Book Co. Inc.

Roy, M. 1979. *Coulsons Valence*, 3rd ed. Oxford: Oxford University Press.

Slater, J.C. 1960. *Quantum Theory of Atomic Structure*, vol. I. New York: McGraw-Hill.

Solved Problems

Problem 1. If Ψ_1 and Ψ_2 are the wave functions for a degenerate energy state E, show that any linear combination $c_1\Psi_1 + c_2\Psi_2$ is also a wave function.

Solution: We can write the wave equation as

$$\frac{\hbar^2}{2m} \cdot \frac{\mathrm{d}^2\Psi}{\mathrm{d}x^2} + (E - V)\Psi = 0$$

Ψ_1 and Ψ_2 are wave functions for the energy E, so we write

$$\frac{\hbar^2}{2m} \cdot \frac{\mathrm{d}^2\Psi}{\mathrm{d}x^2} + (E - V)\Psi_1 = 0 \tag{i}$$

$$\frac{\hbar^2}{2m}\cdot\frac{d^2\Psi}{dx^2}+(E-V)\Psi_2=0 \tag{ii}$$

Multiplying Eqs (i) and (ii) by c_1 and c_2, respectively, and then on adding, we get

$$\frac{\hbar^2}{2m}\left(\frac{d^2c_1\Psi_1}{dx^2}+\frac{d^2c_2\Psi_2}{dx^2}\right)+(E-V)(c_1\Psi_1+c_2\Psi_2)=0$$

$$\frac{\hbar^2}{2m}\cdot\frac{d^2}{dx^2}(c_1\Psi_1+c_2\Psi_2)+(E-V)(c_1\Psi_1+c_2\Psi_2)=0 \tag{iii}$$

It clearly indicates that $(c_1\Psi_1 + c_2\Psi_2)$ is also an Eigen function of the energy E. Answer.

Problem 2. Prove that the bonding and the antibonding MOs of H_2 molecule are orthogonal.

Solution: Given that

$$\Psi_1 = \frac{1}{\sqrt{2(1+S)}}[1s_a+1s_b]\text{ for bonding molecular orbital}$$

$$\Psi_2 = \frac{1}{\sqrt{2(1-S)}}[1s_a-1s_b]\text{ for antibonding molecular orbital}$$

Then, we have to prove that $\int\Psi_1\Psi_2 d\tau = 0$

$$\int\Psi_1\Psi_2\,d\tau \equiv \int\frac{1}{\sqrt{2(1+S)}}[1s_a+1s_b]\times\frac{1}{\sqrt{2(1-S)}}[1s_a-1s_b]d\tau$$

$$=\frac{1}{\sqrt{2(1-S^2)}}(1-1)=0$$

$$\therefore\quad \int\Psi_1\Psi_2 d\tau=0$$

Proved

Problem 3. Find out the normalised ground-state MO wave function for H_2 molecule.

Solution: Given that

$$\Psi_{MO}(1,2)=\Psi_1\Psi_1 \tag{i}$$

$$\left.\begin{aligned}\text{where }\Psi_1 &= c_1 1s_a(1)+c_2 1s_b(1)\\ \Psi_2 &= c_3 1s_a(2)+c_4 1s_b(2)\end{aligned}\right\} \tag{ii}$$

Let us assume that $c_1 = c_2 = c_3 = c_4 = C$ (say)

Now our problem is to find the coefficient C.

Substituting the value of Ψ_1 and Ψ_2 in Eq. (i), we get

$$\Psi_{MO}(1,2)=C^2\left[1s_a(1)+1s_b(1)\right]\left[1s_a(2)+1s_b(2)\right]$$

Under normalisation condition, we may write

$$\int \Psi^*_{\mathrm{MO}}(1,2)\Psi^*_{\mathrm{MO}}(1,2)d\tau \equiv \langle \Psi_{\mathrm{MO}}(1,2) \mid \Psi_{\mathrm{MO}}(1,2)\rangle$$

$$= \langle \Psi_1\Psi_2 \mid \Psi_1\Psi_2\rangle$$

$$= \langle \Psi_1\Psi_1 \mid \Psi_2\Psi_2\rangle$$

$$\langle \Psi_1 \mid \Psi_1\rangle^2 = 1 \qquad \text{(iii)}$$

But $\langle \Psi_1 \mid \Psi_1\rangle = C^2\left[\langle 1s_a(1) \mid 1s_a(1)\rangle + \langle 1s_b(1) \mid 1s_b(1)\rangle + \langle 1s_a(1) \mid 1s_b(1)\rangle + \langle 1s_b(1) \mid 1s_a(1)\rangle\right]$

$$= C^2\left[s_{aa} + s_{bb} + s_{ab} + s_{ba}\right]$$

where $s_{aa} = \langle 1s_a(1) \mid 1s_a(1)\rangle$

$$s_{bb} = \langle 1s_b(1) \mid 1s_b(1)\rangle$$

$$s_{ab} = \langle 1s_a(1) \mid 1s_b(1)\rangle$$

$$s_{ba} = \langle 1s_b(1) \mid 1s_a(1)\rangle$$

where s=overlap integrals.

By definition, $s_{aa} = s_{bb} = 1$ and $s_{ab} = s_{ba} = s$.

$$\therefore \quad \langle \Psi_1 \mid \Psi_1\rangle = c^2(1+1+2s) = 2(1+s)c^2$$

Putting this value in Eq. (iii), we get

$$\left[2(1+s)c^2\right]^2 = 1$$

$$\text{or} \quad \left[2(1+s)c^2\right]^2 = 1$$

$$\therefore \quad C = \frac{1}{\sqrt{2(1+S)}}$$

$$\therefore \quad \Psi_1 = \frac{1}{\sqrt{2(1+S)}}\left[1s_a(1) - 1s_b(1)\right]$$

Similarly, we can show that

$$\Psi_2 = \frac{1}{\sqrt{2(1+S)}}\left[1s_a(1) - 1s_b(1)\right]$$

Answer.

Problem 4. Find the wave function, which corresponds to E, using equation $E_{\pm} = -1 + \frac{J \pm K}{1 \pm S^2}$.

Solution: Given that $E_{\pm} = -1 + \frac{J \pm K}{1 \pm S^2}$,

$$c_1(H_{11} - ES_{11}) + c_2(H_{12} - ES_{12}) = 0$$

Using this equation, one can write

$$c_1(-1 + J - E_-) + c_2(-S^2 + K - E_- S^2) = 0$$

$$\text{or} \quad c_1\left(J - \frac{J-K}{1-S^2}\right) + c_2(+K - \left(\frac{(J-K)S^2}{1-S^2}\right) = 0$$

$$\text{or} \quad c_1\left(\frac{K - JS^2}{1-S^2}\right) + c_2\left(\frac{K - JS^2}{1-S^2}\right) = 0$$

$$\text{or} \quad c_2 = c_1$$

Thus, we can write

$$\Psi_- = c_1(\Psi_1 - \Psi_2)$$

We can evaluate c_1 by the condition that Ψ_- be normalised and

$$\therefore \quad c_1^2(1 - 2S^2 + 1) = 1$$

$$\text{or} \quad c_1^2(2 - 2S^2) = 1$$

$$\text{or} \quad c_1^2 = \frac{1}{(2 - 2S^2)}$$

$$\therefore \quad c_1 = \frac{1}{\sqrt{(2 - 2S^2)}} = \frac{1}{\sqrt{2(1 - S^2)}}$$

$$\therefore \quad \Psi_- = \frac{1}{\sqrt{2(1 - S^2)}} \cdot (\Psi_1 - \Psi_2) \qquad \text{Answer.}$$

It should be kept in mind that Ψ_- is antisymmetric under the interchange of the coordinates of the two electrons.

Problem 5. Find the MOs that correspond to the energies E_+ and E_- when

$$E_\pm = -\frac{1}{2} + \frac{J' \pm K'}{1 \pm S}, \; C_1(H_{AA} - E) + C_2(H_{AB} - ES) = 0$$

$$\& \, C_1(H_{AB} - ES) + C_2(H_{BB} - E) = 0$$

Solution: Given that

$$E_\pm = -\frac{1}{2} + \frac{J' \pm K'}{1 \pm S} \qquad \text{(i)}$$

$$C_1(H_{AA} - E) + C_2(H_{AB} - ES) = 0 \quad \text{(ii)}$$

$$C_1(H_{AB} - ES) + C_2(H_{BB} - E) = 0 \quad \text{(iii)}$$

$$H_{AA} = -\frac{1}{2} + J' \quad \text{(iv)}$$

$$H_{AB} = -\frac{S}{2} + K' \quad \text{(v)}$$

If we put E_+ from the aforementioned equation into either Eq. (ii) or (iii), then we can write

$$C_1\left(J' - \frac{J' + K'}{1+S}\right) + C_2\left(K' - \frac{(J' + K')S}{1+S}\right) = 0$$

$$\text{or,} \quad C_1\left(\frac{J'S + K'}{1+S}\right) + C_2\left(\frac{(K' - J'S)}{1+S}\right) = 0$$

when $C_1 = C_2$.

Then the MO corresponding to $E+$ is given by

$$\Psi_+ = C_1(1s_a + 1s_b)$$

For Ψ_+ to be normalised, we get

$$C_1^2(1 + 2S + 1) = 1$$

$$\text{or} \quad C_1^2(2 + 2S) = 1$$

$$\text{or} \quad C_1^2 = \frac{1}{2 + 2S}$$

$$\therefore \quad C_1 = \frac{1}{\sqrt{2(1+S)}}$$

$$\Psi_+ = \frac{1}{\sqrt{2(1+S)}} 1s_a + 1s_b$$

Now we shall find $\Psi\_$ for which we shall put $E\_$ into

$$C_1(H_{AA} - E) + C_2(H_{AB} - ES) = 0$$

to obtain

$$C_1\left(\frac{(K' + JS')}{1-S}\right) + C_2\left(\frac{(K' - J'S)}{1-S}\right) = 0$$

If $C_2 = -C_1$ and for Ψ_- to be normalised, we may write

$$C_1^2(1 - 2s + 1) = 1$$

$$\text{or} \quad C_1^2 = \frac{1}{(2 - 2S)} \quad \therefore C_1 = \frac{1}{\sqrt{2(1 - S)}}$$

Putting this value, we can write

$$\Psi_- = \frac{1}{\sqrt{2(1 - S)}}(1s_a + 1s_b)$$

This corresponds to the energy E_- for the MO. Answer.

Problem 6. Write down the ground-state electronic configuration of the Be_2, C_2, and Ne_2 using the MO concept and also find the BO.

Solution: On the basis of MO concept, the electronic configurations of the given diatomics are expressed as

- $\text{Be}_2 : \text{KK}\sigma 2s^2\sigma^* 2s^2;\ \text{BO} = \frac{n_b - n_a}{2} = \frac{2 - 2}{2} = 0$
- $C_2 : \text{KK}\sigma 2s^2\sigma^* 2s^2\pi 2p_x^2\pi 2p_y^2; \text{BO} = \frac{6 - 2}{2} = 2$
- $\text{Ne}_2 : \text{KK}\sigma 2s^2\sigma^* 2s^2\pi 2p_x^2\pi 2p_y^2\sigma 2p_z^2\pi^* 2p_x^2\pi^* 2p_y^2\sigma^* 2p_z^2; \text{BO} = \frac{8 - 8}{2} = 0$

From the BO of Ne_2, it is inferred that Ne_2 does not exist. Answer.

Problem 7. Establish that under what conditions the VB wave function and the MO wave function for a diatomic molecule will be identical?

Solution: It should be kept in mind that the VB wave function ignores the ionic term, but the MO wave function considers the ionic contribution.

We know that

$$\Psi_{\text{VB}} = \Psi_{\text{cov}} + \Psi_{\text{ion}} \quad \text{(i)}$$

$$= \underbrace{\Phi_a(1)\Psi_b(2) + \Phi_b(1)\Phi_a(2)}_{\text{Covalent term}} + \underbrace{\lambda[\Phi_a(1)\Phi_a(2) + \Phi_b(1)\Phi_b(2)}_{\text{Ionic term}} \quad \text{(ii)}$$

We can denote this equation in the short form as

$$\Psi_{\text{VB}} = (ab + ba) + \lambda(aa + bb) \quad \text{(iii)}$$

$$\text{Also } V_{\text{MO}} = [\Phi_a(1) + \Phi_b(1)][\Phi_a(2) + \Phi_b(2)] + \lambda'[\Phi_a(1) - \Phi_b(1)][\Phi_a(2) - \Phi_b(2)]$$

$$\equiv (a + b)(a + b) + \lambda'(a - b)(b - a) \quad \text{(iv)}$$

where λ = constant.

Now we have to establish the relationship between λ and λ'.
From Eq. (iii), we can write

$$\Psi_{\text{VB}} = ab + ba + \lambda aa + \lambda bb = 2ab + \lambda(aa + bb) \quad \text{(v)}$$

From Eq. (iv), we can write

$$V_{\text{MO}} = aa + ab + ba + bb + \lambda' aa - \lambda' ab - -\lambda' ba + \lambda' bb$$
$$= (1+\lambda')aa + (1+\lambda')bb + 2(1-\lambda')ab$$
$$= (1-\lambda')\left[\left\{\frac{1+\lambda'}{1-\lambda'}(aa+bb)\right\} + 2ab\right] \quad \text{(vi)}$$

The factor $(1 - \lambda')$ in Eq. (vi) can be treated in the normalisation constant. Hence, $\lambda_{\text{VB}} = \lambda_{\text{cov}}$

$$\text{if} \quad \lambda = \frac{1+\lambda'}{1-\lambda'}$$ Answer.

Problem 8. Write the unnormalised VB wave function for HF with regard to its ground state being formed from the ls orbital of H atom and 2pz orbital of F atom. Assume that HF is (a) purely covalent and (b) purely ionic.

Solution:

a. Electron 1 belongs to ls orbital and electron 2 belongs to 2pz orbital. In other words, we can say that because of the electrons cannot be distinguished, electron 1 may be assumed to be in 2pz and electron 2 in ls orbital.

$$\therefore \ \Psi_{\text{cov}} = 1s(1)2p_z(2) + 1s(2)2p_z(1)$$

b. Both the electrons belong to the same orbital; under such situation

$$\Psi_{\text{ion}} = 2p_z(1)2p_z(2)$$ Answer.

Problem 9. Consider the following MOs:

$$\Psi_1 = \frac{1}{\sqrt{2}}(\Phi_1 + \Phi_2)$$

$$\Psi_2 = \frac{1}{\sqrt{3}}(\Phi_1 + \Phi_2 + \Phi_3) \text{ and}$$

$$\Psi_1 = \frac{1}{2}(\Phi_1 - 2\Phi_2 + \Phi_3)$$

where Φs are AOs that are orthogonal.

a. Which of the aforementioned are normalised?
b. Are Ψ_1 and Ψ_2 mutually orthogonal?
c. Determine if the given MO is normalised

$$\Psi_4 = \frac{1}{\sqrt{2}}(\Psi_1 - \Psi_2)$$

Solution:

a. for MOs to be normalised,

$$\sum_{i=1}^{n} C_i^2 = 1, \text{ where } n = \text{no. of atomic orbital combined}$$

$$\text{For } \Psi_1, \left(\frac{1}{\sqrt{2}}\right)^2 + \left(\frac{1}{\sqrt{2}}\right)^2 = \frac{1}{2} + \frac{1}{2} = 1$$

$$\text{For } \Psi_2, \left(\frac{1}{\sqrt{3}}\right)^2 + \left(\frac{1}{\sqrt{3}}\right)^2 + \left(\frac{1}{\sqrt{3}}\right)^2 = \frac{1}{3} + \frac{1}{3} + \frac{1}{3} = 1$$

$$\text{For } \Psi_3, \left(\frac{1}{2}\right)^2 + (-1)^2 + \left(\frac{1}{2}\right)^2 = \frac{1}{4} + 1 + \frac{1}{4} = \frac{3}{2} \neq 1$$

Hence, Ψ_1 and Ψ_2 are normalised, but Ψ_3 is not normalised. Answer.

b. The condition of orthogonality is

$$\int \Psi_1 \Psi_2 \, d\tau \equiv \Psi_1 \Psi_2 = 0$$

$$\text{Here, } \langle \Psi_1 \mid \Psi_2 \rangle = \frac{1}{\sqrt{2}}(\Phi_1 + \Phi_2) \mid \frac{1}{\sqrt{3}}(\Phi_1 + \Phi_2 + \Phi_2)$$

$$= \frac{1}{\sqrt{6}}\left[\langle \Phi_1 \mid \Phi_1 \rangle + \langle 2\Phi_1 \mid \Phi_2 \rangle + \langle \Phi_1 \mid \Phi_3 \rangle + \langle \Phi_2 \mid \Phi_2 \rangle + \langle \Phi_2 \mid \Phi_3 \rangle\right]$$

The given AOs Φ_1, Φ_2, and Φ_3 are normalised. Suppose that overlap integral $\langle \Phi_i | \Phi_j \rangle = 0$, we obtain

$$\langle \Psi_1 \mid \Psi_2 \rangle = \frac{1}{\sqrt{6}}(1+1) = \frac{2}{\sqrt{6}} \neq 0$$

Therefore, Ψ_1 and Ψ_2 are not orthogonal. Answer.

c. Given that $\Psi_4 = \dfrac{1}{\sqrt{2}}(\Psi_1 - \Psi_2)$

$$\text{or} \quad \Psi_4 = \frac{1}{\sqrt{2}}(\Psi_1 - \Psi_2)$$

$$= \frac{1}{\sqrt{2}}\left[\frac{1}{\sqrt{2}}(\Phi_1 + \Phi_2) - \frac{1}{\sqrt{3}}(\Phi_1 + \Phi_2 + \Phi_3)\right]$$

$$= \left(\frac{1}{2} - \frac{1}{\sqrt{6}}\right)\Phi_1 + \left(\frac{1}{2} - \frac{1}{\sqrt{6}}\right)\Phi_2 - \frac{1}{\sqrt{6}}\Phi_3$$

$$\therefore \quad \sum_{i=1}^{3} C_i^2 = \left(\frac{1}{2}-\frac{1}{\sqrt{6}}\right)^2 + \left(\frac{1}{2}-\frac{1}{\sqrt{6}}\right)^2 + \left(-\frac{1}{\sqrt{6}}\right)^2$$

$$= 2\left(\frac{1}{2}-\frac{1}{\sqrt{6}}\right)^2 + \frac{1}{6}$$

$$= 2\left(\frac{1}{4}+\frac{1}{6}-2\times\frac{1}{2}\times\frac{1}{\sqrt{6}}\right)+\frac{1}{6}$$

$$= \frac{1}{2}+\frac{1}{3}-\frac{2}{\sqrt{6}}+\frac{1}{6}$$

$$= \frac{5}{6}+\frac{1}{6}-\frac{2}{\sqrt{6}}$$

$$= 1-\frac{2}{\sqrt{6}} \neq 1$$

$\therefore$ Φ_4 is not normalised. Answer.

Questions on Concepts

1. What are the important elements of Born approximation? Derive the mathematical equation $T_N(\chi p) + E_e(xp) = E(\chi\rho)$, where the terms have their usual significance.
2. In the case of H_2^+ (hydrogen molecule ion), prove that $E_{\pm} = \dfrac{E_0 \pm H_{ab}}{1 \pm S}$ where the terms have their usual significance.
3. Prove that the overlap integral $\int \Psi_a \Psi_b d\tau = S = e^{-R}[R^3/2 + R + 1]$ in the case of H_2^+ ion, where $R = \dfrac{r_a + r_b}{\mu}$.
4. Evaluate the coulomb integral in the case of H_2^+ ion, i.e., $J = \int \dfrac{\psi_a^2}{r_b} d\tau$ where $\Psi_a = \dfrac{1}{\sqrt{11}} e^{-r_a}$, and arrive at $J = \dfrac{1}{R}\left[1 - e^{-2R}(1+R)\right]$.
5. Evaluate the resonance integral in the case of H_2^+ ion, i.e., $K = \int \dfrac{\psi_a \psi_b}{r_b}$ where the terms have their own significance, and arrive at $K = e^{-R}(1 + R)$.
6. When $\Psi_s = C_1(\Psi_a + \Psi_b)$ and $\Psi_A = C_2(\Psi_a - \Psi_b)$, then how can you prove that $\Psi_S = \dfrac{1}{\sqrt{2(1+S_{ab})}} \cdot (\Psi_a + \Psi_b)$ and $\Psi_A = \dfrac{1}{\sqrt{2(1-S_{ab})}} \cdot (\Psi_a + \Psi_b)$, where the terms have their usual significance.
7. In the case of H_2 molecule, show that if $H_{11} - E = H_{12} - SE$ and $H_{11} - E = -H_{12} + SE$ then $E_1 = 2E_0 + \dfrac{K-J}{S-1}$ and $E_2 = 2E_0 + \dfrac{K-J}{S-1}$, where the terms used are known.
8.
    a. What do you mean by linear combination of atomic orbitals?
    b. How will you find out the molecular wave function, i.e., $\Psi^{\pm} = \dfrac{1}{\sqrt{2}}(\Phi_1 \pm \Phi_2)$.

9. Prove that $\Psi_{\text{MO}} = \dfrac{\sqrt{2\left(1+S^2\right)}}{\sqrt{2(1+S)}}(\Psi_{\text{cov}} + \Psi_{\text{ion}})$ for H_2 molecule.
10. Show from valence bond approach for H_2 molecule, the energy is expressed as $E = \dfrac{H_{11} - H_{12}}{\left(1-S^2\right)}$.
11. Keeping in view the valence bond theory, for H_2 molecule the energy may be expressed as

    $E_{\pm} = \dfrac{H_{11} \pm H_{12}}{1 \pm S^2}$
12. If $\Psi_+ = C_1(\Phi_1 + \Phi_2)$ and $\Psi_- = C_2(\Phi_1 - \Phi_2)$, find the value of C_1 and C_2 under normalisation constant.
13. How will you prove that

$$\Psi_{\text{MO}} - \Psi'_{\text{MO}} = \left[1s_a(1)1s_b(2) + 1s_b(1)1s_b(2)\right]_-$$

    When $R \rightarrow \infty$ and $S = 0$
14. How will you show that the molecular orbital treatment with configuration interaction produces better result with respect to the valence bond treatment?
15. Compare the molecular orbital theory with the valence bond theory.
16. Explain the symmetric and antisymmetric wave functions.
17.
    a. What are the important postulates of symmetric and antisymmetric wave functions?
    b. How will you show that the symmetry behaviour of a wave function does not alter in time?
18. What is the Slater determinant? How does it incorporate Pauli's principle?
19. The Slater atomic orbitals are normalised but not mutually orthogonal. In the Schmidt orthogonalisation procedure, one orbital Ψ is made orthogonal to another orbital Ψ' by forming $\Psi'' = \Psi - C\Psi'$ with $C = \int\Psi^*\Psi' d\tau$ How?
20. Write the explicit expression for the Slater determinant corresponding to the $1s^2 2s^1$ ground state of lithium. Show the antisymmetry of the wave function upon interchange of any two electrons.
21. Show that each successive interchange of the coordinates of any pair of electrons will change the sign but not the magnitude of the wave equation described in the following equation:

$$\Psi = \frac{1}{\sqrt{n!}} \Sigma \pm p(a\alpha)_1 (b\beta)_2 (c\beta)_3 \ldots (n\alpha)_n .$$

22. Mention the important properties of the Slater determinant.
23. Explain the bonding and antibonding orbitals using the equation $\Psi(r_1) = N_{\text{MO}}[a(1) + b(1)]$.
24. The electron density in molecular hydrogen is given by $\int\rho d\tau_1 = 2$.
25. Prove that

$$E_{\text{el}}\left({}^1\psi_1\right) - E_{el}\left({}^3\psi_1\right) = 2K_{12}$$

    where the terms have their usual significance.
26. Discuss the electronic transition in hydrogen molecule.
27. Prove that $M(\psi_0 \cdot {}^1\psi_1) = \dfrac{|R|}{\sqrt{2\left(1-S^2\right)}}$, where the terms have their usual significance.
28.
    a. What do you mean by homopolar diatomics? What do you mean by bonding and antibonding molecular orbitals?

b. Show the formation of bonding molecular orbital and antibonding molecular orbital in the case of H_2 molecule and also give the energy level diagram.

29. How do molecular orbitals form with p valence atomic orbitals? Illustrate the formation of bonding molecular orbital and antibonding molecular orbital.
30. Explain the π and π^* molecular orbital formation.
31. Write down the electronic configuration of H_2, He_2, Li_2, N_2, O_2, and F_2.
32.
   a. What do you mean by heteropolar diatomics?
   b. What are the important conditions which must satisfy for the combinations to occur for the formation of heteronuclear diatomic molecules.
   c. Write down the electronic configuration of HF, NO, BN, and CO.

33.
   a. Using the molecular orbital concept of electronic configuration of molecules, show that O_2 is paramagnetic and evaluate the bond order of the O_2 molecule.
   b. A gas having B2 molecule is found to be paramagnetic. What pattern of molecular orbital must apply in this case?

34. Explain the significance of (a) coulomb integral, (b) exchange integral, and (c) overlap integral.
35. Represent in a diagram contours of equal electron density for bonding and antibonding orbitals of H_2^+ ion.
36. Explain giving diagrams, the combination of (a) s and p_x orbitals and (b) two p orbitals.

Numerical Problems

1. Give the steps involved in deriving $\Psi_1 = C_1(\Psi_a + \Psi_b)$ and $\Psi_2 = C_2(\Psi_a - \Psi_b)$. What will be the normalisation factor if the two nuclei are at infinite distance?

$$\left[\text{Ans.}C_1 = \frac{1}{\sqrt{2+2s}},\quad \Psi_1 = \frac{\Psi_a + \Psi_b}{\sqrt{2+2S}}, C_2 = \frac{1}{\sqrt{2-2S}},\quad \Psi_2 = \frac{\Psi_a + \Psi_b}{\sqrt{2(1+S)}}\right]$$

2. The Heitler–London wave functions for H_2 molecule are

$$\Psi_S = N_S\left[\Psi_a(1)\Psi_b(2) + \Psi_a(2)\Psi_b(1)\right]$$

$$\Psi_a = N_a\left[\Psi_a(1)\Psi_b(2) - \Psi_a(2)\Psi_b(1)\right]$$

Calculate the normalisation constants N_s and N_a. What will be the normalisation factor if the nuclear separation is infinite?

$$\left[\text{Ans.}N_S = \frac{1}{\sqrt{2+2s^2}},\ N_a = \frac{1}{\sqrt{2-2s^2}} \text{ for } s \cdot 0,\ N_S = N_a = \frac{1}{\sqrt{2}}\right]$$

3. Determine the molecular orbitals which correspond to the energies E_+ and E_- in the following equation:

$E_\pm = \dfrac{J' \pm K'}{1 \pm S}$, where $\Delta E_\pm$ = the energy of with respect to a separated proton and a H atom and

$$J' = e^{-2R}\left(1 + \frac{1}{R}\right), K' = \frac{s}{R} - e^{-R}(1+R) \;\&\; S = e^{-R}\left(1 + R + \frac{R^2}{3}\right)$$

[Ans. $\Psi_{\pm} = \frac{1}{\sqrt{2(1 \pm S)}}(1s_a \pm 1s_b)$]

4. Show that the symmetry argument, which we use to show that Ψ_1 does not mix with Ψ_2 and Ψ_3 does not imply that Ψ_1 does not mix with Ψ_4. [Ans. $H_{12} = 0$ and $H_{14} = 0$]
5. Find the ground electronic configuration and bond order of He_2^+ [Ans. $\sigma 1s^2 \cdot \sigma^* 1s^1$, BO = 1/2]
6. Write the electronic configuration of Na_2 and S_2 molecules with the help of molecular orbital concept.

   [Ans. $Na_2\left(KK\sigma 2s^2\sigma^* 2s^2\pi 2p_y^2\pi 2p_z^2\sigma 2p_X^2\pi^* 2p_y^2\pi^* 2p_z^2\sigma^* 2p_X^2\sigma 3s^2\right)$]

   [$S_2\left(KKLL\sigma 3s^2\sigma^* 3s^2\sigma 3p_X^2\pi 3p_y^2\pi 3p_z^2\pi^* 3p_y^1\pi^* 3p_z^1\right]$

7. Prove that the bonding molecular orbital and the antibonding molecular orbital of H_2 are orthogonal where

$$\Psi_1 = \frac{1}{\sqrt{2(1+S)}}(1s_a + 1s_b) \text{ and } \Psi_2 = \frac{1}{\sqrt{2(1-S)}}(1s_a - 1s_b)$$

   [Ans. $\int \Psi_1 \Psi_2 d\tau = 0$]
8. Establish the condition under which valence bond wave function and the molecular orbital wave function for a diatomic become identical. [Ans. $\Psi_{VB} = \Psi_{VB}$ when $\lambda = \frac{1+\lambda'}{1-\lambda'}$]
9. Do you expect that the ground state of He_2^+ would be stable? [Ans. Yes, since the $b \cdot 0 = 1/2$]

11

Multielectronic Systems

In Chapter 9, we computed the solutions of the Schrödinger equation for one- and two-electron systems, which are approximate solutions, but the exact solutions of these and multielectronic system are difficult to obtain. It should be kept in mind that the difficulties are only computational. There is no doubt that various methods have been developed to solve the Schrödinger equation, but they are approximate in nature. Here, we shall take into consideration, in essence, one of them, namely, the method of self-consistent field (SCF), which is more commonly called the Hartree–Fock SCF method. The rigorous approach for getting the Hartree–Fock equations for multielectronic systems is based on the variational principle. Our main purpose here is to estimate the energy of the many-electron system by considering the interaction of all the electrons with the nucleus and the interaction of all the electrons with each other.

11.1 Energy of the Many-Electron System

First of all, we shall know about the SCF. Actually, SCF is a concept used to obtain approximate solutions to multielectronic system in quantum mechanics. The procedure begins with an approximate solution for a particle moving in a single particle potential, which is derived from its average interaction with all other particles. This average interaction is obtained by the wave functions for all other particles. The equation, which describes this average interaction, is solved, and the better solution or improved solution obtained is employed in the calculation of the interaction term. This process is repeated/iterated for the wave function until the wave function and the related energies are not significantly changed in the cycle, self-consistency having been reached.

We want to put forward only the physical meaning of SCF. The spatial distribution of the electrons in an atom alters continuously, but in case of stationary state, the probability density becomes constant. Hence, every electron creates some field, which is supposed to be constant. Then, the motion of every electron can be considered to be independent in the average field of the nucleus and of all other electrons, or in other words, we can say that each electron possesses its wave function and its set of quantum numbers.

Now we are going to find out the energy of the many-electron system. Let us consider a multielectron atom with a closed-shell configuration. The ground state of such type of system can be represented by a single determinant. Let us suppose that this determinant is expressed by

$$\Psi = \frac{1}{\sqrt{N!}}\left|\Phi_a(1)\alpha(1)\Phi_a(2)\beta(2)\Phi_b(3)\alpha(3)\ldots\Phi_n(N)\beta(N)\right| \tag{11.1}$$

where Ψ = wave function

α = spin and β = spin.

The Hamiltonian for the N-electron system may be given by

$$H = \sum_i\left(-\frac{\hbar^2}{2m}\cdot\nabla_i^2 - \frac{ze^2}{4\pi\epsilon_0}\cdot\frac{1}{r_i}\right) + \frac{1}{2}\sum_{ij}\frac{e^2}{4\pi\epsilon_0}\cdot\frac{1}{r_{ij}} \tag{11.2}$$

Considering the a.u., i.e., $\hbar = 1$, $m = 1$, $e = 1$, Eq. (11.2) is written as

$$H = \sum_i \left(-\frac{\nabla_i^2}{2} - \frac{z}{4\pi\epsilon_0} \cdot \frac{1}{r_i} \right) + \frac{1}{2} \sum_{ij} \frac{1}{4\pi\epsilon_0} \cdot \frac{1}{r_{ij}} \tag{11.3}$$

For simplicity sake, we shall avoid writing $4\pi\epsilon_0$. Then aforementioned equation is simply expressed as

$$H = \sum_i \left(-\frac{\nabla_i^2}{2} - \frac{z}{r_i} \right) + \frac{1}{2} \sum_{ij} \frac{1}{r_{ij}} \tag{11.4}$$

where, $\sum_i \left(-\frac{z}{r_i} \right)$ = interaction of all the electrons with the nucleus and

$\sum_{ij} \frac{1}{r_{ij}}$ = interaction of all the electrons with each other.

The factor 1/2 in Eq. (11.4) is put to ensure that the interaction between a pair of electrons is counted only once.

The first term in Eq. (11.4) represents the kinetic energy of N electrons. On using Eqs. (11.1) and (11.4), the wave equation takes the form

$$H\Psi = E\Psi \tag{11.5}$$

If the interaction between the electrons is not taken into consideration, Eq. (11.5) can be separated into N equations, each of which will be a function of only one electron. It is this fact that justifies the use of one electron orbital product wave function. Eq. (11.1) includes a variational parameter to allow the stationary value of the energy to be obtained. The energy related to Eq. (11.5) is obtained in the usual way.

$$E = \int \Psi^* H \Psi d\tau = \int \Psi^* \left(\sum_i -\frac{\nabla_i^2}{2} \right) \Psi d\tau - \int \Psi^* \left(\sum_i \frac{Z}{r_i} \Psi d\tau + \frac{1}{2} \int \Psi^* \sum_{ij} \frac{1}{r_{ij}} \Psi \right) d\tau \tag{11.6}$$

The evaluation of integrals on the RHS of Eq. (11.6) must be taken into consideration separately. Let us consider the first kinetic energy integral

$$T = \frac{1}{N!} \int \left| \Phi_a^*(1)\alpha^*(1) \ldots \Phi_n^*(N)\beta^*(N) \right| \left(\sum_i -\frac{\nabla_i^2}{2} \right) \left| \Phi_a(1)\alpha(1) \ldots \Phi_n(N)\beta(N) \right| d\tau \tag{11.7}$$

As the determinant wave function consists of $N!$ terms, the integral will have $N!^2$ terms. However, it is not essential to evaluate this large number of one-electron integrals.

Now examine the integral with respect to electron (1) in Eq. (11.7) and, in particular, the one with the leading term on the LHS of the integrand. The integral is

$$\frac{1}{N!} \int \Phi_a^*(1)\alpha^*(1) \ldots \Phi_n^*(N)\beta^*(N) \left(-\frac{\nabla_i^2}{2} \right) \times \left| \Phi_a(1)\alpha(1) \ldots \Phi_n(N)\beta(N) \right| d\tau_1 d\tau_2 d\tau_3 - - d\tau_n \tag{11.8}$$

The integration over the other $N-1$ electrons may be factored out, and Eq. (11.8) may be written as

$$\frac{1}{N!} \int \Phi_a^*(1)\alpha^*(1) \left(-\frac{\nabla_i^2}{2} \right) \Phi_a(1)\alpha(1) d\tau_1 \int \Phi_a^*(2)\beta(2) \ldots \Phi_n^*(N)\beta^*(N)(-1)^p$$
$$\times P\left[\Phi_a(2)\beta(2)\Phi_b(3)\alpha(3) \cdots \Phi_n(N)\beta(N) \right] d\tau_2 d\tau_3 - - d\tau_n \tag{11.9}$$

where P = exchange operator.

Among all the terms in the second integral, the only one which will be nonvanishing is the one which is identical to the LHS (with the assumption that orbitals are orthogonal). The factor to multiply the kinetic energy integral in Eq. (11.9) becomes one, because it is only the term on the RHS, which is identical to that present on the left and is nonvanishing.

This also applies to the other terms in the determinantal wave function on the RHS in Eq. (11.7). The value of the integral is a characteristic of the orbital but not of the electron, which in our case becomes a dummy suffix. It should be kept in mind that there will be $N!$ occurrence of the integral $\int \Phi_a^*(x)\left(-\frac{\nabla_x^2}{2}\right)\Phi_a(x)\mathrm{d}v_x$ in Eq. (11.7). This will be the situation for each and every orbital. The value of Eq. (11.7) may be expressed as

$$T = \frac{1}{N!}(T_a + T_a + T_b + T_b + \ldots + T_n) \tag{11.10}$$

which may be expressed simply as

$$T = 2\sum_{i=1}^{n} T_i \tag{11.11}$$

A similar reasoning holds good for the other one-centre integral V_i, which spells out the attraction between electron i and the nucleus. The total electron–nucleus interaction is represented by

$$V = 2\sum_{i} V_i \tag{11.12}$$

where the suffix represents the orbital but not the electron. In expressing the energy, the two one-centre integrals are generally combined together to yield a total one-electron contribution to the energy H_i, which is equal to

$$f_i = T_i + V_i \tag{11.13}$$

and the total one-electron contribution to the energy is

$$f = 2\sum f_i \tag{11.14}$$

We shall now calculate the two-electron integrals in Eq. (11.6), which refer to electron–electron interaction. For this, we shall examine the contribution from the leading term in the LHS, as with the one-electron integrals.

$$\int \Phi_a^*(1)\alpha^*(1)\Phi_a^*(2)\beta^*(2) - - - \Phi_n^*(N)\beta^*(N)\sum_{ij}\frac{1}{r_{ij}}\left|\Phi_a(1)\alpha(1) - - - \Phi_n(N)\beta(N)\right|\mathrm{d}\tau \tag{11.15}$$

In the light of the interaction between electron 1 and electron 2, the relevant term may be expressed as

$$\int \frac{\Phi_a(1)\alpha(1)\Phi_a(2)\beta(2)\Phi_a(1)\alpha(1)\Phi_a(2)\beta(2)}{r_{12}}\mathrm{d}\tau_1\mathrm{d}\tau_2$$
$$\times \int \Phi_b^*(3)\alpha(3)\ldots\Phi_n^*(N)\beta^*(N)(-1)^P\left[\Phi_b(3)\alpha(3)\ldots\Phi_n(N)\beta(N)\right]\mathrm{d}\tau_3\ldots\mathrm{d}\tau_n \tag{11.16}$$

As in the case of one electron, the only contribution to Eq. (11.16) comes from the term in the second integral, which contains both sides of the integral matching. Similarly, there is one such term from each member of the wave function in Eq. (11.15), which is the leading term. Again it is reminded that the value of the said term is characterised by the orbital and not by the electron and the integral is represented by J_{aa}, which is known as *coulomb integral* or *coulombic interaction.*

Now let us examine the contribution from interaction between electron 1 and electron 3 in Eq. (11.15). This refers to the interaction between one electron in orbital Φ_a and one electron present in orbital Φ_b. The corresponding coulomb integral is symbolised by J_{ab}.

A similar J_{ab} will arise from the interaction of electron 1 and electron 4, a third from the interaction between electron 2 and electron 3 and a fourth from electron 2 and electron 4. The total contribution will be $4J_{ab}$, which is due to the coulombic interaction between the two charge clouds.

There is, however, another type of two-electron interaction for which we shall consider the integral similar to the integral present in Eq. (11.16), which consists of $1/r_{13}$ term. This may be written down as a product of two integrals, one over space and the other over spin. This yields

$$\int \frac{\Phi_a^*(1)\Phi_b^*(3)\Phi_a(1)\Phi_b(3)\mathrm{d}v_1\mathrm{d}v_3}{r_{1,3}} \int \alpha^*(1)\alpha^*(3)\alpha(1)\alpha(3)\mathrm{d}\eta_1\mathrm{d}\eta_3 \tag{11.17}$$

On interchanging electrons 1 and 3 on the RHS of each integral, we get

$$\int \frac{\Phi_a^*(1)\Phi_b^*(3)\Phi_a(3)\Phi_b(1)\mathrm{d}v_1\mathrm{d}v_3}{r_{1,3}} \int \alpha^*(1)\alpha^*(3)\alpha(3)\alpha(1)\mathrm{d}\eta_1\mathrm{d}\eta_3 \tag{11.18}$$

The spin integral still integrates out to unity, whereas the space integral gives something, which is known as *resonance integral* and is symbolised by K_{ab}. Its contribution to the energy has a negative sign due to the fact that we interchange two particles in the course of forming the integral, which is in accordance with Pauli's principle that wave function changes sign. There is a further contribution of K_ab, which arises from electron 2 and electron 4. It must be remembered that resonance or exchange integrals come out when the particles concerned have parallel spin; otherwise, the spin integral value will be zero.

In the closed-shell circumstances, the number of exchange integral is equal to half of the coulomb integral. Under this situation, the two-electron contribution to the energy is given by

$$E = H_{ii} + \frac{1}{2}\sum_{ij}\left(4J_{ij} - 2k_{ij}\right) \tag{11.19}$$

$$\text{or,}\quad E = H_{ii} + \sum_{ij}\left(2J_{ij} - K_{ij}\right) \tag{11.20}$$

The total energy of N-electron system is obtained by adding Eqs. (11.14) and (11.18), and hence, we write

$$E = 2\sum_{i} f_i + \sum_{ij}\left(2J_{ij} - K_{ij}\right)$$

which may be finally written as

$$E = H_{ii} + \sum_{ij}\left(2J_{ij} - K_{ij}\right) \tag{11.21}$$

where $2\sum f_i = H_{ii}$.

It is to be noted from the form of integrals J_{ij} and K_{ij} that these are dependent on the distribution of charge. The energy is then the function of the charge distribution, and while estimating the wave function and energy of many-electron system, it is essential to vary the wave function until the energy and charge distribution become self-consistent by the use of variational principle for each cycle. There is no doubt that it is a laborious procedure, and the best wave function, thus, found is still only an approximation to the wave function. It is clear that the better wave function will be found by setting aside one determinant wave function and expressing the ground-state wave function as a linear sum of the determinants:

$$\Psi_0 = \sum_{i=0}^{N} C_i \Phi_i \tag{11.22}$$

(where we express Φ_i for a single determinant)

provided that the configurations chosen have the same symmetry. They will interact with each other, and the final wave function having the lowest energy will be the best approximation to the ground state. For this, an adequate number of determinants are selected. The energy will be lower and hence a better approximation to the ground-state energy as compared with the one determinant approximation.

11.2 Fock Equation and Hartree Equation

Let us consider an operator F for the Fock operator operating on wave function Ψ representing a system of particles, which is orthonormal. According to the basic postulate of the quantum mechanics, the expectation value of an operator is given by

$$\langle F \rangle = \int \frac{\Psi^* F \Psi \mathrm{d}\tau}{\Psi^* \Psi \mathrm{d}\tau} = I(\text{say}) \tag{11.23}$$

where Ψ is a state function.

Completeness of state function is defined as

$$\int \Psi^* \Psi \mathrm{d}\tau = 1 \tag{11.24}$$

It means that the wave function is normalised.

$$\therefore \langle F \rangle = \int \Psi^* F \Psi \mathrm{d}\tau \tag{11.25}$$

Applying variational principle or δ-variation in Eqs. (11.23) and (11.24), we may write

$$\begin{aligned} \delta \langle F \rangle = \delta I = \delta \int \Psi^* F \Psi \mathrm{d}\tau = 0 \\ \text{or,} \quad \int \delta \Psi^* F \Psi \mathrm{d}\tau + \int \Psi^* F \delta \Psi \mathrm{d}\tau = 0 \end{aligned} \tag{11.26}$$

Applying δ-variation on Eq. (11.24), we get

$$\begin{aligned} \delta \int \Psi^* \Psi \mathrm{d}\tau = \delta(1) = 0 \\ \text{or,} \quad \int \delta \Psi^* \Psi \mathrm{d}\tau = \int \Psi^* \delta \Psi \mathrm{d}\tau = 0 \end{aligned} \tag{11.27}$$

Using Lagrange's undetermined multiplier ($\in$) in Eq. (11.27), we obtain

$$\in \int \delta \Psi^* \Psi \mathrm{d}\tau + \in \int \Psi^* \delta \Psi \mathrm{d}\tau = 0 \tag{11.28}$$

Subtracting Eq. (11.28) from Eq. (11.27), we get

$$\int \delta\Psi^*(F-\epsilon)\Psi d\tau + \int \Psi^*(F-\epsilon)\delta\Psi d\tau = 0 \quad (11.29)$$

Selecting $\delta\Psi^*$, $\partial\Psi$, and $\in$ arbitrarily, Eq. (11.29) can be solved, and we shall obtain

$$(F-\in)\Psi = 0 \text{ and } \left(F^*-\epsilon^*\right)\Psi^* = 0$$

Applying Hamiltonian property of the operator

$$F^* = F$$

Then the equations $(F-\in)\Psi = 0$ and $\Psi^*\left(F^*-\in\right) = 0$ become identical, in the form of Schrödinger wave equation, and it may be expressed as

$$(F-\in)\Psi = 0 \quad (11.30)$$

This equation is called the *Fock equation.*

11.2.1 Application in Two-Electron Systems – For Getting Hartree Equation and Energy of Two-Electron System

Let us apply the aforementioned equation in two-electron system by considering that

$$\Psi = \Psi(1,2)$$

and this can be expressed in the form of product as

$$\Psi = \Psi(1)\Psi(2) \quad (11.31)$$

Since Ψ is normalised,

$$\int \Psi^*(1)\Psi(1)d\tau = \int \Psi^*(2)\Psi(2)d\tau = 1 \quad (11.32)$$

$$\text{and} \quad \int \Psi^*\Psi d\tau = \int \Psi^*(1)\Psi^*(2)\Psi(1)\Psi(2)d\tau_1 d\tau_2 = 1 \quad (11.33)$$

Equation (11.30) can be set up in the form if F is treated as Hamiltonian operator for two-electron system. The interaction in two-electron system is shown in Figure 11.1.

The Fock operator and Hamiltonian operator can now be expressed as

$$F = H = H_0(1) + H_0(2) + H'_{12} \quad (11.34)$$

where the terms have their usual significance.

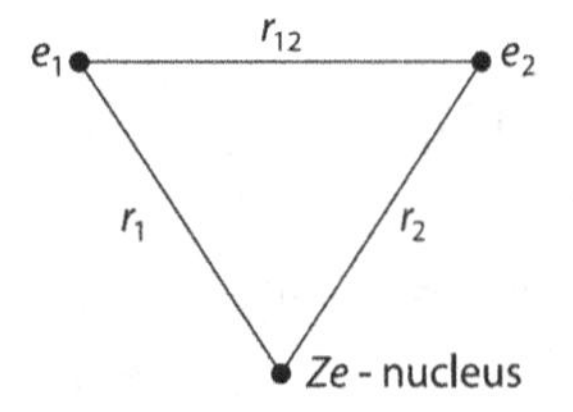

FIGURE 11.1 Interaction between electron–electron and nucleus–electron.

But $(H - \in)\Psi = 0$

$$\therefore \left[H_0(1) + H_0(2) + H'_{12} - \in\right]\Psi = 0 \tag{11.35}$$

$$\text{or,}\left[H_0(1) + H_0(2) + H'_{12} - \in\right]\Psi_1\Psi_2 = 0 \quad \left[\because \Psi = \Psi_1\Psi_2\right] \tag{11.36}$$

Multiplying Eq. (11.36) by Ψ_2^* and integrating over space, we obtain

$$\int \left[\Psi_2^* H_0(1)\Psi_2 + \Psi_2^* H_0(2)\Psi_2 d\tau_2 + \Psi_2^* H'_{12}\Psi_2 - \Psi_2^* \epsilon \Psi_2\right]\Psi_1 d\tau = 0$$

$$\text{or}\left[\int \Psi_2^* H_0(1)\Psi_2 d\tau_2 + \int \Psi_2^* H_0(2)\Psi_2 d\tau_2 + \int \Psi_2^* H'_{12}\Psi_2 d\tau_2 - \int \Psi_2^* \epsilon \Psi_2 d\tau_2\right]\Psi_1 = 0 \tag{11.37}$$

$$\text{or}\left[H_0(1) + \int \Psi_2^* H_0(2)\Psi_2 d\tau_2 + \int \Psi_2^* H'_{12}\Psi_2 d\tau_2 - \epsilon\right]\Psi_1 = 0$$

Similarly, multiplying Eq. (11.36) by Ψ^* and integrating over space, we can write

$$\left[\Psi_1^* H_0(1)\Psi_1 d\tau_1 + H_0(2) + \int \Psi_1^* H'_{12} d\tau_1 - \epsilon\right]\Psi_2 = 0 \tag{11.38}$$

Equations (11.37) and (11.38) are known as *Hartree equation.*

Now for the two-electron system,

$$H_0(1) = -\frac{\hbar^2}{2\mu}\nabla_1^2 - \frac{ze^2}{r_1} \times \frac{1}{4\pi \in_0}$$

$$H_0(2) = -\frac{\hbar^2}{2\mu}\nabla_2^2 - \frac{ze^2}{r_2} \times \frac{1}{4\pi \in_0}$$

$$H'_{12} = V_{12} = \frac{1}{4\pi \in_0} \cdot \frac{e^2}{r_{12}}$$

But in atomic unit system,

$$\hbar = 1, \mu = 1/2, \frac{1}{4\pi \in_0} = 1$$

Then the aforementioned equations will be simply written as

$$H_0(1) = -\nabla_1^2 - \frac{ze^2}{r_1}$$

$$H_0(2) = -\nabla_2^2 - \frac{ze^2}{r_2}$$

$$H'_{12} = \frac{e^2}{r_{12}}$$

Multiplying Eq. (11.37) by Ψ_1^* and integrating over space, we get

$$\int \Psi_1^* H_0(1)\Psi_1 d\tau_1 + \int \Psi_1^* \left(\Psi_2^* H_0(2)\Psi_2 d\tau_2\right)\Psi_1 d\tau_1 + \int \Psi_1^* \left(\int \Psi_2^* H_{12}' \Psi_2 d\tau_2\right)\Psi_1 d\tau_1 = \int \Psi_1^* \in \Psi_1 d\tau_1$$

$$\text{or, } E_0(1) + E_0(2) + \iint \Psi_1^* \Psi_2^* H_{12}' \Psi_2 \Psi_1 \, d\tau_2 \, d\tau_1 = E \tag{11.39}$$

Similarly, multiplying Eq. (11.38) by Ψ_2^* and integrating over space, we get

$$E_0(1) + E_0(2) + \iint \Psi_2^* \Psi_1^* H_{12}' \Psi_2 \Psi_1 \, d\tau_1 \, d\tau_2 = E \tag{11.40}$$

Considering the state function

$$\Phi = \frac{1}{\sqrt{2}} \begin{vmatrix} \Psi_{\alpha_i}(1) & \Psi_{\alpha_j}(2) \\ \Psi_{\beta_i}(1) & \Psi_{\beta_j}(2) \end{vmatrix}$$

the aforementioned equation may be expressed as

$$E = 2\sum_{i=1}^{N} H_{ii} + \sum_{i=1}^{N}\sum_{j=1}^{N}\left(2J_{ij} - K_{ij}\right) \tag{11.41}$$

which is the energy of two-electron system like He and of more than two electrons.

$$J_{ij} = \text{Coulomb integral} = \int \Psi_\alpha(1)\Psi_\beta(2) \cdot \frac{1}{r_{12}} \cdot \Psi_\alpha(1)\Psi_\beta(2) d\tau_1 d\tau_2$$

$$\text{and } K_{ij} = \int \Psi_\alpha(1)\Psi_\beta(2) \cdot \frac{1}{r_{12}} \cdot \Psi_\alpha(2)\Psi_\beta(1) d\tau_1 d\tau_2 = \text{exchange integral}$$

It should be kept in mind that the exchange integral exists only for antisymmetric wave function, but in case of symmetric state, its value becomes zero.

11.3 Hartree and Hartree–Fock Self-Consistent Field Methods

SCF is a very powerful tool for the estimation of ground-state energy Eigen values and Eigen functions of multielectronic systems. The SCF concept was originally expressed in a simple but incorrect form because it did not consider the effects due to exchange symmetry. This form was termed as the Hartree–Fock SCF method. Later on, the Hartree–Fock SCF method was developed, which considered and incorporated the effects of exchange symmetry in the SCF concept. In the present context, we are going to introduce the SCF method in its relation to variation principle.

Another way of representing variational principle is to state that a small change in the trial function Φ, i.e., $\delta\Phi$, should make no alteration in the variational integral. Mathematically, this fact is expressed as

$$\delta\langle\Phi|H|\Phi\rangle = 0 \tag{11.42}$$

It is to be noted that an SCF calculation can be assumed to be a special case of variational technique; it is based on not only the variational principle but also a special model that is on a particular kind of trial Eigen function. The trial Eigen function for a system is expressed as

$$\Phi = \Psi_\alpha(1)\Psi_\beta(2)\ldots\Psi_\nu(N) \tag{11.43}$$

In Eq. (11.43), Ψ_α (1) represents that the electron number one is found in one-electron atomic orbital α. For the two-electron system (like He), the Hamiltonian is expressed as

$$H = \left(\sum_k -\frac{\hbar^2}{2m}\nabla_k^2 - \frac{ze^2}{r_k} \right) + \sum_k \sum_{l>k} \frac{e^2}{r_{kl}} \tag{11.44}$$

$$\text{or } H = H(k) + \sum_k \sum_{l>k} \frac{e^2}{r_{kl}}$$

This equation indicates that the bracket term is equal to $H(k)$, and it also represents the one-electron atomic Hamiltonian for electron k.

The variational integral may be written as

$$\langle \Phi | H | \Phi \rangle = \sum_k \langle \Psi_k(k) | H(k) | \Psi_k(k) \rangle + \sum_K \sum_{l>k} \left\langle \Psi_k(k)\Psi_\lambda(l) \left| \frac{e^2}{r_{kl}} \right| \Psi_k(k)\Psi_\lambda(l) \right\rangle \tag{11.45}$$

Applying the variational principle to the integral in Eq. (11.45), we can write

$$\delta\langle \Phi | H | \Phi \rangle = \sum_k \delta \langle \Psi_k(k) | H(k) | \Psi_k(k) \rangle + \sum_K \sum_{l\neq k} \delta \left\langle \Psi_k(k)\Psi_\lambda(l) \left| \frac{e^2}{r_{kl}} \right| \Psi_k(k)\Psi_\lambda(l) \right\rangle \tag{11.46}$$

Now we shall define an operator $F(k)$ as given in the following:

$$F(k) = H(k) + \sum_{l\neq k} \left\langle \Psi_\lambda(l) \left| \frac{e^2}{r_{kl}} \right| \Psi_\lambda(l) \right\rangle \tag{11.47}$$

where the sum runs over l. The variational principle then opines that

$$\delta \langle \Psi_k(k) | F(k) | \Psi_k(k) \rangle = 0 \tag{11.48}$$

In Expression (11.47), the term

$$\sum_{l\neq k} \left\langle \Psi_\lambda(l) \left| \frac{e^2}{r_{kl}} \right| \Psi_\lambda(l) \right\rangle = \text{expectation value of interelectronic repulsion of the}$$

kth electron with all other electrons

$\Psi_\lambda(l)$ = an approximate Eigen function for electron 1

The aforementioned term is only an expression for the repulsive potential energy experienced by electron k. Eq. (11.48) opines that the best product of one-electron functions should be selected in such a way that one-electron energy mixed to the average interelectronic potential energy is minimised.

The SCF method can be summed up as follows:

1. Select an arbitrary product of one-electron functions as a trial function.
2. Calculate the electrostatic potential of one electron by the use of the aforementioned trial function (say K electron).
3. Find the next best one-electron function for electron k with the help of Eq. (11.48).
4. Continue steps 2 and 3 for all electrons.
5. Then select a new set of one-electron function and repeat steps 2, 3, and 4 until there is no apparent change in the ground-state energy after an iteration.

This method is known as the SCF method due to the fact that the iteration through steps 2 to 5 is continued until the electric field set up by the electron, as estimated in step 2, does not alter or is self-consistent. The result of the SCF method is the best product of one-electron functions, and hence, the Hartree or SCF energy may be estimated easily.

Now examine Eq. (11.47). The Hartree operator $F(k)$ is made up of two terms. The first term is $H(k)$, which is called one-electron energy operator and other one is the coulomb operator, represented by $J(k)$, where

$$J_\lambda(k) = \left\langle \Psi_\lambda(l) \left| \frac{e^2}{r_{kl}} \right| \Psi_\lambda(l) \right\rangle \tag{11.49}$$

$$= \text{Coulomb integral}$$

∴ Equation (11.47) can be written as

$$F(k) = H(k) + \sum_{l \neq k} J_\lambda(k) \tag{11.50}$$

Now the main idea of SCF calculation has been understood. We may apply this idea to a specific model for the electrons. We know that the identical particles show certain characteristic properties due to the fact that these are indistinguishable. In particular, particles consisting of 1/2 integral spin have antisymmetric state functions with regard to the exchange of two identical particles. Particles containing half (1/2) electron spin are fermions and must have antisymmetric state functions. Let us now explore the consequences of exchange symmetry for a two-electron system. The procedure is the Hartree–Fock method.

We are citing an example of antisymmetric two-electron function as given in the following:

$$\Phi = \frac{1}{\sqrt{2}}\left[\Psi_\alpha(1)\Psi_\beta(2) - \Psi_\beta(1)\Psi_\alpha(2)\right]$$

If the trial function to form variational integral similar to Eq. (11.45) is used, we would obtain the following:

$$\langle \Phi | H | \Phi \rangle = \frac{1}{2}\left\langle \Psi_\alpha(1)\Psi_\beta(2) - \Psi_\beta(1)\Psi_\alpha(2) \right| H(1) + H(2) + \frac{e^2}{r_{12}} \left| \ldots \Psi_\alpha(1)\Psi_\beta(2) - \Psi_\beta(1)\Psi_\alpha(2) \right\rangle$$

$$\text{or,}\quad \langle \Phi | H | \Phi \rangle = \frac{1}{\sqrt{2}} \begin{vmatrix} \Psi_\alpha(1) & \Psi_\alpha(2) \\ \Psi_\beta(1) & \Psi_\beta(2) \end{vmatrix} H^* \left| \frac{1}{\sqrt{2}} \right| \begin{matrix} \alpha(1) & \alpha(2) \\ \beta(1) & \beta(2) \end{matrix} \Bigg| \tag{11.51}$$

where $H^* = H(1) + H(2) + H_{12}$.

From Eq. (11.51), we can write down the following possible interactions:

$$\begin{aligned} \langle \Phi | H | \Phi \rangle = {} & \frac{1}{2}\langle \Psi_\alpha(1) | H(1) | \Psi_\alpha(1) \rangle + \frac{1}{2}\langle \Psi_\beta(1) | H(1) | \Psi_\beta(1) \rangle + \frac{1}{2}\langle \Psi_\alpha(2) | H(2) | \Psi_\alpha(2) \rangle \\ & + \frac{1}{2}\langle \Psi_\beta(2) | H(2) | \Psi_\beta(2) \rangle + \frac{1}{2}\left\langle \Psi_\alpha(1)\Psi_\beta(2) \left| \frac{e^2}{r_{12}} \right| \Psi_\alpha(1)\Psi_\beta(2) \right\rangle \\ & - \frac{1}{2}\left\langle \Psi_\beta(1)\Psi_\alpha(2) \left| \frac{e^2}{r_{12}} \right| \Psi_\alpha(1)\Psi_\beta(2) \right\rangle - \frac{1}{2}\left\langle \Psi_\alpha(1)\Psi_\beta(2) \left| \frac{e^2}{r_{12}} \right| \Psi_\beta(1)\Psi_\alpha(2) \right\rangle \\ & + \frac{1}{2}\left\langle \Psi_\beta(1)\Psi_\alpha(2) \left| \frac{e^2}{r_{12}} \right| \Psi_\beta(1)\Psi_\alpha(2) \right\rangle \end{aligned} \tag{11.52}$$

where $\dfrac{e^2}{r_{12}} = H_{12}$.

It is observed that three kinds of integrals appear in Eq. (11.52). The first four terms in Eq. (11.52) represent one-electron integral of the one-electron energy operator. The fifth and eighth terms are coulomb integral or two-electron integrals of coulomb type. Sixth and seventh terms define the exchange operator, $K(k)$ such that

$$\langle K_k(k)|\Psi_\lambda(k)\rangle = \left\langle \Psi_K(l)\left|\frac{e^2}{r_{kl}}\right|\Psi_\lambda(l)\right\rangle|\Psi_\lambda(k)\rangle \tag{11.53}$$

where $\left\langle \Psi_K(l)\left|\dfrac{e^2}{r_{kl}}\right|\Psi_\lambda(l)\right\rangle = \text{exchange integral}$.

We have already talked about coulomb integral earlier, which may be expressed as

$$\langle J_\lambda(k)|\Psi_\lambda(k)\rangle = \left\langle \Psi_K(l)\left|\frac{e^2}{r_{kl}}\right|\Psi_\lambda(l)\right\rangle|\Psi_\lambda(k)\rangle \tag{11.54}$$

where $\left\langle \Psi_K(l)\left|\dfrac{e^2}{r_{kl}}\right|\Psi_\lambda(l)\right\rangle = \text{Coulomb integral} = J_\lambda(k)$.

With the definition of $H(k)$ and $J_\lambda(k)$, Eq. (11.52) may be written as

$$\langle\Phi|H|\Phi\rangle = \sum_k \langle\Psi_K(k)|H(k)|\Psi_K(k)\rangle + \sum_k\sum_{l>k}\langle\Psi_K(k)|J_\lambda(k)|\Psi_K(k)\rangle$$
$$-\sum_k\sum_{l>k}\langle\Psi_K(k)|K_\lambda(k)|\Psi_K(k)\rangle \tag{11.55}$$

$$\text{or,}\ \langle E\rangle = 2\sum_{k=1}^{N} H_{kk} + \sum_{k=1}^{N}\sum_{j=1}^{N}(2J_{kl} - K_{kl}) \tag{11.56}$$

This is the energy of two-electron system or more than two-electron systems.

11.4 Excited State of Helium

The electronic configuration of He is $1s^2$. When He is excited, its electronic configuration may be either $1s^1\,2s^1$ or $1s^1\,2p^1$. Except for the electron exchange, the first one represents nondegenerate, and the second one is threefold degenerate. We shall consider the $1s^1\,2s^1$ excited-state configuration of He. It is a two-electron system, and the different interactions, i.e., electron–electron and nucleus–electron interactions, are shown in Figure 11.2.

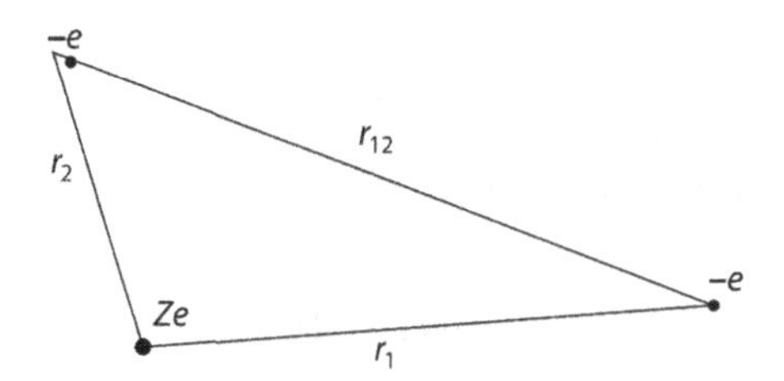

FIGURE 11.2 Interactions in He atom.

Now we will find the energy for the singlet and triplet states, and it will be proved that the triplet state lies lower in energy with respect to the singlet state. The Schrödinger equation for two-electron system (like He) may be written as

$$\left(-\frac{\hbar^2}{2m}\nabla_1^2-\frac{\hbar^2}{2m}\nabla_2^2-\frac{ze^2}{r_1}-\frac{ze^2}{r_2}+\frac{e^2}{r_{12}}\right)\Psi=E\Psi \tag{11.57}$$

$$\text{or,}\quad \left(H_1+H_2+\frac{e^2}{r_{12}}\right)\Psi=E\Psi$$

$$\text{or,}\quad (H_1+H_2+H')\Psi=E\Psi \tag{11.58}$$

where $H'=\frac{e^2}{r_{12}}$ = repulsive energy.

Equation (11.58) for the singlet wave function can be rewritten as given in the following:

$$(H_1+H_2+H')\Psi(1,2)=E_{\text{singlet}}\Psi(1,2) \tag{11.59}$$

where H_1 and H_2 = total energy operators after neglecting electron–electron repulsion and the subscripts 1 and 2 represent the electrons involved.

$$\Psi(1,2)=\text{unknown exact wave function.}$$

It should be kept in mind that the spin part of the wave function is not included due to the fact that the total energy operator does not have spin-dependent terms. Since we are not aware of the exact wave function, we approximate it with the help of the singlet wave function of the following equation:

$$\Psi_{\text{singlet}}=\frac{1}{\sqrt{2}}\left[1s(1)2s(2)+2s(1)1s(2)\right]\frac{1}{\sqrt{2}}\left[\alpha(1)\beta(2)-\beta(1)\alpha(2)\right] \tag{11.60}$$

Since the total energy operator does not include terms that are dependent on spin, Eq. (11.60) will take the form

$$\Psi_{\text{singlet}}=\frac{1}{\sqrt{2}}\left[1s(1)2s(2)+2s(1)1s(2)\right]$$

It is to be remembered that the Ψ_{singlet} is not an Eigen function of the total energy operator. The expectation value for the total energy can be found with the help of the appropriate wave function by multiplying on the left by wave function and integrated over space.

$$\langle E_{\text{singlet}}\rangle=\frac{1}{2}\iint\left[1s(1)2s(2)+2s(1)1s(2)\right](H_1+H_2+H')\times\left[1s(1)2s(2)+2s(1)1s(2)\right]d\tau_1 d\tau_2 \tag{11.61}$$

$$=\frac{1}{2}\int\left[1s(1)2s(2)+2s(1)1s(2)\right]H_1|\left[1s(1)2s(2)+2s(1)1s(2)\right]d\tau_1$$

$$+\frac{1}{2}\int\left[1s(1)2s(2)+2s(1)1s(2)\right]H_2|\left[1s(1)2s(2)+2s(1)1s(2)\right]d\tau_2$$

$$+\frac{1}{2}\iint\left[1s(1)2s(2)+2s(1)1s(2)\right]H'|[1s(1)2s(2)$$

$$+2s(1)1s(2)]d\tau_1 d\tau_2 \tag{11.62}$$

It is clear from Eq. (11.62) that the first integral corresponds to E_{1S} and the second integral corresponds to E_{2s}. Keeping in mind the equation, i.e., $H_i\, 1s\,(i) = E_{1s}\, 1s\,(i)$ and $H_i\, 2s\,(i) = E_{2s}\, 2s\,(i)$, and therefore, Eq. (11.62) may be written as

$$E_{\text{singlet}} = E_{1s} + E_{2s}$$

$$+\frac{1}{2}\iint[1s(1)2s(2)+2s(1)1s(2)]H'[1s(1)2s(2)+2s(1)1s(2)]d\tau_1\,d\tau_2$$

$$= E_{1s} + E_{2s}$$

$$+\frac{1}{2}\iint[1s(1)]^2|H'|[2s(2)]^2\,d\tau_1\,d\tau_2$$

$$+\frac{1}{2}\iint[1s(1)2s(2)]H'[1s(2)2s(1)+]d\tau_1\,d\tau_2$$

$$\text{or, } E_{\text{singlet}} = E_{1s} + E_{2s} + J_{12} + K_{12} \tag{11.63}$$

where $J_{12} = \frac{1}{2}\iint[1s(1)]^2|H'|[2s(2)]^2\,d\tau_1\,d\tau_2$ and

$$K_{12} = \frac{1}{2}\iint[1s(1)2s(2)]H'[1s(2)2s(1)]d\tau_1\,d\tau_2$$

J_{12} is known as the *coulomb integral* and K_{12} is called the *exchange integral* about which we have already learnt.

Similarly, we can calculate the energy for the triplet state, and we can arrive at the following result:

$$E_{\text{triplet}} = E_{1s} + E_{2s} + J_{12} - K_{12} \tag{11.64}$$

$$\text{But } E_{\text{singlet}} - E_{\text{triplet}} = (E_{1s} + E_{2s} + J_{12} + K_{12}) - (E_{1s} + E_{2s} + J_{12} - K_{12}) = 2K_{12} \tag{11.65}$$

Thus, it is clear that the difference between the singlet-state energy and triplet-state energy of He ($1s^1\, 2s^1$) is $2K_{12}$. It means that the singlet-state energy of first excited state of He ($1s^1\, 2s^1$) is greater than that of triplet-state energy, i.e., $E_{\text{singlet}} > E_{\text{triplet}}$, meaning thereby that the triplet state for the first excited state of He lies lower in energy than the singlet states.

It is to be noted that the singlet and triplet wave functions also differ in the degree to which they include electron correlation. It should be mentioned that the electrons avoid each other because of coulomb repulsion. If we assume that the electron 2 goes to electron 1, the spatial part of the singlet wave function may be expressed as

$$\frac{1}{\sqrt{2}}[1s(1)2s(2)+2s(1)1s(2)] \rightarrow \frac{1}{\sqrt{2}}[1s(1)2s(1)+2s(1)1s(1)] = \frac{2}{\sqrt{2}}[1s(1)2s(1)]$$

Since $r_2, \theta_2, \Phi_2 \rightarrow r_1, \theta_1, \Phi_1$, but in case of the spatial part of the triplet wave function,

$$\frac{1}{\sqrt{2}}[1s(1)2s(2)-2s(1)1s(2)] \rightarrow \frac{1}{\sqrt{2}}[1s(1)2s(1)-2s(1)1s(1)] = 0$$

It gives clue to understand that the triplet wave function consists of greater degree of electron correlation as compared with the singlet wave function and because of the fact that the probability of finding both electrons in a given space becomes zero as the electrons approach each other.

One question arises: why is the $E_{\text{singlet}} < E_{\text{triplet}}$? On the basis of the detailed analysis, it can be stated that the electron–electron repulsion is virtually larger in the triplet state than the singlet state. However, it is expected that the electrons are a bit close to the nucleus in the triplet state. The increased electron–nucleus attraction exceeds in influence the electron–electron repulsion, and due to this, the triplet state has a lower energy.

It should be kept in mind that spin influences the energy even though the total energy operator does not include spin term but spin comes into picture in the course of our calculation through antisymmetrisation, which is the requirement of Pauli's exclusion principle.

Finally, it can be concluded that for a given configuration, the spin unpaired state has a lower energy than the paired state.

It is pertinent to note that the singlet state is often referred to as *parahelium*, whereas the triplet state is referred to as *orthohelium*. The energy of orthohelium is found to be less than that of parahelium, which has been experimentally verified. Thus, it is clear that the ground state of He is the parahelium, but in the excited state, both forms exist.

Alternatively

In the first excited state, the electronic configuration of He can be written as $1s^1 2s^1$. It means the two electrons occupy different orbitals. Their wave function may be represented by either $\Psi_{n_1l_1m_1}(r_1)\Psi_{n_2l_2m_2}(r_2)$ or $\Psi_{n_2l_2m_2}(r_1)\Psi_{n_1l_1m_1}(r_2)$, which are represented in short for simplicity sake as a (1) b (2) and b (1) a (2), respectively.

It should be remembered that both wave functions consist of the same energy and their corresponding unperturbed energies will be E_a and E_b, respectively, and it will be $E_a + E_b$ in total. Corresponding to energy E_a, the Hamiltonian will be represented by either H_a or H_1, and corresponding to energy E_b, the Hamiltonian will be H_b or H_2. The perturbation due to the interaction will be $\frac{e^2}{4\pi \epsilon_0 r_{12}}$ or simply $\frac{1}{r_{12}}$ in a.u. To estimate the perturbed energy, we shall apply the perturbation theory appropriate to degenerate states and hence set up the secular determinant. To do this, we require the following matrix element, in which the state 1 is identified by a (1) b (2) and state 2 by b (1) a (2).

$$H_{11} = \left\langle a(1)b(2) \middle| H_1 + H_2 + \frac{1}{r_{12}} \middle| a(1)b(2) \right\rangle = E_a + E_b + J$$

$$H_{22} = H_{11} = E_a + E_b + J$$

$$H_{12} = \left\langle a(1)b(2) \middle| H_1 + H_2 + \frac{1}{r_{12}} \middle| a(2)b(1) \right\rangle$$

$$= E_a + E_b \left\langle a(1)b(2) | a(2)b(1) \right\rangle + \left\langle a(1)b(2) \middle| \frac{1}{r_{12}} \middle| a(2)b(1) \right\rangle = H_{21}$$

The first of the integrals in H_{12} will be equal to zero due to the fact that the orbitals a and b are orthogonal as mentioned in the following.

$$\left\langle a(1)b(2) | a(2)b(1) \right\rangle = \left\langle a(1) | b(1) \right\rangle \left\langle b(2) | a(2) \right\rangle = 0$$

The rest integral is known as the exchange integral, K, i.e.,

$$K = \left\langle a(1)b(2) \middle| \frac{1}{r_{12}} \middle| a(2)b(1) \right\rangle \text{ in a.u.} \tag{11.66}$$

Similar to J, this integral is also positive, and the secular determinant can be framed with the help of above as

$$\begin{vmatrix} H_{11} - ES_{11} & H_{12} - ES_{12} \\ H_{21} - ES_{21} & H_{22} - ES_{22} \end{vmatrix} = \begin{vmatrix} H_{11} - E & H_{12} \\ H_{21} & H_{22} - E \end{vmatrix}$$
$$= \begin{vmatrix} E_a + E_b + J - E & K \\ K & E_a + E_b + J - E \end{vmatrix} = 0 \tag{11.67}$$

or $(E_a + E_b + J - E)^2 = K^2$

$$\therefore \; E_a + E_b + J - E = \pm K$$
$$\therefore \; E = E_a + E_b + J \pm K \tag{11.68}$$

(It should be kept in mind that $S_{11} = S_{22} = 1$, $S_{12} = S_{21} = 0$ because of the orthonormality of the states 1 and 2.)

Thus, the solution of the aforementioned determinant gives the value of the total energy of the first excited state of helium.

The corresponding wave functions with regard to Eq. (11.68) may be expressed as

$$\Psi_{\pm}(1,2) = \frac{1}{\sqrt{2}}\left[a(1)b(2) \pm b(1)a(2)\right] \tag{11.69}$$

or in the form of Ψs, we can write

$$\Psi_{\pm}(r_1, r_2) = \frac{1}{\sqrt{2}}\left[\Psi_{n_1 l_1 m_1}(r_1)\Psi_{n_2 l_2 m_2}(r_2) \pm \Psi_{n_2 l_2 m_2}(r_1)\Psi_{n_1 l_1 m_1}(r_2)\right]$$

where the individual wave functions are representing hydrogenic atomic orbitals having atomic number $z = 2$.

The main feature of the aforementioned result lies in the fact that the degeneracy of the two product functions *a* (1) *b* (2) and *b* (1) *a* (2) is removed due to electron–electron repulsion, and their two linear combinations $\Psi_{\pm}$ differ in energy by $2K$. Since the exchange integral has no classical counterpart, it should be understood as a quantum mechanical correction to the coulomb integral, *J*. However, it is possible to perceive clearly with the mind to the origin of this correction by considering the amplitudes $\Psi_{\pm}$ as one electron approaches the other. It is noticed that $\Psi_{-} = 0$ when $r_1 = r_2$, but $\Psi_{+} \neq 0$. The corresponding $|\Psi_{-}|^2$ and $|\Psi_{+}|^2$, i.e., the corresponding differences in the probability densities are illustrated in Figure 11.3.

It is clear from the figure that $|\Psi_{-}|^2$ becomes zero when the state is described by the wave function Ψ_{-} but in case of Ψ_{+}, there is a small enhancement in the probability that they will be found together. We see in the figure that there is a dip in the probability density $|\Psi_{-}|^2$ when $r_1 = r_2$, and this situation is

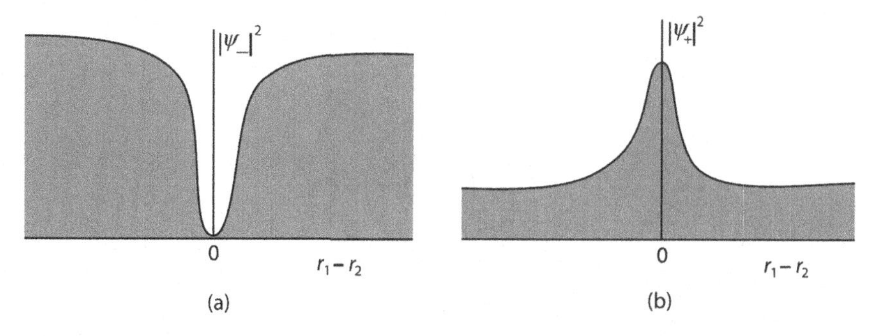

FIGURE 11.3 (a) Formation of Fermi hole by spin correlation; (b) formation of Fermi heap under spin paired condition.

known as *Fermi hole*. It can be concluded from this that due to the existence of Fermi hole, the electrons which occupy Ψ_- tend to discard one another. Hence, Ψ_- will be expected to be lower than Ψ_+ with regard to the average of the electron–electron repulsion energy. It should be kept in mind that in case of Ψ_+, the electrons tend to come closer to each other. Due to this, coulombic potential will be reduced from J to $J - K$ for electrons having wave function Ψ_- and J is increased to $J + K$ for electrons having wave function Ψ_+.

From Figure 11.3, it is clear that a heap is obtained at $r_1 = r_2$ when $|\Psi_+|^2$ is plotted against $r_1 - r_2$. This phenomenon is known as *Fermi heap*.

It is to be noted that the wave function Ψ_- is antisymmetric under interchange of electrons, and we can write

$$\begin{aligned}\Psi_-(2,1) &= \frac{1}{\sqrt{2}}\left[a(2)b(1) - b(2)a(1)\right] \\ &= -\frac{1}{\sqrt{2}}\left[a(1)b(2) - b(1)a(2)\right] \\ &= -\Psi_-(1,2)\end{aligned} \tag{11.70}$$

whereas Ψ_+ is symmetric under electron interchange, i.e.,

$$\begin{aligned}\Psi_+(2,1) &= \frac{1}{\sqrt{2}}\left[a(2)b(1) + b(2)a(1)\right] \\ &= \frac{1}{\sqrt{2}}\left[a(1)b(2) + b(1)a(2)\right] \\ &= \Psi_+(1,2)\end{aligned} \tag{11.71}$$

Thus, we can state that the *Fermi hole* and *Fermi heap* are quantum mechanical phenomena.

11.5 Lithium in the Ground State

Lithium (Li) is a three-electron system (Figure 11.4) in which two electrons remain in the K shell and one electron in the L shell. The ground-state Hamiltonian for lithium may be expressed as

$$H = \left(-\frac{\hbar^2}{2m}\nabla_1^2 - \frac{\hbar^2}{2m}\nabla_2^2 - \frac{\hbar^2}{2m}\nabla_3^2 - \frac{ze^2}{4\pi \epsilon_0 r_1} - \frac{ze^2}{4\pi \epsilon_0 r_2} - \frac{ze^2}{4\pi \epsilon_0 r_3} + \frac{e^2}{4\pi \epsilon_0 r_{12}} + \frac{e^2}{4\pi \epsilon_0 r_{23}} + \frac{e^2}{4\pi \epsilon_0 r_{13}}\right)$$

This equation can be written in general form in a.u. as follows:

$$H = \sum_{i=1}^{3} H^0(i) + H'$$

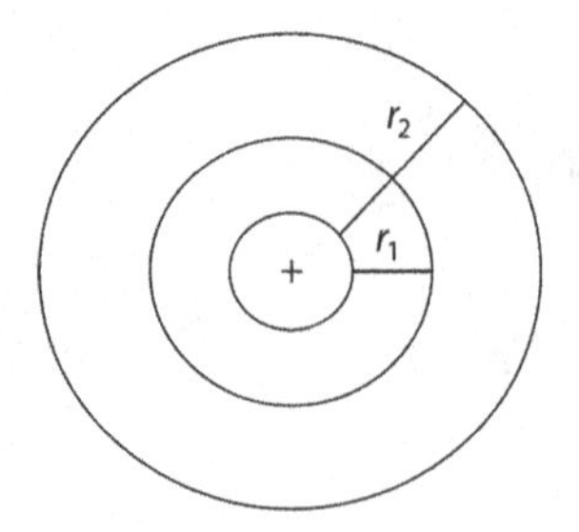

FIGURE 11.4 Lithium atom.

where $H^0(i) = -\nabla_i^2 - \frac{6}{r_i}$, and

$$H' = \sum_{i \neq j}^{3} \frac{2}{r_{ij}}$$

It should be kept in mind that $i = j$ represents self-interaction of electrons and H' is the perturbation term.

$$\therefore\ E = \langle H \rangle = \left\langle \Psi \left| \sum_{i=1}^{3} H^0(i) + H' \right| \Psi \right\rangle$$

The wave function Ψ can be represented by the Slater determinant as

$$\Psi = \frac{1}{\sqrt{3!}} \cdot \begin{vmatrix} 1s\alpha(1) & 1s\alpha(2) & 1s\alpha(3) \\ 1s\beta(1) & 1s\beta(2) & 1s\beta(3) \\ 2s\alpha(1) & 2s\alpha(2) & 2s\alpha(3) \end{vmatrix} \tag{11.72}$$

where the subscripts of α and β in a row are also the subscripts of $1s$ and $2s$.

When the aforementioned determinant is expanded and written out in full, the one-electron integrals may be expressed as given in the following:

$$\begin{aligned} &\frac{1}{6}\Big\langle 1s\alpha(1)1s\beta(2)2s\alpha(3) + 2s\alpha(1)1s\alpha(2)1s\beta(3) + \\ &\quad 1s\beta(1)2s\alpha(2)1s\alpha(3) - 1s\alpha(1)2s\alpha(2)1s\beta(3) \Big| H^0(1) + H^0(2) + H^0(3) \\ &\quad - 1s\beta(1)1s\alpha(2)2s\alpha(3) - 2s\alpha(1)1s\beta(2)1s\alpha(3) \Big| \\ &\qquad\qquad 1s\alpha(1)1s\beta(2)2s\alpha(3) + 2s\alpha(1)1s\alpha(2)1s\beta(3) + \\ &\qquad\qquad 1s\beta(1)2s\alpha(2)1s\alpha(3) - 1s\alpha(1)2s\alpha(2)1s\beta(3) \\ &\qquad\qquad -1s\beta(1)1s\alpha(2)2s\alpha(3) - 2s\alpha(1)1s\beta(2)1s\alpha(3)\Big\rangle \end{aligned} \tag{11.73}$$

Now we shall use the orthogonality conditions for the space and spin functions, and finally we can represent it as

$$\langle u_i\chi_k(1) | u_j\chi_l(1) \rangle = \langle u_i(1) | u_i(1) \rangle \langle \chi_K(1) | \chi_l(1) \rangle = \delta_{ij}\delta_{kl} \tag{11.74}$$

Equation (11.73) can be reduced to

$$\begin{aligned} &\frac{1}{6}\sum_{i=1}^{3}\left[4\langle 1s(i)|H^0(i)|1s(i)\rangle + \langle 2s(i)|H^0(i)|2s(i)\rangle\right] \\ &= 2\,I(1s) + I(2s) \end{aligned} \tag{11.75}$$

It is a fact that the two-electron integrals are very much difficult to solve; however, in order to simplify this calculation, we shall factor out the common spin functions. Thus, Ψ can be expressed as

$$\Psi = \Psi_1\alpha(1)\beta(2)\alpha(3) + \Psi_2\beta(1)\alpha(2)\alpha(3) + \Psi_3\alpha(1)\alpha(2)\beta(3) = \sum_i \Psi_i S_i \tag{11.76}$$

$$\text{where, } \Psi_1 = \frac{1}{\sqrt{3!}}\left[1s(1)1s(2)2s(3) - 2s(1)1s(2)1s(3)\right]$$

$$\Psi_2 = \frac{1}{\sqrt{3!}}\left[1s(1)2s(2)1s(3) - 1s(1)1s(2)2s(3)\right]$$

$$\Psi_3 = \frac{1}{\sqrt{3!}}\left[2s(1)1s(2)1s(3) - 1s(1)2s(2)1s(3)\right]$$

Hence, H' integral will become as follows:

$$\langle\Psi|H'|\Psi\rangle = \left\langle \Psi_1 S_1 + \Psi_2 S_2 + \Psi_3 S_3 \left| \sum_{i<j}^{3} \frac{2}{r_{ij}} \right| \Psi_1 S_1 + \Psi_2 S_2 + \Psi_3 S_3 \right\rangle$$

When orthogonality condition is applied to the spin functions, the result will be

$\langle S_i|S_j\rangle = \delta_{ij}$ and we shall be left with

$$\langle\Psi|H'|\Psi\rangle = \langle\Psi_1|H'|\Psi_1\rangle + \langle\Psi_2|H'|\Psi_2\rangle + \langle\Psi_3|H'|\Psi_3\rangle$$

Now expressing $\langle\Psi_1|H'|\Psi_1\rangle$ in terms of ground-state hydrogen like state function, the integral can be written as

$$\langle\Psi_1|H'|\Psi_1\rangle = \frac{1}{6}\left[\left\langle 1s(1)1s(2)2s(3) - 2s(1)1s(2)1s(3) \left| \frac{2}{r_{12}} + \frac{2}{r_{13}} + \frac{2}{r_{23}} \right| 1s(1)1s(2)2s(3) - 2s(1)1s(2)1s(3) \right\rangle\right]$$

Since all the integrals are equal, i.e., $\langle\Psi_1|H'|\Psi_1\rangle = \langle\Psi_2|H'|\Psi_2\rangle = \langle\Psi_3|H'|\Psi_3\rangle$ (11.77)

corresponding to infractions,

$\langle\Psi_1|H'|\Psi_1\rangle = E'$ can be written as

$$E' = \iint 1s^2(1)2s^2(2)\frac{1}{r_{12}}\,d\tau_1\,d\tau_2 + \iint 1s^2(1)2s^2(2)\frac{1}{r_{12}}\,d\tau_1\,d\tau_2 - \iint 1s(1)2s(2)\frac{1}{r_{12}}1s(2)2s(1)\,d\tau_1\,d\tau_2 \tag{11.78}$$

Now the first integral can be expressed as

$$I = (4\pi)^2 \int_{r_1=0}^{\infty}\int_{r_2=0}^{\infty} u_{1s}^2(r_1)u_{1s}^2(r_2)\cdot\frac{1}{r_{12}} r_1^2 r_2^2\,dr_1 dr_2 \tag{11.79}$$

which can be evaluated based on the ground state of hydrogen-like atom.

$$= (4\pi)^2 \int_{r_1=0}^{\infty} u_{1s}^2(r_1)\left[\int_{t=0}^{r_1} u_{1s}^2(r_2)\frac{r_2^2}{r_1}\,dr_2 + \int_{r_1}^{\infty}\frac{r_2^2}{r_2}u_{1s}^2(r_2)\,dr_2\right] r_1^2\,dr_1$$

where $u_{1s}(r) = \left(\frac{z^3}{\pi}\right)^{1/3} e^{-zr}$ in a.u.

The aforementioned equation can be written as

$$I = (4\pi)^2 \int_{r=0}^{\infty} u_{1s}^2(r_1)[I_1 + I_2] r_1^2 \mathrm{d}r_1 \tag{11.80}$$

where $I_1 = \int_{r=0}^{r_1} \frac{r_2^2}{r_1}\left(\frac{Z^3}{\pi}\right) e^{-2zr_2} \mathrm{d}r_2 = \frac{1}{\pi}\left[-\left(\frac{Z^2}{2} r_1 + \frac{Z}{2} + \frac{1}{4r_1}\right) e^{-2Zr_1} + \frac{1}{4r_1}\right]$

$$I_2 = \int_{r_1}^{\infty} \frac{r_2^2}{r_2}\left(\frac{Z^3}{\pi}\right) e^{-2zr_2} \mathrm{d}r_2 = \frac{1}{\pi}\left[\left(\frac{Z^2 r_1}{2} + \frac{Z}{4}\right) e^{-2Zr_1}\right]$$

$$\therefore \quad I_1 + I_2 = \frac{1}{4\pi}\left[-\left(\frac{Z}{4} + \frac{1}{4r_1}\right) e^{-2Zr_1} + \frac{1}{4r_1}\right]$$

$$\therefore \quad I = (4\pi)^2 \int_{r=0}^{\infty} u_{1s}^2(r_1) \cdot \frac{1}{4\pi}\left[-\left(Z + \frac{1}{r_1}\right) e^{-2Zr_1} + \frac{1}{r_1}\right] r_1^2 \mathrm{d}r_1$$

$$= 4\pi \int_{r=0}^{\infty} \frac{Z^3}{\pi} e^{-2Zr_1}\left\{-\left(Z + \frac{1}{r_1}\right) e^{-2Zr_1} + \frac{1}{r_1}\right\} r_1^2 \mathrm{d}r_1$$

$$= 4Z^3\left[-\int_0^{\infty} Z e^{-4Zr_1} r_1^2 \mathrm{d}r_1 - \int_0^{\infty} \frac{1}{r_1} e^{-4Zr_1} \cdot r_1^2 \mathrm{d}r_1 + \int_0^{\infty} \frac{r_1^2}{r_1} e^{-2Zr_1} \mathrm{d}r_1\right]$$

$$= 4Z^3\left[-\frac{2Z}{(4Z)^3} - \frac{1}{(4Z)^2} + \frac{1}{(2Z)^2}\right] \text{ by using standard integral } \int_0^{\infty} x^n e^{-ax} \mathrm{d}x = \frac{n!}{a^{n+1}}$$

$$= 4Z^3\left[-\frac{2Z}{64Z^3} - \frac{1}{16Z^2} + \frac{1}{4Z^2}\right]$$

$$= 4Z^3\left[-\frac{1}{32Z^2} - \frac{1}{16Z^2} + \frac{1}{4Z^2}\right]$$

$$= 4Z^3 \times \frac{1}{4Z^2}\left[-\frac{1}{8} - \frac{1}{4} + 1\right] \tag{11.81}$$

$$= Z\left[\frac{-1 - 2 + 8}{8}\right] = \frac{5}{8} Z$$

$$\therefore \quad I = \frac{5}{8} Z$$

$\therefore$ E' for the first coulombic integral will be

$$E_1' = \frac{5}{8} Z \frac{e^2}{a_0}$$

$$\therefore \quad \langle \Psi_1 | H' | \Psi_1 \rangle = \frac{5}{8} Z \frac{e^2}{a_0} = J(1s, 1s)$$

The terms $\langle \Psi_2 | H' | \Psi_2 \rangle$ and $\langle \Psi_3 | H' | \Psi_3 \rangle$ give approximately the same results, and they represent coulombic integral and exchange integral respectively, i.e.,

$$\langle \Psi_2 | H' | \Psi_2 \rangle = 2J(1s, 2s) \text{ and}$$

$$\langle \Psi_3 | H' | \Psi_3 \rangle = K(1s, 2s)$$

$$\therefore \text{ Total } E' = J(1s, 1s) + 2J\big((1s, 2s) - K(1s, 2s)\big)$$

$$\text{But } J(1s, 2s) = \frac{17}{81} Z \frac{\acute{e}^2}{a_0},\ K(1s, 2s) = \frac{16}{729} Z \frac{\acute{e}^2}{a_0}$$

$$\therefore E' = \frac{5}{8} Z \frac{\acute{e}^2}{a_0} + 2 \times \frac{17}{81} Z \frac{\acute{e}^2}{a_0} - \frac{16}{729} Z \frac{\acute{e}^2}{a_0}$$

$$= \frac{5}{8} \times 3 \times 2 \left(\frac{\acute{e}^2}{za_0} \right) + \frac{2 \times 17 \times 3 \times 2}{81} \left(\frac{\acute{e}^2}{za_0} \right) - \frac{16 \times 3 \times 2}{729} \left(\frac{\acute{e}^2}{za_0} \right)$$

$$E' = \frac{5}{8} Z \frac{e^2}{\alpha_0} + \frac{17}{81} \acute{e}^2 - \frac{16}{729} Z \acute{e}^2 / a_0$$

$$= 51.0 \text{ eV} + 34.0 \text{ eV} - 1.77 \text{ eV}$$

$$= 85.0 \text{ eV} - 1.77 \text{ eV}$$

$$= 83.23 \text{eV} \tag{11.82}$$

$$\therefore E' = 83.23 \text{ eV} \tag{11.83}$$

Now the ground-state energy E^0 of Li will be calculated by the formula:

$$E^0 = -\left(\frac{1}{n_1^2} + \frac{1}{n_2^2} + \frac{1}{n_3^2} \right) \left(\frac{z^2 \acute{e}^2}{za_0} \right)$$

$$= -\left(\frac{1}{1^2} + \frac{1}{1^2} + \frac{1}{2^2} \right) \left(3^2 \times 13.6 \right) \text{eV}$$

$$= -2.25 \times 9 \times 13.6 \text{eV}$$

$$= -2.25 \times 122.4 \text{eV} = -275.4 \text{ eV}$$

$$\therefore E = E^0 + E^1$$

$$= -275.4 \text{ eV} + 83.23 \text{ eV}$$

$$= -192.17 \text{ eV}$$

But the true ground-state energy of lithium is found to be –203.5 eV, which clearly indicates an error of 5%.

11.6 Atomic Magnets and Magnetic Quantum Numbers

In 1896, Zeeman found that each line in an atomic spectrum was split into component lines when the atoms were placed in a strong magnetic field. The component lines are found to be evenly spaced, and their separation depends on the field strength *H*. In other words, we can say that the separation of evenly spaced lines is proportional to the field strength H. This gives a clue to understand that an electron with a given value of n and l can have states of slightly different energy under the influence of a magnetic field, and these must be denoted by an additional quantum number.

11.6.1 Atomic Magnets

According to the laws of electromagnetism, it can be stated that an electric current having strength i flowing along the path of the perimeter (c) of a circle consisting of area πr^2 will be equivalent to a magnet of moment:

$$M = \frac{\pi r^2 i}{c}(\text{emu}) \tag{11.84}$$

placed at the centre of the circle, as illustrated in Figure 11.5.

We know that electric current has strength i = number of electrons flowing a certain point in unit time × electronic charge.

Suppose that

r = radius of an orbit on which an electron is moving
v = velocity of the electron

The electron will pass a certain point $v/2\pi r$ times s^{-1} and will be equivalent to a current of strength:

$$i = \left(v/2\pi r\right)e \tag{11.85}$$

and the magnetic moment (M) will be

$$M = \frac{\pi r^2}{c} \times ve/2\pi r = \frac{evr}{2c}\text{emu} \tag{11.86}$$

The angular momentum of the electron will be written as

$$P_\Phi = mvr = \frac{lh}{2\pi} \tag{11.87}$$

where l = azimuthal quantum number.

From Eqs. (11.86) and (11.87), we can write

$$M = \frac{evr}{2c} = \frac{emvr}{2mc} = \frac{eP_\Phi}{2mc} = \frac{lh}{2\pi(2mc)} \cdot e = l\mu \tag{11.88}$$

where $\mu = \dfrac{he}{4\pi mc}$.

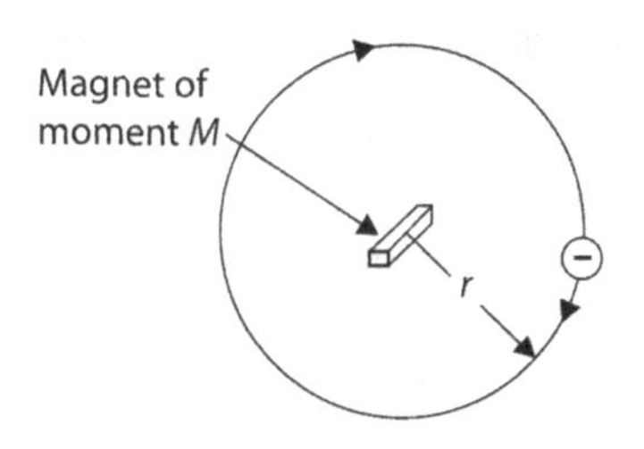

FIGURE 11.5 Magnetic effect exerted by electron in closed orbit having radius r.

The quantity $\frac{he}{4\pi mc}$ is called the Bohr magneton, which is equal to 9.1×10^{-21} emu and 9.247×10^{-24} JT^{-1} (T stands for Tesla).

11.6.2 Magnetic Quantum Number

We now know that the atoms have magnetic moments, and because of this, they tend to orient themselves in a magnetic field. As a result of this, they acquire an additional energy amounting to

$$E_m = -MH \cos\Phi \tag{11.89}$$

where Φ = angle of inclination of the equivalent magnet to the direction of H, i.e., field of strength H.

It is clear that M is quantised, and hence, $\cos \Phi$ must be quantised if E_m is to have only certain values in a certain field. It implies that only certain angles of inclinations are possible, and at these angles, the component of angular momentum P_θ in the direction of field may be expressed as

$$P_\Phi = P_\theta \cos\Phi \tag{11.90}$$

But $P_\Phi = \frac{mh}{2\pi}$ is the requirement for the quantisation of $\cos \Phi$, where m = an integer.

From this relationship, we may write

$$\frac{mh}{2\pi} = P_\theta \cos\Phi = \frac{lh}{2\pi}\cos \Phi \text{ or } \cos \Phi = \frac{mh}{2\pi} \times \frac{2\pi}{lh} = \frac{m}{l} \tag{11.91}$$

But the limiting value of $\cos \Phi = \pm 1$ and hence 'm' will have $(2l + 1)$ values from $+l$ to $-l$, i.e., $l, l - 1, l - 2, \ldots l, 0, -l, \ldots -l$, where m = magnetic quantum number.

11.6.2.1 The Fourth Quantum Number

If the spectral lines in atomic spectra are examined under high resolution carefully and in the absence of external magnetic field, it is observed that many of them have very closely spaced group of lines or multiplets. In case of alkali metals, doublet lines are found having wavelengths 5889.95 and 5895.92 Å. They are found to be separated by 6 Å. In case of alkaline earth metals, singlets and triplets are found, whereas group 3 atoms consist of doublets and quartets.

These observations point to the fact that an electron with quantum number n and l can have states of slightly different energy but not in the presence of magnetic field, and as such, a fourth quantum number is essential to describe the electron fully.

11.6.2.2 Electron Spin

In 1925, Uhlenbeck and Goudsmit put forth an idea that an electron revolves about its own axis and its spin helps to bring about its overall angular momentum of magnitude $\hbar/2$ and associated with it is a magnetic moment of magnitude $he/4\pi mc$. The rotation of the electron about its own axis gives rise to the concept of spinning and the spin angular momentum represented by $p_s = sh/2\pi$, where s is the spin quantum number.

The values of spin quantum number, $s = \pm 1/2$. This corresponds to two possible directions of spin, i.e., one clockwise and another anticlockwise.

It is to be noted that the resultant angular momentum of the electron is found to be the vectorial sum of the orbital and spin angular momenta, i.e., $\frac{jh}{2\pi} = \frac{lh}{2\pi} + \frac{sh}{2\pi}$

where j = resultant or inner quantum number,

or $j = l + s = l \pm 1/2$.

It is, thus, clear that the electron spin increases or decreases the electronic orbital angular momentum and leads to two sublevels having slightly different energy for an orbit of given n and l when $l \neq 0$. Thus, in case of alkali metals having one outer electron, j has the following values:

Orbit Type	→	s	p	d	F
l	→	0	1	2	3
j	→	1/2	3/2, 1/2	5/2, 3/2	7/2, 5/2

It is also to be noted that an electronic transition $s \leftrightarrow p$ will come into picture with the appearance of two closely spaced spectral lines, which originate from the p state having $j = 1/2$ and $j = 3/2$.

11.6.3 Atoms Having Two or More than Two Electrons

It is known to us that the closed inner electron shells do not contribute to the total electronic angular momentum. The total orbital angular momentum (TOAM) is generally represented by L, and it is the vectorial sum of the value of l_1, l_2, l_3, … of the outer electrons, i.e., $L = l_l + l_2 + l_3 + \ldots$ The total spin angular momentum is denoted by S, which is equal to the algebraic sum of the individual spin quantum numbers of s_1, s_2, s_3, …, i.e., $S = s_1 + s_2 + s_3 + \ldots$

We have already stated that the resultant total angular momentum is denoted by the total inner quantum number J. The value of J depends on the value of L and S. J is the vectorial sum of L and S, and it has $(2S + 1)$ values, i.e., $J = L + S,\ L + S - 1,\ L + S - 2,\ \ldots\ |L - S|$.

This gives a clue to understand that each state of total orbital angular momentum L consists of $(2S + 1)$ closely spaced sublevels or a multiplicity of $(2S + 1)$. This also suggests that the transition from this state will lead to the appearance of $2S + 1$ closely spaced spectral lines in the found fine structure.

A group 2 atom has two outermost electrons whose total spin is calculated as follows:

$S = +1/2 - 1/2 = 0$, if the electron spin remains in opposite direction.

$S = +1/2 + 1/2 = 1$, if the electron spin remains parallel.

In the first case, the multiplicity will be $2S + 1 = 2 \times 0 + 1 = 1$, which indicates that the state of the atom is singlet represented by ^{1}M.

In the second case, the multiplicity will be $2S + 1 = 2 \times 1 + 1 = 3$, which denotes that the state of the atom is triplet or ^{3}M.

Similarly, we can calculate the multiplicity in the case of group 3 atoms. In this case, three electrons remain in the outermost orbital, and hence, the atom's multiplicity will be either doublet (^{2}M) or quartet (^{4}M).

11.7 The Gyromagnetic Ratio and the Landé Splitting Factor

Gyromagnetic ratio is defined as the ratio of magnetic moment to angular momentum. This ratio is found to be different for orbital motion and for spin. The value of magnetic moment because of orbital motion is given by

$$M_0 = \sqrt{l(l+1)} \cdot \frac{he}{4\pi mc} \tag{11.92}$$

The value of orbital angular momentum $I\omega$ is given by

$$I\omega = \frac{h}{2\pi}\sqrt{l(l+1)} \tag{11.93}$$

$$\therefore \text{ Orbital gyromagnetic ratio} = \frac{\text{Magnetic moment}}{\text{Angular momentum}}$$

$$= \frac{M_0}{I\omega} = \frac{\sqrt{l(l+1)} \cdot \dfrac{he}{4\pi mc}}{\dfrac{h}{2\pi}\sqrt{l(l+1)}}$$

$$= \sqrt{l(l+1)} \cdot \frac{he}{4\pi mc} \times \frac{2\pi}{h\sqrt{l(l+1)}}$$

$$= \frac{e}{2mc} \tag{11.94}$$

Now, we shall find out the spin gyromagnetic ratio. We know that the angular momentum of spin

$$= \frac{h}{4\pi} \tag{11.95}$$

And the spin magnetic moment $= \dfrac{he}{4\pi mc}$

$$\therefore \text{ the spin gyromagnetic ratio} = \frac{\dfrac{he}{4\pi mc}}{\dfrac{h}{4\pi}}$$

$$= \frac{he}{4\pi mc} \cdot \frac{4\pi}{h}$$

$$= \frac{e}{mc} \tag{11.96}$$

It is thus clear that the spin gyromagnetic ration is twice the orbital gyromagnetic ratio.

11.7.1 Landé 'g' Factor or Splitting Factor

Here, we are going to show the relationship between 'g' and the resultant magnetic moment. For this, let us consider Figure 11.6, which represents the vectorial sum of orbital and spin angular momentum. We have already concluded that the gyromagnetic ratio for spin is twice the gyromagnetic ratio for orbital

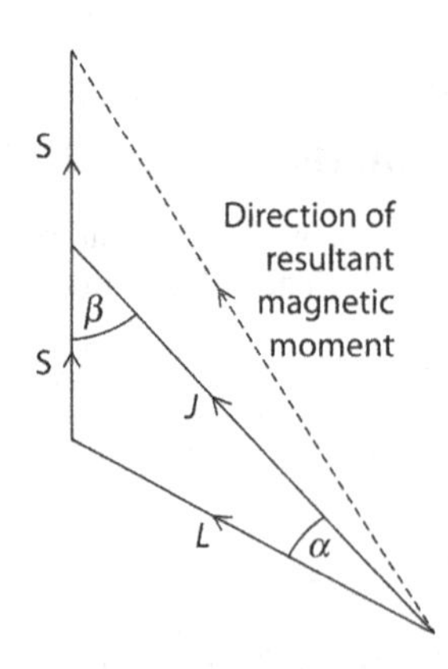

FIGURE 11.6 Relation of direction of resultant magnetic moment to direction of *J*.

motion, and as such, the resultant magnetic moment does not lie in the same direction as *J*, the resultant angular momentum formed by the vectorial sum of *L*, the resultant orbital angular momentum, and *S*, the resultant spin angular momentum. The scalar value of *J* is expressed as

$$|J| = |S|\cos\beta + |L|\cos\alpha \tag{11.97}$$

Let μ = component of the magnetic moment along the direction of J.

$$\mu = \text{gyromagnetic ratio} \times \text{angular momentum}$$

$$= \frac{e}{mc} \times |S|\cos\beta + \frac{e}{2mc} \times |L|\cos\alpha$$

$$= \frac{e}{2mc}\left(2|S|\cos\beta + |L|\cos\alpha\right)$$

$$= \frac{e}{2mc}\left(|J| + |S|\cos\beta\right)$$

$$= \frac{e}{2mc}|J|g, \text{ where, } g = 1 + \frac{|S|\cos\beta}{|J|}$$

$$\therefore \cos\beta = \frac{|S|^2 + |J|^2 - |L|^2}{2|S||J|}$$

$$g = 1 + \frac{|S|^2 + |J|^2 - |L|^2}{2|J|^2} \tag{11.98}$$

$$\therefore g = 1 + \frac{S(S+1) + J(J+1) - L(L+1)}{2J(J+1)} \tag{11.99}$$

This is the expression for Landé 'g' factor.

When $S = 0$ and $J = L$, $g = 1$, but when $L = 0$ and $S = J$, g will be equal to 2, when calculated using Eq. (11.99).

11.7.2 Landé Interval Rule

According to *Landé interval rule*, 'the separation between the two *J* levels of a Russell–Saunders term, brought about by spin–orbit interaction, is proportional to the larger *J* value of the pair'.

We know from definition that

$$\vec{J} = \vec{L} + \vec{S}$$

$$\overrightarrow{J^2} = \overrightarrow{L^2} + \overrightarrow{S^2} + 2 \cdot \vec{L} \cdot \vec{S}$$

$$\therefore 2 \cdot \vec{L} \cdot \vec{S} = \left(\vec{J}^2 - \vec{L}^2 - \vec{S}^2\right)$$

$$\therefore \vec{L} \cdot \vec{S} = \frac{1}{2}\left(\vec{J}^2 - \vec{L}^2 - \vec{S}^2\right)$$

We also know that the Eigen values of the various angular momenta are

$$\overrightarrow{J^2} = J(J+1)\hbar^2; \overrightarrow{L^2} = L(L+1)\hbar^2; \overrightarrow{S^2} = S(S+1)\hbar^2$$

Therefore, the energy of the level characterised by J as a consequence of spin–orbit interaction is given by

$$E_J = \frac{1}{2}\left[J(J+1) - L(L+1) - S(S+1)\right]\hbar^2$$

When $J \rightarrow J - 1$, then energy is given by

$$E_{J-1} = \frac{1}{2}\left[(J-1)(J-1+1) - L(L+1) - S(S+1)\right]\hbar^2$$

$$\therefore\ E_J - E_{J-1} = \Delta E$$

$$= \frac{1}{2}\left[J(J+1) - L(L+1) - S(S+1)\right]\hbar^2 - \frac{1}{2}\left[J(J-1) - L(L+1) - S(S+1)\right]\hbar^2$$

$$= \frac{1}{2}\left[J(J+1) - J(J-1)\right]\hbar^2$$

$$= \frac{1}{2}\left[J^2 + J - J^2 + J\right]\hbar^2$$

$$= \frac{1}{2}\cdot 2J\hbar^2 = J\hbar^2$$

$$\therefore\ \Delta E = J\hbar^2$$

$$\text{or } \Delta E \propto J$$

Hence, the separation between the two J levels is proportional to J. This is called *Landé interval rule.*

11.7.3 Zeeman Effect

When a substance that emits a line spectrum is placed in a strong magnetic field, the single lines are split up into groups of closely spaced lines. This effect is called the *Zeeman effect.* This effect is divided into two classes: (i) normal Zeeman effect and (ii) anomalous Zeeman effect.

In case of the normal Zeeman effect, a single line is split into three if the field is perpendicular to the light path or split up into two lines if the magnetic field is parallel to the light path. In other words, we can say that a spectral line is split into three components under the influence of a strong magnetic field when observed in the direction perpendicular to the magnetic field. The single line is observed to split into two components when viewed in a direction parallel to the magnetic field.

The anomalous Zeeman effect is a complex splitting of the lines into many closely spaced lines. It is called so because of the fact that it does not agree with the classical predictions. This effect is explained by quantum mechanics in terms of electron spin. The anomalous Zeeman effect is observed under the influence of weak magnetic fields.

11.7.3.1 Origin of the Zeeman Effect

We have described the additional energy (E_m) acquired by an electron in an external magnetic field. The value of E_m is given by

$$E_m = -MH\cos\Phi = -l\mu H\cos\Phi$$

$$= -\mu Hm \tag{11.100}$$

where the terms have their usual significance.

A single state of energy

$$E_{n,l} = -\frac{R}{(n-\delta)^2} \quad \text{where, } \delta = \text{quantum defect} \tag{11.101}$$

splits under the influence of the magnetic field into $2l + 1$ levels of total energy

$$E_{n,l,m} = -\frac{R}{(n-\delta)^2} - \mu Hm \tag{11.102}$$

A spectral line having frequency $\overline{\nu_0}$ because of the transition is expressed by

$$\overline{\nu_0} = E_{n_2l_2} - E_{n_1l_1} = R\left\{\frac{1}{(n_1-\delta)^2} - \frac{1}{(n_2-\delta)^2}\right\}\text{cm}^{-1}$$

and is therefore replaced in the presence of a magnetic field by a series of lines consisting of frequency

$$\overline{\nu} = E_{n_2l_2m_2} - E_{n_1l_1m_1} = \overline{\nu_0} + \mu H(m_1 - m_2) \tag{11.103}$$

Now the selection rule allows that m can change only by 0 or ±1, and therefore, the number of possible lines is again limited.

11.7.3.2 The Normal Zeeman Effect

Electrons acquire magnetic moments as a consequence of their orbital and spin angular momenta. We know that in Zeeman effect, the effect of strong magnetic field on atomic spectra is observed. So the orbital and spin angular momenta will interact with magnetic field applied from an outside source, and the consequential shifts in energy should be readily visible or perceivable in the atomic spectrum.

Let us consider the effect of magnetic field on a singlet term, for example, ^1P. In this case, since $S = 0$, the magnetic moment of the considered atom will be only due to the orbital angular momentum.

Suppose B is the magnitude of magnetic field applied in the z-direction. Therefore, the first-order correction in Hamiltonian H' (in place of field strength H used before, we have used B here for removing confusion with Hamiltonian) can be written as

$$H' = -m_z B = -\gamma_e l_z B \tag{11.104a}$$

where $\gamma_e = \frac{e}{2m_e}$ = magnetogyric ratio and

m_e = mass of electron.

$m_Z = \gamma_e m_l \hbar; \quad m_l = l, l-1, \ldots l$

$= -\mu_B m_l$

If many electrons are present, then H' may be written as

$$H' = -M_z B = -\gamma_e(lz_1 + lz_2 + \ldots)B = \mu_B \cdot L_Z B \tag{11.104b}$$

where $\mu_B = -\gamma_e \hbar = \frac{eh}{2m_e}$ = Bohr magneton

Therefore, the first-order correction to energy of ^1P term may be expressed as

$$E' = \langle {}^1PM_L|H'|{}^1PM_L\rangle = -\gamma_e M_L \hbar B = \mu_B M_L B \tag{11.105}$$

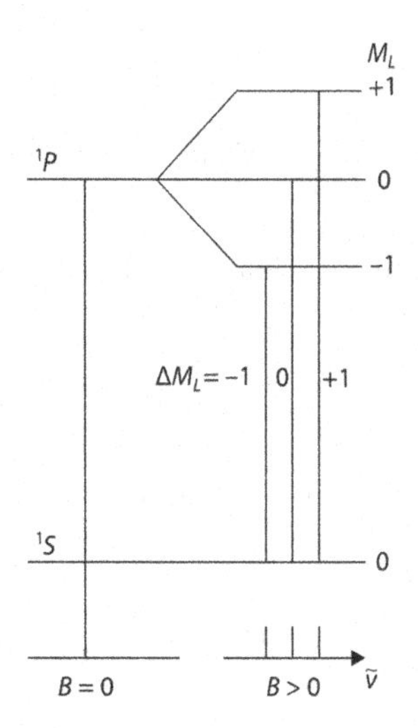

FIGURE 11.7 Splitting of energy levels of an atom in case of normal Zeeman effect, and the splitting of single line into three lines.

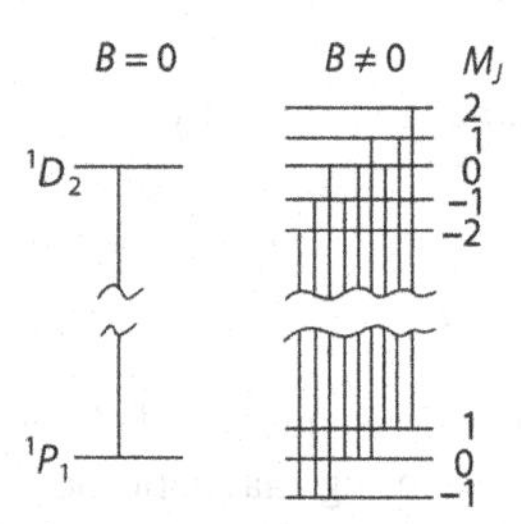

FIGURE 11.8 Normal Zeeman effect of the $^1D_2 \leftarrow {}^1P_1$ transition of Cd.

It should be kept in mind that in case of S term, there is no appearance of orbital angular momentum as well as spin angular momentum; hence, there will be no effect of magnetic field on it. It follows from the aforementioned fact that the transition $^1P \rightarrow {}^1S$ should be split into three limes or three components, as shown in Figure 11.7, with a splitting having the magnitude $\mu_B \cdot B$. It has been noted that the magnetic field splits line by only 0.5 cm^{-1}, which is a very small effect. The splitting of a line in spectra into three lines is known as *normal Zeeman effect*. It is observed only for singlet lines.

It is pertinent to note that for observations made in a direction perpendicular to the magnetic field, the lines having $\Delta M = 0$ are said to be polarised parallel to the applied field. It is known as π line or π components. The lines with $\Delta M = \pm 1$, perpendicular to the field, are circularly polarised. These lines are called *σ-lines* or *σ-components*. It is also important to note that normal Zeeman effect is seen in the absence of spin.

When both the states involved in a transition consist of $J = 0$, both will be split by a field. In case of normal Zeeman effect, the splitting between the adjacent levels is found to be the same for both the states. This is shown for the $^1D_2 \leftarrow {}^1P_1$ (Figure 11.8) transition of Cd. We find that even though nine transitions are actually taking place here, only three spectral lines are observed due to the fact that the three with ΔM_J of −1 consist of the same energy difference, as do the three with ΔM_j of zero and the three with ΔM_j of +1.

11.7.3.3 The Anomalous/Complex Zeeman Effect

The energy in a magnetic field is expressed as

$$E = E_0 - H\mu_H \tag{11.106}$$

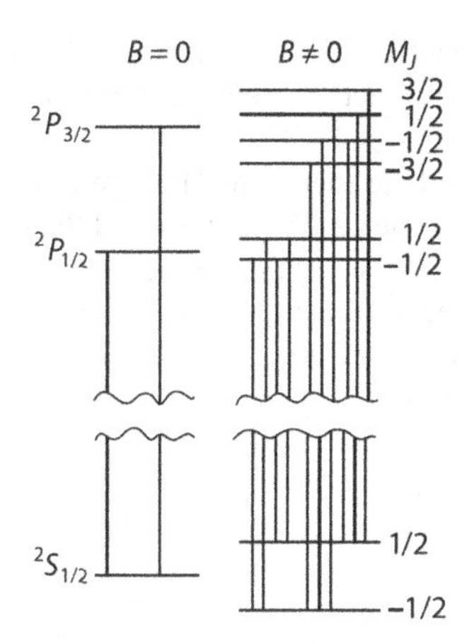

FIGURE 11.9 Anomalous Zeeman effect of the $^2P_{3/2} \leftarrow {}^2S_{1/2}$.

where μ_H = component of the magnetic moment in the direction of the field,
H = magnetic field strength, and
E_0 = energy in the field-free case.

When μ or J is perpendicular to the field direction, $E = E_0$.

In the normal Zeeman effect, we find that the energy in the magnetic field is given by

$$E = E_0 + h\omega_o M \tag{11.107}$$

When $W_0 = \dfrac{1}{2\pi}\,\dfrac{eh}{2\pi mc}$

Here, ω_o is called the Larmor frequency, which may be shown to be the frequency of precession.

For all lines which are not singlets, the so-called *anomalous Zeeman effect* is found. It is observed when the spin angular momentum is nonzero and originates from the unequal splitting of the energy levels in the two terms that cause to participate in the transition. The unequal splitting originates, in turn, from the anomalous magnetic moment of the electron. This effect can be explained only by assuming that the magnitude of the term splitting for a particular field strength is not the same for all terms but differs with values of L and J. We may account this fact by introducing a 'g' factor in Eq. (11.107), after which it becomes

$$E = E_0 + \omega_o M g \tag{11.108}$$

where g = Landé g factor, which is a rational number dependent on J and L.

It is quite clear that even if we keep the aforementioned selection rule $\Delta M = 0, \pm 1$, the number of line components found in a magnetic field will now depend upon the number of term component $(2J + 1)$.

Let us consider the case of D lines of sodium (Na), which corresponds to the transition $^2P_{1/2} \leftarrow {}^2S_{1/2}$ and $^23/2 \leftarrow {}^2S_{1/2}$.

Since M has only two values for terms $^2P_{3/2}$ and $^2S_{1/2}$ and four values for $^2P_{3/2}$, it is clear that with a different values of 'g' for each of the three terms, the number of components of the splitting pattern for one D line of Na will be different from that for the other. In Figure 11.9, we have shown the splitting for $^2P_{3/2} \leftarrow {}^2S_{1/2}$. Two lines are observed in the absence of a field, but ten lines are seen under high resolution under the influence of a magnetic field.

11.8 Stark Effect

The Stark effect is the effect of an external electric field on atomic spectra and is similar to the Zeeman effect in external magnetic field in many ways, but there are some important differences. In the Stark effect, the electric field generates an electric moment in the atom by removing or displacing the electrons with respect to nucleus and consequently polarises the atom. This causes a precession of the resultant angular momentum vector J about the direction of the electric field. L and S remain uncoupled in a very strong electric field.

The splitting of the spectral lines and the energy difference between the subsidiary terms of the atomic states are dependent on the product of the dipole moment and the electric field strength. As the magnitude of dipole moment is proportional to the field strength, the energy difference between the subsidiary terms is found to be proportional to the square of the field strength, i.e., $\Delta E\ \alpha$ (field strength)^2. Therefore, the effect of imposing an external electric field is known as *quadratic Stark effect.* It is to be noted that the resulting spectrum is very complicated.

We have already stated that the Stark effect is identical in some respect to the Zeeman effect, but the Stark effect involves the interaction of matter with an electric field.

Let us suppose that the perturbation operator showing interaction with an electric field of strength E in case of H atom is expressed as

$$V = -E \cdot r \tag{11.109}$$

If $E \rightarrow e$ = magnitude of the electric field

Then,

$$V = -\in r\cos\Phi = -\in z \tag{11.110}$$

It should be kept in mind that the electric field is directed towards the z-axis.

Now, the Stark effect will be examined on the H atom in its $n = 2$ quantum state. This is a fourfold-degenerate state having zero-order wave functions of

$$\Psi_1 = 2s, \Psi_2 = 2p_0, \Psi_3 = 2p_{-1}, \Psi_4 = 2p_1 \tag{11.111}$$

It is to be noted that when $n = 2$, $l = 0$, 1 and therefore, $m_l = 0, 0, -1, +1$. Then the perturbation energy correction will be found by solving the 4×4 secular determinant.

$$\begin{vmatrix} V_{11}-x & V_{12} & V_{13} & V_{14} \\ V_{21} & V_{22}-x & V_{23} & V_{24} \\ V_{31} & V_{32} & V_{33}-x & V_{34} \\ V_{41} & V_{42} & V_{43} & V_{44}-x \end{vmatrix} = 0 \tag{11.112}$$

The matrix elements V_{ij} may be readily estimated, which may be expressed as

$$V_{ij} = -\in \langle \Psi_i | z | \Psi_j \rangle$$

which will vanish unless $\Delta l = \pm 1$. Hence, all the diagonal elements V_{11}, V_{22}, V_{33}, and V_{44} will vanish.

The commutation relation

$$[L_Z, V] = [L_z, -\in_z] = 0 \tag{11.113}$$

It means all the V_{ij} for $\Delta m_l \neq 0$ will be also zero. Hence, V_{12} and V_{21} matrix elements will remain or survive. When you will calculate $V_{12} = V_{21}$, its value will be found to be $3 \in$, i.e.,

$$V_{12} = V_{21} = 3\in (\text{the reader should calculate}) \tag{11.114}$$

And the following will be the roots of the secular determinant:

$$x = -3\in, +3\in, 0 \text{ and } 0 \tag{11.115}$$

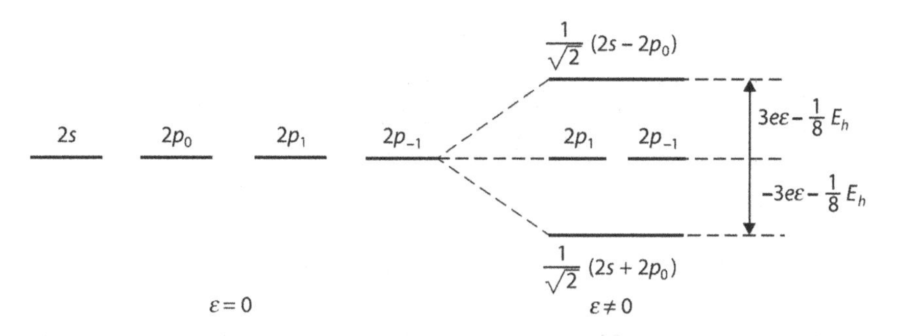

FIGURE 11.10 Stark effect for $n = 2$ for H atom as a result of the electric field.

From this, it is clear that the part of the degeneracy is removed by the electric field perturbation. It should be kept in mind that one level goes up by 3 $\in$, one goes down by 3 $\in$, and the other two remain the same as in the zeroth-order system. This is shown in Figure 11.10.

It is important to note that the real significance of the Stark effect lies in the analogy between the behaviour of the atomic electrons in an external electric field and the behaviour of molecular electron in the axial electric field provided by the two nuclei in the diatomic molecule.

11.9 Coupling of Orbital Angular Momentum

The orbital angular momentum is a vector quantity and the angular momenta of all occupied orbitals in a given atom may be added/combined vectorially to obtain the resultant orbital angular momentum. For example, in a two-electron atom, each electron in an occupied orbital has its own angular momentum. So to get the resultant angular momentum, the individual orbital angular momentum for the two electrons must be combined vectorially. It is important to note that the individual occupied orbital is also affected by the angular momentum of the other occupied orbital. It is because of the fact that they are coupled/combined together by electrostatic forces. These forces are between electrons, and electrons and the nucleus. Each angular momentum acts on the other one and causes precession around the direction of the resultant angular momentum, which is found to be constant in magnitude and direction by the principle of the conservation of angular momentum. This is shown in Figure 11.11.

The resultant orbital angular momentum of the whole occupied orbital system is denoted by vector L_τ, but the value of L for a single electron is given by $L=\sqrt{l(l+1)}\frac{h}{2\pi}$, where l is the azimuthal quantum number. For many-electron systems, the total or resultant orbital angular momentum is equal to $L=\sqrt{L(L+1)}\frac{h}{2\pi}$, where L is the orbital angular momentum.

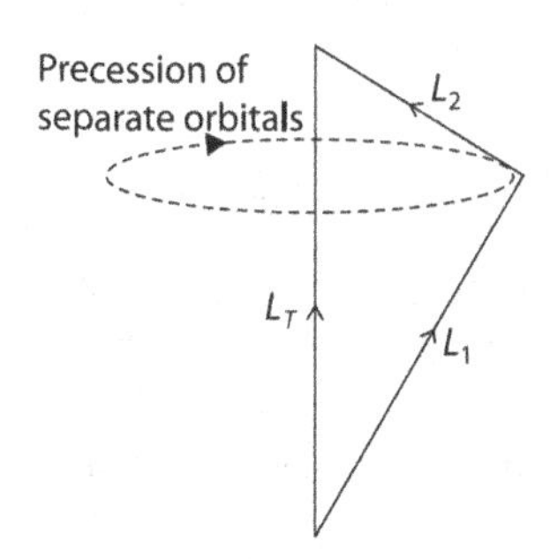

FIGURE 11.11 Combination of orbital angular momentum.

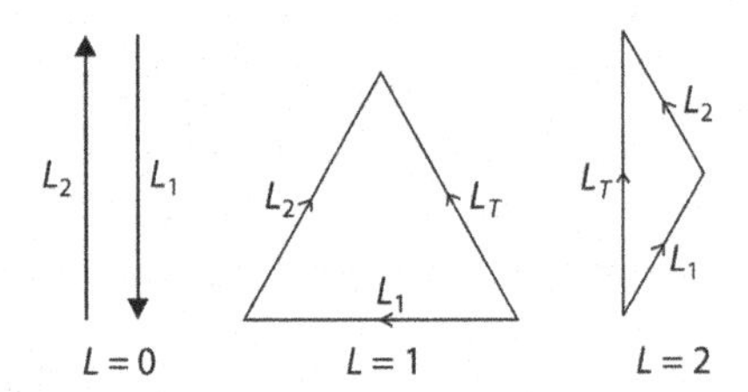

FIGURE 11.12 Vectorial addition of angular momenta of p orbitals for the three possible values of L.

It should be kept in mind that each angular momentum vector makes a constant angle with the direction of L_T; precession of each orbital angular momentum occurs about the direction of L in space. If L is parallel to L_T, it does not precess. For s orbital, $l = 0$, it can be shown empirically from the spectra of inert gases that L_T of completely filled electron shell is zero.

The angular momenta for two p orbitals ($l = 1$) occupied by nonequivalent electrons can couple vectorially in three ways corresponding to $L = 0, 1, 2$ (Figure 11.12).

Since $l = 1$, for two electrons

$$l_1 = 1 \text{ and } l_2 = 1$$

$$\therefore\ L = l_1 + l_2 = 1 + 1 = 2$$

$$= l_1 + l_2 - 1 = 2 - 1 = 1$$

$$= |l_1 - l_2| = |1 - 1| = 0$$

In other words, we can express

$$\sum_i^n l_i = L$$

For the value of $l_1 = l_2$, the angular momentum will be $\sqrt{2}\frac{h}{2\pi}$ and $\sqrt{2}\frac{h}{2\pi}$, respectively, whereas the resultant angular momentum will be $L_T = \sqrt{L(L+1)}\frac{h}{2\pi} = \sqrt{6}\frac{h}{2\pi}, \sqrt{2}\frac{h}{2\pi}$ for $L = 2$ and 1, respectively.

For a d orbital, $l = 2$, and for d^2 electrons,

$$l_1 = 2 \text{ and } l_2 = 2$$

$$\therefore\ L = l_1 + l_2 = 2 + 2 = 4$$

$$= l_1 + l_2 - 1 = 2 + 2 - 1 = 3$$

$$= l_1 + l_2 - 2 = 2 + 2 - 2 = 2$$

$$l_1 + l_2 - 3 = 2 + 2 - 3 = 1$$

$$= |l_1 - l_2| = |2 - 2| = 0$$

Thus, L has five values, i.e., $L = 0, 1, 2, 3$, and 4. These angular momenta of two d orbitals occupied by nonequivalent electrons couple vectorially in five ways corresponding to the five L values. This is shown in Figure 11.13.

Similarly, we can find out the different values of L and Lj in terms of $L = \sqrt{l(l+1)}\frac{h}{2\pi}$ and $L_T = \sqrt{L(L+1)}\frac{h}{2\pi}$ by putting different values of l and L, respectively.

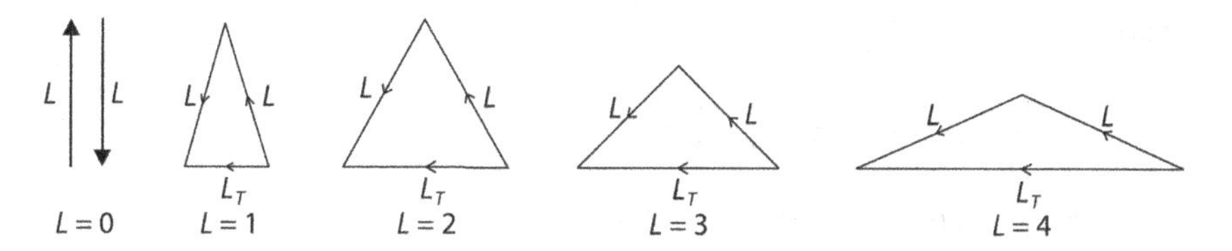

FIGURE 11.13 Vector sum of angular momenta of *d* orbitals for the five possible values of *L* and *LT* is the resultant orbital angular momentum.

11.10 Coupling of Spin Momenta

It has been deduced from spectral studies that in an applied field, the component spins of two electrons in an atom can be combined vectorially in two relative orientations only: one is spinning in the same direction and other is spinning in the opposite direction. The spin quantum number(s) of the individual electrons is(are) combined to yield the resultant spin quantum number(s) of electrons.

$$S = (s_1 + s_2), (s_1 + s_2 - 1), (s_1 + s_2 - 2), \ldots (s_1 - s_2)$$

The vector diagram for the addition/coupling of two total spins is illustrated in Figure 11.14.

The total spin angular momentum vector S is related to the quantum number by the following equation:

$$S = \sqrt{s(s+1)}\frac{h}{2\pi}$$

Now the total angular momentum of spin of one electron is also quantised. Therefore,

$$S = \sqrt{s(s+1)}\frac{h}{2\pi}$$

For two electrons, $S = s_1 + s_2 = +\frac{1}{2} - \frac{1}{2} = 0$

or $S = +\frac{1}{2} + \frac{1}{2} = 1$

Thus, it can be said that when spins of the two electrons are coupled parallel (↑↑), then $S = 1$. But when they are coupled oppositely, then $S = 0$.

For three electrons:

- The coupling ↑↑↑ gives $S = \frac{1}{2} + \frac{1}{2} + \frac{1}{2} = \frac{3}{2}$
- The coupling ↑↑↓ gives $S = \frac{1}{2} + \frac{1}{2} - \frac{1}{2} = \frac{1}{2}$

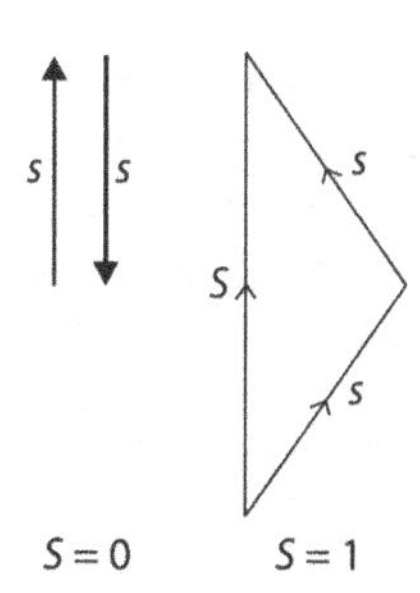

FIGURE 11.14 Vectorial addition of spin angular momenta for two electrons.

For four electrons:

- The coupling ↑↑↑↑ gives $S = \frac{1}{2} + \frac{1}{2} + \frac{1}{2} + \frac{1}{2} = 2$
- The coupling ↑↑↑↓ gives $S = \frac{1}{2} + \frac{1}{2} + \frac{1}{2} - \frac{1}{2} = 1$
- The coupling ↑↑↓↓ gives $S = \frac{1}{2} + \frac{1}{2} - \frac{1}{2} - \frac{1}{2} = 0$

The total spin angular momentum vector S can be calculated using the expression

$$S = \sqrt{S(S+1)}\frac{h}{2\pi}$$

In case of two electrons $S = 0$ or 1, by putting these values in the aforementioned equation, we shall get

$$S = \sqrt{0(0+1)}\frac{h}{2\pi} = 0 \text{ and}$$

$$S = \sqrt{1(1+1)}\frac{h}{2\pi} = \sqrt{2}\frac{h}{2\pi}$$

In the many-electron atom, the resultant spin quantum number S takes an integral number when an even number of electrons are present and a half integral value when an odd number of electrons are present.

11.11 Coupling of Orbital and Spin Angular Momenta

In case of a single electron, it is very much clear that the orbital and spin angular momenta are coupled together by means of their magnetic fields, because there is no electrostatic coupling between them. It has, therefore, been concluded that all the coupling between orbital and spin angular momenta is magnetic. The coupling of angular momenta in the atom may be described theoretically in two possible manners.

- *L–S* or the Russell–Saunders coupling scheme
- *jj*-coupling scheme

We shall consider the two possible ways one by one.

11.11.1 *L-S* or the Russell–Saunders Coupling Scheme

The Russell–Saunders coupling scheme has been named after the astronomers N. Russell and F.A. Saunders, who devoted work to explain the spectra of complex atoms. The *L–S* or Russell–Saunders coupling is an appropriate/adequate way of describing small electronic configurations (atoms with $z < 30$). This is because of the fact that the spin–orbit coupling is weak in comparison with the electrostatic effects. This means that in the central field approximation, the *i*th electrons term in the Hamiltonian H proportional to $\vec{l}_i \cdot \vec{s}_i$ (the so-called orbit–spin effect) is small enough to be negligible. In terms of perturbation theory, this leads to first solving the Schrödinger equation

$$E_{n_i \text{ mean}} \Psi_{i \text{ mean}} = H_{\text{mean}} \Psi_{i \text{ mean}} \tag{11.116}$$

giving unperturbed single particle wave function $\Psi_{n_i l_i m_{li} m_{si}}$. Their angular momenta combine to $\vec{L}$ and $\vec{S}$, and then a finer structure comes up from the coupling of $\vec{L}$ and $\vec{S}$ to a whole-system total angular momentum *J*. It is significant to note that the coupling here can be expressed as vectorial combinations or sums.

$$\vec{L} = \sum_{i=1}^{N} \vec{l_i} \tag{11.117}$$

$$\vec{S} = \sum_{i=1}^{N} \vec{s_i} \tag{11.118}$$

$$\vec{J} = \vec{L} + \vec{S} \tag{11.119}$$

The vectorial combination is illustrated in Figure 11.15.

It is clear from the figure that l_1 and l_2 couple electrostatically to give the resultant orbital angular momentum *L*. Similarly, spin momenta (s_1 and s_2) are coupled electrostatically together to give the resultant spin angular momentum *S*. But *L* and *S* are coupled together magnetically and combined vectorially to give the resultant angular momentum *J*.

The angle between *L* and *S* should be such that *J* has an integral value (or 1/2 integral value if $s = 1/2$ or 3/5 or 5/2…). In general, the variation of *J* may be represented as

$$L + S, L + S - 1, L + S - 2 \ldots L - S + 1, |L - S|$$

This form of coupling is called *L–S coupling or the Russell–Saunders coupling*, which is valid for the lighter elements.

It should be kept in mind that the impact of the vectorial addition together with the fact that only $\vec{L}^2, \vec{S}^2, \vec{J}^2$ and L_z, S_z, J_z are commutating with *H* (and therefore L^2, S^2, J^2, m_L, m_S, and mJ are good quantum numbers) is hardly to be assessed. *L–S* coupling is based on the assumption that the contribution of a single term such as $\vec{l_i} \cdot \vec{s_i}$ is small, but after summing up to $\vec{L}$ and $\vec{S}$, those complete systems related to spin and orbit values converse to each other and then contribute to H. To get the energy contribution, the *bra* and *ket* vectors of an *L–S* coupling state have to act on the term present in the Hamiltonian operator of the system, which is proportional to $\vec{L} \cdot \vec{S}$. This approach is essential for getting the first-order energy correction from $H_{s\text{–}o}$ ('s' for spin and 'o' for orbit) to find a finer structure, which is based on J by the application of perturbation theory.

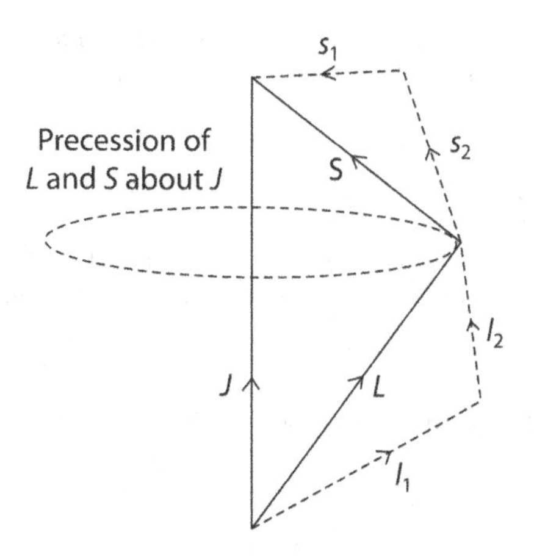

FIGURE 11.15 *L–S* or the Russell–Saunders coupling.

It is to be noted that the term proportional to $^J L \blacksquare\, ^J S$ can be diagonalised giving the same subspace as H_{s-o} may be shown by the Wigner–Eckart theorem. Therefore, we can write

$$\begin{aligned}\langle E_{LS}\rangle &= \langle L,S,J,m_J|H_{LS}|L,S,J,m_J\rangle \propto \left\langle \Psi_{LSJm_J}\left|\vec{L}\cdot\vec{S}\right|\Psi_{LSJm_J}\right\rangle \\ &= \frac{1}{2}\left(\left\langle \Psi_{LSJm_J}\left|\vec{J}^2-\vec{L}^2-\vec{S}^2\right|\Psi_{LSJm_J}\right\rangle\right) \\ &= \frac{1}{2}\left(\left\langle \Psi_{LSJm_J}\left|\left(J(J+1)-L(L+1)-S(S+1)\right)\right|\Psi_{LSJm_J}\right\rangle\right) \\ &= \frac{1}{2}\left(J(J+1)-L(L+1)-S(S+1)\right) \end{aligned} \tag{11.120}$$

assuming that $|\Psi_{LSJm_J}\rangle = |L,S,J,m_J\rangle$ is an orthonormal Eigen base of Hamiltonian H and therefore H_{s-o}.

For a given set of L and S, a large number of values are possible for J. The value of J ranges from $|L + S|$ to $|L - S|$. This is called triangular condition.

$$J = L+S,\ L+S-1,\ L+S-2,\ldots L-S+1, |L-S|. \tag{11.121}$$

This is a very important relation since it is a consequence of the mechanism how angular momenta sum up, and this mechanism governs everything in the course of finding the level structure of neutral atoms as well as ions. The resulting physically possible states are given by a common nomenclature, which is known as the term symbol, i.e.,

$$2S+1_{L_J} \tag{11.122}$$

A term symbol denotes a level (L, S, J), the complete set of quantum numbers, including mJ, is known as a *state*, and the term symbol without giving J is known as a *term*. According to convention, in this notation, L is not described by numbers expressing the Eigen value of the angular momentum operator. Instead, letters are used which are given in the following:

Value of L →	**0**	**1**	**2**	**3**	**4**	**5**	**6**
L in term symbol →	*S*	*P*	*D*	*F*	*G*	*H*	*I*
l_i in orbital configuration →	*s*	*P*	*d*	*f*	*g*	*h*	*i*

This is known as spectroscopic notation because of the relation between the line characteristics ('sharp', 'principal', 'diffuse', and 'fundamental' or 'fine'). The splitting into $2S + 1$ of different ms values leads to the denotation of the 'multiplicity' in the top left corner. The value of m_J is not mentioned in this notation. Nevertheless a state in L-S coupling is characterised by this value, and there is a degeneracy of $2J + 1$ states per value of J.

11.11.2 *jj*-Coupling Scheme

In case of *jj*-coupling scheme, the angular momentum vector $\vec{l}_i$ of a single electron combines with the single electron's spin $\vec{s}_i$, giving rise to the single electron's total angular momentum $\vec{J}$, i.e.,

$$\vec{J} = \vec{l}_i + \vec{s}_i \tag{11.123}$$

After this, all those $\vec{J}_i$ for other electrons are coupled vectorially to give the total angular momentum $\vec{J}$, which can be illustrated as

$$\vec{J} = \sum_{i=1}^{N} \vec{J}_i \tag{11.124}$$

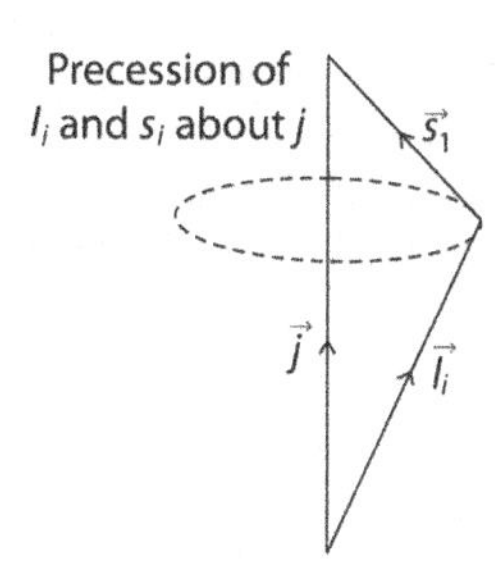

FIGURE 11.16 Coupling of l_i and s_i for a single electron.

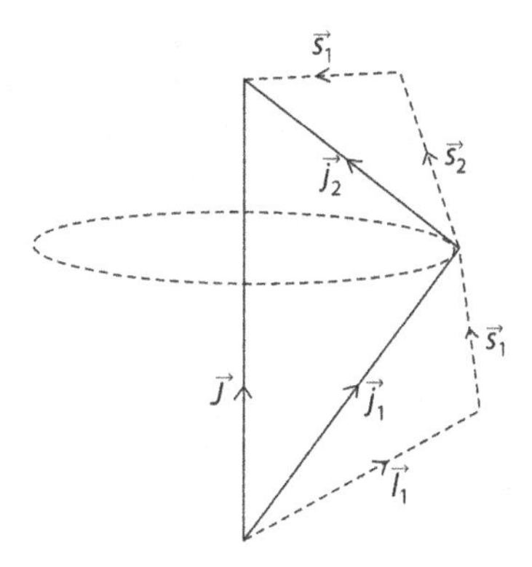

FIGURE 11.17 *jj* coupling.

The vectorial addition of $\vec{l_i}$ and $\vec{s_i}$ to give $\vec{J_i}$ and summation of all j_i to give $\vec{J}$ are illustrated in Figures 11.16 and 11.17, respectively.

This method is found to be applicable to the spectra of heavy elements ($z>30$). One point must be remembered that in case of L–S coupling, the equivalent electrons have common n and l, whereas in case *jj* coupling, the electrons have common n_i, l_i, and j_i and the set of quantum numbers which are important for the coupling are n, l, j, and m_j. The L–S coupling and *jj*-coupling schemes may be compared as follows:

L_1	L_2	-------	$\rightarrow$	L
S_1	S_2	-------	$\rightarrow$	S
$\downarrow$	$\downarrow$			$\downarrow$
J_1	J_2	-------	$\rightarrow$	J

In the L–S scheme of coupling, the orbital and spin angular momenta are added horizontally and then vertically, but in the case *jj*-coupling scheme, orbital and spin angular momenta are summed up vertically and then horizontally.

11.12 Multiplicity and Atomic States

The resultant angular momentum because of electron spin is denoted by the multiplicity. The multiplicity may be defined as *the number of lines in the atomic spectrum that are equivalent to a single line in the absence of spin.* On the basis of experimental observation, an empirical relation for calculating multiplicity is as follows:

Multiplicity $= 2S + 1$
where S = resultant spin quantum number.

The rule of multiplicity is consistent with the values of the multiplicity derived from orbital and spin angular momentum, for example,

$$\begin{aligned} \text{Multiplicity } 1,\quad & S = 0 \\ 2,\quad & S = 1/2 \\ 3,\quad & S = 1 \\ 4,\quad & S = 3/2 \\ 5,\quad & S = 2 \end{aligned}$$

This has been obtained by using the aforementioned empirical formula, multiplicity = $2S + 1$. The multiplicity is expressed as a superscript in front of the capital letter, which indicates the resultant orbital angular momentum, e.g., ^2P. The value of J may be denoted by a subscript at the RHS of the capital letter, for instance, if $J = 3/2$, then $^2P_{3/2}$. Such a symbol indicates the atomic state of the atom under examination.

11.13 Hund's Rule

For finding out the relative orders of the energies of the atomic states, Hund's rules must be used. These rules are mentioned in the following:

- Of the states originating from a given electronic configuration, the lowest in energy will be that which has the highest multiplicity.
- In case of two or more states having same multiplicity, the lowest in energy will be that having the higher L value.
- The term and thus the term symbol having maximum S will have the lowest energy.
- In case where the open shell is less than half-filled, the term symbol which has the lowest J value will have the lowest energy. If the open shell is more than half-filled, the term symbol which has the highest J value will have the lowest energy.

These rules are only valid if the coulomb interaction of the outer electrons is stronger in comparison with the spin–orbit interaction. This is exactly the validity region for L–S coupling but not for *jj* coupling.

11.14 Atomic Terms and Symbols

To be familiar with the terms and symbols, the following procedural methods should be adopted to have the knowledge about their introduction and acceptance.

- Assign the electrons to orbitals with l quantum numbers l_1 and l_2 and find the resultant orbital angular momentum (L), which may take values

$$L = (l_1 + l_2), (l_1 + l_2 - 1), \ldots |l_1 - l_2|$$

- With $s_1 = s_2 = 1/2$, find the resultant spin angular momentum (S), which may take values

$$S = s_1 + s_2,\ s_1 + s_2 - 1$$

- With L values, find the state indicated by the capital letters S, P, D, F, G, …

- Couple the resultant orbital and spin angular momenta to yield a total angular momentum characterised by a quantum number J which may have the values

$$J = L + S,\ L + S - 1, \ldots |L - S|;$$

There will be $2S + 1$ such values for $S < L$ or $2L + 1$ for $L < S$, and this is called the *multiplicity* of the state. The allowed z-components then correspond, for a given J, to $M_J = J, J - 1, \ldots J$.

- Construct the term symbol with $2S + 1$ and J values as

$$2S + 1_{L_J}$$

- Follow Hund's rule, i.e.,
  - The state of maximum spin multiplicity lies lowest.
  - For given multiplicity, the state of maximum L lies lowest.
  - For given S and L, the states of maximum or minimum J lie lowest accordingly, as the subshell is more or less than half-filled.
- Finally, write down the different term symbols and label.

Let us consider a specific example of two electrons p^2, which may be a subshell of C ($1s^2 2s^2 2p^2$). It is clear that $1s^2$ and $2s^2$ will have no influence on the couplings. First of all, we shall calculate L values. For $2p^2$, we can write

$$l_1 = 1 \text{ and } l_2 = 1$$

and $L = l_1 + l_2 = 1 + 1 = 2$

$$= l_1 + l_2 - 1 = 1 + 1 - 1 = 1$$

$$= |l_1 - l_2| = |1 - 1| = 0$$

Thus, $2p^2$ will have $L = 2$, 1, and 0. Then the symbols will be written with the values of L.

$L \rightarrow$	0	1	2
$TS \rightarrow$	S	P	D

Thus, the letter designation for different values of L will be D, P, and S.
Thus the value of resultant spin angular momentum (S) will be calculated.
Since $s_1 = 1/2$ and $s_2 = -1/2$ in case of $2p^2$ electrons,

$$\therefore S = s_1 + s_2 = 1/2 + 1/2 = 1$$

$$= s_1 + s_2 - 1 = 1/2 + 1/2 - 1 = 0$$

Thus, 'S' for $2p^2$ will have 1 and 0 values.

Now $L + S$ values will be calculated with the different values of L and S to get different values of J. The varied possibilities of combinations have been tabulated in Table 11.1.

All theoretically possible terms should not be considered; after all, we have to select only those that should have bearings with the practical aspect of investigation. According to Pauli's exclusion principle and Hund's rule, there are certain limitations on the distribution of quantum numbers and some of the levels are said to be forbidden. The terms in the aforementioned table are selected according to the selection rules, and consequently, we can have the following states only, i.e., P_0, P_1, P_2, D_2, and S_0.

'Multiplicity' due to spin is expressed as $2S + 1$. The results and term symbols are illustrated in Table 11.2.

TABLE 11.1

Various Possibilities of Combinations of $L + S$

$L = 2, 1, 0$	$J = L + S, L + S - 1, L + S - 2 \ldots$	Symbol, Ground State
$L = 2$	1. $J = 3, 2, 0$	D_3, D_2, D_0
Term = D	2. $J = 2, 1, 0$	D_1
$S = 1, 0$		
$L = 1$	1. $J = 2, 1, 0$	P_2, P_1, P_0
Term = P	2. $J = 1, 0, -1$	
$S = 1, 0$		
$L = 0$	1. $J = 1, 0, -2$	S_1, S_0
Term = S	2. $J = 0, -1, -2$	
$S = 1, 0$		

TABLE 11.2

Term Symbol

States	Multiplicity $(2S + 1)$	Term Symbol
D_2	$2S + 1, S = 1, S = 0$	${}^3D_2, {}^1D_2$
P_0	$S = 1, S = 0$	${}^3P_0, {}^1P_0$
P_1	$S = 1, S = 0$	${}^3P_1, {}^1P_0$
P_2	$S = 1, S = 0$	${}^3P_2, {}^1P_2$
S_0	$S = 1, S = 0$	${}^3S_0, {}^1S_0$

We have mentioned Hund's rule earlier, and following the rules, the admissible term symbols will be accepted, i.e., ${}^1S_0$, ${}^3P_0$, ${}^3P_1$, ${}^3P_2$, and ${}^1D_2$.

According to Hund's rule, ${}^3P_0$ is the most stable state.

Now, we want to calculate the degeneracy for p^2 configuration, and we find that ${}^1S_0$: degeneracy of 1

${}^1D_2$: degeneracy of 5

${}^3P_2$: degeneracy of 5

${}^3P_1$: degeneracy of 3

${}^3P_0$: degeneracy of 1

Total: 15 separate possible states.

11.14.1 Terms of Nonequivalent Electrons

Nonequivalent electrons are those that belong to different (n, l) subgroups. In order to find the terms resulting from two nonequivalent electrons, we should first of all find the values of resultant L. In case of one p electron and one d electron, $L = 3, 2, 1$, i.e., two electrons give F, D, and P terms according to the values of $L = 3, 2, 1$. The spins of the two electrons can be either parallel or antiparallel. So, the resultant S values will be 1 and 0, meaning thereby that the singlet and triplet terms will be found. In all, we shall find six terms, namely, ${}^3F$, ${}^3D$, ${}^3P$, ${}^1F$, ${}^1D$, and ${}^1P$. Similarly, two nonequivalent p electrons will yield the terms: ${}^3D$, ${}^3P$, ${}^3S$, ${}^1D$, ${}^1P$, and ${}^1S$.

Suppose, we add a third nonequivalent electron; its l should be combined vectorially to the previously calculated L; and s must similarly be added to S.

Suppose an s electron is added to pd; the L value will remain the same. The possible S values will now be 1/2, 1/2, and 3/2. The possible terms will be ${}^2P$, ${}^2D$, ${}^2F$, ${}^2P$, ${}^2D$, ${}^2F$, ${}^4P$, and ${}^4F$. It should be kept in mind that two different doublet terms for each L will come into picture, because p electrons and d electrons having parallel and antiparallel spins can result in $S = 1/2$ with the addition of s electron. On the other hand, the $S = 3/2$ will be obtained in case where all the three electrons have spins parallel to one another (↑↑↑). The three spin configuration for $s\ p\ d$ will be shown as: ↑↓↑, ↓↑↑, ↑↑↑.

If the third electron added to pd is an f electron ($l = 3$), the possible L values will be 3 + 1, 3 + 2, 3 + 3 (where + denotes vector summation). This will result in the following L values, i.e., 2, 3, 4; 1, 2, 3, 4, 5; 0, 1, 2, 3, 4, 5, 6. The possible S values will be 1/2, 1/2, and 3/2 Thus, the term will be $^4S(1)$,$^4P(2)$, $^4D(3)$, $^4F(3)$, $^4G(3)$, $^4H(2)$, and $^4I(1)$, where the number in the bracket indicates the number of corresponding terms. In the aforementioned case, we assumed that Pauli's principle is not *considered in adding together individual l and s values*. The examples are mentioned in Table 11.3.

11.14.2 Terms of Equivalent Electrons

In case of equivalent electrons, n and l remain the same, but they must at least differ in their values of m_l or m_s. For instance, the two p^2 electrons must have l with the same direction, which yields a D term; m_l is the same for both (–1, 0 or +1), and hence, according to Pauli's principle, the two electrons cannot have both $m_s = +1/2$ or both $m_s = -1/2$, that is, their spin can only be antiparallel for D term giving ^1D only, and ^3D is not possible. Further consideration illustrates that p^2 equivalent electrons give only $^1S\ ^3P\ ^1D$ terms. Similarly, p^3 electrons give $^4S\ ^2P\ ^2D$, which are given in Table 11.4.

Now, we shall mention something about the closed shells. In the shells having a maximum number of equivalent electrons, all the electrons must be antiparallel pairs ($S = 0$) in order to fulfil Pauli's exclusion principle. In such a situation, $L = 0$ because the state can be realised only in one way in a magnetic field,

TABLE 11.3

Terms of Nonequivalent Electrons

Electronic	Configuration
Ss	^1S, ^3S
Sp	^1P, ^3P
Sd	^1D, ^3D
PP	^1S, ^1P, ^1D, ^3S, ^3P, ^3D
Pd	^1P, ^1D, ^1F, ^3P, ^3D, ^3F
Dd	^1S, ^1P, ^1D, ^1F, ^1G, ^3S, ^3P, ^3D, ^3F, ^3G
Sss	^2S, ^2S, ^4S
Ssd	^2D, ^2D, ^4D
Spp	^2S, ^2P, ^2D, ^2S, ^2P, ^2D, ^4S, ^4P, ^4D
Spd	^2P, ^2D, ^2F, ^2P, ^2D, ^2F, ^4P, ^4D, ^4F
PPP	$^2S(2)$, $^2P(6)$, $^2D(4)$, $^2F(2)$, $^4S(1)$, $^4P(3)$, $^4D(2)$, $^4F(1)$
Ppd	$^2S(2)$, $^2P(4)$, $^2D(6)$, $^2F(4)$, $^2G(2)$, $^4S(1)$,$^4P(2)$, $^4D(3)$, $^4F(2)$, $^4G(1)$, $^2S(2)$, $^2P(4)$, $^2D(6)$, $^2F(6)$,$^2G(6)$, $^2H(4)$, $^2I(2)$
Pdf	$^4S(1)$, $^4P(2)$, $^4D(3)$, $^4F(3)$, $^4G(3)$, $^4H(2)$, $^4I(1)$

TABLE 11.4

Terms of Equivalent Electrons

Electronic	Configuration Terms
s^2	^1S
P^2	$^1S\ ^1D\ ^3P$
P^3	$^2P\ ^2D\ ^4S$
p^4	$^1S\ ^1D\ ^3P$
p^5	^2P
p^6	^1S
d^2	$^1S\ ^1D\ ^1G\ ^3P\ ^3F$
d^3	$^2P\ ^2D(2)\ ^2F\ ^2G\ ^2H\ ^4P\ ^4F$

i.e., with $\Sigma l_i = L = 0$. Hence, a closed shell always forms an $^1S_0$ state. The term $^1S_0$ in case of closed shell must result when the shell is divided into two parts, the term types for each part deduced, and the resulting angular momentum vectors combined together. For instance, combining the angular momenta of the terms of p^2 vectorially to the corresponding quantities for p^4 must yield the value for $p\ ^{61}S_0$ state, that is, zero. From this, it is clear that the terms for p^4 will be the same as those for p^2.

11.14.3 Use of *jj* Coupling

We have discussed earlier about the *jj* coupling. One thing is very important in *jj*-coupling scheme that L and S are not specified, and hence, the term symbol loses its importance. Here, we shall find the values of the total angular momentum J, which should be permitted for the electronic configuration $5p\ ^15d^1$.

Coupling of $l_1 = 1$ and $s_1 = 1/2$ gives values for the resultant j_1. Likewise, for j_2. Finally, j_1 and j_2 will couple to give J. We shall use the Clebsch–Gordan series in this case, i.e.,

$$j = j_1 + j_2,\ j_1 + j_2 - 1, j_1 + j_2 - 2, \ldots |j_1 - j_2|$$

Coupling of $l_1 = 1$ and $s_1 = 1/2$ will give $j_1 = 3/2$, 1/2. Coupling of $l_2 = 2$ and $S_2 = 1/2$ will result in $j_2 = 5/2$ and 3/2. Finally, coupling of j_1 and j_2 will yield $J = 4$, 3, 3, 3, 2, 2, 2, 2, 1, 1, 1, and 0. Thus, we can find out the value of J in other cases also using *jj* coupling.

11.15 Slater Rules

Slater has developed a set of empirical rules for estimating the screening effect of the other electrons on a particular electron in an atom. The correction of the nuclear charge for the screening effect is called the screening constant, S. The screening constant (S) is related to the effective nuclear charge (Z^* or s^*) as

$$Z^* = Z - s \tag{11.125}$$

where Z = actual atomic number
Z^* = effective nuclear charge
s = screening constant

Slater empirical rules are summarised in the following to estimate the screening constant.

- The effective principal quantum number n^* is related to the principal quantum number 'n' as given in the following:

n	→	1	2	3	4	5	6
n^*	→	1	2	3	3.7	4	4.2

- The electrons are classified into the following groups and order: (1*s*), (2*s*, 2*p*), (3*s*, 3*p*), (3*d*), (4*s*, 4*p*), (4*d*), (4*f*), (5*s*, 5*p*), etc.
- No contribution from any shell outside the one considered. In other words, the electrons in any group lying above that of a particular group do not contribute to the shielding constant. For example, in $_8O$, the electronic configuration in the group form may be written as ($1s^2$), ($2s^2$, $2p^4$); there is no contribution of the group ($2s^2$, $2p^4$) to the shielding constant of the ($1s^2$) group.
- Each electron other than the particular electron in the same group (except the 1*s* group) contributes a shielding/screening of 0.35 (the 1*s* electron in the (1*s*) group contributes a shielding of 0.30).

- For s and p electrons, the shielding from the immediately underlying shell (i.e., the $[n-1]$ shell) amounts to 0.85 for each electron; the shielding from shells further in (i.e., the $[n-2]$ shell, etc.,) amounts to 1.00 for each electron.
- For d and f electrons, the shielding from all underlying groups is 1.00 for each electron in the underlying group.

Example 1

Let us consider a particular $4s$ electron of 30Zn:

i. The electronic configuration of $_{30}Zn$ is $1s^2\, 2s^2\, 2p^6\, 3s^2\, 3p^6\, 3d^{10}\, 4s^2$.
ii. The group electronic configuration of Zn is $(1s^2)$, $(2s, 2p)^8\, (3s3p)^8\, (3d)^{10}\, (4s^2)$.
iii. a. The contribution of the other electron in the group $4s$ to the screening/shielding is $0.35 \times 1 = 0.35$.
 b. The contribution of the immediately underlying shell ($n = 3$) to the screening is $(10 + 6 + 2)\,(0.85) = 15.30$.
 c. The contribution of the shells before $3s^2$ into the shielding is $(6 + 2 + 2)\,(1.00) = 10$.
 d. Hence, the total screening constant $S = 0.35 + 15.30 + 10.00 = 25.65$.
 e. The effective nuclear charge $Z^* = Z - s$, i.e., $Z^* = 30 - 25.65 = 4.35$.

For a particular $3d$ electron, the $4s$ electrons do not contribute to s. Thus, for a $3d$ electron, $s = (9 \times 0.35) + (6 + 2 + 6 + 2 + 2)\,(1.00) = 21.15$ and $Z^* = 30 - 21.15 = 8.85$.

Example 2

Suppose that we have to estimate the effective nuclear charge for $2s$ electron in carbon ($_6C$). The electronic configuration of C is $1s^2 2s^2 2p^2$. For each electron in the $2s$ or $2p$ group,

$$s = 3\times(0.35) + 2\times 0.85 \text{ as per the rule mentioned earlier.}$$

$$= 1.05 + 1.70 = 2.75$$

$$\therefore Z^* = Z - s = (6 - 2.75) = 3.25$$

Similarly, Z^* for other electron of an atom can be estimated.

11.16 Slater-Type Orbitals

The Slater-type orbitals (STOs) are defined by the following expressions:

$$-\Phi(r,\theta,\phi) = \frac{\left(\dfrac{2\xi}{a_0}\right)^{\frac{n^*+1}{2}}}{\left[(2n^*)!\right]^{1/2}} \cdot r^{n^*-1} \cdot e^{\frac{-\xi r}{a_0}} \cdot Y_l^m(\theta,\Phi) \tag{11.126}$$

where n^*, m, and l denote the usual quantum number, ξ represents the effective nuclear charge, and $Y_l^m(\theta,\Phi)$ is the angular part or the spherical harmonics.

$$-u_{\alpha,n^*,l,m}(r,\theta,\Phi) = \left[(2n^*)!\right]^{-1/2} \cdot (2\alpha)^{\frac{n^*+1}{2}} \cdot r^{n^*-1} \cdot e^{-\alpha r} \cdot Y_l^m(\theta,\Phi) \tag{11.127}$$

where n and α are variational parameters, and l and m are azimuthal and magnetic quantum number, respectively. n^* is called the effective principal quantum number. α is supposed to be Z^*/n^*, i.e., effective nuclear charge/effective principal quantum number.

$$\chi_{\text{STO}} = Nr^{n^*-1} \cdot e^{-\xi r} \cdot Y_l^m(\theta,\Phi) \tag{11.128}$$

where n^* = effective principal quantum number,
ξ = effective nuclear charge,
N = normalisation factor/constant, and
$Y_l^m(\theta,\Phi)$ = spherical harmonics.

All the functions of Eqs. (11.126), (11.127), and (11.128) are same, but they have different notations and represent STOs, which differ in general from hydrogen-like orbitals.

(The hydrogen-like orbitals are defined as

$$\Psi_{nlm}(r,\theta,\Phi) = N_{nl}\rho^l L_{n+1}^{2l+1}(\rho)e^{-2\rho}Y_{1,m}(\theta,\Phi),$$

where $\rho = \dfrac{2zr}{n}$, L = associated Laguerre polynomials, Y_{lm} (θ, Φ) = spherical harmonics, n, l, m = one electron quantum number.)

i. They are complete but not mutually orthogonal.
ii. STOs and hydrogen-like orbitals consist of a different number of nodes, but the STOs have nodeless radial part. The hydrogen-like orbitals consist of $(n - l - 1)$ nodes.

It must be noted that STOs involve some orthogonality; for example, functions having different values of 'l' are orthogonal due to the presence of spherical harmonics part. It should be understood that in Eq. (11.127), if $\alpha = Z/n^*$ and $n^* = 1$, then the resulting STO is just the hydrogen-like 1s orbital.

It should be kept in mind that $Y_l^m(\theta,\Phi)$ are the same spherical harmonics as we find in case of the solution of hydrogen atom. The Slater functions have no spherical nodal surface, but the hydrogen wave functions have. One thing is clear that they are distance dependent.

11.17 Gaussian-Type Orbitals

The Gaussian functions and Gaussian-type orbitals (GTOs) are mathematically expressed as

$$g_{kpq} = Nx^k y^p z^q e^{-\xi r^2} \tag{11.129}$$

where g_{kpq} = Gaussian function,
N = normalisation constant,
kpq = non-negative integers, and
x, y, z = Cartesian coordinates with the origin at nucleus.

The value of (N) is as follows:

$$N = \left(\frac{2\xi}{\pi}\right)^{3/4}\left[\frac{(8\xi)^{k+p+q} \cdot k!p!q!}{(2k)!(2p)!(2q)!}\right]^{1/2} \tag{11.130}$$

where the terms have their usual significance.

The exponential term of Eq. (11.129) is similar to the Gaussian distribution formula and therefore the name. The important fact is this that the GTOs lack the cusp or peak or apex at the nucleus and their nature at long distances also differs from the Slater function. It will be clear from the illustration of Figure 11.18. However, its computational uses make up for their adverse characteristics, and presently, they are increasingly being used.

It is to be noted that the angular dependence of Gaussian functions is expressed by $x^k\, y^p\, z^q$, where $k + p + q = l$, which is one of the quantum numbers, i.e., azimuthal quantum number and therefore an $s, p, d, \ldots$ type function. These types of Gaussian functions are shown in Table 11.5.

If the comparison is made between STOs and GTOs, it is found that r^{n^*-1} term is missing in GTOs with the effect that these are independent of n, i.e., principal quantum number; for example, $2p$ and $3p$ functions are similar.

From Table 11.5, it is clear that there are six types of Gaussians, of which only five are linearly independent. The important computational advantage of Gaussians originates from a mathematical theorem that enunciates that the product of two Gaussians is also a Gaussian. Suppose that the centres of the two Gaussian functions are different; then the product function will be represented at a new centre as illustrated in Figure 11.19.

This helps in calculating the two-electron integrals easily even if they are based on different centres. The other face of the function is that we want a larger basis set to acquire the quality of STO functions,

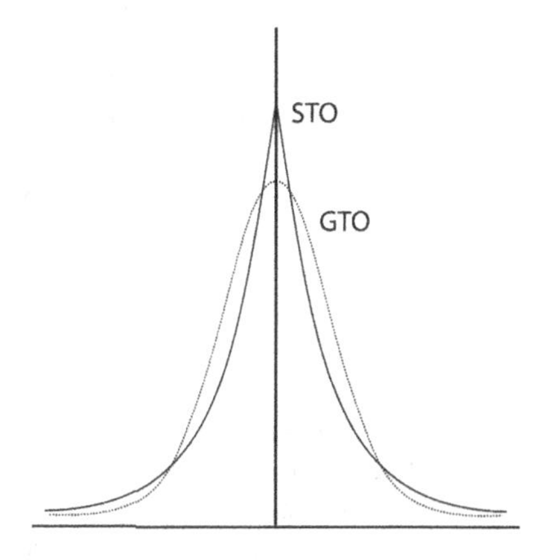

FIGURE 11.18 Comparison of Gaussian-type orbitals (GTOs) and Slater-type orbitals (STOs).

TABLE 11.5

Gaussian-Type Functions

s-Type functions	$k+p+q=0$	$g = Ne^{-\xi r^2}$
p-Type function	$k+p+q=1$	
P_x		$g_x = Nxe^{-\xi r^2}$
P_y		$g_y = Nye^{-\xi r^2}$
p_z		$g_z = Nze^{-\xi r^2}$
d-type function	$k+p+q=2$	
D_{xy}		$g_{xy} = Nxye^{-\xi r^2}$
D_{xz}		$g_{xz} = Nxze^{-\xi r^2}$
D_{yz}		$g_{yz} = Nyze^{-\xi r^2}$
dz^2		$g_z^2 = Nz^2e^{-\xi r^2}$
dx^2		$g_x^2 = Nx^2e^{-\xi r^2}$
dy^2		$g^2 = Ny^2e^{-\xi r^2}$

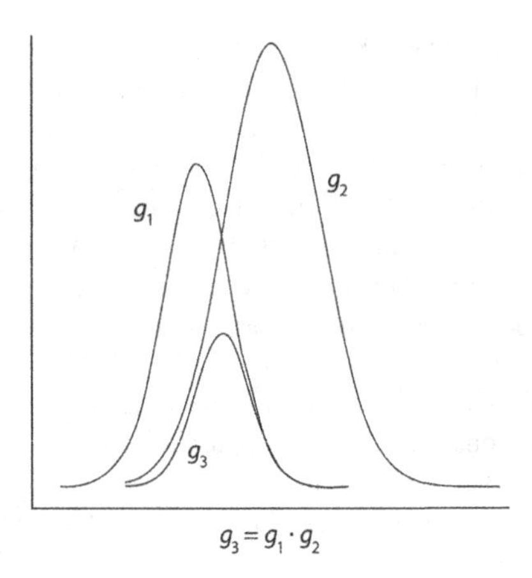

FIGURE 11.19 Product of two Gaussians is a Gaussian.

in general one, which is two to four times greater, but even then it is advised to use the Gaussians in a large number of cases.

It is to be noted that the GTOs constitute a complete set, but they do not show the property of mutual orthogonality. The basic defect of GTOs is that they do a very poor job of denoting the electron probability near the nucleus as well as far away from it. However, this defect may be removed by the use of many GTOs. For example, the linear combination of a large number of GTOs will necessarily replace a single STO. Actually the Gaussian calculations on H atom exhibit this property.

11.17.1 Gaussian Basis Set

It is the requirement of the linear combination of atomic orbital (LCAO) approximation to use the basis set, which should be made up of many well-defined functions centred on each atom. Now, we have to choose the function. The function should be the one that corresponds to the exact solution of H atom, that is, a polynomial in the Cartesian coordinates must be multiplied by a term having exponential in *r*.

It should be kept in mind that the use of these functions was not cost-effective, and the numerical estimations were performed by the use of nodeless STOs defined by the following equation:

$$\phi(r,\theta,\phi) = \frac{\left(\frac{2z^*}{a_0}\right)^{n+1/2}}{\left[(2n^*)\right]^{1/2}} \cdot r^{n^*-1} \cdot e^{\frac{-z^* r}{a_0}} \cdot Y_l^m(\theta,\phi) \tag{11.131}$$

where n^* is the effective principal quantum number and l and m denote the usual quantum number. Z^* is the effective nuclear charge.

Furthermore, researchers showed that the cost of calculations can be made lesser if the atomic orbitals are expanded in terms of Gaussian functions, which is expressed as follows:

$$g_{ijk}(r) = Nx^i y^j z^k e^{-\alpha r^2} \tag{11.132}$$

where i, j, and k are non-negative integers; x, y, z represent the position coordinates measured from the nucleus; and α is the orbital exponent.

If we set $i = j = k = 0$, then an s-type function results, which is called the *zeroth-order Gaussian*.

Also, if any one of i, j, and k is set as 1 and the remaining two are set as 0, then a *p*-type function is generated, which is called the *Gaussian function of first order*. Similarly, a *d*-type function is obtained by all combinations, which yield $i + j + k = 2$, and this type of function is known as *second-order Gaussian*.

Actually with the help of Gaussian functions, integrals are obtained, which are easily estimated. It should be noted that all practical quantum chemical models make use of Gaussian functions to evaluate many difficult integrals with the exception of semi-empirical models. Suppose STOs and Gaussian functions remain radially dependent, even if one cannot say initially that the Gaussian functions are the suitable choice for atomic orbitals. To obtain the solution of this problem, one should approximate the STO by a linear combination of Gaussian functions consisting of various α-values; for example, a best fit to a 1*s*-type STO will be obtained by the use of three Gaussians.

It is observed that the space near the nucleus does not fit up to the mark, but in the bonding region beyond $0.5a_0$, the fit is satisfactory. If we use more Gaussian functions, then the fit near the nucleus will be somewhat improved.

To get the best fit to an atomic orbital, we should not consider the individual Gaussian function as a member of the basis set, but we should construct a normalised linear combination of Gaussian functions having fixed coefficients. The value of such coefficient must be optimised by seeking minimum atom energies. It can also be achieved by the comparison of calculated and experimental values for representative molecules. These linear combinations are known as contracted functions, which become the elements of basis set. It should be kept in mind that although the coefficients in the contracted functions are fixed, the coefficients C_{ij} in equation

$$\left(\Psi_i = \sum_{\mu}^{\text{basic function}} C_{ij}\mu_i\phi_\mu\right)$$

are varied and are optimised in the solution of the Schrödinger equation. In the aforementioned equation,

$$\Psi_i = \sum_{\mu}^{\text{basic funtion}} C_{ij}\mu_i\phi_\mu$$

$C_{ij}\mu_i$ represent the molecular orbital coefficient and ϕ_μ is the basis function.

11.18 Condon–Slater Rules: Evaluation of Matrix Elements

The Hartree–Fock SCF configuration interaction (CI) wave function of an atom/molecule has the appearance of being a Slater determinant or a linear combination of a few Slater determinants. A CI wave function such as $\Psi = \sum C_i\Phi_i$ is a linear combination of various Slater determinants. Therefore, to estimate the energy and the other characteristic properties of atoms and molecules, we have to find the value of the integrals of the form $\langle D'|H|D\rangle$, which is actually the expectation value of energy. D' and D represent the determinants of orthonormal spin orbitals. H denotes the Hamiltonian operator. D' and D may be same or different.

Now let us represent each spin–orbital θ_i as a product of the form $\theta_i = \Phi_i\sigma_i$, where θ_i is the spatial orbital and σ_i is a spin function and σ_i may be either α or β. The Kronecker delta δ_{ij} is represented by the integral $\left(\theta_i(1)\middle|\theta_j(1)\right)$, where $\left(\theta_i(1)\middle|\theta_j(1)\right)$ consists of a sum over the spin coordinate of a particular electron (1) and the integration over its spatial coordinates. Suppose, θ_i and θ_j have unlike spin functions, then equation

$$\sum_{m_s=-1/2}^{1/2} \alpha^*(m_s)\beta(m_s) = 0$$

confirms that θ_i and θ_j are orthogonal, but if θ_i and θ_j consist of same spin function, then their orthogonality will be because of the orthogonality of Φ_i and Φ_j (spatial orbitals).

The Slater determinant for n-electron system is represented as

$$D = \frac{1}{\sqrt{n!}} \begin{vmatrix} \theta_1(1)\theta_2(1)\cdots\theta_n(1) \\ \theta_1(2)\theta_2(2)\cdots\theta_n(2) \\ \cdots \\ \cdots \\ \theta_1(n)\theta_2(n)\cdots\theta_n(n)[0.5pc] \end{vmatrix} \tag{11.133}$$

Suppose, we put $n = 3$, then the Slater determinant of zeroth order will be represented by

$$\Psi^0 = \frac{1}{\sqrt{3!}} \begin{vmatrix} 1s(1)\alpha(1) & 1s(1)\beta(1) & 2s(1)\alpha(1) \\ 1s(2)\alpha(2) & 1s(2)\beta(2) & 2s(2)\alpha(2) \\ 1s(3)\alpha(3) & 1s(3)\beta(3) & 2s(3)\alpha(3) \end{vmatrix} \tag{11.134}$$

D' will have the same form as D, the only difference is that only θ_1, θ_2, θ_3, … θ_n are replaced by θ_1, θ_2, θ_3, … θ_n.

Suppose $D \neq D'$, then the columns of the determinants are arranged in such a way that they have as many left-hand column match as possible. Let the diagonal elements of the determinants be

$$D = \left|1s\ 1\bar{s} \quad 2s\ 2p_z\right| \text{ and } D' = \left|1s1\bar{s} \quad 2p_z\ 3s\right| = -\left|1s\ 1\bar{s} \quad 2p_x 2s\right|$$

so that the first three terms of D and D' match.

The Hamiltonian operator for n electron is given by

$$H = \sum_{i=1}^{n} H_i^o + \sum_{i=1}^{n-1}\sum_{j>1} \frac{1}{r_{ij}} \tag{11.135}$$

where H_i^o = zeroth-order operator or core energy operator

$\sum_{i=1}^{n-1}\sum_{j>1} \frac{1}{r_{ij}}$ = operator for interelectronic repulsion and exchange.

r_{ij} = interelectronic distances

The value of

$$\sum H_i^o = \sum_i \left(\frac{-\nabla_i^2}{2} - \frac{Z}{r_i} \right) \tag{11.136}$$

where $\hbar = m = e = 1$

The integral $\int D'HD\,\mathrm{d}\tau$ or $\langle D'|H|D\rangle$ is expressed as

$$\left\langle D' \left| \sum_{i=1}^{n} H_i \right| D \right\rangle + \left\langle D' \left| \sum_{i=1}^{n-1}\sum_{j>1} \frac{1}{r_{ij}} \right| D \right\rangle \tag{11.137}$$

The first term is one-electron-like system, i.e., $\langle \theta_i | H^0 | \theta_i \rangle$ in view of the orthogonality of orbitals and the second term is a two-electron integral, which is expressed as

TABLE 11.6

Condon–Slater Rules

D and D′ differ by	$\left\langle D'\left\vert\sum_{i=1}^{n}H^0\right\vert D\right\rangle$	$\left\langle D\left\vert\sum_{i=1}^{n-1}\sum_{j>1}\frac{1}{r_{ij}}\right\vert D\right\rangle$
No spin orbitals	$\sum_{i=1}^{n}\langle\theta_i(1)\vert H^0\vert\theta_i(1)\rangle$	$\sum_{i=1}^{n-1}\sum_{j>1}\left[\left\langle\theta_i(1)\theta_j(2)\left\vert\frac{1}{r_{ij}}\right\vert\theta_i(1)\theta_j(2)\right\rangle - \left\langle\theta_i(1)\theta_j(2)\left\vert\frac{1}{r_{ij}}\right\vert\theta_j(1)\theta_i(2)\right\rangle\right]$
One spin orbital $\theta'_n \neq \theta_n$	$\langle\theta'_n(1)\vert H^0\vert\theta_n(1)\rangle$	$\sum_{i=1}^{n-1}\left[\left\langle\theta'_n(1)\theta_j(2)\left\vert\frac{1}{r_{ij}}\right\vert\theta_n(1)\theta_j(2)\right\rangle - \left\langle\theta'_n(1)\theta_j(2)\left\vert\frac{1}{r_{ij}}\right\vert\theta_j(1)\theta_n(2)\right\rangle\right]$
Two spin orbitals $\theta'_n \neq \theta_n$ $\theta'_{n-1} \neq \theta_{n-1}$	0	$\left\langle\theta'_n(1)\theta'_{n-1}(2)\left\vert\frac{1}{r_{ij}}\right\vert\theta_n(1)\theta_{n-1}(2)\right\rangle - \left\langle\theta'_n(1)\theta'_{n-1}(2)\left\vert\frac{1}{r_{ij}}\right\vert\theta_{n-1}(1)\theta_n(2)\right\rangle$
Three or more spin orbitals	0	0

$\left\langle\theta_i(1)\theta_j(2)\left\vert\frac{1}{r_{ij}}\right\vert\theta_r(1)\theta_s(2)\right\rangle$ where θ's are spin orbitals, i.e., $\theta_i(1)=\phi_i(1)\sigma_i(1)$, which is already defined. 1 and 2 represent any two electrons, and i, j, r, and s are subscripts representing the orbitals, which may be same, or partly or wholly different.

For $D' = D$, the Condon–Slater rules take the form as depicted in Table 11.6.

Now, we shall use the value of $\theta_i = \Phi_i\sigma_i$ and apply to the integral $\left(\theta_i(1)\vert H^0\vert\theta_i(1)\right)$, and it can be written as

$$\langle\theta_i(1)\vert H^0\vert\theta_i(1)\rangle = \langle\theta_i(1)\vert H^0\vert\Phi_i(1)\rangle\cdot\langle\sigma_i(1)\vert\sigma_i(1)\rangle$$

$$= \langle\theta_i(1)\vert H^0\vert\Phi_i(1)\rangle \quad [\because \sigma_i \text{ is normalised}]$$

Similarly, the integral

$$\left\langle\theta_i(1)\theta_j(2)\left\vert\frac{1}{r_{ij}}\right\vert\theta_r(1)\theta_s(2)\right\rangle$$
$$= \left\langle\Phi_i(1)\Phi_j(2)\left\vert\frac{1}{r_{ij}}\right\vert\Phi_r(1)\Phi_s(2)\right\rangle\cdot\langle\sigma_i(1)\vert\sigma_r(1)\rangle\cdot\langle\sigma_i(2)\vert\sigma_s(2)\rangle \tag{11.138}$$

It has already been pointed that σ's can be either α or β only, and they are orthonormal.

The integral will vanish otherwise unity. So the integral containing spatial part will be equal to

$$\left\langle\Phi_i(1)\Phi_j(2)\left\vert\frac{1}{r_{ij}}\right\vert\Phi_r(1)\Phi_s(s)\right\rangle \tag{11.139}$$

Then, the integral $\left\langle D' \left| \frac{1}{r_{ij}} \right| D \right\rangle$

$$= \sum_{i=1}^{n-1} \sum_{j>1} \left\langle \Phi_i(1)\Phi_j(2) \left| \frac{1}{r_{ij}} \right| \Phi_i(1)\Phi_j(2) \right\rangle - \delta_{ij} \left\langle \Phi_i(1)\Phi_j(2) \left| \frac{1}{r_{ij}} \right| \Phi_r(1)\Phi_s(2) \right\rangle \quad (11.140)$$

where δ_{ij} = Kronecker delta = 1, if same spin

= 0, if different spin

$$\therefore \langle D|H|D \rangle = \sum_{i=1}^{n} \langle \Phi_i(1)|H^0|\Phi_i(1) \rangle + \sum_{j=1}^{n=1} \sum_{j} (J_{ij} - \delta_{ij} K_{ij}) \quad (11.141)$$

$$\text{where } J_{ij} = \left\langle \Phi_i(1)\Phi_j(2) \left| \frac{1}{r_{ij}} \right| \Phi_i(1)\Phi_j(2) \right\rangle \quad (11.142)$$

$$\text{and} \quad K_{ij} = \left\langle \Phi_i(1)\Phi_j(2) \left| \frac{1}{r_{ij}} \right| \Phi_j(1)\Phi_i(2) \right\rangle \quad (11.143)$$

Let us now consider the $1s^2 2s^2$ configuration, where we have $n = 4$ and the two different spatial orbitals are 1s and 2s, where $\Phi_1 = 1s$ and $\Phi_2 = 2s$.

The spin orbitals are expressed as $\theta_1 = 1s$ and $\theta_2 = 1\overline{s}$, $\theta_3 = 2s$ and $\theta_4 = 2s$.

$$\therefore \sum_{n=1}^{4} \langle \Phi_i(1)|H^0|\Phi_i(1) \rangle = 2\langle 1s|H^0|1s \rangle + 2\langle 2s|H^0|2s \rangle \quad [\because e = 2 \text{ in each orbital}] \quad (11.144)$$

$$\therefore \quad \sum_{i=1}^{3} \sum_{j>1} (J_{ij} - \delta_{ij} K_{ij}) = J_{12} + J_{13} + J_{14} + J_{23} + J_{24} + J_{34} - K_{13} - K_{24}$$

$$= J_{1s1s} + J_{1s2s} + J_{1s2s} + J_{1s2s} + J_{1s2s} - K_{1s2s} - K_{1s2s} \quad (11.145)$$

$$\therefore \quad E = J_{1s1s} + J_{2s2s} + 4J_{1s2s} - 2K_{1s2s}$$

This is the value of energy for $1s^2 2s^2$ configuration of Be.

11.19 Koopman's Theorem

In spite of the fact that we cannot have the total electronic energy by the sum of SCF one electron energies, it is still possible to establish a connection between the $\epsilon_i'^s$ and the physical measurements. The value of

$$\epsilon_i = H_{ii} + \sum_{J=1}^{n} (2J_{ij} - K_{ij}) \quad (11.146)$$

where H_{ii} = average kinetic plus nuclear–electronic attraction energy for the electron in Φ_i ϵ_i = orbital energy or one-electron energy and

$\sum_{J=1}^{n} (2J_{ij} - K_{ij})$ = sum of the coulomb and exchange integrals, which contain all the electronic interaction energy.

If we make certain assumptions, it will be possible to equate the orbital energies and molecular ionisation energies. This fact is related to the important theorem, i.e., *Koopman's theorem*. Koopman has proved that if one electron is removed from Φ_k, a wave function will be obtained, or if one electron is added to the occupied molecular orbital Φj, a Hartree–Fock wave function will be stable with regard to any subsequent variation in Φ_k or Φ_j.

It should be kept in mind that this neglects the question of subsequent variation of all the molecular orbitals Φ with unaltered occupations. It is not essentially true that they remain in optimum condition because of the fact that the potential experienced by them is altered by the removal or addition of an electron. However, Koopman's theorem puts forward a model. The theorem suggests that the wave function is approximated for a positive ion just by removing an electron from one of the occupied Hartree–Fock molecular orbitals for a neutral molecule but without reoptimising any of the molecular orbitals. Let us now work on it, and we shall compare the electronic energies for the two wave functions.

We know that the neutral molecule is assumed to be a closed-shell system and the energy of the system is written as

$$E = \sum_i \left[2H_{ii} + \sum_j \left(2J_{ij} - K_{ij}\right) \right] \tag{11.147}$$

Suppose one electron is removed from Φ_k, then the energy will be expressed as

$$E_k^+ = \sum_{i \neq k} \left[2H_{ii} + \sum_{j \neq k} \left(2J_{ij} - K_{ij}\right) + H_{kk} + \sum_{i \neq k} \left(2J_{ik} - K_{ik}\right) \right] \tag{11.148}$$

In Eq. (11.148), the first term represents the total electronic energy due to all but the unpaired electron in Φ_k.

$$H_{ii} = \text{kinetic and nuclear attraction energies for unpaired electron}$$

The final sum represents the exchange energy and repulsion energy between this electron and all others. The last sum in Eq. (11.148) is actually the void generated in the first sum due to the condition $j = i$. Hence, we can couple these by removing the index restriction and removing the last sum. Finally, we can write

$$E_k^+ = \sum_{i \neq k} \left[2H_{ii} + \sum_j \left(2J_{ij} - K_{ij}\right) + H_{kk} \right] \tag{11.149}$$

We are going to remove the remaining index restriction for comparing E of Eq. (11.147) with Eq. (11.148). This is done by putting i equal to i in the sum and subtracting at the same time the new terms hence produced, which is expressed as

$$E_k^+ = \sum_i \left[2H_{ii} + \sum_j \left(2J_{ij} - K_{ij}\right) - H_{kk} - \sum_j \left(2J_{kj} - K_{kj}\right) \right] \tag{11.150}$$

But on the basis of Eqs. (11.146) and (11.147), we can write

$$E - \epsilon_i = \sum_i \left[2H_{ii} + \sum_j \left(2J_{ij} - K_{ij}\right) - H_{ii} - \sum_{J=1}^{n} \left(2J_{ij} - K_{ij}\right) \right]$$

$$\text{or, } E - \epsilon_i = \sum_i \left[2H_{ii} + \sum_j \left(2J_{ij} - K_{ij}\right) \right] - H_{kk} - \sum_{J=1}^{n} \left(2J_{kj} - K_{kj}\right) \text{ (by putting } i = k \text{ in the last part)} \tag{11.151}$$

Hence, the ionisation energy I_k^0 for ionisation from Φ_k is written as

$$I_k^0 = E_k^+ - E = -\in_k \tag{11.152}$$

This is *Koopman's theorem.* Hence, the negative of the orbital energies of the kth electron is called *Koopman's theorem.*

Alternative Method

We are acquainted with the solutions of the Hartree–Fock equation. Actually, its solutions are orthogonal.

Let us suppose that $u_i(1)$ and $u_k(1)$ are two solutions of the Hartree–Fock equation, and therefore, we can write

$$H_1^0 u_i(1) + \left[\sum_j \int u_j^*(2) V_{12} u_j(2) d^3r_2\right] u_i(1) - \sum_j \left[\int u_j^*(2) V_{12} u_i(2) d^3r_2\right] u_j(1) \tag{11.153}$$
$$= E_i u_i(1)$$

$$H_1^0 u_k(1) + \left[\sum_j \int u_j^*(2) V_{12} u_j(2) d^3r_2\right] u_k(1) - \left[u_j^*(2) V_{12} u_k(2) d^3r_2\right] u_j(1) \tag{11.154}$$
$$= E_k u_k(1)$$

where $V_{12} = e^2/r_{12}$. Multiplying Eq. (11.153) by d $u_k^*(1)$ and then integrating, we get

$$\langle u_k(1)|H_1^0|u_i(1)\rangle + \sum_j \langle u_k(1)u_j(2)|V_{12}|u_i(1)u_j(2)\rangle - \sum_j \langle u_k(1)u_j(2)|V_{12}|u_j(1)u_i(2)\rangle \tag{11.155}$$
$$= E_i \langle u_k(1)u_i(1)\rangle$$

Now, we shall take the complex conjugate of Eq. (11.154) and after multiplying by $u_i(1)$, the equation will be integrated. The result will be that Eq. (11.155) with E_i is replaced by E_k. Thus, we can express

$$E_i \langle u_k(1)|u_i(1)\rangle = E_k \langle u_k(1)|u_i(1)\rangle \tag{11.156}$$

$$\text{or,} \quad \langle u_k|u_i\rangle = 0 \text{ for } E_i \neq E_k \tag{11.157}$$

The total energy for the system is given by

$$E = \langle \Psi|H|\Psi\rangle = \sum_i I_i + \sum_{i<j} [J_{ij} - K_{ij}] \tag{11.158}$$

where $I_i = \langle u_i|H^0|u_i\rangle = \langle \Phi_i|H^0|\Phi_i\rangle$

Φ = spatial function

$$J_{ij} = \langle u_i(1)u_j(2)|V_{12}|u_i(1)u_j(2)\rangle = \langle \Phi_i(1)\Phi_j(2)|V_{12}|\Phi_i(1)\Phi_j(2)\rangle \tag{11.159}$$

$$k_{ij} = \langle u_i(1)u_j(2)|V_{12}|u_j(1)u_i(2)\rangle = \delta\langle m_s^i, m_s^j\rangle(\Phi_i(1)\Phi_j(2)|V_{12}|\Phi_j(1)\Phi_i(2)\rangle$$

Multiplying Eq. (11.153) by u^* (1) and then integrating, we shall obtain

$$E_i = I_i + \sum_i [J_{ij} - K_{ij}] \tag{11.160}$$

$$\text{or, } \sum_i E_i = \sum_i I_i + \sum_{i,j}[J_{ij} - K_{ij}] = \sum_i I_i + 2\sum_{i<j}[J_{ij} - K_{ij}] \tag{11.161}$$

From, Eqs. (11.158) and (11.161), we can write

$$E = \sum_i E_i - \sum_{i<j}[J_{ij} - K_{ij}] \neq \sum_i E_i \tag{11.162}$$

It is clear from this that the total energy E is not the sum of E_i, i.e., single-electron energies.

Let us now consider the difference in energy between a system containing $N+1$ electrons and a system having N electrons. Let the orbitals be labelled as $j = 1, 2, \ldots, n, i$. Then, Eq. (11.158) yields

$$E_{N+1} = \sum_{j=1}^{N} I_j + \sum_{j<k}^{N}[(J_{jk} - K_{jk})] + I_i + \sum_{k}^{N}[(J_{ik} - K_{ik})] \tag{11.163}$$

$$E_{N+1} - E_N = I_i + \sum_{k}^{N}[(J_{ik} - K_{ik})] = E_i \tag{11.164}$$

Equation (11.164) is called *Koopman's theorem.* It strongly suggests the truth that E_r is the energy necessary to remove an electron from the ith orbital.

11.20 Brillouin's Theorem

Let Y' = n-electron wave function with excitation of an electron from the orbital u_k to u'_k. Then,

$$\Psi = (N!)^{-1/2}\det\{u_i, \ldots, u_k \ldots, u_n\}$$

$$\Psi' = (N!)^{-1/2}\det\{u_i, \ldots, u'_k \ldots, u_n\}$$

and

$$\langle \Psi'|H|\Psi'\rangle = \langle u'_k(1)|H_1^0|u_k(1)\rangle + \sum_{p\neq k}\langle u'_k(1)u_p(2)|V_{12}|u_k(1)u_p(2)\rangle - \sum_{p\neq k}\langle u'_k(1)u_p(2)|V_{12}|u_p(2)u_k(1)\rangle$$

$$= \int u'_k(1)\left[H_1^0 + \sum_{i\neq k}\left\{\int u_p^*(2)V_{12}u_p(2)\mathrm{d}^3r_2\right\}u_k(1) - \sum_{p\neq k}\left\{\int u_p^*(2)V_{12}u_k(2)\mathrm{d}^3r_2\right\}u_p(1)\right]d^3r_1 \tag{11.165}$$

For orbitals satisfying the Hartree–Fock equation, i.e.,

$$\left[H_1^0u_a(1) + \sum_{j\neq i}\left\{\int u_j^*(2)(e^2/r_{12})u_j(2)d^3r_2\right\}u_i(1) - \sum_{j\neq i}\left\{u_j^*(2)(e^2/r_{12})u_i(2)d^3r_2\right\}u_i(1)\right]$$

$$= E_iu_i(1)$$

gives

$$\langle \Psi'|H|\Psi'\rangle = \int u_k^*(1)E_ku_k(1)d^3r_1 = 0 \tag{11.166}$$

Thus, the matrix elements between a state described by an n-electron Slater determinant constituted by the Hartree–Fock orbitals and a singly excited-state vanish. This is *Brillouin's theorem.*

11.21 Roothaan's Equations: The Matrix Solution of the Hartree–Fock Equation

The Hartree–Fock method is really simple one, which can be implemented for atoms for solving the Hartree–Fock equations numerically for the spin orbitals. However, the numerical solution for spin orbitals is complex for molecular systems.

Roothaan (1951) suggested to use a known set of basis functions for the purpose of expanding the spin orbitals. We shall illustrate how his suggestion transforms the coupled Hartree–Fock equations into a matrix problem, which can be solved by matrix manipulations.

We are going to start with equation

$$\left[H_1 + \sum (2J_r - K_r)\right]\psi_s(1) = \varepsilon_s \psi_s(1)$$

which is the Hartree–Fock equation for a space orbital and ψ_s is occupied by electrons 1. Here, the spatial function ψ_a (1) is occupied by electron 1, and hence,

$$f_1 \psi_a(1) = \varepsilon_a \psi_\alpha(1) \tag{11.167}$$

where f_1 = Fock operator represented in terms of spatial wave function

Now the aforementioned equation changes to

$$f_1 = H_1 + \sum_u \left[2J_u(1) - K_u(1)\right] \text{ with the change of notation.} \tag{11.168}$$

The coulomb and exchange operators are defined by the following equations completely in terms of spatial coordinates.

$$J_r \psi_s(1) = \left[\int \psi_r^*(2)\left(\frac{e^2}{r_{12}}\right)\psi_r(2) d\tau_2\right]\psi_s(1) \tag{11.169}$$

$$K_r \psi_s(1) = \left[\int \psi_r^*(2)\left(\frac{e^2}{r_{12}}\right)\psi_s(2) d\tau_2\right]\psi_s(1) \tag{11.170}$$

Let θ_j = a set of M basis functions,

ψ_i = spatial wave functions, and

ψ_i may be expressed as a linear combination of the basis functions, which is

$$\psi_i = \sum_{J=1}^{M} C_{ji}\theta_j \tag{11.171}$$

where C_{ji} is an unknown coefficient. It should be understood that from a set of M basis functions, M linearly independent spatial functions can be obtained, and the problem of estimating the wave functions has been transformed to one of calculating the coefficients C_{ji}.

Now, we shall substitute the expansion of equation

$$\psi_i = \sum_{J=1}^{M} C_{ji}\theta_J$$

into equation $f_1\psi_a(1) = \varepsilon_a\psi_a(1)$ and obtain

$$f_1\sum C_{ja}\theta_j(1) = \varepsilon_a C_{ja}\theta_j(1) \tag{11.172}$$

Multiplying Eq. (11.172) by θ_i^* and integrating over r_1, we shall get

$$\sum_{j=1}^{M} C_{ja}\int \theta_i^*(1) f_1\theta_j(1)\mathrm{d}r_1 = \varepsilon_a \sum_{r=1}^{M} C_{ja}\int \theta_i^*(1)\theta_j(1)\mathrm{d}r_1 \tag{11.173}$$

After introducing a more compact notation, the structure of a set of equations becomes clear. In this case, overlap matrix S is introduced with the following elements:

$$S_{ij} = \int \theta_i^*(1)\theta_j(1)\mathrm{d}r_1 \tag{11.174}$$

and the Fock matrix F is represented with elements as

$$F_{ij} = \int \theta_i^*(1) f_1\theta_j(1)\mathrm{d}r_1 \tag{11.175}$$

Then Eq. (11.173) will take the form

$$\sum_{j=1}^{M} F_{ij}C_{ja} = \varepsilon_a \sum_{j=1}^{M} S_{ij}C_{ja} \tag{11.176}$$

where the terms have been written as per Eq. (11.173). Equation (11.176) is a set of M simultaneous equations, which are called *Roothaan's equations*,

Equation (11.176) may be expressed as a single matrix equation as given in the following:

$$FC = Sc\varepsilon \tag{11.177}$$

where C is an $M \times M$ matrix constituted by the elements C_{ja} and ε is an $M \times M$ diagonal matrix having orbital energy ε_a.

It must be noted that Roothaan's equations will have a nontrivial solution only if the secular equation mentioned in the following is satisfied:

$$\det|F - \varepsilon_a S| = 0 \tag{11.178}$$

For its solution, we must adopt the SCF approach, getting with each iteration, a new set of coefficients C_{ja} and continue until a convergence criterion is obtained or reached.

The explicit/expressly stated form of the matrix element F_{ij} is got from Eqs. (11.168), (11.169), and (11.170), and this is

$$F_{ij} = \int \theta_i^*(1)H_1\theta_j(1)\mathrm{d}r_1 + 2j_0\sum_u \int \theta_i^*(1)\psi_u^*(2)\cdot\frac{1}{r_{12}}\psi_u(2)\theta_j(1)\mathrm{d}r_1\mathrm{d}r_2$$
$$- j_0\sum_u \int \theta_i^*(1)\psi_u^*(2)\frac{1}{r_{12}}\theta_j(2)\psi_u(1)\mathrm{d}r_1\mathrm{d}r_2 \tag{11.179}$$

It should be kept in mind that the first term on the RHS is a one-electron integral, which will be represented by h_{ij}. Putting the expansion in Eq. (11.171) gives the following expression for F_{ij}, but completely in terms of integrals over the unknown basis function.

$$F_{ij} = h_{ij} + 2j_0 \sum_{u.l.m} C^*_{lu} C_{mu} \int \theta^*_i(1)\theta^*_j(2) \cdot \frac{1}{r_{12}} \theta_m(2)\theta_j(1) dr_1 dr_2$$

$$- j_0 \sum_{u.l.m} C^*_{lu} C_{mu} \int \theta^*_i(1)\theta^*_j(2) \cdot \frac{1}{r_{12}} \theta_j(2)\theta_m(1) dr_1 dr_2 \quad (11.180)$$

Equation (11.180) can be made easy by introducing the notation given in the following for the two-electron integrals over the basis functions.

$$(ab/cd) = j_0 \int \theta^*_a(1)\theta_b(1)\frac{1}{r_{12}}\theta^*_c(2)\theta_d(2) dr_1 dr_2 \quad (11.181)$$

Equation (11.180) then takes the form

$$F_{ij} = H_{ij} + \sum_{u,l,m} C^*_{lu} C_{mu} \left[2(ij/lm) - (im/lj) \right] \quad (11.182)$$

This equation is usually expressed as

$$F_{ij} = H_{ij} + \sum_{lm} P_{lm} \left[(ij/lm) - \frac{1}{2}(im/ij) \right] \quad (11.183)$$

$$\text{where, } P_{lm} = 2\sum_{u} C^*_{lu} C_{mu} \quad (11.184)$$

The matrix element P_{lm} is called density matrix elements and constitutes total electron density in the region of θ_l and θ_m.

It must be remembered that one-electron matrix element h_{ij} requires to be found out only once due to the fact that it remains unaltered during every iteration. However, p_{lm} is dependent on C_{lu} and C_{mu}, and there is no need to re-evaluate at each iteration.

BIBLIOGRAPHY

Atkins, P.W. 1990. *Physical Chemistry*, 4th ed. New York: Freeman.
Coulson, C.A. 1961. *Valence*, 2nd ed. London: Oxford University Press.
Cowan, R.D. 1981. *The Theory of Atomic Structure and Spectra*. Berkeley: University of California Press.
Hartree, D.R. 1928. 'The wave mechanics of an atom with a non-Coulomb central field.' *Proc. Cambridge Phil. Soc*. 24: 89.
Levine, I.N. 2000. *Quantum Chemistry*, 4th ed. New Jersey: Prentice Hall.
Lowe, J.P. 1993. *Quantum Chemistry*, 2nd ed. New York: Academic Press.
Mulliken, R.S. and W.C. Ermler. 1981. *Polyatomic Molecules*. New York: Academic Press.
Pillar, F.L. 1990. *Elementary Quantum Chemistry*. New York: McGraw-Hill.
Pople, J.A., M. Head-Gordon, D.J. Fox, K. Raghavachari and L.A. Curtiss. 1989. 'Gaussian-1 theory: A general procedure for prediction of molecular energies.' *J. Chem. Phys*. 90: 5622.
Roothaan, C.C.J. 1951. 'New developments in molecular orbital theory.' *Rev. Mod. Phys*. 23: 6989.
Simons, J. and J. Nichols. 1997. *Quantum Mechanics in Chemistry*. New York: Oxford University Press.
Slater, J.C. 1930. 'Note on Hartree's method.' *Phys. Rev*. 35: 210.

Solved Problems

Problem 1. How will you show that $\psi(1, 2) = 1s(1)\alpha(1)1s(2)\beta(2) - 1s(2)\alpha(2)1s(1)\beta(1)$ is an Eigen function of the operator S_z? if so, what will be its Eigen value M_s?

Solution: Given that $\hat{S}_z = \sum_i \hat{S}_z \cdot i$ and $\psi(1,2) = 1s(1)\alpha(1)1s(2)\beta(2) - 1s(2)\alpha(2)1s(1)\beta(1)$

From above

$$\hat{S}_z = \hat{S}_z(1) + \hat{S}_z(2), \text{ where } \hat{S}_z(i) \text{ acts only on electron } i \text{ and } \hat{S}_z(1) = \hat{S}_z(2) = \hbar/2$$

$$\hat{S}_z\Psi(1,2) = \left\{\hat{S}_z(1) + \hat{S}_z(2)\right\}\left\{1s(1)\alpha(1)1s(2)\beta(2) - 1s(2)\alpha(2)1s(1)\beta(1)\right\}$$

or, $\hat{S}_z\Psi(1,2) =$

$$\uparrow \hat{S}_z(1)\left[1s(1)\alpha(1)1s(2)\beta(2) - 1s(2)\alpha(2)1s(1)\beta(1)\right]$$

$$+ \hat{S}_z(2)\left[1s(1)\alpha(1)1s(2)\beta(2) - 1s(2)\alpha(2)1s(1)\beta(1)\right]$$

$$= \hbar/2\left[1s(1)\alpha(1)1s(2)\beta(2)\right] + \frac{\hbar}{2}\left[1s(2)\alpha(2)1s(1)\beta(1)\right]$$

$$+ \hbar/2\left[1s(1)\alpha(1)1s(2)\beta(2)\right] + \frac{\hbar}{2}\left[1s(2)\alpha(2)1s(1)\beta(1)\right]$$

$$= \left(\frac{\hbar}{2} - \frac{\hbar}{2}\right)\left[1s(1)\alpha(1)1s(2)\alpha(2) - 1s(2)\alpha(2)1s(1)\beta(1)\right]$$

$$= 0 \times \Psi(1,2) = 0$$

$$\therefore \hat{S}_z = 0$$

This results shows that the wave function is an Eigen function of $\hat{S}_z$ having $M_s = 0$. Answer.

Problem 2. Find all possible term symbols (without neglecting any due to Pauli's exclusion considerations) for d^2 ground electronic configuration.

Solution: Each d^2 electron has $l = 2$.

$$\therefore L = l_1 + l_2 = 2 + 2 = 4$$

$$= l_1 + l_2 - 1 = 2 + 2 - 1 = 3$$

$$= l_1 + l_2 - 2 = 2 + 2 - 2 = 2$$

$$= l_1 + l_2 - 3 = 2 + 2 - 3 = 1$$

$$|l_1 + l_2| = |2 - 2| = 0$$

Thus, the value of $L = 4$, 3, 2, 1, and 0.

These correspond to possible existence of the following states:

L	0	1	2	3	4
States	S	P	D	F	G

Now, the possible values of $S = \pm 1/2$.

$$\therefore \quad S = s_1 + s_2 = 1/2 + 1/2 = 1$$

$$= s_1 - s_2 = 1/2 - 1/2 = 0$$

Thus, $S \rightarrow 1$ and 0.

$$\therefore \quad \text{Multiplicity} = 2S + 1 = 2 \times 1 + 1 = 3$$

$$= 2S + 1 = 2 \times 0 + 1 = 1$$

Thus, $2S + 1 = 3$ and 1.

Combining the multiplicities with the L values, the possible term symbols will be ^3G, ^1G, ^3F, ^1F, ^3D, ^1D, ^3P, ^1P, ^3S, and ^1S

Pauli's exclusion principle and redundancy will eliminate the ^3S, ^3D, ^3G, ^1P, ^1F term symbols for the atoms, leaving ^1G, ^3F, ^3P, and ^1S Answer.

Problem 3. Find the total number of states in an atom with electronic configuration d^2.

Solution: See Problem 2. We have found the term symbols for d^2 configurations. If we shall include the J values, the following term symbols will come into existence.

$$^1S_0, ^1D_2, ^1G_4, ^3P_2, ^3P_1, ^3P_0, ^3F_4, ^3F_3, s\, ^3F_2$$

The degeneracy will be calculated by $2J + 1$ formula:

Term Symbol	Degeneracy
$^1S_0$	1
$^1D_2$	5
$^1G_4$	9
$^3P_2$	5
$^3P_1$	3
$^3P_0$	1
$^3F_4$	9
$^3F_3$	7
$^3F_2$	5

Thus, the total degeneracy will be 1, 5, 9, 5, 3, 1, 9, 7, and 5 = 45 Answer.

Problem 4. What values of J are consistent with the terms ^2P and ^3D? How many states with different values of J correspond to each?

Solution: The quantum number J can have $J = L + S,\ L + S - 1L + S - 2, \ldots |L - S|$ values. For ^2P term, $L = 1$ and $S = 1/2$. Hence, $J = 3/2$ and $1/2$ and the $J + 1$ values will be $(2 \times 3/2 + 1)$ and $(2 \times 1/2 + 1)$, i.e., 4 and 2 states, respectively.

For ^3D term, $L = 2$ and $S = 1$.

$$\therefore \; J = L + S = 2 + 1 = 3$$

$$L + S - 1 = 2 + 1 - 1 = 2$$

$$|L - S| = |2 - 1| = 1$$

Thus, J has 3, 2, and 1 values.

$$\therefore\ 2J+1=2\times3+1=7$$
$$=2\times2+1=5$$
$$=2\times1+1=3$$

Thus, the degeneracy = 7 + 5 + 3 = 15 Answer.

Problem 5. Find the L, S, and J values for coupling between an electron of p orbital and that of f orbital.

Solution: According to the question, the configuration is p' and f'.

$$\therefore\ \text{Total } S=1/2+1/2=1$$
$$=1/2-1/2=0$$

Thus, S has values 1 and 0.

$$l_1 \text{ for } p1 \text{ and } l_2 \text{ for } f=3$$
$$\therefore\ L=l_1+l_2=1+3=4$$
$$l_1+l_2-1=4-1=3$$
$$|l_1-l_2|=|3-1|=2$$

Thus, L = 4, 3, and 2.

Hence, this corresponds to letters D, F, and G. The states can be written as ^3D_j, ^3F_j, ^3G_j, ^1D_j, ^1F_j, and ^1G_j. The values of J for each value of L can be calculated in the following manner:

i. When $L = 2$, $S = 1$
   $J = 2 + 1 = 3$
   $2 + 1 - 1 = 2$
   $|2 - 1| = 1$

Similarly,

ii. For $L = 2$, S = 0, $J = 2 + 0 = 2$
iii. For $L = 3$, $S = 0$, J = 4, 3, and 2
iv. For $L = 3$, $S = 0$, $J = 3$
v. For $L = 4$, $S = 1$, J = 5, 4, and 3
vi. For $L = 4$, $S = 0$, $J = 4 + 0 = 4$

Therefore, the term symbols will be
$^3D_3$, $^3D_2$, $^3D_1$, $^1D_2$, $^3F_4$, $^3F_3$, $^3F_2$, $^1F_3$, $^3G_5$, $^3G_4$, $^3G_3$, $^1G_4$ Answer.

Problem 6. Find the term symbol for the electronic configuration $2p^1\,3p^1\,3d^1$ system.

Solution: Given that the system is $2p^1\,3p^1$ and $3d^1$.
In the aforementioned system, first consider the configuration $2p^1\,3p^1$. For these two $l_1 = 1$ and $l_2 = 1$.

$$\therefore\ L=l_1+l_2=1+1=2$$
$$l_1+l_2-1=1+1-1=1$$
$$=|l_1-l_2|=|1-1|=0$$

The total spin quantum number will be

$$S = 1/2 + 1/2 = 1$$

$$= 1/2 - 1/2 = 0$$

And the corresponding $2S + 1$ values will be obtained by putting the values of S in $2S + 1$ formula

$$\therefore 2S + 1 = 2 \times 1 + 1 = 3$$

$$= 2 \times 0 + 1 = 1$$

$\therefore$ The term symbol for the coupling of $2p^1\,3p^1$ electrons will be ^3D, ^3P, ^3S and ^1D, ^1P, ^1S.

Further, each of six states will couple with $3d^1$ electron to yield the resultant L and S. For example, $^3D^3d^1$ gives

$$l_1 = 2(\text{for } D), l_2 = 2(\text{for } d^1)$$

$$\therefore \; L = l_l + l_2 = 2 + 2 = 4$$

$$= l_1 + l_2 - 1 = 2 + 2 - 1 = 3$$

$$= l_l + l_2 - 2 = 2 + 2 - 2 = 2$$

$$= l_l + l_2 - 3 = 2 + 2 - 3 = 1$$

$$= |l_l - l_2| - |2 - 2| = 0$$

$$\text{and } \; S = 1 + 1/2 = 3/2, 2S + 1 = 2 \times 2/3 + 1 = 4$$

$$= 1 - 1/2 = 1/2, 2S + 1 = 2 \times 1/2 + 1 = 2$$

The term symbols for coupling for $^3D\,^3d^1$ configuration will be ^4G, ^4F, ^4D, ^4P, ^4S and ^2G, ^2F, ^2D, ^2P, ^2S. For $^3P\,3d^1$, the terms will be ^4F, ^4D, ^4P, ^2F, ^2D, ^2P. Similarly, for the coupling of $^1D\,3d^1$, the terms will be ^2G, ^2F, ^2D, ^2S; for ^1P, $3d^1$, the terms will be ^2F, ^2D, ^2P; and for $^1S\,3d^1$, the term will be ^2D. It will be observed that some of the term symbols occur more than once in different couplings, and they can be coupled to yield

$$^4G,^4F(2),^4D(3),^4P(2),^4S \text{ and}$$

$$^2G(2),^2F(4),^2D(6),^2P(4),^2S(2)$$

Problem 7. Show that a closed-shell electronic configuration is always represented by ^{1}S term.

Solution: In a closed-shell system, for every electron $s = +1/2$ and $s = -1/2$

$$\therefore \; S = +1/2 - 1/2 = 0$$

Similarly, $l_1 = +ml\,s\,l_2 = -ml$

$$\therefore \; L = l_1 + l_2 = \sum m_l = +m_l - m_l = 0$$

$$L = 0$$

$$\therefore \; J = 0 \text{ and } 2S + 1 = 1.$$

Hence, the closed-shell electronic configuration yields only one term ^1S and makes no contribution to L or S. Answer.

Problem 8. Find the Landé 'g' factor for the $^3P_1$ level in the $2p\ 3s$ configuration of the ^6C atom and use this result to predict the splitting of the level when the atom is kept in an external magnetic field of value 0.1 tesla.

Solution: According to question, for the $^3P_1$ state, $S = L = J = 1$, so, the Landé 'g' factor can be calculated by the formula

$$g = 1 + \frac{J(J+1) + S(S+1) - L(L+1)}{2J(J+1)}$$

$$= 1 + \frac{1(1+1) + 1(1+1) - 1(1+1)}{2 \times 1(1+1)}$$

$$= 1 + \frac{2+2-2}{2 \times 2}$$

$$= 1 + \frac{2}{4} = 1 + \frac{1}{2} = \frac{3}{2} = 1.5$$

For $J = 1$, the possible values of $m_j' = -1, 0, +1$, so the level is split into three components, one having same energy and others displaced in energy by

$$\Delta E = \pm \mu_b H g m_j' = \pm 9.3 \times 10^{-24}\ \text{amp m}^2 \times 10^{-1}\ \text{tesla} \times 1.5$$

$$= \pm 1.4 \times 10^{-24}\ \text{joule}$$

$$= \pm 8.7 \times 10^{-6}\,\text{eV}$$ Answer.

Problem 9. Calculate the J values in the following Russell–Saunders terms: ^1P, ^2D, ^5D.

Solution: a. for ^1P

$$\because\ 2S + 1 = 1 \therefore 2S = 1 - 1 = 0 \quad \therefore S = 0$$

But $L = 1$

$$\therefore\ J = L + S,\ L + S - 1 \ldots |L - s| = 1$$

Hence, $^1P \rightarrow {}^1P_1$ Answer.

b. For ^2D

$$2S + 1 = 2 \ \text{ or } \ 2S = 2 - 1 = 1 \quad \therefore\ S = 1/2$$

But $L = 2$

$$\therefore\ J = L + S, L + S - 1 \ldots |L - S|$$

$$= 2 + 1/2 = 5/2$$

$$= 2 + 1/2 - 1 = 3/2$$

$\therefore\ {}^2D \rightarrow {}^2D_{5/2}, {}^2D_{3/2}$ Answer.

c. For ^5D

$$2S+1=5 \text{ or } 2S=5-1=4 \quad \therefore\ S=2$$

$$L=2$$

$$\therefore\ J=L+S=2+2=4$$

$$=L+S-1=2+2-1=3$$

$$=L+S-2=2+2-2=2$$

$$=L+S-3=2+2-3=1$$

$$=|L-S|=|2-2|=0$$

Therefore, $^5D = {}^5D_4, {}^5D_3, {}^5D_2, {}^5D_1$ and $^5D_0$ Answer.

Problem 10. The two nonequivalent electrons with the configuration p^1d^1 are taken into consideration. Assuming the Russell–Saunders coupling scheme, derive the various terms arising from the given configuration and find out the total angular momentum associated with each term.

Solution: For *p* electrons, $l_1 = 1$, and for *d* electrons, $l_2 = 2$.
i.e., $L - l_1 + l_2 - 1 + 2 = 3$

$$l_1+l_2-1=2$$

$$=|l_1-l_2|=|2-1|=1$$

Thus, $L \rightarrow 3, 2, 1$, which denotes the letters *F*, *D*, and *P*, respectively.

Now, for the two nonequivalent electrons, the spin may be either parallel or antiparallel.

$$\text{i.e., } S=s_1+s_2=\frac{1}{2}+\frac{1}{2}=1$$

$$s_1-s_2=\frac{1}{2}-\frac{1}{2}=0$$

Hence, multiplicity $= 2S + 1 = 3$ or 1.

Thus, we find triplet and singlet terms:

$$^3F, {}^3D, {}^3P, {}^1F, {}^1D, {}^1P$$

Now $J=L+S$

$$=3+1=4$$

$$=3+1-1=3$$

$$=|L-S|=|3-1|=2$$

Thus, $J \rightarrow 4, 3, 2$. This is for $S = 1$

For $S=0$ $J=L+S=3+0=3$

$$=L+S-1=3-1=2$$

$$=L+S-2=3-2=1$$

Thus, with these *S* and *J* values, the complete term symbol can be written. Answer.

Problem 11. Verify the Landé interval rule assuming that sublevels corresponding to $J = 5/2$, $7/2$, and $9/2$ are at 225, 560, and 960 cm^{-1}. Suppose that J = 3/2 sublevel is at 0 cm^{-1}.

Solution: According to Landé interval rule,

$$\Delta E = E_J - E_{J-1} = J\hbar^2, \text{ i.e., } \Delta E \propto \text{J}$$

In the present problem,

$$E_{\frac{5}{2}} - E_{\frac{3}{2}} = 225 - 0 = 225 \text{ cm}^{-1}$$

$$= \frac{225}{\frac{5}{2}} = 90 \text{ cm}^{-1}$$

$$E_{\frac{7}{2}} - E_{\frac{5}{2}} = 560 - 225 = 335 \text{ cm}^{-1}$$

$$= \frac{335}{7/2} = 95.7 \text{ cm}^{-1}$$

$$E_{\frac{9}{2}} - E_{\frac{7}{2}} = 960 - 560 = 400 \text{ cm}^{-1}$$

$$= \frac{400}{\frac{9}{2}} = 89 \text{ cm}^{-1}$$

We find that there is a fair consistency in the interval values between the successive pairs of sublevels differing by $\Delta J = 1$. Hence, the interval rule is obeyed. Answer.

Problem 12. Show that $\Psi_1(1, 2) = \Psi(1, 2) + \Psi(2, 1) = 1s\alpha(1)1s\beta(2) + 1s\alpha(2)1s\beta(1)$ is symmetric when the two electrons are interchanged and is not acceptable as an electronic wave function.

Solution: Given that $\Psi_1(1, 2) = \Psi(1, 2) + \Psi(2, 1)$
On interchanging the two electrons, we find that

$$\Psi(1,2) = \Psi(2,1) + \Psi(1,2) = \Psi(1,2)$$

The Ψ is said to be symmetric under the interchange of the two electrons.

But the symmetric wave functions are not acceptable as electronic wave function according to Pauli's exclusion principle. Answer.

Problem 13. Show that the equation
$\text{E} = 2\sum_{j=1}^{N} I_j + \sum_{i=j}^{N}\sum_{j=1}^{N}(2J_{ij} - K_{ij})$ for a helium atom is the same as equation $E = I_1 + I_2 + J_{12}$.

Solution: Let us put $N = 1$ in the equation given in the problem for a helium atom. Applying equation mentioned earlier for E for a helium atom, the equation becomes

$$E = 2I_1 + 2J_{11} - K_{11}$$

From the definition of J_{ij} and K_{ij}, we can write

$$J_{ij} = \iint \Phi_i^*(r_1)\Phi_j^*(r_2)\frac{1}{r_{12}}\Phi_i(r_1)\Phi_j(r_2)\mathrm{d}r_1\,\mathrm{d}r_2 \text{ and } k_{ij}$$

$$= \iint \Phi_i^*(r_1)\Phi_j^*(r_2)\frac{1}{r_{12}}\Phi_i(r_2)\Phi_j(r_1)\mathrm{d}r_1\,\mathrm{d}r_2$$

We observe that $J_{11} = K_{11}$ and hence we have $E = 2I_1 + 2J_{11} - K_{11}$

$$= 2I_1 + 2K_{11} - K_{11}$$
$$= 2I_1 + K_{11}$$
$$= 2I_1 + J_{11}$$

This is exactly $E = I_1 + I_2 + J_{12}$ in different notation. Thus proved. Answer.

Problem 14. Which of the following functions are symmetric?

1. $\Phi(1)\alpha(1)\Phi(2)\alpha(2)$
2. $\Phi(1)\alpha(1)\Phi(2)\beta(2)$
3. $\Phi(1)\beta(1)\Phi(2)\beta(2)$
4. $\Phi(1)\alpha(2)\Phi(2)\beta(1)$

Choose the functions and construct symmetric and antisymmetric wave functions. Illustrate that they are mutually orthogonal.

Solution: According to question,

i. $Y(1,2) = \Phi(1)\alpha(1)\Phi(2)\alpha(2)$
   On interchanging the two electrons, we find that $Y(2,1) = \Phi(2)\alpha(2)\Phi(1)\alpha(1)$
   Thus, $Y(1,2) = Y(2,1)$
ii. $Y(1,2) = \Phi(1)\alpha(1)\Phi(2)\beta(2)$
   On interchanging the two electrons, we find that $Y(2,1) = \Phi(2)\alpha(2)\Phi(1)\alpha(1)$
   $\therefore\ Y(1,2) = Y(2,1)$. Hence, it is not symmetric.
iii. $\Psi(1,2) = \Phi(1)\beta(1)\Phi(2)\beta(2)$

$$\therefore\ \Psi(2,1) = \Phi(2)\beta(2)\Phi(1)\beta(1)$$
$$\therefore\ \Psi(1,2) = \Psi(2,1). \text{ Hence, symmetric.}$$

iv. $\Psi(1,2) = \phi(1)\alpha(2)\phi(2)\beta(1)$

$$\therefore \Psi(2,1) = \phi(2)\alpha(1)\phi(1)\beta(2)$$

$$\therefore\ \Psi(1,2) = \Psi(2,1)$$

Hence, they are not symmetric.

Thus, (i) and (iii) are symmetric.

Now, $\Psi_{\pm} = \frac{1}{\sqrt{2}}\left[\phi(1)\alpha(1)\phi(2)\beta(2) \pm \phi(1)\alpha(2)\phi(2)\beta(1)\right]$

Now we shall check the orthogonality.

$$\int\left[\phi(1)\alpha(1)\phi(2)\beta(2)+\phi(1)\alpha(2)\phi(2)\beta(1)\right]\left[\phi(1)\alpha(1)\phi(2)\beta(2)-\phi(1)\alpha(2)\phi(2)\beta(1)\right]d\tau$$

$$=\left[\int\left[\phi(1)\alpha(1)\phi(2)\beta(2)\right]^2 - \int\left[\phi(1)\alpha(2)\phi(2)\beta(1)\right]^2 d\tau\right]$$

$$=\int\phi(1)^2\,d\tau_1\cdot\int\phi(2)^2\,d\tau_2\cdot\int\alpha(1)^2\,ds_1\cdot\int\beta(2)^2\,ds_2$$

$$-\int\phi(1)^2\,d\tau_1\cdot\int\phi(2)^2\,d\tau_2\cdot\int\alpha(2)^2\,ds_2\cdot\int\beta(1)^2\,ds_1 = 0$$

Problem 15. Estimate the average values of r_1 and r_2 consistent with He atom wave function. $\Psi(1,2) = 1s(1)2s(2)$ and $\Psi(2,1) = 1s(2)2s(1)$. Try to comment on the result.

Solution: According to question, we can write

$$\bar{r}_1 = \frac{\int 1s(1)2s(2)\hat{r}_1 1s(1)2s(2)\mathrm{d}\tau_1\mathrm{d}\tau_2}{\int [1s(1)2s(2)]^2 \mathrm{d}\tau_1\mathrm{d}\tau_2}$$

$$= \frac{1s(1)\hat{r}_1 1s(1)\mathrm{d}\tau_1 \cdot \int 2s(2)\cdot 2s(2)\mathrm{d}\tau_2}{\int [1s(1)]^2 \mathrm{d}\tau_1 \int [2s(2)]^2 \mathrm{d}\tau_2}$$

$$= \frac{\int 1s(1)\hat{r}_1 1s(1)\mathrm{d}\tau_1}{\int [1s(1)]^2 \mathrm{d}\tau_1}$$

Similarly, we can write r_2, i.e.,

$$\bar{r}_2 = \frac{\int 2s(1)\hat{r}_2 2s(2)\mathrm{d}\tau_2}{\int [2s(2)]^2 \mathrm{d}\tau_2}$$

Now we know that

$$1s(1) = \left(\frac{z^3}{\pi}\right)^{1/2} e^{-zr_1} \text{ and } \mathrm{d}\tau_1 = r_1^2 \mathrm{d}r_1 \sin\theta_1 \mathrm{d}\theta_1 \mathrm{d}\Phi_1$$

We obtain

$$\bar{r}_1 = \frac{\int_0^\pi \left(\frac{z^3}{\pi}\right) e^{-2zr_1} r_1^3\, \mathrm{d}r_1 \int_0^\pi \sin\theta_1\, \mathrm{d}\theta_1 \int_0^\pi \mathrm{d}\Phi_1}{\int_0^\pi \left(\frac{z^3}{\pi}\right) e^{-2zr_1} r_1^2\, \mathrm{d}r_1 \int_0^\pi \sin\theta_1\, \mathrm{d}\theta_1 \int_0^\pi \mathrm{d}\Phi_1}$$

$$= \frac{3!}{(2z)^4}\cdot\frac{(2z)^3}{2!} = \frac{3!}{16z^4}\cdot\frac{8z^3}{2!} = \frac{1\cdot 2\cdot 3\ldots 8z^3}{16z^4\cdot 2\cdot 1} = \frac{1\cdot 3}{2\cdot 1\cdot z} = \frac{3}{2z} = \frac{3}{4} \text{ a.u. } (\because\ z = 2)$$

Furthermore, we shall use $2s$ (2), and it will be written as

$$2s(2) = \frac{1}{4}\left(\frac{z^3}{\pi}\right)^{1/2} \cdot (2 - zr_2)e^{-zr_2}$$

We obtain

$$\bar{r}_2 = \frac{\int_0^\infty \left[(2 - zr_2)e^{-zr_2}\right]^2 r_2^2 \, dr_2}{\int_0^\infty \left[e^{-zr_2}(2 - zr_2)\right]^2 r_2^2 \, dr_2}$$

$$= \frac{4\int_0^\infty e^{-2zr_2} r_2^3 dr_2 - 4z\int_0^\infty r_2^4 \cdot e^{-2zr_2} dr_2 + z^2\int_0^\infty r_2^5 \cdot e^{-2zr_2} dr_2}{4\int_0^\infty r_2^2 e^{-2zr_2} r_2^3 dr_2 - 4z\int_0^\infty r_2^3 \cdot e^{-2zr_2} dr_2 + z^2\int_0^\infty r_2^4 \cdot e^{-2zr_2} dr_2}$$

These are standard integrals, and their values can be written as

$$\bar{r}_2 = \frac{\frac{4\times 3!}{(2z)^4} - \frac{4z\times 4!}{(2z)^5} + \frac{z^2\times 3!}{(2z)^6}}{\frac{4\times 2!}{(2z)^3} - \frac{4z\times 3!}{(2z)^4} + \frac{z^2\times 4!}{(2z)^5}} = \frac{6}{z} = 3.0 \text{ a.u.} \quad [\because \; z = 2]$$

After interchange of electron between $1s$ and $2s$, we get

$$\Psi(2,1) = 1s(2)2s(1)$$

$$\therefore \; \bar{r}_1 = \frac{\int 2s(1) r_1 2s(1) d\tau_1}{\int \{2s(1)\}^2 d\tau_1} = 3 \text{ a.u.}$$

$$\text{and } \bar{r}_2 = \frac{\int 1s(2) r_2 1s(2) d\tau_2}{\int \{1s(2)\}^2 d\tau_2} = \frac{3}{2z} = \frac{3}{4} \text{ a.u.} \quad [z = 2]$$

On interchanging, the elections $\bar{r}_1$ and $\bar{r}_2$ of $1s$ and $2s$ electrons are altered. This does not have any sense as the electrons cannot be labeled. Answer.

Problem 16. Show for each row of the Slater determinants for Li, i.e.,

$$\begin{vmatrix} 1s_1\alpha & 1s_1\beta & 1s_1\alpha \\ 1s_2\alpha & 1s_2\beta & 1s_2\alpha \\ 1s_3\alpha & 1s_3\beta & 1s_3\alpha \end{vmatrix} \text{ or } \begin{vmatrix} 1s_1\alpha & 1s_1\beta & 1s_1\beta \\ 1s_2\alpha & 1s_2\beta & 1s_2\beta \\ 1s_3\alpha & 1s_3\beta & 1s_3\beta \end{vmatrix}$$

That the wave function represented by the determinant violates Pauli's exclusion principle.

Solution: We shall list the set of four quantum numbers for each spin orbital, by row as

$$\begin{vmatrix} 1s_1\alpha & 1s_1\beta & 1s_1\alpha \\ 1s_2\alpha & 1s_2\beta & 1s_2\alpha \\ 1s_3\alpha & 1s_3\beta & 1s_3\alpha \end{vmatrix} \begin{vmatrix} (1,0,0,0,1/2) & (1,0,0,0,-1/2) & (1,0,0,0,1/2) \\ (1,0,0,0,1/2) & (1,0,0,0,-1/2) & (1,0,0,0,1/2) \\ (1,0,0,0,1/2) & (1,0,0,0,-1/2) & (1,0,0,0,1/2) \end{vmatrix}$$

$$\begin{vmatrix} 1s_1\alpha & 1s_1\beta & 1s_1\beta \\ 1s_2\alpha & 1s_2\beta & 1s_2\beta \\ 1s_3\alpha & 1s_3\beta & 1s_3\beta \end{vmatrix} \begin{vmatrix} (1,0,0,0,1/2) & (1,0,0,0,-1/2) & (1,0,0,0,-1/2) \\ (1,0,0,0,1/2) & (1,0,0,0,-1/2) & (1,0,0,0,-1/2) \\ (1,0,0,0,1/2) & (1,0,0,0,-1/2) & (1,0,0,0,-1/2) \end{vmatrix}$$

In such case, two of the three entries in each row have the same set of four quantum numbers, and this is why the wave function is not allowed by Pauli's exclusion principle. Answer.

Problem 17. Show by evaluation of the ground-state energy that for Li atom a configuration $1s^3$ is impossible.

Solution: Suppose that $\psi(1,2,3) = 1s(1)1s(2)1s(3)$

$$\therefore\ E_{1s^3} = \int 1s(1)1s(2)1s(3)H1s(1)1s(2)1s(3)\mathrm{d}\tau$$

$$\int [1s(1)1s(2)1s(3)]^2\,\mathrm{d}\tau$$

$$H = H_1 + H_2 + H_3 + \frac{1}{r_{12}} + \frac{1}{r_{23}} + \frac{1}{r_{31}} \text{ (in a.u)}$$

where $H_i = -\frac{1}{2}\nabla_i^2 - \frac{Z}{r_i}$,

where Z is the effective nuclear charge (< 3)

But $E_{1s^3} = 3E_{1s}^{Li+2} + J_{12} + J_{23} + J_{13}$

$$3E_{1s(1)}^{\mathrm{Li+2}} = \int 1s\left[-\frac{\nabla^2}{2} - \frac{3}{r}\right]1s\mathrm{d}\tau$$

$$= \int 1s\left[-\frac{\nabla^2}{2} - \frac{z}{r} - \frac{3-z}{r}\right]1s\mathrm{d}\tau$$

$$= \int 1s\left[-\frac{\nabla^2}{2} - \frac{z}{r}\right]1s\mathrm{d}\tau - \int 1s\left(\frac{3-z}{r}\right)1s\mathrm{d}\tau$$

$$= -\frac{z^2}{2} - (3-z)\int (1s)^2\frac{1}{r}\mathrm{d}\tau$$

But $1s = \left(\frac{z^3}{\pi}\right)^{\frac{1}{2}} e^{-zr}$ and $\mathrm{d}r = r^2\mathrm{d}r\ \sin\theta\ \mathrm{d}\Phi$.

$$\therefore\ E_{1s}^{\mathrm{Li}2^+} = -\frac{z^2}{2} - (3-z)z = \frac{z^2}{2} - 3z$$

We know that

$$J_{12} = J_{23} = J_{13} = \frac{5}{8}z$$

$$\therefore\ E_{1s^3}^{\text{Li}} = 3\left[\frac{z^2}{2} - 3z\right] + 3 \times \frac{5}{8}z$$

even if we substitute $z = 3$, then we get

$$E_{1s^3} = -7.875 \text{ a.u.} = -214\,\text{eV}$$

The exponential value is −202.4 eV, which is lesser than the evaluated value. If $z < 3$ and on the application of variation method, the evaluated energy will be further lowered, which is an absurd value. Therefore, $1s^3$ configuration for Li is impossible. Answer.

Questions on Concepts

1. Calculate the energy of the many-electron system.
2. Prove that $E = H_{li} + \sum_{ij} (2J_{ij} - k_{ij})$, where the terms have their usual significance.
3. Derive the Fock equation.
4. Derive the Hartree equation.
5. What do you mean by the Hartree–Fock self-consistent field (SCF) method?
6. Find $F(k) = H(k) = \sum_{l \neq k} J_{\lambda}(k)$
7. a. Find the expression for excited state of helium.
   b. Why is the $E_{\text{triplet}} < E_{\text{singlet}}$?
   c. What is parahelium?
   d. What do you mean by orthohelium?
8. Prove that $E = E_a + E_b + J \pm K$ for the excited state of He.
9. Explain Fermi hole and Fermi heap.
10. Find an expression for energy of lithium in the ground state.
11. a. What do you mean by Bohr magneton?
    b. Write note on electron spin.
    c. What do you mean by multiplicity?
12. a. What is gyromagnetic ratio?
    b. What is Landé ‘g’ factor?
    c. Prove that $g = l + \frac{s(s+1) + J(J+1) - L(L+1)}{2J(J+1)}$
13. Derive the Landé interval rule.
    or
    Prove that $\Delta E \propto J$.
14. a. What do you mean by Zeemann effect?
    b. What is the origin of Zeemann effect?
15. a. Explain normal Zeemann effect.
    b. What do you mean by π line and σ line?

16. a. What do you mean by the anomalous Zeeman effect?
    b. What is the Stark effect?
17. Explain the coupling of orbital angular momentum.
18. Clearly explain the coupling of spin momenta.
19. Explain *R–S* coupling giving one example.
20. Explain the term symbol and how will you find $^{2s+1}L_J$ for d^2 configuration of an atom?
21. Find the term symbol due to coupling of $2p^1\,3p^1$.
22. a. What do you mean by degeneracy and multiplicity?
    b. Calculate the spectroscopic term symbol for (1) $2p^1\,3d^1$, (2) $2p^1\,3p^1$, and (3) $2p^2\,3d^1$.
23. How will you explain the *jj* coupling? Explain giving example.
24. What is Hund's rule?
25. What are the procedures for finding the terms and symbols?
26. Write notes in terms of nonequivalent electrons.
27. Write notes in terms of equivalent electrons.
28. What is the use of *jj* coupling?
29. a. Write down the Slater rules.
    b. Find the effective nuclear charge of a particular 4s electron of $_{30}Zn$.
30. Write note on Slater-type orbitals.
31. Write notes on Gaussian-type orbitals.
32. Explain Gaussian basis set.
33. How will you find the matrix elements will the help of Condon–Slater rules?
34. Write notes on
    a. Brillouin's theorem
    b. Gaussian basis set
35. a. Write the Hartree–Fock equation and explain the terms involved.
    b. Explain the difference between the Hartree SCF and Hartree–Fock SCF methods.
36. Write short notes on
    a. Koopman's theorem
    b. Brillouin's theorem
37. Derive the Hartree–Fock equation. What are its applications?
38. Prove that $I_k^0 = E_k^+ - E = -\epsilon_k$.
    or

$$\text{Prove that } E_{N+1} - E_N = I_i + \sum_k^N \left[(J_{ik} - K_{ik})\right] = E_i$$

39. Prove that

$$\langle \psi' | H | \psi \rangle = \int u_k^*(1) E_k u_k(1) \mathrm{d}^3 r_1 = 0.$$

40. Derive Roothaan's equation.
    or

    Prove that $F_{ij} = H_{ij} + \sum_{lm} P_{lm}\left[(ij - lm) - \frac{1}{2}(im - lj)\right]$

    What is density matrix?

41. Classify the following functions as symmetric or antisymmetric.
    a. [1*s* (1) 2*s* (2) + 2*s* (1) 1*s* (2)] [α (1) 0 (2) – β (1) α (2)]
    b. [1*s* (1) 2*s* (2) + 2*s* (1) 1*s* (2)] [α (1) α (2)]
    c. [1*s* (1) 2*s* (2) + 2*s* (1) 1*s* (2)] [α (1) 0 (2) – β (1) α (2) + α (1) α (2)]
    d. [1*s* (1) 2*s* (2) + 2*s* (1) 1*s* (2)] [α (1) β (2) + β (1) α (2)]
    e. [1*s* (1) 2*s* (2) – 2*s* (1) 1*s* (2)] [α (1) β (2) + β (1) α (2)]
42. Write the Slater determinant for the ground-state electronic configuration of Be.
43. What atomic terms are possible for the following electronic configurations?
    Which of the possible terms has the lowest energy?
    a. $ns^1\ np^1$ (b) $ns^1\ nd^1$ (c) $ns^2\ np^1$ (d) $ns^1\ np^2$
44. Derive ground-state term symbols for the following:
    a. $s^1 d^5$ (b) f^3 (c) d^2
45. Derive the ground-state term symbols for the following:
    a. Na (b) Na^+ (c) F (d) H
46. What J values are possible for ^6H term?
    Find the number of states associated with each level and illustrate that the total number of states is the same as estimated from the formula $(2S + 1)\ (2L + 1)$.
47. Show that Hamiltonian matrix elements between the Hartree–Fock wave functions ϕ_0 and singly excited determinants are identically zero.
48. Give the levels that arise from the terms given and give the degeneracy of each level.
    a. ^3F (b) ^4D (c) ^2S
49. For the state $^3D3$, give the magnitude of (a) the total electronic orbital angular momentum and (b) total spin angular momentum.
50. Give the symbol for the ground level of the atoms where atomic numbers are less than 10.
51. Write in details in terms of integrals of the Fock operator for a 2*s* orbital of Be.

$$\left[\text{Ans.}\ \left[2s(1) = -\frac{\nabla_1^2}{2} - \frac{Z}{r_1} + 2\left\langle 1s(2)\left|\frac{1}{r_{12}}\right|1s(2)\right\rangle + \left\langle 2s(2)\left|\frac{1}{r_{12}}\right|2s(2)\right\rangle\right.\right.$$

$$\left.\left. - \left[2s^*(1)1s(1)\middle|2s^*(1)2s(1)\right]\left\langle 2s(2)\left|\frac{1}{r_{12}}\right|2s(2)\right\rangle\right]\right]$$

12

Polyatomic Molecules

We are aware of the fact that a molecule is an aggregate of atomic nuclei encircled by electrons. The situation is that the nuclei repel each other but are attracted by electrons, which also repel one another. The aggregate molecule will remain stable only if the sum of the attractive forces between nuclei and electrons just balances the repulsive forces among nuclei and among electrons present.

It is to be noted that we have derived Roothaan's equations and the Hartree–Fock self-consistent field (SCF) methods in Chapter 11 under multielectron system. In this chapter, we shall discuss the matrix form of Roothaan's equations and also deliberate on Roothaan's method for quantum mechanical calculations in one dimension.

In this chapter, we shall also deal with the shapes of molecules. For diatomic molecules, the only structural element is bond length, but in case of polyatomic molecules, both bond lengths and bond angles are important elements, which determine the energy of the molecule. We have studied earlier the nature of bonding in diatomic molecules. We want to extend the discussion of bonding to molecules in which one central atom is joined to other atoms more than one in number. We have discussed while giving idea of the valance bond theory that electrons in a molecule occupy atomic orbitals, and in the course of the bond formation, these orbitals overlap. We also learnt that greater the overlapping, stronger will be the bond formation, and the direction of the bond is estimated by the direction in which the two orbitals overlap as far as practicable. Overlapping of atomic orbitals actually yields a stable bond, provided the atomic orbitals have the same energy and adequate symmetry.

In this chapter, we shall also discuss the shape of molecules for which hybridisation will be dealt with. We shall give the quantum mechanical picture of hybridisation of different kinds.

12.1 Matrix Form of Roothaan's Equations

Roothaan's equations (Eq. 11.176, Chapter 11) are represented by

$$\sum_{j=1}^{M} F_{ij}C_{ja} = \epsilon_a \sum_{j=1}^{M} S_{ij}C_{ja}, \quad \text{where } i = 1, 2, \ldots, M$$

In this equation, the coefficients C_{ja} are related to the MOs ϕ_a and the basis function χ_j by the following equation:

$$\phi_a = \sum_{j=1}^{M} C_{ja}\chi_j \tag{12.1}$$

Now, let us consider that

C = square matrix of order M whose elements belong to C_{ja} which are coefficients,

F = square matrix of order M whose elements are represented by $F_{ij} = \langle \chi_i | \hat{F} | \chi_j \rangle$,

S = square matrix whose elements are denoted by $S_{ij} = \langle \chi_i | \chi_j \rangle$,

$\in$ = square matrix or diagonal square matrix whose diagonal elements belong to $\in_1, \in_2, \ldots \in_M$, which are orbital energies such that $\in_{mi} = \delta_{mi} \in_i$, i.e., elements of $\in$ where δ_{mi} is Kronecker delta, and

χ_j = set of basis functions.

Now, let us suppose that A is an m by n matrix and B is an n by p matrix. The matrix product $C = AB$ will be defined to be the m by p matrix whose elements will be expressed as

$$\begin{aligned} C_{ij} &= x_{i1}y_{1j} + x_{i2}y_{2j} + \ldots + x_{in}\ y_{nj} \\ &= \sum_{k=1}^{n} x_{ik}\ y_{kj} \end{aligned} \tag{12.2}$$

This is known as matrix multiplication rule. When this rule is applied, it will yield (j, i)th elements of the matrix product

$$\begin{aligned} C \in as(C \in)_{ja} &= \sum_m C_{jm} \in_{ma} \\ &= \sum_m C_{jm}\delta_{ma} \in_a = C_{ja} \in_a \end{aligned} \tag{12.3}$$

Therefore, Roothaan's equations will take the form

$$\sum_{j=1}^{M} F_{ij}C_{ja} = \sum_{j=1}^{M} S_{ij}(C \in)_{ja} \tag{12.4}$$

Thus, it can be said that according to matrix multiplication rule, the LHS of Eq. (12.4) is actually the (i, a)th elements of FC, but the RHS is the (i, a)th element of $S(C \in)$. Further, we may express as

$$FC = SC \in \tag{12.5}$$

It is because of the fact that the general element of FC is equal to the general element of $SC \in$. Equation (12.5) can also be expressed as

$$(F - S \in)C = 0 \tag{12.6}$$

This is the matrix representation or form of Roothaan's equations.

12.2 Fock Matrix Elements

For getting the solution of Roothaan's equations, we should first of all express the Fock matrix elements F_{ij} in terms of the basis function χ. The Fock operator is written as

$$\hat{F}(1) = H^{\text{core}}(1) + \sum_{j=1}^{n/2}\left[2\hat{j}_j(1) - k_j(1)\right] \tag{12.7}$$

$$\text{and } F_{ij} = \left\langle \chi_i(1)\middle|\hat{F}(1)\middle|\chi_j(1)\right\rangle$$

$$= \left\langle \chi_i(1)\middle|H^{\text{core}}(1)\middle|\chi_j(1)\right\rangle$$

$$+\sum_{j=1}^{n/2}\left[2\left\langle \chi_i(1)\middle|\hat{j}(1)\middle|\chi_j(1)\right\rangle\right]$$

$$-\left\langle \chi_i(1)\middle|\hat{k}_j(1)\middle|\chi_s(1)\right\rangle\Big] \tag{12.8}$$

The coulomb operator $\hat{J}_j$ and the exchange operator $\hat{k}_j$ have already been introduced, which may be defined as

$$\hat{J}(1)f(1) = f(1)\int |\phi_j(2)|^2 \frac{1}{r_{12}} d\tau_2 \tag{12.9}$$

$$\text{and} \quad \hat{k}_j(1)f(1) = \phi_j(1)\int \frac{\phi_j^*(2)f(2)}{r_{12}} d\tau_2 \tag{12.10}$$

where f = an arbitrary functions and the integrals are definite integrals over all space.

Now, we shall replace f by χ_j in Eq. (12.9), which will be followed by the application of the expansion of Eq. (12.1), which we can write as

$$\hat{j}_j(1)\chi_j(1) = \chi_j(1)\int \frac{\phi_j^*(2)\phi_j(2)}{r_{12}} \cdot d\tau_2$$

$$= \chi_j(1)\sum_t\sum_u c_{ij}^* c_{uj}\int \frac{\chi_t^*(2)\chi_u(2)}{r_{12}} \cdot d\tau_2 \tag{12.11}$$

Multiplying Eq. (12.11) by # and then integrating over electron 1 coordinates yield

$$\left\langle \chi_i(1)\middle|\backslash\hat{J}_j(1)\middle|\chi_j(1)\right\rangle = \sum_t\sum_u c_{ij}^* c_{uj}\int \frac{\chi_i^*(1)\chi_j(1)\chi_t^*(2)\chi_u(2)}{r_{12}} d\tau_1 \cdot d\tau_2$$

$$\text{or} \quad \left\langle \chi_i(1)\middle|\hat{J}_j(1)\middle|\chi_j(1)\right\rangle = \sum_{t=1}^{M}\sum_{u=1}^{M} c_{ij}^* c_{uj}\left(ij/tu\right) \tag{12.12}$$

$$\text{where} \quad \left(ij/tu\right) \equiv \int\int \frac{\chi_i^*(1)\chi_j(1)\chi_t^*(2)\chi_u(2)}{r_{12}} d\tau_1 d\tau_2 \tag{12.13}$$

which is known as two-electron repulsion integral.

In the same way, the replacement of f is done by χ_i in Eq. (12.10), yielding

$$\left\langle \chi_i(1)\left|\hat{k}_j(1)\right|\chi_j(1)\right\rangle = \sum_{t=1}^{M}\sum_{u=1}^{M} c_{ij}^{*}c_{uj}\left(iu/tj\right) \tag{12.14}$$

On substitution of Eqs. (12.14) and (12.12) into Eq. (12.8), and altering the variable order of summation, we obtain the required expression for F_{ij} in the form of integrals over the basis function χ, as mentioned in the following:

$$F_{ij} = H_{ij}^{\text{core}} + \sum_{t=1}^{M}\sum_{u=1}^{M}\sum_{j=1}^{n/2} c_{ij}^{*}c_{uj}\left[2\left(ij/tu\right)-\left(iu/tj\right)\right] \tag{12.15}$$

$$F_{ij} = H_{ij}^{\text{core}} + \sum_{t=1}^{M}\sum_{u=1}^{M} P_{tu}\left[\left(ij/tu\right)-\frac{1}{2}\left(iu/tj\right)\right] \tag{12.16}$$

$$\text{where } P_{tu} \equiv 2\sum_{j=1}^{n/2} c_{ij}^{*}c_{uj},\ t = 1,2,\ldots,M \text{ and } u = 1,2,\ldots,M \tag{12.17}$$

$$H_{ij}^{\text{core}} \equiv \left\langle \chi_i(1)\left|H_{(1)}^{\text{core}}\right|\chi_j(1)\right\rangle \tag{12.18}$$

It should be kept in mind that P_{tu} are also known as density matrix elements.

Furthermore, we shall substitute the expansion of Eq. (12.1), i.e., $\phi_a = \sum c_{ja}\chi_j$ into the equation for the electron probability density, i.e., $\rho(x,y,z) = \sum_j n_j\left|\phi_j\right|^2$ (where $n_j = 0$, 1, or 2) for obtaining the value of p (electron probability density) for a closed-shell molecule which is

$$\rho = 2\sum_{j=1}^{n/2}\phi_i^{*}\phi_j = 2\sum_{i=1}^{M}\sum_{j=1}^{M}\sum_{j=1}^{n/2} c_{ij}^{*}c_{ji}\chi_i^{*}\chi_j = \sum_{i=1}^{M}\sum_{j=1}^{M} p_{ij}\chi_i^{*}\chi_j \tag{12.19}$$

For representing the Hartree–Fock energy in terms of integrals over χ (basis function), we first solve the equation

$$\sum_{i=1}^{n/2} \epsilon_a = \sum_{i=1}^{n/2} H_{ii}^{\text{core}} + \sum_{i=1}^{n/2}\sum_{j=1}^{n/2}\left(2J_{ij} - k_{ij}\right) \tag{12.20}$$

for $\sum_i\sum_j\left(2J_{ij} - k_{ij}\right)$ and the result obtained will be substituted into the following equation, i.e.,

$$E_{HF} = 2\sum_{i=1}^{n/2} \epsilon_a - \sum_{i=1}^{n/2}\sum_{j=1}^{n/2}\left(2J_{ij} - k_{ij}\right) + V_{NN}$$

to get

$$E_{HF} = 2\sum_{i=1}^{n/2} \epsilon_a + \sum_{i=1}^{n/2} H_{ii}^{\text{core}} + V_{NN} \tag{12.21}$$

Using the expansion $\phi_a = \sum^{M} c_{ja}\chi_j$, we shall obtain

$$H_{ii}^{\text{core}} = \left\langle \phi_i \middle| H^{\text{core}} \middle| \phi_i \right\rangle = \sum_i \sum_j c_{ii}^* c_{ji} \left\langle \chi_i \middle| H^{\text{core}} \middle| \chi_j \right\rangle$$
$$= \sum_i \sum_j c_{ii}^* c_{ji} H_{ij}^{\text{core}} \tag{12.22}$$

$$\text{or } E_{HF} = \sum_{i=1}^{n/2} \in_a + \sum_i \sum_j \sum_{i=1}^{n/2} c_{ii}^* c_{ji} H_{ij}^{\text{core}} + V_{NN} \tag{12.23}$$

$$\text{or } E_{HF} = \sum_{i=1}^{n/2} \in_a + \frac{1}{2} \sum_{i=1}^{M} \sum_{j=1}^{M} p_{ij} H_{ij}^{\text{core}} + V_{NN} \tag{12.24}$$

Now, multiplying equation $\hat{F}\phi_a\phi_a$ by ϕ_a^* and then integrating, we shall get

$$\in_a = \left\langle \phi_a \middle| \hat{F} \middle| \phi_a \right\rangle \tag{12.25}$$

Substituting the value of $\phi_a = \sum_{j=1} c_{ja}\chi_j$, we have

$$\in_a = \sum_i \sum_j c_{ia}^* c_{ja} < \chi_j \middle| \hat{F} \middle| \chi_j = \sum_i \sum_j c_{ia}^* c_{ja} F_{ij} \tag{12.26}$$

The first sum in Eq. (12.24) takes the form

$$\sum_a \in_a = \sum_i \sum_j c_{ia}^* c_{ja} F_{ij} = \frac{1}{2} \sum_i \sum_j p_{ij} F_{ij} \tag{12.27}$$

where we have used the definition mentioned in Eq. (12.17). Then, Eq. (12.24) takes the following form:

$$E_{HF} = \frac{1}{2} \sum_{i=1}^{M} \sum_{j=1}^{M} p_{ij} \left(F_{ij} + H_{ij}^{\text{core}} \right) + V_{NN} \tag{12.28}$$

which clearly expresses E_{HF} of a closed-shell molecule in terms of density, Fock, and core Hamiltonian matrix elements evaluated by the use of χ_i as the basis function.

12.3 Roothaan's Method in One Dimension

Quantum mechanical computations in one dimension are of use since these are conceptually and mathematically easily understood. With the suitable guess of basis function, Roothaan's method may be used/applied in computation of self-consistent ground-state energy value of He atom. Roothaan's analytical procedure of calculation has been outlined by Snow and Bills (1981).

The computation presented here and a Hartree numerical SCF computation on the same one-dimensional model have been utilised. The one-dimensional model was presented by Harris and Rioux (1980), considering the following important facts:

- An infinite potential barrier exists at the nucleus.
- The electron nuclear potential energy amounts to $-2/x$ for positive values of x.
- V_{12} is described by a truncated coulombic interaction, which is expressed as $V_{12} = \frac{1}{|x_1 - x_2|} + A$, where V_{12} = electron potential energy. The truncation parameter saves V_{12} from becoming infinite when $x_1 = x_2 \cdot A$ is equal to 0.5 gives a suitable and appropriate value for the He atom's ground-state energy.

The Schrödinger equation for such a model (in atomic units) can be expressed as

$$-\frac{1}{2}\frac{d^2}{dx_1^2} - \frac{2}{x_1} - \frac{1}{2}\frac{d^2}{dx_2^2} - \frac{2}{x_2} + \frac{1}{\left(|x_1 - x_2| + A\right)} - E\phi(1,2) = 0 \tag{12.29}$$

Let us express ϕ (1, 2) as a product of one-electron orbital, i.e., ϕ (1, 2) = ψ (1) ψ (2), which in turn is a linear combination of atomic orbitals

$$\psi(1) = \sum c_j f_j(1) \tag{12.30}$$

and following the method given by Snow and Bills, linear simultaneous equations will result, i.e.,

$$\sum_{j=1} c_j\left(h_{ij} + g_{ij} - \in s_{ij}\right) = 0 \qquad i = 1,2,3, \tag{12.31}$$

The nontrivial solutions may be obtained for Eq. (12.31) only if the determinant of the coefficients is zero.

$$\left|h_{ij} + g_{ij} - \in s_{ij}\right| = 0 \tag{12.32}$$

$$\text{where } h_{ij} = \left\langle f_i \left| -\frac{1}{2}\frac{d^2}{dx^2} - \frac{2}{x_1} \right| f_j \right\rangle \tag{12.33}$$

$$g_{ij} = \sum_k \sum_l c_k c_l \left\langle f_i f_j | V_{12} | f_k f_l \right\rangle$$

$$= \sum_k \sum_l c_k c_l \left\langle ij | kl \right\rangle \tag{12.34}$$

$$s_{ij} = \left\langle f_i | f_j \right\rangle \tag{12.35}$$

For the derivation of Eq. (12.31), we shall assume that ϕ (1, 2) = ψ (1) ψ (2) which it allows us to write Eq. (12.29) as follows:

$$\left(\hat{T}_1 + \hat{V}_{N_1} + \hat{T}_2 + \hat{V}_{N_2} + V_{12}\psi(1)\psi(2) = E\psi(1)\psi(2)\right. \tag{12.36}$$

where the terms represent the kinetic, potential, and interaction energy.

Now, multiplying Eq. (12.36) on the left by ψ (2) followed by integration over the coordinate of electron 2 results in the following after rearrangement.

$$\begin{gathered}\left(\hat{T}_1 + \hat{V}_{N_1} + \langle\psi(2)|V_{12}|\psi(2)\rangle\right)\psi(1) \\ \left(E - \langle T_2\rangle - \langle V_{N2}\rangle\right)\psi(1)\end{gathered} \tag{12.37}$$

where $\langle T_2 \rangle = \left\langle \psi(2) \middle| \hat{T}_2 \middle| \psi(2) \right\rangle$

$$\langle V_{N_2} \rangle = \langle \psi \rangle(2)|V_{N2}|\psi(2)$$

Recognising $\left(E - \langle T_2 \rangle - \langle V_{N_2} \rangle\right)$ = the orbital energy of electron 1, we can write

$$\left(\hat{T}_1 + \hat{V}_{N_1} + \langle \psi(2)|V_{12}|\psi(2) \rangle\right)\psi(1) = E_1\psi(1) \tag{12.38}$$

Equation (12.38) is generally called *Hartree's equation*, and the energy operator on the left is known as the one-electron effective Hamiltonian, H_l^{eff}. Similarly, we can express the Hartree equation for electron 2.

Roothaan solved Eq. (12.38) by representing ψ (1) as a linear combination in a full basis set of functions expressed by Eq. (12.30). Furthermore, we shall substitute Eq. (12.30) into Eq. (12.38), but it should be followed by multiplication on the left by every member of the basis set, and the integration is performed over the coordinate of electron 1, which will give Eq. (12.31).

For the computational purpose, a basis set like

$$\psi = \sum_{j=1} c_j (2J)^{3/2} \cdot xe^{-jx} \tag{12.39}$$

is selected and is truncated after two terms, yielding

$$\psi = c_1 2xe^{-x} + c_2\sqrt{3}2xe^{-2x} \tag{12.40}$$

It should be understood that the first term on the right of Eq. (12.40) is actually the Eigen function for one-dimensional hydrogen atom $\left(H = -\frac{1}{2}\frac{\mathrm{d}^2}{\mathrm{d}x^2} - \frac{1}{x}\right)$. In the first term of the RHS, the exponential parameter is 1, which reflects the fact that the electron experiences the nuclear charge +1.

The second term on the RHS of Eq. (12.40) represents the Eigen function for the one-dimensional helium ion $\left(H = -\frac{1}{2}\frac{\mathrm{d}^2}{\mathrm{d}x^2} - \frac{1}{x}\right)$. The exponential parameter is 2, which reflects that the electron experiences a nuclear charge of +2. Thus, it can be inferred that the electron in the atom experiences an effective nuclear charge between +1 and +2.

This suggests that a linear combination of the hydrogen atom Eigen function and that of He^+ ion Eigen function should serve as a better approximation for the wave function for He atom's electron.

Now, we are able to find the h_{ij} and s_{ij} matrix elements using

$$\int_0^\infty x^n\, e^{-ax}\mathrm{d}x = \frac{n!}{a^{n+1}}.$$

The h_{ij} and s_{ij} values are as follows:

$$h_{11} = -1.5$$

$$h_{12} = h_{21} = -1.6761$$

$$h_{22} = -2.00$$

$$s_{11} = s_{22} = 1.00$$

$$s_{12} = s_{21} = 0.8381$$

These values are verified in the following manner:

$$\text{I}\quad h_{11} = \int_0^\infty 2xe^{-x}|h|2xe^{-x}\mathrm{d}x$$

$$= 4\int_0^\infty xe^{-x}|h|xe^{-x}\mathrm{d}x$$

$$= 4\int_0^\infty xe^{-x}\left\{-\frac{1}{2}\frac{\mathrm{d}^2}{\mathrm{d}x^2}\left(xe^{-x}\right)+\frac{1}{x}\cdot xe^{-x}\right\}\mathrm{d}x$$

$$= -4\int_0^\infty xe^{-x}\left\{\left(\frac{x-2}{2}\right)e^{-x}+2e^{-}x\right\}\mathrm{d}x \qquad \left[\because \frac{\mathrm{d}}{\mathrm{d}x}\left(xe^{-x}\right)=e^{-x}-xe^{-x}\right.$$

$$= -4\int_0^\infty x\left\{\frac{x-2}{2}+2\right\}e^{-2x}\mathrm{d}x \qquad \frac{\mathrm{d}^2}{\mathrm{d}x^2}\left(xe^{-x}\right)=\frac{\mathrm{d}}{\mathrm{d}x}\left(xe^{-x}-xe^{-x}\right)$$

$$= -4\int_0^\infty x\left\{\frac{x-2+4}{2}\right\}e^{-2x}\mathrm{d}x \qquad = -e^{-x}-e^{-x}+xe^{-x} = -2e^{-x}+xe^{-x}$$

$$= -2\int_0^\infty x\{x+2\}e^{-2x}\mathrm{d}x \qquad \left.=(x-2)e^{-x}\right]$$

$$= -2\int_0^\infty x^2e^{-2x}\mathrm{d}x + 2\int_0^\infty xe^{-2x}\mathrm{d}x$$

$$= -2\left[\frac{2!}{(2)^3}+2\cdot\frac{1!}{2^2}\right]$$

$$= \left[\frac{-4}{2^3}-\frac{4}{4}\right] = \left[-\frac{1}{2}-1\right] = -\frac{3}{2} = -1.50 \tag{12.41}$$

Now, we calculate h_{12}

$$\text{II}\quad h_{12} = \int_0^\infty 2xe^{-x}\left[-\frac{1}{2}\frac{\mathrm{d}^2}{\mathrm{d}x^2}-\frac{2}{x}\right]\sqrt{32}xe^{-2x}\mathrm{d}x$$

$$= 2\sqrt{32}\int_0^\infty xe^{-x}\left[-\frac{1}{2}\frac{\mathrm{d}^2}{\mathrm{d}x^2}-\frac{2}{x}\right]\cdot xe^{-2x}\mathrm{d}x$$

$$= 2\sqrt{32}\int_0^\infty xe^{-x}\left[+2(1-x)e^{-2x}-2e^{-2x}\right]\mathrm{d}x$$

$$= -4\sqrt{32}\int_0^\infty x\{1-x-1\}e^{-3x}\mathrm{d}x \tag{12.42}$$

$$= -4\sqrt{32}\int_0^\infty x^2 e^{-3x}\mathrm{d}x$$

$$= -4\sqrt{32}\cdot\frac{2!}{(3)^2} = \frac{-4\sqrt{32}\times 2}{27} = \frac{-8\sqrt{32}}{27} = \frac{-32\sqrt{2}}{27} = -1.6761$$

$$\therefore\ h_{12} = h_{21} = -1.6761$$

$$\left[\text{Note: } \frac{\mathrm{d}^2}{\mathrm{d}x^2}\left(xe^{-2x}\right) = \frac{\mathrm{d}}{\mathrm{d}x}\left\{(1-2x)e^{-2x}\right\}\right.$$

$$\frac{\mathrm{d}^2}{\mathrm{d}x^2}\left(xe^{-2x}\right) = \frac{\mathrm{d}}{\mathrm{d}x}\left\{(1-2x)e^{-2x}\right\}$$

$$= (-2)e^{-2x} + (1-2x)e^{-2x}$$

$$= -2e^{-2x}\left[1+1-2x\right]$$

$$\left.= -4(1-x)e^{-2x}\right]$$

III $$h_{22} = \int_0^\infty \sqrt{32}xe^{-2x}\,|h|\,\sqrt{32}xe^{-2}\mathrm{d}x \tag{12.43}$$

$$= \int_0^\infty \sqrt{32}xe^{-2x}\left[-\frac{1}{2}\frac{\mathrm{d}^2}{\mathrm{d}x^2} - \frac{2}{x}\right]\sqrt{32}xe^{-2}\mathrm{d}x$$

$$= 32\int_0^\infty xe^{-2x}\left[\left(-\frac{1}{2}\right)(-4)(1-x)e^{-2x} - 2e^{-2x}\right]\mathrm{d}x$$

$$= 32\int_0^\infty xe^{-2x}\left[2(1-x)e^{-2x} - 2e^{-2x}\right]\mathrm{d}x$$

$$= 32\int_0^\infty xe^{-2x}\left[2e^{-2x} - 2xe^{-2x} - 2e^{-2x}\right]\mathrm{d}x$$

$$= 32\int_0^\infty xe^{-2x}\left[-2xe^{-2x}\right]\mathrm{d}x$$

$$= -64\int_0^\infty xe^{-2x}\cdot xe^{-2x}\mathrm{d}x$$

$$= -64\int_0^\infty x^2 e^{-4x}\mathrm{d}x \tag{12.44}$$

$$= -64 \cdot \frac{2!}{(4)^3} \qquad \left[\therefore \int_0^\infty x^n e^{-ax} dx = \frac{n!}{(a)^{n+1}}\right]$$

$$= \frac{-64 \times 2}{64} = -2.00$$

IV $s_{12} = \langle f_i / f_j \rangle$ (12.45)

$$= \left\langle 2xe^{-x} \middle| \sqrt{32}xe^{-2x} \right\rangle$$

$$= 2\sqrt{32}\int_0^\infty xe^{-x} xe^{-2x}\mathrm{d}x$$

$$= 2\sqrt{32}\int_0^\infty x^2 e^{-3x}\mathrm{d}x$$

$$= 2\sqrt{32} \cdot \frac{2!}{(3)^3} \qquad \therefore \int_0^\infty x^n e^{-ax} = \frac{n!}{(a)^{n+1}}$$

$$= \frac{2 \times 4\sqrt{2} \times 2}{27} = \frac{16\sqrt{2}}{27} = 0.83805 = s_{21}$$

V $s_{11} = \langle f_i / f_j \rangle = \langle f_i | f_j \rangle$ (12.46)

$$= \int_0^\infty 2xe^{-x} \left| 2xe^{-2x}\mathrm{d}x\right.$$

$$= 4\int_0^\infty x^2 e^{-2x}\mathrm{d}x$$

$$= 4 \cdot \frac{2!}{(3)^3} = 1.00 = s_{22}$$

TABLE 12.1

Successive Values of Coefficients and Energies of Ten Iterations of Roothaan's Equation

Iteration	C_1	C_2	$\in$	V_{ee}	E_{He} (in Hartrees)
1	0.0000	1.0000	−0.8266	1.1529	−2.8266
2	0.2024	0.8242	−0.8556	1.1327	−2.8350
3	0.1527	0.8685	−0.8473	1.1390	−2.8357
4	0.1654	0.8573	−0.8493	1.1385	−2.8357
5	0.1622	0.8602	−0.8488	1.1380	−2.8357
6	0.1630	0.8594	−0.8489	1.1378	−2.8357
7	0.1628	0.8596	−0.8489	1.1378	−2.8357
8	0.1629	0.8596	−0.8489	1.1378	−2.8357
9	0.1628	0.8596	−0.8489	1.1378	−2.8357
10	0.1628	0.8596	−0.8489	1.1378	−2.8357

Thus, we have verified all five integrals.

Now, because of the existence of the truncation parameter in V_{12}, the g_{ij} matrix elements should be computed by numerical method. Putting $A = 0.5$ and applying Simpson's method, one can evaluate the integrals described in Eq. (12.34). On computation, the following values of integrals will be obtained:

$$\langle 11|11\rangle = 0.902392, \langle 12|12\rangle = 0.750754$$

$$\langle 11|12\rangle = 0.785299, \langle 12|22\rangle = 0.924484$$

$$\langle 12|22\rangle = 0.924484, \langle 22|22\rangle = 1.18784$$

Students are advised to prepare a computer programme to check whether the aforementioned values of integrals are correct. After this, the calculation is being done by Snow and Bills procedure. The values of integrals evaluated are tabulated in Table 12.1.

The plot of r^2 vs. x has is given in Figure 12.1, which gives a clear picture of the comparison between the resulting SCF wave function and the basis set function. It is clear from the figure that the ground state energy of 1-D of He atom is equal to twice the one-electron orbital energy minus the electron–electron potential energy, i.e.,

$$E_{\text{He(scf)}} = 2 \in = V_{ee} = -2.8357 \text{ Hartrees} \tag{12.47}$$

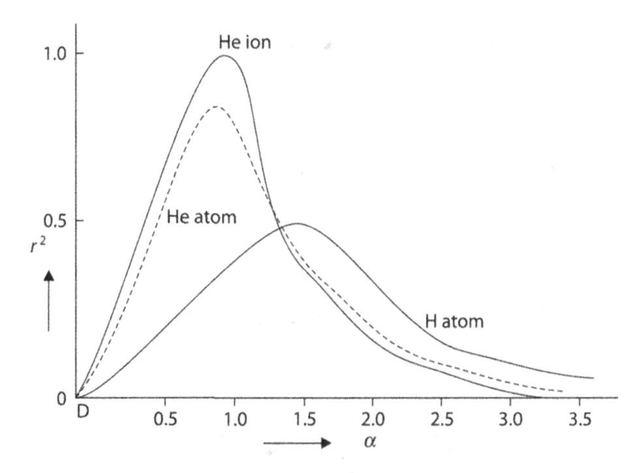

FIGURE 12.1 Comparison of He atom self-consistent field wave function with f_1 and f_2 of the basis set.

Since the h_{22} matrix elements belong to the ground-state energy of He^+ ion, it is really possible to evaluate the ionisation potential of the atom as given in the following:

$$IP = E^{+}_{He} - E_{He(scf)}$$
$$= -2.00 - (-2.8357)$$
$$= 0.8357 \text{ Hartrees}$$

The initial values selected for c_1 and c_2 are 0.0 and 1.0, respectively. This means that both the electrons will remain in the He^+ ion orbital. The values of the coefficients c_1 and c_2 are 0.1628 and 0.8596, respectively. This gives a clue towards understanding that these are making the orbital larger and decreasing the electron–electron potential energy.

12.4 Electronic Energy

The electronic energy is the orbital energy of an electron, which is actually the expectation value of Fock operator F and expressed by

$$E_i = E_i^0 + \sum_{j=1}^{n/2} \left(2J_{ij} - k_{ij}\right)$$

Therefore, the total electronic energy is

$$E = \sum_{i=1} E_i = 2\sum_{i=1}^{n/2} E_i^0 + \sum_{i=1}^{n/2}\sum_{j=1}^{n/2} \left(2J_{ij} - k_{ij}\right) \tag{12.48}$$

We also know that

$$E = \sum 2E_i - \sum_{i=1}^{n/2}\sum_{j=1}^{n/2} \left(2J_{ij} - k_{ij}\right) \tag{12.49}$$

Now, we shall subtract Eq. (12.49) from Eq. (12.48), and the result is obtained as

$$E = \sum_{i=1} E_i = 2\sum_{i=1}^{n/2} E_i^0 + \sum_{i=1}^{n/2}\sum_{j=1}^{n/2} \left(2J_{ij} - k_{ij}\right)$$

$$E = \sum 2E_i - \sum_{i=1}^{n/2}\sum_{j=1}^{n/2} \left(2J_{ij} - k_{ij}\right)$$

$$\underline{\quad - \qquad\qquad + \quad}$$

$$0 = 2\sum_{i=1}^{n/2} E_i^0 - \sum_{j=1}^{n/2} 2E_i + 2\sum_{i=1}^{n/2}\sum_{j=1}^{n/2} \left(2J_{ij} - k_{ij}\right)$$

$$\text{or} \quad 0 = \sum_{i=1}^{n/2} E_i^0 - \sum_{j=1}^{n/2} E_i + 2\sum_{i=1}^{n/2}\sum_{j=1}^{n/2} \left(2J_{ij} - k_{ij}\right)$$

$$\text{or} \quad \sum_{i=1}^{n/2}\sum_{j=1}^{n/2}(2J_{ij}-k_{ij})=\sum_{i=1}^{n/2}E_i-\sum_{i=1}^{n/2}E_i^0 \tag{12.50}$$

Putting the LHS value of Eq. (12.50) in Eq. (12.48), we get

$$E=2\sum_{i=1}^{n/2}E_i^0+\sum_{i=1}^{n/2}E_i-\sum_{i=1}^{n/2}E_i^0$$

$$\text{or } E=\sum_{i=1}^{n/2}E_i^0+\sum_{i=1}^{n/2}E_i \tag{12.51}$$

$$\text{or } E=\sum_{i=1}^{n/2}E_i+\sum_{i}\sum_{j}\sum_{i=1}^{n/2}a_{ii}^*a_{ij}\int\chi_i^*H_i^0\chi_j\mathrm{d}\tau$$

$$\text{or } E=\sum_{1}^{n/2}E_i+\frac{1}{2}\sum_{i}\sum_{j}p_{ij}H_{ij}^0 \tag{12.52}$$

This is the equation for electronic energy.

12.5 Solution of Roothaan's Equation for He Atom

According to Roetti and Clementi, Φ_{1s} atomic orbitals of helium can be expressed as a linear combination of two normalised Slater-type orbital (STO) basis functions

$$\chi_1 \text{ and } \chi_2 \text{ i.e. } \phi_{is}=a_1\chi_1+a_2\chi_2 \tag{12.53}$$

where $\chi_1=2\xi_1^{3/2}e_1^{-\xi_1^r}Y_0^0$

$$\chi_2=2\xi_2^{3/2}e_1^{-\xi_2^r}Y_0^0$$

Here, $Y_0^0=1$, hence χ_1 and χ_2 are written as

$$\left.\begin{aligned}\chi_1&=2\xi_1^{3/2}e^{-\xi_1^r}\\ \text{and } \chi_2&=2\xi_2^{3/2}e^{-\xi_2^r}\end{aligned}\right\} \tag{12.54}$$

The optimum values of ξ_1 and ξ_2 are 1.45 and 2.91, respectively. ξ_1 and ξ_2 are effective atomic numbers.

For the solution of Roothaan's equations, we require the integrals F_{ij}, and s_{ij}. The overlap integrals S_{ij} are

$$s_{11} = \langle \chi_1 | \chi_1 \rangle = 1, \qquad s_{22} = \langle \chi_2 | \chi_2 \rangle = 1$$

$$s_{12} = s_{21} = \langle \chi_1 | \chi_2 \rangle$$

$$= \int_0^\infty 2\xi_1^{3/2} \cdot e^{-\xi_1^r} \cdot 2e^{-}\xi_1^{3/2} \cdot e^{-\xi_2^r} r^2 \mathrm{d}r$$

$$= 4\xi_1^{3/2} \cdot \xi_2^{3/2} \int_0^\infty e^{-\left(-e^{-\xi_1^r + \xi_2^r}\right)} r^2 \mathrm{d}r \tag{12.55}$$

$$\text{or } s_{12} = 4\xi_1^{3/2} \cdot \xi_2^{3/2} \frac{2!}{(\xi_1 + \xi_2)^3} = \frac{8\xi_1^{3/2} \cdot \xi_2^{3/2}}{(\xi_1 + \xi_2)^3} \qquad \left[\because \int_0^\infty x^n e^{-ax} = \frac{n!}{a^{n+1}}\right]$$

Putting the values of ξ_1 and ξ_2, we shall get

$$s_{12} = s_{21} = \frac{8(1.45)^{\frac{3}{2}}(2.91)^{3/2}}{(1.45 + 2.91)^3}$$

$$\text{or } s_{12} = s_{21} = \frac{8(4.2195)^{3/2}}{(4.36)^3} = 8(0.4711)^3$$

$$\text{or } s_{12} = s_{21} = 8 \times 0.1046 = 0.8368$$

Now, we find the value of F_{ij} integrals as described in Eq. (12.16) which are dependent on the values of H_{ij}^{core}, and $(ij/tu) \cdot H^{\text{core}}$ will be obtained using the formula

$$H^{\text{core}} \equiv \frac{-\nabla_1^2}{2} - \sum \frac{z}{r} \tag{12.56}$$

In case of He, H^{core} will be written as:

$$H^{\text{core}} \equiv \frac{-\nabla^2}{2} - \frac{2}{r} = \frac{-\nabla^2}{2} - \frac{\xi}{r} + \frac{(\xi - 2)}{r} \tag{12.57}$$

where the terms have their usual significance.

The integrals H_{ij}^{core} are computed in the similar manner as we have evaluated in the case of He atom by variation method. The values of H_{ij}^{core} will be found as follows:

$$H_{11}^{\text{core}} = \langle \xi_1 | H^{\text{core}} | \chi_1 \rangle$$

$$= \frac{\xi_1^2}{2} + (\xi_1 - 2)\xi_1$$

$$= \frac{1}{2}\xi_1^2 - 2\xi_1 = \frac{1}{2} \times (1.45)^2 - 2(1.45)$$

$$= \left(\frac{1}{2} \times 2.1025 - 2.90\right)$$

$$= (1.0512 - 2.90) = -1.8488 \tag{12.58}$$

Similarly, we shall evaluate H_{22}^{core} by using

$$
\begin{aligned}
H_{22}^{\text{core}} &= \left(\frac{1}{2}\xi_2^2 - 2\xi_2\right)\\
&= \frac{1}{2}(2.91)^2 - 2(2.91)\\
&= \frac{1}{2}(8.4681) - 5.8200\\
&= (4.23405 - 5.82000) = -1.5860
\end{aligned} \tag{12.59}
$$

$$
\begin{aligned}
\text{Now, } H_{12}^{\text{core}} = H_{21}^{\text{core}} &= \left\langle \chi_1 \middle| \hat{H}^{\text{core}} \middle| \chi_2 \right\rangle\\
&= \frac{1}{2}\xi_2^2 s_{12} + \frac{4(\xi_2 - 2)\xi_1^{3/2}\ \xi_2^{3/2}}{(\xi_1 + \xi_2)^2}
\end{aligned} \tag{12.60}
$$

$$
\text{But } s_{12} = \langle \chi_1 | \chi_2 \rangle = \frac{8\xi_1^{3/2}\xi_2^{3/2}}{(\xi_1 + \xi_2)^3}
$$

$$
\therefore\ H_{12}^c = H_{21}^c = \frac{1}{2}\xi_2^2 \times \frac{8\xi_1^{3/2}\ \xi_2^{3/2}}{(\xi_1 + \xi_2)^3} + \frac{4(\xi_2 - 2)\xi_1^{3/2}\ \xi_2^{3/2}}{(\xi_1 + \xi_2)^2} \qquad [\text{where } c \to \text{core}]
$$

$$
\begin{aligned}
&= \xi_1^{3/2}\ \xi_2^{3/2}\frac{(4\xi_1\xi_2 - 8\xi_1\xi_2)}{(\xi_1 + \xi_2)^3}\\
&= \frac{(1.45)^{3/2}(2.91)^{3/2}(4 \times 1.45 \times 2.91 - 8 \times 1.45 - 8 \times 2.91)}{(1.45 + 2.91)^3}\\
&= (75.124738)^{3/2}\frac{(16.878 - 11.6 - 23.28)}{(4.36)^3}\\
&= (75.124738)^{1/2}\frac{(-18.002)}{(4.36)^3}\\
&= (75.124738)^{1/2}\frac{(-18.002)}{82.8818}\\
&= (8.66745)\frac{(-18.002)}{82.8818} = (8.66745)(-0.2172)\\
&= -1.88257 \approx -1.8826
\end{aligned}
$$

We know that (ij/tu) are integrals, which are known as electron repulsion integrals, and such integrals are many in number. These integrals are equal to each other. In the case of real basis functions, we can illustrate

$$
\begin{aligned}
(ij/tu) &= (ji/tu) = (ij/ut) = (ji/ut) = (tu/ij)\\
&= (ut/ij) = (tu/ji) = (ut/ji)
\end{aligned} \tag{12.61}
$$

Now, we can evaluate the integrals (11/11) and (22/22) in the following manner.

Let us denote $f_1 = 2\xi_1^{3/2}e^{-\xi_1 r_2}$ and $f_2 = 2\xi_1^{3/2}e^{-\xi_1 r_2}$ representing electrons at distance r_1 and r_2 from nucleus. Then integral (11/11) can be written as

$$\int_{r_1=0}^{\infty}\int_{r_2=0}^{\infty} |f_1|^2|f_2|^2 \cdot \frac{1}{|r_1 - r_2|} \cdot r_1^2 dr_1 r_2^2 dr_2 \tag{12.62}$$

$$\int_{r_2=0}^{\infty} |f_2|^2 \left[\int_{r_1=0}^{\infty} \frac{1}{|r_1 - r_2|} \cdot |f_1|^2 r_1^2 dr_1 \right] r_2^2 dr_2$$

$$\int_{r_2=0}^{\infty} |f_2|^2 \left[\int_{r_1=0}^{r_2} \frac{r_1^2}{r_2} \cdot 4\xi_1^3 e^{-2\xi_1 r_1} dr_1 + \int_{r_2}^{\infty} \frac{r_1^2}{r_1} \cdot 4\xi_1^3 e^{-2\xi_1 r_1} dr_1 \right] r_2^2 dr_2$$

$$\int_{r_2=0}^{\infty} |f_2|^2 \left[\int_{r_1=0}^{r_2} \frac{4\xi_1^3}{r_2} \cdot r_1^2 \cdot e^{-2\xi_1 r_1} dr_1 + \int_{r_2}^{\infty} 4\xi_1^3 r_1 e^{-2\xi_1 r_1} dr_1 \right] r_2^2 dr_2$$

$$\int_0^{\infty} |f_2|^2 [I_1 + I_2] \cdot r_2^2 dr_2 \cdot 4\xi_1^3 \tag{12.63}$$

$$\text{where, } I_1 = \int_0^{r_2} \frac{r_1^2}{r_2} e^{-2\xi_1 r_1} dr_1 \text{ and } \int_{r_2}^{\infty} r_1 e^{-2\xi_1 r_1} dr_1$$

Performing integration by method of parts, we can express

$$I_1 = \frac{1}{r_2}\int_0^{r_2} r_1^2 e^{-2\xi_1 r_1} dr_1$$

$$= \frac{1}{r_2}\left\{ \left[\frac{r_1^2 e^{-2\xi_1 r_1}}{-2\xi_1} \right]^{r_2} + \frac{2}{2\xi_1}\int_0^{r_2} r_1 e^{-2\xi_1 r_1} dr_1 \right\}$$

$$= \frac{1}{r_2}\left[\frac{r_2^2 e^{-2\xi_1 r_2}}{-2\xi_1} + \frac{1}{\xi_1}\left\{ \frac{r_1 e^{-2\xi_1 r_1}}{-2\xi_1}\Big|_0^{r_2} + \frac{1}{2\xi_1}\int_0^{r_2} 1.e^{-2\xi_1 r_1} dr_1 \right\} \right]$$

$$= \frac{1}{r_2}\left[\frac{r_2^2 e^{-2\xi_1 r_2}}{-2\xi_1} - \frac{1}{2\xi_1^2} r_2 e^{-2\xi_1 r_2} + \frac{1}{2\xi_1^2}\left\{ \frac{e^{-2\xi_1 r_1}}{-2\xi_1}\Big|_0^{r_2} \right\} \right]$$

$$= \frac{1}{r_2}\left[\frac{r_2^2 e^{-2\xi_1 r_2}}{-2\xi_1} - \frac{1}{2\xi_1^2} r_2 e^{-2\xi_1 r_2} - \frac{1}{4\xi_1^3}\left\{ e^{-2\xi_1 r_2} - 1 \right\} \right]$$

$$= \frac{1}{r_2}\left[\left\{-\frac{r_2}{2\xi_1^2} - \frac{1}{4\xi_1^3}\right\}e^{-2\xi_1 r_2} + \frac{1}{4\xi_1^3}\right]$$

$$= -\left[\frac{r_2}{2\xi_1} + \frac{1}{2\xi_1^2} + \frac{1}{4\xi_1^3 r_2}\right]e^{-2\xi_1 r_2} + \frac{1}{4\xi_1^3 r_2} \tag{12.64}$$

Similarly,

$$I_2 = \int_{r_2}^{\infty} r_1 e^{-2\xi_1 r_1} \mathrm{d}r_1$$

$$= \frac{r_1 e^{-2\xi_1 r_1}}{-2\xi_1}\Big|_{r_2}^{\infty} + \frac{1}{2\xi_1}\int_{r_2}^{\infty} e^{-2\xi_1 r_2}\mathrm{d}r_1$$

$$= +\frac{r_2}{2\xi_1}e^{-2\xi_1 r_2} + \frac{1}{2\xi_1}\left\{\frac{e^{-2\xi_1 r_1}}{-2\xi_1}\Big|_{r_2}^{\infty}\right\}$$

$$\frac{r_2}{2\xi_1}e^{-2\xi_1 r_2} - \frac{1}{4\xi_1^2}\left\{e^{-\infty}e^{-2\xi_1 r_2}\right\}$$

$$\frac{r_2}{2\xi_1}e^{-2\xi_1 r_2} + \frac{1}{4\xi_1^2}e^{-2\xi_1 r_2} \tag{12.65}$$

Adding Eqs. (12.64) and (12.65), we get

$$I_1 + I_2 = \left[-\frac{r_2}{2\xi_1} - \frac{1}{2\xi_1^2} - \frac{1}{4\xi^3 r_2} + \frac{r_2}{2\xi_1} + \frac{1}{4\xi_1^2}\right]e^{-2\xi_1 r_2} + \frac{1}{4\xi_1^2 r_2}$$

$$\text{or } \mathrm{I}_1 + \mathrm{I}_2 = \left[-\left(\frac{1}{4\xi_1^2} + \frac{1}{4\xi^3 r_2}\right)e^{-2\xi_1 r_2} + \frac{1}{4\xi_1^2 r_2}\right] \tag{12.66}$$

Substituting this value in Eq. (12.63), we get

$$I = \int_{r_2}^{\infty}|f_2|^2\left[-\left(\frac{1}{4\xi_1^2} + \frac{1}{4\xi_1^3 r_2}e^{-2\xi_1 r_2} + \frac{1}{4\xi_1^3 r_2}\right)\right]\cdot 4\xi_1^3 r_2^2 \mathrm{d}r_2$$

$$= 16\xi_1^6 \int_{r_2=0}^{\infty} e^{-2\xi_1 r_2}\left[-\left(\frac{1}{4\xi_1^2} + \frac{1}{4\xi_1^3 r_2}\right)e^{-2\xi_1 r_2} + \frac{1}{4\xi_1^3 r_2}\cdot r_2^2 \mathrm{d}r_2\right]$$

$$= 16\xi_1^6\left[-\int_0^{\infty}\frac{1}{4\xi_1^2}r_2^2 e^{-4\xi_1 r_2}\mathrm{d}r_2 - \frac{1}{4\xi_1^3}\int_0^{\infty} r_2 e^{-4\xi_1 r_2}\mathrm{d}r_2 + \int_0^{\infty}\frac{r_2 e^{-2\xi_1 r_2}}{4\xi_1^3}\mathrm{d}r_2\right]$$

$$= 16\xi_1^6\left[-\frac{1}{4\xi_1^2}\cdot\frac{2!}{(4\xi_1)^3}-\frac{1}{4\xi_1^3}\cdot\frac{1!}{(4\xi_1)^2}+\frac{1}{4\xi_1^3}\cdot\frac{1}{(2\xi_1)^2}\right]$$

$$= 16\xi_1^6\left[-\frac{1}{128\xi_1^5}-\frac{1}{64\xi_1^5}+\frac{1}{16\xi_1^5}\right]$$

$$= \xi_1\left[-\frac{1}{8}-\frac{1}{4}+1\right] = \xi_1\left[\frac{-1-2+8}{8}\right] = \frac{5}{8}\xi_1$$

$$\therefore\quad \left(\frac{11}{11}\right) = \frac{5}{8}\xi_1 = \frac{5}{8}\times 1.45 = \frac{7.25}{8} = 0.9062$$

Similarly, for integral $(22/22)$, assuming that $f_k = 2\xi_2^{3/2}e^{-2\xi_2 r_1}$ and $f_k = 2\xi_2^{3/2}e^{-2\xi_2 r_2}$ and performing integration like before, we can write

$$(22/22) = \frac{5}{8}\xi_2 = \frac{5}{8}\times 2.91 = 1.81875 \approx 1.8188$$

Similarly, we can find the values of other integrals. Their values are quoted as given in the following:

$$(11/22) = (22/11) = 1.1826$$

$$(12/12) = (21/12) = (12/21) = (21/21) = 0.9536$$

$$(11/12) = (11/21) = (12/11) = (21/11) = 0.9033$$

$$(12/22) = (22/12) = (22/21) = (21/22) = 1.2980$$

For further computational work, we shall guess for the ground-state atomic orbital expansion coefficients c_{ja} as described in Eq. (12.1) with the help of which we can find out the value of P_{tu}, that is, density matrix elements as pointed out in Eq. (12.17). We know that the value of 1s STO is $\frac{27}{16} = 1.6875$ for the optimum value of orbital exponent for He. The value 1.6875 is much closer to $\xi_1 = 1.45$ than $\xi_2 = 2.91$; our expectation is that the coefficient of $\chi_1 > \chi_2$ in equation $\phi_1 = c_{11}\chi_1 + c_{21}\chi_2$.

Let us make a suitable guess for $c_{11}/c_{21} \approx 2$, but a more general method to obtain an initial guess for c_{ja} coefficients is to neglect the electron repulsion integrals in Eq. (12.16) and try to approximate F_{ij} in the secular equation $\det\left(F_{ij} - \epsilon_i\, s_{ij}\right) = 0$ by $H_{ij}^{\text{core}}\left(F_{ij} \approx H_{ij}^{\text{core}}\right)$. Under these circumstances, we will be able to solve the aforementioned secular equation and the equation $\sum_j^M c_{ja}\left(F_{ij} - \epsilon_i\, s_{ij}\right) = 0, \quad i = 1,2,\ldots M.$

This would result $c_{11}/c_{21} \approx 1.5$. It should be kept in mind that to get the real value of coefficients, the normalisation condition, that is, $\int|\phi_1|^2\,d\tau = 1$, will be helpful. For example, we can write

$$c_{21}^2 + c_{11}^2 + 2c_{21}c_{11}s_{12} = 1 \tag{12.67}$$

$$\text{or } c_{21}^2\left[1 + +\frac{c_{11}^2}{c_{21}^2} + \frac{2c_{21}c_{11}}{c_{21}^2}s_{12}\right] = 1$$

$$\text{or } c_{21}^2 = \left[1 + k^2 + 2ks_{12}\right] = 1, \text{ where} \frac{c_{11}}{c_{21}} = k,$$

$$\text{or } c_{21}^2 = \frac{1}{\left[1 + k^2 + 2ks_{12}\right]}$$

$$\therefore \ c_{21} = \frac{1}{\sqrt{1 + k^2 + 2ks_{12}}} \tag{12.68}$$

Putting $k = 2$ and $s_{12} = 0.8368$ in Eq. (12.68), we can write

$$c_{21} = \frac{1}{\sqrt{1 + 2^2 + 2 \times 2 \times 0.8368}} = \frac{1}{\sqrt{1 + 4 + 2 \times 2 \times 0.8368}} = \frac{1}{\sqrt{5 + 3.3472}} = \frac{1}{\sqrt{8.3472}}$$

$$\therefore \ c_{21} = \frac{1}{2.889} = 0.3461 \tag{12.69}$$

$$\therefore \ c_{11} = 2 \times c_{21} = 2 \times 0.3461 = 0.6922 \tag{12.70}$$

Putting $n = 2$ and $M = 2$ in Eq. (12.17), we shall get the density matrix elements.

$$\begin{aligned} P_{11} &= 2c_{11}^*c_{11} = 2 \times 0.6922 \times 0.6922 = 0.9582 \\ P_{12} &= 2c_{11}^*c_{21} = 2 \times 0.6922 \times 0.3461 = 0.4791 \end{aligned} \tag{12.71}$$

$$\text{and } P_{21} = 0.4791$$

$$P_{22} = 2c_{21}^*c_{21} = 2c_{21}^*2 \tag{12.72}$$

$$= 2 \times (0.3461)^2 = 0.2396$$

The next problem is to obtain the Fock matrix elements. This will be done by the use of Eq. (12.16) of this chapter with $M = 2$. Using Eq. (12.61) and also $P_{12} = P_{21}$, for real functions, we shall write the values of F_{11}, F_{12}, and F_{22}.

$$F_{11} = H_{11}^{\text{core}} + \frac{1}{2}P_{11}(11/11) + P_{12}(11/12) + P_{22}\left[(11/12) - \frac{1}{2}(12/21)\right] \tag{12.73}$$

$$F_{12} = F_{21} = H_{12}^{\text{core}} + \frac{1}{2}P_{11}(12/11) + P_{12}\left[\frac{3}{2}(12/12) - \frac{1}{2}(11/22)\right] + \frac{1}{2}P_{22}(12/21) \tag{12.74}$$

$$F_{22} = H_{22}^{\text{core}} + P_{11}\left[(22/11) - \frac{1}{2}(21/12)\right] + P_{12}(22/12) + \frac{1}{2}P_{22}(22/22) \tag{12.75}$$

Substituting the values of H_{ij}^{core} and (ij/tu) integrals, we can write

$$F_{11} = -1.8488 + 0.4531P_{11} + 0.9033P_{12} + 0.7058P_{22} \tag{12.76}$$

$$F_{12} = F_{21} = -1.8488 + 0.4516P_{11} + 0.8391P_{12} + 0.649P_{22} \tag{12.77}$$

$$F_{22} = -1.5860 + 0.7058P_{11} + 1.2980P_{12} + 0.9094P_{22} \tag{12.78}$$

Now, we shall put the values of P_{tu} from Eq. (12.76) into Eq. (12.78) and get the values of F_{ij} which are given in the following.

$$F_{11} = -1.8488 + 0.4531 \times 0.9583 + 0.9033 \times 0.4791 + 0.7058 \times 0.2396$$
$$= -1.8488 + 0.4342 + 0.43277 + 0.1691$$
$$= -1.8488 + 1.030 \quad (12.79)$$

Similarly, from Eq. (12.77), we have

$$F_{11} = F_{21} = -1.8826 + 0.4516P_{11} + 0.8391P_{12} + 0.6490P_{22}$$
$$= -1.8826 + 0.4516 \times 0.9583 + 0.8391 \times 0.4791 + 0.6490 \times 0.2396$$
$$= -1.8826 + 0.43276 + 0.4020 + 0.1555$$
$$= -1.8826 + 0.9902 = -0.8920 \quad (12.80)$$

$$\text{and } F_{22} = -1.5860 + 0.7058P_{11} + 1.2980P_{11} + 0.9094P_{22}$$
$$= -1.5860 + 0.7058 \times 0.9583 + 1.2980 \times 0.4791 + 0.9094 \times 0.2396$$
$$= -1.5860 + 0.6764 + 0.6219 + 0.2179$$
$$= -1.5860 + 1.516 = 0.070 \quad (12.81)$$

Now, we can write the secular determinant as given in the following:

$$\begin{vmatrix} -0.813 - \epsilon_i & -0.892 - 0.8366\ \epsilon_i \\ -0.892 - 0.8366 - \epsilon_i & -0.070 - \epsilon_i \end{vmatrix} = 0$$

This will lead to $\left(-0.813 - \epsilon_i\right)\left(-0.070 - \epsilon_i\right) = \left(-0.892 - 08366\ \epsilon_i\right)\left(-0.892 - 8366\ \epsilon_i\right)$

$$\text{or} \quad 0.3001\ \epsilon_i^2 - 0.609\ \epsilon_i - 0.739 \approx 0 \quad (12.82)$$

$$\therefore \quad \epsilon_i = \frac{+0.609 \underline{+} \sqrt{(0609)^2 + 4(0.3001)(0.739)}}{2 \times 0.3001}$$

$$\therefore \quad \epsilon_i = \frac{+0.609 \underline{+} \sqrt{0.3709 + 0.8871}}{0.6002} = \frac{+0.609 \underline{+} \sqrt{1.258}}{0.6002} = \frac{+0.609 \underline{+} 1.1216}{0.6002}$$

$$\frac{+0.609 + 1.1216}{0.6002} \text{ or } \frac{+0.609 + 1.1216}{0.6002}$$

$$\therefore \ \epsilon_i = -0.8540 \text{ or } +2.8834$$

$$\therefore \ \epsilon_1 = -0.8540 \text{ and } \epsilon_2 = +2.8834 \quad (12.83)$$

Putting the value of ϵ_i, that is, ϵ_1 in Roothaan's equations, we may write

$$c_{11}\left(F_{21} - \epsilon_1\ s_{21}\right) + c_{21}\left(F_{22} - s_{22}\right) \approx 0 \quad (12.84)$$

Putting the required values, we have

$$-0.177c_{11} + 0.784c_{21} = 0$$

$$\text{or } 0.784c_{21} = 0.177c_{11}$$

$$\therefore \frac{c_{11}}{c_{21}} = \frac{0.784}{0.177} = 4.43 \tag{12.85}$$

Now, substituting $k = 4.43$ and $s_{12} = 0.8366$ in the normalisation condition, we shall obtain

$$c_{21} = 0.189 \text{ and } c_{11} = 4.43 \times c_{21} = 0.837$$

Putting these improved values of c_{11} and c_{22} in Eqs. (12.71) to (12.72), we shall obtain $P_{11} = 1.398$, $P_{12} = P_{21} \approx 0.316$, and $P_{22} = 0.071$. The improved values of F_{ij} will be obtained after substitution of P_{tu} values from Eq. (12.76) into Eq. (12.77). On calculation, we shall get the following values of F_{ij}, that is,

$$F_{11} = -0.880,\ F_{12} = F_{21} = -0.940,\ F_{22} = -0.124$$

Finally, the improved secular equation will be expressed as

$$\begin{vmatrix} -0.880 - \epsilon_i & -0.940 - 0.8366\,\epsilon_i \\ -0.940 - 0.8366\,\epsilon_i & -0.124 - \epsilon_i \end{vmatrix} = 0 \tag{12.86}$$

This will give a quadratic equation in E_i as earlier. The roots of the equation will be found to be $E_1 = -0.918$ and $E_2 = 2.810$.

With the help of the improved value of E_i, we shall obtain

$$c_{11} / c_{21} = 4.61 \text{ and } c_{11} = 0.842 \text{ and } c_{21} = 0.183.$$

Now, the another cycle of computation will give

$$P_{11} = 1.148,\ P_{12} = P_{21} = 0.308,\ P_{22} = 0.067$$

$$F_{11} = -0.881,\ F_{12} = F_{21} = -0.940,\ F_{22} = -0.124$$

$$\epsilon_1 = -0.918,\ \epsilon_2 = 2.8092 \tag{12.87}$$

$$c_{11} = 0.842,\ c_{21} = 0.183 \tag{12.88}$$

It is, thus, clear that the last values of c_{21} resemble approximately the values quoted earlier; hence, the computation has converged. Thus, the SCF atomic orbital for the ground state of He with regard to this basis set may be expressed as

$$\phi_1 = 0.842\chi_1 + 0.183\chi_2 \tag{12.89}$$

$$\therefore E_{HF} = -0.918 + \frac{1}{2}\left[1.418(-1.8488) + 2(0.308)(-1.8826) + 0.067(-1.5860) + 0\right]$$

$$= -2.862 \text{ Hartress which is equivalent to} -77.9 \text{ eV.}$$

$$\therefore E_{HF} = -77.9 \text{ eV.} \tag{12.90}$$

Roetti and Clementi have calculated the SCF energy of He atom using $\xi_1 = 1.45363$ and $\xi_2 = 2.91093$ and have found the value of $E_{HF} = -2.8616726$ Hartrees.

12.6 Hybridisation

In order to increase the strength of a bond, it is worth having to find a way of enhancing the overlap integral between the orbitals taking part. The manner to do this can well be understood in terms of one of the fundamental properties of atomic Eigen functions. If ϕ_a and ϕ_b be two Eigen functions of an operator H with Eigen values E_a and E_b, the normalised sum of the Eigen functions will also be an Eigen function of the Hamiltonian having Eigen value $\frac{1}{N}(E_a - E_b)$, where N is the normalisation factor. This follows due to the fact that the Hamiltonian is a linear operator. For finding orbitals that overlap strongly, it is essential to obtain new combination of the orbitals in the valence shell of an atom, which are strongly directional and overlap to a large extent with the orbitals of nearby atoms.

In order to perform this, we mix orbitals having the same valence shell, e.g., s and p orbitals first to give the p orbitals some s character, which enhances the strength of the total outcome of the bonding, and also to give a set of strongly directed bonds. If we look at the electronic configuration of C atom ($1(s^2\ 2s^2\ 2p^2)$, it can be said that the $2s$ electrons are paired and the $2p$ electrons are unpaired, which should be available for bond formation and C atom may be expected to be divalent. But it is a fact that most of the compounds are formed or known in tetravalent state. To achieve this, C atom has to be excited, and due to excitation, one electron from $2s$ orbital will be promoted to $2p$ orbital; under this condition, the valency of C will be four or C will be tetravalent. Thus, one electron remains in s orbital, and three electrons remain in $2p_x$, $2p_y$, and $2p_z$ orbitals, but each of p_x or p_y or p_z will contain one electron.

Since the s orbital and the three p orbitals are not largely different, it is possible to mix them, i.e., make linear combination of these for getting four equivalent hybrid orbitals. The mixing can be performed in three different manners:

1. One s orbital may mix with all three p orbitals to give four hybrid orbitals, and this phenomenon is called sp^3 hybridisation.
2. One s orbital may mix with any two of the p orbitals to yield three sp^2 hybrid orbitals, and this type of mixing is called sp^2 hybridisation.
3. One s orbital may mix with one of the p orbitals to give two sp hybrid orbitals, and this type of mixing is named as sp hybridisation.

We shall deal with the three types of hybridisation quantum mechanically one by one.

12.6.1 sp^3 Hybridisation

Hybridisation can be affected by carrying out a linear combination of different orbitals. Let ψ_1, ψ_2, ψ_3, and ψ_4 denote the resulting bond orbitals. Then these will be written as

$$\left.\begin{aligned}\psi_1 &= a_1\psi_s + b_1\psi_{pz} + c_1\psi_{px} + \mathrm{d}_1\psi_{py}\\ \psi_2 &= a_2\psi_s + b_2\psi_{pz} + c_2\psi_{px} + \mathrm{d}_2\psi_{py}\\ \psi_3 &= a_3\psi_s + b_3\psi_{pz} + c_3\psi_{px} + \mathrm{d}_3\psi_{py}\\ \psi_4 &= a_4\psi_s + b_4\psi_{pz} + c_4\psi_{px} + \mathrm{d}_4\psi_{py}\end{aligned}\right\} \tag{12.91}$$

It should be kept in mind that ψ_1, ψ_2, ... can be normalised and orthogonalised. Let ψ_1 be normalised. Then, we can write

$$\int \psi_1^2 \mathrm{d}\tau = \int \left(a_1\psi_s + b_1\psi_{pz} + c_1\psi_{px} + \mathrm{d}_1\psi_{py}\right)^2 \mathrm{d}\tau$$

$$= \int \left\{a_1^2\psi_s^2 + b_1^2\psi_{ps}^2 + c_1^2\psi_{py}^2 + \mathrm{d}_1^2\psi_{py}^2 + 2a_1c_1\psi_s\psi_{px} + 2a_1\mathrm{d}_1\psi_s\psi_{py}\right.$$

$$\left. + 2b_1c_1\psi_{pz}\psi_{px} + 2b_1\mathrm{d}_1\psi_{pz}\psi_{py} + 2c_1\mathrm{d}_1\psi_{px}\psi_{py}\right\}\mathrm{d}\tau = 1 \tag{12.92}$$

But we see that ψ_s, ψ_{pz}, ψ_{px}, and ψ_{py} are all normalised, and therefore, $\int \psi_s^2 \mathrm{d}\tau$, $\int \psi_{pz}^2 \mathrm{d}\tau$, $\int \psi_{px}^2 \mathrm{d}\tau$, and $\int \psi_{py}^2 \mathrm{d}\tau$, are equal to one and all the pairs $\int \psi_s \psi_{pz} \mathrm{d}\tau$, etc., are orthogonal, and therefore, $\int \psi_s \psi_{pz} \mathrm{d}\tau, = 0$

Thus, we can write

$$a_1^2 + b_1^2 + c_1^2 + d_1^2 = 1 \tag{12.93}$$

Similarly, from the others mentioned in Eq. (12.91), we can write

$$a_i^2 + b_i^2 + c_i^2 + d_i^2 = 1 \tag{12.94}$$

But from orthogonality condition, we have

$$\int \psi_1 \psi_2 \mathrm{d}\tau = 0 = \int \left(a_1\psi_s + b_1\psi_{pz} + c_1\psi_{px} + d_1\psi_{py}\right) \times \left(a_2\psi_s + b_2\psi_{pz} + c_2\psi_{px} + d_2\psi_{py}\right) \tag{12.95}$$

From this, we have

$$a_1a_2 + b_1b_2 + c_1c_2 + d_1d_2 = 0 \tag{12.96}$$

and in general, we can express

$$a_ia_j + b_ib_j + c_ic_j + d_id_j = 0 \tag{12.97}$$

Different combinations will lead to six such equations.

It should be kept in mind that there will be 16 variables (a_1, ... d_4) and only 10 equations (Eqs. 12.93 and 12.97)). Thus, we see that six more equations are required for getting the solution of the problem. These six equations are degree of freedom in the system for which arbitrary values may be assigned. In the present case, we shall assume $c_1 = d_1 = 0$, $d_2 = 0$, $a_2 = a_3 = a_4$, and $d_3 = -d_4$.

i. Let us fix two variables $c_1 = 0$ and $d_1 = 0$. Therefore, ψ_1 will be written as

$$\psi_1 = a_1\psi_s + b_1\psi_{pz}$$

This gives a clue to understand that we are mixing the ψ_s orbital with the ψ_{pz} orbital in the z direction. Hence, we get

$$\psi_1 = a_1\psi_s + b_1\psi_{pz} \tag{12.98}$$

Applying normalisation condition, we obtain

$$a_1^2 + b_1^2 = 1 \text{ or } 1 - a_1^2 = b_1^2$$

$$\text{or } b_1 = \sqrt{\left(1 - a_1^2\right)}$$

$$\therefore \quad \psi_1 = a_1\psi_s + \sqrt{\left(1 - a_1^2\right)\psi_{pz}} \tag{12.99}$$

$$\left.\begin{aligned}\text{But if}\quad \psi_s &= f(r)\\ \psi_{pz} &= \sqrt{3}f(r)\cos\theta\\ \psi_{px} &= \sqrt{3}f(r)\sin\theta\cos\phi\\ \psi_{py} &= \sqrt{3}f(r)\sin\theta\sin\phi\end{aligned}\right\} \tag{12.100}$$

The relative values of the wave functions may be written on the basis of Eq. (12.100) as

$$\left.\begin{aligned}\psi_s &= 1\\ \psi_{pz} &= \sqrt{3}\cos\theta\\ \psi_{px} &= \sqrt{3}\sin\theta\cos\phi\\ \text{and}\qquad \psi_{py} &= \sqrt{3}\sin\theta\sin\phi\end{aligned}\right\}$$

From Eq. (12.99), we have

$$\psi_1 = a_1\psi_s + \sqrt{\left(1-a_1^2\right)}\psi_{pz}$$

For the z direction, $\sin\theta = 0$ and $\cos\theta = 1$.

$$\therefore\quad \psi_1 = a_1\cdot 1 + \sqrt{\left(1-a_1^2\right)\cdot\sqrt{3}\cos\theta}$$

On putting $\cos\theta = 1$, the aforementioned equation changes to

$$\psi_1 = a_1 + \sqrt{\left(1-a_1^2\right)\cdot\sqrt{3}} \tag{12.101}$$

Differentiating this equation with respect to a_1, we get

$$\frac{d\psi_1}{da_1} = 1 - \frac{1}{2}\cdot\frac{\sqrt{3}\cdot 2a_1}{\sqrt{\left(1-a_1^2\right)}} \tag{12.102}$$

For maximum condition, $\frac{d\psi_1}{da_1} = 0$, Hence, Eq. (12.102) will be

$$1 - \frac{1}{2}\cdot\frac{\sqrt{3}\cdot 2a_1}{\sqrt{\left(1-a_1^2\right)}} = 0$$

$$\text{or}\quad 1 - \frac{\sqrt{3}\cdot a_1}{\sqrt{\left(1-a_1^2\right)}} = 0$$

$$\text{or}\quad \frac{\sqrt{3}\cdot a_1}{\sqrt{\left(1-a_1^2\right)}} = 1 \quad\text{or}\quad \left(\frac{\sqrt{3}\cdot a_1}{\sqrt{\left(1-a_1^2\right)}}\right)^2 = 1$$

$$\frac{3a_1^2}{1-a_1^2}=1$$

$$\text{or}\quad 3a_1^2=1-a_1^2$$

$$\text{or}\quad 4a_1^2=1 \quad \therefore\ a_1^2=1/4 \quad \therefore\ a_1=1/2 \tag{12.103}$$

and, therefore,

$$b_1=\sqrt{\left(1-a_1^2\right)}=\sqrt{1-1/4}=\sqrt{\frac{4-1}{4}}=\sqrt{\frac{3}{4}}=\frac{\sqrt{3}}{2} \tag{12.104}$$

Then, the wave function

$$\psi_1=\frac{1}{2}\psi_s+\frac{\sqrt{3}}{2}\psi_{pz} \tag{12.105}$$

$$\begin{aligned}&=\frac{1}{2}f(r)+\frac{\sqrt{3}}{2}\cdot\sqrt{3}f(r)\\&=2f(r)\end{aligned} \tag{12.106}$$

In the absence of hybridisation,

$$\psi_{pz}=\sqrt{3}f(r)\cos\theta=\sqrt{3}f(r)\quad[\because\ \cos\theta=1]$$

Therefore, $\psi_1>\psi_{pz}$, which clearly indicates that the hybridised bond is stronger than the pure bond.

ii. $\psi_2=a_2\psi_s+b_2\psi_{pz}+c_2\psi_{px}+d_2\psi_{py}$

In this case, let $d_2=0$. Hence, we assume that the second bond is formed in the z–x plane.

$$\therefore\quad \psi_2=a_2\psi_s+b_2\psi_{pz}+c_2\psi_{px} \tag{12.107}$$

Applying the normalisation and orthogonality condition,

$$a_2^2+b_2^2+c_2^2=1 \tag{12.108}$$

$$\text{and}\quad a_1a_2+b_1b_2=0 \quad \because\ c_1=0 \tag{12.109}$$

Putting the values of a_1 and b_1 in Eq. (12.109), we have

$$\frac{1}{2}a_2+\frac{\sqrt{3}}{2}b_2=0\ \text{ or }\ b_2=-\frac{a_2}{\sqrt{3}}$$

And from Eq. (12.108), we can write

$$\begin{aligned}c_2^2&=1-a_2^2-b_2^2\\&=1-a_2^2-\frac{a_2^2}{3}=\frac{3-4a_2^2}{3}\end{aligned}$$

$$\therefore \quad c_2 = \left(\frac{3-4a_2^2}{3}\right)^{1/2} \tag{12.110}$$

$$\begin{aligned} \therefore \quad \psi_2 &= a_2\psi_s - \frac{a_2}{\sqrt{3}}\psi_{p_z} + \sqrt{\frac{3-4a_2^2}{3}}\psi_{p_x} \\ a_2 &- \frac{a_2}{\sqrt{3}}\cdot\sqrt{3}\cos\theta + \sqrt{\frac{3-4a_2^2}{3}}\sqrt{3}\sin\theta \end{aligned} \tag{12.111}$$

$$\left[\because \phi = 0 \text{ along } x\right]$$

This function indicates the bond formation in z–x plane inclined at angle θ with respect to z axis. In the present case, a_2 and θ are two unknown constants. To obtain the maximum bond strength, both $\frac{d\psi_2}{da_2} = 0$ and $\frac{d\psi_2}{d\theta_2} = 0$.

Differentiating Eq. (12.111) with respect to a_2, we get $\frac{d\psi_2}{da_2} = 0$, that is

$$\frac{d\psi_2}{da_2} = 1 - \cos\theta - \frac{1}{2}\cdot\frac{1}{\sqrt{3-4a_2^2}}\cdot 8a_2\sin\theta = 0 \tag{12.112}$$

at maximum condition, and differentiating ψ with respect to θ and at maximum condition $\frac{d\psi_2}{d\theta_2} = 0$,

$$\begin{aligned} \therefore \quad & \frac{d\psi_2}{d\theta_2} = a_2\sin\theta + \sqrt{3-4a_2^2}\cdot\cos\theta = 0 \\ \text{or} \quad & a_2\sin\theta - \sqrt{3-4a_2^2}\cdot\cos\theta \end{aligned} \tag{12.113}$$

$$\therefore \quad \sin\theta = \frac{\sqrt{3-4a_2^2}\cdot\cos\theta =}{a_2} \tag{12.114}$$

Substituting this value of sin θ in Eq. (12.112), we shall get

$$1 - \cos\theta + \frac{4a_2}{\sqrt{3-4a_2^2}}\frac{\sqrt{3-4a_2^2}}{a_2}\cdot\cos\theta = 0$$

$$\text{or } 1 - \cos\theta + 4\cos\theta = 0$$

$$\text{or } 1 + 3\cos\theta = 0$$

$$\therefore \quad \cos\theta = -\frac{1}{3} \tag{12.115}$$

$$\therefore \quad \theta = 109°28'$$

Thus, $\theta = 109°28'$ between the two bonds (ψ_1 and ψ_2) indicates the tetrahedral model for carbon atom in its compounds.

Furthermore,
$$\sin\theta = \sqrt{1-\cos^2\theta} = \sqrt{1-\left(\frac{-1}{3}\right)^2}$$
$$= \sqrt{1-\frac{1}{9}} = \sqrt{\frac{8}{9}} = \frac{2\sqrt{2}}{3}$$

Substituting this value in Eq. (12.113), we shall obtain

$$a_2 \cdot \frac{2\sqrt{2}}{3} - \frac{1}{3}\sqrt{3-4a_2^2} = 0$$

$$\text{or } a_2 \cdot \frac{2\sqrt{2}}{3} - \frac{1}{3}\sqrt{3-4a_2^2}$$

$$\text{or } 2\sqrt{2}a_2 = \sqrt{3-4a_2^2}$$

$$\text{or } \left(2\sqrt{2}a_2\right)^2 = \left(\sqrt{3-4a_2^2}\right)^2$$

$$\text{or } 8a_2^2 = 3 - 4a_2^2$$

$$\text{or } 8a_2^2 + 4a_2^2 = 3$$

$$\text{or } 12a_2^2 = 3$$

$$\therefore\ a_2^2 = \frac{3}{12} = \frac{1}{4} \qquad \therefore\ a_2 = \frac{1}{2}$$

$$\text{But } b_2 = \frac{-a_2}{\sqrt{3}} = -\frac{1}{2\sqrt{3}} \qquad \left[\because\ a_2 = \frac{1}{2}\right]$$

From Eq. (12.111), we have

$$\psi_2 = \left[a_2 - \frac{a_2}{\sqrt{3}} \cdot \sqrt{3}\cos\theta + \frac{\sqrt{3-4a_2^2}}{3} \cdot \sqrt{3}\sin\theta\right] f(r)$$

$$= \left[\frac{1}{2} - \frac{1}{2\sqrt{3}} \cdot \sqrt{3}\cos\theta + \frac{\sqrt{3-4a_2^2}}{3} \cdot \sqrt{3}\sin\theta\right] f(r)$$

$$= \left[\frac{1}{2} + \frac{1}{2} \cdot \frac{1}{3} + \frac{\sqrt{3-4\times\frac{1}{4}}}{3} \cdot \sqrt{3} \cdot \frac{2\sqrt{2}}{3}\right] f(r) \qquad \left[\because \cos\theta = -\frac{1}{3} \text{ and } \sin\theta = \frac{2\sqrt{2}}{3}\right]$$

$$= \left(\frac{1}{2} + \frac{1}{6} + \frac{\sqrt{2}}{\sqrt{3}} \cdot \sqrt{3} \cdot \frac{2\sqrt{2}}{3} \right) f(r)$$

$$= \left(\frac{1}{2} + \frac{1}{6} + \frac{4}{3} \right) f(r) = 2f(r)$$

This is in terms of radial wave function. $\therefore \ \psi = \psi_2$. Hence, the bonds are equivalent, and they are making an angle 109°28′ with each other.

iii. The third one is $\psi_3 = a_3\psi_s + b_3\psi_{p_z} + c_3\psi_{p_x} + d_3\psi_{p_y}$.

In the present case, we assume $a_2 = a_3$, but since $a_2 = \frac{1}{2}$, $a_3 = \frac{1}{2}$ so from normalisation, we get

$$a_3^2 + b_3^2 + c_3^2 + \mathrm{d}_3^2 = 1 \tag{12.116}$$

But on the application of orthogonality, we have

$$\left. \begin{aligned} a_1a_3 + b_1b_3 + c_1c_3 = 0 \\ a_2a_3 + b_2b_3 + c_2c_3 = 0 \end{aligned} \right\} \tag{12.117}$$

From Eq. (12.117), we get

$$a_1a_3 + b_1b_3 = 0 \quad (\because \ c_1 = d_1 = 0)$$

$$\text{and } \ a_2a_3 + b_2b_3 + c_2c_3 = 0 \quad (\because \ \mathrm{d}_2 = 0)$$

Substituting $a_1 = 1/2, b_1 = \sqrt{3}/2$ in $a_1a_3 + b_1b_3 = 0$,

and $a_2 = 1/2$ and $\mathrm{b}_2 = -1/\left(2\sqrt{3}\right)$ in $a_2a_3 + b_2b_3 + a_2c_3 = 0$, we shall obtain

$$\frac{1}{2}a_3 + \frac{\sqrt{3}}{2}b_3 = 0$$

$$\text{or } a_3 = -\sqrt{3}b_3$$

$$\text{or } b_3 = -\frac{a_3}{\sqrt{3}} = -\frac{1}{2\sqrt{3}} \tag{12.118}$$

$$\text{and } \frac{1}{2}a_3 + c_2c_3 - \frac{1}{2\sqrt{3}}b_3 = 0$$

$$\text{But } c_2 = \frac{\sqrt{2}}{3}$$

Putting this value, we get

$$\frac{1}{2} \times \frac{1}{2} - \frac{1}{2\sqrt{3}} \cdot b_3 + \sqrt{\frac{2}{3}}c_3 = 0$$

$$\text{or } \frac{1}{4} + \frac{1}{2\sqrt{3}} \times \frac{1}{2\sqrt{3}} + \sqrt{\frac{2}{3}} c_3 = 0$$

$$\text{or } c_3 = -\frac{1}{\sqrt{6}} \tag{12.119}$$

Hence, from the relation $a_3^2 + b_3^2 + c_3^2 + d_3^2 = 1$ and substituting the values, we shall get

$$\frac{1}{4} + \frac{1}{12} + \frac{1}{6} + d_3^2 = 1$$

$$\text{or } \frac{6+2+4}{24} + d_3^2 = 1$$

$$\text{or } \frac{12}{24} + d_3^2 = 1$$

$$d_3^2 = 1 - \frac{1}{2} = \frac{1}{2}$$

$$\therefore \quad d_3 = \frac{1}{\sqrt{2}}$$

Therefore, the wave function can be expressed as

$$\psi_3 = \frac{1}{2}\psi_s - \frac{1}{2\sqrt{3}}\psi_{p_z} - \frac{1}{\sqrt{6}}\psi_{p_x} + \frac{1}{\sqrt{2}}\psi_{p_y} \tag{12.120}$$

Similarly, if we put $a_3 = a_4$ and $d_4 = -d_3$, then ψ_4 can be given by

$$\psi_4 = \frac{1}{2}\psi_s - \frac{1}{2\sqrt{3}}\psi_{p_z} - \frac{1}{\sqrt{6}}\psi_{p_x} + \frac{1}{\sqrt{2}}\psi_{p_y} \tag{12.121}$$

We can find out the values of the Eigen functions ψ_3 and ψ_4. All these may be shown to have the same values, and they will also make the same angle (109°28′) with each other. Hence, the wave functions will have the same electron density, though the bonds formed are in different directions and formed by the linear combination of atomic orbital (LCAO) of *s* and *p* orbitals. The shape of hybrid orbitals (sp^3 hybrid orbitals) is given in Figure 12.2.

Such hybrid orbitals will be four in number, and they will be directed towards the corner of a regular tetrahedron.

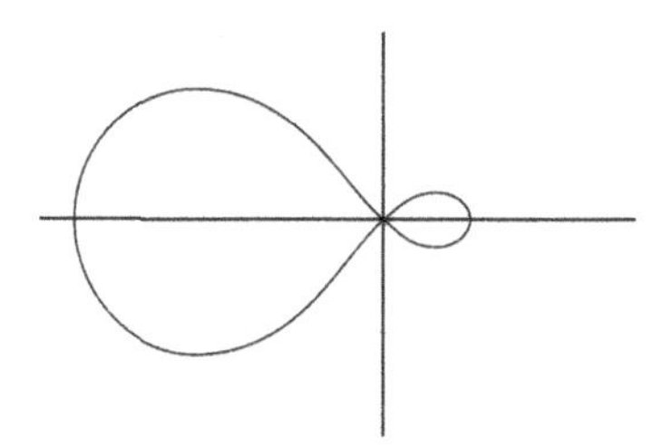

FIGURE 12.2 sp^3 hybrid orbital.

12.6.2 *sp*² Hybridisation

When one *s* orbital and two *p* orbitals combine, three *sp*² hybrid orbitals are formed, and the phenomenon of such formation is called *sp*² hybridisation.

Let us assume that

$$\left.\begin{aligned} \psi_1 &= a_1\psi_s + b_1\psi_{p_x} + c_1\psi_{p_y} \\ \psi_2 &= a_2\psi_s + b_2\psi_{p_x} + c_2\psi_{p_y} \\ \psi_3 &= a_3\psi_s + b_3\psi_{p_x} + c_3\psi_{p_y} \end{aligned}\right\} \quad (12.122)$$

On normalisation, we can write

$$\left.\begin{aligned} a_1^2 + b_1^2 + c_1^2 &= 1 \\ a_2^2 + b_2^2 + c_2^2 &= 1 \\ a_3^2 + b_3^2 + c_3^2 &= 1 \end{aligned}\right\} \quad (12.123)$$

or in general form $a_i^2 + b_i^2 + c_i^2 = 1$. From orthogonality condition, we have

$$\left.\begin{aligned} a_1a_2 + b_1b_2 + c_1c_2 &= 0 \\ a_1a_3 + b_1b_3 + c_1c_3 &= 0 \\ a_2a_3 + b_2b_3 + c_2c_3 &= 0 \end{aligned}\right\} \quad (12.124)$$

Thus, we find that there are nine variables and six equations; so, we have three degrees of freedom. Since the three hybrid orbitals are equivalent, the s orbital is supposed to divide itself equally among them, i.e.,

$a_1^2 + a_2^2 + a_3^2 = \frac{1}{3}$, or $a_1 = a_2 + a_3 = \frac{1}{\sqrt{3}}$, which satisfies only two degrees of freedom.

Let us suppose that $c_1 = 0$ in $\psi_1 = a_1\psi_s + b_1\psi_{p_x} + c_1\psi_{p_y}$ for the purpose of the hybridisation of ψ_s with ψ_{p_x} only. Then,

$$\psi_1 = a_1\psi_s + b_1\psi_{p_x} \quad (12.125)$$

Now, from Eq. (12.125), we can write on normalisation

$$a_1^2 + b_1^2 = 1$$

$$\text{or } b_1^2 = 1 - a_1^2 = 1 - \frac{1}{3} = \frac{2}{3}$$

$$\because\ b_1 = \sqrt{\frac{2}{3}}$$

Hence, from Eq. (12.125), we can write

$$\psi_1 = \frac{1}{\sqrt{3}} f(r) + \sqrt{\frac{2}{3}} \cdot \sqrt{3} f(r) \sin\theta \cos\theta \quad (12.126)$$

But for the x axis, $\theta = 90°$ and $\phi = 0$, and hence, $\psi_1 = \left(\frac{1}{\sqrt{3}} + \sqrt{2}\right) f(r)$.

$$= 2.014\ f(r)$$

Therefore, it may be concluded that $\psi_1 > \psi_{p_x}$, and hence, ψ_1 can finally be expressed

$$\psi_1 = \frac{1}{\sqrt{3}}\psi_s + \sqrt{\frac{2}{3}}\psi_{p_x} \tag{12.127}$$

By orthogonality of ψ_1 and ψ_2, we can express

$$a_1a_2 + b_1b_2 + c_1 + c_2 = 0$$

$$\text{or } \frac{1}{\sqrt{3}} \cdot \frac{1}{\sqrt{3}} + \sqrt{\frac{2}{3}} \cdot b_2 = 0$$

$$\text{or } \frac{1}{3} + \sqrt{\frac{2}{3}} b_2 = 0$$

$$\therefore\ b_2 = -\frac{1}{\sqrt{6}}$$

But by normalisation of ψ_2, we can write $a_2^2 + b_2^2 + c_2^2 = 1$

$$\text{or } \frac{1}{3} + \frac{1}{6} + c_2^2 = 1$$

$$\text{or } \frac{2+1}{6} + c_2^2 = 1$$

$$\text{or } \frac{3}{6} + c_2^2 = 1$$

$$\text{or } \frac{1}{2} + c_2^2 = 1$$

$$\therefore\ c_2^2 = 1 - \frac{1}{2} = \frac{1}{2}$$

$$\therefore\ c_2 = \frac{1}{\sqrt{2}}$$

$$\text{Thus, } \psi_2 = \frac{1}{\sqrt{3}}\psi_s - \frac{1}{\sqrt{6}}\psi_{p_x} + \frac{1}{\sqrt{6}}\psi_{p_y} \tag{12.128}$$

Similarly, considering the orthogonality of ψ_1 and ψ_3, we can write

$$a_1a_3 + b_1b_3 + c_1c_3 = 0 \tag{12.129}$$

and by orthogonality of ψ_2 and ψ_3, one can write

$$a_2a_3 + b_2b_3 + c_2c_3 = 0 \tag{12.130}$$

$$\text{or } \frac{1}{3} + \frac{1}{6} + \frac{1}{\sqrt{2}}c_3 = 0$$

$$\text{or } c_3 = -\frac{1}{\sqrt{2}}$$

Hence, ψ_3 can be written as

$$\psi_3 = \frac{1}{\sqrt{3}}\psi_s - \frac{1}{\sqrt{6}}\psi_{px} - \frac{1}{\sqrt{2}}\psi_{py} \tag{12.131}$$

From the alternative equation of Eq. (12.128), we can write

$$\psi_2 = \left(\frac{1}{\sqrt{3}} - \frac{1}{\sqrt{2}}\cos\phi + \frac{\sqrt{3}}{2}\sin\phi\right)$$

since for xy plane, $\theta = 90°$

Differentiating this equation with respect to ϕ, we get

$$\frac{d\psi_2}{d\phi} = \frac{1}{\sqrt{2}}\sin\phi + \frac{\sqrt{3}}{\sqrt{2}}\cos\phi = 0$$

$$\text{or } \sin\phi + \sqrt{3}\cos\phi$$

$$\text{or } \frac{\sin\phi}{\cos\phi} = -\sqrt{3}$$

$$\text{or } \tan\phi = -\sqrt{3}$$

$\therefore \phi = 120°$, which is the angle between the hybrid orbitals in the XY plane.

In the similar manner, by mixing ψ_s with ψ_{p_x} and ψ_{p_y}, we shall get $\phi = 120°$. It is thus concluded that the hybrids are all inclined at angles of 120° with each other.

It should be kept in mind that sp^2 hybridisation occurs in unsaturated hydrocarbons as in ethylene and also in many boron compounds like BCl_3, BBr_3, BF_3, etc. The structure in such cases is triangular planar.

12.6.3 *sp* Hybridisation

When one *s* orbital and one *p* orbital are mixed, the mixing produces two *sp* hybrid orbitals, and the phenomenon is called *sp* hybridisation.

Let us assume that

$$\left.\begin{aligned} \psi_1 &= a_1\psi_s + b_1\psi_{p_x} \\ \psi_2 &= a_2\psi_s + b_2\psi_{p_x} \end{aligned}\right\} \tag{12.132}$$

By applying normalisation condition, one can write

$$\left.\begin{aligned} a_1^2 + b_1^2 = 1 \\ a_2^2 + b_2^2 = 1 \end{aligned}\right\} \tag{12.133}$$

From orthogonality condition, we can write

$$a_1 a_2 + b_1 b_2 = 0 \tag{12.134}$$

It is, thus, clear that there are four unknown quantities, but three equations are available. We have one degree of freedom, and we are at liberty to assign a particular value to any of these variables a_1, a_2, b_1, and b_2.

Let us consider $b_1 = -b_2$. Such a choice is not arbitrary but has some justification if we remember that ψ_s is being hybridised; when each of the lobes has a coefficient b_1, the other will have $b_2 = -b_1$ as the coefficient.

Under this condition, $b_2 = -b_1$ will lead to $a_1 = a_2$ and $a_1 = b_1$ from Eqs. (12.133) and (12.134).

Hence, $2a_1^2 = 1$ from Eq. (12.133).

$$\therefore\ a_1^2 = \frac{1}{2} \quad \text{or} \quad a_1 = \frac{1}{\sqrt{2}} a_2$$

From Eq. (12.133), we may write

$$2b_1^2 = 1 \quad \therefore\ b_1 = \frac{1}{\sqrt{2}} = -b_2$$

$$\therefore\ b_2 = \frac{-1}{\sqrt{2}}$$

$$\therefore\ \psi_1 = \frac{1}{\sqrt{2}}\psi_s + \frac{1}{\sqrt{2}}\psi_{p_x} \tag{12.135}$$

Similarly, we can write for ψ_2, i.e.,

$$\psi_2 = \frac{1}{\sqrt{2}}\psi_s + \frac{1}{\sqrt{2}}\psi_{p_x} \tag{12.136}$$

In other words, we can express

$$\begin{aligned} \psi_1 &= a_1\psi_s + b_1\psi_{p_x} \\ &= \frac{1}{\sqrt{2}} f(r) + \frac{1}{\sqrt{2}} \cdot \sqrt{3} f(r) \sin\theta \cos\theta \\ &= \left(\frac{1}{\sqrt{2}} + \sqrt{\frac{3}{2}}\right) f(r) \qquad [\because \text{for the } x-\text{axis } \phi = 0° \text{ and } \theta = 90°] \end{aligned}$$

$$\therefore\ \psi_1 = 1.932 f(r) \tag{12.137}$$

Similarly, $\psi_2 = a_2\psi_{px}$

$$= \frac{1}{\sqrt{2}} f(r) - \frac{1}{\sqrt{2}} \cdot \sqrt{3} f(r) \sin\theta \cos\phi$$

$$= \left(\frac{1}{\sqrt{2}} + \sqrt{\frac{3}{2}}\right) f(r) \quad [\because\ \theta = 90° \text{ and } \phi = 180°]$$

$$= 1.932 f(r) \tag{12.138}$$

Thus, $\psi_1 = \psi_2 > \psi_{px}$, and the angle between them is 180°. The *sp* hybrid orbital is illustrated in the Figure 12.3.

The linear structure of ethyne due to *sp* hybridisation is shown in Figure 12.4.

In this case, a triple bond is formed in which one is σ-bond and other two are π bonds.

12.6.4 Hybridisation in H_2O

There are two important conditions of hybridisation scheme: (i) the hybrid orbitals should be mutually orthogonal, and (ii) the hybrid orbitals along the similar bond directions remain equivalent.

Now, we are to consider the formation of hybrid orbitals in H_2O molecule. The bond angle in H–O–H may be predicted by searching the direction in which the two mutually orthogonal and equivalent hybrid orbitals consist of major strengths.

Let us consider that one of the hybrid orbitals represented by ψ_1 lies along the *Z* axis and the other ψ_2 lies in *XZ* plane at an angle θ, which is known as the bond angle, which is shown in Figure 12.5.

Now, let us assume that

$$\left.\begin{aligned} \psi_1 &= a\psi_s + b_1\psi_{pz} \\ \text{and} \quad \psi_2 &= a\psi_s + b_2\psi_{pz} + c_2\psi_{px} \end{aligned}\right\} \tag{12.139}$$

We know that each function is normalised, therefore, we can write

$$a^2 + b_1^2 = 1 \tag{12.140}$$

$$\text{and } a^2 + b_2^2 + c_2^2 = 1 \tag{12.141}$$

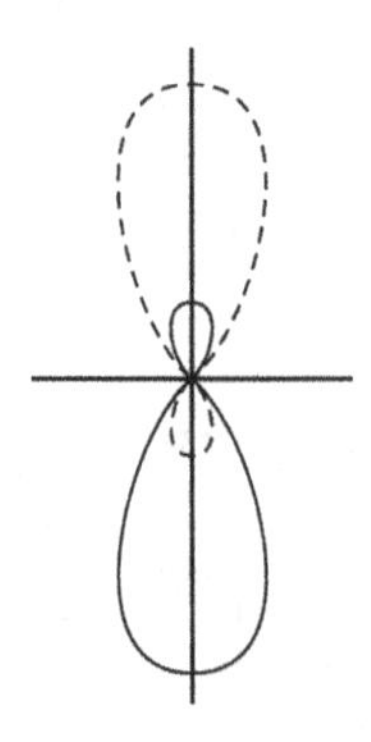

FIGURE 12.3 *sp* hybrid orbitals.

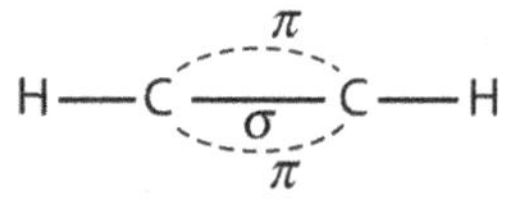

π

H—C—σ—C—H

π

FIGURE 12.4 Linear shape of C_2H_2.

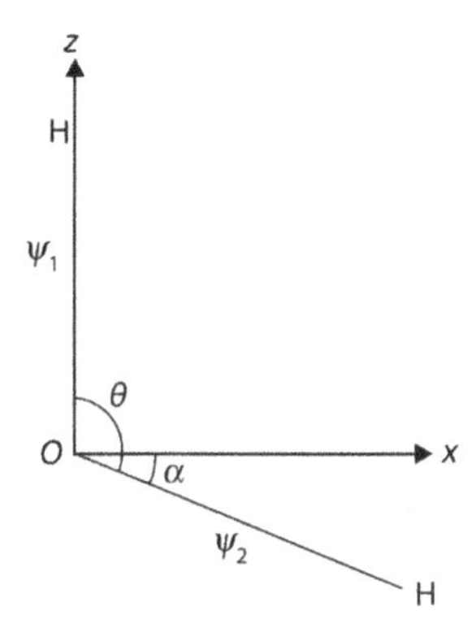

FIGURE 12.5 Hybridisation and bond angle in H_2O.

Since these functions are orthogonal,

$$\langle \psi_1 | \psi_2 \rangle = \langle (a\psi_s + a_1\psi_{pz}) | (a\psi_s + b_2\psi_{pz} + c_2\psi_{px}) \rangle = 0$$

$$\text{or } a^2 + b_1 b_2 = 0 \tag{12.142}$$

From Eq. (12.140), we can write

$$b_1^2 = (1 - a^2)$$

$$\therefore \; b_1 = \sqrt{(1 - a^2)} \tag{12.143}$$

From Eq. (12.142), we have

$$a^2 + b_1 b_2 = 0$$

$$\text{or } b_1 b_2 = -a^2$$

$$\text{or } b_2 \sqrt{1 - a^2} = -a^2 \qquad \left[\because \; b_1 = \sqrt{1 - a^2}\right]$$

$$\therefore \; b_2 = \frac{-a^2}{\sqrt{1 - a^2}} \tag{12.144}$$

From Eq. (12.141), we have

$$a^2 + \frac{a^4}{\left(1-a^2\right)} + c_2^2 = 1$$

$$c_2^2 = 1 - a^2 - \frac{a^4}{1-a^2}$$

$$\text{or} \qquad = \frac{1-a^2-a^2+a^4-a^4}{\left(1-a^2\right)}$$

$$= \frac{\left(1-2a^2\right)}{\left(1-a^2\right)}$$

$$\therefore\ c_2 = \frac{\sqrt{1-2a^2}}{\sqrt{1-a^2}} \tag{12.145}$$

Now, we shall try to find the maximum value of ψ_2, considering the following equation:

$$\phi_2 = a \cdot R(r) + b_2 R(r)\cos\theta + c_2 R(r)\sin\theta\cos\phi \tag{12.146}$$

$$\text{or } f_2 = \frac{\psi_2}{R(r)} = \left\{a + b_2\cos\theta + c_2 R(r)\sin\theta\cos\phi\right\}$$

Putting $\cos\phi = 1$ along the X axis and differentiating f_2 with respect to θ, we get

$$\frac{df_2}{s\theta} = -b_2\sin\theta + c_2\cos\theta = 0$$

$$\therefore\ b_2\sin\theta = c_2\cos\theta$$

$$\text{or } \frac{b_2}{c_2} = \frac{\cos\theta}{\sin\theta} = \frac{\cos\theta}{\sin(90+\alpha)} = \frac{\cos\theta}{\cos\alpha} \tag{12.147}$$

It clear from Eq. (12.147) that $b_2 \propto \cos\theta$ and $c_2 \propto \sin\alpha$, that is, the angle is made by hybrid orbitals ψ_2 with Z axis and X axis, respectively, and in the XZ plane. This will yield directly the bond angle θ, which is given in the following:

$$\cos\theta = b_2 = -\frac{a^2}{\sqrt{1-a^2}}$$

$$\therefore\ \psi_1 = a\psi_s + \left(1-a^2\right)^{1/2}\psi_{pz} \tag{12.148}$$

It gives a clue to understand that the hybrid orbitals ψ has the s character given by a^2 and p character given by $(1-a^2)$, and similar is the situation with ψ_2. With the change of angle from 90° to 180°, the amount of s character (a^2) in hybrid increases from 0 to 1/2 (*sp* hybrid, 180° bond angle). Experimental value of bond angle for H_2O has been found to be 104°28′ and $a^2 = 0.20$, meaning thereby 20% *s* character.

The two hybrid orbitals of O atom in O–H bonding are accordingly

$$\psi_1 = 0.45\psi_s + 0.89\psi_{pz} \tag{12.149}$$

$$\text{and } \psi_2 = 0.45\psi_s + 0.22\psi_{pz} + 0.86\psi_{px} \tag{12.150}$$

It should be kept in mind that Eq. (12.150) does not denote the sp^3 hybrid orbitals. It is really a fact that the sp^3 hybrid orbitals constitute strongest bonds and the tetrahedral molecules are most stable.

12.7 Semi-Empirical Methods

Here, we want to tell the readers about *ab initio* calculations and semi-empirical methods of calculations. The term *ab initio* means, 'beginning from the first principles'. The quantities calculated or evaluated from the first principles are in fact evaluated from the quantum mechanical postulates or axioms and theorems without resorting to a possible source of help.

It should be kept in mind that any advanced knowledge of the experimental results always gives a way to make a suitable choice of initial data for the *ab initio* evaluation, although the actual process of evaluation never resorts to the use of any experimental data in any form.

Thus, if one defines a basis set, calculates/evaluates all the essential integrals, and does an evaluation like the Hartree or Hartree–Fock calculation with or without a subsequent configuration interaction or any other many-body processes, the evaluation is *ab initio.* The selection of the basis set normally depends on our idea of the atomic or molecular system on which the evaluation is carried out. The maximum size of the molecule taken into consideration for an *ab initio* process is mainly found by the capability to compute and process the large number of two-electron integrals while forming the Fock matrix.

The semi-empirical methods are faster and easier than nonempirical methods. Consequently, they are useful for still bigger molecules, but their biggest drawback is that their accuracy is unpredictable. The empirical parameters are calculated on the basis of certain types of experimental data taken from proper types of molecules. Due to their poor predictions, they are often more useful for studying the trends in closely related series of molecules than for evaluating absolute properties of individual molecules.

A semi-empirical method must have a few desirable properties. The approximations involved in a semi-empirical method must not overlook the basic physical interactions, which are responsible for the equilibrium geometry of the molecule. The approximate wave function should be got from a general treatment, which does not need an empirically derived knowledge, such as the covalent or ionic character of a bond. The evaluated results should be easy to interpret. It is important to note that the *ab initio* calculations can be done either for all electrons or only for valence electrons, but the semi-empirical evaluations are done strictly for the valence electrons. In a semi-empirical method, some of the integrals are explicitly evaluated using STO or GTO basis functions. Some of them are neglected and yet some of the integrals are replaced by a set of numerical values found from the observed physical properties like electron affinity, ionisation potential, dissociation energy, etc., of homonuclear diatomic molecules.

12.7.1 Valence Electrons

The electrons present in the outermost electronic configuration of an atom are known as valence electrons. The valence electron process in semi-empirical methods is dependent on the assumption that the nucleus and the 'core electrons' (the total electrons in an atom minus the valence electrons are called the core electrons) in each atom together constitute an atomic core, which can be represented by the effective nuclear charge z^* in the core 'Hamiltonian operator' for one valence electron, i.e.,

$$H^{\text{core}}(1) = \frac{\nabla_1^2}{2} - \sum_n \frac{z^*}{r_1} \tag{12.151}$$

where $H^{\text{core}}(1)$ = Core Hamiltonian operator,
z^* = effective nuclear charge, and
r_1 = distance from the nucleus of an atom to the considered electron.

Equation (12.151) represents the energy of the valence electron 1 in the field of atomic nuclei and the electrons existing in the inner shells. We may then represent the Fock operator in the following manner:

$$\hat{F}(1) = H^{\text{core}}(1) + \sum_{j=1}^{n/2} \left[2\hat{J}_j(1) - \hat{K}_j(1) \right] \qquad (12.152)$$

where n = number of electrons in $\frac{n}{2}$ molecular orbitals.

Now, we shall discuss different semi-empirical methods, which will be helpful in the evaluation of different physical properties of atoms and molecules.

12.7.2 Zero Differential Overlap

The most valuable/significant approximation in the semi-empirical treatments is the *zero differential overlap* (ZDO) approximation. The meaning of ZDO approximation is that if ϕ_μ and ϕ_ν be the two atomic orbitals, then

$$\phi_\mu(r)\phi_\nu(r) = \delta_{\mu\nu} \Big\}_{=1 \text{ for } \mu \neq \nu}^{=1 \text{ for } \mu = \nu} \qquad (12.153)$$

where $\phi_\mu(r)$ and $\phi_\nu(r)$ = two different atomic orbitals belonging to the same atom or two different atoms.

In place of $\phi_\mu(r)$ and $\phi_\nu(r)$, we may also write $\phi_i(r)$ and $\phi_j(r)$, but here, we shall use ϕ_μ and ϕ_ν for convenience.

It is known to us that the ZDO approximation was put forward by Parr in 1952, which implies the following:

$$(\mu\nu \mid \lambda\sigma) = (\mu\mu \mid \lambda\lambda)\delta_{\mu\nu}\delta_{\lambda\sigma} = \gamma_{\mu\lambda}\delta_{\mu\nu}\delta_{\lambda\sigma} \qquad (12.154)$$

where δ represents Kronecker delta. In fact, Eq. (12.154) will may satisfy the weaker condition, that is

$$S_{\mu\nu} = \int \phi_\mu(r)\phi_\nu(r)\,\mathrm{d}v = \delta_{\mu\nu} \qquad (12.155)$$

where $S_{\mu\nu}$ = an integral observed in Roothaan's equations.

The core Hamiltonian integrals may be expressed as

$$H_{\mu\nu}^{\text{core}} = \int \phi_\mu(r) H^{\text{core}} \phi_\nu(r)\,\mathrm{d}v \qquad (12.156)$$

which may consist of overlap distributions, but these are not neglected; rather, they are to be approximated in terms of proper atomic and bond parameters.

Applying the ZDO approximations, the Hartree–Fock–Roothaan equations

$$\left[\sum_\nu F_{\mu\nu}C_{\nu i} = \sum_\nu S_{\mu\nu}C_{\nu i}\varepsilon_i \text{ for } i = 1, 2, \ldots, M \right.$$

or it may be represented as

$\sum_{j=1}^{M} F_{ij}C_{ja} = \varepsilon_a \sum_{j=1}^{M} S_{ij}C_{ja}$, $i = 1,2,\ldots,M$ as we have earlier written for a closed shell may be written as

$$\sum_{v} F_{\mu v}C_{vi} = \varepsilon_i C_{\mu i} \tag{12.157}$$

The elements of the Fock matrix may be represented by the following equation:

$$F_{\mu v} = H_{\mu v} + \sum_{\lambda\sigma} P_{\lambda\sigma}\left[(\mu v \mid \sigma\lambda) - \frac{1}{2}(\mu\lambda \mid \sigma v)\right] \tag{12.158}$$

where $P_{\lambda\sigma}$ = density matrix. And the elements of the Fock matrix may reduce in the following form, that is

$$F_{\mu\mu} = H_{\mu\mu}^{\text{core}} - \frac{1}{2}P_{\mu\mu}(\mu\mu \mid \mu\mu) + \sum_{\lambda} P_{\lambda\lambda}(\mu\mu \mid \lambda\lambda) \text{ and}$$

$$F_{\mu v} = H_{\mu v}^{\text{core}} - \frac{1}{2}P_{\mu v}(\mu\mu \mid vv), \quad \text{for } (\mu \neq v) \tag{12.159}$$

These equations are useful in various semi-empirical techniques or methods.

12.7.3 π_i-Electron Evaluation

According to Pople's π-electron theory, the core Hamiltonian may be expressed as

$$H^{\text{core}} = -\frac{1}{2}\nabla^2 + \sum_{s} V_s \tag{12.160}$$

where V_s = potential energy of a π electron in the field of the σ bonded framework round about the sth atom.

Now, suppose that the s atom contains z_π electrons. Then, we can write

$$\alpha_r = <\phi_r \left| H^{\text{core}} \right| \phi_r \geq w_r - \sum_{s \neq r} z_\pi \gamma_{rs} \tag{12.161}$$

where r = Euler's constant

$$\begin{aligned} \text{and } w_r &= \left\langle \phi_r \left| -\frac{\nabla^2}{2} + V_r \right| \phi_r \right\rangle \\ &= \left\langle \phi_r \left| T_2 + V_2 \right| \phi_r \right\rangle \end{aligned} \tag{12.162}$$

and the matrix element of V_s (as appeared in Eq. 12.160) may be approximated as

$$\left\langle \phi_r \left| V_s \right| \phi_r \right\rangle = -z_\pi \left\langle \phi_r \left| \frac{1}{\left| r - R_s \right|} \right| \phi_r \right\rangle \tag{12.163}$$

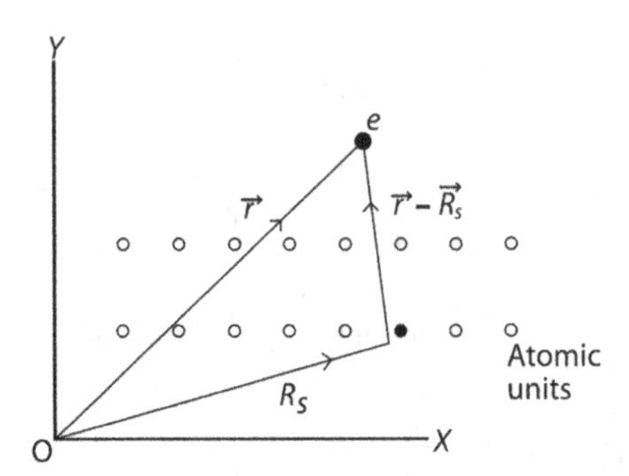

FIGURE 12.6 Representation of $|\vec{r} - R_s|$.

where $\vec{r}$ and $\vec{R}$ may be represented in Figure 12.6.

We know that $V_s = -\dfrac{Z_s e^2}{4\pi \in_0 |r - R_s|}$

But in atomic units

$V_s = -\dfrac{1}{|\vec{r} - \vec{R}_s|}$ because $4\pi \in_0 = 1$, $e = 1$ and $Z_s = 1$.

On the basis of Eq. (12.163), we can write

$$\left.\begin{aligned} \left\langle \phi_r \left| \frac{1}{r - R_S} \right| \phi_r \right\rangle &= \gamma_{rs} \\ \text{or} \quad \left\langle \phi_t \left| \frac{1}{r - R_S} \right| \phi_r \right\rangle &= \gamma_{st} \end{aligned}\right\} \tag{12.164}$$

So that the nuclear interaction integrals and the two electron integrals may be supposed to be at par.

Then the resonance integral may be expressed in a formula as

$$\beta_{rs} = \left\langle \phi_r \left| H^{\text{core}} \right| \phi_s \right\rangle \approx \left\langle \phi_r \left| -\frac{\nabla^2}{2} + V_r + V_s \right| \phi_r \right\rangle \tag{12.165}$$

where β_{rs} = resonance integral, $\langle \phi_r | V_t | \phi_s \rangle$ = integrals with different *r*, *s*, and *t*, which have been neglected by ZDO approximation.

Hence, the elements of the π electron Fock matrix may be given as

$$\left.\begin{aligned} \alpha_{rr}^F &= w_r + \frac{1}{2} P_{rr} \gamma_{rr} + \sum_{s \neq} (P_{ss} - Z_\pi) \gamma_{rs} \\ \text{and} \quad \beta_{rs}^F &= \beta_{rs} - \frac{1}{2} P_{rs} \gamma_{rs} \end{aligned}\right\} \tag{12.166}$$

It is clear that the difference from simple Hückel evaluation is clear. It is obvious that Hückel's π electron theory is dependent on the average values of both and β^F. In the Hückel theory, the resonance integral is limited to the nearest neighbour interaction since it decreases very fast as $|R_r - R_s|$ increases.

12.7.4 Invariance under Transformation

A formal study on the electronic structure of molecules is usually done by the application of LCAO–molecular orbital scheme, that is, by expressing the molecular orbitals as orthonormal LCAOs. The valence orbitals of every atom are linearly transformed among themselves, as given in the following:

$$\phi_v \rightarrow \phi_{v'} = \sum_{\mu} \phi_\mu T_{\mu v}$$
$$= \phi_1 T_{1v} + \phi_2 T_{2v} + \phi_3 T_{3v} + \cdots \quad (12.167)$$

such that the altered or transformed orbitals remain orthogonal to each other, and correspondingly, each molecular orbital can be expressed as a linear combination of altered orbitals. This can be achieved as follows:

The LCAO–molecular orbital scheme may be given by

$$\psi = \phi C \Rightarrow \psi_i = \sum_{\mu} \phi_\mu C_{\mu i}$$
$$= \phi' T^{-1} C \approx \phi' C' \quad (12.168)$$

On the basis of Eq. (12.167), we can write

$$\phi' = \phi T$$
$$\text{or } \phi = \frac{\phi'}{T} = \phi' T^{-1} \quad (12.169)$$

where T^{-1} = inverse of T matrix, which can be used to get

$$\psi = \phi'^{T^{-1}} C = \phi' C' \Rightarrow \psi_i = \sum_{\mu} \phi'_\mu C'_{\mu i} \quad (12.170)$$

It is clear from this that the transformation provides the effect that hybrid orbitals are constituted with the atom's valence orbitals.

We have discussed earlier that $\phi_\mu(r)\phi_v(r) = 0$ for $\mu \neq v$ according to ZDO approximation and does not change as $\phi'_\mu(r)\phi'_v(r) = 0$ for $\mu \neq v$, when μ and v are orbitals on two different atoms, i.e., the neglect of the diatomic differential overlap is valid under any linear transformation.

Consequently, the integrals essential for the Hartree–Fock calculation changes as

$$S_{\mu\upsilon} \rightarrow S'_{\mu\upsilon} \equiv \sum_{\alpha,\beta} T_{\alpha\mu} T_{\beta\upsilon} T_{\beta\upsilon} T_{\alpha\beta}$$

where $\alpha = 1, 2, 3, \ldots,$

$\beta = 1, 2, 3, \ldots,$ and

$\mu = v = 1, 2, 3, \ldots$

$$H_{\mu\upsilon}^{\text{core}} \rightarrow H_{\mu\upsilon}^{\prime\text{core}} = \sum_{\alpha,\beta} T_{\alpha\mu} H_{\alpha\beta}^{\text{core}} \text{ and}$$

$$(\mu\upsilon|\sigma\lambda) \rightarrow (\mu'\upsilon'|\sigma'\lambda') = \sum_{\alpha,\beta,\gamma,\delta} T_{\alpha\mu} T_{\beta\upsilon} T_{\gamma\sigma} T_{\delta\lambda} (\alpha\beta|\gamma\delta) \tag{12.171}$$

It should be kept in mind that in an approximate LCAO–molecular orbital SCF evaluation, it is required that the transformed integrals should not change from the original one with regard to any arbitrary transformation. This fact was first of all pointed by Pople, Santry, and Segal in 1965.

12.7.5 Complete Neglect of Differential Overlap

The method of complete neglect of differential overlap (CNDO) was introduced by Pople and Segal in 1965, which is one of the ZDO-based methods. The ZDO result mentioned in Eq. (12.154) can be made more approximate in order to keep up the invariance under local rotational transformations. The more approximate ZDO result can be expressed as

$$(\mu\upsilon \mid \lambda\sigma) = \gamma_{AB}\delta_{\mu\upsilon}\delta_{\lambda\sigma} \tag{12.172}$$

where ϕ_μ remains on atom A and ϕ_χ on atom B.

The parameter γ_{AB} mainly depends on the behaviour of atoms A and B and also the distance between *A* and *B*, but it has been considered to be independent of clearly described nature of the orbitals represented by ϕ_μ and ϕ_χ. This is illustrated in Figure 12.7.

The ZDO result described in Eq. (12.155) is actually rotationally invariant, and this is why it has been kept. Our next step will be to derive the expression for core Hamiltonian integrals invariant under the transformation.

The core Hamiltonian may be written as

$$H^{\text{core}} = \frac{-\nabla^2}{2} + \sum_A V_A \tag{12.173}$$

where $V_A = \dfrac{-Z_A}{r_A}$

and Z_A = core charge of atom A.

The above yields three kinds of core Hamiltonian integrals, which are mentioned in the following:

$$H_{\mu\mu}^{\text{core}} = u_{\mu\mu} + \sum_{B\neq A} \langle \phi_\mu | V_B | \phi_\mu \rangle \tag{12.174}$$

where ϕ_μ is on the atom A.

$$H_{\mu\upsilon}^{\text{core}} = u_{\mu\upsilon} + \sum_{B\neq A} \langle \phi_\mu | V_B | \phi_\upsilon \rangle \tag{12.175}$$

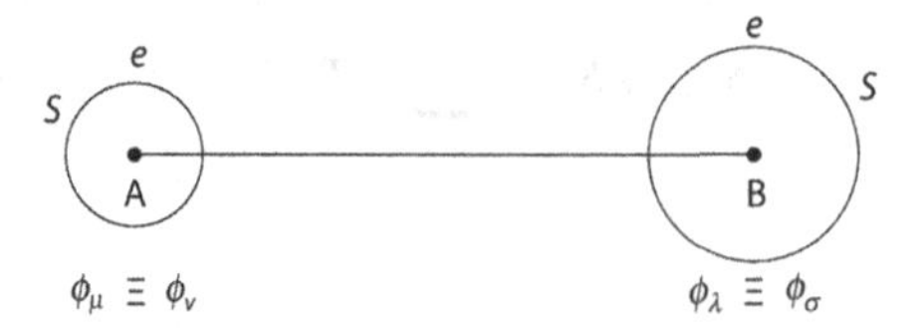

FIGURE 12.7 Invariance of the two-centre two-electron integrals under local rotational transformation.

In this case, ϕ_μ and ϕ_ν are on atom A:

$$H_{\mu\upsilon}^{\text{core}} = \left\langle \phi_\mu \left| -\frac{1}{2}\nabla^2 + V_A + V_B \right| \phi_\upsilon \right\rangle + \sum_{C(\neq A.B)} \langle \phi_\mu | V_C | \phi_\upsilon \rangle \tag{12.176}$$

where ϕ_μ is on atom A and ϕ_ν is on atom B.

And $u = -\frac{1}{2}\nabla^2 + V_A$.

Furthermore, the rotational invariance is found out in steps. The first step is written as

$$\langle \phi_\mu | V_B | \phi_\upsilon \rangle = V_{AB}\delta_{\mu\upsilon} \tag{12.177}$$

Here, it is indicated that ϕ_μ and ϕ_ν are both on atom A.

The $H_{\mu\upsilon}^{\text{core}}$ Hamiltonian can then be expressed as

$$H_{\mu\upsilon}^{\text{core}} = \delta_{\mu\upsilon}\left[U_{\mu\mu} + \sum_{B\neq A} V_{AB} \right] \tag{12.178}$$

The approximation described in Eq. (12.177) is illustrated in Figure 12.8.

In the second step, one should neglect the three-centre integrals, such as

$$\langle \phi_\mu | V_C | \phi_\upsilon \rangle \tag{12.179}$$

The resulting H_2^+-*like* integral can be expressed as:

$$H_{\mu\upsilon}^{\text{core}} = \left\langle \phi_\mu \left| -\frac{1}{2}\nabla^2 + V_A + V_B \right| \phi_\upsilon \right\rangle$$

where ϕ_μ on atom A and ϕ_ν on atom B may be approximated and represented by

$$H_{\mu\upsilon}^{\text{core}} = \beta_{AB}^0 S_{\mu\upsilon} \tag{12.180}$$

where B_{AB}^0 = proportionality constant.

If these approximations are used, one can find that the elements of the Fock matrix given in Eq. (12.159) become simplified forms, i.e.,

$$F_{\mu\mu} = U_{\mu\mu} + \left(P_{AA} - \frac{1}{2}P_{\mu\mu} \right)\gamma_{AA} + \sum_{B\neq A}\left[-Q_B\gamma_{AB} + \left(Z_B\gamma_{AB} + V_{AB} \right) \right] \tag{12.181}$$

$$\text{and } F_{\mu\upsilon} = \beta_{AB}^0 S_{\mu\upsilon} - \frac{1}{2}P_{\mu\upsilon}\gamma_{AB} \tag{12.182}$$

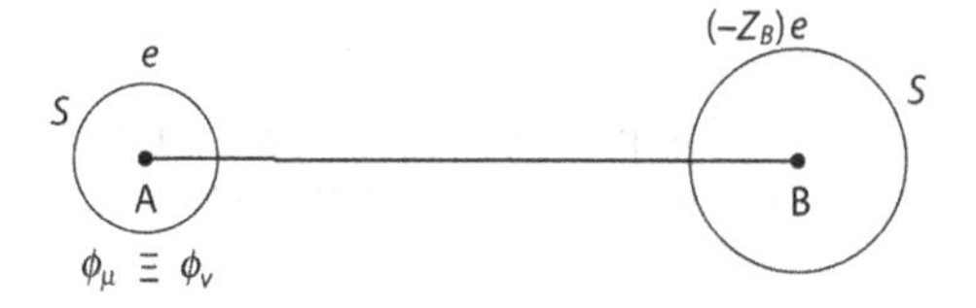

FIGURE 12.8 Approximation described in Eq. (12.177) which keeps the core Hamiltonian matrix invariant.

where P_{AA} = net Mulliken population or Mulliken population matrix on atom A

Q_B = net charge on atom B

$$P_{AA} = \sum_{\mu \in A} P_{\mu\mu} \tag{12.183}$$

$$\text{and } Q_B = Z_B - P_{BB} \tag{12.184}$$

At this stage, we can express the Hartree–Fock total energy as

$$E_{\text{total}}^{HF} = \frac{1}{2}\sum_{\mu\upsilon} P_{\mu\upsilon}\left(H_{\mu\upsilon}^{\text{core}} + F_{\mu\upsilon}\right) + \sum_{A<B} Z_A Z_B R_{AB}^{-1} \tag{12.185}$$

where the terms have their usual meaning.

12.7.6 Parametrisation

There are two systems of parametrisation for CNDO method of calculation. One is CNDO/1 parameter scheme, and the other is CNDO/2 parameter scheme.

The older one is CNDO/1, and the improved version is CNDO/2. In CNDO/1 scheme, the integrals V_{AB} and γ_{AB} are evaluated on the assumption that the charge distribution around each atom is spherically symmetric. The integrals are then given by employing the STOs, which denote the valence s orbitals of the atoms involved. The integrals have been expressed as follows:

$$V_{AB} = -Z_B \int S_A^2(r_1) r_{1B}^{-1} d\upsilon_1 \tag{12.186}$$

$$\text{And } \gamma_{AB} = \int d\upsilon_1 \int d\upsilon_2 s_A^2(r_1) S_B^2(r_2) r_{12}^{-1} \tag{12.187}$$

These integrals are evaluated in the CNDO process by employing the standard exponents, which is made available from Slater's rules. The $U_{\mu\mu}$ terms are computed from the equations of total energy of an atom A and its A^+ ion, the neutral electronic configuration being represented by $2s^i 2p^j$.

$$\begin{aligned} E_A\left(2s^i 2p^i\right) &= iU_{2s,2s} + jU_{2p,2p} + \frac{1}{2}(i+j)(i+j-1)\gamma_{AA}; \\ E_A\left(2s^{i-1} 2p^j\right) &= (i-1)U_{2s,2s} + jU_{2p,2p} + \frac{1}{2}(i+j-1)(i+j-2)\gamma_{AA}; \\ E_A\left(2s^i 2p^{j-1}\right) &= iU_{2s,2s} + (j-1)U_{2p,2p} + \frac{1}{2}(i+j-1)(i+j-2)\gamma_{AA}; \end{aligned} \tag{12.188}$$

These will yield U matrix elements in terms of orbital ionisation potentials I_S and I_P,

$$\begin{aligned} U_{2s,2s} &= -I_s - (i+j-1)\gamma_{AA} = -I_S - (Z_A - 1)\gamma_{AA}, \\ \text{and } U_{2p,2p} &= -I_P - (i+j-1)\gamma_{AA} = -I_P - (Z_A - 1)\gamma_{AA}, \end{aligned} \tag{12.189}$$

We have already stated that β_{AB}^0 (bonding parameter) is considered to be the average of the bonding parameters for the corresponding homonuclear diatomic molecules, i.e.,

$$\beta_{AB}^0 = \frac{1}{2}\left(\beta_{AA}^0 + \beta_{BB}^0\right) \tag{12.190}$$

The overlap integrals, which have been represented by $S_{\mu v}$, are clearly computed, but they are only used to form the off-diagonal elements of the Fock matrix. On the basis of ZDO approximations, the $S_{\mu v}$ can be expressed as

$$S_{\mu v} = \delta_{\mu v} \tag{12.191}$$

and this relationship is still to be used in the Hartree–Fock–Roothaan equation.

The average ionisation potentials required for the formation of $U_{\mu\mu}$ parameters and β^0_{AA} (values of resonance integrals) used in the CNDO/1 scheme/procedure are mentioned in Table 12.2.

The term $(Z_B\gamma_{AB} + V_{AB})$ in $F_{\mu\mu}$ in Eq. (12.181) signifies the effect of penetration of one electron, which stays in the μth orbitals of atom A into the shell of atom B. When V_{AB} and γ_{AB} are computed as different integrals, as described in Eqs. (12.186) and (12.187), it is found that the penetration term leads to an additional stability, and a positive value of bond energy is obtained even when the bond order is zero. This problem is corrected by treating $V_{AB} \approx \gamma_{AB}$, as it was done in case of Pople's π-electron theory.

The equation for V_{AB} can be modified by expressing

$$V_{AB} = -Z_B\gamma_{AB} \tag{12.192}$$

This modification has been shown in Figure 12.9.

This constitutes the background for the improvement in CNDO/2 scheme. The diagonal elements of the Fock matrix are now expressed as

$$F_{\mu\mu} = U_{\mu\mu} + \left(P_{AA} - \frac{1}{2}P_{\mu\mu}\right)\lambda_{AA} - \sum_{B\neq A} Q_B\gamma_{AB} \tag{12.193}$$

The off-diagonal elements $F_{\mu v}$, with $\mu \neq v$, do not change from the second expression mentioned in Eq. (12.182). We can make one more modification in CNDO/2. We have already discussed about CNDO/1,

TABLE 12.2

Average Ionisation Potential and Resonance Integrals in eV for Complete Neglect of Differential Overlap (CNDO)/1 Scheme

			Valence I.P. (eV)		
Element	**Atomic Number**	**Core Charge (Z_c)**	*s*	*P*	β^0_{AA}
H	1	1	13.06		–9
Li	3	1	5.39	3.54	–9
Be	4	2	9.32	5.96	–13
B	5	3	14.05	8.30	–17
C	6	4	19.44	10.67	–21
N	7	5	25.60	13.20	–25
O	8	6	32.40	15.85	–31
F	9	7	40.20	18.66	–39

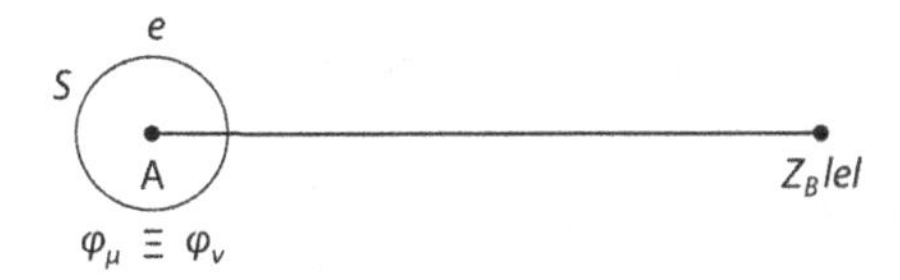

FIGURE 12.9 The Pariser–Parr–Pople type modification of the nuclear integral in complete neglect of differential overlap scheme.

which defines the $U_{\mu\mu}$ matrix elements and conversely the ionisation potential in Eq. (12.189). One may similarly define the electron affinity A_μ by the following mathematical expression:

$$A_\mu = -U_{\mu\mu} - Z_A\gamma_{AA}$$

We are aware of the fact that the chemical characteristics are decided by the tendency of an atom to lose as well as gain an electron. If one assumes the ionisation potential and electron affinity at par for constructing the element $U_{\mu\mu}$, the treatment would give good result of thermochemical properties. Therefore, $U_{\mu\mu}$ may be found out by CNDO/2 scheme by taking the average of orbital ionisation potential and orbital electron affinity, i.e., from the orbital electron negativity.

The initial molecular orbital coefficients needed for the SCF procedure are often obtained from Hückel-type approach, which consists of the solution of ZDO-level SCF Eq. (12.193) with Fock matrix elements:

$$\begin{aligned} F_{\mu\mu} &= U_{\mu\mu} \\ F_{\mu\upsilon} &= \beta^0_{AB} S_{\mu\upsilon} \qquad \mu \neq \upsilon \end{aligned} \tag{12.194}$$

CNDO/2 scheme has been further modified by Santry and Segal in 1967 for including the third row elements of the periodic table. The said authors gave their idea for rewriting the resonance integral as

$$\beta^0_{AB} = \frac{1}{2}K\left(\beta^0_{AA} + \beta^0_{BB}\right) \tag{12.195}$$

They have considered $K = 0.75$ in Eq. (12.195). The quantity β^0_{AA} for homonuclear diatomic of a third row element A can be computed from the corresponding quantity β^0_{xx} for the second row diatomic X_2 by using the formula

$$\beta^0_{AA} = \frac{U_{3s,3s}(A) + U_{3p,3p}(A)}{U_{2s,2s}(X) + U_{2p,2p}(X)} \cdot \beta^0_{xx} \tag{12.196}$$

where the terms have already been introduced.

12.7.7 Intermediate Neglect of Differential Overlap

We have learnt the most important CNDO approximation in Eq. (12.172), which considers only the direct repulsion between the atomic charge densities and neglects all other coulomb integrals. It should be kept in mind that the exchange integrals, both one-centre and two-centre, are fully neglected in the CNDO approach. Hence, the Hartree–Fock total energy computed by CNDO approach for different spin states but having same electronic configuration does not seem to be different. Thus, the states 3P, D, and S of carbon atom, which come from $2s^2\,2p^2$ configuration, are not energetically distinguished in CNDO. CNDO approach is also unable to lead to a net spin density in σ orbitals. This drawback led to intermediate neglect of differential overlap (INDO), wherein the monatomic differential overlap is kept in one-centre two-electron integrals. This idea was given by Pople, Beveridge, and Dobosh in 1967. Due to the retention of the monatomic exchange integrals, the improved treatment resulted, though the theory is still not as correct as the *ab initio* formalism.

Since the only nonvanishing one-centre two-electron integrals are of the kind $(\mu\mu \mid \mu\mu)$, $(\mu\mu \mid vv)$, and $(\mu v \mid \mu v)$, the elements of the Fock matrix in unrestricted INDO scheme are expressed as

$$F^\alpha_{\mu\mu} = U_{\mu\mu} + \sum\left[P_{\lambda\lambda}(\mu\mu \mid \lambda\lambda) - P^\alpha_{\lambda\lambda}(\mu\lambda \mid \mu\lambda)\right] + \sum_{B\neq A}(P_{BB} - Z_B)\gamma_{AB} \quad \text{for } \mu \text{ on atom A.}$$

$$\text{and } F^\alpha_{\mu\upsilon} = \left(2P_{\mu\upsilon} - P^\alpha_{\mu\upsilon}\right)(\mu\upsilon \mid \mu\upsilon) - P^\alpha_{\mu\upsilon}(\mu\mu / \upsilon\upsilon) \tag{12.197}$$

for both μ and v on the atom A.

where $P^{\alpha}_{\mu\upsilon} = (\mu,\ v)$ element of density matrix for electrons having the α spin and
$P^{\alpha}_{\mu\upsilon}$ = corresponding elements of the full density matrix.

The equation $F^{\alpha}_{\mu\upsilon}$ having orbitals belonging to different atoms does not alter from that in the CNDO/2 scheme. The restricted Fock matrix elements can easily be found from these equations.

Let us assume that *ns* and *np* orbitals have the same radial parts. The nonvanishing integrals are obtained as follows:

$$(SS \mid SS) = (SS|XX) = F^0 = \gamma_{AA} \tag{12.198}$$

$$(sX \mid sX) = 1/3G' \tag{12.199}$$

$$(xy \mid xy) = 3/25F^2 \tag{12.200}$$

$$(XX \mid XX) = F^0 + \frac{4}{25}F^2 \tag{12.201}$$

$$\text{and } (xx \mid yy) = F^0 - \frac{2}{25}F^2 \tag{12.202}$$

and so on, where F^0, G', and F^2 are two-electron integrals having the radial parts. The values of these integrals are listed in Table 12.3.

The electronic energy of the elements having configuration $ns^i\,np^j$ can be normally expressed as

$$E = iU_{2s,2s} + jU_{2p,2p} + \frac{1}{2}(i+j)(i+j-1)F^0 - \frac{1}{6}ijG' - \frac{1}{25}j(j-1)F^2 \tag{12.203}$$

Remember that the INDO theory/scheme can be successfully employed for finding out very small energy changes associated with redox reactions, solvation process, and other biochemical reactions in a solution. As the molecular size increases, the INDO computation itself becomes more voluminous.

It is important to note that the INDO scheme is a substantial improvement over CNDO/2 in a problem in which electron spin distribution is significant.

12.7.8 Neglect of Diatomic Differential Overlap

Actually, neglect of diatomic differential overlap (NDDO) approximation neglects differential overlap only when atomic orbitals in question belong to different atoms; thus, dipole–dipole interactions are retained and are represented by the following integrals:

$$\langle s_A(1)s_B(2)|H|P_A(1)P_B(2)\rangle \tag{12.204}$$

TABLE 12.3

Empirical Values of Slater–Condon Parameters

	G'		F^2	
Element	**Rydberg**	**a.u.**	**Rydberg**	**a.u.**
Li	0.184024	0.092012	0.09973	0.049865
Be	0.28140	0.14070	0.17825	0.089125
B	0.39853	0.199265	0.26082	0.13041
C	0.535416	0.267708	0.34744	0.17372
N	0.692058	0.346029	0.43811	0.219055
O	0.86846	0.43423	0.53283	0.266415
F	1.06461	0.532305	0.63160	0.31580

Such kind of integrals are either computed from atomic orbitals or found by empirical methods, but in either case, the invariance conditions of CNDO scheme must be satisfied. Let us suppose that atomic orbitals μ and υ are on atom A and λ and σ are on atom B, where $B = A$. The matrix elements for Roothaan's equations are mentioned in the following:

$$F_{\mu\mu} = U_{\mu\mu} + \sum_{B} V_{\mu\mu,B} + 2\sum_{\upsilon}^{A} R_{\upsilon\upsilon}\left(\langle\mu\upsilon|H|\mu\upsilon\rangle - \frac{1}{2}\langle\mu\mu|H|\upsilon\upsilon\rangle\right)$$
$$+2\sum_{B}\sum_{\lambda\sigma} R_{\lambda\sigma}\langle\mu\lambda|H|\mu\sigma\rangle \tag{12.205}$$

$$F_{\mu\upsilon} \sum_{B} U_{\mu\upsilon,B} + R_{\mu\upsilon}\left(3\langle\mu\mu|H|\upsilon\upsilon\rangle - \langle\mu\upsilon|H|\mu\upsilon\rangle\right) + 2\sum_{B}\sum_{\lambda\sigma} R_{\lambda\sigma}\langle\mu\lambda|H|\mu\sigma\rangle \tag{12.206}$$

$$F_{\mu\lambda} = \beta_{\mu\lambda} - \sum_{\upsilon}^{A}\sum_{\sigma}^{B} R_{\upsilon\sigma}\langle\mu\lambda|H|\mu\sigma\rangle \tag{12.207}$$

This NDDO method was proposed by Pople in 1960. Dewar and his colleagues have introduced a similar modification of NDDO, which they named modified neglect of diatomic overlap (MNDO). The MNDO method has been utilised to compute the heats of formation, ionisation potential/energy, dipole moments, and molecular geometry. Dewar and his coworkers have proposed a large number of 'NDO' approximations such as MINDO, MNDO, and so on.

12.7.9 The Pariser–Parr–Pople Method

This method is a union of two π-electron theories in which one was proposed by Pariser and Parr and the other by Pople. The purpose of the first was to supply a satisfactory description of conjugated systems in the excited states. This theory includes electron correlation and electron repulsion and employs wave functions describing pure spin states. The Pople theory is unable to include electron correlation but produces the π-electron MOs with help of Hartree–Fock–Roothaan SCF method.

Now, we shall describe the mathematical formulation of the Pariser–Parr–Pople (PPP) method. For this, we shall consider the Hamiltonian operator for a conjugated system having $n\pi$ electrons in a.u. which is expressed as

$$H\pi = \sum^{n\pi} H^{c}\pi(i) + \sum_{i=1}^{n\pi}\sum_{j>i} 1/r_{ij} \tag{12.208}$$

where $H^{c}\pi = -1/2\nabla_i^2 + V(i)$ and

$H^{c}\pi(i) =$ the one electron core Hamiltonian operator.

It gives the idea of motion of a π electron in the nuclear field and σ electrons, which are supposed to form a nonpolarised, rigid, and core with positive charge. Now, we shall mention the equations of the wave function and energy of a closed shell having $n\pi$ electrons, respectively, as given in the following:

$$\Psi = |\varphi_1\bar{\varphi_1}\ \varphi_2\bar{\varphi_2}\ldots\varphi_{n/2}\bar{\varphi}_{n/2} \tag{12.209}$$

$$E = 2\sum_{i=1}^{n/2} H_{ii} + \sum_{i}\sum_{j}\left(2j_{ij} - K_{ij}\right) \tag{12.210}$$

where $H_{ii} = \langle\varphi_i(i)H^{c}(i)|\varphi_i(i)\rangle$

$J_{ij} = \langle \varphi^2 i(i) | 1/r_{ij} | \varphi^2 j(j) \rangle$, etc.

Actually, MOs $\{\varphi_i\}$ are solutions of the *HF* equation

$$F(i)_{i(i)} = \in_i \varphi_{i(i)} \tag{12.211}$$

$$\text{Where } F(i) = H^c\pi(i) + \sum_{j=1}^{n/2} (2J_j(i) - K_j(i)) \tag{12.212}$$

Here, J_j = the coulomb operator and K_j = the exchange operator.

Now, we shall express the set of *n*/2 *HF* equations in form of matrix as

$$FC = SC \in \tag{12.213}$$

where C = matrix of the coefficient of $\{\varphi_i\}$ MOs which is represented as combination of MOs $\{X_p\}$.

Taking the help of zero differential overlap (ZDO), it may be assumed that

$$\mathrm{d}(S_{pq}) = \mathrm{d}\int x_{p(i)} x_{q(i)} \mathrm{d}\tau = (x_{p(i)} x_{q(i)}) = 0$$

where $p \neq q$.

This suggest that not only $S_{pq} = 0$ but also $x_{p(i)} x_{q(i)}$ and the multiplication factor $1/rj$ will be zero. Thus, an integral like $(pq|rs) = \langle x_{p(i)} x_{q(i)} | 1/rj | x_{r(j)} \rangle$ becomes $= \delta_{pq}\ \delta_{rs}\ (pp/rr)$.

This is indicative of the fact that all three- and four-centre electron repulsion integrals will amount to zero. Hence, Eq. (12.213) will now take the following form:

$$FC = C \in \tag{12.214}$$

$$\text{where } F_{pp} = \langle \chi_{P(i)} | F(i) \chi_{p(i)} \rangle$$

$$= H_{PP}^C - P_{PP}/2\ \gamma_{pp} + \sum_r P_{rr}\gamma_{pr}$$

$$= H_{PP}^C + P_{PP}/2\gamma_{pp} + \sum_{r \neq p} P_{rr}\gamma_{pr} \tag{12.215}$$

$$F_{pq} = \langle \chi_p(i) | F(i) | \chi_q(i) \rangle$$

$$= H_{pq}^C - F_{pq}\ \gamma_{pq} \tag{12.216}$$

With $\gamma_{pp} = (PP/PP)$ and $\gamma_{pp} = (pp/qq)$.

These expressions can be found out from equation given in the following:

$(H + G)_e = E_e \Psi_e$ with the aid of $(pq / rs) = \delta_{pq}\ \delta_{rs}\ (pp \mid rr)$ approximation.

Furthermore, our aim is to evaluate the integrals occurring in the PPP method.

12.7.9.1 Evaluation of Integrals of Pariser–Parr–Pople Method

In the PPP method, the integrals present are core integrals H_{pp}^C and H_{pq}^C and other integrals are γ_{pp} and γ_{pq} which is known as electron repulsion integrals. The one-electron core Hamiltonian operator in a.u. is expressed as given in the following:

$$H^C(i) = -\nabla_i^2 / 2 + \sum_p V_p(i) = T(i) + \sum V_{\rho(i)} \tag{12.217}$$

where $V_{p(i)}$ = the operator corresponding to the attraction between electron i and atom p in the core.

Let us consider $n\rho$ as the positive charge in a.u. at P, the number of π electrons contributed by the atom P. Thus, n =1 for C, n_p =1 for C, n_p =1 for N in pyridine, 2 for N in pyrrole, etc.

Now, rewriting eq. (12.217) in the following form, i.e.,

$$H^C(i) = T(i) + V_p(i) + \sum_{q \neq p} v_q(i) \tag{12.218}$$

We have

$$\begin{aligned} H_{pp}^c &= \langle \chi_p(i) | H^C(i) | \chi_p(i) \rangle \\ &= \langle \chi_p(i) | T(i) + V_p(i) | \chi_p(i) \rangle \\ &+ \left\langle \chi_p(i) \middle| \sum_{q \neq p} \upsilon_q(i) \middle| \chi_p(i) \right\rangle \end{aligned} \tag{12.219}$$

According to the Goeppert-Mayer approximation the first term of Eq. (12.219) may be written as

$$\langle \chi_p(i) | T(i) + V_p(i) | \chi_p(i) \rangle = -I_p \langle \chi_p(i) | \chi_p(i) \rangle = -I_p \tag{12.220}$$

where I_p = ionization energy of atom P.

It is a fact that the atom exists in a molecule. The ground-state electronic configuration of C is $1s^2\ 2s^2\ 2p^2$, but its valence state electronic configuration in conjugated hydrocarbon is represented as $1s^2\ t_r^1\ t_r^1\ t_r^1\ 2p^1\ \pi$ where t_r = trigonally (sp^2) hybridised atomic orbital.

With the help of Eqs. (12.220) and (12.219), we can write

$$H_{PP}^C = -I_P + \sum_{q \neq p} \langle \chi_p(i) | V_p(i) | \chi_p(i) \rangle \tag{12.221}$$

It must be kept in mind that the integral in Eq. (12.221) represents the attraction of the charge density χ_p^2 to atom q in the core. Let us consider the positive charge n_q at this atom results because of the removal of n_q π electrons of charge density $\chi_q(i)$, and then we may write as

$$\begin{aligned} \langle \chi_p(i) | V_q(i) | \chi_p(i) \rangle &= \langle \chi_p(i) | V_q^o(i) | \chi_p(i) \rangle - n_q \langle \chi_p^2(i) | 1/r_{ij} | \chi_q^2(i) \rangle \\ &= 0 - n_q \gamma_{pq} \end{aligned} \tag{12.222}$$

where the integral containing V_q^o is indicative of the attraction of the charge density $\chi_p^2(i)$ to the neutral atom. It is known as penetration integral whose value is very less that is why it is neglected. At this stage, we are going to substitute Eq. (12.222) in Eq. (12.221), and as a result, we have the following equation:

$$H_c^{pp} = -I_p - \sum_{q \neq p} n_q \gamma_{pq} \tag{12.223}$$

Now the off-diagonal core integral is written in the following:

$$H_{pq}^c = \langle \chi_p(i) | T_i + V_p(i) | \chi_q(i) \rangle + \left\langle \chi_P(i) \middle| \sum_{r \neq p,q} V_r(i) \middle| \chi_q(i) \right\rangle \tag{12.224}$$

Following eq. (12.222), the second integral in RHS of Eq. (12.224) is expressed as

$$\left\langle \chi_p(i) \middle| \sum_{r \neq p,q} V_r(i) \middle| \chi_q(i) \right\rangle = -\sum n_r (pq/rr) \tag{12.225}$$

Applying the ZDO approximation, we have $(pq/rr)=0$ and thus

$$H_{pq}^{C}=\langle\chi_P(i)|T_i+V_p(i)|\chi_q(i)\rangle=\beta_{pq}^{c} \tag{12.226}$$

where β_{pq}^{c} or β is also known as core resonance integral.

J. Linderberg had given a relationship between β, R (bond length), and S (overlap integral) for symmetrically orthogonalized 2 P π AOs, which is

$$\beta=1/R\,\mathrm{d}S/\mathrm{d}R \tag{12.227}$$

This was derived on the basis of commutation relation between linear momentum and the core Hamiltonian operator.

Remember that the overlap integral (S) is related with two Slater-type 2 ρ π. Orbitals having orbital exponent ε which is given in the following:

$$S=e^{-p}\left(1+\rho+2\rho^2/5+\rho^3/15\right) \tag{12.228}$$

where ρ is related to orbital exponent and bond length as

$$\rho=\varepsilon R$$

After substituting Eq. (12.228) into (12.227), we shall obtain

$$\begin{aligned}\beta&=1/R\,\mathrm{d}s/\mathrm{d}R=1/R\,\mathrm{d}\rho/\mathrm{d}R\cdot\mathrm{d}s/\mathrm{d}p\\&=-0.2\varepsilon^2\left(1+\rho+\rho\,2/3\right)e^{-p}\\&=-0.2\varepsilon^2 S(1s,1s)\end{aligned} \tag{12.229}$$

But the value of β thus obtained was underestimated. Later on, a semitheoretical equation was put forward which gave β value in good agreement. The semitheoretical equation is as follows:

$$\beta_{pq}=-s/2(I_P+I_Q)-S^2/4(\varepsilon_P^2+\varepsilon_q^2)-S/2\left[\langle X_{P\,(i)}|\;V_q(i)\;|\;X_P(i)\rangle+\langle X_q(i)\;|V_p(i)|\,x_q(i)\rangle\right] \tag{12.230}$$

where the integrals in Eq. (12.230) is evaluated theoretically for Slater-type orbitals with the help of Roothaan's equation but on assumption that

$$V_t\;(i)=-n_t/r_{it}\,(t=p,q)$$

Remember that for the one-centre integral γ_{pp}, the most widely used formula is given by

$$\gamma_{pp}=I_P-A_P \tag{12.231}$$

where I_P = ionization energy and A_P = electron affinity.

Now, we shall give the simplest equations for γ_{pq} which is given in the following:

$$\gamma_{pq}=1/a+R \tag{12.232}$$

$$\text{and}\;\;\gamma_{pq}=1/\sqrt{a^2+R^2} \tag{12.233}$$

where $a=2/(\gamma_{pp}+\gamma_{qq})$ (in a.u.).

The aforementioned formulae were given by Nishimoto, Metaga, and Ohno.

BIBLIOGRAPHY

Atkins, P. and R. Friedman. 2007. *Molecular Quantum Mechanics*. New York: Oxford University Press.
Goeppert-Mayer, M. and A.L.S. Klar. 1938. *J. Chem. Phys.* 6, 645.
Harriss, D.K. and F. Rioux. 1980. 'A simple Hartree SCF calculation on a one-dimensional model of the He atom.' *J. Chem. Educ.* 57: 491.
Hehre, W.J., L. Schleyer, P.V.R. Radom and J.A. Pople. 1986. *Ab-Initio Molecular Orbital Theory*. New York: Wiley.
Levine, I.N. 2000. *Quantum Chemistry*. Singapore: Pearson Education.
Linderberg, J. 1967. *Chem. Phys. Lett.* 1, 39.
Lowe, J.P. and K.A. Peterson. 2006. *Quantum Chemistry*, 3rd edn. London: Academic Press.
Metaga. 1957. *Z. Phys. Chem.* 13, 140.
Mulliken, R.S., C.A. Ricke, Dorloff and H. Orloff. *J. Chem. Phys.* 17, 1248.
Nishimoto, K. 1964. *Z. Phys. Chem.* 12, 335.
Ohno, K. 1964. *Theor. Chem.* Acta. 2, 219.
Parr, R. 1963. *The Quantum Theory of Molecular Electronic Structure*. New York: W.A. Benjamin Inc.
Pillar, F.L. 1990. *Elementary Quantum Chemistry*. New York: McGraw-Hill.
Pople, J.A and D.L. Beveridge. 1970. *Approximate Molecular Orbital Theory*. New York: McGraw-Hill.
Pople, J.A., D.L. Beveridge, and P.A. Dobosh. 1967. 'Approximate self-consistent molecular orbital theory V: Intermediate neglect of differential overlap.' *J. Chem. Phys.* 47: 2026.
Pople, J.A., D.P. Santry and G.A. Segal. 1965. 'Approximate self-consistent molecular orbital theory I: Invariant procedures.' *J. Chem. Phys.* 43: S129.
Pople, J.A. and G.A. Segal. 1965. 'Approximate self-consistent molecular orbital theory II: Calculations with complete neglect of differential overlap.' *J. Chem. Phys.* 43: 136.
Roothaan, C.C.J. 1951. 'New developments in molecular orbital theory.' *Rev. Mod. Phys.* 23: 6989.
Roothaan, C.J. 1951. *J. Chem. Phys.* 19, 1445.
Snow, R.L and J.L. Bills. 1981. *J. Chem. Educ.* 58: 619.

Solved Problems

Problem 1. How will you show that the three sp^2 hybrid orbitals are orthogonal to one another?

Solution: We know that the two orbitals ψ_1 and ψ_2 will be orthogonal if $\int \psi_1^* \psi_2 \mathrm{d}\tau = 0$. Substituting the values of ψ_1^* and ψ_2 for the first two sp^2 hybrid orbitals, we have

$$\int \psi_1^* \psi_2 \mathrm{d}\tau = \int \left(\frac{1}{3}\psi_s^* + \sqrt{\frac{2}{3}}\psi_{p_x}^* \right)\left(\frac{1}{\sqrt{3}}\psi_s - \frac{1}{\sqrt{6}}\psi_{p_x} + \frac{1}{\sqrt{2}}\psi_{p_x} \right)\mathrm{d}\tau$$

Expanding the product in the aforementioned integrals, we shall get

$$\int \psi_1^* \psi_2 \mathrm{d}\tau = \int \left(\frac{1}{\sqrt{3}}\psi_s^* \cdot \frac{1}{\sqrt{3}}\psi_s - \frac{1}{\sqrt{18}}\psi_s^* \psi_{p_x} + \frac{1}{\sqrt{6}}\psi_s^* \cdot \psi_{p_y} + \sqrt{\frac{2}{3}}\psi_{p_x}^* \cdot \psi_s \right.$$

$$\left. -\sqrt{\frac{2}{18}}\psi_{p_x}^* \cdot \psi_{p_x} + \frac{\sqrt{2}}{3 \times 2}\psi_{p_x}^* \psi_{p_y} \right)\mathrm{d}\tau$$

$$= \int \frac{1}{3}\psi_s^* \cdot \psi_s \mathrm{d}\tau_s - \frac{1}{\sqrt{18}}\int \psi_s^* \cdot \psi_{p_x}\mathrm{d}\tau + \frac{1}{\sqrt{6}}\int \psi_s^* \cdot \psi_{p_y}$$

$$+ \sqrt{\frac{2}{3}}\int \psi_{p_x}^* \cdot \psi_s \mathrm{d}\tau - \frac{1}{3}\int \psi_{p_x}^* \cdot \psi_{p_x}\mathrm{d}\tau + \frac{1}{\sqrt{3}}\psi_{p_x}^* \psi_{p_y}\mathrm{d}\tau$$

$$= \frac{1}{3} - \frac{1}{3} = 0$$

Therefore, Ψ1 and Ψ2 are orthogonal. Answer.

Problem 2. How will you prove that the angle between any two of the sp^2 hybrid orbitals Ψ_1 and Ψ_2 is 0.5°?

Given that $\Psi_1 = 0.45 \cdot 2s + 0.71 \cdot 2p_y + 0.55 \cdot 2p_z$ and $\Psi_2 = 0.45 \cdot 2s - 0.71 \cdot 2p_y + 0.55 \cdot 2p_z$ for water molecule.

Solution: Since $2s$ orbital is spherically symmetric, the angle between Ψ_1 and Ψ_2 will be found out by the relative contributions of $2p_y$ and $2p_z$ orbitals. The following figure will give a clear picture of direction

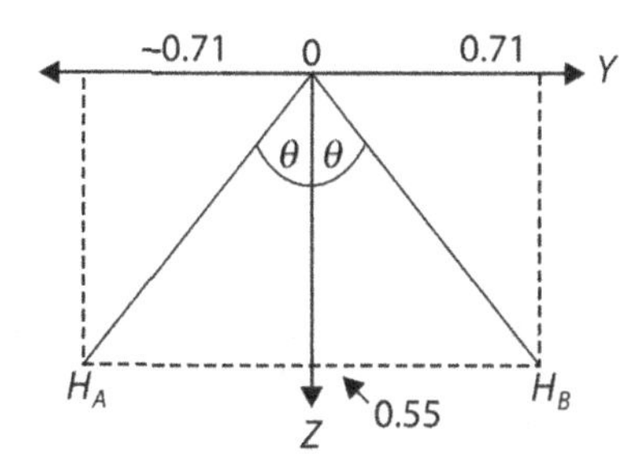

between the two sp^2 hybrid orbitals.

It is to be noted that ψ_1 and ψ_2 may be represented in drawing as vectors whose components will be the coefficient of $2p_y$ and $2p_z$. It should be kept in mind that $2p_y$ and $2p_z$ atomic orbitals are the unit vectors. From the figure, we can write

$\tan\theta = \dfrac{0.71}{0.55} = 1.29$, which corresponds to $\theta = 52.24°$, and hence, $20 \approx 104.5°$. Proved.

Problem 3. The hybrid orbitals ψ_3 are given by the equation

$\psi_3 = \dfrac{1}{\sqrt{3}} 2s - \dfrac{1}{\sqrt{6}} 2p_z - \dfrac{1}{\sqrt{2}} 2p_x$ using the STOs. Show that ψ_3 is directed to 240° from ψ_1.

Solution: The STOs are represented by

$$\psi_{2_s} = \left(\frac{1}{4\pi}\right)^{1/2} R(r)$$

$$\psi_{2p_x} = \left(\frac{3}{4\pi}\right)^{1/2} R(r)\sin\theta\cos\phi$$

$$\psi_{2p_z} = \left(\frac{3}{4\pi}\right)^{1/2} R(r)\cos\phi$$

Putting these values in equation for ψ_3, we shall obtain using $(\phi = 0)$

$$\psi_3 = \frac{R(r)}{(4\pi)^{\frac{1}{2}}}\left[\frac{1}{\sqrt{3}} - \frac{\sqrt{3}}{\sqrt{6}}\cos\theta - \frac{\sqrt{3}}{\sqrt{2}}\sin\theta\right]$$

$$\therefore \quad \frac{d\psi_3}{d\theta} = -\frac{\sqrt{3}}{\sqrt{6}}(-\sin\theta) - \frac{\sqrt{3}}{\sqrt{2}}\cos\theta$$

$$= \left[\frac{\sqrt{3}}{\sqrt{6}}\sin\theta - \frac{\sqrt{3}}{\sqrt{2}}\cos\theta\right]$$

$$= 0 \text{ on maximissing}$$

$$\therefore \quad \frac{\sqrt{3}}{\sqrt{6}}\sin\theta - \frac{\sqrt{3}}{\sqrt{2}}\cos\theta = 0$$

$$\text{or} \quad \frac{\sqrt{3}}{\sqrt{6}}\sin\theta = \frac{\sqrt{3}}{\sqrt{2}}\cos\theta$$

$$\text{or} \quad \frac{\sin\theta}{\cos\theta} = \frac{\sqrt{3}}{\sqrt{2}} \bigg/ \frac{\sqrt{3}}{\sqrt{6}}\cos\theta = \sqrt{3}$$

$$\text{or} \quad \tan\theta = \sqrt{3}$$

$$\therefore \quad \theta = 240°$$

Answer.

Problems 4. Prove that angle between any two of the sp^3 hybrid orbitals is 109°28′.

Solution: It may be shown that the linear combination of three p orbitals $\phi = ap_x + bp_y + cp_z$ can yield another p orbital oriented in a direction depending on the values of a, b, and c.

Let us consider an adequate combination p_1 of the three orbitals in the direction of the first bond. Then the wave function of the hybrid of the first bond can be expressed as

$$\psi_1 = c_1 s + c_2 p_1$$

where c_1 and c_2 are constants.

It is known to us that all the four hybrid orbitals are equivalent, that is, each must have the same amount of s character and the same amount of p character. Hence, each bond should have 25% s character and 75% p character, i.e., 1/4 s character and 3/4 p character.

$$\therefore \psi_1^2 \text{ must contain } \frac{1}{4}s^2 \text{ and } \frac{3}{4}p^2$$

$$\therefore c_1^2 = \frac{1}{4} \quad \therefore c_1 = \frac{1}{2} \quad \text{and } c_2^2 = \frac{3}{4} \quad \therefore c_2 = \frac{\sqrt{3}}{2}.$$

$\therefore$ The hybrid orbital of the first bond will be written as

$$\psi_1 = \frac{1}{2}s + \frac{\sqrt{3}}{2}p_1$$

and $\psi_2 = \frac{1}{2}s + \frac{\sqrt{3}}{2}p_2$

Since ψ_1 and ψ_2 are orthogonal

$$\therefore \langle\psi_1|\psi_2\rangle = \left\langle\left(\frac{1}{2}s + \frac{\sqrt{3}}{2}p_1\right)\middle|\left(\frac{1}{2}s + \frac{\sqrt{3}}{2}p_2\right)\right\rangle = 0$$

$$\text{or} \quad \frac{1}{4}\langle s|s\rangle + \frac{3}{4}\langle p_1|p_2\rangle + \frac{\sqrt{3}}{4}\langle s|p_2\rangle + \frac{\sqrt{3}}{4}\langle p_1|s\rangle = 0$$

$$\text{or} \quad \frac{1}{4} + \frac{3}{4}\langle p_1|p_2\rangle = 0$$

$$\text{or} \quad \frac{1}{4} + \frac{3}{4}\langle p_1|p_2\rangle\cos\theta_{12} = 0 \qquad [\text{writing } p_2 = p_1\cos\theta_{12}]$$

$$\text{or} \quad \frac{1}{4}+\frac{3}{4}\cos\theta_{12} = 0$$

$$\text{or} \quad \frac{3}{4}\cos\theta_{12} = -\frac{1}{4}$$

$$\text{or} \quad \cos\theta_{12} = \frac{-1}{4} / \frac{3}{4} = -\frac{1}{4}\times\frac{4}{3} = -\frac{1}{3}$$

$$\therefore \ \theta_{12} = 109°28'$$

Proved.

Questions on Concepts

1. Express the matrix form of Roothaan's equations.
2. Prove that $E_{HF} = \frac{1}{2}\sum_{i=1}^{M}\sum_{j=1}^{M} P_{ij}\left(F_{ij} + H_{ij}^{\text{core}}\right) + V_{NN}$.
3. Derive $\sum_{j=1} c_j\left(h_{ij} + g_{ij} - \in s_{ij}\right) = 0$ for Roothaan's method in one dimension.
4. Considering $\psi = c_1 2xe^{-x} + c_2\sqrt{32}xe^{-2x}$ and $H = -\frac{1}{2}\frac{d^2}{dx^2} - \frac{2}{x}$, prove that $h_{11} = -1.5$ and $h_{12} = h_{21} = -1.6761$.
5. For one-dimensional Roothaan's method, prove that $s_{12} = \langle f_j \mid f_j \rangle =$ 0.83805.
6. Draw the graph between r^2 vs. χ for comparison of He atom self-consistent field wave function with f_1 and f_2 of the basis set.
7. Prove that the electronic energy of the electron is $E = \sum_{L}^{n} E_i + \sum_i \sum_j P_{ij} H_{ij}^{o}$. where the terms have their usual significance.
8. Suppose $\chi_1 = 2\xi_1^{3/2} e^{-\xi_1 r}$ and $\chi_2 = 2\xi_2^{3/2} e^{-\xi_2 r}$, then find $s_{21} = s_{12} = 0.8368$.
9. Find H_{11}^{core} using $\xi_1 = 1.45$, where $H_{11}^{\text{core}} = \chi_1 \left|H^{\text{core}}\right| \chi_1$.
10. Explain the type of hybridisation in H_2O molecule. Try to find the value of bond angle in H_2O quantum mechanically.
11. Show that the three hybrid bonds in sp^2 hybridisation are inclined to each other by 120°.
12. In *sp* hybridisation, show that the angle between two hybrid bonds is 180°.
13. Prove that the angle between any two of the sp^3 hybrid orbitals is 109°28′.
14. Find ψ_1 and ψ_2 for *sp* hybridisation.
15. Find the values of coefficients of ψ_1 and ψ_2 equations when hybridisation is taking place in H_2O molecule.
16. What do you mean by *ab initio* and semi-empirical methods.
17. Define valence electrons. Write down the equation for $H_{(1)}^{\text{core}}$ and $\hat{F}(1)$ for valence electrons.
18. How does Pariser–Parr–Pople method make the approximation of zero differential overlap while evaluating electron repulsion integrals? [P.U. 2011]
19. a. What is zero differential overlap approximation?
    b. In what sense intermediate neglect of differential overlap (INDO) method is an improvement over complete neglect of differential overlap (CNDO) method?
20. Discuss CNDO theory.
21. Write short notes on INDO theory.

22. Explain invariance under transformation.
23. Prove that $E_{\text{Total}}^{HF} = \frac{1}{2}\sum_{\mu\upsilon} p\mu\upsilon\left(H_{\mu\upsilon}^{\text{core}} + F_{\mu\upsilon}\right) + \sum_{A<B} z_A z_B R_{AB}^{-1}$.
24. Give a clear concept of intermediate neglect of differential overlap.
25. Write a short note on neglect of diatomic differential overlap method.

13

Hückel Molecular Orbital Theory/Method

The basic principle of the Hückel molecular orbital (HMO) theory lies in the fact that it is absolutely necessary a one-electron treatment and the electrons remain in the π orbitals. In other words, we can assume that both the core electrons and the electrons present in the skeleton are regarded as 'frozen'. It is pertinent to note that there is no interaction between σ and π electrons present in the molecule.

Hückel (as early as in 1931) pointed out that it was possible to state the characteristics of conjugated hydrocarbons and polyenes by the quantum mechanical model, which took into consideration only π electrons. The Hückel HMO approximation works best for a class of alternant hydrocarbons (may be aliphatic or aromatic).

The explicit starting point for the derivation of the Hückel method for π electron system is the Eigen value formulation of Schrödinger equation, $H\psi = E\psi$. Hückel applied this equation to molecules, keeping in view that H and Ψ represent molecular Hamiltonian and wave function, respectively.

Multiplying $H\psi = E\psi$ by ψ, we have

$$\psi H\psi = E\psi^2 \tag{13.1}$$

Also multiplying both the sides by the volume element dτ, Eq. (13.1) takes the form

$$\psi H\psi \mathrm{d}\tau = E\psi^2 \mathrm{d}\tau$$

Integrating and rearranging the above equation, we can express this as

$$E = \frac{\int \psi H\psi \mathrm{d}\tau}{\int \psi^2 \mathrm{d}\tau}$$

$$\text{or} \quad E = \frac{\langle H|H|\Psi\rangle}{\langle \Psi|\Psi\rangle} \tag{13.2}$$

This equation represents the energy value/expectation value of energy of the system.

Next, the molecular wave function will be approximated as a linear combination of atomic orbitals (LCAO) by using suitable basis function ϕ_i. Keep in mind that the combination of n basis function will give rise to n molecular orbitals, which can be expressed as

$$\psi = \sum_i C_i\phi_i = C_1\phi_1 + C_2\phi_2 + C_3\phi_3 + ---- C_n\phi_n \tag{13.3}$$

where C_i = variational coefficients

Φ_i = basis functions

Substituting the value of Ψ in Eq. (13.2), we have

$$E = \frac{\left\langle \left(\sum_i C_i\phi_i\right) \middle| H \middle| \left(\sum_j C_j\phi_j\right) \right\rangle}{\left\langle \sum_i \sum_j C_i\phi_i \middle| C_j\phi_j \right\rangle}$$

$$\text{or} \quad E = \frac{\sum_i \sum_j C_iC_j\{\langle\phi_i|H|\phi_j\rangle}{\sum_i \sum_j C_iC_j\{\langle\phi_i|\phi_j\rangle} \tag{13.4}$$

But $\langle\psi_i|H|\psi_j\rangle = H_{ij}$ = Hückel energy integral
and $\langle\psi_i|\psi_j\rangle = S_{ij}$ = overlap integral

Substituting these in the above equation, we shall obtain

$$E = \frac{\sum_i \sum_j C_iC_jH_{ij}}{\sum_i \sum_j C_iC_jS_{ij}} = \frac{N}{D} \tag{13.5}$$

where N = numerator
D = denominator

Differentiating Eq. (13.5) with respect to C_i and on applying the condition of minimum to get minimum value of E, we can write

$$\frac{\partial E}{\partial C_i} = \frac{N'D - D'N}{D^2} \tag{13.6}$$

where N' = first differentiation of N
D' = first differentiation of D

The above equation can also be expressed as

$$\frac{\partial E}{\partial C_i} = \frac{\frac{N'D - D'N}{D}}{D^2/D} = \frac{N' - D'\frac{N}{D}}{D} = \frac{N' - D'E}{D}$$

Under minimum condition, $\frac{\partial E}{\partial C_i} = 0$

$$\therefore \quad \frac{\partial E}{\partial C_i} = \frac{N' - D'E}{D} = 0$$

$$\text{or} \quad N' - D'E = 0 \tag{13.7}$$

From this, we shall get the following set of linear equations:

$$C_1(H_{11} - S_{11}E) + C_2(H_{12} - S_{12}E) + ------ = 0$$

$$C_1(H_{21} - S_{21}E) + C_2(H_{22} - S_{22}E) + ------ = 0$$

These linear equations can be expressed in a more general form as

$$\sum_i C_i \left(H_{ij} - ES_{ij}\right) = 0 (i = 1, 2, 3 \ldots) \tag{13.8}$$

When one proceeds to solve Eq. (13.8), one will find for each root the ratio of the expansion coefficients, and finally, one will get the coefficients uniquely by normalising each orbital. The normalisation condition is expressed as

$$\sum_i \sum_j C_i C_j S_{ij} = 1 \tag{13.9}$$

This is the pertinent points of the principle of the Hückel method.

In 1931, Hückel proposed drastic approximations for energy integral and overlap integral for further simplification of the method.

- $H_{ii} = \alpha$ for all i
  = Coulomb integral, which represents the energy of $2p_z$ electron in each carbon atom having sp^2 hybridisation. α is a negative quantity
- $H_{ij} = \beta$, when C_i and C_j are the nearest neighbour atoms
  = resonance integral, which is a negative quantity
  $H_{ij} = 0$ when $i \neq j$ and C_i and C_j are not the nearest neighbour atoms
- $S_{ij} = \delta_{ij}$ = Kronecker delta, that is, the basis atomic orbitals is orthonormal
  = 0 for $i \neq j$
  1 for $i = j$

In summary, we can put forward the pertinent points of the Hückel drastic assumptions as follows:

1. All overlap integrals are neglected, i.e., $S = 0$.
2. All non-diagonal matrix elements of H are neglected except those connecting the nearest neighbour atoms, i.e., $H_{ij} = \beta$.
3. All diagonal matrix elements of H are set equal, i.e., $H_{ii} = H_{jj} = \alpha$.
4. The carbon atoms of the conjugated system should be labelled, as it will be easier to write the secular equations.
5. Set the secular determinant equal to zero.

Among the above assumptions, the first one is the easiest to motivate. Despite its obvious absurdity, Assumption 2 is reasonable due to the fact that we deal with only one kind of atom. Assumption 3 is also comparatively reasonable.

13.1 Application of the Hückel Molecular Orbital Method to π Systems

Now, we are in a position to apply the HMO method to different π systems, i.e., conjugated polyenes. These systems may be in the straight chain or in cyclic rings. First, we shall apply the HMO method to simple systems, such as ethylene or ethene, allyl system, and butadiene.

13.1.1 Ethylene

Ethylene is a two-carbon system in which there is one π bond between carbon–carbon. The structure of ethylene can be shown as

$$>\underset{1}{C}\overset{\pi}{=}\underset{2}{C}<$$

Each carbon atom is contributing one basis function to the molecular orbital. When the basis function on different atoms will be combined to yield molecular orbital, it will be analogous to the combination of atomic orbitals on the same atom to give hybrid atomic orbitals. When n basis functions are combined, n molecular orbitals will be produced. Applying the LCAO approximation to ethylene, we can write the molecular orbitals as

$$\psi = C_1\phi_1 + C_2\phi_2 \tag{13.10}$$

where ϕ_1 and ϕ_2 = basis functions on carbon atoms 1 and 2, respectively

(In other words, we can say that ϕ_1 denotes the p_z atomic orbital on carbon 1 and ϕ_2 on carbon atom 2.)

C_1 and C_2 = weighting coefficients to be adjusted to have the best ψ

The energy value E has been given in Eq. (13.2) as

$$E = \frac{\langle\psi|H|\psi\rangle}{\langle\psi|\psi\rangle} = \frac{\int \psi H \psi \, d\tau}{\int \psi^2 d\tau}$$

But the energy E of the molecular orbital will be obtained by substituting the value of ψ given in Eq. (13.2).

∴ The energy of the molecular orbital is given by

$$E = \frac{\langle (C_1\phi_1 + C_2\phi_2) H (C_1\phi_1 + C_2\phi_2)\rangle}{\langle (C_1\phi_1 + C_2\phi_2)^2 \rangle} \tag{13.11}$$

$$\text{or } E = \frac{\langle (C_1^2\phi_1 H\phi_1 + C_2^2\phi_2 H\phi_2 + 2C_1C_2\phi_1 H\phi_2)\rangle}{\langle (C_1^2\phi_1^2 + 2C_1C_2\phi_1\phi_2 + C_2^2\phi_2^2)\rangle}$$

$$\text{or } E = \frac{C_1^2\langle\phi_1|H|\phi_1\rangle + C_2^2\langle\phi_2|H|\phi_2\rangle + 2C_1C_1\langle\phi_1|H|\phi_2\rangle}{C_1^2\langle\phi_1|\phi_1\rangle + 2C_1C_2\langle\phi_1|\phi_2\rangle + C_2^2\langle\phi_2|\phi_2\rangle} \tag{13.12}$$

$$\text{or } E = \frac{C_1^2 H_{11} + C_2^2 H_{22} + 2C_1C_2H_{12}}{C_1^2 S_{11} + 2C_1C_2S_{12} + C_2^2 S_{22}}$$

But $H_{11} = H_{22} = \alpha$ = coulomb integral = $\langle\psi_1|H|\psi_1\rangle = \langle\psi_2|H|\psi_2\rangle$

$H_{12} = H_{21} = \beta$ = resonance integral = $\langle\psi_1|H|\psi_2\rangle$

$S_{11} = S_{22} = 1 = \langle\psi_1|\psi_1\rangle = \langle\psi_2|\psi_2\rangle$

$S_{12} = S_{21} = 0 = \langle\psi_1|\psi_2\rangle$

Putting these values in Eq. (13.12), we obtain

$$E = \frac{C_1^2\alpha + C_2^2\alpha + 2C_1C_2\beta}{(C_1^2 + 2C_1C_2S_{12} + C_2^2)}$$

$$= \frac{(C_1^2 + C_2^2)\alpha + 2C_1C_2\beta}{(C_1^2 + 2C_1C_2S_{12} + C_2^2)} \tag{13.13}$$

$$\text{or } E(C_1^2 + 2C_1C_2S_{12} + C_2^2) = (C_1^2 + C_2^2)\alpha + 2C_1C_2\beta$$

Differentiating Eq. (13.13) and applying the condition of minima, i.e., $\frac{\partial E}{\partial C_1} = 0$ and $\frac{\partial E}{\partial C_2} = 0$, we can write

$$\frac{\partial E}{\partial C_1} \cdot \left[C_1^2 + 2C_1C_2S_{12} + C_2^2\right] + E[2C_1 + 2C_2S_{12}] = \frac{\partial}{\partial C_1}\left[\left(C_1^2 + C_2^2\right)\alpha + 2C_1C_2\beta\right] = 0 = 2C_1\alpha + 2C_2\beta$$

$$\text{or} \quad 0 \times \left[C_1^2 + 2C_1C_2S_{12} + C_2^2\right] + E\left[2C_1 + 2C_2S_{12}\right] = 2C_1\alpha + 2C_2\beta$$

$$\text{or} \quad E[2C_1 + 2C_2S_{12}] = 2C_1\alpha + 2C_2\beta$$

$$\text{or} \quad E[C_1 + C_2S_{12}] = C_1\alpha + C_2\beta$$

$$\text{or} \quad C_1(\alpha - E) + C_2(\beta - ES_{12}) = 0 \tag{13.14}$$

Similarly, differentiating Eq. (13.13) with respect to C_2 and putting $\frac{\partial}{\partial C_2} = 0$, we can write

$$\frac{\partial E}{\partial C_2}\left[C_1^2 + 2C_1C_2S_{12} + C_2^2\right] + E\frac{\partial}{\partial C_2}\left[C_1^2 + 2C_1C_2S_{12} + C_2^2\right] = \frac{\partial}{\partial C_2}\left[\left(C_1^2 + C_2^2\right)\alpha + 2C_1C_2\beta\right]$$

$$\text{or} \quad 0 \times \left[C_1^2 + 2C_1C_2S_{12} + C_2^2\right] + E[2C_2 + 2C_1S_{12}] = [2C_2\alpha + 2C_1\beta]$$

$$\text{or} \quad 0 + E[2C_2 + 2C_1S_{12}] = [2C_2\alpha + 2C_1\beta]$$

$$\text{or} \quad E[2C_2 + 2C_1S_{12}] = [2C_2\alpha + 2C_1\beta]$$

$$\text{or} \quad E[C_2 + C_1S_{12}] = [C_2\alpha + C_1\beta]$$

$$\text{or} \quad C_2(\alpha - E) + C_1(\beta - ES_{12}) = 0 \tag{13.15}$$

Since $S_{12} = 0$, Eqs. (13.14) and (13.15) may be expressed as two linear equations.

$$\left.\begin{aligned} C_1(\alpha - E) + C_2\beta &= 0 \\ \text{and} \quad C_2(\alpha - E) + C_1\beta &= 0 \end{aligned}\right\} \tag{13.16}$$

These equations are satisfied if $C_1 = C_2 = 0$. If this is not the situation, then Eq. (13.16) can be expressed as a secular determinant with the following notations:

$$\begin{vmatrix} \alpha - E & \beta \\ \beta & \alpha - E \end{vmatrix} = 0 \tag{13.17}$$

If the notation $\alpha - E/\beta = x$ is introduced, then above determinant can be expressed as

$$\begin{vmatrix} x & 1 \\ 1 & x \end{vmatrix} = 0$$

From this, we have

$$\begin{aligned} x^2 - 1 &= 0 \\ \text{or} \quad x^2 = 1 \quad &\therefore \quad x = \pm 1 \end{aligned} \tag{13.18}$$

Thus, the roots of Eq. (13.18) are $x = +1$ and $x = -1$. Corresponding to these two roots, we shall get two values of E, i.e., E_1 and E_2.

Since $\alpha - E\,\beta = x$, on putting $x = -1$, we have

$$\frac{\alpha - E_1}{\beta} = -1 \text{ or } \alpha - E_1 = -\beta$$

$$\therefore \quad \alpha + \beta = E_1 \text{ (bonding)} \tag{13.19}$$

Putting $x = +1$, we shall get

$$\frac{\alpha - E_2}{\beta} = +1 \text{ or } \alpha - E_2 = \beta$$
$$\therefore \quad \alpha - \beta = E_2\left(\text{antibonding}\right) \tag{13.20}$$

Thus, we obtained two energy levels E_1 and E_2 corresponding to two molecular orbitals ψ_1 and ψ_2. Since β is negative, $E_1 < E_2$. This means that E_1 is the lowest in energy, and therefore, ψ_1MO will represent bonding molecular orbital (BMO) and ψ_2, antibonding molecular orbital. The energy level diagram is represented in Figure 13.1.

The total π-electron energy will be given by

$$E_{\text{total}} = 2E_1 = 2\left(\alpha + \beta\right) \tag{13.21}$$

Since the energy of an electron in a p orbital is α, the energy of $2p_z$ electrons of carbon atom will be 2α. Hence, the energy of π bond will be

$$E_\pi = 2\left(\alpha + \beta\right) - 2\alpha = 2\beta \tag{13.22}$$

13.1.2 Determination of the Hückel Molecular Orbital Coefficients and Molecular Orbitals of Ethylene

For the determination of molecular orbitals, the following two linear equations, i.e., secular equations, will be taken into consideration:

$$C_1\left(\alpha - E\right) + C_2\beta = 0$$

and $C_2(\alpha - E) + C_1\beta = 0$

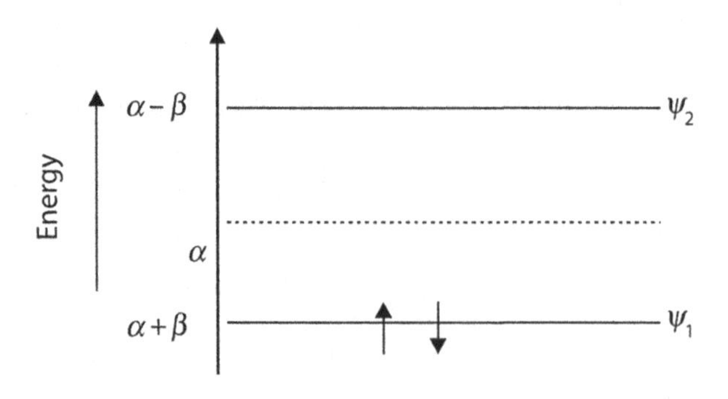

FIGURE 13.1 Bonding and antibonding π orbitals of ethylene molecule.

By dividing these equations by β, we have

$$C_1\frac{(\alpha - E)}{\beta} + \frac{C_2\beta}{\beta} = 0$$

$$\text{and} \quad C_2\frac{(\alpha - E)}{\beta} + \frac{C_1\beta}{\beta} \qquad (13.23)$$

or $C_1x + C_2 = 0$

and $C_2x + C_1 = 0$

Putting the value of $x = -1$ in one of the above secular equations, we obtain

$$C_1(-1) + C_2 = 0$$

$$\text{or} \quad -C_1 + C_2 = 0 \qquad (13.24)$$

$$\therefore \quad C_1 = C_2$$

Since molecular orbitals are normalised, applying the normalisation condition, we get

$$\int \psi^2 d\tau = 1 \quad \text{or} \quad \langle\psi|\psi\rangle = 1$$

Putting the value of ψ, we can write

$$\langle(C_1\phi_1 + C_2\phi_2)|(C_1\phi_1 + C_2\phi_2)\rangle = 1$$

$$\text{or} \quad C_1^2\langle\phi_1|\phi_1\rangle + 2C_1C_2\langle\phi_1|\phi_2\rangle + C_2^2\langle\phi_2|\phi_2\rangle = 1$$

$$\text{or} \quad C_1^2 \cdot 1 + 2C_1C_2 \times 0 + C_2^2 \cdot 1 = 1 \qquad [\because \text{In HMOT } \langle \phi_1|\phi_2\rangle \text{ is neglected}] \qquad (13.25)$$

$$\text{or} \quad C_1^2 + C_2^2 = 1$$

$$\text{But} \quad C_1 = C_2$$

$$\therefore \quad C_1^2 + C_2^2 = 2C_1^2 = 1 \qquad (13.26)$$

$$\text{or} \quad C_1^2 = \frac{1}{2} \quad \therefore \quad C_1 = \frac{1}{\sqrt{2}} = C_2$$

$$\left.\begin{aligned} \therefore \quad \psi_1 &= \frac{1}{\sqrt{2}}(\phi_1 + \phi_2) \\ \text{and} \quad \psi_2 &= \frac{1}{\sqrt{2}}(\phi_1 - \phi_2) \end{aligned}\right\} \qquad (13.27)$$

In these equations, ψ_1 corresponds to E_1 and ψ_2 corresponds to E_2. We call ψ_1 a bonding orbital with two electrons. The ψ_2 is vacant having higher energy than ψ_1 and is known as an antibonding orbital.

13.1.2.1 Graphical Representation: Plots of ψ_1 and ψ_2 vs Distance

When HMO functions ψ_1 and ψ_2 are plotted against various distances along $C_1 - C_2$ axis (in ethylene), graphs are obtained, which are illustrated in Figure 13.2.

It is clear from Figure 13.2a that ψ_1 increases with the distance till it reaches maximum near C_1 and further it decreases and again it increases till it attains maximum corresponding to C_2 and then falls off

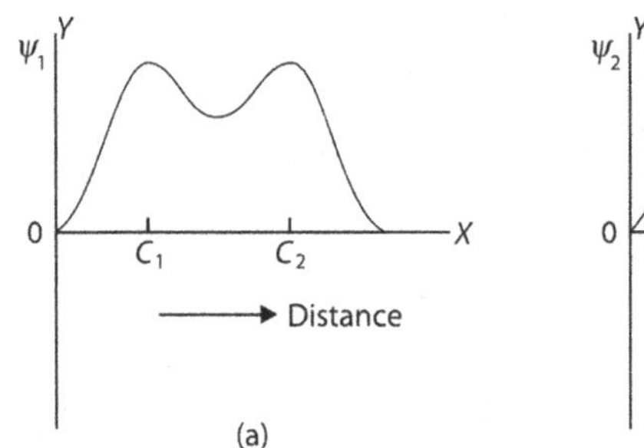

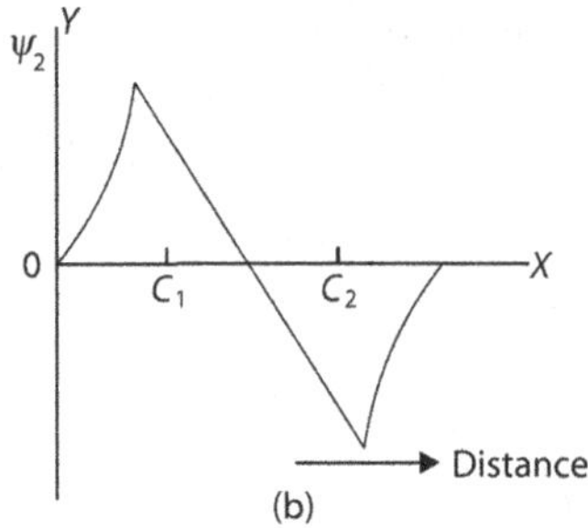

FIGURE 13.2 Plots of ψ_1 and ψ_2 against distance.

and touches the OX axis. Function ψ_2 also increases with the increase of distance and intersect the OX axis and proceeds in the negative direction. It also gives another negative maxima and then increases and cuts the OX axis. It gives a clue to understand that the function ψ_2 changes sign in between C_1 and C_2 axes. The point of intersection is known as *node*.

13.1.2.2 Three-Dimensional Representation

The shapes of HMO may be obtained from 3D plots along different directions. The 3D shapes of bonding and antibonding π molecular orbitals of ethylene are shown in Figure 13.3.

It is clear from the figure that each function changes its sign along the c_1–c_2 axis. The plane consisting of this axis is called the *nodal plane*. It should be kept in mind that the existence of such a plane is the characteristic of π orbitals. It is important to note that Ψ_1 is *ungerade*, whereas Ψ_2 is *gerade*, as Ψ_2 has centre of inversion. It is also to be noted that the '+' and '–' signs denote the signs of the functions Ψ_1 and Ψ_2.

13.1.3 Allyl System

The allyl system is represented by the following structure:

$$\dot{C}H_2-CH=CH_2 \longleftrightarrow \overset{1}{\dot{C}}H=\overset{2}{\dot{C}}H-\overset{3}{\dot{C}}H \longleftrightarrow {}^{H}_{H}\!\!>\dot{C}-\overset{H}{\overset{|}{C^{+}}}-\bar{C}<$$

The allyl radical system is a three-carbon atom system, labelled as above. It has three electrons located in a π orbital framework on three carbons in a linear array. First, we shall set up the system of HMO secular equation. The HMO secular equations are given by

$$\left.\begin{aligned} C_1x + C_2 &= 0 \\ C_2x + C_1 + C_3 &= 0 \\ C_3x + C_2 &= 0 \end{aligned}\right\} \tag{13.28}$$

where $x = \alpha - E\,\beta$.

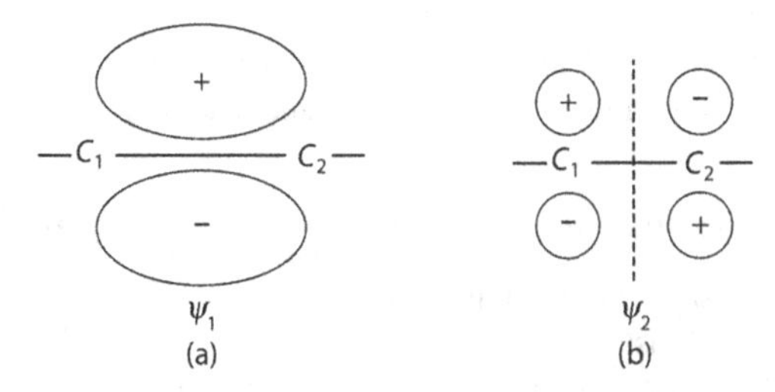

FIGURE 13.3 Shape of (a) bonding and (b) antibonding π molecular orbitals of C_2H_4.

On the basis of HMO secular equations, the secular determinant is set up as

$$\begin{vmatrix} x & 1 & 0 \\ 1 & x & 1 \\ 0 & 1 & x \end{vmatrix} = 0 \tag{13.29}$$

This determinant is expanded as

$$x\begin{vmatrix} x & 1 \\ 1 & x \end{vmatrix} - 1\begin{vmatrix} 1 & 1 \\ 0 & x \end{vmatrix} = 0$$

or $x(x^2 - 1) - 1(x) = (x^3 - x - x) = x^3 - 2x = 0$.

Thus, we get cubic equation

$$x^3 - 2x = 0$$

or $x(x^2 - 2) = 0$

$$\text{or } x\left(x+\sqrt{2}\right)\left(x-\sqrt{2}\right) = 0$$

So, the roots of the equation will be

$$x = 0, x = -\sqrt{2} \text{ and } x = +\sqrt{2} \tag{13.30}$$

When $x = -\sqrt{2}$, then the energy levels will be given by

$$\frac{\alpha - E_1}{\beta} = -\sqrt{2}$$

$$\begin{aligned} &\text{or} \quad \alpha - E_1 = -\sqrt{2}\beta \\ &\text{or} \quad E_1 = \alpha + \sqrt{2}\beta \qquad (\text{BMO}) \end{aligned} \tag{13.31}$$

When $x = +\sqrt{2}$, then the energy level will be given by

$$\frac{\alpha - E_2}{\beta} = +\sqrt{2}$$

$$\begin{aligned} &\text{or} \quad \alpha - E_2 = \sqrt{2}\beta \\ &\text{or} \quad E_2 = \alpha - \sqrt{2}\beta \qquad (\text{antibonding molecular orbital}) \end{aligned} \tag{13.32}$$

When $x = 0$, then the energy level E_3 will be given by

$$\frac{\alpha - E_3}{\beta} = 0$$

or $\alpha - E_3 = 0$

$$\therefore \quad E_3 = \alpha \qquad (\text{Non-bonding molecular orbital, NBMO}) \tag{13.33}$$

It is clear that in the allyl system that the cation has two electrons, the radical has three electrons, and the anion contains four electrons. The energy level diagram for the allyl radical is illustrated in Figure 13.4.

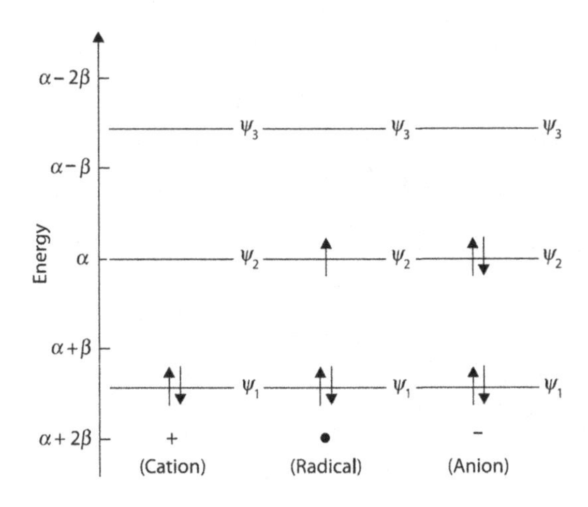

FIGURE 13.4 Energy level diagram for allyl system.

The 2π electrons in the allyl carbonium ion $C_3H_5^+$ are placed in the (BMO) ψ_1 consisting of total π electron energy.

$$E_\pi^+ = 2\left(\alpha + \sqrt{2}\beta\right) = 2\alpha + 2\sqrt{2}\beta \tag{13.34}$$

In the allyl radical, C_3H_5, there are 3π electrons in which the first two electrons go to ψ_1 and the third electron goes to ψ_3 non-bonding molecular orbital (NBMO). The E_π° energy is given by

$$E_\pi^{\circ} = 2\left(\alpha + \sqrt{2}\beta\right) + \alpha = 3\alpha + 2\sqrt{2}\beta \tag{13.35}$$

Among the 4π electrons of allyl anion (C_3H_5), two electrons are accommodated in ψ_1 (BMO) and two in ψ_2 (NBMO).

$$\therefore \quad E_\pi^- = 2\left(\alpha + \sqrt{2}\beta\right) + 2\alpha = 4\alpha + 2\sqrt{2}\beta \tag{13.36}$$

13.1.4 Delocalisation Energy of Allyl System

The resonating structures of allyl cation are represented as follows:

In structure I, the π electrons are localised in the double bond and the molecule has no resonance stabilisation. In structure II, the electrons are delocalised over the total molecular framework.

The delocalisation energy is defined as the energy difference between the localised structure $(E_\pi)_{loc}$ and the energy of actual molecule E_π is called delocalisation energy, that is,

$$DE = \left(E^+\right)_{deloc} = \left(E_\pi\right)_{loc} - E_\pi \tag{13.37}$$

The localised structure I has 2π electrons having energy $2(\alpha + \beta)$, i.e.,

$$\left(E_\pi\right)_{loc} = 2\left(\alpha + \beta\right)$$

But the energy of allyl carbonium ion is

$$E_\pi^+ = \left(2\alpha + 2\sqrt{2}\beta\right)$$

$$\therefore \left(E^{+}\right)_{\text{deloc}} = \left(E_{\pi}\right)_{\text{loc}} - \left(E_{\pi}^{+}\right)$$

$$= 2(\alpha + \beta) - \left(2\alpha + 2\sqrt{2}\beta\right)$$

$$= 2\alpha + 2\beta - 2\alpha - 2\sqrt{2}\beta$$

$$= 2\beta - 2\sqrt{\beta} = \beta\left(2 - 2\sqrt{2}\right)$$

$$= -0.828\beta \tag{13.38}$$

In the case of free radical structure, it resonates between the following structures:

III ⟷ IV

In this case, the delocalisation energy is given by

$$\left(E_{\pi}^{\cdot}\right)_{\text{deloc}} = 2(\alpha + \beta) - \left(2\alpha + 2\sqrt{2}\beta\right)$$

$$= 2\alpha + 2\beta - 2\alpha - 2\sqrt{2}\beta$$

$$= 2\beta - 2\sqrt{2}\beta = -0.828\beta \tag{13.39}$$

In the case of $C_3H_5^-$, the resonance takes place as follows:

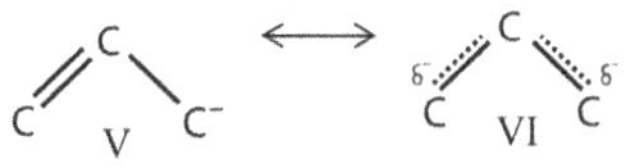

V ⟷ VI

In this case, the delocalisation energy is given by

$$\left(E_{\pi}^{-}\right)_{\text{deloc}} = \left[2(\alpha + \beta) + 2\alpha\right] - \left[2\left(2\alpha + \sqrt{2}\beta\right)\right]$$

$$= \left(2\alpha + 2\beta + 2\alpha - 4\alpha - 2\sqrt{2}\beta\right)$$

$$= \left(4\alpha + 2\beta - 4\alpha - 2\sqrt{2}\beta\right)$$

$$= \beta\left(2 - 2\sqrt{2}\right) = -0.828\beta \tag{13.40}$$

Thus, we find that the allyl cation, radical, and anion have the same value of delocalisation energy.

13.1.5 Determination of the Hückel Molecular Orbital Coefficients and Molecular Orbitals of Allyl System

The HMO secular equations are

$$C_1x + C_2 = 0$$

$$C_2x + C_1 + C_3 = 0$$

$$C_3x + C_2 = 0$$

Substituting, $x = -\sqrt{2}$ in the above equations for obtaining BMO, Ψ_1, we have

$$\left.\begin{aligned} -\sqrt{2}C_1 + C_2 &= 0 \\ -\sqrt{2}C_2 + C_1 + C_3 &= 0 \\ -\sqrt{2}C_3 + C_2 &= 0 \end{aligned}\right\} \tag{13.41}$$

From the first equation, we have

$$C_2 = \sqrt{2}C_1 \tag{13.42}$$

From the third equation,

$$\begin{aligned} -\sqrt{2}C_3 + C_2 &= 0 \\ \therefore\quad C_2 &= \sqrt{2}C_3 \end{aligned} \tag{13.43}$$

From Eqs. (13.42) and (13.43), we get

$$C_1 = C_3 \tag{13.44}$$

For the normalisation condition,

$$\sum_1^3 C_i^2 = 1$$

$$\text{or}\quad C_1^2 + C_2^2 + C_3^2 = 1$$

$$\text{or}\quad C_1^2 + \left(\sqrt{2}c_1\right)^2 + C_1^2 = 1$$

$$\text{or}\quad C_1^2 + 2C_1^2 + C_1^2 = 1$$

$$\text{or}\quad 4C_1^2 = 1 \text{ or } C_1^2 = 1/4$$

$$\therefore\quad C_1 = \frac{1}{2} = C_3;\quad \text{but } C_2 = \sqrt{2}C_3$$

$$\therefore\quad C_2 = \sqrt{2} \times \frac{1}{2} = \frac{1}{\sqrt{2}}$$

Hence, $\Psi_1 = \left(\frac{1}{2}\Phi_1 + \frac{1}{\sqrt{2}}\Phi_2 + \frac{1}{2}\Phi_3\right)$

$$\text{or}\quad \Psi_1 = \frac{1}{2}\left(\Phi_1 + \sqrt{2}\Phi_2 + \Phi_3\right) \tag{13.45}$$

Putting $x = +\sqrt{2}$ in the secular equations, we obtain

$$\left.\begin{aligned} \sqrt{2}C_1 + C_2 &= 0 \\ \sqrt{2}C_2 + C_1 + C_3 &= 0 \\ \sqrt{2}C_3 + C_2 &= 0 \end{aligned}\right\} \tag{13.46}$$

From the first equation of Eq. (13.46), we can write

$$\sqrt{2}C_1 = -C_2 \tag{13.47}$$

From the third equation, we have

$$\sqrt{2}C_3 = -C_2 \tag{13.48}$$

From Eqs. (13.47) and (13.48), we can write

$$C_1 = C_3 \tag{13.49}$$

But for normalisation condition

$$\sum_{1}^{3} C_i^2 = 1$$

$$\left.\begin{aligned} \text{or} \quad & C_1^2 + C_2^2 + C_3^2 = 1 \\ \text{or} \quad & C_1^2 + 2C_1^2 + C_1^2 = 1 \end{aligned}\right. \tag{13.50}$$

$$\text{or} \quad 4C_1^2 = 1 \quad \therefore \quad C_1 = \frac{1}{2} = C_3 \text{ and } C_2 = -\sqrt{2}C_1$$

Therefore, the molecular orbital Ψ_2 for ABMO will be given by

$$\begin{aligned} \Psi_2 &= C_1\Phi_1 + C_2\Phi_2 + C_3\Phi_3 \\ &= \frac{1}{2}\Phi_1 - \frac{\sqrt{2}}{2}\Phi_2 + \frac{1}{2}\Phi_3 \\ &= \frac{1}{2}\left(\Phi_1 - \frac{1}{\sqrt{2}}\Phi_2 + \Phi_3\right) \end{aligned} \tag{13.51}$$

For the root $x = 0$, we shall find the HMO coefficients. The HMO secular equations are

$$C_1x + C_2 = 0$$

$$C_2x + C_1 + C_3 = 0$$

$$C_3x + C_2 = 0$$

Putting $x = 0$ in the first equation for getting NBMO, we have
$C_2 = 0$ (by putting $x = 0$ in the first equation)
Putting $x = 0$ in the second equation, we have

$$C_2 \times 0 + C_1 + C_3 = 0$$

$$\therefore \quad C_1 = -C_3$$

Applying the normalisation condition, we get

$$\sum_{1}^{3} C_i^2 = C_1^2 + C_2^2 + C_3^2 = 1$$

$$\text{or } C_1^2 + 0^2 + C_1^2 = 1$$

$$\text{or } 2C_1^2 = 1 \quad \therefore \quad C_1 = \frac{1}{\sqrt{2}}$$

$$\Psi_3 = (C_1\Phi_1 + C_2\Phi_2 + C_3\Phi_3) \tag{13.52}$$

$$\text{or } \Psi_3 = \left(\frac{1}{\sqrt{2}}\Phi_1, -\frac{1}{\sqrt{2}}\Phi_3\right)$$

$$\text{or } \Psi_3 = \frac{1}{\sqrt{2}}(\Phi_1 - \Phi_3)$$

Thus, the molecular orbitals for the allyl system are

$$\psi_1 = \frac{1}{2}\left(\phi_1 + \sqrt{2}\phi_2 + \phi_3\right) \quad \text{(BMO)}$$

$$\psi_2 = \frac{1}{2}\left(\phi_1 - \sqrt{2}\phi_2 + \phi_3\right) \quad \text{(ABMO)} \tag{13.53}$$

$$\psi_3 = \frac{1}{2}(\phi_1 - \phi_3) \quad \text{(NBMO)}$$

13.1.5.1 Graphical Representation

The plots of ψ_1, ψ_2, and ψ_3 against different points along $C_1 - C_2 - C_3$ axis in the case of allyl system are illustrated in Figure 13.5.

The plots are very clear and one can understand the trend of variation of ψ_1, ψ_2, and ψ_3 with respect to different distances. With respect to point at C_2, both ψ_1 and ψ_3 are observed to be symmetric, while ψ_2 is antisymmetric. In the case of Figure 13.5a, there is no node; in the case of Figure 13.5b, there is one node; and in the case of Figure 13.5c, there are two nodes.

It should be kept in mind that the number of node increases with the increase of energy levels.

13.1.5.2 Three-Dimensional Representation: Plots of ψ_1, ψ_2, and ψ_3 vs Directions

The 3D representation of allyl system is shown in Figure 13.6. It is obtained from the plot of the plots of ψ_1, ψ_0, and ψ_3 in various directions. Its sectional views have been illustrated in Figure 13.6.

It is clear from the figure that dotted lines denote the bases of the nodal planes.

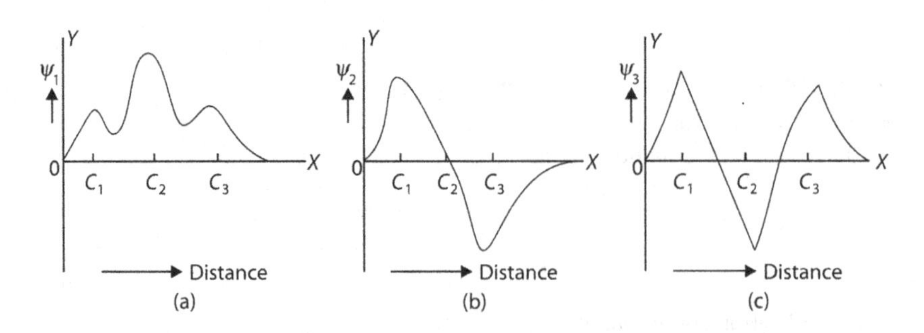

FIGURE 13.5 Plots of three Hückel molecular orbitals in allyl system.

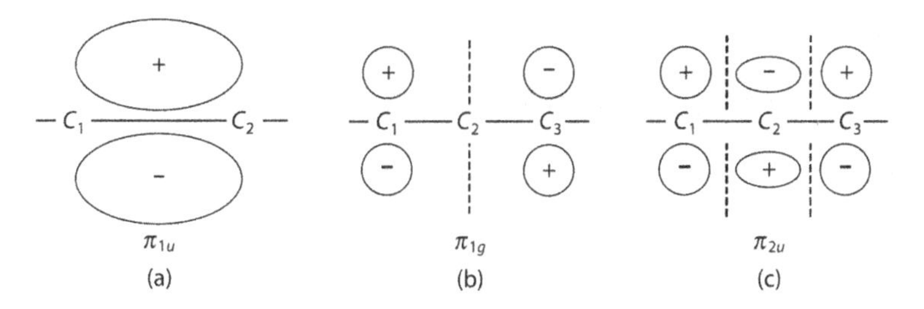

FIGURE 13.6 Three-dimensional representation of Hückel molecular orbitals of allyl system.

13.1.6 Butadiene

Butadiene is a four-carbon atom system, and it has 4π electrons. We shall neglect the electrons forming σ-bonds and only 4π electrons will be taken into consideration. 1–3 Butadiene is structurally represented by

$$>\underset{1}{C}=\underset{2}{\overset{|}{C}}-\underset{3}{\overset{|}{C}}=\underset{4}{C}<$$

The carbon atom of butadiene numbered as 1, 2, 3, and 4, respectively, consists of $2p_z$ orbitals represented by ϕ_1, ϕ_2, ϕ_3, and ϕ_4. The molecular orbital ψ is denoted as usual by LCAO, and the molecular orbital is mathematically expressed as

$$\psi = C_1\phi_1 + C_2\phi_2 + C_3\phi_3 + C_4\phi_4 \tag{13.54}$$

The equation for energy of butadiene is expressed as usual

$$E = \frac{\left\langle (C_1\phi_1 + C_2\phi_2 + C_3\phi_3 + C_4\phi_4) \middle| H \middle| (C_1\phi_1 + C_2\phi_2 + C_3\phi_3 + C_4\phi_4) \right\rangle}{\left\langle (C_1\phi_1 + C_2\phi_2 + C_3\phi_3 + C_4\phi_4)^2 \right\rangle} \tag{13.55}$$

The denominator is given by

$$\begin{aligned}
&\left\langle (C_1\phi_1 + C_2\phi_2 + C_3\phi_3 + C_4\phi_4)^2 \right\rangle \\
&= \left\langle (C_1^2\phi_1^2 + C_2^2\phi_2^2 + C_3^2\phi_3^2 + C_4^2\phi_4^2 \right\rangle + C_1C_2\langle\phi_1|\phi_2\rangle + C_1C_3\langle\phi_1|\phi_3\rangle \\
&+ C_1C_4\langle\phi_1|\phi_4\rangle + C_2C_1\langle\phi_2|\phi_1\rangle + C_2C_3\langle\phi_2|\phi_3\rangle + C_2C_4\langle\phi_2|\phi_4\rangle \\
&+ C_3C_1\langle\phi_3|\phi_1\rangle + C_3C_2\langle\phi_3|\phi_2\rangle + C_3C_4\langle\phi_3|\phi_4\rangle + C_4C_1\langle\phi_4|\phi_1\rangle \\
&+ C_4C_2\langle\phi_4|\phi_2\rangle + C_4C_3\langle\phi_4|\phi_3\rangle
\end{aligned} \tag{13.56}$$

Applying the normalisation condition, we can write

$$\langle\phi_1|\phi_1\rangle = \langle\phi_2|\phi_2\rangle = \langle\phi_3|\phi_3\rangle = \langle\phi_4|\phi_4\rangle = 1 \tag{13.57}$$

We represent the overlap integrals between the two adjacent carbon atoms i and j by S_{ij}, and it amounts to

$$S_{ij} = \langle\phi_i|\phi_j\rangle \tag{13.58}$$

We shall further assume that for the two adjacent atoms

$$S_{12} = S_{21} = S_{23} = S_{32} = S_{34} = S_{43} = S \tag{13.59}$$

And for two non-adjacent carbon atoms, we express

$$S_{13} = S_{31} = S_{14} = S_{41} = S_{24} = S_{42} = 0 \tag{13.60}$$

Thus, the denominator of Eq. (13.55) can be written as

$$D = C_1^2 + C_2^2 + C_3^2 + C_4^2 + 2(C_1C_2 + C_2C_3 + C_3C_4)S \tag{13.61}$$

where D = denominator

Let $S \to 0$ for simplicity sake, then Eq. (13.61) takes the form

$$\begin{aligned} D &= C_1^2 + C_2^2 + C_3^2 + C_4^2 \\ &= \text{denominator of Eq. (13.55)} \end{aligned} \tag{13.62}$$

Now, we shall consider the numerator of Eq. (13.55), which is given by

$$\langle (C_1\phi_1 + C_2\phi_2 + C_3\phi_3 + C_4\phi_4) | H | (C_1\phi_1 + C_2\phi_2 + C_3\phi_3 + C_4\phi_4) \rangle \tag{13.63}$$

$$\begin{aligned} &= C_1^2 \langle \phi_1 H \phi_1 \rangle + C_1C_2 \langle \phi_1 H \phi_2 \rangle + C_1C_3 \langle \phi_1 H \phi_3 \rangle + C_1C_4 \langle \phi_1 H \phi_4 \rangle \\ &+ C_2C_1 \langle \phi_2 H \phi_1 \rangle + C_2^2 \langle \phi_2 H \phi_2 \rangle + C_2C_3 \langle \phi_2 H \phi_3 \rangle + C_2C_4 \langle \phi_2 H \phi_4 \rangle \\ &+ C_3C_1 \langle \phi_3 H \phi_1 \rangle + C_3C_2 \langle \phi_3 H \phi_2 \rangle + C_3^2 \langle \phi_3 H \phi_3 \rangle + C_3C_4 \langle \phi_4 H \phi_4 \rangle \\ &+ C_4C_1 \langle \phi_4 H \phi_1 \rangle + C_4C_2 \langle \phi_4 H \phi_2 \rangle + C_4C_3 \langle \phi_4 H \phi_3 \rangle + C_4^2 \langle \phi_4 H \phi_4 \rangle \end{aligned} \tag{13.64}$$

It is known to us that

$$\langle \phi_1 H \phi_1 \rangle = \langle \phi_2 H \phi_2 \rangle = \langle \phi_3 H \phi_3 \rangle = \langle \phi_4 H \phi_4 \rangle = \alpha \tag{13.65}$$

$$\text{Further, } \langle \phi_1 H \phi_2 \rangle = \beta_{12} = \langle \phi_2 H \phi_1 \rangle = \beta_{21} \tag{13.66}$$

$$\beta_{12} = \beta_{21} = \beta_{23} = \beta_{32} = \beta_{34} = \beta_{43} = \beta \tag{13.67}$$

$$\text{and} \quad \beta_{13} = \beta_{31} = \beta_{14} = \beta_{41} = \beta_{24} = \beta_{42} = 0 \tag{13.68}$$

for non-adjacent carbon atoms.

Thus, the numerator (N) of Eq. (13.55) can be written as

$$N = \alpha\left(C_1^2 + C_2^2 + C_3^2 + C_4^2\right) + 2\beta(C_1C_2 + C_2C_3 + C_3C_4) \tag{13.69}$$

Hence, Eq. (13.55) can be expressed as

$$E = \frac{\alpha\left(C_1^2 + C_2^2 + C_3^2 + C_4^2\right) + 2\beta(C_1C_2 + C_2C_3 + C_3C_4)}{\left(C_1^2 + C_2^2 + C_3^2 + C_4^2\right)} \tag{13.70}$$

$$\text{or} \quad E\left(C_1^2 + C_2^2 + C_3^2 + C_4^2\right) = \alpha\left(C_1^2 + C_2^2 + C_3^2 + C_4^2\right) + 2\beta(C_1C_2 + C_2C_3 + C_3C_4) \tag{13.71}$$

Differentiating this equation with respect to C_1, we shall obtain

$$\frac{\delta E}{\delta C_1}\left(C_1^2 + C_2^2 + C_3^2 + C_4^2\right) + 2C_1E = 2C_1\alpha + 2C_2\beta \tag{13.72}$$

For minimum condition, $\frac{\delta E}{\delta C_1} = 0.$

Therefore, Eq. (13.72) takes the form

$$2C_1E = 2C_1\alpha + 2C_2\beta$$

or $C_1E = C_1\alpha + C_2\beta$

$$\text{or} \quad C_1(\alpha - E) + C_2\beta = 0 \tag{13.73}$$

Similarly, differentiating Eq. (13.71) with respect to C_2, C_3, and C_4, respectively, and putting

$$\frac{\delta E}{\delta C_2} = 0, \frac{\delta E}{\delta C_3} = 0, \text{ and } \frac{\delta E}{\delta C_4} = 0, \text{ we obtain}$$

$$2C_2E = 2C_1\beta + 2C_3\beta + 2C_2\alpha \tag{13.74}$$

$$\text{or} \quad C_2E = \beta C_1 + \beta C_3 + C_2\alpha$$

$$\text{or} \quad C_2(\alpha - E) + C_1\beta + C_3\beta = 0$$

Similarly, others will be given by

$$C_3(\alpha - E) + C_2\beta + C_4\beta = 0 \tag{13.75}$$

$$\text{and} \quad C_4(\alpha - E) + C_3\beta = 0 \tag{13.76}$$

From Eqs. (13.73–13.76), we can write the secular determinant as

$$\begin{vmatrix} \alpha - E & \beta & 0 & 0 \\ \beta & \alpha - E & \beta & 0 \\ 0 & \beta & \alpha - E & \beta \\ 0 & 0 & \beta & \alpha - E \end{vmatrix} = 0 \tag{13.77}$$

Dividing all the terms in the determinant by β and substituting $\alpha - E/\beta = x$, we shall get

$$\begin{vmatrix} x & 1 & 0 & 0 \\ 1 & x & 1 & 0 \\ 0 & 1 & x & 1 \\ 0 & 0 & 1 & x \end{vmatrix} = 0 \tag{13.78}$$

The expression of the determinant yields

$$x\begin{vmatrix} x & 1 & 0 \\ 1 & x & 1 \\ 0 & 1 & x \end{vmatrix} - 1\begin{vmatrix} 1 & 1 & 0 \\ 0 & x & 1 \\ 0 & 1 & x \end{vmatrix} = 0$$

$$\text{or} \quad x\left[x\left(x^2 - 1\right) - 1(x)\right] - 1\left(x^2 - 1\right) = 0$$

$$\text{or} \quad x\left[\left(x^3 - x\right) - x\right] - x^2 + 1 = 0$$

$$\text{or} \quad x^4 - x^2 - x^2 - x^2 + 1 = 0$$

$$\text{or} \quad x^4 - 3x^2 + 1 = 0$$

$$y^2 - 3y + 1 = 0 \tag{13.79}$$

Putting $x^2 = y$, Eq. (13.79) takes the form

$$y = \frac{+3 \pm \sqrt{-3^2 - 4 \times 1 \times 1}}{2} = \frac{3 \pm \sqrt{9-4}}{2} = \frac{3 \pm \sqrt{5}}{2}$$
$$\therefore \quad y = \frac{3+\sqrt{5}}{2} \text{ and } \frac{3-\sqrt{5}}{2} \tag{13.80}$$

This is a quadratic equation, and the value of y will be

$$y = \frac{+3 \pm \sqrt{-3^2 - 4 \times 1 \times 1}}{2} = \frac{3 \pm \sqrt{9-4}}{2} = \frac{3 \pm \sqrt{5}}{2}$$
$$\therefore \quad y = \frac{3+\sqrt{5}}{2} \text{ and } \frac{3-\sqrt{5}}{2} \tag{13.81}$$

or $y = 2.618$ and 0.382.

When $y = 2.618 = x^2$

$$\therefore \quad x = \pm\sqrt{2.618} = \pm 1.618 \tag{13.82}$$

When $y = 0.382 = x^2$

$$\therefore \quad x = \pm(0.382)^{1/2} = \pm 0.618 \tag{13.83}$$

Then the four roots of the equation are

$$x = +1.618, -1.618, +0.618, \quad \text{and} \quad -0.618$$

When $x = -1.618$, we get

$$\frac{\alpha - E_1}{\beta} = -1.618$$

or $\alpha - E_1 = -1.618\beta$

$$\therefore \quad E_1 = \alpha + 1.618\beta \quad [\text{BMO}] \tag{13.84}$$

When $x = -0.618$, we have

$$\frac{\alpha - E_2}{\beta} = -0.618$$
$$\therefore \quad E_2 = \alpha + 0.618 \qquad [\text{BMO}] \tag{13.85}$$

When $x = +0.618$, we get

$$\frac{\alpha - E_3}{\beta} = +0.618$$

or $\alpha - E_3 = 0.618\beta$

$$\therefore \quad E_3 = \alpha - 0.618\beta \quad [\text{ABMO}] \tag{13.86}$$

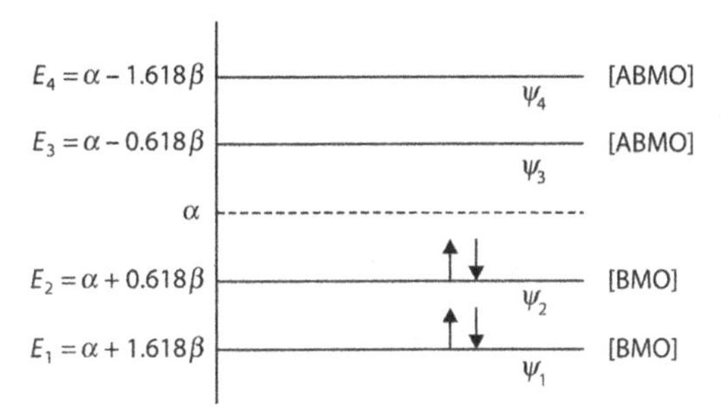

FIGURE 13.7 Hückel molecular orbital energy levels in butadiene.

When $x = +1.618$, we get

$$\frac{\alpha - E_4}{\beta} = 1.618$$
$$\therefore \quad E_4 = \alpha - 1.618\beta \quad \text{(ABMO)} \tag{13.87}$$

With the help of four energy values for butadiene molecule, we shall construct the energy level diagram, which is illustrated in Figure 13.7.

13.1.7 Delocalisation Energy of Butadiene

According to Pauli's exclusion principle, the 4π electrons of butadiene are accommodated in BMOs ψ_1 and ψ_2. Hence, the π electron energy will be given by

$$E_\pi = 2(\alpha + 1.618\beta) + 2(\alpha + 0.618\beta) = 4\alpha + 4.472\beta \tag{13.88}$$

The butadiene molecule resonates between localised structure, which contains two conjugated double bonds and a delocalised structure.

But

$$(E_\pi)_{\text{deloc}} = (E_\pi)_{\text{loc}} - E_\pi \tag{13.89}$$

$$\text{and} \quad (E_\pi)_{\text{loc}} = 2(2\alpha + 2\beta)$$
$$\therefore \quad (E_\pi)_{\text{deloc}} = 2(2\alpha + 2\beta) - (4\alpha + 4.472\beta)$$
$$= 4\alpha + 4\beta - 4\alpha - 4.472\beta$$
$$= -0.472\beta \tag{13.90}$$

This is the value of delocalisation energy of butadiene.

13.1.8 Hückel Molecular Orbital Coefficients and Molecular Orbitals

From Eqs. (13.73–13.76), the secular equations are written in terms of x as

$$\left.\begin{aligned} C_1x + C_2 &= 0 \\ C_1 + C_2x + C_3 &= 0 \\ C_2 + C_3x + C_4 &= 0 \\ C_3 + C_4x &= 0 \end{aligned}\right\} \tag{13.91}$$

Putting $x = -1.618$ for BMO ψ_1, we have

$$\left.\begin{aligned} -1.618C_1 + C_2 &= 0 \\ C_1 - 1.618C_2 + C_3 &= 0 \\ C_2 - 1.618C_3 + C_4 &= 0 \\ C_3 - 1.618C_4 &= 0 \end{aligned}\right\} \tag{13.92}$$

From the first equation,

$$C_2 = 1.618C_1 \approx 1.62C_1$$

Putting the value of C_2 in the second equation, we get

$$C_1 - 1.618(1.618C_1) + C_3 = 0$$

$$\text{or} \quad C_1 - 2.618C_1 + C_3 = 0$$

$$\text{or} \quad C_1(1 - 2.618) + C_3 = 0$$

$$\text{or} \quad -1.618C_1 + C_3 = 0$$

$$\therefore \quad C_3 = 1.618C_1 \approx 1.62C_1$$

Putting the value of C_3 in the fourth equation, we get

$$C_3 - 1.618C_4 = 0$$

or $1.62C_1 - 1.62C_4 - 0$
$\therefore C_1 = C_4$

We also know that the normalisation condition is

$$\sum_1^4 C_i^2 = 1$$

$$\text{or} \quad C_1^2 + C_2^2 + C_3^2 + C_4^2 = 1$$

$$\text{or} \quad C_1^2 + (1.62)^2 C_1^2 + (1.62C_1)^2 + C_1^2 = 1$$

$$\text{or} \quad C_1^2 + 2.62C_1^2 + 2.62C_1^2 + C_1^2 = 1$$

$$\text{or} \quad 7.24C_1^2 = 1 \quad \therefore \quad C_1^2 = \frac{1}{7.24} \quad \therefore \quad C_1 = \sqrt{\frac{1}{7.24}}$$

$$\therefore \quad C_1 = 0.372$$

$$\because \quad C_1 = C_4 \therefore C_4 = 0.372$$

$$\because \quad C_2 = 1.62C_1 = 1.62 \times 0.372 = 0.602$$

$$\therefore \quad C_3 = 1.62C_1$$

$$\therefore \quad C_3 = 1.62 \times 0.372 = 0.602$$

Therefore, the resulting MOs are given by

$$\begin{aligned} \psi_1 &= C_1\phi_1 + C_2\phi_2 + C_3\phi_3 + C_4\phi_4 \\ \text{or} \quad \psi_1 &= 0.372\phi_1 + 0.602\phi_2 + 0.602\phi_3 + 0.372\phi_4 \\ \text{or} \quad \psi_1 &= 0.372(\phi_1 + \phi_4) + 0.602(\phi_2 + \phi_3) \end{aligned} \tag{13.93}$$

Similarly, putting the value of $x = -0.618$, $+0.618$, and 1.618, respectively, the values of ψ_2, ψ_3, and ψ_4 will be evaluated. They will be

$$\psi_2 = 0.602(\phi_1 - \phi_4) + 0.372(\phi_2 - \phi_3) \tag{13.94}$$

$$\psi_3 = 0.602(\phi_1 + \phi_4) - 0.372(\phi_2 + \phi_3) \tag{13.95}$$

$$\psi_4 = 0.372(\phi_1 - \phi_2) - 0.602(\phi_2 - \phi_4) \tag{13.96}$$

Equations (13.93–13.96) are the molecular orbitals of butadiene.

13.1.8.1 Graphical Representation

Figure 13.8 gives the diagrams obtained when graphs of ψ_1, ψ_2, ψ_3, and ψ_4 are plotted against various points along $C_1 - C_2 - C_3 - C_4$ axis in the case of butadiene.

It is clear from the figure that ψ_1 has no node, ψ_2 has one node, ψ_3 has two nodes, and ψ_4 consists of three nodes along the bond axis.

13.1.8.2 Three-Dimensional Representation

The four sections of HMOs along a vertical plane are illustrated in Figure 13.9.

13.2 Application of the Hückel Method to Some Cyclic Polyenes

We have applied the Hückel method for the determination of energy level and HMO coefficients in the case of straight chain polyenes.

Here, we are going to apply the HMO method in the case of cyclic polyenes. The cyclic polyenes may be three-membered rings, four-membered rings, five-membered rings, six-membered rings, etc.

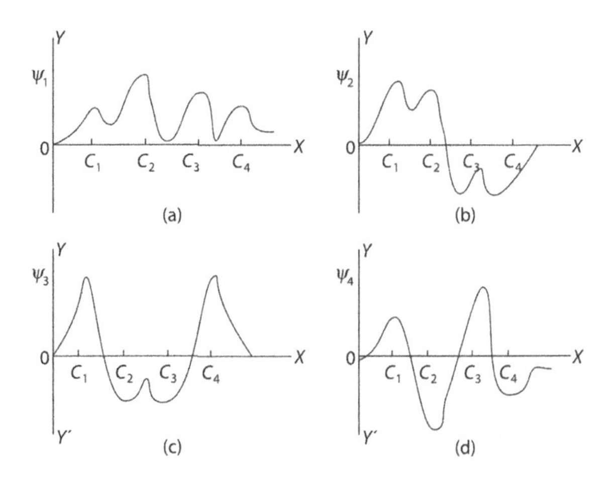

FIGURE 13.8 Four Hückel molecular orbitals of butadiene along the bond axis.

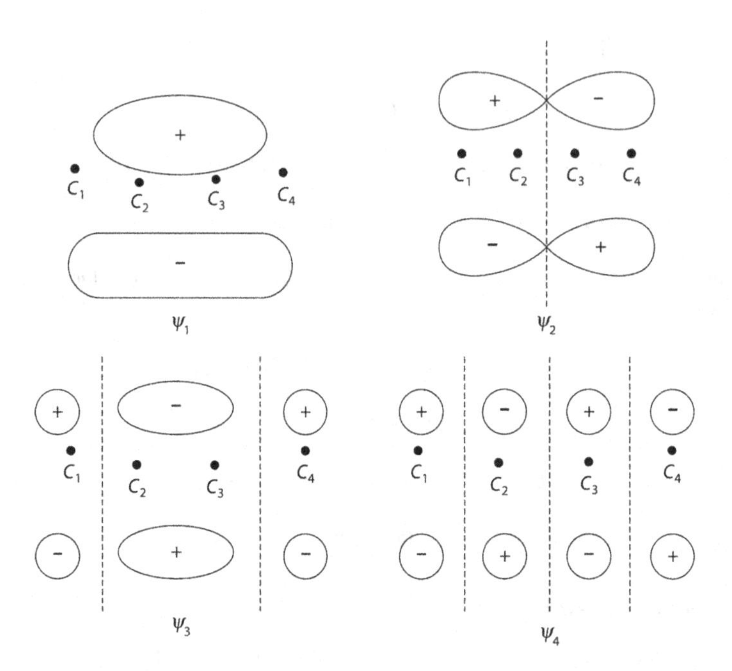

FIGURE 13.9 Four Hückel molecular orbitals of butadiene.

13.2.1 Cyclopropenyl System

The cyclopropenyl system is a triangular ring having π electrons. Cyclopropenyl radical has 3π electrons; cyclopropenyl cation has 2π electrons, whereas cyclopropenyl anion has 4π electrons. They only differ in their delocalisation energies.

The structural formula for the propenyl systems is given in Figure 13.10.

The carbon atoms of the system have been numbered. Let the HMO function for the above system be

$$\psi = C_1\phi_1 + C_2\phi_2 + C_3\phi_3 \tag{13.97}$$

The secular equations for cyclopropenyl system is given by

$$\left.\begin{array}{l} C_1x + C_2 + C_3 = 0 \\ C_2x + C_3 + C_1 = 0 \\ C_3x + C_1 + C_2 = 0 \end{array}\right\} \tag{13.98}$$

From these secular equations, the secular determinant may be constructed as follows:

$$\begin{vmatrix} x & 1 & 1 \\ 1 & x & 1 \\ 1 & 1 & x \end{vmatrix} = 0 \tag{13.99}$$

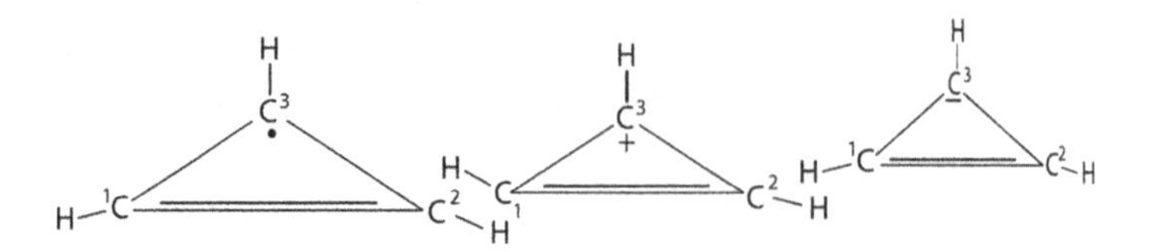

FIGURE 13.10 Cyclopropenyl system.

where $x = \dfrac{\alpha - E}{\beta}$

Expansion of Eq. (13.99) gives

$$x\begin{vmatrix} x & 1 \\ 1 & x \end{vmatrix} - 1\begin{vmatrix} 1 & 1 \\ 1 & x \end{vmatrix} + 1\begin{vmatrix} 1 & x \\ 1 & 1 \end{vmatrix} = 0$$

$$\text{or } x\left(x^2 - 1\right) - 1(x - 1) + 1(1 - x) = 0$$

$$\text{or } x^3 - x - x + 1 + 1 - x = 0 \tag{13.100}$$

$$\text{or } x^3 - 3x + 2 = 0$$

$$\text{or } (x + 2)\left(x^2 - 2x + 1\right) = 0$$

$$\text{or } (x + 2)(x - 1)^2 = 0$$

Therefore, the roots of the above equation will be –2, +1, +1, i.e., $x_1 = -2$, $x_2 = 1$, $x_3 = 1$. Putting $x_1 = -2$, we have

$$\frac{\alpha - E_1}{\beta} = -2$$

$$\text{or } \alpha - E_1 = -2\beta \tag{13.101}$$

$$\therefore \; E_1 = \alpha + 2\beta$$

Putting, $x_2 = 1$, we have

$$\frac{\alpha - E_2}{\beta} = 1$$

$$\text{or } \alpha - E_2 = \beta \tag{13.102}$$

$$\text{or } E_2 = \alpha - \beta$$

$$\text{similarly, } E_3 = \alpha - \beta \tag{13.103}$$

Thus, the energy level diagram of cyclopropenyl system is illustrated in Figure 13.11.

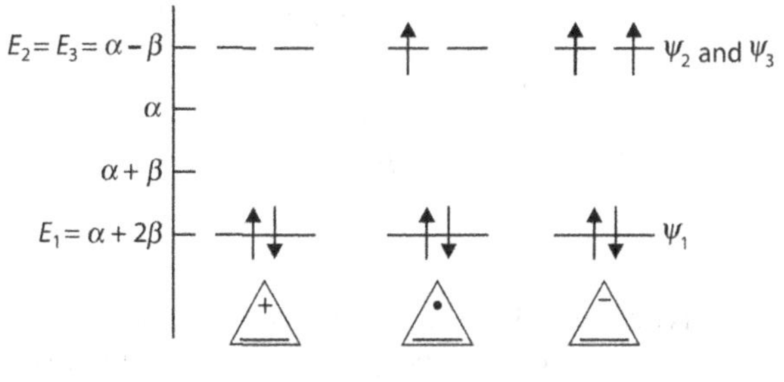

FIGURE 13.11 π electrons in cyclopropenyl system.

13.2.2 Delocalisation of Cyclopropenyl System

I. The π electron energy of cyclopropenyl cation $\left(\begin{matrix} CH \\ \| \\ CH \end{matrix} > \overset{+}{C}H\right)$ is estimated as follows:

$$E_\pi = 2(\alpha + 2\beta) = 2\alpha + 4\beta$$

$$\text{and} \quad DE = (2\alpha + 4\beta) - (2\alpha + 2\beta)$$

$$= 2\alpha + 4\beta - 2\alpha - 2\beta = 2\beta$$

II. The π electron energy of cyclopropenyl radical $\left(\begin{matrix} CH \\ \| \\ CH \end{matrix} > \dot{C}H\right)$ is

$$E_\pi = 2(\alpha + 2\beta) + (\alpha - \beta) = 2\alpha + 4\beta + \alpha - \beta$$

$$= 2\alpha + 4\beta + \alpha - \beta$$

$$= 3\alpha + 3\beta$$

And the energy of one isolated double bond having 2π electrons is $(2\alpha + 2\beta)$. Therefore, the delocalisation energy will be

$$DE = (3\alpha + 3\beta) - (2\alpha + 2\beta + \alpha)$$

$$= 3\alpha + 3\beta - 3\alpha - 2\beta$$

$$= \beta$$

III. The π electron energy of cyclopropenyl anion $\left(\begin{matrix} CH \\ \| \\ CH \end{matrix} > \bar{C}H\right)$ is given by

$$E_\pi = 2[\alpha + 2\beta] + 2(\alpha - \beta)$$

$$= 2\alpha + 4\beta + 2\alpha - 2\beta$$

$$= 4\alpha + 2\beta$$

And the delocalisation energy will be given by

$$DE = [2(\alpha + 2\beta) + 2(\alpha - \beta)] - [2(\alpha + \beta) + 2\alpha]$$

$$= 2\alpha + 4\beta + 2\alpha - 2\beta - 2\alpha - 2\beta - 2\alpha$$

$$= 4\alpha + 4\beta - 4\alpha - 4\beta$$

$$= 0$$

From the above calculation, it is clear that the delocalisation energy of cyclopropenyl cation is the highest, which gives a clue to understand that cyclopropenyl cation has the greatest stability. Delocalisation energy of cyclopropenyl anion is the lowest; hence, it is less stable.

13.2.3 Hückel Molecular Orbital Coefficients and Molecular Orbitals

In the case of cyclopropenyl system, the values of the roots of Eq. (13.100) are found to be 2, +1, and +1.
The secular equations are

$$C_1x + C_2 + C_3 = 0$$

$$C_2x + C_3 + C_1 = 0$$

$$C_3x + C_1 + C_2 = 0$$

Putting $x = -2$, we have

$$\left.\begin{aligned} -2C_1 + C_2 + C_3 = 0 \\ -2C_2 + C_3 + C_1 = 0 \\ -2C_3 + C_1 + C_2 = 0 \end{aligned}\right\} \quad (13.104)$$

From Eq. (13.104), we can write

$$C_2 + C_3 = 2C_1$$

$$C_3 + C_1 = 2C_2$$

$$\text{and} \quad C_1 + C_2 = 2C_3$$

Considering the second and third equations, we can write

$$\begin{aligned} & C_1 + C_3 = 2C_2 \\ & C_1 + C_2 = 2C_3 \\ & - \quad - \quad - \\ \hline & C_3 - C_2 = 2C_2 - 2C_3 \\ \text{or} \quad & 3C_3 = 3C_2 \quad \therefore \quad C_2 = C_3 \end{aligned} \quad (13.105)$$

From the third equation, we have

$$\begin{aligned} & C_1 + C_2 = 2C_3 \\ \text{or} \quad & C_1 + C_2 = 2C_2 \quad [\because \; C_2 = C_3] \\ \text{or} \quad & C_1 = 2C_2 - C_2 = C_2 \\ \therefore \quad & C_1 = C_2 \end{aligned} \quad (13.106)$$

From Eqs. (13.105) and (13.106), we can write

$$C_1 = C_2 = C_3 \quad (13.107)$$

Applying the normalisation condition, i.e., $\sum_{1}^{3} C_i^2 = 1$

$$\begin{aligned} \text{or} \quad & C_1^2 + C_2^2 + C_3^2 = 1 \\ \text{or} \quad & 3C_1^2 = 1 \quad \therefore \quad C_1^2 = 1/3 \\ \therefore \quad & C_1 = 1/\sqrt{3} = C_2 = C_3 \end{aligned} \quad (13.108)$$

Therefore, HMO ψ_1 can be expressed as

$$\begin{aligned}\psi_1 &= C_1\phi_1 + C_2\phi_2 + C_3\phi_3 \\ &= \frac{1}{\sqrt{3}}(\phi_1 + \phi_2 + \phi_3)\end{aligned} \tag{13.109}$$

For getting ψ_2 and ψ_3, we select $x \pm 1$, and when this is put in secular equations, we shall get $C_1 + C_2 + C_3 = 0$ (such equation will be three in number)

$$\text{And } C_1^2 + C_2^2 + C_3^2 = 0$$

It is clear that we have three unknowns but two equations. Under this condition, let us choose

$$C_1 = -C_2 \text{ and } C_3 = 0 \tag{13.110}$$

Applying normalisation condition, we have

$$\begin{aligned}&C_1^2 + C_2^2 + C_3^2 = 1 \\ \text{or} \quad &C_1^2 + C_1^2 + 0^2 = 1 \quad [\because \; C_1 = -C_2] \\ \text{or} \quad &2C_1^2 = 1 \quad \therefore \quad C_1 = \frac{1}{\sqrt{2}} \\ \therefore \quad &\psi_2 = \frac{1}{\sqrt{2}}\phi_1 - \frac{1}{\sqrt{2}}\phi_2\end{aligned} \tag{13.111}$$

For finding the expression for ψ_3, we shall choose $C_1 = \frac{1}{\sqrt{2}}$, $C_3 = -\frac{1}{\sqrt{2}}$, and $C_2 = 0$, and hence, ψ_3 can be expressed as

$$\psi_3 = \frac{1}{\sqrt{2}}\phi_1 - \frac{1}{\sqrt{2}}\phi_3 \tag{13.112}$$

13.2.4 Cyclobutadiene

Cyclobutadiene is a square system consisting of 4π electrons. If we consider that this system is perfectly square, we can inscribe the square in a circle as given in Figure 13.12.

The carbon atoms in the cyclobutadiene are labelled as 1, 2, 3, and 4. The LCAO-molecular orbital or HMO of cyclobutadiene may be written in the following form:

$$\psi = \sum_{i=1}^{4} C_i\phi_i = C_1\phi_1 + C_2\phi_2 + C_3\phi_3 + C_4\phi_4 \tag{13.113}$$

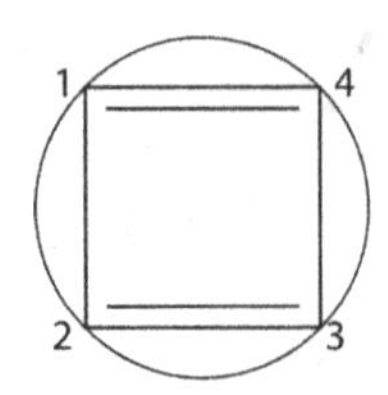

FIGURE 13.12 *Cyclobutadiene.*

The secular equations for cyclobutadiene are given by

$$\left.\begin{aligned} C_1x + C_2 + C_4 &= 0 \\ C_2x + C_1 + C_3 &= 0 \\ C_3x + C_2 + C_4 &= 0 \\ C_4x + C_1 + C_3 &= 0 \end{aligned}\right\} \tag{13.114}$$

The secular determinant in terms of x can be expressed as

$$\begin{vmatrix} x & 1 & 0 & 1 \\ 1 & x & 1 & 0 \\ 0 & 1 & x & 1 \\ 1 & 0 & 1 & x \end{vmatrix} = 0 \tag{13.115}$$

The Laplace expansion of this determinant gives the following:

$$x\begin{vmatrix} x & 1 & 0 \\ 1 & x & 1 \\ 0 & 1 & x \end{vmatrix} - 1\begin{vmatrix} 1 & 1 & 0 \\ 0 & x & 1 \\ 1 & 1 & x \end{vmatrix} - 1\begin{vmatrix} 1 & x & 0 \\ 0 & 1 & x \\ 1 & 0 & 1 \end{vmatrix} = 0$$

$$\text{or} \quad x\left[x\left(x^2-1\right)-1(x)\right]-1\left[1\left(x^2-1\right)-1(-1)\right]$$

$$-1\left[1(1)-x(-x)+1(-1)\right]=0$$

$$\text{or} \quad x\left[x^3-x-x\right]-1\left[x^2-1+1\right]-1\left[1+x^2-1\right]=0$$

$$\text{or} \quad x^4-x^2-x^2-x^2+1-1-1-x^2+1=0$$

$$\text{or} \quad x^4-4x^2=0$$

$$\text{or} \quad x^2\left(x^2-4\right)=0$$

$$\text{or} \quad x^2(x+2)(x-2)=0$$

Therefore, the four roots obtained are $x = 0, 0, -2$, and $+2$.

It may be written as $x_1 = -2$, $x_1 = 0$, $x_3 = 0$, and $x_4 = +2$. Corresponding to these roots, four energy values are obtained.

When $x_1 = -2$, then

$$\frac{\alpha - E_1}{\beta} = -2,$$

$$\text{or} \quad \alpha - E = -2\beta \tag{13.116}$$

$$\text{or} \quad E_1 = \alpha + 2\beta \qquad [\text{BMO}]$$

when $x_2 = 0$, E_2 and E_3 will be

$$\therefore \quad E_2 = \alpha = E_3 \quad [\text{Degenerate pair, NBMO}] \tag{13.117}$$

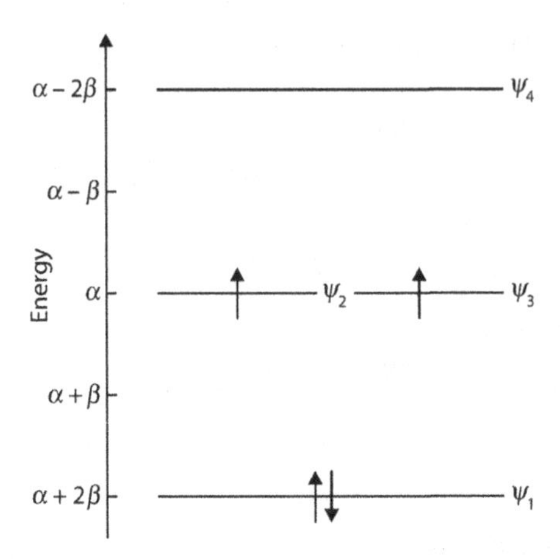

FIGURE 13.13 HMO energy levels in cyclobutadiene.

When $x_4 = +2$, then

$$\frac{\alpha - E_4}{\beta} = 2$$

$$\text{or} \quad \alpha - E_4 = 2\beta \tag{13.118}$$

$$\therefore \quad E_4 = \alpha - 2\beta \quad [\text{ABMO}]$$

On the basis of these four values of energy, the energy level diagram for cyclobutadiene can be constructed as shown in Figure 13.13.

13.2.5 Delocalisation Energy of Cyclobutadiene

$E\pi$ value of cyclobutadiene is given by

$$E_\pi = 2(\alpha + 2\beta) + \alpha + \alpha = 4\alpha + 4\beta$$

$$(E_\pi)_{\text{loc}} = (2\alpha + 2\beta) + (2\alpha + 2\beta) = 4\alpha + 4\beta$$

Therefore, the delocalisation energy of cyclobutadiene is obtained as follows:

$$\begin{aligned} DE &= (E_\pi)_{\text{loc}} - E_\pi \\ &= (4\alpha + 4\beta) - (4\alpha + 4\beta) = 0 \end{aligned} \tag{13.119}$$

Thus we see that delocalisation energy $DE = 0$, and hence, there is no additional stability of cyclobutadiene. Due to understability, this compound is difficult to synthesise.

13.2.6 Hückel Molecular Orbital Coefficient and Molecular Orbitals

The secular equations are

$$\begin{aligned} C_1x + C_2 + C_4 &= 0 \\ C_1 + C_2x + C_3 &= 0 \\ C_2 + C_3x + C_4 &= 0 \\ C_1 + C_3 + C_4x &= 0 \end{aligned}$$

Putting $x = -2$, the above equations take the form

$$\left.\begin{aligned} -2C_1 + C_2 + C_4 &= 0 \\ C_1 - 2C_2 + C_3 &= 0 \\ C_2 - 2C_3 + C_4 &= 0 \\ C_1 + C_3 - 2C_4 &= 0 \end{aligned}\right\} \tag{13.120}$$

Now, from the first and third equations of Eq. (13.120), we can write

$$\begin{aligned} -2C_1 + C_2 + C_4 &= 0 \\ -2C_3 + C_2 + C_4 &= 0 \\ + \quad - \quad - & \end{aligned}$$

On subtraction, $-2C_1 + 2C_3 = 0$.

$$\therefore \quad C_1 = C_3 \tag{13.121}$$

From the second and fourth equations of Eq. (13.120), we can write

$$\begin{aligned} C_1 - 2C_2 + C_3 &= 0 \\ C_1 + C_3 - 2C_4 &= 0 \\ - \quad - \quad + & \\ \hline 2C_4 - 2C_2 &= 0 \end{aligned} \tag{13.122}$$

$$\therefore \quad C_2 = C_4$$

From the third equation of Eq. (13.120), we have

$$C_2 - 2C_3 + C_4 = 0$$

$$\text{or} \quad C_4 - 2C_3 + C_4 = 0 \qquad [\therefore \quad C_2 = C_4]$$

$$\left.\begin{aligned} \text{or} \quad 2C_4 - 2C_3 &= 0 \\ \therefore \quad C_3 &= C_4 \end{aligned}\right. \tag{13.123}$$

From Eqs. (13.121–13.123), we can conclude that

$$C_1 = C_2 = C_3 = C_4 \tag{13.124}$$

Applying normalisation condition, we get

$$\sum_{i=1}^{4} C_i^2 = 1 \quad \text{or } C_1^2 + C_2^2 + C_3^2 + C_4^2 = 1$$

$$\text{or} \quad 4C_1^2 = 1 \quad \therefore \quad C_1^2 = 1/4 \tag{13.125}$$

$$\therefore \quad C_1 = 1/2 = C_2 = C_3 = C_4$$

Therefore, HMO, ψ_1 can be written as

$$\psi_1 = \frac{1}{2}(\phi_1 + \phi_2 + \phi_3 + \phi_4). \tag{13.126}$$

Putting $x - 0$ in Eq. (13.120), we get

$$\left.\begin{aligned} C_2 + C_4 &= 0 \\ C_1 + C_3 &= 0 \\ C_2 + C_4 &= 0 \\ C_1 + C_3 &= 0 \end{aligned}\right\} \tag{13.127}$$

Here, we have to choose any value for the coefficients and also normalisation condition. So, we get

$$C_1 = -C_3$$

Let $C_2 = 0$ and $C_4 = 0$.

In normalisation condition,

$$\therefore \quad \sum_{i=1}^{4} C_i^2 = 1$$

$$\text{or} \quad C_1^2 + C_2^2 + C_3^2 + C_4^2 = 1 \tag{13.128}$$

$$\text{or} \quad C_1^2 + 0 + (-C_1)^2 + 0 = 1$$

$$\text{or} \quad C_1^2 + C_1^2 = 1 \quad \text{or} \quad 2C_1^2 = 1 \quad \therefore \quad C_1 = \frac{1}{\sqrt{2}} \quad \text{and} \quad C_3 = -\frac{1}{\sqrt{2}}$$

$$\therefore \quad \psi_2 = \frac{1}{\sqrt{2}}(\phi_1 - \phi_3) \tag{13.129}$$

Again, we can write from the above that $C_2 + C_4 = 0$.

$\therefore$ $C_2 = -C_4$ and we may choose $C_1 = 0$ and $C_3 = 0$.

Now applying normalisation condition, we have

$$C_1^2 + C_2^2 + C_3^2 + C_4^2 = 1$$

$$\text{or} \quad 0^2 + C_2^2 + 0^2 + (-C_2)^2 = 1$$

$$\text{or} \quad 2C_2^2 = 1 \quad \therefore \quad C_2^2 = \frac{1}{2} \quad \therefore \quad C_2 = \frac{1}{\sqrt{2}} \quad \text{and} \quad C_4 = -\frac{1}{\sqrt{2}} \tag{13.130}$$

$$\therefore \quad \psi_3 = \frac{1}{\sqrt{2}}(\phi_2 - \phi_4) \tag{13.131}$$

Finally, when $x = 2$, then using the method of elimination, we obtain

$$C_1 = 1/2, C_2 = -1/2, C_3 = 1/2, C_4 = -1/2$$

$$\text{Hence, HMO}, \psi_4 = \frac{1}{2}(\phi_1 - \phi_2 + \phi_3 - \phi_4) \tag{13.132}$$

Thus, ψ_1, ψ_2, ψ_3, and ψ_4 molecular orbitals are determined.

13.2.7 Cyclopentadienyl System

It is a five-membered ring system consisting of 5π electrons. This system includes cyclopentadienyl radical having 5π electrons and cyclopentadienyl cation having 6π electrons. The structures of these cyclopentadienyl system are given as follows:

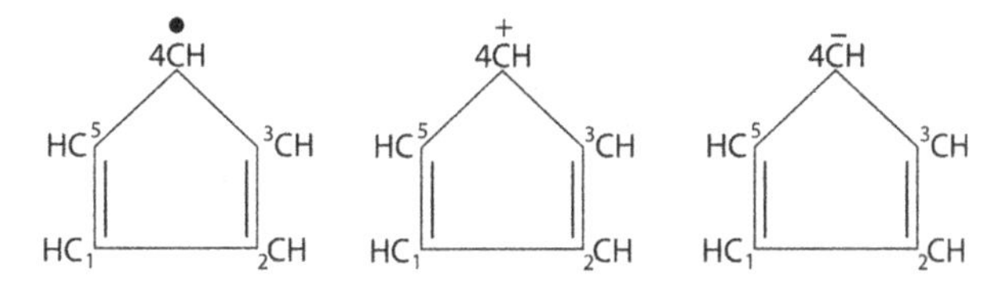

The cyclopentadienyl system has been numbered. Let the HMO function for this system be

$$\psi_1 = C_1\phi_1 + C_2\phi_2 + C_3\phi_3 + C_4\phi_4 + C_5\phi_5$$

$$\text{or} \quad \psi_1 = \sum_{i=1}^{5} C_i\phi_i \tag{13.133}$$

The secular equations for the system are expressed as

$$\left.\begin{aligned} C_1x + C_2 + C_5 &= 0 \\ C_2x + C_1 + C_3 &= 0 \\ C_3x + C_2 + C_4 &= 0 \\ C_4x + C_3 + C_5 &= 0 \\ C_5x + C_1 + C_4 &= 0 \end{aligned}\right\} \tag{13.134}$$

From the equations, the secular determinant can be written in terms of x as follows:

$$\begin{vmatrix} x & 1 & 0 & 0 & 1 \\ 1 & x & 1 & 0 & 0 \\ 0 & 1 & x & 1 & 0 \\ 0 & 0 & 1 & x & 1 \\ 1 & 0 & 0 & 1 & x \end{vmatrix} = 0 \tag{13.135}$$

where $x = \dfrac{\alpha - E}{\beta}$

Laplace expansion of the above determinant will give

$$x\begin{vmatrix} x & 1 & 0 & 0 \\ 1 & x & 1 & 0 \\ 0 & 1 & x & 1 \\ 0 & 0 & 1 & x \end{vmatrix} - 1\begin{vmatrix} 1 & 1 & 0 & 0 \\ 0 & x & 1 & 0 \\ 0 & 1 & x & 1 \\ 1 & 0 & 1 & x \end{vmatrix} + 1\begin{vmatrix} 1 & x & 1 & 0 \\ 0 & 1 & x & 1 \\ 0 & 0 & 1 & x \\ 1 & 0 & 0 & 1 \end{vmatrix} = 0$$

This will further reduce to

$$x^2\begin{vmatrix} x & 1 & 0 \\ 1 & x & 1 \\ 0 & 1 & x \end{vmatrix} - x\begin{vmatrix} 1 & 1 & 0 \\ 0 & x & 1 \\ 0 & 1 & x \end{vmatrix} - 1\begin{vmatrix} x & 1 & 0 \\ 1 & x & 1 \\ 0 & 1 & x \end{vmatrix} + 1\begin{vmatrix} 0 & 1 & 0 \\ 0 & x & 1 \\ 1 & 1 & x \end{vmatrix}$$

$$+1\begin{vmatrix} 1 & x & 1 \\ 0 & 1 & x \\ 0 & 0 & 1 \end{vmatrix} - x\begin{vmatrix} 0 & x & 1 \\ 0 & 1 & x \\ 1 & 0 & 1 \end{vmatrix} + 1\begin{vmatrix} 0 & 1 & 1 \\ 0 & 0 & x \\ 1 & 0 & 1 \end{vmatrix} = 0$$

The expansion of above determinant gives the following:

$$x^3\left(x^2-1\right)-x^2(x)-x\left[\left(x^2-1\right)-1(0)\right]-1\left[x\left(x^2-1\right)-(x)\right]+1\left[-1(-1)\right]$$

$$+1\left[1(1)\right]-x\left[-x(-x)-1\right]+1\left[0(1\times 0-0\times x)-1(0-x)+1(0)\right]=0$$

$$=x^3\left(x^2-1\right)-x^2(x)-x^3+x-\left(x^3-x-x\right)+1+1-x\left(x^2-1\right)+x=0$$

$$=x^5-x^3-x^3-x^3+x-x^3+2x+2-x^3+x+x=0$$

$$=x^5-5x^3+5x+2=0$$

Thus, the above determinants give a polynomial on expansion, which is

$$x^5-5x^3+5x+2=0 \tag{13.136}$$

When we put $x = -2$, Eq. (13.136) is satisfied; therefore, –2 will be one of the roots of the said equation. On dividing the above equation by $(x + 2)$, the remaining polynomial will be

$$x^4-2x^3-x^2+2x+1=0$$

Let $f(x) - x^4 - 2x^3 - x^2 + 2x + 1 = 0$

Put $y = 2x$

$\therefore\ x = y/2$

Putting $x = y/2$ in the above equation, it takes the form

$$f(y)=\frac{y^4}{16}-\frac{2y^3}{8}-\frac{y^2}{4}+\frac{2y}{2}+1=0$$

Multiplying both the sides by 16, we get

$$f(y)=y^4-4y^3-4y^2+16y+16=0$$

Again, put $z - y - 3 = 2x - 3$, this is done for diminishing the roots by 3.

$$\therefore\ f(z)=(z+3)^4-4(z+3)^3-4(z+3)^2+16(z+3)+16=0$$

$$=\left(z^2+6z+9\right)\left(z^2+6z+9\right)-4(z+3)\left(z^2+6z+9\right)-4\left(z^2+6z+9\right)+16z+64=0$$

$$=z^4+8z^3+14z^2-8z+1=0$$

This will be equivalent to

$$f(z)=\left(z^2+pz+q\right)\left(z^2+pz+q_1\right)=0$$

$$=z^4+2pz^3+\left(p^2+q+q_1\right)z^2+p(q+q_1)z+qq_1=0$$

Comparing the z containing equation, we get

$$2p=8\Rightarrow p=4$$

$$\left(p^2+q+q_1\right)=14\Rightarrow q+q_1=-2$$

$$p(q+q_1)=4(-2)=-8 \quad \text{and}$$

$$qq_1=1 \quad \therefore \quad q=\frac{1}{q_1}$$

Putting $q=\frac{1}{q_1}$ in $q+q_1=-2$, we get $\frac{1}{q_1}+q_1=-2$ or $1+q_1^2=-2q_1$

$$1+2q_1+q_1^2=0$$

$$\therefore \quad q_1=\frac{-2\pm\sqrt{4-4}}{2}$$

$$\therefore \quad q_1=-1$$

i.e., the equations will become

$$f(z)=(z^2+4z-1)(z^2+4z-1)=0$$

When $z^2+4z-1=0$

$$\therefore \quad z=\frac{-4\pm\sqrt{16+4}}{2}=\frac{-4\pm2\sqrt{5}}{2}=-2\pm\sqrt{5}$$

$$\text{or} \quad 2x-3=-2\pm\sqrt{5}$$

$$\text{or} \quad 2x=1\pm\sqrt{5}$$

$$\therefore \quad x=\frac{1\pm\sqrt{5}}{2}$$

Therefore, the roots will be $x=\frac{1+\sqrt{5}}{2},\frac{1-\sqrt{5}}{2}$

$$\text{or} \quad x=\frac{1+2.2361}{2} \quad \text{and} \quad \frac{1-2.2361}{2}$$

$$=+1.618(\text{Twice}) \text{ and } \frac{-1.2361}{2}=-0.618(\text{Twice})$$

Thus, we get five roots and these roots are

$$x=-2, x=-0.618(\text{Twice}) \text{ and } x=+1.618(\text{Twice})$$

Putting $x=-2$, we shall get

$$\frac{\alpha-E_1}{\beta}=-2 \quad \text{or} \quad \alpha-E_1=-2\beta$$

$$\therefore \quad E_1=\alpha+2\beta \tag{13.137}$$

When $x=-0.618$

$$\therefore \quad \frac{\alpha-E_2}{\beta}=-0.618 \text{ or } \alpha-E_2=-0.618\beta \tag{13.138}$$

$$\therefore \quad E_2=\alpha+0.618\beta=E_3 \text{ (Degenerate pair)}$$

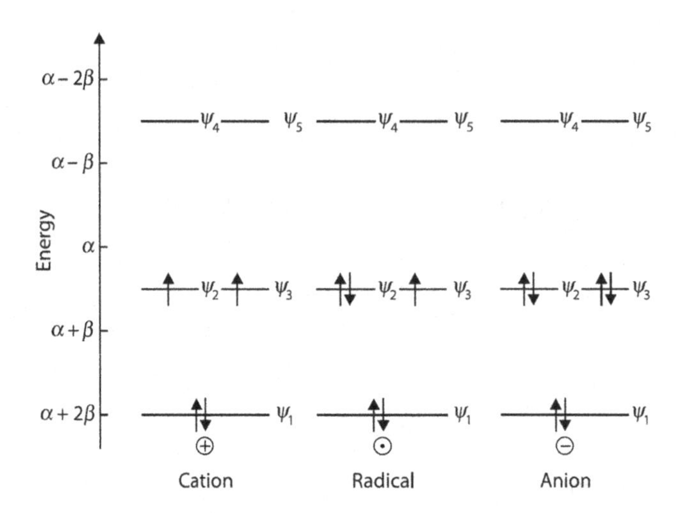

FIGURE 13.14 Energy level diagram of cyclopentadienyl systems.

When $x = 1.618$,

$$\therefore \quad \frac{\alpha - E_4}{\beta} = 1.618 \text{ or } \alpha - E_4 = 1.618\beta$$
$$\therefore \quad E_4 = \alpha - 1.618\beta = E_5 \left(\text{Degenrate pair}\right) \tag{13.139}$$

Thus, to sum up

$$\left.\begin{aligned} E_1 &= \alpha + 2\beta \\ E_2 &= E_3 = \alpha + 0.618\beta \\ E_4 &= E_5 = \alpha - 1.618\beta \end{aligned}\right\} \tag{13.140}$$

From these values of energy, the energy level diagram can be constructed, as given in Figure. 13.14.

13.2.8 Delocalisation Energy of Cyclopentadienyl Systems

- The π electron energy of cyclopentadienyl system (cation) consisting of 4π electrons is given by

$$E_\pi = \left[2(\alpha + 2\beta) + 2(\alpha + 0.618\beta)\right] = 4\alpha + 5.236\beta$$

  and DE = $(4\alpha + 5.236\beta) - 4(\alpha + \beta) = 1.236\beta$
- The cyclopentadienyl radical contains 5π electrons and E_π is given by

$$E_\pi = 2(\alpha + 2\beta) + 3(\alpha + 0.618\beta) = 5\alpha + 5.854\beta$$

  and DE = $(5\alpha + 5.854\beta) - [4(\alpha + \beta) + \alpha] = 1.854\beta$
- The cyclopentadienyl anion contains 6π electrons. Therefore,

$$E_\pi = 2(\alpha + 2\beta) + 4(\alpha + 0.618\beta) = 6\alpha + 6.472\beta$$

$$\therefore \quad \text{DE} = (6\alpha + 6.472\beta) - \left[4(\alpha + \beta) + 2\alpha\right] = 2.472\beta$$

From the values of delocalisation energy of cyclopentadienyl system, it is observed that cyclopentadienyl anion has the highest value of delocalisation energy, and hence, its stability will be greatest.

13.2.9 Hückel Molecular Orbital Coefficient and Molecular Orbitals

Substituting the value of $x = -2$ in Eq. (13.134), we obtain

$$\left.\begin{aligned} -2C_1 + C_2 + C_5 &= 0 \\ C_1 - 2C_2 + C_3 &= 0 \\ C_2 - 2C_3 + C_4 &= 0 \\ C_3 - 2C_4 + C_5 &= 0 \\ C_1 + C_4 - 2C_5 &= 0 \end{aligned}\right\} \tag{13.141}$$

From the above equations, we can write

$$C_2 + C_5 = 2C_1 \quad \text{or} \quad C_2 = 2C_1 - C_5$$

$$-2C_2 + C_3 = -C_1 \quad \text{or} \quad C_3 = 2C_2 - C_1$$

$$C_4 = (2C_5 - C_1)$$

$$C_2 = 2C_3 - C_4$$

$$C_3 = 2C_4 - C_5$$

$$\begin{aligned} \text{or} \quad C_2 &= 2(2C_4 - C_5) - C_4 \\ &= 4C_4 - 2C_5 - C_4 \\ &= 3C_4 - 2C_5 \\ &= 3(2C_5 - C_1) - 2C_5 \\ &= 6C_5 - 3C_1 - 2C_5 \\ &= 4C_5 - 3C_1 \end{aligned}$$

But, $C_2 = 2C_1 - C_5$.

Equating the values of C2, we get

$$2C_1 - C_5 = 4C_5 - 3C_1$$

$$\text{or} \quad 5C_1 = 5C_5 \tag{13.142}$$

$$\therefore \quad C_1 = C_5$$

$\because C_2 = 2C_1 - C_5 = 2C_1 - 2C_1$

$$\therefore \quad C_2 = C_1 \tag{13.143}$$

But $C_3 = 2\,C_1 - C_1 = 2\,C_1 - C_1 = C_1$

$$\therefore C_3 = C_1 \tag{13.144}$$

$$\left.\begin{aligned} C_4 &= 2(C_5) - C_1 = 2C_5 - C_1 = 2C_5 - C_1 = C_1 \\ \therefore \quad C_4 &= C_1. \end{aligned}\right. \tag{13.145}$$

Thus, with the help of Eqs. (13.142), (13.143), and (13.145), we can write

$$C_1 = C_2 = C_3 = C_4 = C_5. \tag{13.146}$$

Applying normalisation condition, we may write

$$\sum_{i=1}^{5} C_i^2 = 1$$

$$\text{or} \quad C_1^2 + C_2^2 + C_3^2 + C_4^2 + C_5^2 = 1$$

$$\text{or} \quad 5C_1^2 = 1 \quad \therefore \quad C_1 = \frac{1}{\sqrt{5}}$$

$$\text{or} \quad C_1 = C_2 = C_3 = C_4 = C_5 = \frac{1}{\sqrt{5}} \tag{13.147}$$

$$\text{Hence,} \quad \psi_1 = \frac{1}{\sqrt{5}}(\phi_1 + \phi_2 + \phi_3 + \phi_4 + \phi_5)$$

Now, we shall find ψ_2. For finding ψ_2, it will be convenient to set $C_5 = 0$. Putting $C_5 = 0$ in Eq. (13.134), we obtain

$$\left.\begin{aligned} C_1x + C_2 &= 0 \\ C_2x + C_1 + C_3 &= 0 \\ C_3x + C_2 &= 0 \\ C_4x + C_3 &= 0 \\ \text{and} \quad C_1 + C_4 &= 0 \end{aligned}\right\} \tag{13.148}$$

Putting $x = -0.618$, we write

$$\left.\begin{aligned} -0.618C_1 + C_2 &= 0 \\ -0.618C_2 + C_1 + C_3 &= 0 \\ -0.618C_3 + C_2 &= 0 \\ -0.618C_4 + C_3 &= 0 \\ C_1 + C_4 &= 0 \end{aligned}\right\} \tag{13.149}$$

Form first equation of Eq. (13.149), we have

$$C_2 = +0.618C_1$$

Putting this value in the second equation, we get

$$-0.618(0.618C_1) + C_1 + C_3 = 0$$

$$\text{or} \quad -0.618 \times 0.618C_1 + C_1 + C_3 = 0$$

$$\text{or} \quad -0.3819C_1 + C_1 + C_3 = 0$$

$$\text{or} \quad (1 - 0.3819)C_1 + C_3 = 0$$

$$\text{or} \quad 0.618C_1 = -C_3$$

$$\therefore \quad C_3 = -0.618C_1 \tag{13.150}$$

$$\text{and} \quad C_4 = -C_1$$

Applying normalisation condition, we have

$$C_1^2 + C_2^2 + C_3^2 + C_4^2 = 1$$

$$\text{or} \quad C_1^2 + (0.618C_1)^2 + (-0.618C_1)^2 + C_1^2 = 1$$

$$\text{or} \quad C_1^2 + 0.3819C_1^2 + 0.3819C_1^2 + C_1^2 = 1$$

$$\text{or} \quad C_1^2 + 0.7638C_1^2 + C_1^2 = 1$$

$$\text{or} \quad 2.7638C_1^2 = 1 \tag{13.150}$$

$$\text{or} \quad C_1^2 = \frac{1}{2.7638} = 0.3618$$

$$\therefore \quad C_1 = (0.3618)^{1/2} = 0.6015$$

$$\therefore \quad C_2 = 0.618C_1 = 0.618 \times 0.6015 = 0.372$$

$$\therefore \quad C_3 = -0.618C_1 = -0.618 \times 0.6015 = -0.372$$

and $C_4 = -C_1 = -0.6015$.

Therefore, the state function ψ_2 molecular orbitals is expressed as

$$\psi_2 = C_1\phi_1 + C_2\phi_2 + C_3\phi_3 + C_4\phi_4$$

$$\text{or} \quad \psi_2 = 0.6015\phi_1 + 0.372\phi_2 - 0.372\phi_3 - 0.6015\phi_4 \tag{13.151}$$

$$\text{or} \quad \psi_2 = 0.6015(\phi_1 - \phi_4) + 0.372(\phi_2 - \phi_3)$$

Putting $x = +1.619$ in Eq. (13.134) and making $C_5 = 0$, the equation is expressed as

$$1.618C_1 + C_2 = 0$$

$$1.618C_2 + C_1 + C_3 = 0$$

$$1.618C_3 + C_2 + C_4 = 0$$

$$C_1 + C_4 = 0$$

$$\because \quad 1.618C_1 + C_2 = 0 \quad \therefore \quad C_2 = -1.618C_1$$

and $1.618C_2 + C_1 + C_3 = 0$

or $(1.618)(-1.618C_1) + C_1 + C_3 = 0$

or $-2.6179C_1 + C_1 + C_3 = 0$

or $-1.6179 + C_3 = 0 \therefore C_3 \approx 1.618C_1$

and $C_4 = -C_1$.

Applying normalisation condition, we obtain

$$C_1^2 + (-1.6179C_1)^2 + (1.6179C_1)^2 + (-C_1)^2 = 1$$

or $$C_1^2 + 2.6176C_1^2 + 2.6176C_1^2 + C_1^2 = 1$$

or $$2C_1^2 + 5.2352C_1^2 = 1$$

or $$7.2352C_1^2 = 1$$

$$\therefore \quad C_1^2 = \frac{1}{7.2352} = 0.1382$$

$$\therefore \quad C_1 = (0.1382)^{1/2} = 0.37175 \approx 0.372$$

$$\therefore \quad C_2 = -1.618C_1 = (-1.618)(0.372) = -0.6018 \approx -0.602$$

$$\therefore \quad C_3 = 1.618C_1 = (1.618)(0.372) =\approx 0.602$$

and $$C_4 = -C_1 = -0.372$$

Therefore, ψ_3 molecular orbital is expressed as

$$\begin{aligned}\psi_3 &= C_1\phi_1 + C_2\phi_2 + C_3\phi_3 + C_4\phi_4 \\ &= 0.372\phi_1 - 0.602\phi_2 + 0.602\phi_3 - 0.372\phi_4 \\ &= 0.372(\phi_1 - \phi_4) - 0.602(\phi_2 - \phi_3).\end{aligned} \tag{13.152}$$

By similar methods, we can arrive at the following:

$$\psi_4 = 0.196(\phi_1 + \phi_4) - 0.512(\phi_2 + \phi_3) + 0.633\phi_5$$

and $$\psi_5 = 0.512(\phi_1 + \phi_4) - 0.196(\phi_2 + \phi_3) - 0.633\phi_5$$

13.2.10 Benzene

Benzene is a hexagonal cyclic ring containing 6π electrons. Its structural formula is shown in Figure 13.15.

The carbon atoms of benzene ring have been numbered from 1 to 6. Let us suppose that the HMO function for the benzene is represented by

$$\psi = \sum_{i=1}^{6} C_i\phi_i$$

where the terms have their usual meaning.

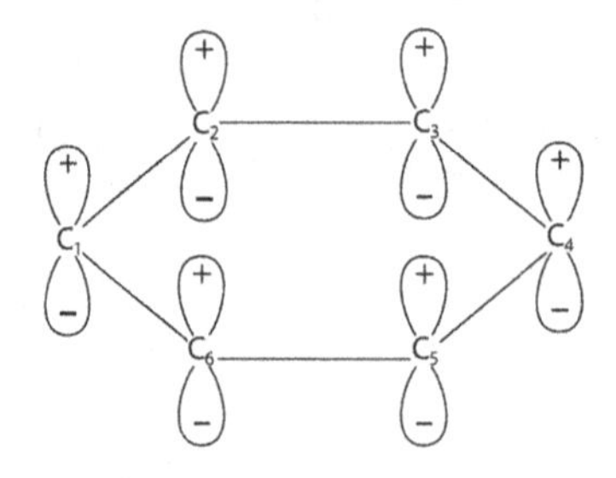

FIGURE 13.15 Structure of benzene.

The secular equations can be written as

$$\left.\begin{aligned} C_1x + C_2 + C_6 &= 0 \\ C_1 + C_2x + C_3 &= 0 \\ C_2 + C_3x + C_4 &= 0 \\ C_3 + C_4x + C_5 &= 0 \\ C_4 + C_5x + C_6 &= 0 \\ C_1 + C_5 + C_6x &= 0 \end{aligned}\right\} \tag{13.153}$$

From these secular equations, the secular determinant can be constructed in terms of dimensionless variable x as follows:

$$\begin{vmatrix} x & 1 & 0 & 0 & 0 & 1 \\ 1 & x & 1 & 0 & 0 & 0 \\ 0 & 1 & x & 1 & 0 & 0 \\ 0 & 0 & 1 & x & 1 & 0 \\ 0 & 0 & 0 & 1 & x & 1 \\ 1 & 0 & 0 & 0 & 1 & x \end{vmatrix} = 0 \tag{13.154}$$

The Laplace expansion of the earlier determinant gives the following:

$$x\begin{vmatrix} x & 1 & 0 & 0 & 0 \\ 1 & x & 1 & 0 & 0 \\ 0 & 1 & x & 1 & 0 \\ 0 & 0 & 1 & x & 1 \\ 0 & 0 & 0 & 1 & x \end{vmatrix} - 1\begin{vmatrix} 1 & 1 & 0 & 0 & 0 \\ 0 & x & 1 & 0 & 0 \\ 0 & 1 & x & 1 & 0 \\ 0 & 0 & 1 & x & 1 \\ 1 & 0 & 0 & 1 & x \end{vmatrix} - 1\begin{vmatrix} 1 & 1 & 0 & 0 & 0 \\ 0 & x & 1 & 0 & 0 \\ 0 & 1 & x & 1 & 0 \\ 0 & 0 & 1 & x & 1 \\ 1 & 0 & 0 & 1 & x \end{vmatrix}$$

$$\Rightarrow x^2\begin{vmatrix} x & 1 & 0 & 0 \\ 1 & x & 1 & 0 \\ 0 & 1 & x & 1 \\ 0 & 0 & 1 & x \end{vmatrix} - x\begin{vmatrix} 1 & 1 & 0 & 0 \\ 0 & x & 1 & 0 \\ 0 & 1 & x & 1 \\ 0 & 0 & 1 & x \end{vmatrix} - 1\begin{vmatrix} x & 1 & 0 & 0 \\ 1 & x & 1 & 0 \\ 0 & 1 & x & 1 \\ 0 & 0 & 1 & x \end{vmatrix} - 1\begin{vmatrix} x & 1 & 0 & 0 \\ 1 & x & 1 & 0 \\ 0 & 1 & x & 1 \\ 0 & 0 & 1 & x \end{vmatrix}$$

$$-1\begin{vmatrix} 0 & 1 & 1 & 0 \\ 0 & 0 & x & 1 \\ 0 & 0 & 1 & x \\ 1 & 0 & 0 & 1 \end{vmatrix} + x\begin{vmatrix} 0 & x & 1 & 0 \\ 0 & 1 & x & 1 \\ 0 & 0 & 1 & x \\ 1 & 0 & 0 & 1 \end{vmatrix} - 1\begin{vmatrix} 1 & 0 & 1 & 0 \\ 0 & 0 & x & 1 \\ 0 & 0 & 1 & x \\ 1 & 0 & 0 & 1 \end{vmatrix} = 0$$

$$\Rightarrow x^3\begin{vmatrix} x & 1 & 0 \\ 1 & x & 1 \\ 0 & 1 & x \end{vmatrix} - x^2\begin{vmatrix} 1 & 1 & 0 \\ 0 & x & 1 \\ 0 & 1 & x \end{vmatrix} - x\begin{vmatrix} x & 1 & 0 \\ 1 & x & 1 \\ 0 & 1 & x \end{vmatrix} + x\begin{vmatrix} 0 & 1 & 0 \\ 0 & x & 1 \\ 0 & 1 & x \end{vmatrix} - x\begin{vmatrix} x & 1 & 0 \\ 1 & x & 1 \\ 0 & 1 & x \end{vmatrix}$$

$$+1\begin{vmatrix} 1 & 1 & 0 \\ 0 & x & 1 \\ 0 & 1 & x \end{vmatrix} - 1\begin{vmatrix} 0 & 1 & 0 \\ 0 & x & 1 \\ 1 & 1 & x \end{vmatrix} - 1\begin{vmatrix} 1 & x & 1 \\ 0 & 1 & x \\ 0 & 0 & 1 \end{vmatrix} + x\begin{vmatrix} 0 & x & 1 \\ 0 & 1 & x \\ 0 & 0 & 1 \end{vmatrix} - 1\begin{vmatrix} 0 & 1 & 1 \\ 0 & 0 & x \\ 0 & 0 & 1 \end{vmatrix}$$

$$-x^2\begin{vmatrix} 0 & x & 1 \\ 0 & 1 & x \\ 1 & 0 & 1 \end{vmatrix} + x\begin{vmatrix} 0 & 1 & 1 \\ 0 & 0 & x \\ 1 & 0 & 1 \end{vmatrix} + 1\begin{vmatrix} 0 & x & 1 \\ 0 & 1 & x \\ 1 & 0 & 1 \end{vmatrix} - 1\begin{vmatrix} 0 & 0 & 1 \\ 0 & 0 & x \\ 1 & 0 & 1 \end{vmatrix} = 0$$

$$\Rightarrow x^4\left(x^2-1\right)-x^3(x)-x^2\left(x^2-1\right)+0-x\left(x^2-1\right)+x\cdot x+0-x\left(x^2-1\right)+x\cdot x+\left(x^2-1\right)$$
$$-0-1-1+0+x\times 0-1(0)-x^3(x)+x^2+x^2+0+0+x^2-1-0=0$$
$$\Rightarrow x^6-x^4-x^4-x^4+x^2-x^4+x^2+x^2-x^4+x^2+x^2+x^2-1-1-1+0+x\times 0-1(0) \quad (13.155)$$
$$-x^3(x)+x^2+x^2+0+0+x^2-1=0$$
$$\Rightarrow x^6-6x^4+9x^2-4=0$$

$$\Rightarrow (x+2)\left(x^5-2x^4-2x^3+4x^2+x-2\right)=0$$
$$\Rightarrow (x+2)(x-2)\left(x^4-2x^2+1\right)=0$$
$$\Rightarrow (x+2)(x-2)\left(x^2-1\right)^2=0$$

$$\Rightarrow (x+2)(x-2)\left(x^2-1\right)\left(x^2-1\right)=0$$
$$\Rightarrow (x+2)(x-2)(x+1)(x-1)(x+1)(x-1)=0 \quad (13.156)$$

Therefore, the roots of the polynomial will be
$x = -2, +2, -1, +1, -1, +1$. It means $x = -2$, -1(twice), $+1$(twice), and $+2$.
Hence, the energy levels will be expressed as follows:

$$\text{When, } x=-2; E_1=\alpha+2\beta \qquad (BMO)$$
$$x=-1; E_2=E_3=\frac{\alpha-E}{\beta}=-1 \text{ or } E_2=\alpha+\beta \qquad \text{(Degenerate pair; BMO)}$$
$$x=+1; E_4=E_5=\alpha-\beta \qquad \text{(Degenerate pair; ABMO)} \quad (13.157)$$
$$x=+2; E_6=\alpha-2\beta$$

Form the above values of benzene, the MO energy level diagram can be illustrated as shown in Figure 13.16.

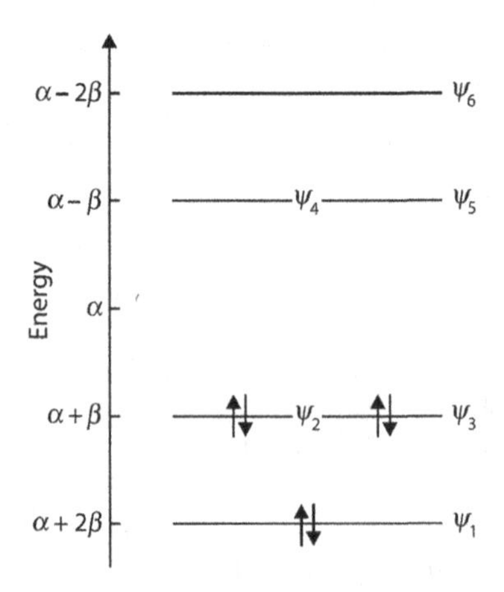

FIGURE 13.16 Hückel molecular orbital energy level diagram of benzene.

13.2.11 Delocalisation Energy of Benzene

Benzene or cyclohexatriene contains 6π electrons. The π electron energy of benzene will be given by

$$E_\pi = 2(\alpha + 2\beta) + 4(\alpha + \beta) = 6\alpha + 8\beta$$

The delocalisation energy is given by

$$(E_\pi)_{\text{deloc}} = (6\alpha + 8\beta) - 6(\alpha + \beta) = 2\beta$$

It is clear that π electron energy is $(6\alpha + 8\beta)$, and $6(\alpha + \beta)$ denotes the energy of benzene because of the presence of three isolated double bonds.

13.2.12 Hückel Molecular Orbital Coefficients and Molecular Orbitals

To obtain the values of the HMO coefficients, we refer to the original set of six simultaneous equations [Eq. (13.153)], which are

$$C_1x + C_2 + C_6 = 0$$

$$C_2x + C_1 + C_3 = 0$$

$$C_3x + C_2 + C_4 = 0$$

$$C_4x + C_3 + C_5 = 0$$

$$C_5x + C_4 + C_6 = 0$$

$$C_6x + C_1 + C_5 = 0$$

For the value of $x = -2$, we shall find the values of the coefficients. For this, we shall substitute $x = -2$ in the above equations. The above equation takes the form

$$-2C_1 + C_2 + C_6 = 0$$

$$-2C_2 + C_1 + C_3 = 0$$

$$-2C_3 + C_2 + C_4 = 0$$

$$-2C_4 + C_3 + C_5 = 0$$

$$-2C_5 + C_4 + C_6 = 0$$

$$-2C_6 + C_1 + C_5 = 0$$

It is further expressed as

$$C_2 + C_6 = 2C_1 \tag{13.158}$$

$$C_1 + C_3 = 2C_2 \tag{13.159}$$

$$C_2 + C_4 = 2C_3 \tag{13.160}$$

$$C_3 + C_5 = 2C_4 \tag{13.161}$$

$$C_4 + C_6 = 2C_5 \tag{13.162}$$

$$C_1 + C_5 = 2C_6 \tag{13.163}$$

From Eqs. (13.158) and (13.162), we can write

$$\begin{array}{r} C_2 + C_6 = 2C_1 \\ C_4 + C_6 = 2C_5 \\ - \quad - \quad - \\ \hline C_2 - C_4 = 2C_1 - 2C_5 \end{array}$$

Adding Eq. (13.160), we shall obtain

$$\begin{array}{l} C_2 - C_4 = 2C_1 - 2C_5 \\ C_2 + C_4 = 2C_3 \\ \hline 2C_2 = 2C_1 + 2C_3 - 2C_5 \end{array}$$

$$\text{or} \quad C_1 + C_3 = 2C_1 + 2C_3 - 2C_5 \quad [\because 2C_2 = C_1 + C_3]$$

$$\text{or} \quad C_1 + C_3 - 2C_5 = 0$$

$$\text{or} \quad C_1 + C_3 - 2C_5 = 0$$

But $C_1 + C_3 = 2C_2$

$$\therefore \quad 2C_2 = 2C_5 \quad \therefore \quad C_2 = C_5$$

From Eq. (13.158), we have

$$\begin{array}{r} C_2 + C_6 = 2C_1 \\ C_5 + C_6 = 2C_1 \\ C_1 + C_5 = 2C_6 \\ - \quad - \quad - \\ \hline C_6 - C_1 = 2C_1 - 2C_6 \end{array}$$

or $3C_6 = 3C_1$ or, $C_1 = C_6$

From Eq. (13.163), we have

$$C_1 + C_5 = 2C_6 = 2C_1$$

$$\text{or} \quad C_5 = 2C_1 - C_1 = C_1 \quad \therefore \quad C_1 = C_5$$

From Eq. (13.162), we have

$$C_4 + C_6 = 2C_5$$

$$\text{or,} \quad C_4 + C_1 = 2C_1$$

$$\text{or,} \quad C_4 = 2C_1 - C_1 = C_1$$

$$\therefore \quad C_4 = C_1$$

From Eq. (13.160), we can write

$$C_2 + C_4 = 2C_3$$

$$\text{or,} \quad C_2 + C_4 = 2C_3$$

$$\text{or,} \quad 2C_1 = 2C_3 \quad \therefore \quad C_1 = C_3$$

Now, we have

$$C_2 = C_5, C_1 = C_6, C_1 = C_5, C_1 = C_3, \quad \text{and } C_1 = C_4$$

From these, it is clear that

$$C_1 = C_2 = C_3 = C_4 = C_5 = C_6 \tag{13.164}$$

Applying normalisation condition, we can write

$$\sum_{i=1}^{6} C_i^2 = 1 = C_1^2 + C_2^2 + C_3^2 + C_4^2 + C_5^2 + C_6^2 = 1$$

$$\text{or} \quad 6C_1^2 = 1 \quad \text{or,} \quad C_1^2 = \frac{1}{6} \quad \therefore \quad C_1 = \frac{1}{\sqrt{6}} \tag{13.165}$$

Hence, the resulting molecular orbitals are

$$\psi_1 = C_1\phi_1 + C_2\phi_2 + C_3\phi_3 + C_4\phi_4 + C_5\phi_5 + C_6\phi_6$$

$$\text{or} \quad \psi_1 = \frac{1}{\sqrt{6}}[\phi_1 + \phi_2 + \phi_3 + \phi_4 + \phi_5 + \phi_6][\because C_1 = C_2 = C_3 = C_4 = C_5 = C_6] \tag{13.166}$$

For the case $x = -1$, we must solve the following set of equations:

$$\left.\begin{aligned} -C_1 + C_2 + C_3 &= 0 \\ C_1 - C_2 + C_3 &= 0 \\ C_2 - C_3 + C_4 &= 0 \\ C_3 - C_4 + C_5 &= 0 \\ C_4 - C_5 + C_6 &= 0 \\ C_1 + C_5 - C_6 &= 0 \end{aligned}\right\} \tag{13.167}$$

We at once note that these equations are not linearly independent. For example, the first equation can be found by adding the fourth and fifth equations and subtracting the second equation.

$$C_3 - C_4 + C_5 = 0$$

$$C_4 - C_5 + C_6 = 0$$

$$\overline{\quad C_3 + C_6 = 0 \quad}$$

$$C_1 - C_2 + C_3 = 0$$

$$- \quad + \quad -$$

on subtraction $C_1 - C_2 + C_6 = 0$

which is the first equation of Eq. (13.167).

It is, thus, clear that it is possible to get four linearly independent equations among the constant C_i by adding the first and second equations, the third and fourth equations, the fifth and sixth equations, and considering one of the six original ones, say the second, as a fourth relationship.

Adding the first and second equations, we get

$$\begin{array}{l} C_1 + C_2 + C_6 = 0 \\ C_1 - C_2 + C_3 = 0 \\ \hline C_3 + C_6 = 0 \end{array}$$

$$\text{or} \quad C_3 = -C_6 \tag{13.168}$$

Adding the third and fourth equations, we get

$$\begin{array}{c} C_2 - C_3 + C_4 = 0 \\ C_3 - C_4 + C_5 = 0 \\ \hline C_2 + C_5 = 0 \end{array}$$

$$\text{or} \quad C_2 = -C_5 \tag{13.169}$$

Adding the fifth and sixth equations, we get

$$\begin{array}{c} C_4 - C_5 + C_6 = 0 \\ C_1 + C_5 - C_6 = 0 \\ \hline C_1 = -C_4 \end{array} \tag{13.170}$$

$$\text{and} \quad C_1 - C_2 + C_3 = 0 \tag{13.171}$$

Since there are four equations and six unknown quantities, we may arbitrarily assign unknown values to two of the six. Let us consider

$$C_2 = C_5 = 0, \text{ then we shall find}$$

$$C_1 = -C_3 = -C_4 = C_6$$

Applying normalisation condition, we have

$$C_1^2 + C_3^2 + C_4^2 + C_6^2 = 1$$

$$\text{or} \quad C_1^2 + (-C_1)^2 + (-C_1)^2 + C_1^2 = 1$$

$$\text{or} \quad C_1^2 + C_1^2 + C_1^2 + C_1^2 = 1$$

$$4C_1^2 = 1$$

$$C_1^2 = \frac{1}{4}$$

$$C_1 = \frac{1}{2}$$

This implies that

$$C_1 = \frac{1}{2}, C_3 = -\frac{1}{2}, C_4 = -\frac{1}{2} \text{ and } C_6 = \frac{1}{2}.$$

Now we are able to write ψ_2 MOs as

$$\psi_2 = C_1\phi_1 + C_3\phi_3 + C_4\phi_4 + C_6\phi_6$$

$$\text{or} \quad \psi_2 = C_1\phi_1 + C_3\phi_3 + C_4\phi_4 + C_6\phi_6 \tag{13.172}$$

$$\text{or} \quad \psi_2 = \frac{1}{2}(\phi_1 - \phi_3 - \phi_4 + \phi_6)$$

We have also got second $x = -1$. Putting $x = -1$ in Eq. (13.153), we get the following:

$$-C_1 + C_2 + C_6 = 0$$
$$C_1 - C_2 + C_3 = 0$$
$$C_2 - C_3 + C_4 = 0$$
$$C_3 - C_4 + C_5 = 0$$
$$C_4 - C_5 + C_6 = 0$$
$$C_1 + C_5 - C_6 = 0$$

Putting $C_2 = C_5 = 0$ like above, we get

$$\begin{aligned} -C_1 + C_6 &= 0 \\ C_1 + C_3 &= 0 \\ -C_3 + C_4 &= 0 \\ C_3 - C_4 &= 0 \\ C_4 + C_6 &= 0 \\ \text{and} \quad C_1 - C_6 &= 0 \end{aligned} \tag{13.173}$$

From the above, we obtain

$$c_1 + c_3 = 0 \quad \therefore \quad c_1 = -c_3 \tag{13.174}$$

Adding $c_3 - c_4 = 0$

$$\begin{aligned} \text{and} \quad & \underline{c_4 + c_6 = 0} \\ & c_3 + c_6 = 0 \quad \therefore \quad c_3 = -c_6 \end{aligned} \tag{13.175}$$

Since the second $x = -1$ also corresponds to the molecular orbital with equal energy, we may require the wave functions for the two orbitals (corresponding to $x = -1$ and $x = -1$) to be mutually orthogonal. We shall solve the case of the second $x = -1$ by invoking or calling on this orthogonality, because in the absence of this condition, we obtain the same result as we have found already for the first $x = -1$ case. Therefore, we can write the integral $\int \psi_2\psi_3 d\tau = 0$ and

$$\int c_1(\phi_1 - \phi_3 - \phi_4 + \phi_6)(c_1\phi_1 + c_2\phi_2 + c_3\phi_3 + c_4\phi_4 + c_5\phi_5 + c_6\phi_6)d\tau = 0 \tag{13.176}$$

$$\text{or } c_1c_1\langle\phi_1|\phi_1\rangle + c_1c_2\langle\phi_1|\phi_2\rangle + c_1c_3\langle\phi_1|\phi_3\rangle + c_1c_4\langle\phi_1|\phi_4\rangle + c_1c_5\langle\phi_1|\phi_5\rangle$$
$$+ c_1c_6\langle\phi_1|\phi_6\rangle - c_1^2\langle\phi_3|\phi_1\rangle - c_1c_2\langle\phi_3|\phi_2\rangle - c_1c_3\langle\phi_3|\phi_3\rangle - c_1c_4\langle\phi_3|\phi_4\rangle$$
$$- c_1c_5\langle\phi_3|\phi_5\rangle - c_1c_6\langle\phi_3|\phi_6\rangle - c_1^2\langle\phi_4|\phi_1\rangle - c_1c_2\langle\phi_4|\phi_2\rangle - c_1c_3\langle\phi_4|\phi_3\rangle$$
$$- c_1c_4\langle\phi_4|\phi_4\rangle - c_1c_5\langle\phi_4|\phi_5\rangle - c_1c_6\langle\phi_4|\phi_6\rangle + c_1^2\langle\phi_6|\phi_1\rangle + c_1c_2\langle\phi_6|\phi_2\rangle$$
$$+ c_1c_3\langle\phi_6|\phi_3\rangle + c_1c_4\langle\phi_6|\phi_4\rangle + c_1c_5\langle\phi_6|\phi_5\rangle + c_1c_6\langle\phi_6|\phi_6\rangle = 0$$

This will yield

$$\begin{aligned} & c_1c_1 - c_1c_3 - c_1c_4 + c_1c_6 = 0 \\ \text{or} \quad & c_1(c_1 - c_3 - c_4 + c_6) = 0 \\ \text{or} \quad & c_1 - c_3 - c_4 + c_6 = 0 \\ & c_1 + c_6 = c_3 + c_4 \end{aligned} \tag{13.177}$$

Now, we shall use Eqs. (13.174), (13.175), and (13.177) for getting c_1, c_3, c_4, and c_6. Putting $c_3 = -c_1$ and $c_4 = -c_6$ from the above equation and substituting in Eq. (13.177), we shall obtain

$$\begin{aligned} & c_1 - c_4 = -c_1 + c_4 \\ \text{or} \quad & 2c_1 = 2c_4 \quad \therefore \quad c_1 = c_4 \end{aligned} \tag{13.178}$$

Further applying the normalisation condition, i.e., $\int \psi_i \psi_i d\tau = 1$, we get

$$\begin{aligned} & C_1^2 + C_3^2 + C_4^2 + C_6^2 = 1 \\ \text{or} \quad & C_1^2 + (-C_1)^2 + C_1^2 + (-C_1)^2 = 1 \\ \text{or} \quad & C_1^2 + C_1^2 + C_1^2 + C_1^2 = 1 \\ \text{or} \quad & 4C_1^2 = 1 \\ \text{or} \quad & C_1^2 = \frac{1}{4} \\ \text{or} \quad & C_1 = \frac{1}{2} \end{aligned}$$

Thus, the MOs ψ_3 will be expressed as

$$\begin{aligned} & \psi_3 = C_1\phi_1 + C_3\phi_3 + C_4\phi_4 + C_6\phi_6 \\ & \psi_3 = C_1\phi_1 - C_1\phi_3 + C_1\phi_4 - C_1\phi_6 \\ \text{or} \quad & \psi_3 = C_1(\phi_1 - \phi_3 + \phi_4 - \phi_6) \\ \text{or} \quad & \psi_3 = \frac{1}{2}(\phi_1 - \phi_3 + \phi_4 - \phi_6) \end{aligned} \tag{13.179}$$

By similar method, we can find out the other ψMOS.

$$\psi_4 = \frac{1}{\sqrt{12}}(\phi_1 - \phi_2 - 2\phi_3 - \phi_4 + \phi_5 + 2\phi_6)$$

$$\psi_5 = \frac{1}{\sqrt{12}}(\phi_1 + \phi_2 - 2\phi_3 + \phi_4 + \phi_5 - 2\phi_6)$$

$$\text{and } \psi_6 = \frac{1}{\sqrt{6}}(\phi_1 - \phi_2 + \phi_3 - \phi_4 + \phi_5 - \phi_6)$$

13.2.13 Graphical Representation of Molecular Orbitals in Benzene

The graphical representation of molecular orbital in benzene is given in Figure 13.17.

It is clear from the graphical representation of benzene that the energy increases with the increase in the number of nodal planes.

13.3 Electron Density

The electron density is actually the probability of finding a π electron on an atom in a conjugated system. It is mathematically expressed as

$$q_r = \sum_r n_{ij} c_{jr}^2 \tag{13.180}$$

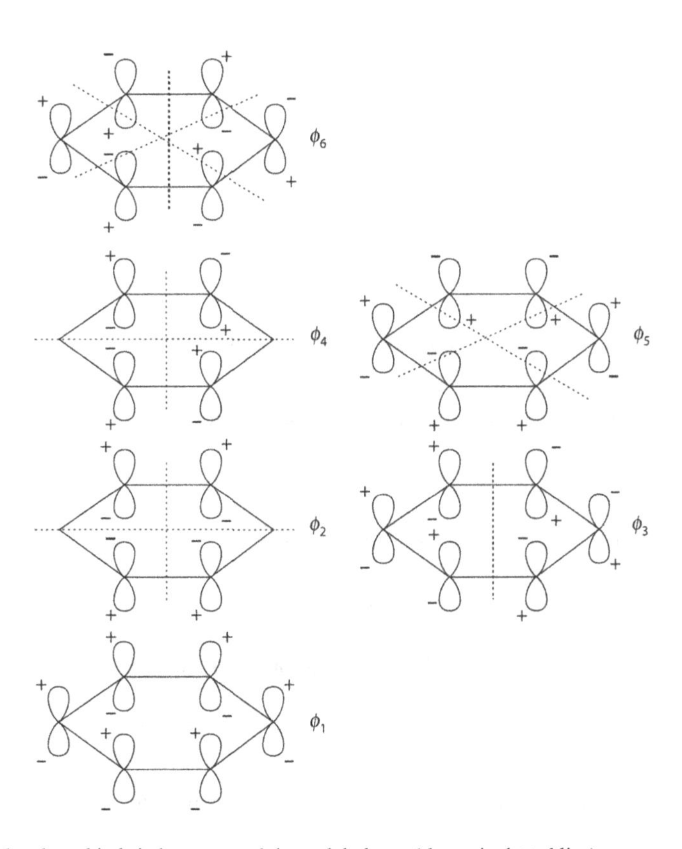

FIGURE 13.17 Molecular orbitals in benzene and the nodal planes (shown in dotted line).

where q_r = total electron density at an atom; is the sum of electron densities contributed by each atom in each molecular orbital

C_{jr} = coefficient of atom r in the jth molecular orbital

j = electron related to a particular carbon

n_{ij} = number of electron in an orbital

Now, we shall take interest to estimate the electron density at different carbon atoms in the molecules.

13.3.1 Ethylene

Ethylene contains only 2π electrons, and these occupy the molecular orbital. The ethylene system is represented as follows:

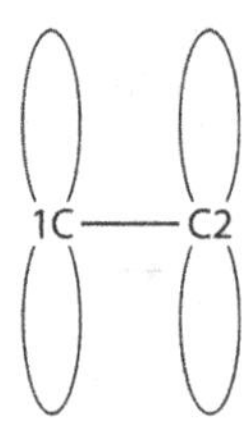

In the case of ethylene/ethane,

$$q_1 = n_1 c_{11}^2 \left(\text{using } q_r = \sum_r n_{ij} c_{jr}^2 \right)$$

where n_1 = number of electron in the first molecular orbital

But $n_1 = 2$,

$$\therefore \quad q_1 = n_1 c_{11}^2 = 2\left(\frac{1}{\sqrt{2}}\right)^2 = 2 \times \frac{1}{2} = 1, \quad \left[\because \quad c_{11} = \frac{1}{\sqrt{2}}\right]$$

Similarly, the electron density at carbon atom 2 is given by

$$q_2 = n_2 = n_2 c_{22}^2$$

$$q_2 = 2\left(\frac{1}{\sqrt{2}}\right) = 1, \quad \left[\because \quad \text{Number of electrons in the carbon atom 2 is also 2}\right]$$

Thus, we observe that the π electron density on the two carbon atoms of ethylene is equal to unity.

13.3.2 Butadiene

Butadiene is represented by the following structural formula:

$$\overset{1}{C}H_2 = \overset{2}{C}H - \overset{3}{C}H = \overset{4}{C}H_2$$

In the case of butadiene, the total electron density at carbon atom 1 is given by

$$q_r = \sum_r n_{ij} c_{jr}^2 = 2\left(c_{11}^2 + c_{12}^2\right)$$

= number of electrons in an orbital (square of coefficient of the first atomic orbital

+ square of coefficient of the second atomic orbital)

$$q_1 = 2\left(C_{21}^2 + C_{22}^2\right) \text{ [By using Eq. (13.180)]}$$

$$= 2\left\{(0.372)^2 + (0.602)^2\right\} \quad [\because C_{11} = 0.372 \text{ and } C_{12} = 0.602]$$

$$= 2\{0.138+0.362\} = 2(0.500) = 1$$

$$\therefore q_1 = 1$$

Similarly, we may calculate the electron density q_2, q_3, and q_4 on the carbon atoms 2, 3, and 4, respectively. They have been estimated as follows:

$$q_2 = 2\left(C_{21}^2 + C_{22}^2\right)$$

$$= 2\left\{(0.602)^2 + (0.372)^2\right\} = 1$$

$$q_3 = 2\left(C_{31}^2 + C_{32}^2\right)$$

$$= 2\left\{(0.602)^2 + (-0.372)^2\right\} = 1$$

$$q_4 = 2\left(C_{41}^2 + C_{42}^2\right)$$

$$= 2\left\{(0.372)^2 + (-0.602)^2\right\} = 1$$

Thus, in the case of butadiene, it is clear that the electron densities of all the carbon atoms are equal, and they amount to unity. It should be kept in mind that this result will be found for all the polyenes.

13.3.3 Benzene

In the case of benzene, the three molecular orbitals are occupied by two electrons in the ground state; hence, the electron density at carbon atom 1 of benzene may be given by

$$q_1 = 2\left(C_{11}^2 + C_{12}^2 + C_{13}^2\right)$$

$$\text{or} \quad q_1 = 2\left[\left(\frac{1}{\sqrt{6}}\right)^2 + \left(\frac{1}{\sqrt{12}}\right)^2 + \left(\frac{1}{2}\right)^2\right]$$

$$\text{or} \quad q_1 = 2\left[\frac{1}{6} + \frac{1}{12} + \frac{1}{4}\right] = 2\left[\frac{4+2+6}{24}\right]$$

$$\text{or} \quad q_1 = \frac{2\times 12}{24} = 1$$

Similarly, the electron density at carbon atom 2 is expressed as

$$q_2 = 2\left[C_{21}^2 + C_{22}^2 + C_{23}^2\right]$$

$$= 2\left[\left(\frac{1}{\sqrt{6}}\right)^2 + \left(\frac{1}{\sqrt{12}}\right)^2 + \left(\frac{1}{2}\right)^2\right]$$

$$= 2\left[\frac{1}{6} + \frac{1}{12} + \frac{1}{4}\right]$$

$$= 2\left[\frac{4+2+6}{24}\right] = \frac{2\times 12}{24} = 1$$

The electron density at carbon atom 3 is given by

$$q_3 = 2\left[C_{31}^2 + C_{32}^2 + C_{33}^2\right]$$

or $$q_3 = 2\left[\left(\frac{1}{\sqrt{6}}\right)^2 + \left(\frac{1}{\sqrt{12}}\right)^2 + \left(\frac{1}{2}\right)^2\right]$$

or $$q_4 = 2\left[\frac{1}{6} + \frac{1}{12} + \frac{1}{4}\right] = \frac{2 \times 12}{24} = 1$$

Further, we can show that $q_4 = q_5 = q_6 = 1$.

Hence, $q_1 = q_2 = q_3 = q_4 = q_5 = q_6 = 1$. It is, thus, inferred that the π electron densities at all the carbon atoms of benzene are equal and they amount to unity.

13.4 Bond Order

The concept of π bond order was put forward by Coulson in 1939. This concept has proved to be of immense value in the electronic interpretation of organic chemistry. It is mathematically defined as

$$P_{rs} = \sum n_j c_{jr} c_{js} \tag{13.181}$$

where $p_{rs} = \pi$ bond between carbons r and s

n_j = number of electrons in the jth molecular orbital

Since in the case of polyenes, the bond order is associated with the π electrons only; it may simply be called π bond order. Now we shall find the π bond order of some of the polyenes.

13.4.1 Ethylene

Using the formula contained in Eq. (13.181), the partial bond order associated with the molecular orbital $\psi_1 = \frac{1}{\sqrt{2}}(\Phi_1 + \Phi_2)$ is given by $P'_{12} = \frac{1}{\sqrt{2}} \cdot \frac{1}{\sqrt{2}} = \frac{1}{2}$.

Similarly, the partial bond order associated with the antibonding molecular orbital is given by

$$P'_{12} = \frac{1}{\sqrt{2}} \cdot \left(-\frac{1}{\sqrt{2}}\right) = -\frac{1}{2}$$

From the values of P_{12} and P'_{12}, it is inferred that the positive value of partial bond order strengthens the π bond, whereas a negative value weakens the same.

In the ground state of ethene, the 2π electrons are associated with ψ_1, the total π bond order

$$P_{12} = 2 \times 1/2 = 1$$

13.4.2 Butadiene

Butadiene is structurally written as $H_2C = CH - CH = CH_2$. It contains 4π electrons, which are occupied in the two bonding orbitals.

$$\psi_1 = 0.3717\Phi_1 + 0.6015\Phi_2 + 0.6015\Phi_3 + 0.3717\Phi_4$$

and $$\psi_2 = 0.6015\Phi_1 + 0.3717\Phi_2 - 0.3717\Phi_3 - 0.6015\Phi_4$$

On this basis, we shall obtain the following values for the partial bond orders between different carbon atoms:

$$P'_{12} = 0.3717 \times 0.6015 = 0.224$$

$$P'_{34} = 0.6015 \times 0.6015 = 0.362$$

$$P'_{3,4} = 0.6015 \times 0.3717 = 0.224$$

We observe that $P'_{2.3} > P'_{12} = P'_{3,4} > 0$, which gives a clue to understand that the contribution of electrons belongs to ψ_1, which gives strength to the central bond.

Similarly, for ψ_2 molecular orbital, we shall get

$$P'_{12} = 0.6015 \times 0.3717 = 0.224$$

$$P'_{2,3} = (+0.3717) \times (-0.3717) = -0.138$$

$$P'_{3,4} = (-0.3717) \times (0.6015) = 0.224$$

Thus, we can write $P'_{12} = P'_{3,4} > P'_{2,3} < 0$.

This indicates that electrons belonging to ψ_2 strengthen the end bonds but not the central bond. Hence, the total bond order in butadiene will be

$$P''_{12} = 2P_{12} + 2P'_{12} = 2 \times 0.224 + 2 \times 0.224$$

$$= 0.448 + 0.448$$

$$P''_{23} = 2P_{23} - 2P'_{23} = 2 \times 0.362 - 2 \times 0.138$$

$$= 0.724 - 0.276$$

$$= 0.448$$

where 2 denotes the number of electrons in each orbital. We can represent the bond orders as

$$H_2\underset{1}{C}\underline{0.896}\underset{2}{C}H\underline{0.448}\underset{3}{C}H\underline{0.896}\underset{4}{C}H_2$$

Sum of the π bond orders will be

$$S_x = 0.896 + 0.896 + 0.448$$

$$= 2.240$$

Thus, it can be said that the total π bond is greater than 2 by 0.240.

13.4.3 Benzene

Considering the bonding orbitals of benzene, we can estimate the partial bond order and π bond order as follows:

$$P^1_{12} = \frac{1}{\sqrt{6}} \cdot \frac{1}{\sqrt{6}} = \frac{1}{6}$$

$$P^2_{12} = \frac{2}{\sqrt{12}} \cdot \frac{2}{\sqrt{12}} = \frac{1}{6}$$

$$P_{12}^3 = 0$$

$$P_{23}^1 = \frac{1}{\sqrt{6}} \cdot \frac{1}{\sqrt{6}} = \frac{1}{6}$$

$$P_{23}^2 = \frac{1}{\sqrt{12}} \cdot \frac{1}{\sqrt{12}} = \frac{1}{12}$$

$$P_{23}^3 = \frac{1}{\sqrt{4}} \cdot \frac{1}{\sqrt{4}} = \frac{1}{4}$$

Now π bond order

$$P_{12} = 2\left(\frac{1}{6} + \frac{1}{6}\right) + 0 = \frac{2}{3} = 0.667$$

and $$P_{1,2} = P_{2,3} = P_{2,4} = P_{4,5}$$

$$P_{2,3} = 2\left(\frac{1}{6} - \frac{1}{12} + \frac{1}{4}\right) = \frac{2}{3} = 0.667$$

Thus, $P_{1,2} = P_{2,3} = P_{3,4} = P_{3,4}$.

It is, thus, clear that all the six bonds in benzene are equivalent.

13.5 Free Valence

The concept of free valence (F_r) was introduced by Coulson. It is defined as the difference between the maximum bond orders (N_{max}) around a carbon atom and the total estimated bond orders (N_{rs}), i.e.,

$$F_r = N_{max} - N_{rs} \tag{13.182}$$

$$= N_{max} - \sum P_{rs} \tag{13.183}$$

where F_r = free valence

N_{max} = maximum bond order

N_{rs} = sum of the orders of all bonds joining atoms r and s

ΣP_{rs} = total bond orders, where the summation extends over all the bonds about atom r

It should be noted that N_{max} depends on the types of bonds involved. Values of N_{max} have been given in Table 13.1.

TABLE 13.1

Values of N_{max}

Types of Carbon Atom	N_{max}
Primary	4
Secondary	$3+\sqrt{2}$
Tertiary	$3+\sqrt{3}$
π bond	1.732

13.5.1 Ethylene

In the case of ethylene, the value of $N_{rs} = 1 + 1 + 1 + 1 = 4$, and the value of N_{max} for unsaturated carbon compound is 4.732.

$$\therefore \quad F_r = 4.732 - 4 = 0.732$$

13.5.2 Butadiene

For the butadiene $\left(H_2\overset{1}{C}=\overset{2}{C}H-\overset{3}{C}H=\overset{4}{C}H_2\right)$ π bond structure, it has already been shown that the bond order between the atoms 1 and 2 is 0.896, between the atoms 2 and 3 is 0.448, and between the atoms 3 and 4 is 0.896.

$\therefore$ The free valence on carbon atom 1 will be

$$F_1 = 1.732 - 0.896 = 0.836$$

The value of F_r at carbon atoms 2 and 3 will be

$$F_2 \text{ or } F_3 = 1.732 - (0.836 + 0.488)$$

$$= 1.732 - 1.284 = 0.448$$

The free valence on atom 4 will be same as on carbon atom 1 and it will amount to 0.836.

13.5.3 Benzene

In the case of benzene, all six positions are equivalent. It is clear from the structure of benzene that each carbon atom is linked to two other carbon atoms by partial double bonds. For carbon atom 1, the total π bond order is given by $N_r = P_{12} + P_{16} = \frac{4}{3} = 1.333$, and hence, the free valence

$$F_1 = \sqrt{3} - 4/3 = (1.732 - 1.333) = 0.399 \approx 0.40$$

Finally, we can write that for benzene

$$F_1 = F_2 = F_3 = F_4 = F_5 = F_6 = 0.40$$

One thing is very important that the concept of free valence is able to provide the quantum mechanical interpretation of Thiele's theory of partial valence. It should also be noted that positions of high free valence are prone to attack by homolytic reagents. In other words, we can write that the magnitude of free valence at a given position of benzene or a molecule determines the extent of possibility of attack by free radicals on that particular position.

13.6 Generalised Treatment of the Hückel Molecular Orbital Theory to Open-Chain Conjugated System

The Hückel molecular orbital has been generalised for open-chain conjugated π-electron system.

Let us consider the non-cyclic determinant as follows:

$$\begin{vmatrix} x & 1 & 0 & 0 & 0 & ----- & 0 & 0 & 0 \\ 1 & x & 1 & 0 & 0 & ----- & 0 & 0 & 0 \\ 0 & 1 & x & 1 & 0 & ----- & 0 & 0 & 0 \\ 0 & 0 & 1 & x & 1 & ----- & 0 & 0 & 0 \\ - & - & - & - & - & ----- & - & - & - \\ 0 & 0 & 0 & 0 & 0 & 0 & 0 & 1 & x \end{vmatrix} = 0 \tag{13.184}$$

This is $r \times r$ determinant. It may be denoted by D_r. Excepting for the first and last rows, all but three elements in each row vanish. First, we want to prove that

$$D_r = xD_{r-1} - D_{r-2}$$

After expanding Eq. (13.184), the determinant takes the following form:

$$x\begin{vmatrix} (1) & (2) & (3) & (4) & ---- & (r-3) & (r-2) & (r-1) \\ x & 1 & 0 & 0 & ---- & 0 & 0 & 0 \\ 1 & x & 1 & 0 & ---- & 0 & 0 & 0 \\ 0 & 1 & x & 1 & ---- & 0 & 0 & 0 \\ - & - & - & - & ---- & - & - & - \\ 0 & 0 & 0 & 0 & ---- & 0 & 1 & x \end{vmatrix} = 0$$

$$-1\begin{vmatrix} 1 & 1 & 0 & 0 & ---- & 0 & 0 & 0 \\ 0 & x & 1 & 0 & ---- & 0 & 0 & 0 \\ 0 & 1 & x & 1 & ---- & 0 & 0 & 0 \\ 0 & 0 & 0 & 0 & ---- & 0 & 1 & x \end{vmatrix} = 0 \tag{13.185}$$

Now subtracting the first column from the second in the second determinant of order (r–1), we observe that Eq. (13.185) may be expressed as

$$\begin{aligned} xD_{r-1} - D_{r-2} &= 0 \\ \therefore \quad D_r &= xD_{r-1} - D_{r-2} \end{aligned} \tag{13.186}$$

If $x = 2\cos\theta = e^{i\theta} + e^{\ i\theta}$, then

$$D_r = Ae^{ir\theta} + Be^{-ir\theta}$$

where A and B depend on θ but independent of r.

$$\begin{aligned} \text{Now, } xD_r &= \left(e^{i\theta} + e^{-i\theta}\right)\left(Ae^{+ir\theta} + Be^{-ir\theta}\right) \\ &= Ae^{i(r+1)\theta} + Be^{-i(r+1)\theta} + Ae^{i(r-1)\theta} + Be^{-i(r-1)\theta} \end{aligned} \tag{13.187}$$

If $D_r = Ae^{ir\theta} + Be^{-ir\theta}$, then a similar relation holds true for all values of r. Therefore, from Eq. (13.187), one may write

$$\begin{aligned} xD_r &= D_{r+1} + D_{r-1} \\ \text{or} \quad D_{r+1} &= xD_r - D_{r-1} \end{aligned}$$

This relation is compatible or in harmony with Eq. (13.184), which we have proved. Therefore, the relation $D_r = Ae^{ir\theta} + Be^{-ir\theta}$ is valid for all the values of r.

Further, we shall prove that

$$A = \frac{e^{i\theta}}{e^{i\theta} - e^{-i\theta}} \text{ and } B = \frac{-e^{-i\theta}}{e^{i\theta} - e^{-i\theta}} \tag{13.188}$$

From Eq. (13.184), we can write

$$D_1 = |x| = x \text{ and } D_2 = \begin{vmatrix} x & 1 \\ 1 & x \end{vmatrix} = x^2 - 1$$

From equation $e^{ir\theta} + B^{-ir\theta}$, we have

$$D_1 = Ae^{i\theta} + Be^{-i\theta} \text{ and } D_2 = Ae^{i2\theta} + Be^{-i2\theta} \tag{13.189}$$

But from equation $2\cos\theta = e^{i\theta} + e^{-i\theta}$ and Eq. (13.189), we can write

$$xD_1 = \left(e^{i\theta} + e^{-i\theta}\right)\left(Ae^{i\theta} + Be^{-i\theta}\right)$$

$$= Ae^{i2\theta} + B + A + e^{-i2\theta} = D_2 + A + B$$

From above, $D_1 = x$ and $D_2 = x^2 - 1$

We find $x^2 = x^2 - 1 + A + B$

$$\therefore \quad A + B = 1 \tag{13.190}$$

We are aware that $Ae^{+i\theta} + Be^{-i\theta} = D_1 = x = e^{i\theta} + e^{-i\theta}$

$$\text{or} \quad Ae^{i\theta} + (1 - A)e^{-i\theta} = e^{i\theta} + e^{-i\theta}$$

$$\text{or} \quad A\left(e^{i\theta} - e^{-i\theta}\right) + e^{-i\theta} = e^{i\theta} + e^{-i\theta}$$

$$\text{or} \quad A\left(e^{i\theta} - e^{-i\theta}\right) + e^{-i\theta} = e^{i\theta} + e^{-i\theta}$$

$$\text{or} \quad A\left(e^{i\theta} - e^{-i\theta}\right) = e^{i\theta}$$

$$\therefore \quad A = \frac{e^{i\theta}}{e^{i\theta} - e^{-i\theta}}$$

Similarly, applying Eq. (13.190), we shall obtain

$$B = -\frac{e^{-i\theta}}{e^{i\theta} - e^{-i\theta}}$$

Putting the values of A and B in equation $D_r = Ae^{ir\theta} + Be^{-ir\theta}$, we get

$$D_r = \frac{e^{i\theta}}{e^{i\theta} - e^{-i\theta}} \cdot e^{ir\theta} - \frac{e^{-i\theta}}{e^{i\theta} - e^{-i\theta}} \cdot e^{-ir\theta}$$

$$= \frac{e^{i(r+1)\theta}}{e^{i\theta} - e^{-i\theta}} + \frac{e^{-i(r+1)\theta}}{e^{i\theta} - e^{-i\theta}}$$

$$= \frac{\cos(r+1)\theta + i\sin(r+1)\theta - \cos(r+1)\theta + i\sin(r+1)\theta}{\left(e^{i\theta} - e^{-i\theta}\right)}$$

$$= \frac{2i\sin(r+1)\theta}{(\cos\theta + i\sin\theta - \cos\theta + i\sin\theta)} \tag{13.191}$$

$$= \frac{2i\sin(r+1)\theta}{2i\sin\theta} = \frac{\sin(r+1)\theta}{\sin\theta}$$

$$\therefore D_r = \frac{\sin(r+1)\theta}{\sin\theta}$$

Now, we find the general solution of Eq. (13.184) by putting $D_r = 0$. Thus,

$$\frac{\sin(r+1)\theta}{\sin\theta} = 0$$

$$\text{or} \quad \sin(r+1)\theta = 0$$

$$\text{or} \quad (r+1)\theta = n\pi$$

$$\therefore \quad \theta = \frac{n\pi}{(r+1)}$$

$$\text{But} \quad x = 2\cos\theta$$

$$\therefore \quad x = 2\cos\frac{n\pi}{(r+1)}, \text{ where } n = 1,2,3\ldots \tag{13.192}$$

This general solution may be applied to obtain the root x of a secular equation and the energy levels, as done previously. In Eq. (13.192), r = the number of π electrons.

13.6.1 Ethylene

In ethylene ($> C = C <$) $r = 2\pi$ electrons. We have

$$x = 2\cos\frac{n\pi}{(r+1)}$$

Putting $n = 1$ and $r = 2$, we shall get

$$x = 2\cos\frac{\pi}{3} = 2\cos 60^\circ = 2\times\frac{1}{2} = 1$$

When $n = 2$ and $r = 2$,

$$x = 2\cos\frac{2\pi}{3} = 2\cos 120^\circ = -2\times\frac{1}{2} = -1$$

Therefore, the energy levels will be calculated as follows:

When $x = 1$, but $x = \dfrac{\alpha - E_1}{\beta}$

$$\frac{\alpha - E_1}{\beta} = 1 \text{ or } \alpha - E_1 = \beta$$

$$\therefore \quad E_1 = \alpha - \beta \qquad [\text{ABMO}]$$

When $x = -1$, then $\dfrac{\alpha - E_2}{\beta} = -1$

$$\text{or } E_2 = \alpha + \beta[\text{BMO}]$$

13.6.2 Butadiene

First of all, we shall find the roots of the secular equation of butadiene $\left(\overset{1}{C}H_2 = \overset{2}{C}H - \overset{3}{C}H = \overset{4}{C}H_2\right)$ with the formula

$$x = 2\cos\frac{n\pi}{(r+1)} \quad [\text{Putting } n = 1]$$

$$x_1 = 2\cos\frac{\pi}{4+1} = 2\cos\frac{\pi}{5} \qquad [\because r = 4 \text{ for butadiene} = \text{no. of } \pi \text{ electrons}]$$

$$= 2\cos 36^\circ = 2 \times 0.809 = 1.618$$

Putting $n = 2$ and $r = 4$, we get

$$x_2 = 2\cos\frac{2\pi}{4+1} = 2\cos\frac{2\pi}{5} = 2\cos 72^\circ = 2 \times 0.309 = 0.618$$

Putting $n = 3$ and $r = 4$, we get

$$x_3 = 2\cos\frac{n\pi}{(r+1)} = 2\cos\frac{3\pi}{5} = 2\cos 108^\circ = 2(-0.309) = -0.618$$

Also by putting $n = 4$ and $r = 4$, we obtain the fourth root, i.e.,

$$x_4 = 2\cos\frac{n\pi}{(r+1)} = 2\cos\frac{4\pi}{5}\pi = 2\cos 144^\circ = 2(-0.809) = -1.618$$

Thus, the four energy levels will be written as

$$E_1 = \alpha + 1.618\beta$$

$$E_2 = \alpha + 0.618\beta$$

$$E_3 = \alpha - 0.618\beta$$

$$\text{and} \quad E_4 = \alpha - 1.618\beta$$

From these, the energy level diagram can be constructed as done earlier.

13.7 Generalised Treatment of the Hückel Molecular Orbital Theory to Cyclic Polyenes

Hückel's treatment of cyclic polyenes satisfactorily explains some peculiar aromatic properties. Here, we consider the $2p_z$ atomic orbitals (i.e., p_π orbitals) of the different carbon atoms of the cyclic polyene systems as the basis orbitals. The one-electron molecular orbital is expressed as

$$\psi_k = \sum_{j=1}^{r} C_{kj}\Phi_j \tag{13.193}$$

where r = number of carbon atoms in the molecule, and each atom contributes one p_π orbital for the formation of molecular orbital.

The expected energies of molecular orbitals are found by solving $r \times r$ secular determinant. Hückel made the following approximations for solving $r \times r$ determinant:

- All overlap integral $S_{ij} = 0$ but $S_{ii} = 1$
- $H_{ij} = 0$ if i and j are not bonded but
  $H_{ij} = \beta$ if i and j are bonded
- $H_{ii} = \alpha$ = the energy of an electron in a p_π orbital of carbon atom.

Consider the following secular determinant:

$$\begin{vmatrix} \alpha - E & \beta & 0 & 0 & 0 & --- & 0 & \beta \\ \beta & \alpha - E & \beta & 0 & 0 & --- & 0 & 0 \\ 0 & \beta & \alpha - E & \beta & 0 & --- & 0 & 0 \\ 0 & 0 & \beta & \alpha - E & \beta & --- & 0 & 0 \\ --- & --- & --- & --- & --- & --- & --- & \\ 0 & 0 & 0 & 0 & 0 & --- & \beta & \alpha - E \end{vmatrix} = 0 \tag{13.194}$$

Dividing throughout by β and putting $\frac{\alpha - E}{\beta} = x$, the above determinant takes the following form:

$$\begin{vmatrix} x & 1 & 0 & 0 & 0 & ----- & 0 & 1 \\ 1 & x & 1 & 0 & 0 & ----- & 0 & 0 \\ 0 & 1 & x & 1 & 0 & ----- & 0 & 0 \\ 0 & 0 & 1 & x & 1 & ----- & 0 & 0 \\ --- & --- & --- & --- & --- & --- & --- & --- \\ 1 & 0 & 0 & 0 & 0 & r-1 & 1 & x \end{vmatrix} = 0 \tag{13.195}$$

It should be kept in mind that this determinant is different from the non-cyclic determinant D_r given in Section 13.6. Here, we shall denote C_r as the cyclic determinant. It is to be noted that in all the rows, all but three elements vanish. A cyclic determinant C_r of the order r is formed from r quantities, $x_1, x_2, \ldots, x_r$ by arranging them in different rows as follows:

$$\begin{vmatrix} x_1 & x_2 & x_3 & x_4 & ----- & x_{r-1} & x_r \\ x_r & x_1 & x_2 & x_3 & ----- & x_{r-2} & x_{r-1} \\ x_{r-1} & x_r & x_1 & x_2 & ----- & & x_{r-2} \\ --- & --- & --- & --- & --- & --- & --- \\ x_2 & x_3 & x_4 & x_5 & ----- & x_r & x_1 \end{vmatrix} \tag{13.196}$$

Equation (13.195) can be expressed as

$$C_1 \cdot 0 + C_2 \cdot 0 + C_3 \cdot 0 + \cdots + C_{n-1} \cdot 1 + C_n \cdot x + C_{n+1} \cdot 1 + \cdots + C_r \cdot 0 = 0$$

or in simple way

$$C_{n-1} + C_n \cdot x + C_{n+1} = 0 \tag{13.197}$$

where n–1 and $n + 1$ are successive atoms of the ring. In a ring of r carbon atoms, the nth atom will also be the $(r + n)$th atom such that

$$C_n = C_{r+n} \tag{13.198}$$

Therefore, we have to search a solution periodic in *r*. For this, let us suppose that

$$C_n = e^{in\theta} \tag{13.199}$$

$$\text{But}\, e^{in\theta} = e^{i(n+r)\theta} = e^{ir\theta} \cdot e^{in\theta} \qquad \left[\because\ C_n = C_{r+n}\right]$$

$$\therefore\ e^{ir\theta} = 1 \tag{13.200}$$

$$\text{But}\, e^{ir\theta} = \cos r\theta + i\sin r\theta$$

$$\therefore\ \cos r\theta + i\sin r\theta = 1 \tag{13.201}$$

We know that cos $(\pm 2k\pi) = 1$, where $k = 0,1,2,\ldots$

From this fact, the real part of Eq. (13.201), we obtain

$$\cos r\theta = 1 \text{ or } r\theta = 2k\pi, \text{ where}, k = 0_1, \pm 1, \pm 2\ldots$$

$$\therefore\ \theta = \frac{2k\pi}{r}, \text{ where } k = 0, \pm 1, \pm 2\ldots \tag{13.202}$$

Putting Eq. (13.199) in Eq. (13.197), we get

$$e^{i(n-1)\theta} + e^{in\theta}x + e^{i(n-1)\theta} = 0$$

Dividing this equation by $e^{in\theta}$, we shall obtain

$$\begin{aligned} & e^{-i\theta} + x + e^{i\theta} = 0 \\ \text{or}\quad & x = -\left[e^{-i\theta} + e^{i\theta}\right] = -2\cos\theta \end{aligned} \tag{13.203}$$

Putting the value of $\theta =$ in Eq. (13.203), we shall get

$$x = -2\cos\frac{k\pi}{r} \tag{13.204}$$

This equation will yield the possible roots of the determinant mentioned in Eq. (13.195), and consequently, we shall be able to find the values of energy, and hence, the energy level diagram can be constructed.

13.7.1 Cyclopropenyl Radical

The cyclopropenyl radical has the following structural formula:

CH
|| •CH
CH

It contains three π electrons and $r = 3$. Applying the formula for obtaining roots, i.e., $x = -2\cos\frac{2k\pi}{r}$, we shall put the values of *k* and *r*.

When $k = 0$, we have

$$x_1 = -2\cos\frac{2 \times 0 \cdot \pi}{r} = -2\cos 0 = -2$$

$$\therefore\ \frac{\alpha - E_1}{\beta} = -2$$

or, $\alpha - E_1 = \alpha + 2\beta$.

When $k = +1$,

$$x_2 = -2\cos\frac{2\pi}{3} = -2\cos 120^{\circ} = -2\times\left(-\frac{1}{2}\right) = +1$$

$$\therefore \quad \frac{\alpha - E_2}{\beta} = +1$$

$$\text{or} \quad \alpha - E_2 = \beta$$

$$\therefore \quad E_2 = \alpha - \beta$$

13.7.2 Cyclobutadiene

The structural formula for cyclobutadiene is given by

□ or
$^{1}CH—C^{4}H$
$\| \quad |$
$^{2}CH—C_{3}H$

It has 4π electrons, and hence, $r = 4$.

When $k = 0$, we have

$$x_1 = -2\cos\frac{2k\pi}{r} = -2\cos\frac{2\times 0\cdot\pi}{4} = -2\cos 0 = -2$$

$$\therefore \quad -2 = \frac{\alpha - E_1}{\beta} \text{ or } \alpha - E_1 = -2\beta$$

or $E_1 = \alpha + 2\beta$

When $k = 1$,

$$x_2 = -2\cos\frac{2\times 1\cdot\pi}{4} = -2\cos\frac{2\pi}{4} = -2\cos 90 = 0$$

$$\therefore \quad 0 = \frac{\alpha - E_2}{\beta}$$

$$\text{or} \quad \alpha - E_2 = 0$$

$$\therefore \quad E_2 = \alpha$$

When $k = 2$,

$$x_3 = -2\cos\frac{2.2\pi}{4} = -2\cos\frac{4\pi}{4} = -2\cos\pi = +2$$

$$\therefore \quad \frac{\alpha - E_3}{\beta} = x \quad \text{or} \quad 2 = \frac{\alpha - E_3}{\beta} \quad \text{or} \quad \alpha - E_3 = +2\beta$$

or $E_3 = \alpha - 2\beta$.

13.7.3 Cyclopentadienyl Radical

The structural formula of cyclopentadienyl radical is given by

$^{2}CH=CH^{1}$
$|$ $\quad \cdot CH^{5}$
$_{3}CH=CH_{4}$

It contains 5π electrons. We have $r = 5$. Now putting different values of k and also $r = 5$ in the equation $x = -2\cos\frac{2k\pi}{r}$, we shall get different values of the roots.

When $k = 0$,

$$x_1 = -2\cos\frac{2\times 0\pi}{5} = -2\cos 0 = -2$$

$$\therefore \quad \frac{\alpha - E_1}{\beta} = -2 \quad \text{or} \quad \alpha - E_1 = -2\beta$$

$$\therefore \quad E_1 = \alpha + 2\beta$$

When $k = 1$,

$$x_2 = -2\cos\frac{2\times 1\times \pi}{5} = -2\cos\frac{2\pi}{5} = -2\cos 72 = -2\times 0.309 = -0.618$$

$$\therefore \quad \frac{\alpha - E_2}{\beta} = -0.618 \quad \text{or} \quad \alpha - E_2 = -0.618\beta$$

$$\therefore \quad E_2 = \alpha + 0.618\beta$$

When $k = +2$,

$$x_3 = -2\cos\frac{2\times 2\cdot \pi}{5} = -2\cos\frac{4\pi}{5} = -2\cos 144 = -2\times(-0.809) = +1.618$$

$$\therefore \quad \frac{\alpha - E_3}{\beta} = +1.618 \quad \text{or} \quad \alpha - E_3 = +1.618\beta$$

$$\therefore \quad E_3 = \alpha - 1.618\beta$$

13.7.4 Benzene

Benzene contains 6π electrons, i.e., $r = 6$. Putting different values of k, we shall get six roots.

When $k = 0$,

$$x_1 = -2\cos\frac{2k\pi}{r} = -2\cos\frac{2\times 0\times \pi}{6}$$

$$\therefore \quad \frac{\alpha - E_1}{\beta} = -2 \text{ or } \alpha - E_1 = -2\beta$$

$$\therefore \quad E_1 = \alpha + 2\beta$$

When $k = 1$,

$$x_2 = -2\cos\frac{2\times 1\times \pi}{6} = -2\cos\frac{2\pi}{6} = -2\cos 60^\circ = -1$$

$$\therefore \quad \frac{\alpha - E_2}{\beta} = -1 \quad \text{or} \quad \alpha - E_2 = -\beta$$

$$\therefore \quad E_2 = \alpha + \beta$$

When $k = 2$, then

$$x_3 = -2\cos\frac{2\times 2\pi}{6} = -2\cos\frac{4\pi}{6} = -2\cos 120^\circ = 1$$

$$\therefore \quad \frac{\alpha - E_3}{\beta} = 1 \quad \text{or} \quad \alpha - E_3 = \beta$$

$$\therefore \quad E_3 = \alpha - \beta$$

When $k = 3$, then

$$x_4 = -2\cos\frac{2\times 3\times \pi}{6} = -2\cos\pi = +2$$

$$\therefore \quad \frac{\alpha - E_4}{\beta} = 2 \quad \text{or} \quad \alpha - E_4 = 2\beta$$

$$\therefore \quad E_4 = \alpha - 2\beta$$

Thus, we can find the energy values of different cyclic polyenes.

13.8 Extended Hückel Theory

The extended Hückel theory (abbreviated as EHT) was pioneered by Wolfsberg and Helmholz (1952) and later developed by Hoffmann (1963, 1964). Actually, this theory is not restricted to the electrons in π orbitals. It also considers the molecules that are not planar and that do not essentially have conjugated systems of bonds.

The EHT starts with the approximation that all the valence shell electrons or valence electrons are involved in the treatment, but inner shell electrons are treated separately. The fact is that a minimal set of basis functions is utilised, which consists of all atomic orbitals in the valence shell of each atom. Further, the basis set is not assumed to be orthogonal, and all overlap integrals are estimated on an analytical basis but electron-electron repulsive energy is neglected. The inclusion of overlap integral imparts some kind of dependence on geometry to the EH method. For example, for a molecule like ethane (C_2H_6), the EHT basis set contains $2s$, $2p_x$, $2p_y$, and $2p_z$ atomic orbitals of the two-carbon atom and the 1s atomic orbitals of six H atoms.

The process is based on the variational principle. The Hamiltonian for the valence electron is supposed to be the sum of effective one-electron Hamiltonians, which are not expressly denoted.

$$H_{\text{Val}} = \sum_i H_{\text{eff}}(i) \tag{13.205}$$

where $H_{\text{eff}}(\text{i})$ is not clearly specified. The molecular orbitals are approximated as a linear combination of the valence atomic orbitals Φ_r of the atoms and mathematically expressed as

$$\Psi_r = \sum_r C_{ri}\Phi_r \tag{13.206}$$

The atomic orbitals used are generally Slater-type orbitals (STOs), which include fixed orbital exponents found from Slater's rules. Equation (13.205) is the simplified version of Hamiltonian, and it can be separated into many one-electron problems.

$$H_{\text{eff}}(i)\Psi_i = e_i\Psi_i$$

$$\text{and} \quad E_{\text{val}} = \sum_i e_i \tag{13.207}$$

Application of variation principle to the linear trial function mentioned in Eq. (13.206) will yield secular equation and also equations for molecular orbital coefficients.

$$\det\left(H_{rs} - e_i s_{rs}\right) = 0 \tag{13.208}$$

$$\text{and} \quad \sum_s \left[\left(H_{rs}^{\text{eff}} - e_i s_{rs}\right) C_{si}\right] = 0, \text{ where, } r = 1,2,3,\ldots \tag{13.209}$$

All these are at par with the simple HMO theory. The elements with $r = s$ (the diagonal elements) are set equal to valence state ionisation energy (VSIE) of the considered orbital; Wolfsberg and Helmholz tried to approximate the off-diagonal elements by taking the mean of the VSIE of each orbital times a fixed constant times the overlap integral. The equation relating to this is given by

$$H_{rs}^{\text{eff}} = \frac{1}{2} k\left(H_{rr}^{\text{eff}} + H_{ss}^{\text{eff}}\right) s_{rs} \tag{13.210}$$

The value of the constant k varies from 1 to 3 but a value of 1.75 is assumed to be common. Ballhausen and Gray (1964) used geometric mean instead of arithmetic mean in taking the approximate value of Hrs.

We know that since H_{rr}^{eff} and H_{SS}^{eff} are found to be negative, H_{rs}^{eff} will be negative. When the H_{rs}^{eff} and s_{rs} integrals are estimated or computed, the secular equation will be solved. It must be kept in mind that the EHT is employed in finding the conformation of the lowest energy by repeated estimations with different conformations. These computations are sometimes found to be more sophisticated.

13.8.1 Hetero Atom Substitutions

The EHT also allows the substitution of hetero atom in the conjugated system. The Hückel method requires modification after substitution of hetero atom. It is modified by putting different values of α_x, β_{cx}, and β_{xx}, where X is the hetero atom in the conjugated system. By carrying out such modifications, different values of orbital energies are obtained. The linear expansion for π orbital now consists of $2p_z$ orbital of the hetero atom X in addition to the $2p_z$ orbital of the carbon atoms. It must be kept in mind that the most significant difference is generated by the presence of $2p_z$ electrons of the hetero atom. The nitrogen atom remains in sp^2 hybridised state, and its $2p_z$ orbital contains two electrons and three σ orbitals. Oxygen in the similar state consists of two $2p_z$ electrons and two σ orbitals, but a pair of electrons exists in the non-bonding orbital, which is oriented in the direction of third sp^2 hybrid orbital.

It is to be noted that the value of $(\beta_{CN} - \beta)$ is very small, and due to this, a very small second order effect is introduced on the charge distribution. Therefore, we can employ the same parameter β for conjugated molecules having one or two carbon atoms substituted by N or nitrogen containing groups. The parameter αx is approximately expressed by the following equation:

$$\alpha_x = \alpha + k_x \beta \tag{13.211}$$

In the case of N, k_x has been taken equal to 0.5.

13.8.2 General Improvement

The Hückel computation can be improved if a non-zero overlap integral for two p_z orbitals is brought as neighbouring atoms, and we can express the integral as

$$S_{i,i\pm1} = S \neq 0 \tag{13.212}$$

We can obtain the following integral if we express each $2p_z$ orbital by a single STO.

$$S = \frac{\xi_{2p}^5}{\pi} \int z^2 \exp\left[-\xi_{2p}\left|r - \frac{R}{2}\right| - \xi_{2p}\left|r + \frac{R}{2}\right|\right] d\tau \tag{13.213}$$

$$= \frac{1}{\pi} \int z^2 \exp^- \left[-\left| r - \xi_{2p} \frac{R}{2} \right| - \left| r + \xi_{2p} \frac{R}{2} \right| \right] d\tau \qquad (13.214)$$

The exponent of the STO has been taken as = 3.25. The absolute magnitude of the argument of the exponential function in the integral in Eq. (13.212) is found to be nearly $\xi_{2p}R$, and this is the least value, which for $R \sim 2.5$ a.u. approximates to be 8.1. It must be noted that the argument increases as z increases. Thus, it is important to note that the value of S is found to be small for the π electron operation; and results obtained on the basis of Eq. (13.212) have slight difference from those computed with $S_{i,\, i\pm 1} = 0$. The integral $S \rightarrow 0.1$ for σ orbitals.

We are going to cite some examples in which EHT is applied.

13.8.3 Extended Hückel Theory Applied to Pyrrole

Pyrrole is a heterocyclic compound consisting of the hetero atom N. There is a lone pair of electrons on the N atom. It should be noted that N in pyrrole donates both the lone pairs of electrons to the ring. This factor makes necessary changes in the value of coulomb integral α_N associated with the hetero atom. It is well known that the C–N bond is more polar than the C–C bond, and this is why the electron is more stable in the C–N bond than in the C–C bond. Consequently, the resonance integral (β) also needs to be changed. Pauling and Wheland have proposed a good relation, which shows good agreement with experimental value, i.e.,

$$\alpha_N(2) = \alpha + 1.5\beta \qquad (13.215)$$

where $\alpha_N(2)$ = Coulomb integral associated with nitrogen atom and (2) indicates that N atom donates two electrons to the pyrrole ring.

This structural formula of pyrrole is given as

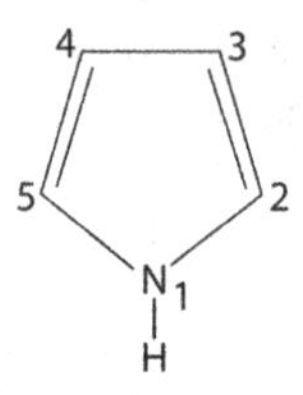

The nitrogen and carbon atoms in pyrrole have been numbered from 1 to 5, but starting from N atom. Let us suppose that the HMO function for the pyrrole is represented by

$$\psi = \sum_{i=1}^{5} C_i \phi_i \qquad (13.216)$$

where the terms have their usual significance. The secular equations for pyrrole can be written as

$$\left.\begin{aligned} &(x+1.5)C_1 + C_2 + C_5 = 0 \\ &xC_2 + C_3 + C_1 = 0 \\ &xC_3 + C_2 + C_4 = 0 \\ &xC_4 + C_3 + C_5 = 0 \\ &xC_5 + C_1 + C_4 = 0 \end{aligned}\right\} \qquad (13.217)$$

The secular determinant can be expressed on the basis of Eq. (13.217) as

$$\begin{vmatrix} x+1.5 & 1 & 0 & 0 & 1 \\ 1 & x & 1 & 0 & 0 \\ 0 & 1 & x & 1 & 0 \\ 0 & 0 & 1 & x & 1 \\ 1 & 0 & 0 & 1 & x \end{vmatrix} = 0 \tag{13.218}$$

where $x = \dfrac{\alpha - E}{\beta}$

The Laplace expansion of the above determinant gives

$$(x+1.5)\begin{vmatrix} x & 1 & 0 & 0 \\ 1 & x & 1 & 0 \\ 0 & 1 & x & 1 \\ 0 & 0 & 1 & x \end{vmatrix} - 1\begin{vmatrix} 1 & 1 & 0 & 0 \\ 0 & x & 1 & 0 \\ 0 & 1 & x & 1 \\ 1 & 0 & 1 & x \end{vmatrix} + 1\begin{vmatrix} 1 & x & 1 & 0 \\ 0 & 1 & x & 1 \\ 0 & 0 & 1 & x \\ 1 & 0 & 0 & 1 \end{vmatrix} = 0$$

$$\Rightarrow (x+1.5)(x)\begin{vmatrix} x & 1 & 0 \\ 1 & x & 1 \\ 0 & 1 & x \end{vmatrix} - x\begin{vmatrix} 1 & 1 & 0 \\ 0 & x & 1 \\ 0 & 1 & x \end{vmatrix} - 1\begin{vmatrix} x & 1 & 0 \\ 1 & x & 1 \\ 0 & 1 & x \end{vmatrix} + 1\begin{vmatrix} 0 & 1 & 0 \\ 0 & x & 1 \\ 1 & 1 & x \end{vmatrix}$$

$$+1\times 1\begin{vmatrix} 1 & x & 1 \\ 0 & 1 & x \\ 0 & 0 & 1 \end{vmatrix} - x\begin{vmatrix} 0 & x & 1 \\ 0 & 1 & x \\ 1 & 0 & 1 \end{vmatrix} + 1\begin{vmatrix} 0 & 1 & 1 \\ 0 & 0 & x \\ 1 & 0 & 1 \end{vmatrix} = 0$$

Expansion of all these determinants gives the following polynomial:

$$x^5 + 1.5x^4 - 5x^3 - 4.5x^2 + 5x + 3.5 = 0 \tag{13.219}$$

The roots of the above polynomials are given as follows:

$$x = -2.55, -1.15, -0.618, +1.20 \text{ and } +1.62$$

Therefore, the energy levels may be expressed as

$$\left.\begin{aligned} x_1 &= -2.55, & E_1 &= \alpha + 2.55\beta \\ x_2 &= -1.15, & E_2 &= \alpha + 1.15\beta \\ x_3 &= -0.618, & E_3 &= \alpha + 0.618\beta \\ x_4 &= +1.20, & E_4 &= \alpha - 1.20\beta \\ x_5 &= +1.62, & E_5 &= \alpha - 1.62\beta \end{aligned}\right\} \tag{13.220}$$

The energy level diagram can be constructed as given in Figure 13.18, which illustrates the filling up molecular orbitals.

13.8.4 Delocalisation Energy of Pyrrole

The total π electron energy is given by

$$\begin{aligned} E_\pi &= 2(\alpha + 2.55\beta) + 2(\alpha + 1.15\beta) + 2(\alpha + 0.618\beta) = 6\alpha + 8.62\beta \\ \therefore \quad (E_\pi)_{\text{deloc}} &= \left[2(\alpha + 1.5\beta) + 4(\alpha + \beta) - (6\alpha + 8.62\beta)\right] \\ &= -1.62\beta \end{aligned} \tag{13.221}$$

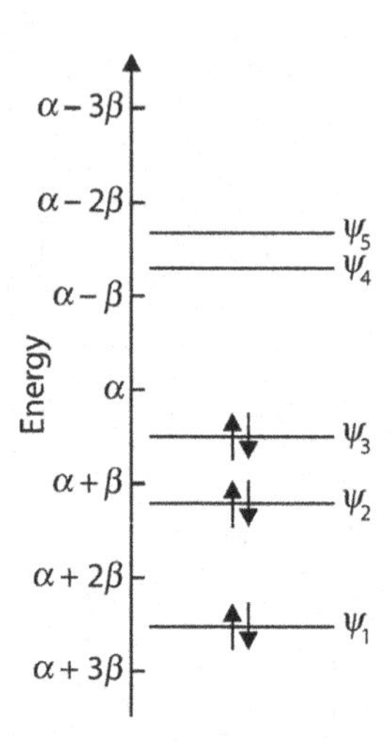

FIGURE 13.18 The energy level diagram of pyrrole.

13.8.5 Hückel Molecular Orbital Coefficients and Molecular Orbitals

The HMO coefficients can be found out by Gaussian elimination and the normalisation condition. The reader should find out the values of c_1, c_2, c_3, c_4, and c_5 using the Gaussian elimination method. The values of the molecular orbitals will be

$$\psi_1 = 0.749\phi_1 + 0.393(\phi_2 + \phi_5) + 0.254(\phi_3 + \phi_4)$$

$$\psi_2 = 0.503\phi_1 + 0.089(\phi_2 + \phi_5) - 0.605(\phi_3 + \phi_4)$$

$$\psi_3 = 0.602(\phi_2 - \phi_5) + 0.372(\phi_3 - \phi_4)$$

$$\psi_4 = 0.430\phi_1 - 0.580(\phi_2 + \phi_5) + 0.267(\phi_3 + \phi_4)$$

and $$\psi_5 = 0.372(\phi_2 - \phi_5) - 0.602(\phi_3 - \phi_4)$$

13.8.6 Pyridine

Pyridine is a simple heterocyclic compound having 6π electrons. It most closely resembles benzene in structure expect that a nitrogen atom replaces a C–H group. Since the value of electronegativity of 'N' (x = 3.5) is more than that of 'C' (x = 3.0), the values of the coulomb's integral α and the resonance integral β need modification. Since the electronegativity difference between 'N' and 'C' is 0.5, α will be modified to (α+0.5).

The modification in the resonance integral will be done on the consideration of the bond parameter, k. The bond parameter, k, is expressed as

$$k = \mathrm{E(C=N)} - \mathrm{E(C-N)}/\mathrm{E(C=C)} - \mathrm{E(C-C)}$$

where E's represent the empirical bond energies. The bond energies of C=N, C–N, C=C, and C–C are 612, 289, 612, and 345 KJ/mol, respectively. Putting these values in the above equation, we get

$$K = (612 - 289)\mathrm{KJ/mol}/(612 - 345)\mathrm{KJ/mol} = 1.21$$

which is rounded off to 1.2. Therefore, the resonance integral β will be modified as 1.2 β.

The structural formula of pyridine is expressed as

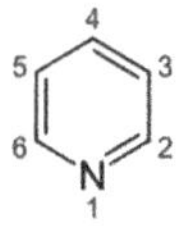

The numbering of elements in pyridine has been done from 1 to 6 starting from nitrogen (N). Now, we are in a position to write down the secular equation for pyridine as follows:

$$(X+0.5)C_1 + 1.2\ C_2 + 1.2\ C_6 = 0$$

$$C_2X + 1.2C_1 + C_3 = 0$$

$$C_3X + C_4 + C_2 = 0$$

$$C_4X + C_3 + C_5 = 0$$

$$C_5X + C_4 + C_6 = 0$$

$$C_6X + 1.2C_1 + C_5 = 0$$

On this basis, the secular determinant for pyridine may be written as follows:

$$\begin{vmatrix} x+0.5 & 1.2 & 0 & 0 & 0 & 1.2 \\ 1.2 & x & 1 & 0 & 0 & 0 \\ 0 & 1 & x & 1 & 0 & 0 \\ 0 & 0 & 1 & x & 1 & 0 \\ 0 & 0 & 0 & 1 & x & 1 \end{vmatrix} = 0$$

Since the manual solution of this 6 × 6 determinant is tedious, the polynomial of this determinant has been found out by Gaussian elimination or by matrix calculator, which is

$$X^6 + \tfrac{1}{2}x^5 - 172/25\,x^4 - 2x^3 + 291/25\,x^2 + 3/2\,x - 144/25 = 0$$

$$Or\ 50X^6 + 25X^5 - 344\,X^4 - 100X^3 + 582X^2 + 75X - 288 = 0$$

and the six roots of this polynomial have been determined by the above same procedure. The six roots are shown as follows:

1.0000, – 1.0000, – 2.2300, – 1.2390, 0.9850, and 2.0540

Hence, the energy levels are

$$X = -2.2300\text{: } E_1 = +2.2300\,\beta \cdot \left[\ \alpha - \frac{E}{\beta} = x\right]$$

$$X = -1.2390\ :\ E_2 = \alpha + 1.2390\ \beta$$

$$X = -1.0000\ :\ E_3 = \alpha + 1.0000\beta$$

$$X = 2.0540\ :\ E_4 = \alpha - 2.0540\ \beta$$

$$X = 1.0000 \ : \ E_5 = \alpha - 1.0000\ \beta$$

$$X = 0.9850 \ : \ E_6 = \alpha - 0.9850\ \beta$$

The energy level diagram of pyridine and filling up of the MOs are shown as follows:

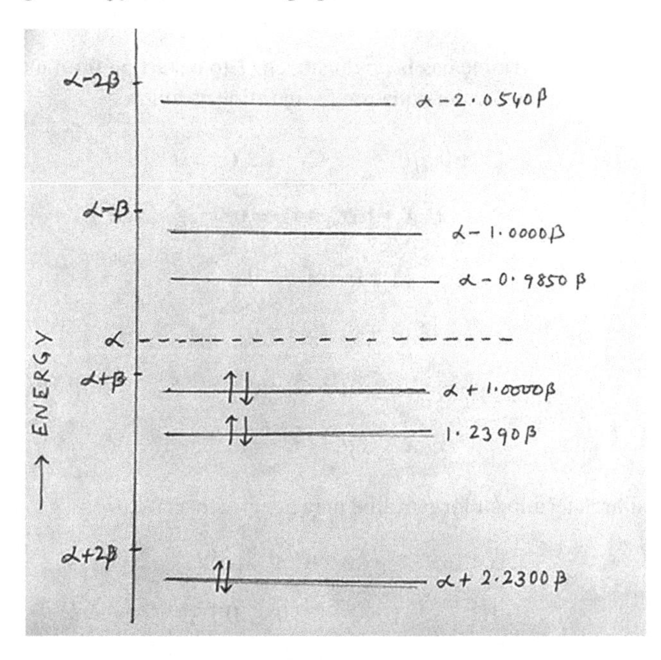

There are 6π electrons in pyridine in the lowest three molecular orbitals. The total π electronic energy in pyridine is

$$E\pi = 2(\alpha + 2.2300\beta) + 2(\alpha + 1.2390\beta) + 2(\alpha + 1.0000\beta)$$

$$= 6\alpha + 8.938\beta$$

Therefore, the delocalisation energy in pyridine is

$$\text{DE or } (E\pi) = E\pi(\text{pyridine}) - 3E\pi(C_2H_4)$$

$$= (6\alpha + 8.938\beta) - 3(2\alpha + 2\beta)$$

$$= 2.938\beta$$

Resonance energy of benzene is 2β and that of pyridine is 2.938β so pyridine has 0.938 *β more resonance energy*. Therefore, pyridine is more stable than benzene. All the six MOs of pyridine are non-degenerate.

13.8.7 Hückel Molecular Orbital Coefficients and Molecular Orbitals

In the case of pyridine, there are six secular equations consisting of six HMO coefficients, i.e., $C_1(C_N)$, C_2, C_3, C_4, C_5, and C_6. It is very difficult to handle the secular equations for getting the values of C_1, C_2, C_3, ...,C_6. Therefore, we have found the HMO coefficients using Gaussian elimination method.

It is important to note that as per symmetric reflection in the plane, one can write $C_2 = C_6$ and $C_3 = C_5$. However, according to the antisymmetric reflection in the plane, the carbon atoms 2, 3, 5, and 6 may be considered equivalent due the same value of α, then one may write:

$$C_2 = C_3 = -C_5 = -C_6 \text{ and}$$

$$C_2 = -C_3 = C_5 = -C_6$$

On the basis of Gaussian elimination method, the HMO coefficients corresponding to orbital energies for pyridine may be seen ahead.

HMO Coefficients of Pyridine

x_i	C_1	C_2	C_3	C_4	C_5	C_6
$x_1 = -2.230$	0.589	0.442	0.309	0.269	0.309	0.442
$x_2 = -1.239$	–0.483	–0.149	0.396	0.639	0.396	–0.149
$x_3 = -1.000$	0.000	–0.500	–0.500	0.000	0.500	0.500
$x_4 = 0.985$	0.506	–0.313	–0.299	0.607	–0.299	–0.313
$x_5 = 1.000$	0.000	–0.500	0.500	0.000	–0.500	0.500
$x_6 = 2.054$	–0.404	0.430	–0.398	0.388	–0.398	0.430

Wave functions in order of increasing energy have been written as follows:

$x_1 = -2.230$	$\Psi 1 = 0.589\ \varphi_1 + 0.442\ \varphi_2 + 0.309\ \varphi_3 + 0.269\ \varphi_4 + 0.309\ \varphi_5 + 0.442\ \varphi_6$
$x_2 = -1.239$	$\Psi 2 = -0.483\ \varphi_1 - 0.149(\varphi_2 + \varphi_6) + 0.396\ (\varphi_3 + \varphi_5) + 0.639\ \varphi_4$
$x_3 = -1.000$	$\Psi 3 = -0.500\ (\varphi_2 + \varphi_3) + 0.500\ (\varphi_5 + \varphi_6)$
$x_4 = 0.985$	$\Psi 4 = 0.506\ \varphi_1 - 0.313\ (\varphi_2 + \varphi_6) - 0.299\ (\varphi_3 + \varphi_5) + 0.607\ \varphi_4$
$x_5 = 1.000$	$\Psi 5 = -0.500\ (\varphi_2 + \varphi_5) + 0.500\ (\varphi_3 + \varphi_6)$
$x_6 = 2.054$	$\Psi 6 = -0.404\ \varphi_1 + 0.430\ (\varphi_2 + \varphi_6) - 0.398\ (\varphi_3 + \varphi_5) + 0.388\ \varphi_4$

13.8.8 Electron Density

The electron density at 'N' atom in pyridine may be given by

$$\begin{aligned} q_1 &= 2\ (C^2{}_{11} + C^2{}_{12} + C^2{}_{13}) \\ &= 2\ [(0.589)^2 + 2(-0.483)^2 + 2(0.0)^2] \\ &= 2\ (0.347 + 0.233) \\ &= (0.694 + 0.466) \\ &= 1.16 \end{aligned}$$

Similarly, the electron density at carbon 2 may be written as

$$\begin{aligned} Q_2 &= 2\ (C^2{}_{21} + C^2{}_{22} + C^2{}_{23}) \\ &= 2[(0.442)^2 + (-0.149)^2 + (0.500)^2] \\ &= 2(0.195364) + 2(0.0222) + 2(0.25) \\ &= (0.390728 + 0.0444 + 0.5000) \\ Q_2 &= q_6 = 0.835 \end{aligned}$$

Similarly, the reader is advised to calculate the electron density at other carbon atoms of pyridine. The other values of electron densities at other carbon atoms may be found to be $q_3 = q_5 \approx 1.004$. $q_4 = 0.961$.

13.8.9 Bond Order

The partial bond orders between different atoms in pyridine are found as follows:

$$P_{12} = 0.589 \times 0.442 = 0.260$$
$$P_{13} = 0.442 \times 0.309 = 0.137$$
$$P_{34} = 0.309 \times 0.269 = 0.083$$

Similarly, one can find the other bond orders.

13.8.10 HMO Treatment to Naphthalene

Naphthalene is an aromatic hydrocarbon having 10 π electrons. In this aromatic hydrocarbon, two benzene rings are fused. The structural formula of naphthalene and the numbering of its carbon atoms have been shown as follows:

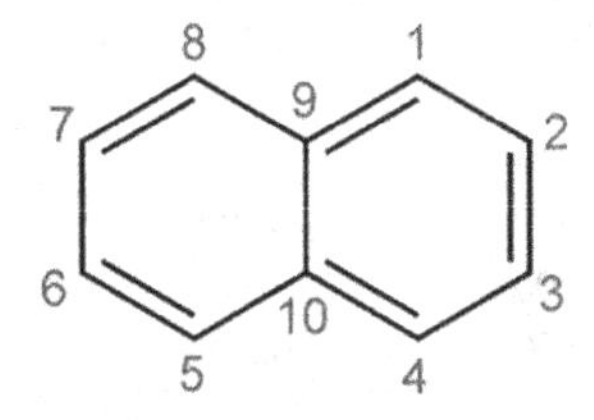

Now, we are in a position to write down the secular equations of naphthalene and they are expressed as follows:

$$
\begin{aligned}
&C_1x + C_8 + C_9 = 0\\
&C_2x + C_3 + C_9 = 0\\
&C_3x + C_2 + C_4 = 0\\
&C_4x + C_3 + C_5 = 0\\
&C_5x + C_4 + C_{10} = 0\\
&C_6x + C_7 + C_{10} = 0\\
&C_7x + C_6 + C_8 = 0\\
&C_8x + C_1 + C_7 = 0\\
&C_9x + C_1 + C_2 + C_{10} = 0\\
&C_{10}x + C_5 + C_6 + C_9 = 0
\end{aligned}
$$

The Hückel secular determinant is set up with help of secular equations as follows:

$$
\begin{vmatrix}
x & 0 & 0 & 0 & 0 & 0 & 0 & 1 & 1 & 0\\
0 & x & 1 & 0 & 0 & 0 & 0 & 0 & 1 & 0\\
0 & 1 & x & 1 & 0 & 0 & 0 & 0 & 0 & 0\\
0 & 0 & 1 & x & 1 & 0 & 0 & 0 & 0 & 0\\
0 & 0 & 0 & 1 & x & 0 & 0 & 0 & 0 & 1\\
0 & 0 & 0 & 0 & 0 & x & 1 & 0 & 0 & 1\\
0 & 0 & 0 & 0 & 0 & 1 & x & 1 & 0 & 0\\
1 & 0 & 0 & 0 & 0 & 0 & 1 & x & 0 & 0\\
1 & 0 & 0 & 0 & 0 & 0 & 1 & x & 0 & 0\\
1 & 1 & 0 & 0 & 0 & 0 & 0 & 0 & x & 1\\
0 & 0 & 0 & 0 & 1 & 1 & 0 & 0 & 1 & x
\end{vmatrix} = 0
$$

The polynomial of this determinant is obtained by Gaussian elimination method, and it is expressed as

$$X^{10} - 11X^8 + 41X^6 - 65X^4 + 43X^2 - 9 = 0$$

The roots of this polynomial obtained by matrix calculator are given as follows:

$X_1 = -2.3028$, $X_2 = -1.6180$, $X_3 = -1.3028$
$X_4 = -1.0000$, $X_5 = -0.6180$, $X_6 = 2.3028$
$X_7 = 1.6180$, $X_8 = 1.3028$, $X_9 = 1.0000$.
and $X_{10} = 0.6180$.

Hence the energy levels of naphthalene are

$X_1 = -2.3028$: $E_1 = \propto + 2.3028\,\beta$
$X_2 = -1.6180$: $E_2 = \propto + 1.6180\,\beta$
$X_3 = -1.3028$: $E_3 = \propto + 1.3028\,\beta$
$X_4 = -1.0000$: $E_4 = \propto + 1.0000\,\beta$
$X_5 = -0.6180$: $E_5 = \propto + 0.6180\,\beta$
$X_6 = 2.3028$: $E_6 = \propto - 2.3028\,\beta$
$X_7 = 1.6180$: $E_7 = \propto - 1.6180\,\beta$
$X_8 = 1.3028$: $E_8 = \propto - 1.3028\,\beta$
$X_9 = 1.0000$: $E_9 = \propto - 1.0000\,\beta$
$X_{10} = 0.6180$: $E_{10} = \propto - 0.6180\,\beta$

The energy levels and filling up of the MOs of naphthalene are shown as follows:

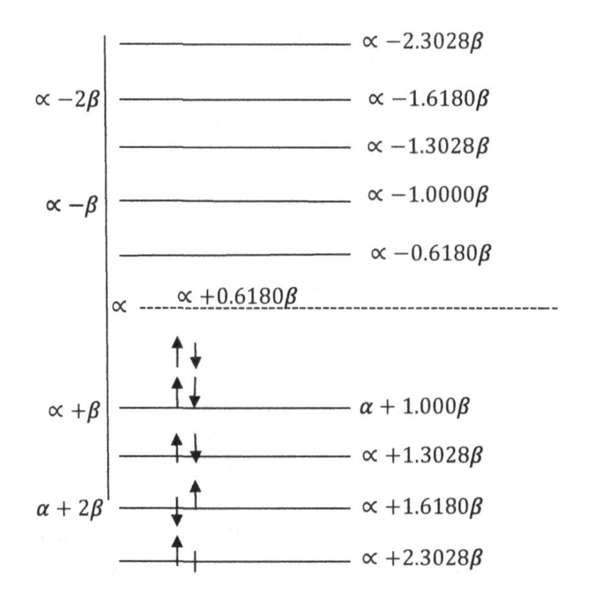

If MO diagram of naphthalene is considered, it will be noticed that it consists of a single low-energy orbital which holds two electrons.

Calculation of $E\pi$ and DE of naphthalene:

The total electronic energy in naphthalene is calculated as

$$E\pi = 2(\alpha + 2.3028\beta) + 2(\alpha + 1.6180\beta) + 2(\alpha + 1.3028\beta) + 2(\alpha + 1.0000\beta) + 2(\alpha + 0.6180\beta)$$

$$= 10\alpha + 13.6832\beta$$

Therefore, delocalisation energy (DE) = $(10\,\alpha + 13.6832\,\beta) - 10\,(\alpha + \beta)$

$$= 3.6832\beta$$

13.8.11 Hückel Molecular Orbital Coefficients and Molecular Orbitals

It has been shown earlier that there are ten secular equations consisting of ten orbital coefficients right from C_1, C_2, C_3, C_4, ... to C_{10}. It is very tedious to handle the said equations manually for finding the HMO coefficients. Fortunately, there are now available a number of microcomputer programmes with the help of which it will be easier to get the HMO coefficients and hence molecular orbitals by using different energy values (x). At this stage, we shall find the values of C_1, C_2, C_3, C_4,... C_{10} by using only bonding molecular orbital's (i.e., with different negative or positive values of x). We have obtained the different values of HMO coefficients using Gaussian elimination process. The values of HMO coefficients have been mentioned in the following table.

Values of HMO Coefficients Corresponding to Orbital Energies for Naphthalene

x_i	C_1	C_2	C_3	C_4	C_5	C_6	C_7	C_8	C_9	C_{10}
−2.3028	0.3006	0.2307	0.2307	0.3006	0.3006	0.2307	0.2307	0.3006	0.4614	0.4614
−1.6180	0.2629	0.4253	0.4253	0.2629	−0.2629	−0.4253	0.4253	−0.2629	0.0000	0.0000
−1.3028	0.3996	0.1735	−0.1735	−0.3996	−0.3996	−0.1735	0.1735	−0.3996	0.3470	−0.3470
−10000	0.0000	−0.4082	−0.4082	0.0000	0.0000	−0.4082	−0.4082	0.0000	0.4082	0.4082
−0.6180	0.4253	0.2629	−0.2629	−0.4253	0.4253	0.2629	−0.2629	−0.4253	0.0000	0.0000
0.6180	0.4253	−0.2629	−0.2629	0.4253	−0.4253	0.2629	0.2629	0.4253	0.0000	0.0000
1.0000	0.0000	0.4082	−0.4082	0.0000	0.0000	−0.4082	0.4082	0.0000	−0.4082	0.4082
1.3028	0.3996	−0.1735	−0.1735	0.3996	0.3996	−0.1735	0.1735	−0.3996	−0.3470	0.3470
1.6180	0.2629	−0.4253	0.4253	−0.2629	0.2629	−0.4253	0.4253	−2.2629	0.0000	0.0000
2.3028	0.3006	−0.2307	0.2307	−0.3006	−0.3006	0.2307	−0.2307	0.3006	−0.4614	0.4614

Now, we shall express the Hückel molecular orbitals corresponding to each value of x and with the help of different values of HMO coefficients (C_i).

$x_1 = -2.303$	$\Psi_1 = 0.3006\,(\varphi_1 + \varphi_4 + \varphi_5 + \varphi_8) + 0.2307(\varphi_2 + \varphi_3 + \varphi_6 + \varphi_7) + 0.4614(\varphi_9 + \varphi_{10})$
$x_2 = -1.6180$	$\Psi_2 = 0.2629\,(\varphi_1 + \varphi_4 - \varphi_5 - \varphi_8) + 0.4253(\varphi_2 + \varphi_3 - \varphi_6 - \varphi_7)$
$x_3 = -1.3028$	$\Psi_3 = 0.3996\,(\varphi_1 - \varphi_4 - \varphi_5 + \varphi_8) + 0.1735\,(\varphi_2 - \varphi_3 - \varphi_6 + \varphi_7) + 0.3470(\varphi_9 - \varphi_{10})$
$x_4 = -1.0000$	$\Psi_4 = -0.4082\,(\varphi_2 + \varphi_3 + \varphi_6 + \varphi_7) + 0.4082\,(\varphi_9 + \varphi_{10})$
$x_5 = -0.6180$	$\Psi_5 = 0.4253(\varphi_1 - \varphi_4 + \varphi_5 - \varphi_8) + 0.2629\,(\varphi_2 - \varphi_3 + \varphi_6 - \varphi_7)$
$x_6 = 0.6180$	$\Psi_6 = 0.4253\,(\varphi_1 + \varphi_4 - \varphi_5 - \varphi_8) - 0.2629\,(\varphi_2 + \varphi_3 - \varphi_6 - \varphi_7)$
$x_7 = 1.0000$	$\Psi_7 = 0.4082\,(\varphi_2 - \varphi_3 - \varphi_6 + \varphi_7) - 0.4082\,(\varphi_9 - \varphi_{10})$
$x_8 = 1.3028$	$\Psi_8 = 0.3996\,(\varphi_1 + \varphi_4 + \varphi_5 + \varphi_8) - 0.1735\,(\varphi_2 + \varphi_3 + \varphi_6 + \varphi_7) - 0.3470(\varphi_9 + \varphi_{10})$
$x_9 = 1.6180$	$\Psi_9 = 0.2629\,(\varphi_1 - \varphi_4 + \varphi_5 - \varphi_8) - 0.4253\,(\varphi_2 - \varphi_3 + \varphi_6 - \varphi_7)$
$x_{10} = 2.3028$	$\Psi_{10} = 0.3006\,(\varphi_1 - \varphi_4 - \varphi_5 + \varphi_8) - 0.2307\,(\varphi_2 - \varphi_3 - \varphi_6 + \varphi_7) - 0.4614(\varphi_9 - \varphi_{10})$

REFERENCES

Coulson, C.A. 1939. 'The electronic structure of some polyenes and aromatic molecules VII: Bonds of fractional order by the molecular orbital method.' *Proc. R. Soc. Lond.* 169A: 413.
Hoffmann, R. 1963. 'An extended Hückel theory I: Hydrocarbons.' *J. Chem. Phys.* 39: 1397.
Hoffmann, R. 1964. 'An extended Hückel theory IV: Carbonium ions.' *J. Chem. Phys.* 40: 2245, 2474, 2480.
Wolfsberg, M. and L. Helmholtz. 1952. 'The spectra and electronic structure of the tetrahedral ions MnO_{4-}, CrO_{4-}, and ClO_{4-}.' *J. Chem. Phys.* 20: 837

BIBLIOGRAPHY

Ballhausm, C.J. and H.B. Gray. 1964. *Molecular Orbital Theory.* New York: W.A. Benjamin Inc., 118.
Coulson, C.A. 1947. 'The theory of the structure of free radicals.' *Discuss, Faraday Soc.* 2: 9.
Dewar, M.J.S. 1969. *The Molecular Orbital Theory of Organic Chemistry.* New York: McGraw-Hill.
Goodisman, J. 1977. *Contemporary Quantum Chemistry.* New York, London: Plenum.
Hückel, E. 1931. 'Quantentheoretische beiträge zum benzolproblem.' *Zeitschrift/ur Physik* 70: 204.
Lowe, J.P. and K.A. Peterson. 2006. *Quantum Chemistry*, 3rd ed. London: Academic Press.
Pilar, F.L. 1990. *Elementary Quantum Chemistry.* New York: McGraw-Hill.
Richards, W.G. and J.A. Horsley. 1970. *Ab-initio Molecular Orbital Calculations for Chemists.* Oxford: Clarendon Press.
Roberts, J.D. 1961. *Notes on Molecular Orbital Theory of Conjugated Systems.* New York: W.A. Benjamin Inc.
Salem, L. 1966. *Molecular Orbital Theory of Conjugated Systems.* New York: W.A. Benjamin Inc.

Solved Problems

Problem 1. Using the HMO theory, how will you show that the triangular state of H_3^+ is more stable than the linear state of H_3^+?

Solution: The linear state of H_3^+ is represented as $H_1 - H_2 - H_3^+$

The secular equation is written as

$$C_1x + C_2 = 0$$

$$C_2x + C_2 + C_3 = 0$$

$$C_3x + C_2 = 0$$

On the basis of the above, the secular determinant can be expressed as

$$\begin{vmatrix} x & 1 & 0 \\ 1 & x & 1 \\ 0 & 1 & x \end{vmatrix} = 0$$

The Laplace expansion of this determinant will give

$$x\begin{vmatrix} x & 1 \\ 1 & x \end{vmatrix} - \begin{vmatrix} 1 & 1 \\ 0 & x \end{vmatrix} = 0$$

$$\text{or} \quad x(x^2 - 1) - 1(x) = 0$$

$$\text{or} \quad x^3 - x - x = 0$$

$$\text{or} \quad x^3 - 2x = 0$$

$$\therefore \quad x = 0 \quad \text{or,} \quad x^2 - 2 = 0$$

When $x^2 - 2 = 0$, $x^2 = 2$

$$\therefore \quad x = \pm\sqrt{2}$$

Therefore, the three roots are 0, $\sqrt{2}$, and $-\sqrt{2}$.

Hence, the Hückel energy levels are

$$x_1 = -\sqrt{2};\ E_1 = \alpha + \sqrt{2}\beta \text{ [BMO]}$$

$$x_1 = 0;\ E_2 = \alpha \text{ [NBMO]}$$

$$x_3 = +\sqrt{2};\quad E_3 = \alpha - \sqrt{2}\beta \text{ [ABMO]}$$

$\therefore$ Total Hückel's energy of linear H_3^+ is

$$E_{H_3^+} = 2(\alpha + \sqrt{2}\beta) = 2\alpha + 2\sqrt{2}\beta$$

Now we shall consider the triangular structure of H_3^+. The structure of H_3^+ in the triangular state is represented by

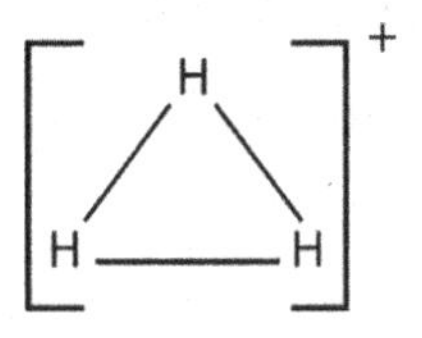

The secular equation is given by

$$C_1x + C_2 + C_3 = 0$$

$$C_2x + C_1 + C_3 = 0$$

$$C_3x + C_1 + C_2 = 0$$

The secular determinant on the basis of above secular equations is represented by

$$\begin{vmatrix} x & 1 & 1 \\ 1 & x & 1 \\ 1 & 1 & x \end{vmatrix} = 0$$

This when expanded gives the following polynomial, i.e.,

$$x^3 - 3x + 2 = 0$$

$$\text{or} \quad (x+2)(x^2 - 2x + 1) = 0$$

$$\text{or} \quad (x+2)(x-1)^2 = 0$$

Therefore, the roots will be –2, 1, and 1.

Hence, Hückel's energy levels are

$$x_1 = -2; \quad E_1 = \alpha + 2\beta$$

$$x_2 = x_3 = 1; \quad E_2 = E_3 = \alpha - \beta \quad \text{(Degenerate pair)}$$

Therefore, the Hückel energy of H_3^+ in the triangular state is given by

$$E_{H_3^+} = 2(\alpha + 2\beta) = 2\alpha + 4\beta.$$

We find that the energy for $E_{H_3^+}$ in triangle form is lower than that in linear state. Therefore, the triangular form of H_3^+ is more stable. Answer.

Problem 2. Find the energy levels of bicyclobutadiene using the HMO theory.

Solution: The structure of bicyclobutadiene is represented by

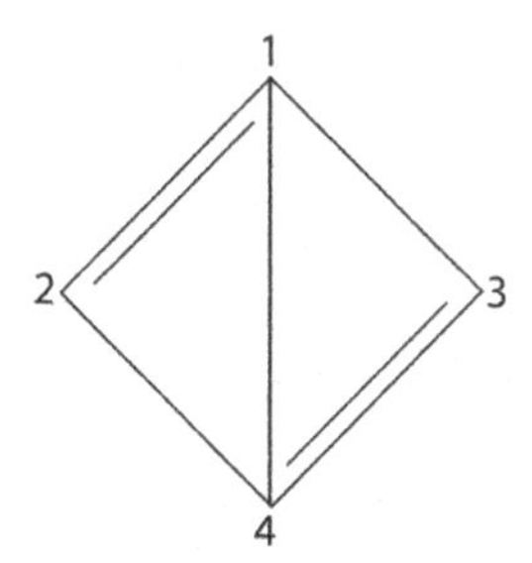

The carbon atoms have been numbered and now the secular equation can be written as

$$C_1x + C_2 + C_3 + C_4 = 0$$

$$C_2x + C_1 + C_4 = 0$$

$$C_3x + C_1 + C_4 = 0$$

$$C_4x + C_2 + C_1 + C_3 = 0$$

On the basis of this, the secular determinant can be expressed as

$$\begin{vmatrix} x & 1 & 1 & 1 \\ 1 & x & 0 & 1 \\ 1 & 0 & x & 1 \\ 1 & 1 & 1 & x \end{vmatrix} = 0$$

Using the Laplace expansion, we can write

$$\Rightarrow x\begin{vmatrix} x & 0 & 1 \\ 0 & x & 1 \\ 1 & 1 & x \end{vmatrix} - 1\begin{vmatrix} 1 & 0 & 1 \\ 1 & x & 1 \\ 1 & 1 & x \end{vmatrix} + 1\begin{vmatrix} 1 & x & 1 \\ 1 & 0 & 1 \\ 1 & 1 & x \end{vmatrix} - 1\begin{vmatrix} 1 & x & 0 \\ 1 & 0 & x \\ 1 & 1 & 1 \end{vmatrix} = 0$$

$$\Rightarrow x\left[x\begin{vmatrix} x & 1 \\ 1 & x \end{vmatrix} + 1\begin{vmatrix} 0 & x \\ 1 & 1 \end{vmatrix}\right] - \left[1\left(x^2 - 1\right) + 1(1 - x)\right] + 1\left[-1 - x(x - 1) + 1(1)\right]$$

$$-1\left[(1 - x) - x(1 - x)\right] = 0$$

$$\Rightarrow x\left[x\left(x^2 - 1\right) + (-x)\right] - \left[x^2 - 1 + 1 - x\right] + \left[-1 - x^2 + x + 1\right] - 1\left[-x - x + x^2\right] = 0$$

$$\Rightarrow x\left[x^3 - x - x\right] - \left[x^2 - x\right] + \left[-x^2 + x\right] - 1\left[-2x + x^2\right] = 0$$

$$\Rightarrow \left(x^4 - 2x^2 - x^2 + x - x^2 + x + 2x - x^2\right) = 0$$

$$\Rightarrow x^4 - 5x^2 + 4x = 0$$

$$\Rightarrow x\left(x^3 - 5x + 4\right) = 0$$

Putting $x = 1$ in the bracket, we get zero; so, $(x - 1)$ is also one of the factors.

$$\therefore \quad x(x - 1)\left(x^2 + x - 4\right) = 0$$

$\therefore$ The two roots are $x = 0$ and $x = 1$. But $x^2 + x - 4 = 0$ is a quadratic equation,

$$x = \frac{-1 \pm \sqrt{1^2 + 4 \times 4}}{2} = \frac{-1 \pm \sqrt{17}}{2}$$

i.e., x is either $\dfrac{-1 + \sqrt{17}}{2}$ or $\dfrac{-1 - \sqrt{17}}{2}$

$$\therefore \quad x = \frac{-1+4.123}{2} = 1.5675 \text{ and}$$

$$x = \frac{-1-14.123}{2} = \frac{-5.123}{2} = -2.5615$$

Thus, the four roots are

$$x_1 = 0, \quad x_3 = 1.5615$$

$$x_2 = 1 \text{ and } x_4 = -2.5615$$

$$\text{For } x_1 = 0; \; E_1 = \alpha$$

$$x_2 = 1; \; E_2 = \alpha - \beta$$

$$x_3 = 1.5615; \; E_3 = \alpha - 1.5615\beta$$

$$\text{and } x_4 = -2.5615; \; E_4 = \alpha + 2.5615\beta$$

$$\therefore E_\pi = 2(\alpha + 2.5615\beta)2\alpha = 4\alpha + 5.123\beta$$

$$\therefore \text{DE} = 4\alpha + 5.123\beta - 2(2\alpha + 2\beta)$$

$$= 1.123\beta$$

where $(2\alpha + 2\beta)$ = energy of ethene/ethylene. Answer.

Problem 3. Using the HMO approximation, find the energy of trimethylene methane and draw the energy level diagram.

Solution: The structural formula of trimethylene methane is

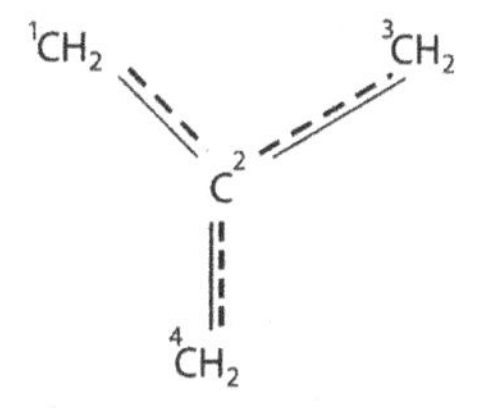

The carbon atoms have been numbered as given above. The secular equations can be written as

$$C_1x + C_2 = 0$$

$$C_2x + C_1 + C_3 + C_4 = 0$$

$$C_3x + C_2 = 0$$

$$C_4x + C_2 = 0$$

On the basis of these equations, the secular determinant can be expressed as

$$\begin{vmatrix} x & 1 & 0 & 0 \\ 1 & x & 1 & 1 \\ 0 & 1 & x & 0 \\ 0 & 1 & 0 & x \end{vmatrix} = 0$$

The Laplace expansion yields

$$x\begin{vmatrix} x & 1 & 1 \\ 1 & x & 0 \\ 1 & 0 & x \end{vmatrix} - 1\begin{vmatrix} 1 & 1 & 1 \\ 0 & x & 0 \\ 0 & 0 & x \end{vmatrix} = 0$$

$$\Rightarrow x\left[x\left(x^2\right) - 1(x) + 1(-x)\right] - x^2 = 0$$

$$\Rightarrow x\left[x^3 - x - x\right] - x^2 = 0$$

$$\Rightarrow x^4 - 2x^2 - x^2 = 0$$

$$\Rightarrow x^4 - 3x^2 = 0$$

$$\Rightarrow x^2\left(x^2 - 3\right) = 0$$

Thus, $x^2 = 0 \therefore x = 0, 0$

and $x^2 = 3 \therefore x = \pm\sqrt{3}$

Hence, the roots are $x_1 = +\sqrt{3}$, $x_2 = 0, x_3 = 0$ and $x_4 = -\sqrt{3}$

Hence, the energy levels are

$$E_1 = \alpha + \sqrt{3}\beta,\ E_2 = E_3 = \alpha \text{ and } E_4 = \alpha - \sqrt{3}\beta$$

The energy level diagram of trimethylene methane will be

$\alpha - \sqrt{3}\beta$ —— E_4

α —↑— E_2 —↑— E_3

$\alpha + \sqrt{3}\beta$ —↑↓— E_1

It is, thus, clear that the two of the 4π electrons are put in the first energy level. The E_2 and E_3 energy levels are two-fold degenerate in which the electrons are unpaired.

Thus, $E_\pi = 2\left(\alpha + \sqrt{3}\beta\right) + 2\alpha = 4\alpha + 2\sqrt{3}\beta$ Answer

Problem 4. Find the HMO orbital energies of methylene cyclopropene by a graphical solution of its polynomial equation $x^4 - 4x^2 - 2x + 1 = 0$.

Solution: Putting various values of x in equation $x^4 - 4x^2 - 2x + 1 = 0$, $f(x)$ is obtained, and $f(x)$ vs x graph is plotted.

x	f(x)	x	f(x)	x	f(x)	x	f(x)
2.2	0.665	0.9	–3.384	–0.2	1.242	–1.3	–0.316
2.0	–3.000	0.8	–2.550	–0.3	1.248	–1.4	–0.198
1.9	–4.210	0.6	–1.510	–0.4	1.180	–1.5	0.062
1.7	–5.608	0.4	–0.414	–0.6	0.889	–1.6	0.514
1.5	–5.938	0.2	0.440	–0.8	0.449	–1.7	1.192
1.4	–5.798	0.1	0.760	–0.9	0.216	–1.9	3.390
1.3	–5.504	0.0	1.000	–1.0	0.000	–2.0	5.000
1.0	–4.000	–0.1	1.160	–1.2	–0.286	–	–

On the basis of these data, a graph between f(x) against × is plotted, which has been illustrated in Figure 13.19.

From the graph, it is clear that the curve cuts the X-axis at four points, which are the roots of the polynomial, and the roots are

$$x = 2.17, 0.31, -1.00 \text{ and } -1.48$$

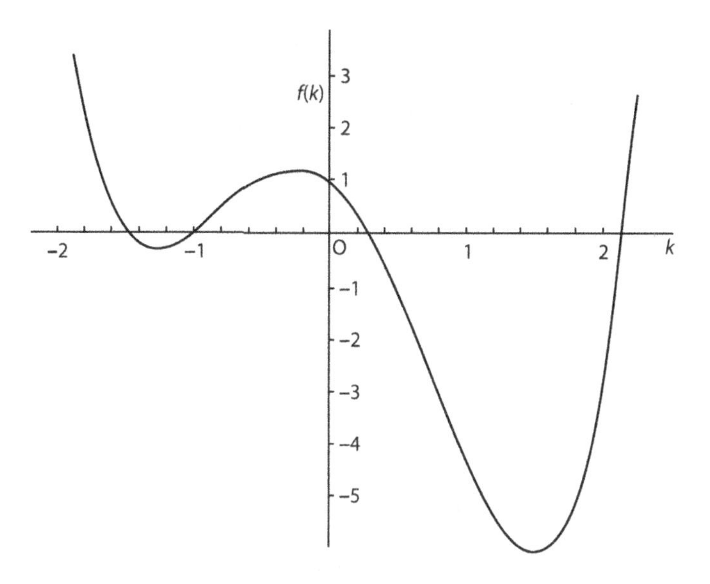

FIGURE 13.19 Graphical solution of the secular polynomial for methylene cyclopropene.

Now, we shall find the HMO orbital energies with different roots of x.

$$x_1 = 2.17,\ E_1 = \alpha - 2.17\beta$$

$$x_2 = 0.31,\ E_2 = \alpha - 0.31\beta$$

$$x_3 = -1.00,\ E_3 = \alpha + \beta$$

$$x_4 = -1.48,\ E_4 = \alpha + 1.48\beta.$$

Answer.

Problem 5. Set up the system of secular equations, secular determinant of 2-azabutadiene, having empirical parameter $\alpha_N = \alpha + 0.5\beta$. Also find the polynomial from the expansion of the determinant.

Solution: The structural formula of 2-azabutadiene is given by

$$H_2 - \overset{1}{C} = \overset{2}{N} - \overset{3}{C}H = \overset{4}{C}H_2$$

The secular equation can be written as

$$C_1 x + C_2 = 0$$

$$(C_2 + 0.5)x + C_1 + C_3 = 0$$

$$C_3 x + C_2 + C_4 = 0$$

$$C_4 x + C_3 = 0$$

On the basis of these equations, the secular determinant is expressed as

$$\begin{vmatrix} x & 1 & 0 & 0 \\ 1 & x+.5 & 1 & 0 \\ 0 & 1 & x & 1 \\ 0 & 0 & 1 & x \end{vmatrix} = 0$$

We shall expand this determinant by Laplace procedure, and thus, we can write

$$x\begin{vmatrix} x+.5 & 1 & 0 \\ 1 & x & 1 \\ 0 & 1 & x \end{vmatrix} - 1\begin{vmatrix} 1 & 1 & 0 \\ 0 & x & 1 \\ 0 & 1 & x \end{vmatrix} = 0$$

$$\Rightarrow x\left[(x+.5)\begin{vmatrix} x & 1 \\ 1 & x \end{vmatrix} - 1\begin{vmatrix} 1 & 1 \\ 0 & x \end{vmatrix}\right] - 1\left[1\begin{vmatrix} x & 1 \\ 1 & x \end{vmatrix} - 1\begin{vmatrix} 0 & 1 \\ 0 & x \end{vmatrix}\right] = 0$$

$$\Rightarrow x\left[(x+.5)(x^2-1) - 1(x)\right] - 1\left[(x^2-1\right] = 0$$

$$\Rightarrow \left[x(x+.5)(x^2-1) - x^2\right] - x^2 + 1 = 0$$

$$\Rightarrow (x^2+0.5x)(x^2-1) - x^2 - x^2 + 1 = 0$$

$$\Rightarrow x^4 + 0.5x^3 - x^2 - 0.5x - 2x^2 + 1 = 0$$

$$\Rightarrow x^4 + 0.5x^3 - 3x^2 - 0.5x + 1 = 0$$

This is the polynomial. Answer.

Problem 6. Establish the secular polynomial for 1,2-diazabutadiene ($\alpha_N = \alpha + 0.5\beta$).

Solution: Given that

$$\alpha_N = \alpha + 0.5\beta$$

$$\beta_{CN} = \beta$$

$$\beta_{NN} = \beta.$$

The Skeleton representation of 1,2-diazabutadiene is given by

$$\overset{1}{N}-\overset{2}{N}-\overset{3}{C}-\overset{4}{C}$$

The secular equations are given by

$$C_1(x+0.5) + C_2 = 0$$

$$C_2(x+0.5) + C_1 + C_3 = 0$$

$$xC_3 + C_2 + C_4 = 0$$

$$xC_4 + C_3 = 0$$

On the basis of these equations, the determinant is given by

$$\begin{vmatrix} x+.5 & 1 & 0 & 0 \\ 1 & x+.5 & 1 & 0 \\ 0 & 1 & x & 1 \\ 0 & 0 & 1 & x \end{vmatrix} = 0$$

Laplace expansion of this determinant gives

$$(x+.5)\begin{vmatrix} x+.5 & 1 & 0 \\ 1 & x & 1 \\ 0 & 1 & x \end{vmatrix} - 1\begin{vmatrix} 1 & 1 & 0 \\ 0 & x & 1 \\ 0 & 1 & x \end{vmatrix} = 0$$

$$\Rightarrow (x+.5)\left[(x+.5)\begin{vmatrix} x & 1 \\ 1 & x \end{vmatrix} - 1\begin{vmatrix} 1 & 1 \\ 0 & x \end{vmatrix}\right] - 1\left[1\begin{vmatrix} x & 1 \\ 1 & x \end{vmatrix} - 1\begin{vmatrix} 0 & 1 \\ 0 & x \end{vmatrix}\right] = 0$$

$$\Rightarrow (x+.5)(x+.5)\left(x^2-1\right) - (x+.5)(x) - 1\left(x^2-1\right) = 0$$

$$\Rightarrow (x+.5)^2\left(x^2-1\right) - x^2 - 0.5x - x^2 + 1 = 0$$

$$\Rightarrow \left(x^2+x+.25\right)\left(x^2-1\right) - x^2 - 0.5x - x^2 + 1 = 0$$

$$\Rightarrow x^4 + x^3 + 0.25x^2 - x^2 - x - 0.25 - x^2 - 0.5x - x^2 + 1 = 0$$

This is the polynomial. Answer.

Questions on Concepts

1. What are the pertinent points of Hückel's drastic assumptions?
2. Using the (HMO) theory, find the energy of the following allylic systems: $C_3H_5^+$, $C_3H_5^-$ and $C_3H_5^{\cdot}$ [Ans. $E_1 = \alpha + \sqrt{2}\beta, E_2 = \alpha$ and $E_3 = \alpha - \sqrt{2}\beta$]
3. Use the HMO theory to calculate the energy of trimethylene methane and construct the energy level diagram.
4. Calculate using the HMO theory, whether the linear $[H-H-H]^+$ or the triangular state is more stable

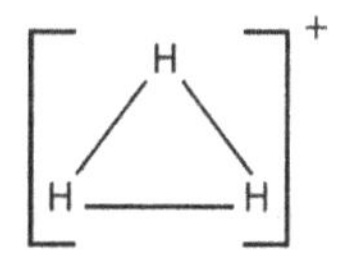

5. Using the HMO theory, set up and solve the secular determinantal equation for ethylene. Also construct the energy level diagram of C_2H_4.
6. Use the HMO theory to set up and solve the Hückel secular determinantal equation for allyl system, which consists of the allyl radical and the related cation and anion. Find the delocalisation energy of the three species and also construct the energy level diagram. After finding the HMO coefficients, also find the molecular wave functions.
7. (a) Apply the HMO theory to butadiene for finding the secular determinant, polynomials, and its roots. After obtaining the roots, find the energy levels and energy level diagrams. (b) How will you determine the molecular orbitals of butadiene?
8. Write down Hückel's secular equation and solve it for hexatriene.
9. Write out Hückel's π-electron determinant for the following molecules, in terms of appropriate α and 0.
   a. Ethyl amine ($CH_2 = CH - NH_2$) by considering the N as part of the π system.
   b. Cyclopropenyl radical.
   c. Benzene.

10. Set up secular determinant for pyridine on the basis of HMO theory.
11. Estimate the π-electron energy level and the total π-electron energy of bicyclobutadiene.

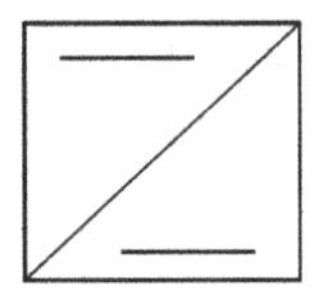

12. The total π electron energy of naphthalene is $E_\pi = 10\alpha + 13.68\beta$. Find the delocalisation energy of naphthalene.
13. Try to set up but do not try to solve, the HMOT determinantal equation for naphthalene.
14. Give the graphical and 3D representations of allyl system. Indicate the nodal planes.
15. Apply HMOT to cyclobutadiene and find the energy levels. Also construct the energy level diagram.
16. Find the HMO coefficients and molecular orbitals of cyclobutadiene.
17. Find the HMO coefficients of cyclopentadienyl radical and then write down the molecular wave functions.
18. a. Set up and solve the determinantal equation for benzene.
    b. Find the energy levels and also its delocalisation energy.
19. Find HMO coefficients and ψ_1 and ψ_2 for benzene.
20. Give the graphical representation of benzene and locate the nodal planes.
21. Find the electron density of the following:
    (a) C_6H_6 (b) C_2H_4 (c) butadiene
22. Estimate the bond order of the following
    (a) C_2H_4 (b) $CH_2 = CH - CH = CH_2$ (c) C_6H_6.
23. What do you mean by free valence? Find the free valence of C_2H_4, butadiene, and benzene.
24. Give the generalised treatment of HMO theory to open-chain conjugated system.
25. Apply the generalised treatment of HMOT to (a) ethylene, (b) butadiene, and (c) benzene.
26. Prove that $x = 2\cos\frac{n\pi}{r+1}$ for open-chain conjugated system.
27. Give the generalised treatment of HMOT to cyclic polyenes.
28. Prove that $x = -2\cos\frac{2k\pi}{r}$, where the terms have their usual significance.
29. Apply the generalised treatment of HMOT to cyclopropenyl radical, cyclobutadiene, and cyclopentadienyl radical.
30. Find all roots of benzene by using $x = -2\cos\frac{2k\pi}{r}$ formula.
31. Write a brief account of the extended Hückel theory (EHT) or Hückel theory.
32. Apply EHT to pyrrole.
33. Set up and solve the Hückel secular determinantal equation for butadiene. Construct the molecular orbital diagram for this system and calculate its delocalisation energy. [P.U. 2010]
34. Use the HMO approximation to obtain (a) Bond orders; (b) free valence for butadiene. [P.U. 2010]
35. (a) Find the HMO of benzene; (b) calculate the π-electronic charge on each carbon atom and the total bond orders in benzene. [P.U. 2010]
36. Using the HMO theory, set up and solve the secular determinantal equation for cyclopropene. Construct the molecular orbital diagram for this system and also determine its molecular orbitals. [P.U. 2010]

37. a. What are the assumptions involved in the Hückel molecular orbital method?
    b. How does extended HMO method differ from the HMO method? [P.U. 2007]
38. Set up and solve Hückel secular equation for pyridine molecule.
39. Is there any need to modify the coulombs integral, and the resonance integral? If yes, why and how?
40. Draw the energy level diagram of pyridine with bonding orbital energies -2.2300β, -1.2390β, and 1.0000β.
41. a. How will you prove that pyridine is more stable than benzene?
    b. Calculate the E and delocalisation energy of pyridine if the bonding orbital energies are $-2.2300\,\beta$, $-1.2390\,\beta$, and $-1.0000\,\beta$
42. Find the electron density of 'N' in pyridine if the Ψ_1, Ψ_2, and Ψ_3 are given
    $\Psi_1 = 0.589\,\varphi_1 + 0.442\,\varphi_2 + 0.309\,\varphi_3 + 0.269\,\varphi_4 + 0.309\,\varphi_5 + 0.442\,\varphi_6$
    $\Psi_2 = -0.483\,\varphi_1 - 0.149\,(\varphi_2 + \varphi_6) + 0.396\,(\varphi_3 + \varphi_5) + 0.639\,\varphi_4$
    $\Psi_3 = -0.500\,(\varphi_2 + \varphi_3) + 0.500\,(\varphi_5 + \varphi_6)$
43. Set up and solve Hückel secular equation for naphthalene molecule.
44. Find the $E\pi$ and delocalisation energy of naphthalene.

14

Density Functional Theory

Density functional theory (DFT) is one of the tools of computational chemistry. The density functional calculations are often said to be DFT calculations. It is just like ab initio and semi-empirical calculations based on the Schrödinger equation. It must be kept in mind that DFT does not compute a wave function but it directly gives a method to obtain the electron distribution or electron density function. Therefore, it may be said that the DFT is not based on the wave function, but it is based on the electron density or charge density or electron density function. It is represented or designated by $\rho(x, y, z)$. The fact is that it is probability per unit volume, which is expressed as $\rho(x,y,z) = \frac{P}{\mathrm{d}x \cdot \mathrm{d}y\mathrm{d}z'}$, where P is the probability of finding an electron in a volume element $\mathrm{d}x{\cdot}\mathrm{d}y{\cdot}\mathrm{d}z$. Note that $\rho(x, y, z)$ is purely a number. It will be clear from the following representation:

$$\rho(x,y,z) = \frac{P}{\mathrm{d}x \cdot \mathrm{d}y\mathrm{d}z} \cdot \mathrm{d}x \cdot \mathrm{d}y \cdot \mathrm{d}z = P \approx \text{pure number}$$

The unit of ρ can be logically expressed as Vol^{-1}. Therefore, ρ can be mathematically represented by

$$\rho = \sum_{i=1}^{n} n_i |\psi_i|^2$$

But for closed shell molecules $n_i = 2$

$$\therefore \quad \rho = 2\sum_{i=1}^{n} |\psi_i|^2 \cdot \frac{\mathrm{d}x \cdot \mathrm{d}y \cdot \mathrm{d}z}{\mathrm{d}x \cdot \mathrm{d}y \cdot \mathrm{d}z} = 2\sum_{i=1}^{n} |\psi_i|^2 \tag{14.1}$$

We have used the words function and functional above so we shall explain these two mathematical words one by one.

14.1 Function

A quantity 'y' *is said to be a function of another quantity* 'x', written as $y = f(x)$, if a change in one produces a change in the other. Thus, in the statement $y = 3x^2 + 5x$, i.e., $f(x) = 3x^2 + 5x$, y is a function of x and a change in the value of x produces a change in the value of y. If $x = 1$, then $f(1) = 3 \times 1^2 + 5 \times 1 = 8$, if $x = 2$, then $f(2) = 3 \times 2^2 + 5 \times 2 = 22$. Thus, it is clear that the value of 'y' is dependent on the value of x.

The function can also be defined in the following manner:

A function $f(x)$ is a mathematical rule, which connects a number with every value of x for which the function f is defined. For example, the function $f(x) = x^3 + 1$ connects the number 9 with $x = 2$, i.e., when we put $x = 2$, $f(2) = 9$; similarly, for $x = 3$ $f(3) = (3)^2 + 1 = 28$. Thus, this pretty much explains what is a function.

14.2 Functional

The word functional is included in the DFT. A functional is a mathematical rule by which a function is transformed into a number. For example,

$$f(x) = x^2 \int_0^3 f(x)\mathrm{d}x \rightarrow \left.\frac{x^3}{3}\right]_0^3 \rightarrow \frac{(3)^3}{3} = 9 \tag{14.2}$$

From this example, it is clear that the functional $\int_0^3 f(x)\mathrm{d}x$ transforms the function x^2 into the number 9.

Similarly,

$$f(x) = x^3 \int_0^2 f(x)\mathrm{d}x \rightarrow \left.\frac{x^4}{4}\right]_0^2 = 4 \tag{14.3}$$

It means the functional $\int_0^2 f(x)\mathrm{d}x$ transforms the function x^2 into the number 4. We denote the fact that the integral is a functional of $f(x)$ by writing down as

$$\int_0^2 f(x)\mathrm{d}x = F[f(x)] \tag{14.4}$$

This gives a clue to state that a functional is a function of a 'definite' function (as noted above).

The above facts can also be understood in the following way.

Let us consider a definite integral from $x = a$ to $x = b$, which is denoted as

$$I = \int_a^b f(x)\mathrm{d}x \tag{14.5}$$

The value of I depends on the choice of the integrand function f and is a functional of f. It is to be noted that the definite integral having definite limits would be a different functional because of the fact that it will give rise to a different value for the same function. We may express the functional relationship as

$$I = I[f] \tag{14.6}$$

where $I[......]$ stands for the functional, which in the present case means taking the definite integral within x = a to $x = b$ limits of the function, which is independent. Taking an example, let

$$I = \int_0^2 [x^3\mathrm{d}x] \approx I[......]$$

= functional

$$= \left[\frac{x^4}{4}\right]_0^2 = 4$$

This helps distinguish between a function and a functional clearly.

14.3 Hohenberg–Kohn Theorem

14.3.1 Theorem 1

For *non-degenerate state, any ground state property of a molecule is a functional of the ground state electron density function.*

For example, for energy

$$E_0 = f\left[\rho_0(r)\right] = E_v\left[\rho_0(r)\right] \tag{14.7}$$

It may be deduced from this that the ground state electronic energy of a molecule is a functional of p_o. The Hohenberg–Kohn (H–K) theorem was enunciated in 1964.

Proof: Let ψ_o = ground state wave function of a molecule having n electrons

H = electronic Hamiltonian

$$= -\frac{h^2}{2m}\sum \nabla_i^2 - \sum_\alpha \sum_i \frac{z_\alpha e^2}{r_{i\alpha}} + \sum_i \sum_{i>j} \frac{e^2}{r_{ij}} \tag{14.8}$$

Equation (14.8) in a.u. takes the following form:

$$H = -\frac{1}{2}\sum \nabla_i^2 + \sum_{i=1}^{n} V(r_i) \sum_i \sum_{i>j} \frac{1}{r_{ij}} \tag{14.9}$$

where $V(r_i) = -\sum_\alpha \frac{z_\alpha}{r_{i\alpha}}$ = external potential

= potential energy of interaction between electron i and the nuclei.

It is called the external potential because of the fact that it is generated by charges external to the system of electron. In the case of the system of non-degenerate ground state, the external potential and the number of electrons present in the system are found out by the ground state electron probability density or electron density $\rho(r)$.

The equation, which is related to the number of electrons and $p(r)$, is given by

$$\rho_0(r) = n \int_{\text{allspace}} \cdots \int \left|\psi(r_1, r_2) - m_s\right|^2 \mathrm{d}r_2 \ldots \mathrm{d}r_n \tag{14.10}$$

Applying the normalisation condition, the above equation can be written as

$$\rho_0(r)\mathrm{d}r = n = \text{number of electrons} \tag{14.11}$$

We have stated above that $\rho_o(r)$ determines the external potential $V(r_i)$, but we shall see that this fact is false and V_a and V_b are two external potentials, which differ by more than a constant. It is also supposed that each yields ρ_o (ground state electron density).

Let H_a and H_b = n electron Hamiltonians

$V_a(r_i)$ and $V_b(r_i)$ = external potentials corresponding to H_a and H_b

where $V_a(r_i)$ and $V_b(r_i)$ are not necessarily given by $V(r_i) = -\sum_\alpha \frac{z_\alpha}{r_{i\alpha}}$

$\psi_{o,a}$ and $\psi_{o,b}$ = ground state different wave functions, which are normalised and differ by more than one additive constant

$E_{o,a}$ and $E_{o,b}$ = ground state energy corresponding to H_a and H_b, respectively

In the case of non-degenerate ground state, there will be only one normalised function, that is, ψ_o, which will give rise to the exact ground state energy E_o. When ψ_o is used as a trial variation function on the basis of variation theorem, we can safely say that the use of any normalised well-behaved function that differs from ψ_o will cause the variation integral greater than E_o. It is mathematically expressed as

$$\langle \phi | H | \phi \rangle > E_o \tag{14.12}$$

if $\phi \neq \phi_o$ the ground state is non-degenerate.

Therefore, we shall use $\psi_{0,b}$ as the trial function having Hamiltonian H_a, which will yield the following expression:

$$\begin{aligned} E_{o,a} &< \langle \psi_{0,b} | H_a | \psi_{0,b} \rangle = \langle \psi_{0,b} | H_a + H_b - H_b | \psi_{0,b} \rangle \\ &= < \langle \psi_{0,b} | H_a - H_b | \psi_{0,b} \rangle + \langle \psi_{0,b} | H_b | \psi_{0,b} \rangle \end{aligned} \tag{14.13}$$

It must be kept in mind that H_a and H_b differ only in their external potentials V_a and V_b.

$$H_a - H_b = \sum_{i=1}^{n} \left[V_a(r_i) - V_b(r_i) \right] \tag{14.14}$$

And we shall write

$$E_{0,a} < \psi_{0,b} \left| \sum_{i=1}^{n} \left[V_a(r_i) - V_b(r_i) \right] \right| \psi_{0,b} + E_{0,b} \tag{14.15}$$

It is to be noted that $V_a(r_i)$ and $V_b(r_i)$ are one-electron operators, and if a function of the spatial co-ordinates x_i, y_i, z_i of electron i be considered, then we can write

$$\int \psi^* \sum_{i=1}^{n} B(r) \psi \, d\tau = \int \rho(r) B(r) dr \tag{14.16}$$

where $B(r)$ = a function of the special co-ordinates x_i, y_i, z_i of electron i

Using the Eq. (14.16), we can write Eq. (14.15) as

$$E_{0,a} < \int \rho_{0,b}(r) \left[V_a(r) - V_b(r) \right] dr + E_{0,b} \tag{14.17}$$

Because the integration has been done over $\psi_{0,b}$, we obtain $\rho_{o,b}$ corresponding to $\psi_{o,b}$.

If a and b are interchanged, then the above equation will take the form

$$E_{0,b} < \int \rho_{0,a}(r) \left[V_b(r) - V_a(r) \right] dr + E_{0,a} \tag{14.18}$$

But by hypothesis, the two different wave functions yield the same electron density, i.e., $\rho_{o,a} = \rho_{o,b}$.

Now putting $\rho_{o,a} = \rho_{0,b}$ and adding Eqs. (14.17) and (14.18), we shall get

$$E_{0,a} + E_{0,b} < E_{0,b} + E_{0,a} \tag{14.19}$$

which is false.

It gives a clue to understand that two different external potentials, which generate the same ground state electron density, must be false. This is the initial assumption we have made. This implies that ρ_0 is able to compute the external potential as well as the number of n electrons present in the system. From ρ_0, we will be also able to find out the molecular electronic Hamiltonian (H_{el}), ground state wave function, energy, and other properties.

$$\text{Thus, } E_0 = f\left[\rho_0(r)\right] \approx E_0 = E_\upsilon\left[\rho_0(r)\right] \tag{14.20}$$

where v indicates that E_o depends on the external potential $V(r)$. The mathematical expression described in Eq. (14.20) can be stated as

'The ground state energy E_o is a functional of the function $\rho_o(r)$.'

From Eq. (14.8), it is clear that the electronic Hamiltonian is the sum of electronic kinetic energy terms, electron nuclear attraction, and electron–electron repulsions, and therefore, considering averages of Eq. (14.8) for the ground state, the energy can be expressed mathematically as

$$E = \bar{T} + \bar{V}_{ne} + \bar{V}_{ee} \tag{14.21}$$

The RHS of Eq. (14.21) represents the molecular properties determined by ground state electronic wave function, which is determined by $p_o(r)$. Therefore, it can be stated that each of the averages in Eq. (14.21) is a functional of p_o.

From Eq. (14.8), we can write

$$\hat{V}_{ne} = \sum_i^n V(r_i) \tag{14.22}$$

where $V(r_i) = -\sum_\alpha \frac{z_\alpha}{r_{i\alpha}}$ in a.u.

$$\therefore \quad \hat{V}_{ne} = \left\langle \psi_0 \mid \sum_{i=1}^n V(r_i) \mid \psi_0 \right\rangle$$

$$\text{or} \quad \hat{V}_{ne} = \int \rho_0(r) V(r) \mathrm{d}r \tag{14.23}$$

In Eq. (14.23), $V(r)$ is equal to nuclear attraction potential energy function for a given electron, which is located at a point r. Thus, $\bar{V}_{ee}[\rho_0]$ becomes known but $\bar{T}[\rho_0]$ and $\bar{V}_{ee}$ are not known. Thus,

$$E_0 = E_\upsilon[\rho_0] = \int \rho_0(r)V(r)\mathrm{d}r + \bar{T}[\rho_0] + \bar{V}_{ee}[\rho_0]$$

$$= \int \rho_0(r)V(r)\mathrm{d}r + f[\rho_0] \tag{14.24}$$

where $f[\rho_0] = \bar{T}[\rho_0] + \bar{V}_{ee}[\rho_0]$ and $f[\rho_0]$ do not depend on the external potential. Thus, Eq. (14.24) is unable to estimate a practical way to compute E_o with the aid of ρ_o due to the fact that $f[\rho_o]$ is not known.

14.3.2 Theorem 2

For every trial density function $\rho_{tr}(r)$, which satisfies $\int \rho_{tr}(r)\mathrm{d}r = n$ and $\rho_{tr}(r) \geq 0$ for all 'r', the following inequality holds:

$$E_\upsilon[\rho_0] \leq E_\upsilon[\rho_{tr}]'' \tag{14.25}$$

where E_v = energy functional in ground state electron density in Eq. (14.24).

Proof: Let us suppose that

ρ_{tr} = trial density function which is non-negative

ρ_{tr} satisfies the two conditions, i.e., $\rho_{tr}(r) \geq 0$ and $\int \rho_{tr}(r)\mathrm{d}r = n$ for all r. We have discussed above that ρ_{tr} determines V_{tr} and ψ_{tr}. (This fact is only true if there exists an external potential, which will yield an antisymmetric wave function that corresponds to ρ_{tr}, which is a trial density function.) Under this condition, ρ_{tr} is called v-representable. It should be noted that all $\rho_{tr}^{\prime s}$ may not be v-representable, though this fact does not make any difference in the application of DFT.

Levy has tried to reformulate the H–K theorem in which the need of v-representability is eliminated. Let us suppose that

ψ_{tr} = trial wave function
ρ_{tr} = trial density function for a molecule
H = molecular Hamiltonian

With these parameters and applying variation theorem, we can express

$$\langle \psi_{tr}|H|\psi_{tr}\rangle = \left\langle \psi_{tr} \mid \hat{T} + \hat{V}_{ee} + \sum_{i=1}^{n} V(r_i) \mid \psi_{tr} \right\rangle \geq E_0 = E_\upsilon[\rho_0] \tag{14.26}$$

Now we are going to use the fact that the average kinetic energy and potential energy are functional of the electron density. Using Eq. (14.23) and further replacement of ψ_o by ψ_{tr}, Eq. (14.26) is written as

$$\bar{T}[\rho_{tr}] + \bar{V}_{ee}[\rho_{tr}] + \int \rho_{tr} V(r) \mathrm{d}r \geq E_\upsilon[\rho_0] \tag{14.27}$$

It should be kept in mind that the functional $\bar{T}$ and $\bar{V}_{ee}$ are same as described in Eqs. (14.24) and (14.26), although ρ_o and ρ_{tr} differ. The LHS of Eq. (14.27) differs from Eq. (14.24) in the fact that in one there is ρ_o term and in the other ρ_{tr} term. After putting ρ_{tr} in place of ρ_o in Eq. (14.27), we get

$$E_\upsilon[\rho_{tr}] \geq E_\upsilon[\rho_0] \tag{14.28}$$

which indicates that any trial density cannot yield a lower ground state energy than the true ground state energy. Later on, *Levy* proved these theorems for degenerate systems also.

14.3.3 Alternative Proof of Hohenberg–Kohn Theorems

14.3.3.1 Theorem 1

For a non-degenerate ground state ψ_g, the external potential energy function V *is a unique functional of the single particle density $\rho_g'(r)$.*

Proof: The theorem is proved by reductio ad absurdum. The reductio ad absurdum refers to a method of proving the falsity of a premise (previous statement from which another is inferred) by showing that the logical consequence or result is absurd.

Let V and V' be two different external potentials for all electrons of a system corresponding to two Hamiltonians H and H', respectively.

ψ_g and ψ_g' = ground state functions
E_g and E_g' = ground state energies, which are unique functional of ρ_g', i.e., electron density

Also suppose that ψ_g and ψ_g' match to the same particle density $\rho'(r)$. It must be kept in mind that ψ_g and ψ_g' are different from each other because of the fact that they are the Eigen functions of different operators H and H', respectively.

We all know that

$$\left.\begin{aligned} H &= T + u + V \\ \text{and} \quad H' &= T + u + V' \end{aligned}\right\} \tag{14.29}$$

where T = kinetic energy operator
U = two-electron interaction energy operator

We can write

$$H' = T + U + V$$
$$H = T + U + V$$
$$- \quad - \quad - \quad -$$

On subtraction, $(H' - H) = (V' - V)$

$$\left.\begin{aligned} \therefore \quad & H' = H + (V' - V) \\ \text{Similarly} \quad & H = H' + (V - V') \end{aligned}\right\} \tag{14.30}$$

This shows the relationship between H and H'. From variation theorem, we can express

$$E'_g = \int \psi_g'^{*} H' \psi'_g \mathrm{d}\tau < \int \psi_g^{*} H' \psi_g \mathrm{d}\tau \tag{14.31}$$

But from Eq. (14.30), one can write

$$\int \psi_g^{*} H' \psi_g \mathrm{d}\tau = E_g + \int \rho'_g(r)\left[V'(r) - V(r)\right]\mathrm{d}^3 r \tag{14.32}$$

where $V'(r)$ and $V(r)$ = external potentials

Therefore, from Eqs. (14.31) and (14.32), we can write

$$E'_g < E_g + \int \rho'_g(r)\left[V'(r) - V(r)\right]\mathrm{d}^3 r \tag{14.33}$$

Starting with E_g, one can write an equation exactly similarly to Eq. (14.33)

$$E_g < E'_g + \int \rho'_g(r)\left[V(r) - V'(r)\right]\mathrm{d}^3 r \tag{14.34}$$

Adding Eqs. (14.33) and (14.34), we shall get

$$E'_g + E_g < E_g + E'_g \tag{14.35}$$

which is false. Thus, it is concluded that either $V = V'$, in which case ψ_g and ψ'_g may be degenerate with less than sign replaced by equal to sign in the above equations, or V and V' are different and ψ_g and ψ'_g do not match to $\rho'_g(r)$. Hence, ψ_g and the external potential V are unique functional of electron density $\rho'_g(r)$. Q.E.D.

14.3.3.2 Theorem 2

For a given external potential V, the correct single particle density $\rho'_g(r)$ minimises the ground state energy E_g, which is a unique functional of $\rho'_g(r)$.

Proof: We are aware of the fact that ψ_g is a unique functional of $\rho_g'(r)$, and hence, all ground state expectation values will be functionals of $\rho_g'(r)$. The expectation value of kinetic energy (*T*) and the expectation value of two-electron interaction energy (*U*) are universal functionals of $\rho_g'(r)$. Thus, we may express

$$T\left[\rho_g'(r)\right] = \int \psi_g^* T \psi_g \mathrm{d}\tau \tag{14.36}$$

$$U\left[\rho_g'(r)\right] = \int \psi_g^* U \psi_g \mathrm{d}\tau \tag{14.37}$$

Adding the LHS of Eqs. (14.36) and (14.37), we obtain

$$T\left[\rho_g'(r)\right] + U\left[\rho_g'(r)\right] = f\left[\rho_g'(r)\right] \tag{14.38}$$

On addition of the LHS of Eqs. (14.36) and (14.37), a new function is obtained, which is named as the universal functional *f*.

Let the energy expectation value of *H* be written as E_v. The ground state energy will be expressed as

$$E_\upsilon\left[\rho_g'(r)\right] = f\left[\rho_g'(r)\right] + \int \rho_g'(r)\cdot V(r)\mathrm{d}^3r$$

According to the variation theorem, for another function ψ,

$$E_\upsilon\left[\psi\right] = f\left[\rho'(r)\right] + \int \rho'(r)\cdot V(r)\mathrm{d}^3r > E_\upsilon\left[\rho_g'(r)\right]$$

Thus, the minimum property of $E_v\left[\rho_g'(r)\right]$ is set up. Q.E.D.

14.4 Kohn–Sham Energy

In finding out the Kohn–Sham energy, we shall take into account the energy of a molecule as a deviation from an ideal energy. We have already discussed about the unknown functional. The unknown functional consists of a small discrepancy, the approximation of which is the prime problem. In the case of an ideal system, the electrons do not interact and the ground state electron density ρ_r is found to be the same as in the case of real ground state system, which is ρ_o. Thus, we can write $\rho_r = \rho_o$. We are aware of the fact that the ground state electronic energy of the real molecule is equal to the sum of the electronic kinetic energy, the nucleus–electron attraction potential energy, and the electron–electron repulsion energy, which are functional of ground state electron density. We can express all these energies in the form of expectation values as

$$E_0 = \left\langle T\left[\rho_0\right]\right\rangle + \left\langle V_{ne}\left[\rho_0\right]\right\rangle + \left\langle V_{ee}\left[\rho_0\right]\right\rangle \tag{14.39}$$

where the terms have already been explained. The value of the middle term of Eq. (14.39) may be written as

$$\left\langle V_{ne}\left[\rho_0\right]\right\rangle = \sum_{i=1}^{2n} \sum_{\text{necleiA}} -\frac{Z_A}{r_{iA}} = \sum_{r_{iA}}^{2n} V(r_i) \tag{14.40}$$

where $V(r_i)$ = external potential. The *n–e* potential energy is the sum of overall 2*n* electrons.

The density function ρ can be introduced into V_{ne} by using

$$\int \psi \sum_{i=1}^{2n} f(r_i)\psi \mathrm{d}\tau = \int \rho(r) f(r)\mathrm{d}r \tag{14.41}$$

where $f(r_i)$ = a function of the co-ordinates of $2n$ electrons of a system; ψ = total wave function

From Eqs. (14.40) and (14.41), calling the concept of expectation value $\langle V_{ne} \rangle = \langle \psi \mid V_{ne} \mid \psi \rangle$.

We obtain

$$\langle V_{ne} \rangle = \int \rho_0(r)V(r)\mathrm{d}r \tag{14.42}$$

Putting this value in Eq. (14.39), we obtain

$$E_0 = \int \rho_0(r)V(r)\mathrm{d}r + \langle T[\rho_0] \rangle + \langle V_{ee}[\rho_0] \rangle \tag{14.43}$$

In this equation, the functionals in $\langle T[\rho_0] \rangle$ and $\langle V_{ee}[\rho_0] \rangle$ are not known; therefore, the above equation for energy cannot be used.

For the use of Eq. (14.43), *Kohn* and *Sham* put forward an idea of a system of non-interacting electrons. Let us consider that $\Delta \langle T[\rho_0] \rangle$ be treated as the deviation of the real kinetic energy from that of the reference system:

$$\therefore \quad \Delta\langle T[\rho_0] \rangle \equiv \langle T[\rho_0] \rangle - \langle T[\rho_0] \rangle \tag{14.44}$$

Now, we shall consider $\Delta\langle V_{ee} \rangle$ as the deviation of the electron–electron repulsion energy from charge cloud coulomb repulsion energy. This repulsion energy will be considered as equivalent to the sum of the repulsion energies for pairs of infinitesimal volume elements $\rho(r_1)\mathrm{d}r_1$ and $\rho(r_2)\mathrm{d}r_2$ separated at a distance r_{12} multiplied by half. It clearly gives a clue to understand that $r_1|r_2$ and again $r_2|r_1$ repulsion energy have not been taken into consideration. It must be kept in mind that the sum of infinitesimals will be an integral, which has been given as follows:

$$\Delta\langle V_{ee}[\rho_0] \rangle = \langle V_{ee}[\rho_0] \rangle - \frac{1}{2}\int\int \frac{\rho_0(r_1)\rho_0(r_2)}{r_{12}}\mathrm{d}r_1\mathrm{d}r_2 \tag{14.45}$$

Making use of Eqs. (14.43–14.45), we can write

$$E_0 = \int \rho_0(r)V(r)\mathrm{d}r + \langle T_r[\rho_0] \rangle + \frac{1}{2}\int\int \frac{\rho_0(r_1)\rho_0(r_2)}{r_{12}}\mathrm{d}r_1\mathrm{d}r_2 + \Delta\langle T[\rho_0] \rangle + \Delta\langle V_{ee}[\rho_0] \rangle \tag{14.46}$$

The sum of the last two terms is called the *exchange-correlation energy functional*, and it is represented by E_{xc}, i.e.,

$$E_{xc}[\rho_0] = \Delta\langle T[\rho_0] \rangle + \Delta\langle V_{ee}[\rho_0] \rangle \tag{14.47}$$

Putting this value in Eq. (14.46), we obtain

$$E_0 = \int \rho_0(r)V(r)\mathrm{d}r + \langle T_r[\rho_0] \rangle + \frac{1}{2}\int\int \frac{\rho_0(r_1)\rho_0(r_2)}{r_{12}}\mathrm{d}r_1\mathrm{d}r_2 + E_{xc}(\rho_0) \tag{14.48}$$

This equation consists of four terms.

The first term is actually the integral of the density multiplied by the external potential, which can be written as

$$\int \rho_0(r)V(r)\mathrm{d}r = \int \left[\rho_0(r_1)\sum_{\text{nucleA}} -\frac{Z_A}{r_{1A}}\right]\mathrm{d}r_1$$
$$= -\sum_{\text{nuclieA}} Z_A \int \frac{\rho_0(r_1)}{r_{1A}}\mathrm{d}r_1 \tag{14.49}$$

Knowing the value of ρ_o, the value of integral can be computed.

The second term is $(T_r[\rho_o]])$, which can be expressed as

$$\langle T_r[\rho_0]\rangle = \langle \psi_r \rangle \left| \sum_{i=1}^{2n} -\frac{\nabla_i^2}{2} \right| \psi_r \tag{14.50}$$

It must be remembered that ψ_r can be expressed exactly in the form of a single Slater determinant of occupied spin molecular orbitals. Thus, for four electrons, ψ_r can be written as

$$\psi_r = \frac{1}{\sqrt{4!}} \begin{vmatrix} \psi_1^{ks}(1)\ \alpha(1)\psi_1^{ks}(1)\ \beta(1)\psi_2^{ks}(1)\ \alpha(1)\psi_2^{ks}(1)\ \beta(1) \\ \psi_1^{ks}(2)\ \alpha(2)\psi_1^{ks}(2)\ \beta(2)\psi_2^{ks}(2)\ \alpha(2)\psi_2^{ks}(2)\ \beta(2) \\ \psi_1^{ks}(3)\ \alpha(3)\psi_1^{ks}(3)\ \beta(3)\psi_2^{ks}(3)\ \alpha(3)\psi_2^{ks}(3)\ \beta(3) \\ \psi_1^{ks}(4)\ \alpha(4)\psi_1^{ks}(4)\ \beta(4)\psi_2^{ks}(4)\ \alpha(4)\psi_2^{ks}(4)\ \beta(4) \end{vmatrix} \tag{14.51}$$

In this determinant, there are 16 spin orbitals, and these are called *KS spin orbitals* of the reference system. Equation (14.50) can be written in terms of *KS* orbitals as follows:

$$\langle T_r[\rho_0]\rangle = -\frac{1}{2}\sum_{i=1}^{2n} \langle \psi_1^{ks}(1)|\nabla_1^2|\psi_1^{ks}(1)\rangle \tag{14.52}$$

The integrals are readily computed.

The third term in Eq. (14.48) will be found out if ρ_o is known. The fourth term is $E_{xc}[\rho_o]$, for which some method should be invented by careful thought. Finally, the energy equation may be written as

$$E_0 = -\sum_{\text{nucileiA}} Z_A \int \frac{\rho_0(r_1)}{r_{1A}}\mathrm{d}r_1 - \frac{1}{2}\sum_{i=1}^{2n} \langle \psi_1^{ks}(1)|\nabla_1^2|\psi_1^{ks}(1)\rangle$$
$$+\frac{1}{2}\int\int \frac{\rho_0(r_1)\rho_0(r_2)}{r_{12}}\mathrm{d}r_1\mathrm{d}r_2 + E_{xc}[\rho_0] \tag{14.53}$$

The error due to E_{xc} $[\rho_o]$ will be relatively small.

14.5 Kohn–Sham Equations

We are acquainted with the fact that the ground state energy of a system can be expressed as

$$E_0 = F[\rho] + \int \rho(\vec{r})V_{Ne}\mathrm{d}\vec{r} \tag{14.54}$$

where $F[\rho]$ = universal functional, which is equal to the sum of the kinetic energy, the classical coulomb interaction, and the non-classical portion.

It can be mathematically written as

$$F[\rho] = T[\rho] + J[\rho] + E_{ncl}[\rho] \tag{14.55}$$

In the RHS, only $J[\rho]$ is known. The main problem lies in obtaining expressions for $T[\rho]$ and $E_{ncl}[\rho]$. In 1965, Kohn and Sham gave some approach to solve the problem. They suggested to compute the exact kinetic energy of a non-interacting reference system with the same density as the real, interacting one. The mathematical formulation of the above fact can be written as

$$T_s = -\frac{1}{2}\sum_i^n \langle \psi_i | \nabla^2 | \psi_i \rangle \cdot \rho_s(\vec{r}) = \sum_i^n \sum_s |\psi_i(\vec{r}, s)|^2 = \rho(\vec{r}) \tag{14.56}$$

where ψ_i represents the orbitals of the non-interacting system. It may be said that T_s is not equal to the true kinetic energy of the system. Kohn and Sham accounted for the above by giving an expression for the functional $F[\rho]$, which is

$$F[\rho] = T_s[\rho] + J[\rho] + E_{xc}[\rho] \tag{14.57}$$

where E_{xc} = exchange-correlation energy

$T_s[\rho]$ = dominant part of the true kinetic energy $T[\rho]$

From Eq. (14.57), we can write

$$E_{xc}[\rho] = F[\rho] - T_s[\rho] - J[\rho] \tag{14.58}$$

But $F[\rho] = T[\rho] + J[\rho] + E_{ncl}[\rho]$

Putting the value of F [p] in Eq. (14.58), we can write

$$E_{xc}[\rho] = T[\rho] + J[\rho] + E_{ncl}[\rho] - T_s[\rho] - J[\rho]$$

$$= (T[\rho] - T_s[\rho]) + E_{ncl}[\rho]$$

We know that $E_{ncl}[\rho] = (E_{ee}[\rho] - J[\rho])$. Putting the value of E_{ncl} in the above equation, we get

$$E_{xc}[\rho] = (T[\rho] - T_s[\rho]) + (E_{ee}[\rho] - J[\rho]) \tag{14.59}$$

It should be remembered the exchange and correlation energy, $E_{xc}[\rho]$ is the functional in which everything is unknown. At this stage, the question arises that how can we uniquely find the orbitals in our non-interacting reference system? It may be stated in other words that how can we define a potential V_s such that it will provide with a Slater determinant, which describes the character of the same density as the real system? For the solution of the problem, we are going to write down an expression for the energy of the interacting system as

$$E[\rho] = F[\rho] + E_{Ne}[\rho] \tag{14.60}$$

$$E[\rho] = T_s[\rho] + J[\rho] + E_{xc}[\rho] + E_{Ne}[\rho] \tag{14.61}$$

This equation may further be expressed as

$$E[\rho]=T_s[\rho]+\frac{1}{2}\int\int\frac{\rho(\vec{r}_1)\rho(\vec{r}_2)}{r_{12}}\mathrm{d}r_1\mathrm{d}r_2+E_{xc}[\rho]+\int V_{Ne}\rho[\vec{r}]\mathrm{d}\vec{r}$$

$$=-\frac{1}{2}\sum_i^n\langle\psi_i|\nabla^2|\psi_i\rangle+\frac{1}{2}\sum_i^n\sum_j^n\iint|\psi_i(\vec{r}_1)|^2.\frac{1}{r_{12}}|\psi_j(\vec{r}_2)|^2\,\mathrm{d}r_1\,\mathrm{d}r_2+E_{xc}[\rho] \tag{14.62}$$

$$-\sum_i^n\int\sum_A^m\frac{Z_A}{r_{1A}}|\psi_i(\vec{r}_1)|^2\,\mathrm{d}\vec{r}_1$$

For E_{xc}, no explicit form can be given in the above equation. At this point, we shall apply the variational principle. One may ask that what condition must the orbitals $\{\psi_i\}$ fulfil in order to minimise the above energy expression under the constraint $\langle\psi_i|\psi_j\rangle\rangle=\delta_{ij}$.

The result obtained as a consequence constitutes K–S equations, which are written as

$$\left(-\frac{1}{2}\nabla^2+\left[\int\frac{\rho(\vec{r}_2)}{r_{12}}+V_{xc}(\vec{r}_2)-\sum_A^m\frac{Z_A}{r_{1A}}\right]\right)\psi_i$$

$$=\left(-\frac{1}{2}\nabla^2+V_s(\vec{r}_1)\right)\psi_i=\epsilon_i\,\psi_i \tag{14.63}$$

where $V_s(\vec{r}_1)=\int\frac{\rho(\vec{r}_2)}{r_{12}}\mathrm{d}\vec{r}_2+V_{xc}(\vec{r}_1)-\sum_A^m\frac{Z_A}{r_{1A}}$. The K–S equation (Eq. (14.63)) can also be represented as

$$\hat{h}^{ks}(1)\psi_i^{ks}(1)=\epsilon_i^{ks}\,\psi_i^{ks}(1) \tag{14.64}$$

14.5.1 Comments

- It is to be kept in mind that V_s depends on the density, and hence, K–S equations have to be solved iteratively.
- V_{xc} is defined as the functional derivative of E_{xc}, with respect to ρ and can be represented as $V_{xc}=\frac{\delta\,E_{xc}}{\delta\rho}$.
- Suppose the exact forms of E_{xc} and V_{xc} are known, then the Kohn–Sham strategy would give the exact energy.

14.6 Local Density Approximation

The word/term 'local' in local density approximation (LDA) has been used due to the fact that for any point only the conditions (electron density) at that point are taken into consideration. The assumption of LDA approximation is that the density $[\rho\,(r)]$ is varying very slowly with the position, and the inhomogeneous density of a solid or a molecule may be computed with the aid of the homogeneous electron gas functional.

Actually, the LDA is the basis of all approximate exchange-correlation functional. The idea of a uniform electron gas is the characteristic of this model. This is indeed a system in which electrons move on a positive background charge distribution in such a way that the total ensemble becomes neutral.

Keeping in mind the central idea of LDA, we can express the E_{xc} in the following form:

$$E_{xc}^{\mathrm{LDA}}[\rho]=\int\rho(\vec{r})\,\epsilon_{xc}\left(\rho(\vec{r})\right)\mathrm{d}\vec{r} \tag{14.65}$$

where the terms have their usual significance and $\in_{xc}\left(\rho(\vec{r})\right)$ = exchange-correlation energy per particle of a uniform electron gas of density $\rho(\vec{r})$.

This energy per particle is weighted with the probability or density $\rho(\vec{r})$ that there is an electron at this position. We can further write $\in_{xc}\left(\rho(\vec{r})\right)$ as a sum of exchange and correlation contributions as follows:

$$\in_{xc}\left(\rho(\vec{r})\right)=\in_{x}\left(\rho(\vec{r})\right)+\in_{c}\left(\rho(\vec{r})\right) \tag{14.66}$$

where $\in_{x}\left(\rho(\vec{r})\right)$ = exchange energy of an electron in a uniform electron gas of a particular density.

The value of $\in_{x}\left(\rho(\vec{r})\right)$ was first of all derived by Bloch and Dirac in the later part of the 1920s.

The value for $\in_{x}\left(\rho(\vec{r})\right)$ is given by

$$\in_{x}\left(\rho(\vec{r})\right)=-\frac{3}{4}\left(\frac{3\rho(\vec{r})}{\pi}\right)^{1/3} \tag{14.67}$$

It is clear that no such explicit expression is known for $\in_{c}\left(\rho(\vec{r})\right)$. However, highly accurate numerical quantum Monte–Carlo simulations of the uniform electron gas exist in the literature.

14.6.1 Comments on LDA

- The accuracy of the LDA for the exchange energy is found within 10%.
- When we find the ionisation energies of the atoms, dissociation energies of molecules, etc., their accuracy is found to be 10–20%, but the calculation of bond lengths of molecules has been found with an accuracy of ~2%.
- The moderate accuracy obtained by LDA is virtually insufficient for the most applications in chemistry.
- LDA may fail in systems, such as heavy fermions.

14.6.2 Application of the LDA

The following are the main applications of the LDA:

- In the determination of the lattice constant and bulk modulus of semi-conductor materials and insulators.
- Since the crystal volume V is accurately estimated, density is also obtained.

The bulk modulus is expressed as

$$B=-V\left(\frac{\partial\rho}{\partial V}\right)=-V\left(\frac{\partial^2 E}{\partial V^2}\right)$$

It is to be noted that the bulk moduli are computed by changing the lattice parameters and plotting the energy as a function of *V*. The curvature at the minimum of the *E*(*V*) plot is found to be proportional to the lattice constant.

14.6.3 Electron Gas

What is electron gas?

Jellium is supposed to be a hypothetical electrically neutral and infinite volume system. It comprises an infinite number of interacting electrons. The important feature of these electrons is to change their position into every part of space whose positive charge is frequently and uniformly spread. It is to be noted that the number of electrons per unit volume in the jellium has been found to be a constant value ρ, which is not zero ($\rho \neq 0$). The electron in jellium constitutes a uniform electron gas.

14.6.4 The Local Spin Density Approximation

Let us consider that electrons have α and β spins present in the homogenous electron gas and are assigned by ψ_α^{ks} and ψ_β^{ks}, which are spatial KS orbitals respectively.

Let ρ^α = Electron density function of electron having spin α from ψ_α^{ks} orbital and
ρ^β = electron density function of electron having spin β from ψ_β^{ks} orbital.

If LDA is worked out in detail, better results than LDA are found. This 'unrestricted' LDA has been named as local spin density approximation (LSDA), handle system with one or more unpaired electrons, for example, radicals and system in which electrons become unpaired like molecules far away from their equilibrium geometries even for simple molecules it seems more willing to forgive towards the use of E_{xc} functionals which is inexact.

In the case of species whose all the electrons are paired, the LSDA becomes equivalent to LDA, i.e., LSDA $\cong$ LDA. Keep in mind that like $E_X{}^{LDA}$ and $v_{XC}{}^{LDA}$ (its functional derivative of $E_X{}^{LDA}$), $E_{XC}{}^{LSDA}$ and $v_{XC}{}^{LSDA}$ can be accurately estimated. It has been found that LSDA geometries, frequencies, and electron distribution characteristics tend to be reasonably appreciable but dissociation energies have been found very poor. If uniform electron gas type LSDA calculation is replaced by an approach involving electron density and its gradient both, then calculations are found to be more accurate.

14.6.5 Generalised Gradient Approximation or Gradient Correlated Functional

Since the electron density in an atom or molecule changes largely from one place to another place, it is not a matter of surprise that electron gas model has serious drawbacks. Nowadays, most DFT calculations utilise exchange-correlation energy functions E_{xc}, which not only involve the LSDA but also use electron density and its gradient or slope (i.e., first derivative w.r.t position). These functional are often called 'Generalised Gradient Approximation' (GGA). They are also known as non-local functionals. The term non-local signifies the fact estimating the gradient of $\rho(r)$ at a point equals to the value of ρ an infinitesimal distance beyond the local point of the coordinate r. Since the gradient is equal to change in ρ over an infinitesimal distance beyond the local point of the coordinate r, $d\rho/dr$ is the change in ρ divided by change in r. While referring to gradient correlated functionals, the term non-local should be avoided.

The exchange-correlation energy functional is expressed as $E_{xc} = E_x + E_c$, i.e., sum of the exchange energy functional and correlation energy functional, both are negative $|E_X| >> |E_C|$. The value of $E_x = -30.19$ Hartrees and that of $E_c = -0.72$ Hartress for 'Ar' atom if calculated on the basis of HF method. It is a fact that gradient corrections have shown more effective when applied to exchange energy functional and an advancement in practical DFT.

Becke 88 functional is actually 'a new and largely improved functional for exchange energy'. Another example is the Gill 1996 (G96) functional which is the gradient correlated exchange energy functional. Lee–Yang–Parr and Perdew 1986 (P86) functionals are the examples of gradient corrected correlation energy functionals. These functionals are normally used with Gaussian type functions having e^{-r^2} basis functions for representing KS orbitals.

A calculation with B_{88} for E_x, LYP for E_c, and the basis set G-31G*

may be designed as B_{88} LYP |G-31* other possible combinations of E_x and E_c and basis set G–31 G*. Other possible combinations of E_x and E_c and basis set are expressed as G_{96LYP} |6–31+G* and G96 P86|6–311G*, etc.

14.6.6 Meta-Generalised Gradient Approximation

It is known to us that the functionals which use the first derivative of the electron density ρ is called GGA functionals. It may be mathematically represented as

$$E_{xc}^{GGA}\left[\rho^\alpha \rho^\beta\right] = \int f\left(\rho^\alpha(r), \rho^\beta(r)\ \nabla\rho^\alpha(r)\nabla\rho^\beta(r)\right) dr$$

where f = some function of the spin densities and their gradients. The GGA functional may further be improved by using the second derivative of electron density ρ and/or a quantity known as the kinetic energy density, τ (defined below). Such functionals are known as meta-generalised gradient approximation (mGGA) functionals and have the following form:

$$E_{XC}^{MGGA}\left[\rho^{\alpha}\rho^{\beta}\right]=\int f\left(\rho^{\alpha}\rho^{\beta}\nabla\rho^{\beta}\nabla^{2}\rho^{\alpha}\nabla^{2}\rho^{\beta}\tau_{\alpha}\tau_{\beta}\right)dr$$

The kinetic energy density $\tau(r)$ is found by summing up the squares of the gradients of the Kohn–Sham MOs, which are represented as

$$\tau_{(r)}=1/2\sum_{i=1}^{\text{Occupied}}\left|\nabla\varphi_{i}^{ks}(r)\right|^{2}$$

This varies with ρ in the same way as the Laplacian of ρ.

It should be kept in mind that mGGA DFT calculations need more time than GGA estimations and give somewhat better results as compared to GGA calculations. Becke's $B_{95\,\text{meta}}$ GGA correlation functional having two parameters whose values when introduced to atomic correlation energy data is often used.

14.6.7 Hybrid Functionals

Hybrid functionals are those functionals which increase the value of the DFT exchange-correlation energy with a term estimated with the aid of HF theory. According to the HF theory, we have the following expression for the electronic energy:

$$E=2\sum_{i=1}^{n}Hii+\sum_{i=1}^{n}\cdot\sum_{j=1}^{n}(2\,J_{ij}-K_{ij})$$

where the sums are considered over 'n' occupied spatial orbitals. Removing the core energy (first term, which is only KE and nucleus attraction energy and coulomb potential energy (having coulomb integral J) from the above equation, we shall be left with exchange energy) having only the double sum of exchange integral k which is given as follows:

$$E_{x}=-\sum_{i=1}^{n}\sum_{j=1}k_{ij} \tag{14.68}$$

Putting the KS orbitals into equation (14.68), we can write the equation for HF–exchange energy as follows:

$$E_{x}^{HF}=-\sum_{i=1}^{n}\sum_{j=1}^{n}\left\langle\psi_{i}^{ks}(1)\psi_{j}^{ks}(2)\middle|1/r_{ij}\middle|\psi_{i}^{ks}(2)\psi_{i}^{ks}(1)\right\rangle \tag{14.69}$$

Since the wave function of non-interacting electron can be represented by KS slater determinant, E_x^{HF} will represent the exact exchange energy for a system having non-interacting electrons whose electron density equals to that of real system. Keep in mind that Becke gave B3LYP hybrid GGA functional which is widely used functional. Becke proposed the B_{97} hybrid GGA functional which is found to be an improvement on B3LYP, B3PW91, and BIB95.

The exchange energy is then expressed as

$$E_{XC}=E_{X}^{GGA}+C_{X}E_{X}^{exact}+E_{c}^{GGA} \tag{14.70}$$

where c_x = a parameter

E_X^{GGA} and E_c^{GGA} = certain GGA functionals, which contain three and six parameters, respectively.

14.6.8 Time-Dependent DFT

We are aware of the following equation, i.e.,

$$-\hbar^2/2m_c\Delta_1^2\left\{-\hbar^2/2mc\nabla_1^2 - J_0\sum_{i=1}^{N} z1/r11 + J_0\int\frac{\rho(r_2)}{r_{12}}dr_2 + V_{XC}(r_1)\right\}\varphi(r_1)\} = E_i\psi_i(r_1) \quad (14.71)$$

The Kohn–Sham time-dependent DFT equation may be expressed, keeping in view the above equation, as follows:

$$\left\{-\hbar^2/2mc\nabla_1^2 - J_0\sum_{i=1}^{n} z1/r11 + J_0\int\frac{\rho(r_2)}{r_{12}}dr_2 + V_{ext}(t) + V_{XC(r,t)}\right\}\psi_i(r,t)\} = i\hbar d/dt\psi_{i(r,t)} \quad (14.72)$$

$$\text{and } \rho(r,t) = \sum_{i=1}^{n}\left|\psi_{i(r,t)}\right|^2 \quad (14.73)$$

where V_{ext} = external potential (e.g., the oscillating electromagnetic field).

Exchange-correlation potential, KS orbitals, and density are all time dependent. It is to be noted that the dependence of $V_{xc}(r,t)$ on ρ (r,t) does not match with the functional dependence of the time-independent density.

14.6.9 Application of Density Functional Theory

The following are the applications of DFT:

1. In finding out the geometries of molecules.
2. In estimating the quantities related to kinetics and thermodynamics.
3. It helps in finding out the bond orders, charges, and dipole moments, which are characteristic properties arising from electron distribution.
4. It helps in estimating, ionisation energies, electron affinity, electronegativity, hardness, softness, etc.

Below are the shortcomings:

1. Since the functional are only approximate, DFT used nowadays is not variational.
2. DFT is not accurate as compared to the highest level ab initio methods.
3. It has limitations with regard to band gaps and optical properties and strongly correlated systems.
4. DFT has semi-empirical character which is not a fundamental feature of basic method.
5. DFT at present is mainly a ground state theory, although the methods of applying it for excited states have been developed.

BIBLIOGRAPHY

Capelle, K. 2003. *J. Chem. Phys* 119: 1285.
Chermette, H. 1999. *J. Comp. Chem.* 20: 129.
Dunning, T.H. Jr. 1989. *J. Chem Phys.* 90: 1007.
Gilbert, T.L. 1975. *Phys. Rev B.* 12: 2111.
Gorling, A. 2000. *Phys. Rev. Lett.* 85: 4229.
Hehre, W.J. and Lou, L. 1997. *A Guide to Density Functional Calculations in Spartan.* Irvine CA: Wave function Inc.
Hohenberg, P. and Kohn, W. 1964. *Phys Rev. B.* 136: 864.

Jensen, F. 1999. *Introduction to Computational Chemistry*. New York: Wiley, 180.
Johnson, E.R. and Becke, A.D. 2005. *J. Chem. Phys.* 123: 024101.
Joulbert D. (Ed), *Density Functionals: Theory and Applications* (springer lecture notes in physics vol 500, 1998).
Koch, W. and Holthausen, M.C. 2001. *A Chemists Guide to Density Functional Theory*. Weinheim: Wiley-VCH.
Lammert, P.E. 1975. *J. Chem. Phys.* 125: 2111.
Lendvay, G. 1994. *J. Phys. Chem.* 98: 6098.
Lundin, U. and Eriksson, O. 2021. *Int .J. Quantum. Chem.* 81: 247.
Maitra, N.T., Burke, K. and Woodward, C. 2002. *Phys. Rev. Letts.* 89: 023002.
Muskar, J., Wander, A. and Harrison, N.M. 2001. *Chem. Lett.* 342: 397.
Parr, R.G. and Yang, W. 1995. *Annu. Rev Phys Chem.* 46: 701.
Parr, R.G. and Yang, W. 1955. Rev. Phys. Chem. 46: 701.
Parr, R.G. and Yang, W. 1989. *Density Functional Theory of Atoms and Molecules*. New York: Oxford University Press.
Seminario, J.M. and Politzer, P. 1995. *Density-Functional Theory: A Tool for Chemistry*. Amsterdam: Elsevier.
Seminario, J.M. and Politzer, P. 1995. *Modern Density Functional Theory; A Tool for Chemistry*. Amasterdam: Elsevier.
Sham, L.J. 1986. *Int. J. Quant. Chem.* 19: 49195.
Sutton, A.P. 1993. *Electronic Structure of Materials*. Oxford: Clarendon Press.
Tao, J. and Perdew, J.P. 2005. *J. Chem. Phys.* 122: 114102.
Truong, T.N., Duncan, W.T. and Bell, R.L. 1996. *Chemical Applications of Density Functional Theory*. Washington, DC: American Chemical Society.
Vydrow, O.A and Scuseria, G.E. 2004. *J. Chem. Phys* 121: 8187.

Questions on Concepts

1. What do you mean by function and functional? Explain giving examples.
2. Prove the Hohenberg–Kohn Theorem 1.
3. Prove the Hohenberg–Kohn Theorem 2.
4. Derive expression for Kohn–Sham energy.
5. Derive Kohn–Sham equations. Give some comments.
6. Write short notes on
   a. LDA
   b. v-Representable
   c. Applications of LDA
7. Explain LDA and give an equation for $\in_x \left(\rho(\vec{r})\right)$. Also comment on LDA.
8. a. What are the basic assumptions of the DFT?
   b. What is the essential difference between the wave function theory and the DFT?
9. State and prove Hohenberg–Kohn theorems. Do they determine the Hamiltonian operator?
10. Given that $f(x) = x^2$. Find the value of functional $\int_0^3 f(x)\mathrm{d}x$. [Ans. 9]
11. A functional is a function of a function. Explore the concept of a function of a functional.
12. If the wave function of a molecule is given, is it possible to estimate the electron density function? How?
13. What is v-representability?
    [Hint: A density $\rho(r)$ is said to be v-representable if it is associated with an antisymmetric ground state wave function of some Hamiltonian H distinguished by a special external potential $V_{\text{ext}}(r, R)$. It is to be mentioned that not all densities are v-representable. There is no general way to say if a density is v-representable.]

14. What is N-representability?

    [Hint: A density $\rho(r)$ is N-representable if it is associated with an antisymmetric wave function, not essentially a ground state wave function associated with a Hamiltonian H which requires]

$$\rho(r) \geq 0, \quad \int \rho(r)\mathrm{d}r = N_e$$

15. What do you mean by electron gas? Explain LSDA. How is it different from LDA?
16. Explain GGA? How is it better than LSDA?
17. Write a short note on mGGA. Do you think that mGGA is better than GGA calculation?
18. What are hybrid functionals? Write a short note on hybrid functionals?
19. Write down the time-dependent DFT equation and explain the terms involved there in?
20. Mention some merits and shortcomings of DFT?

Appendix I

Useful Integrals

1. $$\int_0^\infty e^{-ax^2}\,dx = \left(\frac{\pi}{4a}\right)^{1/2}$$

2. $$\int_0^\infty x e^{-ax^2}\,dx = \frac{1}{2a}$$

3. $$\int_0^\infty x^2 e^{-ax^2}\,dx = \frac{\sqrt{\pi}}{4}\cdot\frac{1}{a^{3/2}}$$

4. $$\int_0^\infty x^3 e^{-ax^2}\,dx = \frac{1}{2a^2}$$

5. $$\int_0^\infty x^4 e^{-ax^2}\,dx = \frac{3\sqrt{\pi}}{8}\cdot\frac{1}{a^{5/2}}$$

6. $$\int_0^\infty x^5 e^{-ax^2}\,dx = \frac{1}{a^3}$$

7. $$\int_0^\infty x^6 e^{-ax^2}\,dx = \frac{15\sqrt{\pi}}{16}\cdot\frac{1}{a^{7/2}}$$

8. $$\int_0^\infty x^n e^{-ax}\,dx = \frac{n!}{a^{n+1}}\ (a > 0,\ n \text{ positive integers})$$

9. $$\int_0^\infty x^{2n} e^{-ax^2}\,dx = \frac{1.3.5\ldots(2n-1)}{2^{n+1}a^n}\cdot\frac{\sqrt{\pi}}{a}\ (a > 0,\ n \text{ positive integers})$$

10. $$\int_0^\infty x^{2n} e^{-ax^2}\,dx = \frac{n!}{2a^{n+1}}\ (a > 0,\ n \text{ positive integers})$$

11. $$\int_{-0}^{+\infty} x^n e^{-ax^2}\,dx = 0 \quad \text{if } n \text{ is odd}$$

12. $$\int_{0}^{\infty} \frac{x dx}{e^x - 1} = \frac{\pi^2}{6}$$

13. $$\int_{0}^{\infty} \frac{x^3 dx}{e^x - 1} = \frac{\pi^4}{15}$$

14. $$\int_{0}^{\infty} e^{-ax} \cos bx dx = \frac{a}{a^2 + b^2}, \quad a > 0$$

15. $$\int_{0}^{\infty} e^{-ax} \cos bx dx = \frac{a}{a^2 + b^2}, \quad a > 0$$

16. $$\int_{0}^{\infty} \cos\, bx\; e^{-a^2x^2} dx = \frac{\sqrt{\pi}}{2a}, e^{-b^2/4a^2}, \quad a > 0$$

17. $$\int_{0}^{\infty} e^{-x^2} dx = 0.157\ \pi^{1/2}$$

18. $$\int_{0}^{\infty} x^n e^{-x/a}\,dx = n!\ a^{n+1}$$

19. $$\int_{-a}^{a} \sin^2\left(\frac{n\pi x}{2a}\right) \int_{-a}^{a} \cos^2\left(\frac{n\pi x}{2a}\right) = a$$

20. $$\int_{-a}^{a} x^2 \sin^2\left(\frac{n\pi x}{2a}\right) dx = \frac{1}{3} a^3 \left[1 - 6(-1)^2 / n^2\pi^2\right]$$

21. $$\int_{-\infty}^{\infty} e^{-a^2x^2 \pm bx}\,dx = \left(\frac{\pi^{1/2}}{a}\right) e^{b^2/4a^2}$$

22. $$\int_{-\infty}^{\infty} x^{-2} \sin^2 x dx = \pi$$

23. $$\int x e^{ax}\,dx = \frac{ax - 1}{a^2} \cdot e^{ax}$$

24. $$\int x^2 e^{ax}\,dx = \left(\frac{x^2}{a} - \frac{2x}{a^2} + \frac{2}{a^3}\right) \cdot e^{ax}$$

25. $\int x^n e^{ax} dx = e^{ax} \sum_{r=0}^{n} (-1)^r \cdot \frac{n! x^{n-r}}{(n-r)! a^{r+1}}$

26. $\int x^n \sin(ax) dx = -\frac{x^n \cos(ax)}{a} + \frac{n}{a} \int x^{n-1} \cos(ax) dx$

27. $\int x^n \cos(ax) dx = -\frac{x^n \sin(ax)}{a} + \frac{n}{a} \int x^{n-1} \sin(ax) dx$

28. $\int_b^\infty x^n e^{-ax} dx = -\frac{n! e^{-ab}}{a^{n+1}} \left[1 + ab + \frac{(ab)^2}{2!} + \ldots + \frac{(ab)^n}{n!} \right]$

29. $\int_0^\pi \sin^2 mx dx = \int_0^\pi \cos^2 mx dx = \frac{\pi}{2}$

30. $\int_0^\infty \frac{\sin x}{\sqrt{x}} dx = \int_0^\infty \frac{\cos x}{\sqrt{x}} dx = \sqrt{\frac{\pi}{2}}$

31. $\int \left(\sin^2 ax \right) dx = \frac{1}{2} x - \frac{1}{4a} \sin 2ax + c$

32. $\int \left(\cos^2 ax \right) dx = \frac{1}{2} x - \frac{1}{4a} \sin 2ax + c$

33. $\int \left(x^2 \sin^2 ax \right) dx = \frac{1}{6} x^3 - \left(\frac{1}{4a} x^2 - \frac{1}{8a^3} \right) \sin 2ax + \frac{1}{4a^2} \cdot x \cos 2ax + c$

34. $\int \left(x^2 \cos^2 ax \right) dx = \frac{1}{6} x^3 - \left(\frac{x^2}{4a} - \frac{1}{8a^3} \right) \sin 2ax + \frac{1}{4a^2} \cdot x \cos 2ax + c$

35. $\int x^m e^{ax} dx = \frac{x^{me^{ax}}}{a} - \frac{m}{a} \int x^{m-1} e^{ax} dx + c$

36. $\int \frac{e^{ax}}{x^m} dx = -\frac{1}{m-1} \cdot \frac{e^{ax}}{x^{m-1}} + \frac{a}{m-1} \int \frac{e^{ax}}{x^{m-1}} dx + c$

Appendix II

Mathematical Relationships

1. $\sin(x \pm y) = \sin x \cos y \pm \cos x \sin y$

2. $\cos(x \pm y) = \cos x \cos y + \sin x \sin y$

3. $\cos 2x = 1 - 2\sin^2 x = 2\cos^2 x - 1$

4. $\sin 2x = 2\sin x \cos x$

5. $\cos(x - y) - \cos(x + y) = 2\sin x \sin y$

6. $\cos(x - y) + \cos(x + y) = 2\cos x \cos y$

7. $\rightarrow \sin(x + y) + \sin(x - y) = 2\sin a \cos \beta$

8. $e^{\pm i\theta} = \cos\theta \pm i\sin\theta$

9. $\cos\theta = \dfrac{e^{i\theta} + e^{-i\theta}}{2}$

10. $\sin\theta = \dfrac{e^{i\theta} - e^{-i\theta}}{2i}$

11. $e^{x} = 1 - x + \dfrac{x^2}{2!} + \dfrac{x^3}{3!} + \ldots$

12. $\sin x = x - \dfrac{x^3}{3!} + \dfrac{x^5}{5!} - \dfrac{x^7}{7!} + \ldots$

13. $\cos x = 1 - \dfrac{x^2}{2!} + \dfrac{x^4}{4!} - \dfrac{x^6}{6!} + \ldots$

14. $(1 \pm x)^n = 1 \pm nx + \dfrac{n(n-1)x^2}{2!} \pm \dfrac{n(n-1)(n-2)x^3}{3!} + \ldots, \qquad x^2 < 1$

15. $\dfrac{1}{1-x} = 1 + x + x^2 + x^3 + x^4 + \ldots, \qquad x^2 < 1$

16. $\ln(1+x) = x - \dfrac{x^2}{2} + \dfrac{x^3}{3} - \dfrac{x^4}{4} + \ldots, \quad -1 < x \le 1$

Appendix III

Physical Constants

Constant	Symbol	Value
Avogadro number	N	6.022×10^{23} mol^{-1}
Bohr radius	a_0	5.292×10^{-11} m
Bohr magneton	μ_B	9.2741×10^{-24} Jτ^{-1}
Boltzmann constant	K	1.381×10^{-23} JK^{-1}
Electronic charge	E	1.602×10^{-19} c
Electronic rest mass	m_e	9.109×10^{-31} kg
Faraday constant	F	9.649×10^{4} c mol^{-1}
Gas constant	R	8.314 JK^{-1} mol^{-1}
Planck constant	h	6.626×10^{-34} Js
Proton charge	e	1.602×10^{-19} c
Proton rest mass	m_p	1.6726×10^{-27} kg
Rydberg constant	R_∞	2.1799×10^{-23} J or 1.097373 cm^{-1}
Stefan–Boltzmann constant	Σ	5.67032×10^{-18} Jm^{-2} K^{-4} s^{-1}
Speed of light	C	3.00×10^{8} ms^{-1}

Conversion Factors for Energy

Quantity	Joule	kj mol^{-1}	eV	a.u.	cm^{-1}	H_z
1 Joule	1	6.022×10^{20}	6.242×10^{18}	2.2939×10^{17}	5.035×10^{22}	1.509×10^{33}
1 a.u.	4.359×10^{-18}	2625	27.21	1	2.195×10^{b}	6.580×10^{lb}
1 cm^{-1}	1.986×10^{-23}	1.196×10^{-2}	1.240×10^{-4}	4.556×10^{-6}	1	2.998×10^{10}
1H$_Z$	6.626×10^{-34}	3.990×10^{-13}	4.136×10^{-15}	1.520×10^{-16}	3.336×10^{-11}	1
1 kjmol^{-1}	1.661×10^{-21}	1	1.036×10^{-2}	3.089×10^{-4}	83.60	2.506×10^{12}

Equivalent Units

Unit	Symbol	Equivalent Unit (s)
Angstrom	A	10^{-10} m
Hertz	Hz	s^{-1}
Joule	J	Nm, Kg m^{2} S^{-2}
Litre	í	Dm^{3}, 10^{-3} m^{3}
Newton	N	Kgm S^{-2}
Pascal	Pa	Nm^{-2}kg m^{-1} S^{-2}
Volt	V	JCr^{-1}, kg m^{2} S^{-3} A^{-1}

Model Question Papers

Set – I

1. a. If $A = \begin{pmatrix} 2 & 3 & 1 \\ 0 & -1 & 5 \end{pmatrix}$ and $B = \begin{pmatrix} 1 & 2 & -1 \\ 0 & -1 & 3 \end{pmatrix}$

Find 2*A* – 2*B*

Hint: According to the properties of matrices, we can write

$$A = \begin{pmatrix} 2 & 3 & 1 \\ 0 & -1 & 5 \end{pmatrix} \text{ and } B = \begin{pmatrix} 1 & 2 & -1 \\ 0 & -1 & 3 \end{pmatrix}$$

$$\therefore\ 2A = \begin{pmatrix} 4 & 6 & 2 \\ 0 & -2 & 10 \end{pmatrix} \quad \therefore\ 2B = \begin{pmatrix} 2 & 4 & -2 \\ 0 & -2 & 6 \end{pmatrix}$$

$$\therefore\ 2A - 2B = \begin{pmatrix} 2 & 2 & 4 \\ 0 & 0 & 4 \end{pmatrix}$$

b. Show that

$$A = \begin{pmatrix} -2 & 1 & 3 \\ 0 & -1 & 1 \\ 1 & 2 & 0 \end{pmatrix}$$ is a non-singular matrix. Find adj A and A^{-1}

Hint $A = \begin{pmatrix} -2 & 1 & 3 \\ 0 & -1 & 1 \\ 1 & 2 & 0 \end{pmatrix}$

$$\therefore\ \text{adj } A = \begin{pmatrix} A_{11} & A_{12} & A_{13} \\ A_{21} & A_{22} & A_{23} \\ A_{31} & A_{32} & A_{33} \end{pmatrix}$$

where A_{ij}'s are the co-factors of matrix element.

$$\therefore\ A_{11} = \begin{pmatrix} -1 & 1 \\ 2 & 0 \end{pmatrix} = -2$$

$$A_{12} = \begin{pmatrix} 0 & 1 \\ 1 & 0 \end{pmatrix} = -1, \text{ and so on}$$

$$\therefore\ A^{-1} = \frac{\text{adj } A}{|A|} \text{ where } |A| \text{ is determinant } A.$$

c. If A is a Hermitian matrix, show that iA is skew-Hermitian.

[Hint: A matrix is skew Hermitian if $a_{ij} = -\bar{a}_{ji}$, where $\bar{a}_{ji}$ is conjugate of a_{ij} and $a_{ij} = -\bar{a}_{ji}$ for Hermitian, then iA is its conjugate and A is real and iA is pure imaginary.]

d. For an electron wave, the wave function $\psi = Ae\dfrac{2\pi ix}{\lambda}$ show that $p_x = \pm\cdot\dfrac{h}{2\pi i}\dfrac{\mathrm{d}}{\mathrm{d}x}$

Hint: Given that $\psi = Ae^{\pm}\dfrac{2\pi ix}{\lambda} = Ae^{\pm iks}$

where, $k = 2\pi/\lambda$

According to de-Broglie relation

$$\lambda p = h$$

$$\therefore\ p = h/\lambda$$

Differentiating ψ w.r.t. x, we get

$$\frac{\mathrm{d}\psi}{\mathrm{d}x} = \pm ik\psi = \pm\frac{i2\pi}{\lambda}\psi = \pm\frac{2\pi ip}{h}\psi$$

$$\text{or}\quad \pm\frac{h}{2\pi i}\cdot\frac{\mathrm{d}}{\mathrm{d}x}\psi = p\psi$$

$$\therefore\ p_x = \pm\frac{h}{2\pi i}\cdot\frac{\mathrm{d}}{\mathrm{d}x}$$

2. Set up and solve the Hückel secular determinantal equation for butadiene. Construct the molecular orbital diagram for this system and calculate the delocalisation energy. [P.U. 2010]

   [Hint: It is fully described in Chapter 13.]

3. a. Derive Hohenberg-Kohn theorem.

   b. Explain the Kohn-Sham equation. [P.U. 2010]

   [Hint: These have been given in Chapter 14.]

4. How does extended molecular orbital method differ from the Hückel molecular orbital method? [P.U. 2011]

   [Hint: It has been mentioned in Chapter 13.]

5. Find the ground state energy for the one-dimensional harmonic oscillator

$$H = -\frac{h}{2m}\cdot\frac{\mathrm{d}^2}{\mathrm{d}x^2} + \frac{1}{2}mw^2x^2$$

using the Gaussian trial function, $\psi(x) = A\exp(-bx^2)$,

where b' is constant.

Hint: Given trial function $\psi(x) = A\exp(-bx^2)$. The value of A will be found out by normalisation of the function i.e., $\int A\exp(-bx^2)\cdot A\exp(-bx^2)\mathrm{d}x = 1$

$$\text{or } A^2\int_{-\infty}^{-\infty}\exp(-2bx^2)\mathrm{d}x = 1$$

and therefore A will be found out. The value of

$$A = \left(\frac{2b}{\pi}\right)^{1/4}$$

we know that $\langle H \rangle = \langle T \rangle + \langle V \rangle$
and it is given that:

$$H = -\frac{h^2}{2m} \cdot \frac{d^2}{dx^2} + \frac{1}{2} mw^2 x^2$$

the expectation value is

$$\langle T \rangle = \int_{-\infty}^{+\infty} \psi^* \left[-\frac{h^2}{2m} \cdot \frac{d^2}{dx^2} \right] \psi \, dx$$

$$\therefore \ \langle T \rangle = \int_{-\infty}^{+\infty} A \exp\left(-bx^2\right) \cdot \left(\frac{-h^2}{2m} \cdot \frac{d^2}{dx^2} \right) A e^{-bx^2} dx$$

$$= \frac{h^2 b}{2m}$$

Again $\langle V \rangle = 1/2 mw^2 |A|^2 \int \exp\left(-2bx^2\right) x^2 dx$

$$= \frac{mw^2}{8b}$$

$$\therefore \ \langle H \rangle = \frac{h^2 b}{2m} + \frac{mw^2}{8b}$$

According to the theory

$$E_{gs} \leq \langle \psi | H | \psi \rangle = \langle H \rangle$$

The value of $\langle H \rangle$ exceeds E_{gs} for any b. After minimizing, we shall get

$$\frac{d}{db} \langle H \rangle = \frac{h^2}{2m} - \frac{mw^2}{b^2}$$

$$\therefore \ b = \frac{mw}{2h}$$

Putting the value of $\langle H \rangle$, we get

$$\langle H \rangle_{\min} = \frac{h^2}{2m} \cdot \frac{mw}{2h^2} + \frac{mw^2}{8} + \frac{2h}{mw}$$

6. How will you show that the 'variation' and perturbation methods are the powerful techniques to obtain approximate Eigen values and Eigen functions?
   [Hint: It has been fully described in Chapter 9.]
7. a. Derive an expression for time-independent Schrödinger wave equation.
   [Hint: It has been derived in Chapter 6.]
   b. Use the HMO theory and show whether the linear state (H–H–H⁺) or the triangular state H–H–H (triangle) of H_3^+ is more stable?
   Hint: For linear geometry of H_3^+, the secular equation is expressed as

$$c_1x + c_2 = 0$$
$$c_2x + c_1 + c_3 = 0$$
$$c_3x + c_2 = 0$$

and the corresponding secular determinant is written as

$$\begin{vmatrix} x & 1 & 0 \\ 1 & x & 1 \\ 0 & 1 & x \end{vmatrix} = 0$$

On solving, we get

$$x = 0, +\sqrt{2} \text{ and } -\sqrt{2}$$

$$\therefore\ E_{H_3^+} = 2\alpha + 2\sqrt{2}\beta$$

In case triangular state, the secular equations will be

$$c_1x + c_2 + c_3 = 0$$
$$c_2x + c_1 + c_3 = 0$$
$$c_3x + c_1 + c_2 = 0$$

and the corresponding determinant will be

$$\begin{vmatrix} x & 1 & 1 \\ 1 & x & 1 \\ 1 & 1 & x \end{vmatrix} = 0$$

and the polynomial equation will be

$$x^3 - 3x + 2 = 0$$

The roots will be $x = 1, 1, -2$

$$\therefore\ E_{H_3^+} = 2\alpha + 4\beta$$

Therefore, $E_{H_3^+}$ (triangular state) will be more stable than $E_{H_3^+}$ (linear state).

8. Write notes on the following
   a. Brillouin's theorem [P.U. 2010]
      [Hint: It is given in Chapter 11.]
   b. Gaussian basis set [P.U. 2010]
      [Hint: Mentioned in Chapter 11.]
9. How will you show that the molecular orbital treatment with configuration interaction produces better results with respect to the valence bond treatment?
   [Hint: It is fully described in Chapter 10.]
10. Derive $\sum_{j=1} c_j\left(h_{ij} + g_{ij} - \in s_{ij}\right) = 0$ for Roothaan's method in one dimension.
   [Hint: Consult Chapter 12.]

Set – II

1. a. Find the transpose of the following matrix
[Hint: While obtaining the transpose of a matrix A, the row is changed into column.]

$$\text{Here } A = \begin{pmatrix} 1 & 4 & 10 & 2 & 9 & 2 \\ 12 & 2 & 7 & 8 & 11 & 6 \\ 19 & 3 & 0 & 5 & 21 & 1 \end{pmatrix}$$

$$\therefore\ A^T = \begin{pmatrix} 1 & 12 & 19 \\ 4 & 2 & 3 \\ 10 & 7 & 0 \\ 2 & 8 & 5 \\ 9 & 11 & 21 \\ 2 & 6 & 1 \end{pmatrix}$$

b. Given two matrices are:

$$A = \begin{pmatrix} 1/2 & 0 \\ 0 & 1/2 \end{pmatrix} \text{ and } B = \begin{pmatrix} -1/2 & 0 \\ 0 & 1/2 \end{pmatrix}$$

Form the matrix $R = BA - AB$.

$$\text{Hint: Given that } A = \begin{pmatrix} 1/2 & 0 \\ 0 & 1/2 \end{pmatrix} \text{ and } B = \begin{bmatrix} -1/2 & 0 \\ 0 & 1/2 \end{bmatrix}$$

$$\therefore\ AB = \begin{pmatrix} -1/4 & 0 \\ 0 & 1/4 \end{pmatrix},\ BA = \begin{pmatrix} 1/4 & 0 \\ 0 & 1/4 \end{pmatrix}$$

$$\therefore\ R = BA - AB = \begin{pmatrix} 0 & 0 \\ 0 & 0 \end{pmatrix}$$

c. Show that the square of the expectation value of energy is equal to the expectation value of square of energy.

Hint: We have to show that $\{\langle\psi|E|\psi\rangle\} = \langle\psi \mid E^2 \mid \psi\rangle$, where ψ is a normalised wave function.

We know that $\int \psi^*\psi \ d\tau = 1$

$$\therefore\ \langle E\rangle\langle E\rangle = \langle\psi^*|E|\psi\rangle\langle\psi^*|E|\psi\rangle$$

$$= \langle\psi^*|E^2|\psi\rangle,\ \text{since}\langle\psi\rangle\ \langle\psi\rangle = 1]$$

d. The two Eigen functions of the operator $\dfrac{d^2}{d\Phi^2}$ are $e^{+im\Phi}$ and $e^{-im\Phi}$. Show that any linear combination of these two Eigen functions is also an Eigen function of the operator $\dfrac{d^2}{d\Phi^2}$.

Hint: Let $\Psi = c_1 e^{im\Phi} + c_2 e^{-im\Phi}$ be linear combination of two state functions.

Then $\dfrac{d\psi}{d\Phi} = (im)c_1 e^{im\Phi} - (im)c_2 e^{-im\Phi}$

$$\therefore \ \frac{d^2\varphi}{d\Phi^2} = (im)(im)c_1 e^{im\Phi} - (im)(-im)c_2 e^{-im\Phi}$$

$$= i^2 m^2 c_1 e^{im\Phi} + i^2 m^2 c_2 e^{-im\Phi}$$

$$= i^2 m^2 [c_1 e^{im\Phi} + c_2 e^{-im\Phi}]$$

$$= -m^2 \psi$$

which is an Eigen value problem.]

2. a. State and prove the variation theorem.

Hint: The state function will be stationary with respect to its variation in parameters, i.e., $\dfrac{d\psi}{dx} = 0$, ψ will have its extreme value. The expectation value of energy is expressed as

$$\langle E \rangle = \frac{\langle \psi | \hat{H} | \psi \rangle}{\langle \psi | \psi \rangle}, \ \hat{H} \text{ is Hamiltonian operator.}$$

Variational principle requires $\delta\langle E \rangle = 0$, which is Schrödinger equation. Considering variation in ψ, we get

$$\langle E \rangle \langle \psi | \psi \rangle = \langle \psi | \hat{H} | \psi \rangle$$

$$\text{or } \ \delta\{\langle E \rangle \langle \psi | \psi \rangle\} = \delta\{\langle \psi | \hat{H} | \psi \rangle\}$$

$$\text{or } \ \delta\langle E \rangle \langle \psi | \psi \rangle + \langle E \rangle \delta\{< \psi | \psi >\} = \langle \delta\psi | \hat{H} | \psi \rangle + \langle \psi | \hat{H} | \delta\psi \rangle$$

$$\text{or } \ \langle \psi | \psi \rangle \delta\langle E \rangle = \langle \delta\psi | \hat{H} \langle E \rangle | \psi \rangle + \langle \psi | \hat{H} - \langle E \rangle | \delta\psi \rangle$$

$$\text{or } \ \langle \delta\psi | \hat{H} - \langle E \rangle | \psi \rangle = 0 \text{ , for an arbitrary } \psi$$

This is possible if and only if

$$\left(\hat{H} - \langle E \rangle | \psi \rangle\right) = 0$$

or $\hat{H}\psi = E\psi$, which is a Schrödinger equation.

b. Use the variation principle to estimate the ground state energy of a harmonic oscillator using the trial function $\phi(x) = \dfrac{1}{1+\beta x^2}$

The Hamiltonian of the Harmonic oscillator is $\hat{H} = -\dfrac{h}{2\mu} \cdot \dfrac{d^2}{dx^2} + 1/2kx^2$.

Hint: The trial state function is given by $\Phi(x) = \dfrac{1}{1+\beta x^2}$. According to question, Hamiltonian operator for harmonic oscillator is

$$\hat{H} = -\frac{h}{2\mu}\frac{d^2}{dx^2} + \frac{1}{2}kx^2$$

But $E = \langle \psi | \hat{H} | \psi \rangle$, Putting Φ (*x*) in place of ψ, we get

$$E = \int_{-\infty}^{+\infty} \left(\frac{1}{1+\beta x^2} \right) \left(\frac{-h^2}{2\mu} \cdot \frac{d^2}{dx^2} + \frac{1}{2} kx^2 \right) \left(\frac{1}{1+\beta x^2} \right) dx$$

$$= \int_{-\infty}^{+\infty} \Phi(x) \left(\frac{-h^2}{2\mu} \cdot \frac{d^2\Phi(x)}{dx^2} + \frac{1}{2} kx^2 \Phi(x) dx \right)$$

$$= \int_{-\infty}^{+\infty} \left(\frac{1}{1+\beta x^2} \right) [\ldots]$$

The integral can be solved for ground state energy by using the standard integration

$$\int_{-\infty}^{+\infty} \frac{dx}{(a^2 + x^2)^n} = \frac{(2n-3)!!}{(2n-2)!!} \left(\frac{\pi}{a^{2n-1}} \right), \quad [a > 0, n = 2,3,\ldots]$$

$$\int_{-\infty}^{+\infty} \frac{x^2 dx}{(a^2 + x^2)^2} = \frac{2}{a} \int_0^{\pi/2} \sin^2\theta \, d\theta = \frac{\pi}{2a}$$

$$\int_{-\infty}^{+\infty} \frac{x^2 dx}{(a^2 + x^2)^4} = \frac{2}{a^5} \int_0^{\pi/2} \sin^2\theta \cos^4\theta \, d\theta = \frac{\pi}{16a^5}$$

c. Find the probability density $|\psi|^2$ of a particle if its wave function is given as $\psi = Ae^{ix}$. Explain the significance of the result.

[Hint: Particle's state function is given by $\psi = Ae^{ix}$

$$\therefore |\psi|^2 = \psi^* \psi = (Ae^{ix})(Ae^{-ix})$$

$$= A^2$$

Therefore, the probability will depend on parameter *A*]

3. Write down Schrödinger wave equation for hydrogen atom in cartesian as well as spherical polar co-ordinates. Separate it into radial and angular equations. Describe a procedure for solving radial equation and find its first three wave functions.

[P.U. 2006]

Hint: Schrödinger wave equation in cartesian form is given by

$\nabla^2\psi + \frac{2\mu}{h^2}(E - V) = 0$, which can be transformed into spherical polar coordinate in the following form:

$$\frac{1}{r^2} \frac{d}{dr} \left(r^2 \frac{d\psi}{dr} \right) + \frac{1}{r^2 \sin\theta} \cdot \frac{d}{d\theta} \left(\sin\theta \frac{d\psi}{d\theta} \right) + \frac{1}{r^2 \sin^2\theta} \cdot \frac{d^2\psi}{d\Phi^2} + \frac{2\mu}{h^2} \{E - V(r)\} \psi = 0$$

Setting $\psi = R(r) Y_l^m(\theta, \Phi)$, the equation can be separated into an angular part and a radial part. Radial part of wave equation will be expressed as

$$\frac{1}{r^2} \frac{d}{dr} \left(r^2 \frac{dR}{dr} \right) + \left[\frac{2\mu}{h^2} \left\{ E - V(r) - \frac{\lambda}{r^2} \right\} \right] R = 0$$

From the angular part of wave function, $\lambda = l(l+1)$ and taking potential $V(r) = -Ze^2/r$, the radial part of the equation becomes

$$\frac{1}{r^2}\frac{\mathrm{d}}{\mathrm{d}r}\left(r^2\frac{\mathrm{d}R}{\mathrm{d}r}\right)+\frac{2\mu}{h^2}\left[E+\frac{Ze^2}{r}-\frac{l(l+1)h^2}{2\mu r^2}\right]R=0$$

Now transforming by introducing dimensionless variable, $\rho = \alpha r$, the above equation can be expressed as

$$\frac{1}{\rho^2}\frac{\mathrm{d}}{\mathrm{d}\rho}\left(\rho^2\frac{\mathrm{d}}{R}\mathrm{d}\rho\right)+\left[\frac{\lambda'}{\rho}-\frac{1}{4}-\frac{l(l+1)}{\rho^2}\right]R=0$$

where, $\alpha^2 = \dfrac{8\mu|E|}{h^2}$, $\lambda' = \dfrac{2\mu ze^2}{\alpha h^2}$

The wave equation can be solved by introducing $R(\rho) = F(\rho)e^{-\rho/2}$,

where, $F(\rho) = \sum_{v=0}^{\infty} a_v \rho^{spv}$ where, s is a parameter,

Differentiating and substituting in equation, the solution can be obtained as

$$R(r) = Ce^{-\rho/2}\cdot\rho^l(l)L_{n+1}^{2l+1}(\rho)$$

where, C = normalisation constant
$L_{n+1}^{2l+1}(\rho)$ = Laguerre polynomial
After normalisation, we have

$$R(r)=\sqrt{\left[\left(\frac{2Z}{na_0}\right)^3\frac{(n-l-1)!}{2n\{(n+1)!\}^3}\right]}\exp\left[\frac{-Zr}{na_0}\right]\left(\frac{2Zr}{na_0}\right)^l\cdot L_{n+1}^{2l+1}\left(\frac{2Zr}{na_0}\right)$$

For H atom, $Z = 1$, $A_0 = \dfrac{h^2}{\mu e^2}$, first Bohr radius.

The first three solutions of radial part are given as

$$\psi_{100} = \frac{1}{\sqrt{\pi}}\frac{1}{a_0^{3/2}}\exp\left(\frac{-r}{2a_0}\right)$$

$$\psi_{200} = \frac{1}{4\sqrt{2\pi}}\frac{1}{a_0^{3/2}}\left(2-\frac{r}{a_0}\right)\exp\left(\frac{-r}{2a_0}\right)$$

$$\psi_{300} = \frac{1}{81\sqrt{3\pi}}\frac{1}{a_0^{3/2}}\left(27-18\rho+2\rho^2\right)e^{\frac{-\rho}{2}}.$$

4. a. What is zero differential overlap approximation? [P.U.]
   [Hint: Consult Chapter 12.]

   b. Derive ground state term symbols of the following
   (a) 2p 3p 3d system (b) V, (c) C_r
   [Hint: It is mentioned in Chapter 11.]

   c. Show by evaluation of the ground state energy that for Li atom, a configuration $1s^3$ is impossible.
   [Hint: It has been solved in Chapter 11.]

5. a. Discuss the Condon–Slater rules.
   [Hint: It is fully described in Chapter 11.]
   b. Write the Hartree–Fock equation and explain the terms involved
   [Hint: See Chapter 11.]
   c. Explain the difference between Hartree-SCF and Hartree–Fock SCF method.
   [Hint: It is given in Chapter 11.]
6. What is density functional theorem? Discuss the treatment of chemical concepts with density functional theory.
   [Hint: Consult Chapter 14.]
7. Use the HMO approximation to obtain
   a. bond orders.
   b. charge distribution.
   c. Free valence for butadiene. [P.U. 2011]
   [Hint: These have been calculated in Chapter 13.]
8. Considering $\psi = c_1 2xe^{-x} + c_2\sqrt{32}xe^{-2x}$ and $H = -\frac{1}{2}\frac{d^2}{dx^2} - \frac{2}{x}$ prove that $h_{11} = -1.5$ and $h_{12} = h_{21} = -1.6761$.
   [Hint: It has been proved in Chapter 12.]
9. Prove that the electronic energy of the electron is

$$E = \sum^{n/L} E_i + \frac{1}{2}\sum_i\sum_j P_{ij}H_{ij}^0$$

   where the terms have their usual significance.
   [Hint: It has been proved in Chapter 12.]
10. Prove that the angle between any two of the sp^3 hybrid orbitals is 109°28′
    [It has been solved in Chapter 12 under Section 'Solved Problems'.]
    or
    Derive Hermite Differential equation. [P.U. 2007]
    [Hint: Schrödinger wave equation for one-dimensional harmonic oscillator is

$$\frac{d^2\psi}{dx^2} + \frac{2m}{h^2}\left(E - \frac{1}{2}kx^2\right)\psi = 0$$

    Introducing $\xi = \alpha x$, such that $\frac{d^2\psi}{dx^2} = \alpha^2\frac{d^2\psi}{d\xi^2}$

    The above equation becomes

$$\frac{d^2\psi}{d\xi^2} + \left[\frac{2mE}{h^2\alpha^2} - \frac{mk}{\alpha^4 h^2}\xi^2\right]\psi = 0$$

    Choosing $\frac{mk}{\alpha^4 h^2} = 1 \Rightarrow = \alpha = \left(\frac{mk}{h^2}\right)^{1/4}$

    and $\frac{2mE}{h^2\alpha^2} = \lambda$, the equation can be transformed into

$$\frac{d^2\psi}{d\xi^2} + \left(\lambda - \xi^2\right)\psi = 0$$

    which is the Hermite differential equation.]

Set – III

1. a. What is an operator? Explain linear and Hermitian operators with examples.

[Hint: An operator is a mathematical operation like multiplication, division, addition, subtraction, differentiation and integration, etc., which operates on certain function $f(x)$ and changes its value.]

Linear operator: An operator $\hat{\alpha}$ is a linear operator if for arbitrary operands

$$\psi_1(x) \text{ and } \psi_2(x), \text{ we get}$$

$$\hat{\alpha}(\psi_1 + \psi_2) = \hat{\alpha}\psi_1 + \alpha\hat{\psi}_2$$

and for operator $\hat{C}$

$$\hat{c}(\psi_1 + \psi_2) = \hat{c}\psi_1 + \hat{c}\psi_2$$

and for an arbitrary constant C

$$\hat{\alpha}c = c\hat{\alpha}$$

Hermitian operator (self adjoint)

An operator $\hat{H}$ is said to be Hermitian if it operates on any two well-behaved functions $f(x)$ and $g(x)$ vanishing at infinity and satisfying the equation

$$\int_{-\infty}^{+\infty} g^*(x)\left[\hat{H}f(x)\right] = \int_{-\infty}^{+\infty}\left[\hat{H}g(x)\right]^* \cdot f(x)\mathrm{d}x$$

$$\text{or} \quad \int_{-\infty}^{+\infty} \psi^* \hat{H}\psi \mathrm{d}x = \int_{-\infty}^{+\infty} \psi\left(\hat{H}\psi\right)^* \mathrm{d}x \quad \text{for an arbitrary } \psi$$

b. Show that Eigen functions of a Hermitian operator corresponding to different Eigen values are orthogonal.

[Hint: Let X_1 and X_2 be Eigen functions belonging to operator $\hat{H}$. The Eigen values are λ_1 and λ_2.

$$\therefore \hat{H}X_1 = \lambda_1 X_1, \hat{H}X_2 = \lambda_1 X_2$$

$$\therefore X_2\hat{H}H_1 = \lambda_1 X_2 X_1 \text{ and } \lambda_1\hat{H}X_2 = \lambda_2 X_1 X_2$$

$$\Rightarrow \left(X_2\hat{H}H_1 - X_1\hat{H}X_2\right) = (\lambda_1 - \lambda_2)X_1X_2$$

For a Hermitian operator

$$\left(\hat{H}X_1X_2\right) - \left(X_1\hat{H}X_2\right)$$

$$\therefore \left(X_2\hat{H}X_1\right) - \left(\hat{H}X_2X_1\right) = 0$$

As $\lambda_1 - \lambda_2 \neq 0$

$\therefore$ $(X_1, X_2) = 0$, i.e., Eigen functions X_1 and X_2 are orthogonal.]

c. Given operator $\frac{\mathrm{d}}{\mathrm{d}r}$ with a set of Eigen function exp $(-ar)$ ('a' being real and positive) and real Eigen values. Is the operator necessarily Hermitian?

[Hint: Let the operator $\frac{\mathrm{d}}{\mathrm{d}r} = \hat{H}$ and Eigen function $\psi = \exp(-ar)$

$$\therefore\ \hat{H}\psi = \frac{\mathrm{d}}{\mathrm{d}r}\left(e^{-ar}\right) = -ae^{-ar}$$

$$\therefore\ \left(\hat{H}\psi, \psi\right) = -ae^{-2ar}$$

and $\left(\psi,\ \hat{H}\psi\right) = -ae^{-2\alpha r}$, this shows that $\left(\psi,\ \hat{H}\psi\right) = \left(\hat{H}\psi,\ \psi\right)$, hence the operator is necessarily Hermitian]

2. a. Explain perturbation theory and apply it in the derivation of energy and wave function for He atom. [P.U. 2009]

[Hint: The perturbation theory is defined as, 'changes in energy levels and Eigen functions of a system takes place when a small disturbance in the system is applied'. The importance may be seen in space-dependent potential, or changes in Hamiltonian on application of electric field and magnetic field. Stark effect and Zeeman effect may be observed through this phenomenon. Perturbation principle demands change in Hamiltonian, energy Eigen values, and Eigen function by small quantity from equilibrium state, i.e.,

$$\hat{H} = \hat{H}_0 + \lambda H'$$

$$E = E_0 + \lambda E'$$

$$\psi = \psi_0 + \lambda \psi'$$

The Schrödinger wave equation holds for equilibrium and perturbed state both, i.e.,

$$\hat{H}\psi = E\psi$$

$$\hat{H}_0\psi_0 = E_0\psi_0$$

From the above equation, we can write

$$\left(\hat{H}_0 + \lambda \hat{H}'\right)\left(\psi_0 + \lambda\psi'\right) = \left(E_0 + \lambda E'\right)\left(\psi_0 + \lambda\psi'\right)$$

$$\Rightarrow \hat{H}_0\psi_0 + \lambda\hat{H}_0\psi' + \lambda\hat{H}'\psi_0 + \lambda^2\hat{H}'\psi' = E_0\psi_0 + \lambda E_0\psi' + \lambda E'\psi_0 + \lambda^2 E'\psi'$$

$$\Rightarrow \lambda_0\left(\hat{H}_0\psi_0 - E_0\psi_0\right) + \lambda\left(\hat{H}_0\psi' - E_0\psi'\right) + \lambda\left(\hat{H}'\psi_0 + E'\psi_0\right)$$

$$+ \lambda^2\left(\hat{H}'\psi' - E'\psi'\right) = 0$$

From these, we may write

$$H_0\psi_0 = E_0\psi_0$$

$$\left(\psi_0, \psi'\right) = 0$$

$$(\psi_0,\psi_0)=1$$

$\therefore \quad E' = \dfrac{\langle \psi_0 | \hat{H} | \psi' \rangle}{\langle \psi_0 | \psi' \rangle}$ can be obtained; and similarly ψ' can be obtained. These have been illustrated in Chapter 9 in detail.]

First order perturbation

Energy and wave function for He-atom

First order perturbation equation is $\left(H_0 - E_n^0\right)\psi_n' = (E_n' - H')\psi_n'$

Energy Eigen value is

$E_n' = \langle \psi_n^0 | H' | \psi_n^0 \rangle$ and the wave function may be written as

$$\psi_n' = \psi_n^0 + \lambda\psi'$$

$$= \psi_n^0 + \lambda \sum_{k\neq n} \frac{\langle \psi_n^0 | H' | \psi_n^0 \rangle}{\left(E_n^0 - E_k^0\right)} \psi_k^0$$

He atom has two electrons and a nucleus, which is shown as

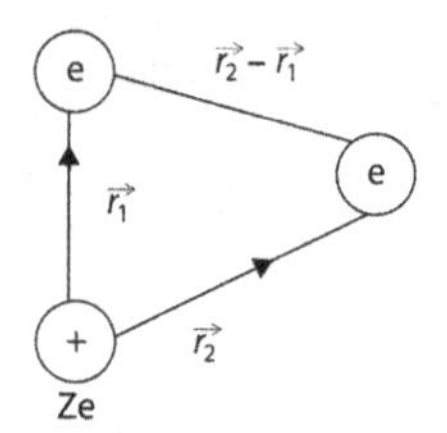

The atom is conceived as having +*Ze* charge on nucleus and two electrons at distances $\vec{r}_1$ and $\vec{r}_2$.

$\therefore$ Hamiltonian of the system is given by

$$H = \frac{-h^2}{2\mu}\left(\nabla_1^2 + \nabla_2^2\right) - \left(\frac{1}{r_1} + \frac{1}{r_2}\right) Ze^2 + \frac{e^2}{|\vec{r}_1 - \vec{r}_2|}$$

where, μ = Reduced mass

or $H = H_0 + H'$

where, $H_0 = \dfrac{-h^2}{2\mu}\left(\nabla_1^2 + \nabla_2^2\right) - \left(\dfrac{1}{r_1} + \dfrac{1}{r_2}\right) Ze^2$

$$H' = \frac{e^2}{|\vec{r}_1 - \vec{r}_2|}$$

The wave function $\phi \cdot (\vec{r}_1 - \vec{r}_2) = \psi_1(\vec{r}_1)\psi_2(\vec{r}_2)$

where, ψ_1 and ψ_2 are hydrogen-like wave functions.

We know that

$$\psi(r) = \sqrt{Z^3/\pi}\, \exp(-zr),\ z = Ze^2/h^2$$

unperturbed Eigen value and Eigen function are

$$E_0 = -\frac{Z^2h^2}{\mu}$$

$$\phi_0 = \frac{Z^3}{\pi}\exp\left[-Z(r_1 + r_2)\right]$$

First order perturbation energy,

$$E'_m = \langle \psi_m | H' | \psi_m \rangle$$

or $E'_m = \dfrac{Z^6 e^2}{\pi^2} \displaystyle\iint \frac{\exp\left[-2Z(r_1 + r_2)\right]}{|\vec{r}_1 - \vec{r}_2|} dr_1^3 \cdot dr_2^3$

After evaluation, $E'_m = \dfrac{5}{8} Ze^2$

$$\therefore \quad \text{Ground state energy} = \left(\frac{Z^2 h^2}{\mu} - \frac{5}{8} Ze^2 \right)$$

[N.B. for the details, consult Chapter 9.]

3. Write notes on

   a. Projection operator
      [Hint: Consult Chapter 4.]

   b. Function and functional
      [Hint: It is given in Chapter 14.]

   c. Local density approximation
      [Hint: For this consult Chapter 14.]

4. Derive an expression for energy of a free particle in a three-dimensional box.
   [Hint: It is fully derived in Chapter 7.]

5. Find the F and T equations for a rigid rotor and solve them.
   [Hint: It has been given in Chapter 7.]

6. Show that for H atom. 1s and 2s orbitals are orthogonal.
   [Hint: Consult Chapter 8.]

7. Using HMO theory, set up and solve the secular determinantal equation for cyclopropene. Construct the molecular diagram for this system and also determine its molecular orbitals.
   [Answered in Chapter 13.] [P.U. 2010]

8. a. Briefly state the various postulates of quantum mechanics.
      [Hint: It has been given in Chapter 5.]

   b. Show that if A and B have simultaneous Eigen vectors (i.e., both diagonalised by the same similarity transformation), then A and B commute.
      [Hint: Let $|\psi_i>$ be a simultaneous Eigen vectors of operators A and B.

$$\therefore \quad A|\psi_i> = a_i|\psi_i>,\ B|\psi_i> = b_i|\psi_i>$$

$$\therefore \quad |\psi\rangle = \sum_{i=1}^{n} c_i \,|\psi_i>$$

$$\therefore \quad AB|\psi\rangle = \sum_{i=1}^{n} c_i \ AB|\psi_i\rangle$$

$$= \sum_{i=1}^{n} c_i \ a_i b_i |\psi_i\rangle$$

$$\text{Similarly, } BA|\psi\rangle = \sum_{i=1}^{n} c_i \; BA|\psi_i\rangle$$

$$= \sum_{i=1}^{n} c_i \; b_i a_i |\psi_i\rangle$$

Where, c_i, a_i and b_i are numbers

$$\therefore \quad (AB - BA)|\psi\rangle = 0$$

$\therefore AB - BA = 0$, i.e., A and B commute.]

c. Express the matrix

$$Z = \begin{pmatrix} 3+i2 & 4+i3 \\ 5+i4 & 3-i5 \end{pmatrix} \text{ in the X + iY form}$$

$$\text{[Hint: } Z = \begin{pmatrix} 3+i2 & 4+i3 \\ 5+i4 & 3-i5 \end{pmatrix}$$

$$= \begin{pmatrix} 3 & 4 \\ 5 & 3 \end{pmatrix} + i\begin{pmatrix} 2 & 3 \\ 4 & -5 \end{pmatrix} \text{ which is of X + iY form]}$$

9. Find the value of coefficients of ψ_1 and ψ_2 equations, when hybridisation has taken place in H_2O molecule.

   [Hint: It is fully described in Chapter 12.]

10. State and prove the Hohenberg–Kohn theorem 2.

    [Hint: It is stated and proved in Chapter 14.]

Glossary

Ab Initio**:** Beginning from the first principles.

Acceptable Wave Function: Single-valued, nowhere infinite, continuous, piecewise continuous, and first derivative function.

Antisymmetric Wave Function: The interchange of any pair of particles alters the sign of the wave function.

Basis Function: A set of functions from which any other function in the same space can be built.

Black Body: A perfect absorber and emitter of radiation.

Bohr Radius: It is the maximum distance (r_{max}) in $r^2\psi^2$ vs. **r** graph.

Bra: It is similar to the complex conjugate of the wave function for a state (symbol $<|$).

Compton Effect: When X-rays are scattered by solids, the frequency of scattered rays is lower than that of the incident rays. This phenomenon is termed as the Compton effect.

Cyclobutadiene: A cyclic square system consisting of 4π electrons.

Degenerate States: Independent states consisting of equal value of the state defining property (i.e., energy).

Degree: The degree of the highest derivative in a differential equation.

Delocalisation Energy: The energy difference between localised structure and the actual molecule.

Density Functional Theory (DFT): A tool of computational chemistry.

Determinant: A quantity represented by a square array implying expansion in terms of elements of the array.

Diagonal Matrix: A square matrix whose all elements are zero except the diagonal element.

Dynamical Variable: Any property of a system of interest.

Eigen Function: Function satisfying an Eigen value equation.

Eigen Value Equation: The equation that contains on one side an operator operating on a function yielding a constant times the unaltered function, i.e., $\hat{A}f = af$.

Eigen Value: Operator operating on function yields a constant times the unchanged function. The constant is the Eigen value.

Expectation Value: The average value of a property in some state in quantum mechanics.

Fermi Heap: The heap in the probability density $|\psi_+|^2$ vs. $(r_1 - r_2)$ graph.

Fermi Hole: The dip in the probability density $|\psi_-|^2$ vs. $r_1 - r_2$ graph.

Free Valence: The difference between the maximum bond orders around a carbon atom and total estimated bond orders.

Function: A quantity y is called a function of another quantity x if a change in one produces a change in another, i.e., $y = f(x)$.

Functional: A mathematical rule by which a function is transformed into a number.

Gerade: A German word meaning even.

Group Velocity: The velocity of wave group is the group velocity.

Gyromagnetic Ratio: Ratio of magnetic moment to angular momentum.

Heisenberg's Uncertainty Principle: It is not possible to specify both the position and the momentum of a particle simultaneously.

Hermitian Matrix: A matrix that is equal to its conjugate transpose.

Hermitian Operator: A linear operator say $\hat{A}$ for which $\langle \Psi | \hat{A} | \Phi \rangle$ equals $\langle \hat{A}\Psi | \Phi \rangle$, where Ψ and Φ are two quadratically integrable functions.

Heteropolar Diatomics: The bond formed between dissimilar atoms.

Homopolar Diatomics: The bond formed between identical atoms.

j.j. Coupling: It is a scheme in which the angular momentum vector $\underset{li}{\rightarrow}$ of a single electron unites with the single electron spin $\underset{Si}{\rightarrow}$ yielding the single electrons total angular momentum $\underset{j}{\rightarrow}$.

Ket: It is similar to the wave function for a state (symbol $|>|$).
Kronecker Delta Function: A function, δ_{ij}, which is equal to zero if $i \neq j$ and unity if $i = j$.
Landés Interval Rule: The separation between the two J levels of an R–S term, brought about by spin orbit interaction, is proportional to the larger J value of the pair.
Matrix: An ordered array of elements (real or complex), the operation of which is subject to certain rules.
Multiplicity: Number of lines in the atomic spectrum that are equivalent to a single line in the absence of spin.
Node: The position where the wave function equals zero or A surface on which a wave function is zero.
Normalized Functions: Functions for which $\langle\psi|\psi\rangle = 1$ or $\int \psi^* \psi \, d\tau = 1$.
Null Matrix: A matrix having all elements zero.
Observable: Any dynamical property of a system of interest.
Operator: A symbol of mathematical procedure, which changes one function into another.
Order: The order of the highest derivative in the differential equation.
Orthogonal Functions: Function for which $\langle\Psi|\Phi\rangle = 0$.
Orthogonal Matrix: A matrix in which inverse is equal to transpose.
Orthonormal: The property of being both orthogonal and normalised.
Phase Velocity: The wave velocity is the phase velocity.
Photoelectric effect: When light of appropriate frequency falls on a clean metal surface, electrons are ejected. This phenomenon is known as photoelectric effect.
Projection Operator: An operator $P \in (V, V)$ is called projection operator if $P^2 = P$.
Quantum Theory: The radiation is emitted and absorbed by a black body in discrete amounts, which is known as quanta and is equal to $E = hv$.
Quantum: The packet of energy $E = hv$.
R–S Coupling: It is an adequate way of describing small electronic configurations.
Rayleigh–Jeans Radiation Law: This is mathematically defined as $E_\lambda = 8\pi kT/\lambda^4$.
Rigid Rotor: A rigid body rotating about a fixed axis.
Row Matrix: Matrix consisting of only one row and any number of columns.
Rydberg Constant: The constant factor having value $109{,}677.6\,\text{cm}^{-1}$.
Scalar Matrix: A diagonal matrix whose diagonal elements are equal.
Screening Constant: The correction of the nuclear charge for the screening effect.
Self-consistent Field: Concept used to obtain approximate solutions to multi-electronic systems in quantum mechanics.
Square Matrix: An $m \times n$ matrix having $m = n$.
Stark Effect: The energy levels are shifted when an atom is placed in a uniform electric field and as a result, splitting of spectral lines occurs.
Symmetric Matrix: A square matrix having $a_{ij} = a_{ji}$.
Symmetric Wave Function: The interchange of any pair of particles among its arguments does not change the wave function.
Term Symbol: A term symbol denotes a level (L, S, J), the complete set of quantum numbers.
Term: An expression fully specifying the angular momentum $st\vec{a}te$ of an atom or a molecule.
Threshold Frequency: The minimum frequency (v_0) of light required to emit electrons from the metal surface when the light strikes the metal surface.
Ungerade: The German word meaning odd.
Unit Matrix: A square matrix having its each diagonal element 1 and off-diagonal element zero.
Valence Electrons: The electrons present in the outermost electronic configuration of an atom.
Wien Radiation Law: It is mathematically defined as $\lambda_{max}\, T = 2.898 \times 10^{-3} m.K.\ or\ E_\lambda = \frac{a}{\lambda^5} e^{-\beta/\lambda T}$
Zeeman Effect: It is the effect in which a substance emitting a line spectrum by a strong magnetic field is split up into group of closely spaced lines.
Zero-Point Energy: Minimum energy of a particle in one-dimensional box. i.e., $E = h^2/8ma^2$.

Index

Printed in the United States
By Bookmasters